Hardy Martingales

This book presents the probabilistic methods around Hardy martingales for an audience interested in applications to complex, harmonic, and functional analysis. Building on the work of Bourgain, Garling, Jones, Maurey, Pisier, and Varopoulos, it discusses in detail those martingale spaces that reflect characteristic qualities of complex analytic functions. Its particular themes are holomorphic random variables on Wiener space, and Hardy martingales on the infinite torus product, and numerous deep applications to the geometry and classification of complex Banach spaces, e.g., the SL^∞ estimates for Doob's projection operator, the embedding of L^1 into L^1/H^1, the isomorphic classification theorem for the polydisk algebras, or the real variables characterization of Banach spaces with the analytic Radon Nikodym property. Due to the inclusion of key background material on stochastic analysis and Banach space theory, it's suitable for a wide spectrum of researchers and graduate students working in classical and functional analysis.

Paul F.X. Müller is Professor at Johannes Kepler University Linz, Austria. He is the author of more than 50 papers in complex, harmonic and functional analysis, and of the monograph *Isomorphisms between H^1 spaces* (Springer, 2005).

NEW MATHEMATICAL MONOGRAPHS

All the titles listed below can be obtained from good booksellers or from Cambridge University Press. For a complete series listing visit www.cambridge.org/mathematics.

1. M. Cabanes and M. Enguehard *Representation Theory of Finite Reductive Groups*
2. J. B. Garnett and D. E. Marshall *Harmonic Measure*
3. P. Cohn *Free Ideal Rings and Localization in General Rings*
4. E. Bombieri and W. Gubler *Heights in Diophantine Geometry*
5. Y. J. Ionin and M. S. Shrikhande *Combinatorics of Symmetric Designs*
6. S. Berhanu, P. D. Cordaro and J. Hounie *An Introduction to Involutive Structures*
7. A. Shlapentokh *Hilbert's Tenth Problem*
8. G. Michler *Theory of Finite Simple Groups I*
9. A. Baker and G. Wüstholz *Logarithmic Forms and Diophantine Geometry*
10. P. Kronheimer and T. Mrowka *Monopoles and Three-Manifolds*
11. B. Bekka, P. de la Harpe and A. Valette *Kazhdan's Property (T)*
12. J. Neisendorfer *Algebraic Methods in Unstable Homotopy Theory*
13. M. Grandis *Directed Algebraic Topology*
14. G. Michler *Theory of Finite Simple Groups II*
15. R. Schertz *Complex Multiplication*
16. S. Bloch *Lectures on Algebraic Cycles (2nd Edition)*
17. B. Conrad, O. Gabber and G. Prasad *Pseudo-reductive Groups*
18. T. Downarowicz *Entropy in Dynamical Systems*
19. C. Simpson *Homotopy Theory of Higher Categories*
20. E. Fricain and J. Mashreghi *The Theory of H(b) Spaces I*
21. E. Fricain and J. Mashreghi *The Theory of H(b) Spaces II*
22. J. Goubault-Larrecq *Non-Hausdorff Topology and Domain Theory*
23. J. Śniatycki *Differential Geometry of Singular Spaces and Reduction of Symmetry*
24. E. Riehl *Categorical Homotopy Theory*
25. B. A. Munson and I. Volić *Cubical Homotopy Theory*
26. B. Conrad, O. Gabber and G. Prasad *Pseudo-reductive Groups (2nd Edition)*
27. J. Heinonen, P. Koskela, N. Shanmugalingam and J. T. Tyson *Sobolev Spaces on Metric Measure Spaces*
28. Y.-G. Oh *Symplectic Topology and Floer Homology I*
29. Y.-G. Oh *Symplectic Topology and Floer Homology II*
30. A. Bobrowski *Convergence of One-Parameter Operator Semigroups*
31. K. Costello and O. Gwilliam *Factorization Algebras in Quantum Field Theory I*
32. J.-H. Evertse and K. Győry *Discriminant Equations in Diophantine Number Theory*
33. G. Friedman *Singular Intersection Homology*
34. S. Schwede *Global Homotopy Theory*
35. M. Dickmann, N. Schwartz and M. Tressl *Spectral Spaces*
36. A. Baernstein II *Symmetrization in Analysis*
37. A. Defant, D. García, M. Maestre and P. Sevilla-Peris *Dirichlet Series and Holomorphic Functions in High Dimensions*
38. N. Th. Varopoulos *Potential Theory and Geometry on Lie Groups*
39. D. Arnal and B. Currey *Representations of Solvable Lie Groups*
40. M. A. Hill, M. J. Hopkins and D. C. Ravenel *Equivariant Stable Homotopy Theory and the Kervaire Invariant Problem*
41. K. Costello and O. Gwilliam *Factorization Algebras in Quantum Field Theory II*
42. S. Kumar *Conformal Blocks, Generalized Theta Functions and the Verlinde Formula*

Hardy Martingales

Stochastic Holomorphy, L^1-Embeddings, and Isomorphic Invariants

PAUL F.X. MÜLLER
Johannes Kepler University Linz

CAMBRIDGE
UNIVERSITY PRESS

CAMBRIDGE
UNIVERSITY PRESS

University Printing House, Cambridge CB2 8BS, United Kingdom

One Liberty Plaza, 20th Floor, New York, NY 10006, USA

477 Williamstown Road, Port Melbourne, VIC 3207, Australia

314–321, 3rd Floor, Plot 3, Splendor Forum, Jasola District Centre, New Delhi – 110025, India

103 Penang Road, #05–06/07, Visioncrest Commercial, Singapore 238467

Cambridge University Press is part of the University of Cambridge.

It furthers the University's mission by disseminating knowledge in the pursuit of education, learning, and research at the highest international levels of excellence.

www.cambridge.org
Information on this title: www.cambridge.org/9781108838672
DOI: 10.1017/9781108976015

First published 2022

A catalogue record for this publication is available from the British Library.

ISBN 978-1-108-83867-2 Hardback

To Joanna

Contents

Preface

In this book we present probabilistic methods developed for applications to complex and functional analysis. We will study, in depth, spaces of martingales that reflect characteristic qualities of holomorphic functions; specifically,

(i) the space of integrable Hardy martingales $H^1(\mathbb{T}^{\mathbb{N}})$ defined by restrictions on the support of their Fourier coefficients;
(ii) the space of holomorphic random variables $H^1(\Omega)$ on Wiener space, characterized by their Itô-integral representation.

Stochastic Holomorphy

The interplay between (nonconstant) holomorphic functions f and complex Brownian motion (z_t) goes back to the work of Paul Lévy who noted that $f(z_t)$ is the path of a complex Brownian motion (Lévy, 1948). More precisely, $f(z_t)$ is distributionally indistinguishable from $z_{\beta(t)}$ where

$$\beta(t) = \int_0^t |f'(z_s)|^2 ds.$$

Itô and McKean (1965) presented a proof of Picard's theorem (asserting that a nonconstant analytic function on $\mathbb{C}$ omits at most one value) by applying Lévy's result to the universal covering map onto $\mathbb{C}\backslash\{0, 1\}$.

Being more specific, we let $f\colon \mathbb{D} \to \mathbb{C}$ be analytic and bounded where $\mathbb{D} = \{z \in \mathbb{C} : |z| < 1\}$. We let τ denote the exit time of (z_t) from $\mathbb{D}$. The process $(f(z_t) : t < \tau)$ may be expanded by Itô integrals,

$$f(z_t) = \int_0^t f'(z_s)dz_s, \qquad t < \tau, \tag{0.0.1}$$

and hence forms a Brownian martingale. Doob (1953) proved martingale convergence theorems, showing that $\lim_{t\to\tau} f(z_t)$ exists almost surely, and developed the tools (conditioned Brownian motion) by which martingale convergence is transformed into radial limits such that

$$\lim_{r\to 1} f(r\zeta) \quad \text{exists for almost every} \quad \zeta \in \mathbb{T},$$

where $\mathbb{T} = \{z \in \mathbb{C} : |z| = 1\}$. Thus Fatou's theorem and Privalov's theorem were among the first results in complex analysis obtained by stochastic methods. Many years later Burkholder et al. (1971) showed that

$$\mathbb{E}|f(z_\tau)| \le C\mathbb{E} \sup_{s<\tau} |\Re f(z_s)|,$$

which in turn, by Doob's conditioned Brownian motion, gives rise to the real variable characterization of the Hardy space $H^1(\mathbb{T})$.

The Itô integral representation (0.0.1) provides the model for an intrinsically *stochastic* concept of *holomorphy*. We say that $F_t \colon \Omega \to \mathbb{C}$ is a holomorphic martingale on Wiener space $(\Omega, (\mathcal{F}_t), \mathbb{P})$ if there exists a complex-valued, $(\mathcal{F}_t)$ adapted process (X_t) satisfying

$$F_t = F_0 + \int_0^t X_s dz_s.$$

A holomorphic random variable is just the limit $F = \lim_{t\to\infty} F_t$ of an equiintegrable holomorphic martingale.

In applications to analysis, it is important that holomorphic martingales form a linear space and that they are stable under stopping times, pointwise multiplication, and composition with analytic functions. Varopoulos (1980, 1981) found a key to profound probabilistic methods in complex analysis by analyzing the action of Doob's conditional expectation operator,

$$N(F)(e^{i\theta}) = \mathbb{E}\left(F | z_\tau = e^{i\theta}\right),$$

on a holomorphic random variable F. It consists in linking the space of p-integrable holomorphic random variables $H^p(\Omega)$ to the corresponding Hardy space $H^p(\mathbb{T})$ by means of the commutative diagram,

$$\begin{array}{ccc} H^p(\mathbb{T}) & \xrightarrow{\ \mathrm{Id}\ } & H^p(\mathbb{T}) \\ & {\scriptstyle M}\searrow \quad \nearrow{\scriptstyle N} & \\ & H^p(\Omega) & \end{array},$$

where $Mf = f(z_\tau)$. Chapter 1 develops central ideas of stochastic holomorphy together with their applications to the Marcinkiewicz decomposition and to complex convexity estimates for $H^p(\mathbb{T})$ and for the Banach space $SL^\infty(\mathbb{T})$ of harmonic functions with uniformly bounded square functions.

Hardy Martingales and Banach Spaces

Let X be a complex Banach space. We turn now to vector-valued martingales on the infinite torus product $(\mathbb{T}^{\mathbb{N}}, (\mathcal{F}_k), \mathbb{P})$, where $\mathbb{P}$ is the Haar probability measure, and $(\mathcal{F}_k)$ the canonical filtration on $\mathbb{T}^{\mathbb{N}}$. Following Garling (1988), we say that an $(\mathcal{F}_k)$ martingale $(F_k : \mathbb{T}^{\mathbb{N}} \to X)$ is a *Hardy martingale* if conditioned on $\mathcal{F}_{k-1}$,

$$F_k - F_{k-1} \text{ defines an element in } H_0^1(\mathbb{T}, X).$$

This book is on Hardy martingales, their roots in the fields of stochastic, complex, and harmonic analysis, and their numerous, deep applications to the geometry and classification of complex Banach spaces.

We begin our review with the role vector-valued Hardy martingales play in the general study of Banach spaces and their isomorphic invariants. We single out the analytic Radon–Nikodym property (aRNP) of a complex Banach space X. A vector-valued analytic function $f: \mathbb{D} \to X$ may be defined simply by demanding that $x^*(f): \mathbb{D} \to \mathbb{C}$ be analytic for each x^* in the dual space X^*. This condition implies the validity of the Cauchy integral formula for f, and hence its expansion into norm convergent power series with coefficients in X. It was observed very early on that the validity of Fatou's theorem is no longer assured, for any Banach space X. If for every bounded analytic $f: \mathbb{D} \to X$,

$$\lim_{r \to 1} f(r\zeta) \quad \text{exists in } X \text{ for almost every} \quad \zeta \in \mathbb{T},$$

then we say that X satisfies the aRNP. Heins (1969), for instance, proved that the aRNP holds true for the Banach spaces $\ell^p, 1 \leq p < \infty$, but not for c_0. Bukhvalov and Danilevich (1982) showed that it is satisfied for L^1, and, more generally, for any Banach lattice not containing c_0.

From the point of view of Banach space theory, such a phenomenon requires clarification. Considerable effort has gone into the determination of the geometric, probabilistic, and real variable characterization of those Banach spaces satisfying the aRNP. In this book, we present definitive characterizations of aRNP due to Ghoussoub, Lindenstrauss, and Maurey (Ghoussoub et al. [1989]; Ghoussoub and Maurey [1989]). While their conditions are expressed in purely analytic terms (the existence of small pluri-subharmonic slices and the pluri-subharmonic variational principle), the method of proof is a probabilistic one, exploiting vector-valued Hardy martingales on the infinite torus product.

Having pluri-subharmonic characterizations at hand, Ghoussoub, Maurey and Schachermayer were able to prove that $\mathrm{J}_*\mathrm{T}$, the predual of the James tree space, satisfies the aRNP (Ghoussoub and Maurey, 1989; Ghoussoub et al., 1989).

Above, we considered Hardy martingales and radial limits of analytic functions ranging in a general Banach space X. We now turn to describing the significance of Hardy martingales in the analysis of concrete Banach spaces of analytic functions. The following theorems were instrumental in establishing the importance of Hardy martingales in problems of hard analysis.

(i) Maurey (1980) started the proof that $H^1(\mathbb{T})$ is isomorphic to dyadic-H^1 by discretizing the process $(f(z_t): t < \tau)$, with $f \in H^1(\mathbb{T})$. The result of the discretization is a Hardy martingale (F_k) with uniformly bounded increments satisfying

$$\mathbb{E}(\sup |F_n|) \leq C \sup_n \mathbb{E}|F_n| \quad \text{and} \quad |\Delta F_n| \leq \epsilon \int_{\mathbb{T}} |f| dm.$$

Maurey's discretization yields that any Hardy martingale may be regarded a subsequence of a Hardy martingale *with predictable increments* and almost identical L^1-norm.

(ii) The L^1 three-space problem asks if, for any subspace X of L^1, either X or L^1/X contain a subspace isomorphic to L^1. Pełczyński pointed out the importance of the case $X = H^1(\mathbb{T})$, which was resolved by Bourgain (1983a) in terms of Hardy martingales on the infinite torus product.

(iii) Bourgain's theorem (Bourgain, 1983a) asserts that L^1 is isomorphic to a subspace of L^1/H^1. In the converse direction, we have Pisier's (Davis et al., 1984) result that L^1/H^1 is not finitely represented in L^1. Pisier exhibited a uniformly bounded Hardy martingale $(G_n: \mathbb{T}^{\mathbb{N}} \to L^1/H^1)$ for which

$$\mathbb{E}\|\Delta G_n\|_{L^1/H^1_0} \geq 1,$$

for $n \in \mathbb{N}$, sharply contrasting with martingale cotype 2 estimates valid for Hardy martingales ranging in L^1. Since (G_n) is a sequence of Riesz products on $\mathbb{T}^{\mathbb{N}}$, an important link is hereby established between Hardy martingales and harmonic analysis on compact abelian groups.

(iv) Bourgain (1984b), connecting Pisier's analysis of vector-valued Riesz products and cotype 2 estimates for $A^*(\mathbb{T})$, showed that $A^*(\mathbb{T} \times \mathbb{T})$ is not finitely represented in $A^*(\mathbb{T})$. Multi-indexed Hardy martingales play a central role in the ensuing isomorphic classification theorem for polydisk algebras.

(v) Garling (1991) obtained the unconditional convergence of martingales by verifying it directly for Hardy martingales using Yudin's Lemma, and showing that the general result follows from the special case and the boundedness of the Riesz projection. Thus Garling's study determines the position of Hardy martingales within the field of harmonic analyis.

This book contains detailed proofs of the theorems listed above and provides a thorough investigation of complex Banach spaces vis-a-vis Hardy martingales. We investigate their convergence (aRNP), unconditional convergence (aUMD), the notions of Hardy martingale cotype and complex uniform convexity.

In writing the book, I intended to present elements of stochastic analysis and Banach space theory to the classical complex analyst – at the same time, I wanted the book to be useful for the functional analyst interested in Banach spaces of analytic functions and/or probability.

Contents

Chapter 1 presents elements of stochastic analysis, emphasizing Itô calculus, Doob's embedding and projection, holomorphic martingales, stopping time decompositions, and SL^∞ estimates by means of conditioned Brownian motion.

Chapter 2 gives an overview of Hardy martingales, highlighting examples originating in harmonic analysis on $\mathbb{T}^{\mathbb{N}}$, and complex Brownian martingales. We obtain central norm estimates for projection operators onto Hardy martingales and spaces of homogeneous polynomials, prove L^1 estimates for the martingale square function, and, by nonlinear teleskoping, we prove the Davis and Garsia inequality for Hardy martingales.

Chapter 3 is devoted to proving that L^1 embeds into $L^1(\mathbb{T})/H^1_0(\mathbb{T})$. The heart of the matter is the corresponding result for the quotient space formed by martingales on the infinite torus product

$$L^1(\mathbb{T}^{\mathbb{N}})/H^1_0(\mathbb{T}^{\mathbb{N}}).$$

Once this is established, the transfer methods developed by Bourgain, Bonami, and Meyer yield the embedding of L^1 into $L^1(\mathbb{T})/H^1_0(\mathbb{T})$.

Chapter 4 begins with Kwapien's representation of operators on L^0 and some of its consequences. We work through the examples of L^1-subspaces (other than H^1) for which a positive solution to the L^1 three-space problem is known. These are the reflexive subspaces of L^1, subspaces isomorphic to ℓ^1, and the complemented subspaces. We present Talagrand's space $X \subset L^1$ and prove that neither X nor L^1/X contain a copy of L^1, contrasting the Johnson–Maurey–Schechtman construction of a normalized weakly null sequence (φ_i) in L^1 that fails to contain an unconditional subsequence.

Chapter 5 presents the proof of the dimension conjecture for polydisk algebras and investigates basic isomorphic invariants of complex Banach spaces induced by Hardy martingales: the analytic Radon–Nikodym property, the analytic UMD property, and Hardy martingale cotype.

Chapter 6 continues the analysis of the projection operators introduced in Chapter 2. We prove Rosenthal's L^p theorem and utilize it to show that these operators are bounded on $L^p(Y)$ where Y is a reflexive subspace of L^1 and $1 < p < \infty$.

Chapter 7 is a case by case study of the isomorphic invariants associated with Hardy martingales, covering L^1 quotients by reflexive spaces, the trace class, iterated $L^p(L^q)$-spaces, the James tree space, and L^1/H^1. The following results are presented:

(i) L^1 quotients by reflexive spaces are of cotype 2 (Kislyakov, Pisier) and satisfy aUMD (Bourgain and Davis).
(ii) The trace class S^1 satisfies Hardy martingale cotype-2 estimates and fails aUMD (Haagerup and Pisier).
(iii) The iterated $L^p(L^q)$-spaces are super-reflexive Banach lattices, with an unconditional basis and if $p \neq q$, fail aUMD (Qiu).
(iv) The predual of the James tree space satisfies the aRNP (Ghoussoub, Maurey, and Schachermayer).
(v) L^1/H^1 is of cotype 2 (Bourgain). Fails the aRNP.

Prerequisites. For the most part our prerequisites are contained in the books by Dieudonné (1968a,b) and Rudin (1974, 1991) and in the first chapters of Durrett (1984) and Wojtaszczyk (1991). There are instances, however, mainly in Chapter 7, where we rely on more specialized sources.

Acknowledgments

While writing this book, I benefited from having access to several unpublished sources dating back to the 1980s. These are the notes by Pallares and Kouba (1986–1987), recording the seminar talks at Groupe de Travail sur les Espaces de Banach directed by Maurey and Pisier; the lecture notes by Pisier entitled "H^p-spaces of Banach space valued functions" (1991, unpublished); Garling's exposition of holomorphic martingales (Garling, D. J. H. $L^1(\Omega)/H^1(\Omega)$ H is a Grothendieck space of cotype 2. pages 1–18, 1992, unpublished.), and detailed seminar notes by Schmuckenschläger on complex uniform convexity (1993/4, unpublished).

I presented several parts of Chapters 1, 2, 3, and 5 at Texas A&M University during SUMIRFAS meetings, at the Mathematical Institute of the Polish Academy of Sciences (Warszawa and Gdansk), and at Yale University. It is my pleasure to thank S. Dilworth, W. Hensgen, A. Kamont, K. Kazaniecki, W. B. Johnson, P. W. Jones, G. Pisier, G. Racher, M. Rzeszut, G. Schechtman, Th. Schlumprecht, M. Wojciechowski, and P. Wojtaszczyk for most valuable feedback and discussions.

In Linz, the topics of this book provided the framework for research seminars, conducted jointly with R. Lechner, M. Passenbrunner, and J. Penteker, who were very helpful in suggesting improvements to the exposition and in correcting actual errors.

I am very grateful to Alexander Lindenberger and David I. Müller for their assistance in proofreading and for producing the figures in the text. I especially thank W. Schachermayer, M. Schmuckenschläger, Ch. Stegall, and P. Yuditskii for unfailing advice and generous help.

Finally, I express my appreciation to the Austrian Science Foundation (FWF) for support by means of FWF-projects P20166, P22549, P23987, and P28352, P344214 and I-5231.

1

Stochastic Holomorphy

Here, we cover basic results in stochastic holomorphy that form the foundation of our work in later chapters of this book. The main focus lies on the space of p-integrable holomorphic random variables, $H^p(\Omega)$, and on $SL^\infty(\Omega)$, consisting of complex Brownian martingales with uniformly bounded quadratic variation process.

Sections 1.2 and 1.3 contain estimates for the martingale maximal function, the quadratic variation process, and the Marcinkiewicz decomposition for holomorphic random variables in $H^1(\Omega)$. We investigate the stochastic Hilbert transform $\mathcal{H}$, the Doob projection N, and the martingale embedding operator M in considerable detail in Sections 1.4–1.6. The relation

$$\mathrm{Id} = NM$$

forms the basis for most applications of stochastic holomorphy to complex analysis. It is exploited in Section 1.7 and throughout the course of this book.

Finally, in Section 1.8 we prove that Doob's projection preserves square function estimates and that

$$N\colon SL^\infty(\Omega) \to SL^\infty(\mathbb{T})$$

forms a bounded linear surjection, inverting the action of the martingale embedding operator $M\colon SL^\infty(\mathbb{T}) \to SL^\infty(\Omega)$.

1.1 Preliminaries

Our preliminary section forms a listing of standard concepts in real, complex, and stochastic analysis, all of which are accessible by means of graduate-level textbooks. The following topics are reviewed in the subsections below:

(i) Conditional expectation, discrete time martingales, almost sure convergence, uniform integrability and convergence in L^1, Doob's inequalities, Davis decomposition.

(ii) Complex Brownian motion, Brownian martingales, stochastic integrals, Itô's formula in complex format.

(iii) Harmonic and subharmonic functions in the unit disk, the boundary convergence theorems of Fatou and Littlewood, the Poisson kernel and Kakutani's theorem, Green's function, and the occupation time formula for Brownian motion.

(iv) Convolution operators, the de La Valle Poussin kernels, the Fejer kernels, Hardy spaces in the unit disk, the Riesz factorisation, the F. and M. Riesz theorem, the Hilbert transform, the predual of $H^1(\mathbb{T})$, and polydisk algebras.

(v) Weak topology, reflexive Banach spaces, uniform integrability, weak compactness in L^1, and the Dunford–Pettis theorem.

1.1.1 Conditional Expectation and Martingales

In this section, we fix notation and recall basic concepts of measure and integration such as conditional expectation, filtered measure spaces, martingale sequences, uniform integrability, Doob's convergence theorems, and maximal inequalities. Basic references include the books by Kahane (1985) and Neveu (1975).

Measure distance: Let $(F, \mathcal{F}, \mu)$ be a finite measure space. We denote the linear space of $\mathcal{F}$-measurable scalar-valued functions by $\mathcal{L}^0 = \mathcal{L}^0(F, \mathcal{F}, \mu)$. We put

$$\mathcal{N} = \mathcal{N}(F, \mathcal{F}, \mu) = \left\{ f \in \mathcal{L}^0 : \mu(\{f \neq 0\}) = 0 \right\}.$$

Given $f, g \in \mathcal{L}^0$, we define the measure distance by

$$d_\mu(f, g) = \inf\{\epsilon > 0 : \mu(\{\omega \in F : |f(\omega) - g(\omega)| > \epsilon\}) < \epsilon\}.$$

We have $d_\mu(f, g) = 0$ if and only if $f - g \in \mathcal{N}$. The quotient space

$$L^0(F, \mathcal{F}, \mu) = \mathcal{L}^0 / \mathcal{N},$$

equipped with the distance induced by d_μ, becomes a complete metric space. When convenient (and feasible) we suppress the explicit dependence on the measure space $(F, \mathcal{F}, \mu)$ and compress our notation to $L^0 = L^0(F, \mathcal{F}, \mu)$.

Continuous operators on L^0: Suppose we are given a second measure space (S, Σ, λ), defining $L^0(S, \Sigma, \lambda)$ of equivalence classes of Σ-measurable functions $f: S \to \mathbb{R}$. Consider now a linear operator,

$$T: L^0(F, \mathcal{F}, \mu) \to L^0(S, \Sigma, \lambda).$$

The continuity of T, with respect to the metrics d_μ and d_λ, is equivalent to the following condition: For any $\epsilon > 0$ there exists $\delta > 0$ such that

$$\mu(\{\omega \in F: |f(\omega)| > \delta\}) < \delta \quad \text{implies} \quad \lambda(\{s \in S: |(Tf)(s)| > \epsilon\}) < \epsilon,$$

for any $f \in L^0(F, \mathcal{F}, \mu)$.

Lebesgue spaces: For $1 \leq p < \infty$, we let $\mathcal{L}^p(F, \mathcal{F}, \mu)$ denote the space of $\mathcal{F}$-measurable, p-integrable, scalar-valued functions equipped with the seminorm

$$\|f\|_p = \left(\int_F |f|^p d\mu \right)^{1/p}.$$

For $1 \leq p < \infty$, the quotient space

$$L^p(F, \mathcal{F}, \mu) = \mathcal{L}^p(F, \mathcal{F}, \mu)/N,$$

equipped with its canonical quotient norm, forms a Banach space called the Lebesgue space of p-integrable functions. Similarly, we form the Banach space

$$L^\infty(F, \mathcal{F}, \mu) = \mathcal{L}^\infty(F, \mathcal{F}, \mu)/N,$$

where $\mathcal{L}^\infty(F, \mathcal{F}, \mu)$ denotes the space of $\mathcal{F}$-measurable functions that are μ-essentially bounded. The space $\mathcal{L}^\infty(F, \mathcal{F}, \mu)$ is equipped with the seminorm, given by the essential supremum,

$$\|f\|_\infty = \mu - \text{ess sup}_{\omega \in F} |f(\omega)| = \inf\{t > 0: \mu\{|f| > t\} = 0\}.$$

We frequently shorten the notation of $L^p(F, \mathcal{F}, \mu)$ to $L^p(F)$ and further to L^p, when the context allows us to do so.

Conditional Expectation

We fix a finite measure space $(F, \mathcal{F}, \mu)$ and a sub-sigma-algebra $\mathcal{G}$ of $\mathcal{F}$. For $f \in L^1(F, \mathcal{F}, \mu)$, we consider the set function

$$\lambda(A) = \int_A f d\mu, \quad A \in \mathcal{G}.$$

In view of Lebesgue's theorem on dominated convergence, $\lambda: \mathcal{G} \to \mathbb{R}$ is a finite signed measure on the measurable space $(F, \mathcal{G})$. Moreover, by the absolute continuity of the Lebesgue integral, for any sequence $A_n \in \mathcal{G}$ satisfying

$\lim_{n\to\infty}\mu(A_n) = 0$, we obtain $\lim_{n\to\infty}\lambda(A_n) = 0$. The Radon–Nikodym theorem asserts that there exists a $\mathcal{G}$-measurable function $g\colon F \to \mathbb{R}$ such that

$$\lambda(A) = \int_A g d\mu, \quad A \in \mathcal{G}, \quad \text{and} \quad \int_F |g| d\mu \leq \int_F |f| d\mu.$$

Moreover, up to $\mathcal{G}$-measurable sets of vanishing μ-measure, g is uniquely determined. We say that g is the conditional expectation of f with respect to $\mathcal{G}$ and write

$$\mathbb{E}(f|\mathcal{G}) = g.$$

Clearly, taking the conditional expectation defines a linear operation, and $f \geq 0$ implies $\mathbb{E}(f|\mathcal{G}) \geq 0$. Moreover, $\mathbb{E}(1_F|\mathcal{G}) = 1_F$ and

$$\|\mathbb{E}(f|\mathcal{G})\|_p \leq \|f\|_p,$$

for $f \in L^p(F)$, $1 \leq p \leq \infty$. We frequently use the following properties:

(i) If $f, h \in L^1(F)$ and h is $\mathcal{G}$-measurable, then $\mathbb{E}(hf|\mathcal{G}) = h\mathbb{E}(f|\mathcal{G})$.
(ii) If $\mathcal{G}_1$ is a sub-sigma-algebra of $\mathcal{G}$, then $\mathbb{E}(\mathbb{E}(f|\mathcal{G})|\mathcal{G}_1) = \mathbb{E}(f|\mathcal{G}_1)$ for $f \in L^1(F)$.

Let $(\mathcal{G}_k)_{k=0}^{\infty}$ be a sequence of sub-sigma-algebras of $\mathcal{F}$ satisfying $\mathcal{G}_{k-1} \subseteq \mathcal{G}_k \subseteq \mathcal{F}$ for $k \in \mathbb{N}$, and $\mathcal{G}_0 = \{\emptyset, F\}$. Let $\mathcal{G}$ denote the sigma-algebra generated by $\bigcup \mathcal{G}_k$, and let $\mu_{\mathcal{G}}$ be the restriction of μ to $\mathcal{G}$. We then say that $(F, (\mathcal{G}_k), \mu_{\mathcal{G}})$ forms a filtered finite measure space.

The following far-reaching theorem addresses convergence of the conditional expectations $\mathbb{E}(f|\mathcal{G}_k)$. As it turns out, for $f \in L^1$, we have convergence in the L^1-norm and point-wise convergence almost everywhere.

Theorem 1.1.1 *For $f \in L^1(F)$, set $g = \mathbb{E}(f|\mathcal{G})$ and $f_k = \mathbb{E}(f|\mathcal{G}_k)$. Then*

$$\lim_{k\to\infty} \int_F |g - f_k| d\mu = 0,$$

and there exists $E \subset F$ satisfying $\mu(E) = 0$ such that $g(\omega) = \lim_{k\to\infty} f_k(\omega)$, for $\omega \in F\setminus E$.

Examples

The canonical product filtration on the infinite torus product forms a concrete realization of a filtered probability space. We explicitly describe the dyadic filtration embedded in the infinite torus product.

Example 1.1.1 (The product filtration on $\mathbb{T}^{\mathbb{N}}$) Let $\mathbb{T} = \{e^{i\theta} : \theta \in [0, 2\pi[\}$ be equipped with the Lebesgue sigma-algebra and normalized angular measure, denoted dm. Thus m is the Haar measure on $\mathbb{T}$, i.e., the unique rotation invariant probability measure on the sigma-algebra of Lebesgue measurable subsets of $\mathbb{T}$. We denote by

$$\mathbb{T}^{\mathbb{N}} = \left\{(z_i)_{i=1}^{\infty} : z_i \in \mathbb{T}\right\}$$

its countable product equipped with the product sigma-algebra and normalized product Haar measure, denoted $\mathbb{P}$. We denote by $\mathcal{F}_k$ the sigma-algebra on $\mathbb{T}^{\mathbb{N}}$ generated by the cylinder sets

$$\left\{\left(A_1, \ldots, A_k, \mathbb{T}^{\mathbb{N}}\right)\right\},$$

where A_i, $i \leq k$, are measurable subsets of $\mathbb{T}$. A measurable function F defined on $\mathbb{T}^{\mathbb{N}}$ is measurable with respect to $\mathcal{F}_k$ if it depends only on the first k variables of $\mathbb{T}^{\mathbb{N}}$. The conditional expectation with respect to the sigma-algebra $\mathcal{F}_k$ acts as integration with respect to the variable z_i, where $i \geq k+1$. Explicitly, if $f \in L^1(\mathbb{T}^{\mathbb{N}})$, then for almost every $x \in \mathbb{T}^k$ we have

$$\mathbb{E}(f|\mathcal{F}_k)(x) = \int_{\mathbb{T}^{\mathbb{N}}} f(x, z) d\mathbb{P}(z). \tag{1.1.1}$$

The filtered probability space $(\mathbb{T}^{\mathbb{N}}, (\mathcal{F}_k), \mathbb{P})$ is our preferred framework for discussing discrete-time martingales.

Example 1.1.2 (The dyadic filtration on $\mathbb{T}^{\mathbb{N}}$) We define the independent Rademacher functions $\sigma_k : \mathbb{T}^{\mathbb{N}} \to \{-1, 1\}$, by

$$\sigma_k(z) = \operatorname{sign}(\cos_k(z)),$$

where $z = (z_k)$ and $\cos_k(z) = \Re z_k$. Set $\mathcal{D}_k$ to be the finite algebra of subsets in $\mathbb{T}^{\mathbb{N}}$ generated by $\sigma_1, \ldots, \sigma_k$. Let $\mathcal{D}$ be the sigma-algebra generated by $\bigcup \mathcal{D}_k$. Denoting $\mathbb{P}_{\mathcal{D}}$ to be the restriction of $\mathbb{P}$ to $\mathcal{D}$, we obtain the filtered probability space $(\mathbb{T}^{\mathbb{N}}, (\mathcal{D}_k), \mathbb{P}_{\mathcal{D}})$.

For $1 \leq p \leq \infty$, the space $L^p(\mathbb{T}^{\mathbb{N}}, \mathcal{D}, \mathbb{P}_{\mathcal{D}})$ is closed in $L^p(\mathbb{T}^{\mathbb{N}}, \mathcal{F}, \mathbb{P})$. The conditional expectation operator

$$\mathbb{E}(\cdot \,|\, \mathcal{D}) : L^p(\mathbb{T}^{\mathbb{N}}, \mathcal{F}, \mathbb{P}) \to L^p(\mathbb{T}^{\mathbb{N}}, \mathcal{D}, \mathbb{P}_{\mathcal{D}})$$

is a surjective, idempotent contraction for $1 \leq p \leq \infty$.

Example 1.1.3 (The dyadic filtration in $[0, 1[$) We say that $I \subseteq [0, 1[$ is a dyadic interval if there exists $n \in \mathbb{N} \cup \{0\}$ and $1 \leq i \leq 2^n$ such that $I = [(i-1)/2^n, i/2^n[$. We let $\mathcal{I}$ denote the collection of all dyadic intervals contained in the unit interval $[0,1[$. For $I \in \mathcal{I}$ there exist uniquely determined $I_1, I_2 \in \mathcal{I}$ such that $I_1 \cap I_2 = \emptyset$ and $I_1 \cup I_2 = I$. In that case $|I_1| = |I_2| = |I|/2$. If $\inf I_1 = \inf I$ we say that I_1 is the left half of I, and

$$h_I = 1_{I_1} - 1_{I_2}, \tag{1.1.2}$$

is called the L^∞-normalized Haar function, supported on I. Let $\mathcal{I}_k = \{I \in \mathcal{I} : |I| = 2^{-k}\}$ and let $\mathcal{G}_k = \sigma\{\mathcal{I}_k\}$ denote the sigma-algebra generated by $\mathcal{I}_k$. For $f \in L^1[0,1[$, we have

$$\mathbb{E}(f|\mathcal{G}_k) = \sum_{I \in \mathcal{I}_k} \int_I f \frac{dt}{|I|} 1_I, \tag{1.1.3}$$

and

$$\mathbb{E}(f|\mathcal{G}_{k+1}) - \mathbb{E}(f|\mathcal{G}_k) = \sum_{I \in \mathcal{I}_k} \int_I f h_I \frac{dt}{|I|} h_I. \tag{1.1.4}$$

With $([0,1[, \mathcal{L}, \lambda)$ denoting Lebesgue's measure space, a filtered probability space is formed by $([0,1[, (\mathcal{G}_k), \lambda)$.

Dyadic intervals $\mathcal{I}$ are canonically ordered as follows. If $I, J \in \mathcal{I}$ and $|I| \leq |J|$ then $I \prec J$; if $|I| = |J|$ and $\inf I < \inf J$ then $I \prec J$. The canonical order on $\mathcal{I}$ is often called the lexicographic order for obvious reasons.

Martingale Convergence

We fix again a finite measure space $(F, \mathcal{F}, \mu)$, and let $(\mathcal{G}_k)$ be a sequence of increasing sub-sigma-algebras of $\mathcal{F}$, satisfying $\mathcal{G}_0 = \{\emptyset, F\}$. We say that a sequence of μ-integrable functions (f_k) is a $(\mathcal{G}_k)$-martingale if

$$f_k = \mathbb{E}(f_{k+1}|\mathcal{G}_k), \quad k \in \mathbb{N}. \tag{1.1.5}$$

We write $\Delta f_k = f_k - f_{k-1}$ and say that (Δf_k) is the martingale difference sequence of (f_k). Should there exist $f \in L^1(F)$ such that $\mathbb{E}(f|\mathcal{G}_k) = f_k$, we say that the martingale (f_k) is closed by f.

The following theorems, due to Doob, assert that the martingale condition (1.1.5) imposes considerable constraints on the size and oscillations of an L^1-bounded sequence (f_k). In view of Theorem 1.1.1, a closed martingale converges in the norm of $L^1(F)$, and hence forms a uniformly integrable subset of $L^1(F)$. We recall that $\{f_k : k \in \mathbb{N}\}$ is (defined to be) uniformly integrable in $L^1(F)$, if

$$\lim_{t \to \infty} \sup_{k \in \mathbb{N}} \int_{\{|f_k| > t\}} |f_k| d\mu = 0.$$

The following theorem provides the converse implication, and thus sheds light on the relation between general martingales and closed ones.

Theorem 1.1.2 (Doob's martingale convergence theorem) *For any $(\mathcal{G}_k)$-martingale sequence (f_k), the following conditions are equivalent:*

(i) *There exists* $f \in L^1(F)$ *such that* $f_k = \mathbb{E}(f|\mathcal{G}_k)$ *for* $k \in \mathbb{N}$.
(ii) *The set* $\{f_k : k \in \mathbb{N}\}$ *is uniformly integrable in* $L^1(F)$.
(iii) *There exists* $f \in L^1(F)$ *such that* $\lim_{k\to\infty} \|f - f_k\|_1$.

We next address the question of almost sure convergence in the context of general submartingales. A sequence of μ-integrable functions (f_k) is a $(\mathcal{G}_k)$ submartingale if f_k is $\mathcal{G}_k$-measurable and

$$f_k \leq \mathbb{E}(f_{k+1}|\mathcal{G}_k), \quad k \in \mathbb{N}.$$

Theorem 1.1.3 *For any* $L^1(F)$*-bounded submartingale* (f_k)*, there exists* $E \subset F$ *with* $\mu(E) = 0$ *such that*

$$f(\omega) = \lim_{k\to\infty} f_k(\omega),$$

exists for $\omega \in F \backslash E$. *If* $\{f_k\}$ *is a uniformly integrable subset of* $L^1(F)$*, then*

$$\lim_{k\to\infty} \|f - f_k\|_1.$$

Summarizing Theorems 1.1.2 and 1.1.3, every L^1-bounded martingale converges almost surely; however its convergence in the L^1-norm requires uniform integrability. By contrast, for $p > 1$, every L^p-bounded martingale is norm-convergent in L^p. This assertion is a direct consequence of Doob's maximal inequalities stated in Theorem 1.1.4.

Theorem 1.1.4 *Let* (f_k) *be an* $L^1(F)$*-bounded* $(\mathcal{G}_k)$*-martingale. Then*

$$t\mu\left\{\max_{k\leq n} |f_k| > t\right\} \leq \int_F |f_n| d\mu, \tag{1.1.6}$$

where $t > 0$, $n \in \mathbb{N}$. *If* $p > 1$ *and* (f_k) *is an* $L^p(F)$*-bounded* $(\mathcal{G}_k)$*-martingale then*

$$\int_F \max_{k\leq n} |f_k|^p d\mu \leq \left(\frac{p}{p-1}\right)^p \int_F |f_n|^p d\mu, \tag{1.1.7}$$

for $n \in \mathbb{N}$. *Inequalities* (1.1.6) *and* (1.1.7) *hold true if* (f_n) *is a nonnegative submartingale.*

In Burkholder's classical L^p-inequality, maximal functions are replaced by martingale square functions.

Theorem 1.1.5 (Burkholder's theorem) *For* $1 < p < \infty$, *there exist* $c_p > 0$ *and* $C_p < \infty$ *such that for any* $L^p(F)$*-bounded* $(\mathcal{G}_k)$*-martingale* (f_k) *then*

$$c_p^p \int_F |f_n|^p d\mu \leq \int_F \left(|f_0|^2 + \sum_{k=1}^{n} |f_k - f_{k-1}|^2\right)^{p/2} d\mu \leq C_p^p \int_F |f_n|^p d\mu, \tag{1.1.8}$$

where $f_0 = \mathbb{E}(f_n|\mathcal{G}_0)$ *and* $n \in \mathbb{N}$.

As $p \to 1$, the constants in Theorem 1.1.5 satisfy $c_p \to 0$ and $C_p \to \infty$. The limiting case, $p = 1$, in Doob's maximal inequality (1.1.7) and Burkholder's square function inequality (1.1.8) is captured by the following result from Davis (1970).

Theorem 1.1.6 (Davis's theorem) *For any $L^1(F)$-bounded $(\mathcal{G}_k)$-martingale (f_k),*

$$c_1 \int_F \max_{k\leq n} |f_k| d\mu \leq \int_F S(f_n) d\mu \leq C_1 \int_F \max_{k\leq n} |f_k| d\mu,$$

where $f_0 = \mathbb{E}(f_n|\mathcal{G}_0)$, $n \in \mathbb{N}$, and

$$S(f_n) = \left(|f_0|^2 + \sum_{k=1}^{n} |f_k - f_{k-1}|^2 \right)^{1/2}, \tag{1.1.9}$$

the constants $c_1 > 0$ and $C_1 < \infty$ are independent of the martingale and the underlying filtration.

The basic tool invented by Davis (1970) in the proof of Theorem 1.1.6 is referred to as the Davis decomposition of martingales.

Theorem 1.1.7 *For any $(\mathcal{G}_k)$-martingale (f_k), there exist $(\mathcal{G}_k)$-martingale sequences (g_k) and (b_k) satisfying*

$$f_n = g_n + b_n, \quad n \in \mathbb{N}, \tag{1.1.10}$$

$$|\Delta g_n| \leq 8 \max_{k\leq n-1} |f_k|, \quad n \in \mathbb{N}, \tag{1.1.11}$$

and

$$\sum_{k=1}^{n} \|\Delta b_k\|_{L^1(\mu)} \leq 8\| \max_{k\leq n} |f_k|\|_{L^1(\mu)}, \quad n \in \mathbb{N}, \tag{1.1.12}$$

where $\Delta b_k = b_k - b_{k-1}$ and $\Delta g_k = g_k - g_{k-1}$.

For the specific constants in Theorem 1.1.7, we refer to Garsia (1973, Theorem III 3.5).

Martingale Spaces

Given a filtered measure space $(F, (\mathcal{G}_k), \mu)$, we define the martingale Hardy space $H^1(F, (\mathcal{G}_k), \mu)$ to consist of those $(\mathcal{G}_k)$-martingales $f = (f_k)$ for which

$$\|(f_k)\|_{H^1(F,(\mathcal{G}_k),\mu)} = \|S(f)\|_{L^1(F,\mu)} < \infty,$$

where

$$S(f) = \left(|f_0|^2 + \sum_{k=1}^{\infty} |f_k - f_{k-1}|^2 \right)^{1/2}.$$

The conditional square function of a $(\mathcal{G}_k)$-martingale $f = (f_k)$ is defined by

$$S_{\mathrm{cd}}(f) = \left(|f_0|^2 + \sum \mathbb{E}_{k-1}|f_k - f_{k-1}|^2\right)^{1/2}.$$

It gives rise to the space of predictable martingales $\mathcal{P}(F, (\mathcal{G}_k), \mu)$, which consists of those $(\mathcal{G}_k)$-martingales $f = (f_k)$ for which

$$\left\|(f_k)\right\|_{\mathcal{P}(F,(\mathcal{G}_k),\mu)} = \left\|S_{\mathrm{cd}}(f)\right\|_{L^1(F,\mu)} < \infty. \tag{1.1.13}$$

In Section 2.5 we will prove the Burkholder–Gundy inequality, which asserts that

$$\left\|(f_k)\right\|_{H^1(F,(\mathcal{G}_k),\mu)} \leq 2\left\|(f_k)\right\|_{\mathcal{P}(F,(\mathcal{G}_k),\mu)}. \tag{1.1.14}$$

Martingales in the Hardy space $H^1(F, (\mathcal{G}_k), \mu)$ and in $\mathcal{P}(F, (\mathcal{G}_k), \mu)$ converge almost surely and in L^1; they may thus be identified with their almost sure limits.

Theorem 1.1.8 *Each martingale $(f_k) \in H^1(F, (\mathcal{G}_k), \mu)$ is closed. That is, there exists $f \in L^1(F, \mu)$, such that*

$$\lim_{k\to\infty} \left\|f_k - f\right\|_{L^1} = 0,$$

and $f_k = \mathbb{E}(f|\mathcal{G}_k)$, for $k \in \mathbb{N}$. The same conclusion holds true for martingales $(f_k) \in \mathcal{P}(F, (\mathcal{G}_k), \mu)$.

Proof By Davis's theorem (Theorem 1.1.6), each martingale $(f_k) \in H^1(F, (\mathcal{G}_k), \mu)$ is a uniformly integrable subset of $L^1(F, \mu)$. Hence Doob's martingale convergence theorem (Theorem 1.1.2) shows that the sequence (f_k) converges in $L^1(F, \mu)$, and that there exists $f \in L^1(F, \mu)$, measurable with respect to $\mathcal{G} = \sigma(\bigcup \mathcal{G}_k)$, such that $f_k = \mathbb{E}(f|\mathcal{G}_k)$, for $k \in \mathbb{N}$.

By Inequality (1.1.14) any martingale in $\mathcal{P}(F, (\mathcal{G}_k), \mu)$ is contained in $H^1(F, (\mathcal{G}_k), \mu)$. Hence, the conclusion of the theorem holds for $\mathcal{P}(F, (\mathcal{G}_k), \mu)$. □

Operators of Weak Type (1:1)

The space $L^{1,\infty} = L^{1,\infty}(F, \mathcal{F}, \mu)$ consists of those $\mathcal{F}$-measurable $f\colon F \to \mathbb{C}$ for which

$$\|f\|_{1,\infty} = \sup_{t>0} t\mu\{|f| > t\} < \infty.$$

We say that a sublinear operator $T\colon L^1(F, \mathcal{F}, \mu) \to L^{1,\infty}(F, \mathcal{F}, \mu)$ is of weak type (1:1) if there exists $C < \infty$ such that

$$\|T(f)\|_{1,\infty} \leq C\|f\|_1, \quad f \in L^1. \tag{1.1.15}$$

We repeatedly use that a sublinear operator T is of weak type (1:1) if and only if for any $f \in L^1(F, \mathcal{F}, \mu)$ and $\varphi \in L^\infty(F, \mathcal{F}, \mu)$ with $\|\varphi\|_\infty \leq 1$, we have

$$\int_F |T(f)|^{1/2} |\varphi| d\mu \leq C_1 \left(\int_F |f| d\mu \cdot \int_F |\varphi| d\mu \right)^{1/2}, \qquad (1.1.16)$$

for some $C_1 < \infty$. We refer to Wojtaszczyk's book (1991, Lemma III.I.11) for the equivalence between Inequalities (1.1.15) and (1.1.16).

1.1.2 Brownian Martingales

We review basic results on Brownian martingales, stopping times, martingale convergence theorems, stochastic integrals, complex Brownian motion, and Itô's formula and some of its numerous consequences. These concepts are covered, for example, in the books by Revuz and Yor (1991), Bass (1995), and Durrett (1984).

Brownian Motion

Let $(\Omega, \mathcal{F}, \mathbb{P})$ be a fixed probability space. Let $\mathbb{R}_0^+ = \{t \in \mathbb{R} : t \geq 0\}$. A map $X \colon \Omega \times \mathbb{R}_0^+ \to \mathbb{R}$ defines uniquely maps

$$X_t \colon \Omega \to \mathbb{R}, \qquad \omega \to X(\omega, t),$$

for $t \geq 0$. We say that $X = (X_t : 0 \leq t < \infty)$ forms a real-valued stochastic process if each of the maps X_t is $\mathcal{F} - \mathcal{B}$ measurable where $\mathcal{B}$ denotes the Borel sigma-algebra on $\mathbb{R}$. Similarly we define complex-valued processes, and processes indexed by $\mathbb{N}_0$.

A real stochastic process $(x_t : 0 \leq t < \infty)$ on a probability space $(\Omega, \mathcal{F}, \mathbb{P})$ is called one-dimensional Brownian motion (or simply Brownian motion) if

(i) $x_0(\omega) = 0$ for almost every $\omega \in \Omega$.
(ii) For $t \in \mathbb{R}_0^+$ and $h > 0$, the increment $x_{t+h} - x_t$ is independent of the process up to time t, $(x_s : 0 < s \leq t)$, and Gaussian distributed with mean 0 and variance h; that is, for every measurable $A \subseteq \mathbb{R}$,

$$\mathbb{P}(\{x_{t+h} - x_t \in A\}) = \int_A e^{-x^2/2h} \frac{dx}{\sqrt{2\pi h}}.$$

(iii) The function $t \to x_t(\omega)$ is continuous, for almost every $\omega \in \Omega$.

We let $\mathcal{F}_t$ denote the completion, with respect to $(\Omega, \mathcal{F}, \mathbb{P})$, of the sigma-algebra generated by $(x_s : 0 < s \leq t)$. We have (see for instance Bass [1995, p. 17–18]) that

$$\mathcal{F}_t = \bigcap_{\epsilon > 0} \mathcal{F}_{t+\epsilon}, \qquad (1.1.17)$$

and for each $s > 0$, the process $(x_{s+t} - x_s : t > 0)$ is independent of $\mathcal{F}_s$. We say that a stochastic process

$$(F_t \colon \Omega \to \mathbb{C} : 0 \le t < \infty)$$

is $(\mathcal{F}_t)$-adapted if F_t is $\mathcal{F}_t$-measurable for $t \ge 0$. An adapted process $(F_t : 0 \le t < \infty)$ is an $(\mathcal{F}_t)$-martingale if $\mathbb{E}|F_t| < \infty$ for $t > 0$, and

$$F_s = \mathbb{E}(F_t | \mathcal{F}_s)$$

for $0 < s < t < \infty$. (It is a submartingale if $F_s \le \mathbb{E}(F_t | \mathcal{F}_s)$.)

The first examples of $(\mathcal{F}_t)$-martingales are Brownian motion (x_t) itself, and the process $(x_t^2 - t : t > 0)$. More generally, for any choice of $0 < s_0 < s_1$, the processes

$$u_t = x_{t \wedge s_1} - x_{t \wedge s_0} \quad \text{and} \quad u_t^2 - (t \wedge s_1 - t \wedge s_0), \qquad 0 < t < \infty, \tag{1.1.18}$$

form $(\mathcal{F}_t)$-martingales. Each $F \in L^1(\Omega)$ determines an $(\mathcal{F}_t)$-martingale by

$$F_t = \mathbb{E}(F | \mathcal{F}_t). \tag{1.1.19}$$

In this section we collect concepts of martingale theory, including stopping times and optional stopping, Doob's maximal inequalities, Itô integrals and Itô calculus. With these tools at our disposal, we will be able to conclude that for (1.1.19),

$$\lim_{t \to \infty} F_t(\omega) = F(\omega),$$

for almost every $\omega \in \Omega$, that

$$\lim_{t \to \infty} \mathbb{E}|F_t - F| = 0,$$

and that the maps $t \to F_t(\omega)$ (called trajectories) are continuous for almost every $\omega \in \Omega$.

Stopping Times

Let $\mathcal{F}_\infty$ denote the sigma-algebra generated by $\{\mathcal{F}_t : t > 0\}$. An $\mathcal{F}_\infty$-measurable map $T \colon \Omega \to \mathbb{R}_0^+ \cup \{\infty\}$ is called an $(\mathcal{F}_t)$-stopping time if

$$\{T \le t\} \in \mathcal{F}_t, \tag{1.1.20}$$

for any $t > 0$. The stopping time sigma-algebra generated by T is defined to be

$$\mathcal{F}_T = \{A \in \mathcal{F}_\infty : A \cap \{T \le t\} \in \mathcal{F}_t\}. \tag{1.1.21}$$

The stopping time T is $\mathcal{F}_T$-measurable, hence $\{T = \infty\} \in \mathcal{F}_T$. For *fixed* $t > 0$, we define $T \wedge t \colon \Omega \to [0, t]$ by

$$T(\omega) \wedge t = \min\{T(\omega), t\},$$

which is in turn a stopping time, and we let $\mathcal{F}_{T\wedge t}$ denote the stopping time sigma-algebra generated by $T \wedge t$. Fix an $(\mathcal{F}_s)$-adapted process $(X_s : s \geq 0)$. We define the sampled random variable (*singular*) $X_{T\wedge t}\colon \Omega \to \mathbb{R}$ by putting

$$X_{T\wedge t}(\omega) = X_{T(\omega)\wedge t}(\omega).$$

Note that $X_{T\wedge t}$ is an $\mathcal{F}_{T\wedge t}$-measurable random variable. Finally, stringing together the sampled random variables we obtain the so-called stopped process,

$$(X_{T\wedge t} : t > 0).$$

The following theorem, a basic result due to Doob, clarifies the relation between sampling, stopping, and taking conditional expectations.

Theorem 1.1.9 (Optional stopping theorem) *Let T be an $(\mathcal{F}_t)$-stopping time, and let $(X_t : t > 0)$ be a continuous $(\mathcal{F}_t)$-martingale. Then the stopped process $(X_{T\wedge t} : t > 0)$ forms an $(\mathcal{F}_t)$-martingale, that is,*

$$X_{T\wedge s} = \mathbb{E}(X_{T\wedge t}|\mathcal{F}_s), \qquad 0 \leq s \leq t < \infty.$$

If there exists $X \in L^1$ satisfying

$$\lim_{t\to\infty} \mathbb{E}|X - X_t| = 0, \tag{1.1.22}$$

then the sampled random variable

$$X_T(\omega) = \begin{cases} X(\omega) & \mathit{for} \quad \omega \in \{T = \infty\}; \\ X_{T(\omega)}(\omega) & \mathit{for} \quad \omega \in \{T < \infty\}, \end{cases}$$

satisfies

$$X_T = \mathbb{E}(X|\mathcal{F}_T). \tag{1.1.23}$$

Moreover,

$$X_{T\wedge t} = \mathbb{E}(X|\mathcal{F}_{T\wedge t}), \quad t > 0, \tag{1.1.24}$$

and

$$\lim_{t\to\infty} \mathbb{E}|X_{T\wedge t} - \mathbb{E}(X|\mathcal{F}_T)| = 0. \tag{1.1.25}$$

The assertions of Theorem 1.1.9 are usually presented as two separate theorems, referred to as Doob's optional stopping theorem, and Doob's optional sampling theorem, respectively. Stretching the established conventions to some degree, we refer to Theorem 1.1.9 as the "optional stopping theorem."

Martingale Convergence (Continued)

With Theorem 1.1.9 in mind, we now isolate an intrinsic criterion for the L^1-convergence of an L^1-bounded submartingale, and recall the notion of uniform integrability. A subset V of $L^1(\Omega)$ is called uniformly integrable (or equi-integrable) if

$$\lim_{c\to\infty} \sup_{v\in V} \int_{\{|v|>c\}} |v|\, d\mathbb{P} = 0.$$

Any L^1-convergent sequence is uniformly integrable, and if a sequence (x_n) converges in measure, then it is L^1-convergent if and only if $\{x_n : n \in \mathbb{N}\}$ is a uniformly integrable subset of L^1. The following submartingale convergence theorem collects basic properties of L^1-bounded martingales and submartingales.

Theorem 1.1.10 (Submartingale convergence theorem) *If $(X_t : t > 0)$ is an L^1-bounded, continuous $(\mathcal{F}_t)$-submartingale, then*

$$X = \lim_{t\to\infty} X_t$$

exists $\mathbb{P}$-almost everywhere on Ω. The uniform integrability of $\{X_t : t > 0\}$ in $L^1(\Omega, \mathbb{P})$ is equivalent to

$$\lim_{t\to\infty} \mathbb{E}|X - X_t| = 0.$$

In the special case where $(X_t : t > 0)$ is an L^1-bounded $(\mathcal{F}_t)$-martingale (and not merely a submartingale) then uniform integrability of $\{X_t : t > 0\}$ is equivalent to

$$X_t = \mathbb{E}(X|\mathcal{F}_t)$$

for $t > 0$.

Paraphrasing the submartingale convergence theorem (Theorem 1.1.10), any continuous, L^1-bounded submartingale is convergent almost everywhere and its convergence in L^1 is equivalent to uniform integrability.

We now turn to Doob's maximal inequalities. They allow us to conclude that for $p > 1$, any L^p-bounded martingale is L^p-convergent – this being the main distinction between martingales bounded in L^1 and L^p $(p > 1)$.

Theorem 1.1.11 (Doob's theorem) *Let $(X_t : t > 0)$ be a continuous $(\mathcal{F}_t)$-martingale. Then for any $\epsilon > 0$, and $1 \le p < \infty$,*

$$\mathbb{P}\left\{\sup_{t>0} |X_t| > \epsilon\right\} \le \epsilon^{-p} \sup_{t>0} \mathbb{E}|X_t|^p. \tag{1.1.26}$$

If $1 < p < \infty$, *then*

$$\mathbb{E}\left(\sup_{t>0} |X_t|^p\right) \leq C_p^p \sup_{t>0} \mathbb{E}|X_t|^p, \tag{1.1.27}$$

where $C_p = p/(p-1)$. *The estimates* (1.1.26) *and* (1.1.27) *hold true if* $(X_t : t > 0)$ *is a right continuous nonnegative submartingale.*

Doob's inequality combined with the submartingale convergence theorem imply that for $p > 1$, an L^p-bounded, continuous $(\mathcal{F}_t)$-martingale $(X_t : t > 0)$ is L^p-convergent. Indeed, $X = \lim_{t\to\infty} X_t$ exists $\mathbb{P}$-almost everywhere on Ω, by the first part of Theorem 1.1.10. Since $|X| \leq \sup_{t>0} |X_t|$, Doob's inequality (1.1.27) gives $X \in L^p$ and, moreover,

$$\mathbb{E}|X|^p + \mathbb{E} \sup_{t>0} |X_t|^p \leq 2C_p^p \sup_{t>0} \mathbb{E}|X_t|^p < \infty.$$

Hence, $(|X - X_t|^p)$ is an L^1-bounded, uniformly integrable submartingale satisfying $\lim_{t\to\infty} |X - X_t|^p = 0$ almost surely. Thus,

$$\lim_{t\to\infty} \mathbb{E}|X - X_t|^p = 0,$$

as claimed. As $p > 1$ we clearly have $\lim_{t\to\infty} \mathbb{E}|X - X_t| = 0$. Moreover, if $(X_t : t > 0)$ is a martingale, then Theorem 1.1.10 implies that $X_t = \mathbb{E}(X|\mathcal{F}_t)$ for any $t > 0$.

Stochastic Integrals

We review the notion of Itô's stochastic integral and state its basic properties. We proceed in two basic steps starting with elementary step functions. The construction exploits the estimates of Doob's martingale inequality (1.1.26).

Step 1: Let $n \in \mathbb{N}$, fix $0 < s_0 < s_1 < \cdots < s_n$ and square-integrable, $\mathcal{F}_{s_i}$-measurable, real-valued random variables $(Y_i : 0 \leq i \leq n-1)$. For $s > 0$ we define the stochastic process $(X_s : s > 0)$ by

$$X_s = \sum_{i=1}^{n} Y_{i-1} 1_{[s_{i-1}, s_i[}(s),$$

and reserve the term "elementary step functions" for any such process. For $t > 0$, define

$$\int_0^t X_s dx_s = \sum_{i=1}^{n} Y_{i-1}(x_{t\wedge s_i} - x_{t\wedge s_{i-1}}). \tag{1.1.28}$$

We call (1.1.28) the stochastic integral of $(X_s : s > 0)$ with respect to Brownian motion $(x_s : s > 0)$. By the continuity of Brownian motion, the maps $t \to F_t(\omega)$ are continuous for almost every $\omega \in \Omega$. By (1.1.18), the processes

$$F_t = \int_0^t X_s dx_s \quad \text{and} \quad F_t^2 - \int_0^t X_s^2 ds, \qquad 0 < t < \infty, \tag{1.1.29}$$

form $(\mathcal{F}_t)$-martingales, and hence

$$\lim_{t\to\infty} \mathbb{E} F_t^2 = \mathbb{E} \int_0^\infty X_s^2 ds. \tag{1.1.30}$$

Step 2: Next, we extend the definition to $(\mathcal{F}_s)$-measurable integrands $(X_s : s > 0)$ satisfying

$$\mathbb{E} \int_0^\infty X_s^2 ds < \infty. \tag{1.1.31}$$

If Inequality (1.1.31) holds true, there exists a sequence of real-valued elementary step functions $\left(X_s^{(n)} : s > 0\right)$ such that

$$\lim_{n\to\infty} \mathbb{E} \int_0^\infty \left(X_s - X_s^{(n)}\right)^2 ds = 0. \tag{1.1.32}$$

Using the construction of Step 1, the real-valued elementary step functions $\left(X_s^{(n)} : s > 0\right)$ define the stochastic integrals

$$F_t^{(n)} = \int_0^t X_s^{(n)} dx_s,$$

for any $t > 0$. Recall that $\left(F_t^{(n)} : t > 0\right)$ forms a martingale for each fixed $n \in \mathbb{N}$, and that the maps $t \to F_t^{(n)}(\omega)$ are continuous for $n \in \mathbb{N}$ and almost every $\omega \in \Omega$. Fix $m, n \in \mathbb{N}$. The maximal estimate (1.1.26) applied with $p = 2$ to $\left(F_t^{(n)} - F_t^{(m)} : t > 0\right)$ gives

$$\mathbb{P}\left\{\sup_{t>0} \left|F_t^{(n)} - F_t^{(m)}\right| > \epsilon\right\} \le \epsilon^{-2} \sup_{t>0} \mathbb{E}\left|F_t^{(n)} - F_t^{(m)}\right|^2, \qquad n, m \in \mathbb{N}. \tag{1.1.33}$$

Fix $t > 0$. By (1.1.30) and (1.1.32), the sequence $\left(F_t^{(n)} : n \in \mathbb{N}\right)$ satisfies Cauchy's condition in L^2 uniformly for $t > 0$. Hence, there exists $F_t \in L^2$, such that

$$\lim_{n\to\infty} \mathbb{E}\left|F_t - F_t^{(n)}\right|^2 = 0, \quad \text{and} \quad \mathbb{E} F_t^2 = \mathbb{E} \int_0^t X_s^2 ds. \tag{1.1.34}$$

We define $\int_0^t X_s dx_s := F_t$. By Inequality (1.1.33), the maps $t \to F_t(\omega)$ are continuous for almost every $\omega \in \Omega$. By (1.1.31) and (1.1.34), the process $(F_t : t > 0)$ forms an L^2-bounded $(\mathcal{F}_t)$-martingale. By Doob's theorem (Theorem 1.1.11) there exists $F \in L^2(\Omega, \mathbb{P})$ such that

$$\lim_{t\to\infty} \mathbb{E}|F - F_t|^2 = 0.$$

Finally, we define F to be the stochastic integral of $(X_s : s > 0)$ with respect to Brownian motion, and denote

$$\int_0^\infty X_s dx_s := F. \tag{1.1.35}$$

Remark: The processes $F_t = \int_0^t X_s dx_s$ and $F_t^2 - \int_0^t X_s^2 ds$ form L^1-convergent $(\mathcal{F}_t)$-martingales. Let T be an $(\mathcal{F}_t)$-stopping time and let $\mathcal{F}_T$ denote the associated stopping time sigma-algebra. Then

$$(1_{\{s\leq T\}} X_s : s > 0)$$

is an $(\mathcal{F}_s)$-adapted process, and the optional stopping theorem (Theorem 1.1.9) asserts that

$$\mathbb{E}(F|\mathcal{F}_T) = F_T = \int_0^T X_s dx_s = \int_0^\infty 1_{\{s\leq T\}} X_s dx_s, \tag{1.1.36}$$

and that

$$F_{t\wedge T} = \int_0^{t\wedge T} X_s dx_s \quad \text{and} \quad F_{t\wedge T}^2 - \int_0^{t\wedge T} X_s^2 ds, \qquad t \geq 0, \tag{1.1.37}$$

form $(\mathcal{F}_t)$-martingales satisfying $\lim_{t\to\infty} \mathbb{E}F_{t\wedge T}^2 = \mathbb{E}F_T^2$.

Complex Brownian Motion

Let $(x_t : t > 0)$ and $(y_t : t > 0)$ be two independent copies of (one-dimensional) Brownian motion started at $0 \in \mathbb{R}$, on a probability space $(\Omega, \mathcal{F}, \mathbb{P})$. We form complex Brownian motions $(z_t : t > 0)$ and $(\overline{z}_t : t > 0)$ by putting

$$z_t = x_t + iy_t, \quad \textit{and} \quad \overline{z}_t = x_t - iy_t, \qquad t \geq 0.$$

Define $\mathcal{F}_t$ to be the completion (with respect to $\mathbb{P}$) of the sigma-algebra generated by $(z_s : 0 \leq s \leq t)$. Since $x_t^2 - t$, $y_t^2 - t$, and $x_t y_t$ form $(\mathcal{F}_t)$-martingales we infer that

$$z_t\overline{z}_t - 2t, \quad z_t z_t, \quad \textit{and} \quad \overline{z}_t\overline{z}_t, \qquad t \geq 0,$$

are $(\mathcal{F}_t)$-martingales as well. We call $(\Omega, (\mathcal{F}_t), \mathbb{P})$ Wiener's filtered probability space. Next we review Itô stochastic integrals and Itô's representation theorems for complex Brownian motion. Let $(Y_t : t > 0)$ be a complex-valued, square-integrable, and $(\mathcal{F}_t)$-adapted process satisfying

$$\mathbb{E}\int_0^\infty |Y_s|^2 ds < \infty.$$

Put $u_t = \Re Y_t$ and $v_t = \Im Y_t$ and set

$$\int_0^\infty Y_s dz_s = \int_0^\infty u_s dx_s - v_s dy_s + i\int_0^\infty u_s dy_s + v_s dx_s,$$

and

$$\int_0^\infty Y_s d\bar{z}_s = \int_0^\infty u_s dx_s + v_s dy_s + i \int_0^\infty v_s dx_s - u_s dy_s,$$

where the real stochastic integrals on the right-hand side are taken with respect to the independent copies of (one-dimensional) Brownian motion $(x_t : t > 0)$ and $(y_t : t > 0)$. We begin by recording the integration by parts formula for stochastic integrals with respect to complex Brownian motion.

Theorem 1.1.12 *Let (F_t) and (G_t) be square-integrable complex-valued martingales with Itô integral representation*

$$F_t = F_0 + \int_0^t X_s dz_s + Y_s d\bar{z}_s, \qquad G_t = G_0 + \int_0^t U_s dz_s + V_s d\bar{z}_s, \tag{1.1.38}$$

where $F_0, G_0 \in \mathbb{C}$. Then,

$$F_t G_t - F_0 G_0 = \int_0^t (F_s U_s + G_s X_s) dz_s + (F_s V_s + G_s Y_s) d\bar{z}_s + \langle F, G \rangle_t, \tag{1.1.39}$$

where

$$\langle F, G \rangle_t = 2 \int_0^t (U_s Y_s + V_s X_s) ds, \tag{1.1.40}$$

is of bounded variation, called the covariance process of F and G.

Remarks: The covariance formula for stochastic integrals and the integral representation of the quadratic variation process are a consequence, respectively a special case, of Theorem 1.1.12.

(i) Taking expectations in (1.1.39) and (1.1.40) gives the covariance formulas,

$$\mathbb{E}(F_t G_t) - F_0 G_0 = 2\mathbb{E} \int_0^t (U_s Y_s + V_s X_s) ds, \qquad t > 0. \tag{1.1.41}$$

(ii) Let $(C_s : s > 0)$ be an $(\mathcal{F}_s)$-adapted process, and assume that $V_t = \int_0^t C_s ds$ is of bounded variation over each finite interval. Let (F_t) be given by (1.1.38). Then integration by parts assumes the form,

$$V_t F_t = \int_0^t V_s (X_s dz_s + Y_s d\bar{z}_s) + \int_0^t C_s F_s ds, \quad t > 0, \tag{1.1.42}$$

and yields the covariance formula $\mathbb{E}(V_t F_t) = \mathbb{E}\left(\int_0^t C_s F_s ds\right)$.

(iii) Let $F \in L^2(\Omega)$ be complex valued. We say that $\langle F\rangle_t = \langle F, \overline{F}\rangle_t$, is the quadratic variation process of F, and let

$$\langle F\rangle = \lim_{t\to\infty}\langle F\rangle_t. \tag{1.1.43}$$

As $(t \to \langle F\rangle_t)$ is increasing, the (finite or infinite) limit (1.1.43) is well defined. We say that $\langle F\rangle$ is the quadratic variation of F.

With respect to complex Brownian motion, Itô's representation theorem reads as follows:

Theorem 1.1.13 (Itô's theorem) *Let $F \in L^2(\Omega)$ be complex-valued. There exist complex-valued $(\mathcal{F}_t)$-adapted processes $(X_t : t > 0)$ and $(Y_t : t > 0)$ satisfying*

$$\mathbb{E}\int_0^\infty |X_s|^2 + |Y_s|^2 ds < \infty,$$

such that

$$F - \mathbb{E}F = \int_0^\infty X_s dz_s + Y_s d\overline{z}_s.$$

The processes $(X_t : t > 0)$ and $(Y_t : t > 0)$ are almost surely uniquely determined. Moreover,

$$\mathbb{E}(F|\mathcal{F}_t) - \mathbb{E}F = \int_0^t X_s dz_s + Y_s d\overline{z}_s,$$

for $t \geq 0$, and the trajectories $t \to \mathbb{E}(F|\mathcal{F}_t)(\omega)$ are continuous for almost every $\omega \in \Omega$.

Remarks: Since $L^2(\Omega)$ is dense in $L^1(\Omega)$, Itô's theorem extends routinely to $L^1(\Omega)$. The following two observations contain the information needed for the extension.

(i) Let $F \in L^1(\Omega)$ and put $F_t = \mathbb{E}(F|\mathcal{F}_t)$. We claim that Itô's theorem (Theorem 1.1.13) implies that the trajectory $t \to F_t(\omega)$ is continuous for almost every $\omega \in \Omega$. Indeed, since L^2 is dense in L^1, we may choose a sequence $F^{(n)} \in L^2$ such that $\mathbb{E}|F - F^{(n)}| \longrightarrow 0$ as $n \to \infty$. By Itô's theorem, $t \to F_t^{(n)}(\omega)$ is continuous for $n \in \mathbb{N}$ and almost all $\omega \in \Omega$. Invoking Doob's maximal inequality (1.1.26) gives

$$\mathbb{P}\left(\left\{\sup_{t>0}\left|F_t^{(m)} - F_t^{(n)}\right| > \epsilon\right\}\right) \leq \epsilon^{-1}\mathbb{E}\left|F^{(m)} - F^{(n)}\right|, \qquad m, n \in \mathbb{N}. \tag{1.1.44}$$

Hence, in view of the Borel–Cantelli Lemma, there exist $n(k) \uparrow \infty$ such that the maps $t \to F_t^{(n(k))}(\omega)$ converge uniformly (in t), for almost all $\omega \in \Omega$. On the other hand, $F_t = \lim_{n\to\infty} F_t^{(n)}$ with convergence in L^1. Consequently, $t \to F_t(\omega)$ is continuous for almost all $\omega \in \Omega$.

(ii) Let $F \in L^1(\Omega)$ be real valued. Then $F_t = \mathbb{E}(F|\mathcal{F}_t)$ is continuous almost surely. Fix $n \in \mathbb{N}$ and define the stopping times $T(n) = \inf\{t > 0 : |F_t| > n\}$. Clearly, we have

$$\lim_{n\to\infty} \|F - F_{T(n)}\|_{L^1(\Omega)} = 0. \tag{1.1.45}$$

By the existence and uniqueness assertions of Theorem 1.1.13, there exist $(\mathcal{F}_t)$-adapted processes $(X_s : s > 0)$ and $(Y_s : s > 0)$ satisfying

$$F_{T(n)} = F_0 + \int_0^{T(n)} X_s dz_s + Y_s d\overline{z}_s.$$

Taking into account that $t \to F_t(\omega)$ is continuous, the definition of the stopping time $T(n)$ gives $\mathbb{E}\int_0^{T(n)} |X_s|^2 + |Y_s|^2 ds \le n^2$. By Equation (1.1.45), we have

$$F - F_0 = \lim_{n\to\infty} \int_0^{T(n)} X_s dz_s + Y_s d\overline{z}_s \quad \left(= \int_0^{\infty} X_s dz_s + Y_s d\overline{z}_s\right), \tag{1.1.46}$$

with convergence in L^1. We call Equation (1.1.46) the stochastic integral representation of F. Similarly, the quadratic variation process of F is defined by

$$\langle F\rangle_t = \lim_{n\to\infty} \langle F_{t\wedge T(n)}\rangle \quad \text{and} \quad \langle F\rangle = \lim_{t\to\infty} \langle F\rangle_t. \tag{1.1.47}$$

Itô's Formula

Let (F_t) be an $(\mathcal{F}_t)$-martingale and let $f: \mathbb{C} \to \mathbb{C}$ be twice continuously differentiable. Itô's formula displays the process $(f(F_t) : t > 0)$ as an $(\mathcal{F}_t)$-martingale plus an $(\mathcal{F}_t)$-adapted process of bounded variation. We begin by recalling the complex first-order partial differential operators,

$$\partial_z = \frac{1}{2}(\partial_x - i\partial_y) \quad \text{and} \quad \overline{\partial_z} = \frac{1}{2}(\partial_x + i\partial_y).$$

We assume that there exist complex-valued adapted processes (X_t) and (Y_t) such that (F_t) is represented by stochastic integrals,

$$F_t = F_0 + \int_0^t X_s dz_s + Y_s d\overline{z}_s, \qquad t > 0. \tag{1.1.48}$$

We write the content of Equation (1.1.48) as

$$dF_s = X_s dz_s + Y_s d\overline{z}_s \quad \text{and} \quad d\overline{F_s} = \overline{Y_s} dz_s + \overline{X_s} d\overline{z}_s.$$

Theorem 1.1.14 *Let $f: \mathbb{C} \to \mathbb{C}$ be twice continuously differentiable. Let (F_t) be given by the Itô integrals* (1.1.48). *Then,*

$$f(F_t) = f(F_0) + \int_0^t \partial_z f(F_s) dF_s + \overline{\partial_z} f(F_s) d\overline{F_s} + \frac{1}{2}\int_0^t \partial_z\partial_z f(F_s) d\langle F, F\rangle_s$$
$$+ \int_0^t \partial_z\overline{\partial_z} f(F_s) d\langle F, \overline{F}\rangle_s + \frac{1}{2}\int_0^t \overline{\partial_z\partial_z} f(F_s) d\langle \overline{F}, \overline{F}\rangle_s.$$

We recall that by equation (1.1.40), the covariance processes arising in Itô's formula, are given explicitly as follows:

$$d\langle F,F\rangle_s = 4X_sY_s ds,\quad d\langle \overline{F},\overline{F}\rangle_s = 4\overline{X_s}\cdot\overline{Y_s}ds,\quad d\langle F,\overline{F}\rangle_s = 2\left(X_s\overline{X_s}+Y_s\overline{Y_s}\right)ds.$$

We apply Itô's formula to special cases of the stochastic integrals (1.1.48). The following list makes explicit the resulting simplifications in Theorem 1.1.14:

(i) If $Y_s = 0$ for $s > 0$, then $d\langle F,\overline{F}\rangle_s = 2|X_s|^2 ds$ and $\langle F,F\rangle_s = \langle \overline{F},\overline{F}\rangle_s = 0$. Hence,

$$f(F_t) = f(F_0) + \int_0^t \partial_z f(F_s)dF_s + \overline{\partial}_z f(F_s)d\overline{F_s} + 2\int_0^t \partial_z\overline{\partial}_z f(F_s)|X_s|^2 ds.$$

(ii) If $Y_s = 0$ and $X_s = 1$ for $s > 0$, then F_s is just complex Brownian motion z_s. Since $\langle z,\overline{z}\rangle_s = 2s$ and $\langle z,z\rangle_s = \langle \overline{z},\overline{z}\rangle_s = 0$, we have

$$f(z_t) = f(0) + \int_0^t \partial_z f(z_s)dz_s + \overline{\partial}_z f(z_s)d\overline{z}_s + 2\int_0^t \left(\overline{\partial}_z\partial_z f\right)(z_s)ds. \quad (1.1.49)$$

(iii) If $Y_s = 0$ for $s > 0$ and f is analytic, then $\langle F,F\rangle_s = \langle \overline{F},\overline{F}\rangle_s = 0$ and $\overline{\partial}_z f(F_s) = 0$. Hence,

$$f(F_t) = f(F_0) + \int_0^t \partial_z f(F_s)dF_s. \quad (1.1.50)$$

(iv) If $Y_s = \overline{X_s}$ for $s > 0$, then by (1.1.48) F_s is real valued, and we have $d\langle F,F\rangle_s = d\langle \overline{F},\overline{F}\rangle_s = d\langle F,\overline{F}\rangle_s = |2X_s|^2 ds$. Moreover, since F_s is real valued we have $dF_s = d\overline{F_s}$ and hence,

$$\partial_z f(F_s)dF_s + \overline{\partial}_z f(F_s)d\overline{F_s} = (\partial_x f)(F_s)dF_s.$$

Taking into account that $\partial_z\partial_z + 2\partial_z\overline{\partial}_z + \overline{\partial_z\partial_z} = \partial_{x,x}$, Itô's formula gives,

$$f(F_t) = f(F_0) + \int_0^t \partial_x f(F_s)\left(X_s dz_s + \overline{X_s}d\overline{z}_s\right) + \frac{1}{2}\int_0^t \partial_{x,x} f(F_s)|2X_s|^2 ds.$$

(v) If X_s and Y_s are real valued, and $X_s = Y_s$ for $s > 0$, Theorem 1.1.14 gives Itô's formula for one-dimensional Brownian motion. Indeed, under this assumption, (1.1.48) reduces to

$$F_t = F_0 + \int_0^t (2X_s)dx_s, \quad (1.1.51)$$

and $\left(X_s dz_s + \overline{X_s} d\overline{z}_s\right) = (2X_s)dx_s$, where x_s denotes one-dimensional Brownian motion. Thus we obtain Itô's formula for real-valued stochastic integrals (1.1.51) driven by one-dimensional Brownian motion,

$$f(F_t) = f(F_0) + \int_0^t \partial_x f(F_s)(2X_s)dx_s + \frac{1}{2}\int_0^t \partial_{x,x} f(F_s)|2X_s|^2 ds.$$

Finally, we recall Itô's formula applied to real-valued processes "driven by complex Brownian motion."

Theorem 1.1.15 *Let $R = (R_t)$ be real valued, given by the stochastic integrals*

$$R_t - R_0 = \int_0^t Y_s dz_s + \overline{Y_s} d\overline{z}_s + 2\int_0^t C_s ds, \tag{1.1.52}$$

where (Y_s) is adapted, and (C_s) is real valued, adapted such that $t \to \int_0^t C_s ds$ is of bounded variation. Let $g\colon \mathbb{R} \to \mathbb{R}$ be twice continuously differentiable and assume that $g(R_t)$ is equi-integrable. Then,

$$\begin{aligned} g(R_t) - g(R_0) &= \int_0^t g'(R_s)\left(Y_s dz_s + \overline{Y_s} d\overline{z}_s\right) \\ &\quad + 2\int_0^t \left(g'(R_s)C_s + g''(R_s)|Y_s|^2\right) ds. \end{aligned} \tag{1.1.53}$$

The first integral on the right-hand side of (1.1.53) *is a Brownian martingale and the second integral is of bounded variation (over each finite interval).*

Remark: We call the term $2\int_0^t C_s ds$ the drift term of the process $(R_t : t > 0)$ defined by (1.1.52). Similarily, $2\int_0^t \left(g'(R_s)C_s + g''(R_s)|Y_s|^2\right) ds$ is the drift term of $(g(R_t) : t > 0)$ in (1.1.53).

Paul Lévy's Theorem

Lévy's theorem asserts that complex Brownian motion $(z_t\colon t > 0)$ is conformally invariant. Specifically – with probability one – the composition of the process $(z_t(\omega)\colon t > 0)$ with an entire function is again a complex Brownian motion path; more precisely, it is a time change of Brownian motion.

Let (z_t) be complex Brownian motion in $\mathbb{C}$, and put $\tau = \inf\{t > 0 : |z_t| > 1\}$. Let $\varphi\colon \mathbb{D} \to \mathbb{C}$ be analytic so that $\varphi(0) = 0$. Define

$$s(t) = \int_0^t |\varphi'(z_r)|^2 dr, \qquad t < \tau.$$

Note that $s(t)$ is increasing and differentiable with $ds(t) = |\varphi'(z_t)|^2 dt$. Put $\sigma = s(\tau)$ and let $\alpha(s)$ be the inverse of $s(t)$. Thus $\alpha(s) = t$ if and only if $s = s(t)$ and $s < \sigma$.

Theorem 1.1.16 (Lévy's theorem) *The process* $(w_s = \varphi(z_{\alpha(s)}): s < \sigma)$ *is Brownian motion, confined to* $\mathcal{L} = \varphi(\mathbb{D})$*, and stopped at* σ*.*

The book by Bass (1995, Chapter V) is our basic reference for the conformal invariance of complex Brownian motion and Lévy's theorem.

1.1.3 Harmonic and Subharmonic Functions

Let U be an open subset of $\mathbb{C}$. A function $u\colon U \to \mathbb{R}$ is harmonic if it has continuous partial derivatives of order ≤ 2, and $\Delta u(z) = 0$, where $\Delta = \partial_{x,x} + \partial_{y,y}$. Harmonic functions satisfy the mean value property and are in fact characterized by it. Let $u\colon U \to \mathbb{R}$ be a continuous function on an open subset $U \subseteq \mathbb{C}$. Then u is harmonic if and only if for any $z \in U$ there exists $r_0 = r_0(z) > 0$ such that,

$$u(z) = \int_{\mathbb{T}} u(z + r\zeta)dm(\zeta), \quad \text{for} \quad r < r(z).$$

The characterization of harmonic functions in terms of the mean value property gives rise to a convenient definition of vector-valued harmonic functions.

Poisson Kernels and Kakutani's Theorem

Let $\mathbb{D} = \{z \in \mathbb{C}: |z < 1\}$ and $\mathbb{T} = \{\zeta \in \mathbb{C}: |\zeta| = 1\}$. We let m denote the unique rotation invariant probability measure on $\mathbb{T}$. Let $(z_t)_{t>0}$ denote the complex Brownian motion on Wiener space $(\Omega, \mathbb{P})$, normalized by $\langle z_t, \overline{z_t}\rangle = 2t$, $\langle z_t, z_t\rangle = 0$, and started at $z_0 = 0$. Set

$$\tau(\omega) = \inf\left\{t > 0 : z_t(\omega) \in \mathbb{C}\backslash\overline{\mathbb{D}}\right\}.$$

Kakutani's theorem asserts that stopped complex Brownian motion $z_\tau : \Omega \to \mathbb{T}$ is uniformly distributed over $\mathbb{T}$, that is

$$\mathbb{P}(\{z_\tau \in I\}) = m(I), \tag{1.1.54}$$

for every measurable $I \subseteq \mathbb{T}$. Given $z \in \mathbb{D}$, let

$$P(z, \zeta) = (1 - |z|^2)/(|z - \zeta|^2), \quad \zeta \in \mathbb{T}, \tag{1.1.55}$$

and

$$\tau_z(\omega) = \inf\left\{t > 0 : z + z_t(\omega) \in \mathbb{C}\backslash\overline{\mathbb{D}}\right\}.$$

Kakutani's theorem combined with Lévy's theorem (Brownian motion is conformally invariant) asserts that for any measurable $I \subseteq \mathbb{T}$,

$$\mathbb{P}\left(\{\omega \in \Omega : z_{\tau_z(\omega)}(\omega) \in I\}\right) = \int_I P(z, \zeta)dm(\zeta), \tag{1.1.56}$$

where dm denotes the normalized Haar measure on $\mathbb{T} = \{z \in \mathbb{C} : |z| = 1\}$. Subsequently, $P(z, \zeta)$ is called the Poisson kernel for $\mathbb{D}$.

Green's Function

Let $U \subset \mathbb{C}$ be bounded, open, and connected, and let $w \in U$. We assume that U has a C^2-boundary, denoted ∂U. We let $G_U(\cdot, w)\colon U \to \mathbb{R} \cup \{\infty\}$ denote the Green's function of U with pole at $w \in U$, normalized and defined uniquely by the following two conditions:

(i) $G_U(z, w) - \frac{1}{\pi} \ln \frac{1}{|z-w|}$ is harmonic in U.
(ii) $\lim_{z \to \zeta} G_U(z, w) = 0$ for $\zeta \in \partial U$.

For instance, $(1/\pi) \ln(1/|z|)$ is the Green's function of $\mathbb{D}$ with pole at $0 \in \mathbb{D}$.

The identity (1.1.57) gives the connection between complex Brownian motion, started at $w \in U$, and Green's function of U with pole at $w \in U$. Let $dA(z)$ denote the area measure in $\mathbb{C}$, and let $\tau_{U,w} = \inf\left\{t > 0 : w + z_t \in \mathbb{C} \backslash \overline{U}\right\}$. Then

$$\mathbb{E} \int_0^{\tau_{U,w}} f(w + z_t)dt = \iint_U f(z) G_U(z, w) dA(z), \tag{1.1.57}$$

for any bounded $f\colon U \to \mathbb{R}$. Equation (1.1.57) is called the "occupation time formula" for Brownian motion. See Bass (1995) or Durrett (1984).

Radial Limits of Harmonic Functions

In this section we review the harmonic extension and its reverse, the boundary behavior of harmonic functions.

Let $\mathbb{D} = \{z \in \mathbb{C}\colon |z| < 1\}$. Let $0 \leq r < 1$, $\zeta \in \mathbb{T}$, and define

$$p_r(\zeta) = \sum_{n=-\infty}^{\infty} r^{|n|} \zeta^n.$$

For $w \in \mathbb{T}$ and $z = rw$, the Fourier expansion of the Poisson kernel $\zeta \to P(z, \zeta)$ defined in (1.1.55) yields

$$P(z, \zeta) = p_r(w \cdot \overline{\zeta}), \qquad \zeta \in \mathbb{T}. \tag{1.1.58}$$

Given $f \in L^p(\mathbb{T})$ and $1 \leq p \leq \infty$, we define

$$P_r f(w) = \int_{\mathbb{T}} p_r(w \cdot \overline{\zeta}) f(\zeta) dm(\zeta), \tag{1.1.59}$$

and put $u(rw) = P_r f(w)$. Thus defined, $u\colon \mathbb{D} \to \mathbb{R}$ is a harmonic function, called the harmonic extension of $f\colon \mathbb{T} \to \mathbb{C}$ to the unit disk. Minkowski's inequality, i.e., the triangle inequality in the space L_p, yields

$$\|P_r f\|_p \leq \|f\|_p,$$

and if $f \geq 0$ then $P_r f \geq 0$. Theorem 1.1.17 asserts that radial limits of the harmonic extension exist almost surely, and give back f.

Theorem 1.1.17 *Let $1 \leq p < \infty$ and let $u\colon \mathbb{D} \to \mathbb{R}$ be harmonic. The following conditions are equivalent:*

(i) *There exists $f \in L^p(\mathbb{T})$ such that $u(rw) = P_r(f)(w)$ for $0 < r < 1$ and $w \in \mathbb{T}$.*

(ii) *There exists $f \in L^p(\mathbb{T})$ such that $\lim_{r\to 1} \int_{\mathbb{T}} |u(rw) - f(w)|^p dm(w) = 0$.*

Fatou's theorem gives the existence of radial limits, almost everywhere for L^1-bounded harmonic functions in the unit disk. The precise meaning of L^1-boundedness is specified by Inequality (1.1.60) below.

Theorem 1.1.18 (Fatou's theorem) *Let $u\colon \mathbb{D} \to \mathbb{R}$ be a harmonic function satisfying*

$$\sup_{0<r<1} \int_{\mathbb{T}} |u(r\zeta)| dm(\zeta) < \infty. \tag{1.1.60}$$

There exists $E_u \subset \mathbb{T}$ with $m(E_u) = 0$ such that $\lim_{r\to 1} u(r\zeta)$ exists for $\zeta \in \mathbb{T}\backslash E_u$.

Sketch of proof The first step in the proof of Theorem 1.1.18 consists of showing that the estimates (1.1.60) imply that the harmonic function $u\colon \mathbb{D} \to \mathbb{R}$ is represented as Poisson integral,

$$u(z) = \int_{\mathbb{T}} P(z, w) d\mu(w), \qquad z \in \mathbb{D}, \tag{1.1.61}$$

where μ is a finite signed Borel measure, defined on $\mathbb{T}$. The measure μ arises as weak* cluster point of the measures

$$d\mu_r(\zeta) = u(r\zeta) dm(\zeta), \qquad 0 < r < 1.$$

The second step of the proof concerns the existence of radial limits for the Poisson integrals (1.1.61). Let $d\mu = f dm + d\nu$ be the Lebesgue decomposition of μ where $f \in L^1(\mathbb{T})$ and where $d\nu$ is a measure, singular with respect to dm. The main step in the proof of Fatou's theorem (Theorem 1.1.18) exhibits a measurable set $E_u \subset \mathbb{T}$ with $m(E_u) = 0$ such that

$$\lim_{r\to 1} u(r\zeta) = f(\zeta), \qquad \zeta \in \mathbb{T}\backslash E_u.$$

Radial Limits of Subharmonic Functions

Here, we review basic theorems by Littlewood on the boundary behavior of subharmonic functions in the unit disk.

Let U be an open subset of $\mathbb{C}$. A function $v\colon U \to \mathbb{R} \cup \{-\infty\}$ is subharmonic if it is upper semicontinuous and if for $w \in U$ there exists $\rho = \rho(w) > 0$ such that

$$v(w) \leq \int_{\mathbb{T}} v(w + r\zeta)dm(\zeta), \quad \text{for} \quad 0 \leq r < \rho.$$

Recall that $v\colon U \to \mathbb{R} \cup \{-\infty\}$ is upper semicontinuous if $\{w \in U : u(w) < t\}$ is an open subset of U for any $t > 0$. (The notion of upper semicontinuity carries over to any $\mathbb{R} \cup \{-\infty\}$ valued function defined on a topological space T.) Our sources for harmonic and subharmonic functions are the books by Garnett (1981, Chapter II), Koosis (1980) and Ransford (1995).

Example 1.1.4 The following list contains important classes of subharmonic functions that arise frequently in this book. We fix an open set $U \subseteq \mathbb{C}$.

(i) Let $v\colon U \to \mathbb{R}$ have continuous partial derivatives of order ≤ 2. Then v is subharmonic if and only if $\Delta v(z) \geq 0$ for $z \in U$.
(ii) In view of Itô's formula (1.1.49), composing a smooth subharmonic function $v\colon U \to \mathbb{R}$ and complex Brownian motion (z_t) gives rise to a submartingale. More precisely, if $z_0 \in U$, and if $\tau_U = \inf\{t > 0, z_0 + z_t \in \mathbb{C}\backslash U\}$, then $(v(z_0 + z(t)) : t < \tau_U)$ forms a submartingale on Wiener space $(\Omega, (\mathcal{F}_t), \mathbb{P})$.
(iii) Let $f\colon U \to \mathbb{C}$ be an analytic function. Then $\log|f|$ and $|f|^\alpha$ for $0 \leq \alpha < \infty$ are subharmonic functions on U.
(iv) If $u\colon U \to \mathbb{C}$ is merely harmonic on U, then all we can say is that u^p is subharmonic for $1 \leq p < \infty$. More generally, if $v\colon U \to \mathbb{R} \cup \{-\infty\}$ is sub-harmonic and if $g\colon \mathbb{R} \cup \{-\infty\} \to \mathbb{R}$ is convex and increasing then $g \circ v$ is a subharmonic function on U.
(v) Assume that for the domain U there exists the Green's function $G_U(z, w)$. Then for any $w \in U$,

$$z \mapsto -G_U(z, w), \qquad z \in U,$$

defines a subharmonic function on U, taking values in $[-\infty, 0)$. This applies, in particular, to the unit disk $\mathbb{D}$ where the Green's function is given explicitly by $G_{\mathbb{D}}(z, w) = \frac{1}{\pi} \log\left|\frac{1-\bar{z}w}{z-w}\right|$ where $z, w \in \mathbb{D}$.

The boundary behavior of subharmonic functions in the unit disk is described by Theorem 1.1.19, which is due to Littlewood (1928).

Theorem 1.1.19 (Littlewood's theorem) *Let $v\colon \mathbb{D} \to \mathbb{R}$ be a subharmonic function satisfying*

$$\sup_{0<r<1} \int_{\mathbb{T}} |v(r\zeta)| dm(\zeta) < \infty.$$

There exists $E_v \subset \mathbb{T}$ with $m(E_u) = 0$ such that $\lim_{r \to 1} v(r\zeta)$ *exist for $\zeta \in \mathbb{T}\backslash E_v$.*

Littlewood (1928) first proves that there exist a harmonic function $u\colon \mathbb{D} \to \mathbb{R}$ satisfying Inequality (1.1.60) and a nonnegative Borel measure μ on $\mathbb{D}$, such that

$$v(z) = u(z) - \int_{\mathbb{D}} G_{\mathbb{D}}(z, w)d\mu(w), \tag{1.1.62}$$

and

$$\int_{\mathbb{D}} (1 - |w|)d\mu(w) < \infty.$$

Setting $b(z) = u(z) - v(z)$ for $z \in \mathbb{D}$, Littlewood obtains a measurable set $E_b \subset \mathbb{T}$ with $m(E_b) = 0$ such that

$$\lim_{r\to 1} b(z) = 0,$$

for $\zeta \in \mathbb{T}\backslash E_b$. By Fatou's theorem (Theorem 1.1.18) there exists $E_u \subset \mathbb{T}$ with $m(E_u) = 0$ such that $\lim_{r\to 1} u(r\zeta)$ exists for $\zeta \in \mathbb{T}\backslash E_u$. Hence, for $E_v = (E_u \cup E_b)$ the conclusion of Theorem 1.1.19 holds true.

We refer to Tsuji (1975, Theorem IV.34, p. 172) for the detailed presentation of Littlewood's theorem (Theorem 1.1.19). See also exercise 19 of chapter II in Garnett (1981). The obvious parallel between those results and the submartingale convergence theorem (Theorem 1.1.10) was first exhibited in Doob's fundamental paper on Brownian motion (1954).

1.1.4 Kernels, Operators, and Spaces

Let $\mathbb{T} = \{\zeta \in \mathbb{C} : |\zeta| = 1\}$. We let m denote the unique rotation invariant probability measure defined on the sigma-algebra of Lebesgue measurable subsets of $\mathbb{T}$. (Thus m is the Haar measure on $\mathbb{T}$.) We review basic integral kernels arising in complex and harmonic analysis, discuss the Hardy spaces $H^p(\mathbb{T})$, the disk algebra $A(\mathbb{T})$, its dual space, and its annihilator within $M(\mathbb{T})$, the space of signed Radon measures on $\mathbb{T}$.

Fejer kernels: Closely related to the Poisson kernels are the Fejer kernels. For $n \in \mathbb{N}$, $\zeta \in \mathbb{T}$, we define the Fejer kernel by,

$$F_n(\zeta) = \sum_{m=-n}^{n} \frac{1 - |m|}{1 + n}\zeta^m. \tag{1.1.63}$$

Each Fejer kernel is a positive function on $\mathbb{T}$ (Katznelson [1968, Lemma I.2.5]), and $\int_{\mathbb{T}} F_n dm = 1$ by construction. Hence, the convolution operator,

$$F_n * f(w) = \int_{\mathbb{T}} F_n(w \cdot \overline{\zeta})f(\zeta)dm(\zeta), \quad w \in \mathbb{T}, \tag{1.1.64}$$

is a contraction on $L^p(\mathbb{T})$ for $1 \leq p \leq \infty$. Moreover, Fejer's theorem (Katznelson [1968, Section I.2]) asserts that

$$\lim_{n\to\infty} \|F_n * f - f\|_p = 0, \tag{1.1.65}$$

for any $f \in L^p(\mathbb{T})$ and $1 \leq p < \infty$.

de La Vallee Poussin kernel: We say that $V_n(\zeta) = 2F_{2n+1}(\zeta) - F_n(\zeta)$ is the de La Vallee Poussin kernel. Clearly, the assertion (1.1.65) of Fejer's theorem carries over to the sequence of convolution operators,

$$(V_n f)(w) = \int_{\mathbb{T}} f(w) V_n\left(w\overline{\zeta}\right) dm(\zeta), \tag{1.1.66}$$

where dm denotes the normalized Haar measure on $\mathbb{T}$. Its associated Fourier multiplier is given by

$$q(k) = \begin{cases} 0, & |k| \geq 2n; \\ \text{linear}, & n < |k| \leq 2n; \\ 1, & |k| \leq n. \end{cases}$$

Indeed, $\widehat{V_n}(k) = q(k)$, for $k \in \mathbb{N}$.

The Hilbert transform: Let H denote the Hilbert transform, given initially as a Fourier multipler on $L^2(\mathbb{T})$ by

$$H(y^n) = (-i)\text{sign}(n)y^n,$$

for $y \in \mathbb{T}$. Clearly, the Hilbert transform defines a contraction on $L^2(\mathbb{T})$. Let $f : \mathbb{T} \to \mathbb{R}$ be square integrable. If $u : \mathbb{D} \to \mathbb{R}$ denotes the harmonic extension of f and $v : \mathbb{D} \to \mathbb{R}$ the harmonic extension of $H(f)$, then $u + iv$ is analytic in $\mathbb{D}$.

By the theorem of Riesz (see Koosis [1980, Chapter V]), the Hilbert transform extends to a bounded operator on $L^p(\mathbb{T})$, $1 < p < \infty$, and

$$\|H(f)\|_{L^p(\mathbb{T})} = A(p)\|f\|_{L^p(\mathbb{T})},$$

where $A(p) \leq p^2 A/(p-1)$. The limiting case $p = 1$ is covered by Kolmogorov's theorem asserting the weak-type (1:1) estimates

$$\sup_{t>0} tm\{|H(f)| > t\} \leq A_1 \|f\|_{L^1(\mathbb{T})}. \tag{1.1.67}$$

Again, we refer to Koosis [1980, Chapter V] for Kolmogorov's theorem. The Hilbert transform does not extend to a bounded operator on $L^1(\mathbb{T})$. In this sense, Kolmogorov's weak-type (1:1) may not be improved.

Next, we isolate an important special class of L^1-functions for which their Hilbert transform is again integrable. Let $z \in \mathbb{T}$, $0 < \delta \leq 2$, and put $I = \{\zeta \in \mathbb{T} : |\zeta - z| < \delta\}$. We say that a measurable $a_I : \mathbb{T} \to \mathbb{R}$ is an atom (associated to the interval I) if the following three conditions hold true:

$$\operatorname{supp} a_I \subseteq I, \quad \|a_I\|_{L^\infty(\mathbb{T})} \leq \frac{1}{\operatorname{diam} I}, \quad \int_{\mathbb{T}} a_I dm = 0. \tag{1.1.68}$$

There exits a constant $C_1 < \infty$ such that for any atom a_I satisfying (1.1.68),

$$\int_{\mathbb{T}} |H(a_I)| dm \leq C_1. \tag{1.1.69}$$

Consequently, if f is a convex combination of atoms, then $f + iH(f) \in L^1(\mathbb{T})$ with $\|f + iH(f)\|_1 \leq C_1$. The deep theorem of Fefferman asserts that the converse implication holds as well (see García-Cuerva and Rubio de Francia [1985]).

Theorem 1.1.20 *Assume that $f+iH(f) \in L^1(\mathbb{T})$. Then there exists a sequence of atoms (a_n) and a sequence of scalars (λ_n) such that*

$$\sum_{n=1}^{\infty} |\lambda_n| \leq A_1 \|f + iH(f)\|_{L^1(\mathbb{T})} \quad \text{and} \quad f + iH(f) = \lambda_0 + \sum_{n=1}^{\infty} \lambda_n(a_n + iH(a_n)), \tag{1.1.70}$$

where convergence holds in $L^1(\mathbb{T})$, and where $A_1 < \infty$ is independent of $f + iH(f)$.

Hardy spaces $H^p(\mathbb{T})$: Consider first $1 \leq p \leq \infty$. The space $H^p(\mathbb{T})$ consists of those complex-valued $f \in L^p(\mathbb{T})$ for which

$$\int_{\mathbb{T}} f(\zeta)\zeta^n dm(\zeta) = 0, \quad n \in \mathbb{N}. \tag{1.1.71}$$

Thus defined, $H^p(\mathbb{T})$ is a closed subspace of $L^p(\mathbb{T})$, and it is a Banach space when equipped with the $L^p(\mathbb{T})$ norm. Hardy spaces are characterized in terms of the harmonic extension to the unit disk $\mathbb{D}$. Indeed, by the Fourier expansion of the Poisson kernels (1.1.58), if $f \in H^p(\mathbb{T})$, then its Poisson integrals,

$$F(z) = \int_{\mathbb{T}} P(z, \zeta) f(\zeta) dm(\zeta),$$

define an analytic function $F : \mathbb{D} \to \mathbb{C}$. The converse implication holds true as well. If the Poisson integrals of $f \in L^p(\mathbb{T})$ give rise to an analytic function in $\mathbb{D}$, then $f \in H^p(\mathbb{T})$.

We denote by $H^p_0(\mathbb{T})$ the space of those $f \in H^p(\mathbb{T})$ for which $\int_{\mathbb{T}} f dm = 0$. By (1.1.71), the space $H^\infty(\mathbb{T})$ is the annihilator of $H^1_0(\mathbb{T})$,

$$H^\infty(\mathbb{T}) = \left\{ f \in L^\infty(\mathbb{T}) : \int_{\mathbb{T}} f(\zeta) g(\zeta) dm(\zeta) = 0, \quad g \in H^1_0(\mathbb{T}) \right\}, \tag{1.1.72}$$

yielding the duality relation

$$(L^1(\mathbb{T})/H_0^1(\mathbb{T}))^* = H^\infty(\mathbb{T}). \tag{1.1.73}$$

Similarily, we find

$$(H^1(\mathbb{T}))^* = L^\infty(\mathbb{T})/H_0^\infty(\mathbb{T}). \tag{1.1.74}$$

We will also consider $0 < p < 1$, in which case the Hardy space $H^p(\mathbb{T})$ is defined to be the $L^p(\mathbb{T})$-closure of $\text{span}\{z^n : n \in \mathbb{N}, z \in \mathbb{T}\}$. If $0 < p < 1$, $H^p(\mathbb{T})$ is a complete quasi metric space, and

$$\|f + g\|_p^p \leq \|f\|_p^p + \|g\|_p^p, \quad f, g \in H^p(\mathbb{T}). \tag{1.1.75}$$

Riesz factorization: The classical F. Riesz factorization theorem for $H^p(\mathbb{T})$ $(0 < p < \infty)$ asserts that for $h \in H^p(\mathbb{T})$ there exist $f, g \in H^{2p}(\mathbb{T})$ such that $h = f \cdot g$ and

$$\int_{\mathbb{T}} |h|^p dm = \left(\int_{\mathbb{T}} |f|^{2p} dm \int_{\mathbb{T}} |g|^{2p} dm \right)^{1/2}. \tag{1.1.76}$$

See García-Cuerva and Rubio de Francia (1985, Corollary I.3.4), or Koosis (1980, Chapter IV, Section D).

A predual of $H^1(\mathbb{T})$: Here, we review the F. and M. Riesz theorem, which asserts that $H^1(\mathbb{T})$ is a dual sub-space of $L^1(\mathbb{T})$. We use Koosis (1980, Chapter VII.A) and Pełczyński (1977, Chapter 1) as basic sources.

The disk algebra $A(\mathbb{T})$ consists of those complex-valued $f \in C(\mathbb{T})$ for which

$$\int_{\mathbb{T}} f(\zeta)\zeta^n dm(\zeta) = 0, \quad n \in \mathbb{N}. \tag{1.1.77}$$

Equipped with the norm $\|f\|_\infty = \sup_{\zeta \in \mathbb{T}} |f(\zeta)|$, the disk algebra is a closed subspace of $C(\mathbb{T})$. By the Riesz representation theorem, $C(\mathbb{T})^* = M(\mathbb{T})$, the space of finite, complex-valued Borel meaures on $\mathbb{T}$. Elementary functional analysis now yields the identity

$$(C(\mathbb{T})/A(\mathbb{T}))^* = A(\mathbb{T})^\perp, \tag{1.1.78}$$

where

$$A(\mathbb{T})^\perp = \left\{ \mu \in C(\mathbb{T})^* : \int_{\mathbb{T}} f d\mu = 0, f \in A(\mathbb{T}) \right\} \tag{1.1.79}$$

is the annihilator of $A(\mathbb{T})$ in $C(\mathbb{T})^*$. The F. and M. Riesz theorem identifies the $A(\mathbb{T})^\perp$. In short, it asserts that

$$A(\mathbb{T})^\perp = H_0^1(\mathbb{T}),$$

with equality of norms, where $H_0^1(\mathbb{T}) = \left\{ f \in H^1(\mathbb{T}) : \int f dm = 0 \right\}$.

Theorem 1.1.21 (F. and M. Riesz theorem) *Let $\mu \in C(\mathbb{T})^*$. If*

$$\int_{\mathbb{T}} \zeta^n d\mu(\zeta) = 0, \quad for \quad n \in \mathbb{N} \cup \{0\},$$

then μ is absolutely continuous with respect to the Lebesgue measure on $\mathbb{T}$. Moreover, there exists a uniquely defined $f \in H_0^1(\mathbb{T})$ such that $d\mu = f dm$, and $\|\mu\| = \|f\|_{L^1(\mathbb{T})}$.

In view of (1.1.78) and (1.1.79), Theorem 1.1.21 asserts that $H_0^1(\mathbb{T})$ is a separable dual space, and specifically that

$$(C(\mathbb{T})/A(\mathbb{T}))^* = H_0^1(\mathbb{T}). \tag{1.1.80}$$

By the Hahn–Banach theorem, we identify the dual of the disk algebra with a quotient space as $A(\mathbb{T})^* = C(\mathbb{T})^*/A(\mathbb{T})^\perp$. As $A(\mathbb{T})^\perp = H_0^1(\mathbb{T})$, we obtain

$$A(\mathbb{T})^* = L^1/H_0^1 \oplus M_{\text{sing}},$$

where M_{sing} denotes the space of finite, complex Borel measures that are singular with respect to the Lebesgue measure on $\mathbb{T}$.

Polydisk algebras: Let $d \in \mathbb{N}$, $z \in \mathbb{T}^d$ and $n \in \mathbb{Z}^d$ where $z = (z_1, \dots, z_d)$ and $n = (n_1, \dots, n_d)$. We put $z^n = \prod_{j=1}^d z_j^{n_j}$ and define the n-th Fourier coefficient of $f \in C\left(\mathbb{T}^d\right)$ by

$$\widehat{f}(n) = \int_{\mathbb{T}^d} f(z) z^{-n} dm_d(z),$$

where m_d denotes the normalized Haar measure in $\mathbb{T}^d$. The polydisk algebra is defined as

$$A(\mathbb{T}^d) = \left\{f \in C(\mathbb{T}^d) : \ \widehat{f}(n) = 0, \quad n \in \mathbb{Z}^d \backslash \mathbb{N}_0^d\right\},$$

equipped with the uniform norm on $\mathbb{T}^d$. Clearly $A(\mathbb{T}^d)$ is the closure in $C(\mathbb{T}^d)$, (with the uniform norm) of the analytic polynomials span $\left\{z^n : n \in \mathbb{N}_0^d, z \in \mathbb{T}^d\right\}$.

By the F. and M. Riesz theorem (Theorem 1.1.21), $L^1/H_0^1 \subset A^*(\mathbb{T})$. Let $f_j \in L^1/H_0^1$, let $s \in \mathbb{T}^d$ with $s = (s_j)$, and define

$$f(s) = f_1(s_1) \cdots f_d(s_d).$$

Then,

$$\|f\|_{A^*(\mathbb{T}^d)} = \|f_1\|_{L^1/H_0^1} \cdots \|f_d\|_{L^1/H_0^1}.$$

The space $H^\infty(\mathbb{T}^d)$ consists of all bounded analytic functions in $\mathbb{D}^d$.

Ball algebras: For $z \in \mathbb{C}^n$ with $z = (z_1, \dots, z_n)$ put $|z|^2 = |z_1|^2 + \cdots + |z_n|^2$. Let $\mathbb{B}_n = \{z \in \mathbb{C}^n : |z| < 1\}$. The ball algebra $A(\mathbb{B}_n)$ consists of all $f \in C(\overline{\mathbb{B}_n})$ that are analytic in $\mathbb{B}_n$. We let $H^\infty(\mathbb{B}_n)$ denote the space of all bounded analytic functions in $\mathbb{B}_n$.

The trace class S^1: Here, we review the definition of the trace class and list its most elementary properties. This paragraph is based on Wojtaszczyk (1991, Chapter III.G).

Let K be a separable complex Hilbert space, and let A be a compact operator on K. Then A^*A is self-adjoint, compact, and positive semidefinite. By the spectral theorem, there exists an orthonormal sequence $\{v_j\}$ in K, and a sequence of non-negative, non increasing scalars $\{\mu_j\}$ converging to zero such that $A^*Ax = \sum \mu_j \langle x, v_j \rangle v_j$. We put $\lambda_j(|A|) = \mu_j^{1/2}$ and define the nonnegative selfadjoint operator $|A|$ on K by setting

$$|A|(x) = \sum_{j=1}^{\infty} \lambda_j(|A|)\langle x, v_j \rangle v_j.$$

Thus $\lambda_j(|A|)$ are the eigenvalues of $|A|$, called the singular values of A. The Schmidt representation theorem for A asserts that there exists a further orthonormal sequence $\{w_j\}$ in K, so that

$$A(x) = \sum_{j=1}^{\infty} \lambda_j(|A|)\langle x, v_j \rangle w_j,$$

for $x \in K$. If we let U denote the linear extension of the map $U(v_j) = w_j$, then the Schmidt representation may be restated as $A = U|A|$. Since U is a rotation and $|A|$ is nonnegative, $U|A|$ is referred to as the polar decomposition of A.

We denote by $S^\infty(K)$ the space of all compact operators A on K equipped with the norm

$$\|A\|_{S^\infty} = \max \lambda_j(|A|).$$

By the Spectral theorem, the norm defined on $S^\infty(K)$ is just the usual operator norm. The Hilbert–Schmidt class $S^2(K) \subseteq S^\infty(K)$ consists of those compact operators for which the singular values are square summable. We put

$$\|A\|_{S^2} = \left(\sum_{j=1}^{\infty} \lambda_j(|A|)^2 \right)^{1/2}.$$

Clearly $S^2(K)$ is a Hilbert space. Finally, we define the trace class $S^1(K) \subseteq S^2(K)$ to consist of those compact operators with summable singular values, and put

$$\|A\|_{S^1} = \sum_{j=1}^{\infty} \lambda_j(|A|).$$

In view of the Schmidt representation, the trace class $S^1(K)$ coincides with the space of all nuclear operators on the Hilbert space K, and the nuclear norm coincides with the trace norm. For any two Hilbert–Schmidt operators $A_1, A_2 \in S^2(K)$, their product is in the trace class, $A_1 \cdot A_2 \in S^1(K)$, and conversely for $A \in S^1(K)$ there exist factoring $A_1, A_2 \in S^2(K)$ such that $A = A_1 \cdot A_2$.

We assume now that the underlying Hilbert space K is finite dimensional. Then the trace of any linear operator on K is well defined. The trace duality formula provides the link betwen the trace class $S^1(K)$, and $S^\infty(K)$ as follows:

$$\|A\|_{S^1} = \max\{|\mathrm{tr}(B^*A)|\}, \tag{1.1.81}$$

where the maximum is taken over those linear operators B on K for which $\|B\|_{S^\infty} \leq 1$. The converse estimate holds as well, and

$$\|B\|_{S^\infty} = \max\{|\mathrm{tr}(B^*A)| : \|A\|_{S^1} \leq 1\}.$$

We now let $\{e_1, \ldots, e_n, \ldots\}$ be an orthonormal basis in the separable Hilbert space K. Then,

$$e_{i,j} = e_i \otimes e_j, \qquad i, j \in \mathbb{N},$$

ordered along the layers $\mathcal{L}_k = \{(i, j) : \max\{i, j\} = k\}$, forms a Schauder basis in $S^1(K)$. Given $A \in S^1(K)$, we write $A = (a_{i,j})$ if $A = \sum a_{i,j} e_{i,j}$. We have

$$\|A\|_{S^1} = \sup_{n \in \mathbb{N}} \left\| \sum_{i,j=1}^{n} a_{i,j} e_{i,j} \right\|_{S^1}.$$

With respect to the norm in S^1, the tensor product basis $e_{i,j} = e_i \otimes e_j$ is symmetric and unconditional in the following *restricted* sense: For $(\epsilon_i) \in \{1, -1\}^{\mathbb{N}}$, for any choice of bijections $\sigma, \tau : \mathbb{N} \to \mathbb{N}$, and any $n \in \mathbb{N}$,

$$\left\| \sum_{i,j=1}^{n} a_{i,j} e_{i,j} \right\|_{S^1} = \left\| \sum_{i,j=1}^{n} \epsilon_i \epsilon_j a_{i,j} e_{\sigma(i),\tau(j)} \right\|_{S^1}. \tag{1.1.82}$$

For a thorough discussion of the trace class we refer to Wojtaszczyk (1991, Chapter III.G) and Kwapień and Pełczyński (1970).

1.1.5 Weak Topology

We briefly review the weak topology in a Banach space X, the concept of weak compactness, and the connection between weak compactness in L^1 and uniform integrability. We use Wojtaszczyk (1991, Sections II.A and III. C.) as basic reference.

Let X be a Banach space, and let X^* denote its dual space, the space of continuous linear functionals on X. We define the $\sigma(X, X^*)$-topology on X by decreeing that for each $x \in X$, its neighborhood basis is given by the sets

$$\mathcal{U}(x, \epsilon, y_1, \ldots, y_n) = \left\{ v \in X : \max_{1 \le i \le n} |y_i(x) - y_i(v)| < \epsilon \right\},$$

where $y_i \in X^*, i \le n, n \in \mathbb{N}, \epsilon > 0$. We use the term weak-topology on X to refer to the $\sigma(X, X^*)$-topology.

Dunford–Pettis theorem: We say that a subset $Z \subseteq X$ is relatively weakly compact if the closure of Z in the weak-topology is compact, again with respect to the weak-topology on X.

Let $(F, \mathcal{F}, \mu)$ be a finite measure space and let V be a subset of $L^1 = L^1(\Omega, \mathcal{F}, \mu)$. We say that V is uniformly integrable in L^1 if

$$\lim_{t \to \infty} \sup_{y \in V} \int_\Omega 1_{\{|y| > t\}} |y| d\mu = 0.$$

Theorem 1.1.22 (Dunford–Pettis) *For every $V \subset L^1$, the following conditions are equivalent:*

(i) *V is relatively $\sigma(L^1, L^\infty)$ compact.*
(ii) *V is uniformly integrable in L^1.*

Reflexive spaces: Let X^{**} denote the dual space of X^*. We define the operator $I: X \to X^{**}$ by putting

$$I(x)(y) = y(x), \qquad x \in X, \quad y \in X^*.$$

The Hahn–Banach Theorem yields that I is well defined, $\|I(x)\|_{X^{**}} = \|x\|_X$, and that I is an injective operator. We say that X is a reflexive Banach space if I is a surjective operator.

Theorem 1.1.23 *Let X be a Banach space, let R be a reflexive subspace of X, and let $Z \subset R$. If*

$$\sup_{z \in Z} \|z\|_X < \infty,$$

then Z is a relatively weakly compact subset of X.

Alaoglu's theorem: Reversing the roles played by the spaces X and X^* we obtain the $\sigma(X^*, X)$-topology on X^*. Given $y \in X^*$, we define the basis of its $\sigma(X^*, X)$ neighborhoods to be

$$\mathcal{U}(y, \epsilon, x_1, \ldots, x_n) = \left\{ w \in X^* : \max_{1 \le i \le n} |x_i(y) - x_i(w)| < \epsilon \right\},$$

where $x_i \in X, i \le n, n \in \mathbb{N}, \epsilon > 0$. The $\sigma(X^*, X)$-topology is called the weak-*topology on X^*.

Theorem 1.1.24 (Alaoglu's theorem) *For any Banach space X, the $\sigma(X^*, X)$-topology restricted to the norm closed unit ball of X^* is compact.*

Recall that a Banach space is called separable if it contains a norm-dense, countable subset. Let B_X and B_{X^*} denote the norm closed unit balls of X and X^* respectively. Then,

(i) If X^* is separable the $\sigma(X, X^*)$-topology restricted to B_X is metrizable.
(ii) If X is separable, the $\sigma(X^*, X)$-topology restricted to B_{X^*} is metrizable.

1.1.6 Schauder Basis

A sequence $\{x_k\}$ in a Banach space X is called a Schauder basis if for each $x \in X$ there exists a uniquely determined sequence of scalars $a_k = a_k(x)$ such that

$$\lim_{n \to \infty} \left\| x - \sum_{k=1}^{n} a_k x_k \right\|_X = 0. \tag{1.1.83}$$

In this case, as proved by Banach, there exists $K < \infty$ such that

$$\sup_{n \in \mathbb{N}} \left\| \sum_{k=1}^{n} a_k x_k \right\|_X \le K \|x\|_X, \quad x \in X. \tag{1.1.84}$$

We denote by $\mathrm{bc}_X\{x_k\}$ (basis constant) the infimum over those $K < \infty$ satisfying Inequality (1.1.84). Exploiting the uniqueness in the partial sum expansion (1.1.83) and the norm estimates (1.1.84), we find that the biorthogonal functionals $x_k^* \in X^*$, defined by $x_k^*(x_n) = 0$ if $k \neq n$, and $x_k^*(x_k) = 1$, satisfy

$$\sup_{k \in \mathbb{N}} \|x_k^*\|_{X^*} \le \mathrm{bc}_X\{x_k\}.$$

A Schauder basis $\{x_k\}$ is called unconditional, in a Banach space X, if there exists $K < \infty$ such that

$$\sup_{n \in \mathbb{N}} \sup_{(\varepsilon_k) \in \{-1,+1\}} \left\| \sum_{k=1}^{n} a_k \varepsilon_k x_k \right\|_X \le K \|x\|_X, \tag{1.1.85}$$

for any $x \in X$. We denote by $\mathrm{ubc}_X\{x_k\}$ (unconditional basis constant) the infimum over those $K < \infty$ satisfying Inequality (1.1.85). Finally, we define

$$\mathrm{ubc}\{X\} = \inf \{\mathrm{ubc}_X\{x_k\}\}, \tag{1.1.86}$$

where the infimum is taken over all unconditional Schauder bases $\{x_k\}$ in X.

The Haar system: Illustrating these concepts in the context of the Lebesgue spaces $L^p[0,1[$, $1 \le p < \infty$, we turn to the Haar system $\{h_I : I \in \mathcal{I}\}$ in lexicographic order (see equation (1.1.2) for its definition).

(i) By equations (1.1.3) and (1.1.4), the Haar system is a Schauder basis for $L^1[0,1[$ with $\mathrm{bc}_{L^1}\{h_I\} = 1$. However, $\mathrm{ubc}_{L_1}\{h_I\} = \infty$, and hence the Haar system is not unconditional in $L^1[0,1[$. To see this, fix $n \in \mathbb{N}$, $I_n = [0, 2^{-n}[$, put $f = 1_{I_n}/|I_n|$, let $I_k = [0, 2^{-k}[$, $(1 \le k \le n)$, and set

$$a_k = \int f h_{I_k} \frac{dt}{|I_k|}.$$

By equations (1.1.3) and (1.1.4), we have $f = 1_{[0,1[} + \sum_{k=1}^{n-1} a_k h_{I_k}$, $\|f\|_1 = 1$, and yet, as is easily observed,

$$\left\| \sum_{k=1}^{n-1} (-1)^k a_k h_{I_k} \right\|_{L^1} \ge c \ln n. \tag{1.1.87}$$

(ii) We have $\mathrm{ubc}\{L^1\} = \infty$ and Pełczyński's theorem (see Theorem 4.2.12) asserting that $L^1[0,1[$ is not even isomorphic to a subspace of a Banach space with an unconditional basis.

(iii) By Burkholder's theorem (Theorem 1.1.5, for $1 < p < \infty$, the Haar system is an unconditional Schauder basis in $L^p[0,1[$, with $\mathrm{bc}_{L^p}\{h_I\} = 1$ and $\mathrm{ubc}_{L^p}\{h_I\} \sim p^2/(p-1)$.

For basic information on the Haar system, its subsequences, and rearrangements, we refer to Müller (2005), Novikov and Semenov (1997), and Penteker (2015).

1.2 Holomorphic Martingales

We recall that the space $H^p(\mathbb{T})$ consists of those $f \in L^p(\mathbb{T})$ for which the harmonic extension to $\mathbb{D}$ is an analytic function. In Section 1.1.2, we introduced complex Brownian motion (z_t) and put $\tau = \inf\{t > 0 : |z_t| > 1\}$. If $f \in H^p(\mathbb{T})$, its harmonic extension satisfies $\overline{\partial_z} f(z) = 0$, for $z \in \mathbb{D}$. Hence, by Itô's formula (1.1.50) putting $F = f(z_T)$ gives

$$F = \mathbb{E}F + \int_0^\infty 1_{\{t\leq T\}} \partial_z f(z_s) dz_s. \tag{1.2.1}$$

The representation (1.2.1) is the model for defining holomorphic random variables on Wiener's filtered probability space $(\Omega, (\mathcal{F}_t), \mathbb{P})$.

An integrable $F\colon \Omega \to \mathbb{C}$ is called a holomorphic random variable if there exists a complex-valued, adapted process (X_s) such that

$$F = F_0 + \int_0^\infty X_s dz_s. \tag{1.2.2}$$

The subspace of $L^p(\Omega)$ consisting of holomorphic random variables is denoted $H^p(\Omega)$. For a given $F \in H^p(\Omega)$ and

$$F_t = \mathbb{E}(F|\mathcal{F}_t), \tag{1.2.3}$$

we call (F_t) the holomorphic martingale associated to F. Combining (1.2.2) and (1.2.3) gives

$$F_t = F_0 + \int_0^t X_s dz_s.$$

It is important to point out that holomorphic random variables are stable under the following operations:

- stopping times
- pointwise multiplication
- composition with analytic functions.

We first show stability under stopping times. Indeed, given an $(\mathcal{F}_t)$ stopping time $\rho\colon \Omega \to \mathbb{R}^+$ and its generated stopping time sigma-algebra $\mathcal{F}_\rho$, then by Theorem 1.1.9, optional stopping yields,

$$\mathbb{E}(F|\mathcal{F}_\rho) = F_0 + \int_0^\rho X_s dz_s = F_0 + \int_0^\infty 1_{\{s<\rho\}} X_s dz_s. \tag{1.2.4}$$

As the process $(1_{\{s<\rho\}} X_s)$ is adapted to the filtration $(\mathcal{F}_t)$, the right-hand side of (1.2.4) is a holomorphic random variable. Thus holomorphic random variables are stable under stopping times. Next, we turn to pointwise multiplication. If $F, G \in L^2(\Omega)$, are holomorphic random variables with Itô integrals,

$$F = F_0 + \int_0^\infty X_s dz_s, \quad \text{and} \quad G = G_0 + \int_0^\infty Y_s dz_s,$$

then the integration by parts formula stated in Theorem 1.1.12 gives

$$FG = F_0 G_0 + \int_0^\infty (F_s Y_s + G_s X_s) dz_s. \tag{1.2.5}$$

Hence, FG is a holomorphic random variable, and the product $(F_t G_t)$ is a holomorphic martingale satisfying

$$F_t G_t = \mathbb{E}(FG|\mathcal{F}_t), \quad t > 0.$$

Finally, holomorphy is preserved under composition with analytic functions. We let $f\colon \mathbb{C} \to \mathbb{C}$ be analytic and assume that F is a holomorphic random variable such that the composition $f(F)$ is integrable. Then Itô's formula gives

$$f(F) = f(F_0) + \int_0^\infty \partial_z f(F_s) dF_s. \tag{1.2.6}$$

Hence, $f(F)$ is a holomorphic random variable and $f(F_t)$ a holomorphic martingale satisfying

$$f(F_t) = \mathbb{E}(f(F)|\mathcal{F}_t), \quad t > 0.$$

In summary, the space of holomorphic random variables is stable under the following operations: Stopping times, pointwise multiplication, composition with analytic functions.

1.2.1 Submartingale Estimates

In this section, we begin with a version of Itô's formula suitably adapted to processes $\left(g(|F_t|^2)\right)$ where (F_t) is a holomorphic martingale. By judiciously choosing testing functions g, we will obtain integral estimates for the maximal function $\sup |F_t|$, and the quadratic variation $\langle F_t \rangle^{1/2}$. The section closes with complex convexity estimates for $H^1(\Omega)$ derived from Theorem 1.2.1.

Theorem 1.2.1 *Let $F \in H^1(\Omega)$, with Itô integral $F = F_0 + \int_0^\infty X_s dz_s$, and assume that $g\colon \mathbb{R}^+ \to \mathbb{R}$ be twice continously differentiable. If $g(|F_t|^2)$ is equi-integrable, then*

$$g(|F_t|^2) - 2\int_0^t g'(|F_s|^2)|X_s|^2 ds - 2\int_0^t g''(|F_s|^2)\left|\overline{F_s} X_s\right|^2 ds \tag{1.2.7}$$

defines an $(\mathcal{F}_t)$-martingale and coincides with the stochastic integrals

$$g(|F_0|^2) + \int_0^t g'(|F_s|^2)\left(\overline{F_s} X_s dz_s + F_s \overline{X_s} d\overline{z_s}\right). \tag{1.2.8}$$

If

$$g'(r) + g''(r) r \geq 0, \qquad \text{for} \quad r > 0, \tag{1.2.9}$$

then $\left(g(|F_t|^2)\right)$ forms a submartingle.

Proof We use Theorem 1.1.12 (integration by parts) to obtain the following integral representation:

$$|F_t|^2 = F_t\overline{F_t} = |F_0|^2 + \int_0^t \overline{F_s}X_s dz_s + F_s\overline{X_s}d\overline{z}_s + 2\int_0^t |X_s|^2 ds. \tag{1.2.10}$$

Itô's formula (1.1.53) applied to (1.2.10) yields that

$$g(|F_t|^2) - 2\int_0^t g'(|F_s|^2)|X_s|^2 ds - 2\int_0^t g''(|F_s|^2)\left|\overline{F_s}X_s\right|^2 ds \tag{1.2.11}$$

coincides with

$$g(|F_0|^2) + \int_0^t g'(|F_s|^2)\left(\overline{F_s}X_s dz_s + F_s\overline{X_s}d\overline{z}_s\right). \tag{1.2.12}$$

We verify Inequality (1.2.9) by observing that the density of the drift term of $g(|F_t|^2)$ is given by

$$2|X_s|^2\left(g'(|F_s|^2) + g''(|F_s|^2)|F_s|^2\right),$$

hence Inequality (1.2.9) implies that $g(|F_t|^2)$ is a submartingale. □

We derive significant structural estimates for holomorphic martingales by exploiting Theorem 1.2.1. First, using

$$g(t) = t^{\alpha/2},$$

with $\alpha > 0$, we obtain Garling's maximal function estimates (1.2.18), and also the square function estimate (1.2.21). Second, considering

$$g(t) = (1+t)^{1/2},$$

and applying Theorem 1.2.1, we derive complex convexity estimates (1.2.32) for $H^1(\Omega)$.

Theorem 1.2.2 *Let $F \in H^1(\Omega)$, with Itô integral $F = F_0 + \int_0^\infty X_s dz_s$. For $0 < \alpha \le 1$, the process $(|F_t|^\alpha)$ is a submartingale, the drift term of $|F_t|^\alpha$ is given by*

$$\frac{\alpha^2}{2}\int_0^t |X_s|^2|F_s|^{\alpha-2}ds, \tag{1.2.13}$$

and

$$\mathbb{E}|F_t|^\alpha = \mathbb{E}\left[|F_0|^\alpha + \frac{\alpha^2}{2}\int_0^t |X_s|^2|F_s|^{\alpha-2}ds\right]. \tag{1.2.14}$$

Consequently, $|\mathbb{E}F_t|^\alpha \le \mathbb{E}|F_t|^\alpha$ for $t > 0$.

Proof Let $g(t) = t^{\alpha/2}$ and consider the process $|F_t|^\alpha = g\left(|F_t|^2\right)$. By Itô's formula (1.2.8), its drift is given as

$$\int_0^t 2g'(|F_s|^2)|X_s|^2 + 2g''(|F_s|^2)\left|\overline{F_s}X_s\right|^2 ds. \tag{1.2.15}$$

Inserting $g'(t) = \frac{\alpha}{2}t^{\alpha/2-1}$ and $g''(t) = \frac{\alpha}{2}\left(\frac{\alpha}{2}-1\right)t^{\alpha/2-2}$, the integrand (1.2.15) coincides with

$$\alpha|F_s|^{\alpha-2}|X_s|^2 + \alpha\left(\frac{\alpha}{2}-1\right)|F_s|^{\alpha-4}|F_s|^2|X_s|^2. \tag{1.2.16}$$

By arithmetic, the term (1.2.16) equals $(\alpha^2/2)|F_s|^{\alpha-2}|X_s|^2$. This proves the assertions (1.2.13) and (1.2.14). □

Remark: We record the specialization of Theorem 1.2.2 to holomorphic random variables of the form $F = f(z_\tau)$, where $f \in H^1(\mathbb{T})$ and $\tau = \inf\{t > 0 : |z_t| > 1\}$. As z_τ is uniformly distributed over $\mathbb{T}$, we have $F \in H^1(\Omega)$ and $\mathbb{E}F = \int f dm$. By the mean value property, we have $\int_{\mathbb{T}} f dm = f(0)$, hence Theorem 1.2.2 gives

$$\left|\int_{\mathbb{T}} f dm\right|^\alpha < \int_{\mathbb{T}} |f|^\alpha dm. \tag{1.2.17}$$

The assertions of Theorem 1.2.2, that is assertions (1.2.13) and (1.2.14), give rise to significant estimates for holomorphic martingales. We begin by merging the submartingale property, established in Theorem 1.2.2, with Doob's inequality. Recall that for $1 < p \leq \infty$, a continuous submartingale (G_t) in $L^p(\Omega)$ satisfies the maximal function estimate

$$\mathbb{E}\left(\sup_{t>0}|G_t|\right)^p \leq C_p^p \sup_{t>0}\mathbb{E}(|G_t|^p),$$

where $C_p = p/(p-1)$. For holomorphic martingales, use (1.2.13) to extrapolate Doob's inequality to the range $0 < p \leq 1$, and prove Garling's estimate (1.2.18).

Theorem 1.2.3 (Garling's theorem) *Let $F \in H^1(\Omega)$, and $0 < \alpha \leq 1$. Then $|\mathbb{E}(F)|^\alpha \leq \mathbb{E}(|F|^\alpha)$, and*

$$\mathbb{E}(\sup|F_t|^\alpha) \leq e\mathbb{E}(|F|^\alpha). \tag{1.2.18}$$

Proof Let $p > 1$. By Theorem 1.2.2, $(|F_t|^{\alpha/p})$ is a submartingale. Applying the Doob's maximal inequality in $L^p(\Omega)$ gives

$$\mathbb{E}(\sup|F_t|^\alpha) \leq C_p^p\mathbb{E}(|F|^\alpha). \tag{1.2.19}$$

Note that the left-hand side of Inequality (1.2.19) does not depend on p. Since $C_p = p/(p-1)$, and $e = \inf\{(p/(p-1))^p : 1 < p < \infty\}$, we get Inequality (1.2.18). □

Remark: First, recall Theorem 1.1.10, which asserts that for any L^1-bounded martingale (F_t) the pointwise limit $F = \lim_{t\to\infty} F_t$ exists almost surely (but not necessarily in L^1). However, if its stochastic integral representation is of the form

$$F_t = \int_0^t X_s dz_s,$$

then Theorem 1.2.3 shows that $\mathbb{E}\sup_{t>0}|F_t| < \infty$. Hence, $\{F_t : t > 0\}$ is a uniformly integrable subset of L^1, and Theorem 1.1.10 implies that

$$\lim_{t\to\infty} \mathbb{E}|F_t - F| = 0.$$

The second consequence of Theorem 1.2.2 exploits assertion (1.2.14). Let $F \in H^1(\Omega)$ with Itô integral representation $F = F_0 + \int X_s dz_s$. We recall that its quadratic variation is defined by equation (1.1.47). Next, we obtain an interpolatory estimate for the quadratic variation of a holomorphic martingale in terms of its maximal function and the limiting holomorphic random variable.

Theorem 1.2.4 *Let $F \in H^1(\Omega)$. Let $0 < \alpha \leq 1$ and*

$$S_{(\alpha)}(F) = \left(|F_0|^2 + (\alpha/4)\langle F\rangle\right)^{1/2}.$$

Then

$$\mathbb{E}\left(S^\alpha_{(\alpha)}(F)\right) \leq \left(\mathbb{E}\sup_{0<t<\infty}|F_t|^\alpha\right)^{1-\alpha/2} (\mathbb{E}|F|^\alpha)^{\alpha/2}, \tag{1.2.20}$$

and

$$\mathbb{E}\left(S^\alpha_{(\alpha)}(F)\right) \leq C_\alpha \mathbb{E}|F|^\alpha, \tag{1.2.21}$$

where $C_\alpha = (e^{1-\alpha/2})$.

Proof Let $F \in H^1(\Omega)$ and $F_t = F_0 + \int_0^t X_s dz_s$, where $F_t = \mathbb{E}(F|\mathcal{F}_t)$. Observe first the simple pointwise estimate,

$$|F_0|^2 + \frac{\alpha}{2}\int_0^\infty |X_s|^2 ds \leq \sup_{t>0}|F_t|^{2-\alpha}\left(|F_0|^\alpha + \frac{\alpha}{2}\int_0^\infty |X_s|^2|F_s|^{\alpha-2}ds\right). \tag{1.2.22}$$

Raising both sides of (1.2.22) to the power $\alpha/2$, taking expectations, and applying Hölder's inequality with conjugate indices $2/(2-\alpha)$ and $2/\alpha$ to its right-hand side gives

$$\mathbb{E}\left(S^\alpha_{(\alpha)}(F)\right) \leq \left(\mathbb{E}\sup_t|F^\alpha_t|\right)^{1-\alpha/2}\left(|F_0|^\alpha + \frac{\alpha}{2}\mathbb{E}\int_0^\infty |X_s|^2|F_s|^{\alpha-2}ds\right)^{\alpha/2}.$$

Finally, invoking Theorem 1.2.2 and rewriting accordingly the second factor on the right-hand side gives the interpolatory estimates (1.2.20). Combining Inequality (1.2.20) with Inequality (1.2.18) gives Inequality (1.2.21). □

Remark: Theorem 1.3.1 (the Burkholder–Davis–Gundy inequality) combined with Theorem 1.2.3 (Garling's theorem) gives Inequality (1.2.21), albeit with a constant that is worse than $C_\alpha = (e^{1-\alpha/2})$. Finally, we remark that the proof of Theorem 1.2.4 shows that there exists $C_p \le C\sqrt{p}$ such that

$$\mathbb{E}\left(|F_0|^2 + \langle F\rangle\right)^{p/2} \le C_p^p \mathbb{E}|F|^p,$$

for $1 \le p < \infty$ and $F \in H^p(\Omega)$.

1.2.2 Complex Convexity Estimates

We begin by presenting sharp complex uniform convexity estimates obtained inependently by Bonami (1970, Chapter III, Theorem 7), Weissler (1980) and Davis, Garling, and Tomczak-Jaegermann (1984). More recently, Aleksandrov (2007) obtained an elegant analytic proof. A real variable, n-dimensional extension was obtained by Ivanisvili et al. (2020). For ease of comparison, we restate those using complex Brownian motion z_τ stopped at $\tau = \inf\{t > 0 : |z_t| > 1\}$.

Theorem 1.2.5 *For any $x, y \in \mathbb{C}$,*

$$\left(|x|^2 + \frac{1}{2}|y|^2\right)^{1/2} \le \mathbb{E}|x + yz_\tau|, \tag{1.2.23}$$

and $\frac{1}{2}$ is the best (=largest) constant for which this inequality holds true.

Proof The proof given here is due Schmuckenschläger (personal communication, 1994). By rescaling and rotation, it suffices that we prove Inequality (1.2.23) with $x = 1$ and $y > 0$. Applying Theorem 1.2.2 with $F = 1 + \int_0^\tau y dz_t$ gives

$$\mathbb{E}|1 + yz_\tau| = 1 + \frac{1}{2}\mathbb{E}\int_0^\tau y^2|1 + yz_t|^{-1}dt. \tag{1.2.24}$$

The proof proceeds by obtaining suitable lower estimates for the right-hand side of (1.2.24).

Step 1: Let $dA(z)$ denote the area measure on $\mathbb{D}$ and let $G(z) = (1/\pi)\ln(1/|z|)$ denote Green's function of the unit disk with pole at $0 \in \mathbb{D}$. For $y > 0$, put

$$a(y) = 1 + \frac{1}{2}\iint_{\mathbb{D}} y^2|1 + yz|^{-1}G(z)dA(z).$$

Note that $a(y)$ coincides with the right-hand side of (1.2.24). On the other hand, by an elementary calculation, we have $a'(0) = 0$ and

$$a''(0) = \iint_{\mathbb{D}} G(z)dA(z) = 2\int_0^1 r\ln\left(\frac{1}{r}\right)dr = \frac{1}{2}. \tag{1.2.25}$$

Step 2: Let $y > 0$ and put

$$h(y) = 1 + \frac{1}{2}\iint_{\mathbb{D}} y^2\left(1 + |yz|^2\right)^{-1/2} G(z)dA(z).$$

Here, we show that $a(y) \geq h(y)$. To this end, fix $z \in \mathbb{D}$ and let $w(\zeta) = (1 + yz\zeta)$ for $\zeta \in \mathbb{T}$. Clearly, we have

$$\int_{\mathbb{T}} |w|^2 dm = \left(1 + |yz|^2\right).$$

Putting $f(s) = s^{-1/2}$ defines a convex function on $\mathbb{R}^+$. Hence, by Jensen's inequality,

$$\int_{\mathbb{T}} |1 + yz\zeta|^{-1} dm(\zeta) = \int_{\mathbb{T}} f\left(|w|^2\right) dm \geq f\left(\int_{\mathbb{T}} |w|^2 dm\right) = \left(1 + |yz|^2\right)^{-1/2}.$$

Invoking rotational invariance, we find that

$$\begin{aligned} \iint_{\mathbb{D}} |1 + yz|^{-1} G(z)dA(z) &= \iint_{\mathbb{D}}\int_{\mathbb{T}} |1 + yz\zeta|^{-1} dm(\zeta)G(z)dA(z) \\ &\geq \iint_{\mathbb{D}} \left(1 + |yz|^2\right)^{-1/2} G(z)dA(z). \end{aligned} \tag{1.2.26}$$

It remains to note that (1.2.26) implies $a(y) \geq h(y)$.

Step 3: Here, we evaluate and simplify the integral defining $h(y)$. By rewriting it in polar coordinates and integrating by parts we obtain

$$\begin{aligned} h(y) &= 1 + \int_0^1 y^2 r(1 + y^2 r^2)^{-1/2} \ln\left(\frac{1}{r}\right) dr \\ &= (1 + y^2)^{1/2} - \ln\left(1 + (1 + y^2)^{1/2}\right) + \ln 2. \end{aligned} \tag{1.2.27}$$

Step 4: A quick calculation using (1.2.27) shows that

$$h'(y) = y / \left(1 + (1 + y^2)^{1/2}\right). \tag{1.2.28}$$

Put $g(y) = \left(1 + \frac{1}{2}y^2\right)^{1/2}$. Then $g(0) = 1$ and

$$g'(y) = \frac{y}{2}\left(1 + \frac{1}{2}y^2\right)^{-1/2}. \tag{1.2.29}$$

Comparing (1.2.29) and (1.2.28) we obtain $h'(y) \geq g'(y)$, for $y > 0$. Since $h(0) = g(0)$, we find

$$h(y) \geq g(y).$$

Recalling that we proved $a(y) \geq h(y)$ and retracing our steps gives Inequality (1.2.23).

Step 5: We finally show that $(1/2)$ is the largest constant for which Inequality (1.2.23) holds true. To this end, fix $\delta > 0$ and put

$$b(y) = (1 + \delta^2 y^2)^{1/2}.$$

Recall the definition of $a(y)$ and assume that

$$a(y) \geq b(y), \tag{1.2.30}$$

for $y > 0$. Since $a(0) = b(0)$ and $a'(0) = b'(0)$, assumption (1.2.30) implies that $a''(0) \geq b''(0)$. We have $b''(0) = \delta^2$ by a short calculation, and $a''(0) = (1/2)$ by equation (1.2.25). Summing up, we have obtained $(1/2) \geq \delta^2$. □

Remark: In Section 2.5, Proposition 2.5.7, we will obtain the following extension of Inequality (1.2.23) due to Haagerup and Pisier (1989). For any $f \in H^1(\mathbb{T})$,

$$\left(|f(0)|^2 + \frac{1}{2}\left(\int_{\mathbb{T}} |f - f(0)| dm\right)^2\right)^{1/2} \leq \int_{\mathbb{T}} |f| dm. \tag{1.2.31}$$

Since Theorem 1.2.5 asserts that $1/2$ is the largest constant for which Inequality (1.2.23) holds true, we conclude that the best constant in Inequality (1.2.31) equals $1/2$.

The next theorem presents a complex convexity estimate, where the special holomorphic random variable (yz_τ) is replaced by any $F \in H^1(\Omega)$ satisfying $\mathbb{E}F = 0$. The proof below uses Theorem 1.2.1 with $g(t) = (1 + t)^{1/2}$. We point out that the best constant β in Theorem 1.2.6 is undetermined.

Theorem 1.2.6 *Let $F \in H^1(\Omega)$ with $\mathbb{E}F = 0$ and $z \in \mathbb{C}$. Then for $\beta^2 \leq 1/6$,*

$$\mathbb{E}\left(|z|^2 + \beta^2 |F|^2\right)^{1/2} \leq \mathbb{E}(|z + F|). \tag{1.2.32}$$

Proof Let $F \in H^1(\Omega)$, with $\mathbb{E}F = 0$ and stochastic integral representation $F = \int_0^\infty X_s dz_s$. By rotating and scaling it suffices to consider $z = 1$.

Step 1: Put $F_t = \mathbb{E}(F|\mathcal{F}_t)$. We verify first that $\left((1 + |F_t|^2)^{1/2}\right)$ is a submartingale and that for $0 < t < \infty$,

$$\mathbb{E}(1 + |F_t|^2)^{1/2} \leq 1 + \mathbb{E}\int_0^t (1 + |F_s|^2)^{-1/2} |X_s|^2 ds. \tag{1.2.33}$$

With $g(t) = (1 + t)^{1/2}$ we have $(1 + |F_t|^2)^{1/2} = g(|F_t|^2)$. By Formula (1.2.8) in Theorem 1.2.1, the drift term of the process $g(|F_t|^2)$ is given by

$$\int_0^t 2g'(|F_s|^2)|X_s|^2 + 2g''(|F_s|^2)|\overline{F_s} X_s|^2 ds. \tag{1.2.34}$$

Taking into account that $2g'(t) = (1+t)^{-1/2}$ and $4g''(t) = -(1+t)^{-3/2}$ it is easy to see that the density in (1.2.34) equals

$$(1+|F_s|^2)^{-1/2}|X_s|^2 - \frac{1}{2}(1+|F_s|^2)^{-3/2}|F_s|^2|X_s|^2. \tag{1.2.35}$$

By arithmetic, (1.2.35) reduces to

$$\frac{1}{2}(1+|F_s|^2)^{-1/2}|X_s|^2\left(2-(1+|F_s|^2)^{-1}|F_s|^2\right). \tag{1.2.36}$$

Since $(1+|F_s|^2)^{-1}|F_s|^2 \leq 1$, the process $(1+|F_t|^2)^{1/2}$ is a submartingale and by integrating the terms in (1.2.36) we obtain the estimate (1.2.33).

Step 2: The basic estimate (1.2.33) and a simple scaling argument show that for any $\beta > 0$,

$$\mathbb{E}\left(1+\beta^2|F|^2\right)^{1/2} \leq 1+\beta^2\mathbb{E}\int_0^\infty \left(1+\beta^2|F_s|^2\right)^{-1/2}|X_s|^2 ds. \tag{1.2.37}$$

Now fix $\beta^2 \leq 1/6$ and $w \in \mathbb{C}$. Then $\beta^2\left(1+\beta^2|w|^2\right)^{-1/2} \leq (1/2)|1+w|^{-1}$, by elementary arithmetic. Inserting $w = F_s$, and multiplying both sides by $|X_s|^2$ gives

$$\beta^2\left(1+\beta^2|F_s|^2\right)^{-1/2}|X_s|^2 \leq (1/2)|1+F_s|^{-1}|X_s|^2. \tag{1.2.38}$$

Comparing the right-hand sides of Inequalities (1.2.37) and (1.2.38) gives

$$\mathbb{E}\left(1+\beta^2|F|^2\right)^{1/2} \leq 1+(1/2)\mathbb{E}\int_0^\infty |X_s|^2|1+F_s|^{-1}ds, \tag{1.2.39}$$

for $\beta^2 \leq 1/6$. By Theorem 1.2.2, the right-hand side of Inequality (1.2.39) equals $\mathbb{E}|1+F|$. □

Let $h \in H^1(\mathbb{T})$ and $F = h(z_\tau)$. As $F \in H^1(\Omega)$, the estimate (1.2.32) of Theorem 1.2.6 translates into the following complex convexity estimate for $H^1(\mathbb{T})$, due to Bourgain (1983a),

$$\int_{\mathbb{T}}\left(|h_0|^2+\beta^2|h-h_0|^2\right)^{1/2}dm \leq \int_{\mathbb{T}}|h|dm, \tag{1.2.40}$$

where $h_0 = \int_{\mathbb{T}} hdm$. We next present Bourgain's original proof of Inequality (1.2.40), based on equation (1.1.76), the Riesz factorization for $H^1(\mathbb{T})$. The best value for $\beta > 0$ in Inequality (1.2.40) is unknown.

Proposition 1.2.7 *For $h \in H^1(\mathbb{T})$ with $h_0 = \int_{\mathbb{T}} hdm$, and $\kappa^2 \leq 1/9$, then*

$$\int_{\mathbb{T}}\left(|h_0|^2+\kappa^2|h-h_0|^2\right)^{1/2}dm \leq \int_{\mathbb{T}}|h|dm.$$

Proof Let $h \in H^1(\mathbb{T})$. By equation (1.1.76) (Riesz factorization) there exist $f, g \in H^2(\mathbb{T})$ so that with $f_0 = \int f dm$, and $g_0 = \int g dm$, the following factorization holds:

$$h = f \cdot g, \quad h_0 = f_0 \cdot g_0, \quad \left(\int_{\mathbb{T}} |h| dm \right)^2 = \int_{\mathbb{T}} |f|^2 dm \int_{\mathbb{T}} |g|^2 dm. \tag{1.2.41}$$

Note that by orthogonality (Pythagoras' theorem), we have

$$\int_{\mathbb{T}} |f|^2 dm = |f_0|^2 + \int_{\mathbb{T}} |f - f_0|^2 dm. \tag{1.2.42}$$

Using $h = f \cdot g$ and $h_0 = f_0 \cdot g_0$, we write

$$h - h_0 = (f - f_0)g_0 + f_0(g - g_0) + (f - f_0)(g - g_0).$$

This identity translates directly into the vector identity

$$\begin{pmatrix} h - h_0 \\ 3h_0 \end{pmatrix} = \begin{pmatrix} (f - f_0)g_0 \\ h_0 \end{pmatrix} + \begin{pmatrix} f_0(g - g_0) \\ h_0 \end{pmatrix} + \begin{pmatrix} (f - f_0)(g - g_0) \\ h_0 \end{pmatrix}. \tag{1.2.43}$$

As $h_0 = f_0 \cdot g_0$, we evaluate the Eucledian length of the first term on the right-hand side as

$$\left| \begin{pmatrix} (f - f_0)g_0 \\ h_0 \end{pmatrix} \right| = (|f_0|^2 + |f - f_0|^2)^{1/2} |g_0|,$$

similarly for the second term. By the Cauchy Schwarz inequality, followed by (1.2.42) and the factorization identity (1.2.41), we get

$$\int_{\mathbb{T}} \left(|f_0|^2 + |f - f_0|^2 \right)^{1/2} |g_0| dm \leq \int_{\mathbb{T}} |h| dm, \tag{1.2.44}$$

and by symmetry,

$$\int_{\mathbb{T}} \left(|g_0|^2 + |g - g_0|^2 \right)^{1/2} |f_0| dm \leq \int_{\mathbb{T}} |h| dm., \tag{1.2.45}$$

Next estimate routinely the third term of the vector identity (1.2.43) as

$$\left| \begin{pmatrix} (f - f_0)(g - g_0) \\ h_0 \end{pmatrix} \right| \leq \left(|g_0|^2 + |g - g_0|^2 \right)^{1/2} \left(|f_0|^2 + |f - f_0|^2 \right)^{1/2}.$$

With the Cauchy–Schwarz inequality and equations (1.2.42), and (1.2.41), one obtains

$$\int_{\mathbb{T}} \left(|g_0|^2 + |g - g_0|^2 \right)^{1/2} \left(|f_0|^2 + |f - f_0|^2 \right)^{1/2} dm \leq \int_{\mathbb{T}} |h| dm. \tag{1.2.46}$$

Since

$$\left| \begin{pmatrix} h - h_0 \\ 3h_0 \end{pmatrix} \right| = \left(9|h_0|^2 + |h - h_0|^2 \right)^{1/2},$$

by the triangle inequality in $\mathbb{R}^2$ applied to the vector identity (1.2.43) and by adding the three estimates (1.2.44)–(1.2.46), we get

$$\int_{\mathbb{T}} \left(9|h_0|^2 + |h - h_0|^2\right)^{1/2} dm \leq 3 \int_{\mathbb{T}} |h| dm.$$

The factor of 3 in the right-hand side determines the size of κ in the complex convexity estimate. Finally, dividing by 3 gives the result. □

1.2.3 Stopping Time Decompositions

In this section we systematically exploit that holomorphic martingales are stable under stopping times. We obtain the Marcinkiewicz decomposition for the pair $(H^p(\Omega), H^\infty(\Omega))$ and connect it to complex convexity estimates for $H^1(\Omega)$, obtaining a significant strengthening of Theorem 1.2.6.

Marcinkiewicz Decompositions

Our first theorem asserts that for $F \in H^p(\Omega)$ and $\lambda > 0$ there exists $G \in H^\infty(\Omega)$ satisfying $|G| \leq \lambda$ and

$$\mathbb{E}|F - G| \leq C^p \lambda^{1-p} \mathbb{E}|F|^p.$$

We will obtain G from F by conditional expectation with respect to a suitably chosen stopping-time sigma-algebra.

Theorem 1.2.8 (Marcinkiewicz decomposition theorem) *Let* $1 \leq p < \infty$, $F \in H^p(\Omega)$ *with* $\mathbb{E}(F) = 0$, *and* $\lambda > 0$. *Let*

$$\rho = \inf\{t >.0 : |F_t| > \lambda\}.$$

Put $A = \{\rho < \infty\}$ *and* $G = \mathbb{E}(F|\mathcal{F}_\rho)$. *Then* $G \in H^\infty(\Omega)$ *satisfies*

$$|G| \leq \lambda, \quad \mathbb{E}|F - G| \leq 2\mathbb{E}(1_A|F|) \quad \textit{and} \quad \lambda\mathbb{P}(A) \leq \mathbb{E}(1_A|F|). \tag{1.2.47}$$

Moreover

$$\mathbb{P}(A) \leq 2C_1^p \lambda^{-p} \mathbb{E}|F|^p \quad \textit{and} \quad \mathbb{E}|F - G| \leq C_2^p \lambda^{1-p} \mathbb{E}|F|^p, \tag{1.2.48}$$

where $C_2^p = 2C_1^{p-1}$.

Proof Let $F \in H^p(\Omega)$ with $F = \int_0^\infty X_s dz_s$. Since $\mathbb{E}(F|\mathcal{F}_\rho) = \int_0^\rho X_s dz_s$, we obtain a holomorpohic random variable by putting $G = \mathbb{E}(F|\mathcal{F}_\rho)$. Moreover, as $\{\rho < \infty\} = \{\sup_t |F_t| > \lambda\}$ we have $|G| \leq \lambda$, hence $G \in H^\infty(\Omega)$. The difference

$$F - G = \int_\rho^\infty X_s dz_s,$$

is obviously supported in $A = \{\rho < \infty\}$. Hence $\mathbb{E}|F - G| = \mathbb{E}(1_A|F - G|)$. We verify next that

$$\mathbb{E}(1_A|F - G|) \leq 2\mathbb{E}(1_A|F|).$$

To this end, $\mathbb{E}_\rho$ denotes the conditional expectation with respect to $\mathcal{F}_\rho$. By definition, A is measurable with respect to $\mathcal{F}_\rho$. Hence $1_A|\mathbb{E}_\rho(F)| \leq \mathbb{E}_\rho(1_A|F|)$, and taking expectations gives

$$\mathbb{E}(1_A|G|) \leq \mathbb{E}(\mathbb{E}_\rho(1_A|F|)) = \mathbb{E}(1_A|F|). \tag{1.2.49}$$

By the triangle inequality, $\mathbb{E}(1_A|F - G|) \leq 2\mathbb{E}(1_A|F|)$. This gives the first estimate in (1.2.47). Note that $|G(\omega)| = \lambda$, for $\omega \in A$. Equivalently, $1_A|G| = 1_A\lambda$, and hence

$$\mathbb{E}(1_A|G|) = \lambda\mathbb{P}(A). \tag{1.2.50}$$

Combining (1.2.49) and (1.2.50) gives the last estimate in (1.2.47). Next we turn to proving (1.2.48). By Garling's estimate, there exists $C_1 > 0$ independent of p such that

$$\mathbb{E}\left(\sup_t |F_t|^p\right) \leq C_1^p\mathbb{E}(|F|^p), \qquad 1 \leq p < \infty.$$

Since $A = \{\sup_t : |F_t| > \lambda\}$, by Chebyshev's inequality,

$$\mathbb{P}(A) \leq \lambda^{-p}\mathbb{E}\left(\sup_t |F_t|^p\right) < C_1^p\lambda^{-p}\mathbb{E}(|F|^p). \tag{1.2.51}$$

On the other hand, using (1.2.47) and Hölder's inequality gives

$$\mathbb{E}(|F - G|) \leq 2\mathbb{E}(1_A|F|) \leq 2(\mathbb{P}(A))^{1-1/p}(\mathbb{E}(|F|^p))^{1/p}. \tag{1.2.52}$$

It remains to use the upper bound (1.2.51) in Inequality (1.2.52) to finally obtain the right-hand side of (1.2.48). □

The space $SL^\infty(\Omega)$: We defined the quadratic variation process of $F \in L^2(\Omega)$ to be the covariance process of F and $\overline{F}$. (See Theorem 1.1.12 and the remarks thereafter.) We let $\langle F\rangle_t = \langle F, \overline{F}\rangle_t$, and set

$$\langle F\rangle = \lim_{t\to\infty}\langle F\rangle_t.$$

Since $(t \to \langle F\rangle_t)$ is nondecreasing, the (finite or infinite) limit above is well defined. We refer to $\langle F\rangle$ as the quadratic variation of F. Let the space $SL^\infty(\Omega) = \left\{F \in L^2(\Omega) : \langle F\rangle \in L^\infty(\Omega)\right\}$ be equipped with the norm

$$\|F\|_{SL^\infty(\Omega)} = |\mathbb{E}(F)| + \left\|\langle F\rangle^{1/2}\right\|_\infty .$$

Next, we turn to the Marcinkiewicz decomposition for the pair of spaces $H^1(\Omega)$ and $H^\infty(\Omega) \cap SL^\infty(\Omega)$, improving Theorem 1.2.8 as follows: Given $F \in H^p(\Omega)$, there exists $G \in H^\infty(\Omega)$ satisfying simultaneously $|G| \le \lambda$ and $\langle G\rangle^{1/2} \le \lambda$, while still approximating F in the L^1-norm

$$\mathbb{E}|F - G| \le C^p \lambda^{1-p} \mathbb{E}|F|^p.$$

This is achieved by simply imposing further restrictions on the stopping time used in the proof of Theorem 1.2.8.

Theorem 1.2.9 *Let* $1 < p < \infty$, $F \in H^p(\Omega)$ *with* $\mathbb{E}F = 0$, *and* $\lambda > 0$. *Put*

$$\rho = \inf\left\{t > 0 : |F_t| > \lambda \text{ or } \langle F\rangle_t^{1/2} > \lambda\right\} \quad \textit{and} \quad G = \mathbb{E}(F|\mathcal{F}_\rho).$$

Then $G \in H^\infty(\Omega)$ *and satisfies the following estimates:*

$$|G| \le \lambda, \quad \langle G\rangle^{1/2} \le \lambda, \textit{ and } \mathbb{E}|F - G| \le C\lambda^{1-p}\mathbb{E}|F|^p.$$

Proof Fix $1 < p < \infty$ and $F \in H^1(\Omega)$ with $\mathbb{E}F = 0$, and let $\lambda > 0$. Since $H^1(\Omega)$ is closed under stopping times and $|G| = |F_\rho| \le \lambda$, we have $G \in H^\infty(\Omega)$. Moreover, $\langle G\rangle = \langle F\rangle_\rho \le \lambda^2$. We next turn to proving the estimate for the error integral $\mathbb{E}|F - G|$. To this end, put

$$A = \{\rho < \infty\}, \quad A_1 = \left\{\sup_{t>0} |F_t| > \lambda\right\} \quad \text{and} \quad A_2 = \left\{\langle F\rangle^{1/2} > \lambda\right\}.$$

Since $\mathbb{E}F = 0$ and $\langle F\rangle = \sup_{t>0}\langle F\rangle_t$, we have $A \subseteq A_1 \cup A_2$. By construction, $\operatorname{supp}(F - G) \subseteq A$ and hence

$$\mathbb{E}|F - G| \le \mathbb{E}\left(1_{A_1}|F - G|\right) + \mathbb{E}\left(1_{A_2}|F - G|\right). \tag{1.2.53}$$

We have $\mathbb{P}(A_2) \le \lambda^{-p}\mathbb{E}\langle F\rangle^{p/2}$ and $\mathbb{P}(A_1) \le \lambda^{-p}\mathbb{E}\left(\sup_{t>0}|F_t|^p\right)$. We use Hölder's inequality on the right-hand side of Inequality (1.2.53) and invoke (1.2.21) and Doob's maximal function estimate. Taking into account that $\mathbb{E}|F - G|^p \le 2^p\mathbb{E}|F|^p$ finishes the proof. □

Complex Convexity Estimates (Continued)

Here we further develop the stopping time decompositions introduced in Theorem 1.2.8, and strengthen the complex convexity estimates of Theorem 1.2.6.

Theorem 1.2.10 (Complex convexity theorem) *Let* $F \in H^1(\Omega)$, *with* $\mathbb{E}F = 0$, *and* $z \in \mathbb{C}$. *Put*

$$\rho = \inf\{t > 0 : |F_t| > 12|z|\}, \quad \textit{and} \quad G = \mathbb{E}(F|\mathcal{F}_\rho).$$

Then $G \in H^\infty(\Omega)$ *satisfies* $|G| \le 12|z|$, *and*

$$\left(|z|^2 + \kappa^2\mathbb{E}|G|^2\right)^{1/2} + \kappa\mathbb{E}|F - G| \le \mathbb{E}|z + F|,$$

where $\kappa \le 1/60$.

Theorem 1.2.10 goes back to Bourgain's work (Bourgain, 1983a). Its proof proceeds by establishing a lower estimate for the term on the left-hand side of Inequality (1.2.32).

Theorem 1.2.11 *Let $F \in H^1(\Omega)$, with $\mathbb{E}F = 0$, and let $\beta, \lambda > 0$. Put*

$$\rho = \inf\{t > 0 : |F_t| > 2\lambda/\beta\}, \qquad \text{and} \qquad G = \mathbb{E}(F|\mathcal{F}_\rho).$$

Then $G \in H^\infty(\Omega)$ with $|G| \leq 2\lambda/\beta$, and

$$\left(\lambda^2 + \kappa^2\beta^2\mathbb{E}|G|^2\right)^{1/2} + \kappa\beta\mathbb{E}|F - G| \leq \mathbb{E}\left(\lambda^2 + \beta^2|F|^2\right)^{1/2}, \tag{1.2.54}$$

whenever $\kappa \leq 1/10$.

Proof By scaling and rotating we may assume that $\beta = \lambda = 1$. We put $A = \{\rho < \infty\}$, and let $B = \Omega \backslash A$. By Theorem 1.2.8, we have $G \in H^\infty(\Omega)$,

$$|G| \leq 2, \quad \mathbb{E}(1_A|F|) \geq 2\mathbb{P}(A), \quad 2\mathbb{E}(1_A|F|) \geq \mathbb{E}|F - G|. \tag{1.2.55}$$

By averaging, (1.2.55) gives $\mathbb{E}(1_A|F|) \geq \mathbb{P}(A) + (1/4)\mathbb{E}|F - G|$ and hence,

$$\mathbb{E}1_A(1 + |F|^2)^{1/2} \geq \mathbb{P}(A) + (1/4)\mathbb{E}|F - G|. \tag{1.2.56}$$

We clearly have

$$\mathbb{E}1_B(1 + |F|^2)^{1/2} \geq \mathbb{P}(B). \tag{1.2.57}$$

Since $\mathbb{P}(A) + \mathbb{P}(B) = 1$, addding the estimates (1.2.56) and (1.2.57) gives

$$\mathbb{E}(1 + |F|^2)^{1/2} \geq 1 + \frac{1}{4}\mathbb{E}|F - G|. \tag{1.2.58}$$

Next, use that $(1+x)^{1/2} \geq 1+x/3$ for $0 < x \leq 1$, write $F = G+(F-G)$, apply the triangle inequality, and invoke $|G| \leq 2$. This gives the pointwise estimate

$$(1 + |F|^2)^{1/2} \geq 1 + \frac{1}{12}|G|^2 - \frac{1}{2}|F - G|.$$

By integrating, we find

$$\mathbb{E}(1 + |F|^2)^{1/2} \geq 1 + \frac{1}{12}\mathbb{E}|G|^2 - \frac{1}{2}\mathbb{E}|F - G|. \tag{1.2.59}$$

It remains to form a weighted average of Inequalities (1.2.58) and (1.2.59), and to take the square root. Thus, for $\kappa = 1/10$ we obtain

$$\mathbb{E}(1 + |F|^2)^{1/2} \geq \left(1 + \kappa^2\mathbb{E}|G|^2\right)^{1/2} + \kappa\mathbb{E}|F - G|.$$ □

Proof of Theorem 1.2.10: Apply Theorem 1.2.11 to F with $\beta^2 = 1/6$ and $\lambda = |z|$. Then use Theorem 1.2.6.

1.3 Extrapolation

Let $1 < p < \infty$, and let (F_t) be an L^p-bounded, continuous $(\mathcal{F}_t)$-submartingale. For $t \in [0, \infty]$, we consider the maximal functions $F_t^* = \sup_{0<s<t} |F_s|$, and put $F^* = F_\infty^*$. Recall that Doob's inequality asserts that

$$\mathbb{E}(F_t^{*p}) \le C_p^p \sup_{0<s<t} \mathbb{E}|F_s|^p, \tag{1.3.1}$$

where $C_p = p/(p-1)$. In this section we prove the Burkholder–Davis–Gundy estimates (Burkholder, Davis, and Gundy [1972]) by extrapolating (adequately modified versions of) Doob's inequality across the entire scale of L^α-spaces, where $0 < \alpha < \infty$.

Theorem 1.3.1 (The Burkholder–Davis–Gundy inequalities) *Let (F_t) be an $(\mathcal{F}_t)$-martingale with $F_0 = 0$. Let $0 < \alpha < 2$. If $\mathbb{E}(F^*)^\alpha < \infty$, then*

$$\mathbb{E}\langle F\rangle^{\alpha/2} \le C_a \mathbb{E}(F^*)^\alpha. \tag{1.3.2}$$

Conversely, if $\mathbb{E}\langle F\rangle^{\alpha/2} < \infty$, then

$$\mathbb{E}(F^*)^\alpha \le C_a \mathbb{E}\langle F\rangle^{\alpha/2}, \tag{1.3.3}$$

where $C_a \le C/(2-\alpha)$.

Proof We present the proof first under the a priori condition that (F_t) is square integrable. The general case in Inequality (1.3.2) is obtained by considering the sequence of stopping times

$$S(n) = \inf\{t > 0 : |F_t| > n\},$$

applying Inequality (1.3.2) to the stopped martingale $t \to F_{t\wedge S(n)}$, and finally using that

$$\lim_{n\to\infty} \mathbb{E} \sup_{t>0} |F_t - F_{t\wedge S(n)}|^\alpha = 0.$$

Extending the validity of Inequality (1.3.3) to martingales satisfying $\mathbb{E}\langle F\rangle^{\alpha/2} < \infty$, we define stopping times

$$R(n) = \inf\{t > 0 : \langle F\rangle_t > n\},$$

and apply Inequality (1.3.3) to the stopped martingales $t \to F_{t\wedge R(n)}$. For any fixed $x > 0$, the proof of Theorem 1.3.1 examines and compares the stopping times

$$R = \inf\{t > 0 : \langle F_t\rangle > x\} \quad \text{and} \quad S = \inf\left\{t > 0 : |F_t|^2 > x\right\}, \tag{1.3.4}$$

yielding the relative distributional estimates (1.3.6) and (1.3.11).

Proof of equation (1.3.2): We begin with an overview of the basic steps. Starting with Itô's formula, we obtain first the L^2 estimate,

$$\mathbb{E}\langle F_T\rangle \le \mathbb{E}(F_T^*)^2, \tag{1.3.5}$$

for any stopping time $T : \Omega \to [0,\infty]$. In the second step we will examine the stopping times R, S, defined in (1.3.4), and obtain

$$\mathbb{P}(\{\langle F\rangle > x\}) \le (C+1)\mathbb{P}\left(\{F^{*2} > x\}\right) + \frac{C}{x}\mathbb{E}\left(F^{*2}1_{\{(F^*)^2\le x\}}\right). \tag{1.3.6}$$

Finally, we apply the Marcinkiewicz integration trick to show that the distributional estimates (1.3.6) imply

$$\mathbb{E}\left(\langle F_T\rangle^{\alpha/2}\right) \le A_\alpha \mathbb{E}\left(F_T^*\right)^\alpha, \qquad 0 < \alpha < 2, \tag{1.3.7}$$

where $A_\alpha \le C/(2-\alpha)$. We now turn to presenting the details of the proof.

Step 1: Fix an L^2-bounded $(\mathcal{F}_t)$-martingale (F_t), and an $(\mathcal{F}_t)$-stopping time T. As, by Theorem 1.1.13,

$$|F_{T\wedge t}|^2 - \langle F_{T\wedge t}\rangle,$$

is an $(\mathcal{F}_t)$-martingale we have $\mathbb{E}\langle F_{T\wedge t}\rangle = \mathbb{E}F^2_{T\wedge t} \le \mathbb{E}\left(F_T^{*2}\right)$. Taking the limit $t \to \infty$ on the left-hand side gives Inequality (1.3.5).

Step 2: Fix $x > 0$ and consider the stopping times R, S defined in (1.3.4). We have $\{S = \infty\} = \left\{(F^*)^2 < x\right\}$ and $\{R < \infty\} = \{\langle F\rangle > x\}$. Hence, with $T = S \wedge R$,

$$\left\{\langle F\rangle > x, (F^*)^2 \le x\right\} = \{S = \infty, R < \infty\} \subseteq \{\langle F_T\rangle > x\}.$$

Taking expectations and invoking Inequality (1.3.5) gives

$$\mathbb{P}\left\{\langle F\rangle > x, (F^*)^2 \le x\right\} \le \frac{1}{x}\mathbb{E}\langle F_T\rangle \le \frac{C}{x}\mathbb{E}\left(F_T^{*2}\right). \tag{1.3.8}$$

We have the following pointwise upper bound for $\left(F_T^*\right)^2$,

$$\begin{aligned}(F_T^*)^2 &= (F_T^*)^2\, 1_{\{S<\infty\}} + (F_T^*)^2\, 1_{\{S=\infty\}}\\ &\le x1_{\{S<\infty\}} + (F_T^*)^2\, 1_{\{S=\infty\}}\\ &\le x1_{\{(F^*)^2>x\}} + (F^*)^2 1_{\{(F^*)^2\le x\}},\end{aligned}$$

In view of Inequality (1.3.8) we obtain,

$$\begin{aligned}\mathbb{P}\{\langle F\rangle > x\} &\le \mathbb{P}\left\{(F^*)^2 \ge x\right\} + \mathbb{P}\left\{\langle F\rangle > x, (F^*)^2 \le x\right\}\\ &\le \mathbb{P}\left\{(F^*)^2 \ge x\right\} + \frac{C}{x}\mathbb{E}\left(F_T^{*2}\right)\\ &\le (1+C)\mathbb{P}\left\{(F^*)^2 \ge x\right\} + \frac{C}{x}\mathbb{E}\left((F^*)^2 1_{\{(F^*)^2\le x\}}\right).\end{aligned} \tag{1.3.9}$$

Step 3: Here, we draw the consequences of the distributional estimates (1.3.9) using the Marcinkiewicz trick. We fix $0 < \alpha < 1$, apply Fubini's theorem, and calculate to obtain

$$\int_0^\infty \mathbb{E}\left((F^*)^2 1_{\{(F^*)^2 \le x\}}\right) x^{\alpha/2-2} dx = \mathbb{E}\left(\int_{(F^*)^2}^\infty x^{\alpha/2-2} dx (F^*)^2\right) = \frac{2}{2-\alpha}\mathbb{E}((F^*)^\alpha).$$

Since $\mathbb{E}\langle F\rangle^{\alpha/2} = \alpha/2 \int_0^\infty \mathbb{P}\{\langle F\rangle > x\} x^{\alpha/2-1} dx$, we routinely finish the proof of Inequality (1.3.2) by combining (1.3.9) and the above identity.

Proof of Inequality (1.3.3): We give an overview of the three basic steps yielding the proof of Inequality (1.3.3). First we show that for any stopping time $T : \Omega \to \mathbb{R}_0^+ \cup \{\infty\}$,

$$\mathbb{E}\,(F_T^*)^2 \le C_2^2 \mathbb{E}\langle F_T\rangle. \tag{1.3.10}$$

To get Inequality (1.3.10) we use Doob's inequality for $p = 2$. Second, we examine the stopping times R, S, and verify the distributional estimates,

$$\mathbb{P}\left(\left\{(F^*)^2 > x\right\}\right) \le (C_2^2 + 1)\mathbb{P}(\{\langle F\rangle > x\}) + \frac{C_2^2}{x}\mathbb{E}\left(1_{\{\langle F\rangle \le x\}}\langle F\rangle\right). \tag{1.3.11}$$

Finally, we employ the Marcinkiewicz integration trick to show that Inequality (1.3.11) implies that

$$\mathbb{E}\,(F_T^*)^\alpha \le A_a \mathbb{E}\langle F_T\rangle^{\alpha/2}, \qquad 0 < \alpha < 2. \tag{1.3.12}$$

Step 1: Fix an L^2-bounded $(\mathcal{F}_t)$-martingale (F_t). Let T be an $(\mathcal{F}_t)$-stopping time. As

$$|F_{T\wedge t}|^2 - \langle F_{T\wedge t}\rangle,$$

is an $(\mathcal{F}_t)$-martingale we have $\mathbb{E}F_{T\wedge t}^2 = \mathbb{E}\langle F_{T\wedge t}\rangle$. Invoking Doob's inequality with $p = 2$ gives Inequality (1.3.10).

Step 2: Fix $x > 0$ and consider the stopping times R, S defined in (1.3.4). Clearly, we have $\{S < \infty\} = \left\{(F^*)^2 > x\right\}$ and $\{R = \infty\} = \{\langle F\rangle \le x\}$. A moment's reflection reveals that with $T = S \wedge R$,

$$\{S < \infty, R = \infty\} \subseteq \left\{(F_T^*)^2 > x\right\}. \tag{1.3.13}$$

Hence, by (1.3.13), Chebyshev's inequality, and Inequality (1.3.10) we obtain

$$\mathbb{P}\left\{(F^*)^2 > x, \langle F\rangle \le x\right\} \le \mathbb{P}\left\{(F_T^*)^2 > x\right\} \le \frac{1}{x}\mathbb{E}\,(F_T^*)^2 \le \frac{C}{x}\mathbb{E}\langle F_T\rangle. \tag{1.3.14}$$

Next, observe the pointwise upper bound for $\langle F_T\rangle$,

$$\begin{aligned}\langle F_T\rangle &= \langle F_T\rangle 1_{\{R<\infty\}} + \langle F_T\rangle 1_{\{R=\infty\}}\\ &\le x 1_{\{R<\infty\}} + \langle F_T\rangle 1_{\{R=\infty\}}\\ &\le x 1_{\{\langle F\rangle\ge x\}} + \langle F\rangle 1_{\{\langle F\rangle\le x\}}.\end{aligned} \tag{1.3.15}$$

Taking expectations and invoking Inequality (1.3.14) we obtain

$$\begin{aligned}\mathbb{P}\left\{(F^*)^2 > x\right\} &\le \mathbb{P}\{\langle F\rangle \ge x\} + \mathbb{P}\left\{(F^*)^2 > x, \langle F\rangle \le x\right\}\\ &\le \mathbb{P}\{\langle F\rangle \ge x\} + \frac{C}{x}\mathbb{E}\langle F_T\rangle\\ &\le (1+C)\mathbb{P}\{\langle F\rangle \ge x\} + \frac{C}{x}\mathbb{E}\left(\langle F\rangle 1_{\{\langle F\rangle\le x\}}\right).\end{aligned} \tag{1.3.16}$$

Step 3: Here, we exploit Inequality (1.3.16) and use the Marcinkiewicz integration trick. We fix $0 < \alpha < 2$, apply Fubini's theorem, and calculate to obtain

$$\int_0^\infty \mathbb{E}\left(\langle F\rangle 1_{\{\langle F\rangle\le x\}}\right) x^{\alpha/2-2}dx = \mathbb{E}\left(\int_{\langle F\rangle}^\infty x^{\alpha/2-2}dx\langle F\rangle\right) = \frac{2}{2-\alpha}\mathbb{E}\left(\langle F\rangle^{\alpha/2-1}\langle F\rangle\right).$$

We routinely finish the proof of Inequality (1.3.3) by combining Inequality (1.3.16) and the above identity. Tracing back the constants in the calculation gives

$$A_\alpha \le \left(\alpha(1+C) + \frac{c\alpha}{2-\alpha}\right),$$

where $C = C_2^2$ is the constant in Doob's inequality. □

We continue with Burkholder's weak type (1:1) estimate for the maximal function of martingale transforms.

Theorem 1.3.2 *Let (F_t) and (G_t) be $(\mathcal{F}_t)$-martingales satisfying $F_0 = G_0 = 0$. If $\sup_{t>0}\mathbb{E}|F_t| < \infty$, and*

$$\langle F_t\rangle \le \langle G_t\rangle, \quad t > 0,$$

then

$$x\mathbb{P}\{G^* > x\} \le 2\sup_{t>0}\mathbb{E}|F_t|, \quad x > 0.$$

Proof In the first part of the proof we work under the a priori restriction that (F_t) is bounded in L^2 and use that then the martingale (F_t) is uniformly integrable. Fix $x > 0$ and note that

$$\{G^* > x\} \subseteq \{G^* > x, F^* \le x\} \cup \{F^* > x\}.$$

Doob's maximal inequality (1.1.26) gives $x\mathbb{P}\{F^* > x\} \le \sup_{t>0}\mathbb{E}|F_t|$. Thus the theorem follows from the measure estimate $x\mathbb{P}\{G^* > x, F^* \le x\} \le \mathbb{E}|F|$.

Step 1: Consider the stopping time,

$$T = \inf\{t > 0 \colon |F_t| > x\}.$$

We have $\{F^* \leq x\} = \{T = \infty\}$. Hence, $\{G^* > x, F^* \leq x\} \subseteq \left\{G_T^* > x\right\}$, and, in view of Doob's inequality (1.1.26), we obtain

$$\mathbb{P}\{G^* > x, F^* \leq x\} \leq \mathbb{P}\{G_T^* > x\} \leq x^{-2}\mathbb{E}(G_T)^2. \tag{1.3.17}$$

Now we invoke the hypothesis $\langle F_T \rangle \leq \langle G_T \rangle$, use that $|F_T| \leq x$, and that the optional stopping theorem (Theorem 1.1.9) – applied to the uniformly integrable martingale (F_t) – yields $\mathbb{E}|F_T| \leq \mathbb{E}|F|$ where $F = \lim F_t$. This gives,

$$\mathbb{E}(G_T)^2 = \mathbb{E}\langle G_T \rangle \leq \mathbb{E}\langle F_T \rangle = \mathbb{E}|F_T|^2 \leq x\mathbb{E}|F_T| \leq x\mathbb{E}|F|. \tag{1.3.18}$$

Combining Inequalities (1.3.18) and (1.3.17) gives $x\mathbb{P}\{G^* > x, F^* \leq x\} \leq \mathbb{E}|F|$, which finishes the proof under the assumption that (F_t) is bounded in L^2.

Step 2: To eliminate this restriction we again use the reducing stopping times $S(n) = \inf\{t > 0 : |F_t| > n\}$, apply the first part of the proof to each of the L^2-bounded martingales $(F_{S(n)\wedge t})$, and pass to the limit $n \to \infty$. Thus, we finally obtain,

$$x\mathbb{P}\{G^* > x\} = \sup_n x\mathbb{P}\left\{G_{S(n)}^* > x\right\} \leq 2\sup_n \sup_{t>0} \mathbb{E}|F_{S(n)\wedge t}| = 2\sup_{t>0} \mathbb{E}|F_t|,$$

for $(\mathcal{F}_t)$-martingales (F_t) and (G_t) satisfying $\sup_{t>0} \mathbb{E}|F_t| < \infty$, $\langle F_t \rangle \leq \langle G_t \rangle$ and $F_0 = G_0 = 0$. □

1.4 Stochastic Hilbert Transforms

Let $R = (R_t)$ be a real-valued, square-integrable martingale on Wiener space with stochastic integral representation $R_t = R_0 + \int_0^t Y_s dz_s + \overline{Y_s} d\overline{z}_s$. We define the stochastic Hilbert transform of R by putting

$$\mathcal{H}R = i \int_0^\infty \overline{Y_s} d\overline{z}_s - Y_s dz_s. \tag{1.4.1}$$

Note that $\mathcal{H}R$ is again real valued and $\mathcal{H}^2 R = R - \mathbb{E}R$. Moreover, for any stopping time $T \colon \Omega \to \mathbb{R}_0^+ \cup \infty$, we have the identity

$$(\mathcal{H}R)_T = \mathcal{H}(R_T).$$

Consider $R = u(z_\tau)$ where $u \in L^2(\mathbb{T})$ is real valued. Itô's formula connects the martingale operator $\mathcal{H}$ to the classical Hilbert transform H by the identity

$\mathcal{H}R = (Hu)(z_\tau)$. Indeed, put $v = Hu$ and denote by u and v the respective harmonic extensions of u and v to the unit disk. Since $u + iv$ is analytic, we have $\overline{\partial}u = -i\overline{\partial}\, v$ and $\partial u = i\partial v$. By Itô's formula, $u(z_\tau) = u(0) + \int_0^\tau \partial u(z_s)dz_s + \overline{\partial}u(z_s)d\overline{z}_s$, and consequently,

$$\begin{aligned} v(z_\tau) &= \int_0^\tau \partial v(z_s)dz_s + \overline{\partial}v(z_s)d\overline{z}_s \\ &= i\int_0^\tau \overline{\partial}u(z_s)d\overline{z}_s - \partial u(z_s)dz_s \\ &= \mathcal{H}u(z_\tau). \end{aligned}$$

Theorem 1.4.1 *Let $R \in L^2(\Omega)$ be real valued with $\mathbb{E}R = 0$. Then $\mathbb{E}(R^2) = \mathbb{E}((\mathcal{H}R)^2)$ and $R + i\mathcal{H}R \in H^2(\Omega)$. Moreover, the quadratic variation processes of R and $\mathcal{H}R$ coincide.*

Proof Let $R \in L^2(\Omega)$ have stochastic integral representation $R = \int_0^\infty Y_s dz_s + \overline{Y_s}d\overline{z}_s$. Theorem 1.1.12 yields $\mathbb{E}(R^2) = 4\mathbb{E}\int_0^\infty \overline{Y_s}Y_s ds$. Similarly, invoking (1.4.1) gives

$$\mathbb{E}\left((\mathcal{H}R)^2\right) = 4\mathbb{E}\int_0^\infty \overline{Y_s}Y_s ds.$$

Comparing the defining integrals for R and $\mathcal{H}R$ shows that $R + i\mathcal{H}R = 2\int Y dz$. Hence, $R + i\mathcal{H}R \in H^2(\Omega)$. Finally, by (1.4.1), the covariance formula shows that

$$\langle \mathcal{H}R\rangle_t = 4\int_0^t \overline{Y_s}Y_s ds = \langle R\rangle_t. \qquad \square$$

Now we are ready for our first main theorem on the stochastic Hilbert transform. It rests on Theorem 1.3.1 (the Burkholder–Davis–Gundy inequalities) and Theorem 1.3.2 (the weak type (1:1) estimates for the maximal function of martingale transforms).

Theorem 1.4.2 *There exists $C > 0$ such that for any real-valued $R \in L^1(\Omega)$ satisfying $\mathbb{E}R^* < \infty$, the following estimates hold true:*

$$C^{-1}\mathbb{E}(|R| + |\mathcal{H}(R)|) \le \mathbb{E}\sup_t |R_t| \le C\mathbb{E}(|R| + |\mathcal{H}(R)|), \tag{1.4.2}$$

and,

$$\sup_{\lambda \ge 0} \lambda\mathbb{P}\left\{\sup_t |\mathcal{H}(R)_t| > \lambda\right\} \le C\mathbb{E}|R|. \tag{1.4.3}$$

Proof Assume first that R is square integrable and $\mathbb{E}R = 0$. Since R and $\mathcal{H}R$ are real valued, and since $R + i\mathcal{H}R$ is a holomorphic random variable, applying Garling's estimate gives

$$\mathbb{E}\sup_t |R_t| \leq \mathbb{E}\sup_t |R_t + i\mathcal{H}R_t| \leq e\mathbb{E}|R + i\mathcal{H}R| \leq e\mathbb{E}(|R| + |\mathcal{H}(R)|).$$

Conversely, invoking the Burkholder–Davis–Gundy inequality (1.3.3) gives

$$\mathbb{E}(|R| + |\mathcal{H}(R)|) \leq \sqrt{2}\mathbb{E}\sup_t |R_t + i\mathcal{H}R_t| \leq C\mathbb{E}(\langle R\rangle + \langle \mathcal{H}R\rangle)^{1/2}.$$

Taking into account that $\langle \mathcal{H}R\rangle = \langle R\rangle$ and applying the Burkholder–Davis–Gundy inequality (1.3.2) we find

$$\mathbb{E}(\langle R\rangle + \langle \mathcal{H}R\rangle)^{1/2} \leq \sqrt{2}\mathbb{E}\langle R\rangle^{1/2} \leq C\mathbb{E}\sup_t |R_t|.$$

To get rid of the a priori assumption that R is square integrable, we fix R with $\mathbb{E}\sup_t |R_t| < \infty$, let $n \in \mathbb{N}$, and consider the stopping times $T(n) = \inf\{t > 0 : |R_t| > n\}$, where $R_t = \mathbb{E}(R|\mathcal{F}_t)$. Apply the result to the stopped martingale $t \to R_{t\wedge T(n)}$, use that

$$\lim_{n\to\infty} \mathbb{E}\sup_{t\to\infty} |R_{t\wedge T(n)} - R_t| = 0,$$

and pass to the limit $n \to \infty$. Finally, as $\langle \mathcal{H}(R)_t\rangle = \langle R_t\rangle$, the weak type estimate 1.4.3 follows directly from Theorem 1.3.2. □

As the stochastic Hilbert transform $\mathcal{H}$ preserves the quadratic variation process, $\langle R\rangle = \langle \mathcal{H}R\rangle$, Theorem 1.3.1 (the Burkholder–Davis–Gundy inequalities) and Doob's inequality imply that $\mathcal{H}$ extends to a bounded operator on $L^p(\Omega)$ for $1 < p < \infty$. Hence, the orthogonal projection,

$$\frac{1}{2}(\mathrm{Id} + i\mathcal{H}) \colon L^2(\Omega) \to H^2(\Omega),$$

extends boundedly to $L^p(\Omega)$, generating a large space of holomorphic random variables in $L^p(\Omega)$, $1 < p < \infty$. The next theorem provides constructions with $L^\infty(\Omega)$ estimates.

Theorem 1.4.3 *Let $R \in L^\infty(\Omega)$ be real valued and $F = \exp(R + i\mathcal{H}(R))$. Then $F \in H^\infty(\Omega)$, with $|F| = \exp(R)$ and $\mathbb{E}F = \exp\mathbb{E}(R)$.*

Proof By Theorem 1.4.1, $G = R + i\mathcal{H}(R) \in H^2(\Omega)$ and by hypothesis $|F| = \exp(R)$ is bounded. Using Itô's formula, $\exp(G) = \exp(G_0) + \int \exp(G_s)dz_s$. Hence, the composition $\exp(G)$ defines a bounded holomorphic random variable, so that $F \in H^\infty(\Omega)$ and

$$F_t = \exp(R_t + i(\mathcal{H}R)_t), \quad 0 \leq t < \infty. \tag{1.4.4}$$

□

Theorem 1.4.4 *Let $\lambda > 0$. Let $F \in H^2(\Omega)$ and $R \in L^2(\Omega)$ be real valued such that*

$$F = \exp(R + i\mathcal{H}R).$$

Put $A = \{|F| > \lambda\}$ and $Z = R1_{\Omega\setminus A} + (\ln\lambda)1_A$. Then $G = \exp(Z + i\mathcal{H}Z)$ is a bounded holomorphic variable satisfying

$$|G| \le \lambda \quad \textit{and} \quad \mathbb{E}|F - G| \le \mathbb{E}|F|^2/\lambda. \tag{1.4.5}$$

Moreover, for $S = G/F$,

$$\mathbb{E}|1 - S|^2 \le 2(1 - \mathbb{E}(S)) \le 2\mathbb{E}(1_A \ln(|F|/\lambda)) \le (2/\lambda)\mathbb{E}(1_A|F|). \tag{1.4.6}$$

Proof By homogeneity it suffices to consider the case $\mathbb{E}|F|^2 = 1$. By construction, $G \in H^\infty(\Omega)$ and $|G| \le \lambda$. Moreover, $|G| \le |F|$. Hence (G/F) is contained in the unit ball of $H^\infty(\Omega)$, such that by (1.4.4),

$$\mathbb{E}(G/F) = \exp\mathbb{E}\ln(|G/F|). \tag{1.4.7}$$

As $\mathbb{E}(G/F) \in \mathbb{R}$, we find by arithmetic that

$$\mathbb{E}|1 - G/F|^2 = 2(1 - \mathbb{E}(G/F)) - \mathbb{E}(1 - |G/F|^2). \tag{1.4.8}$$

Using (1.4.7) and unwinding the construction of G, we have

$$\mathbb{E}(G/F) = \exp\mathbb{E}(Z - R) = \exp -\mathbb{E}(\ln(|F|/\lambda)1_A).$$

The elementary inequalities $1 - \exp(-t) \le t$, $\ln a < a$, and $\ln a < a^2/2$ for $a \ge 1$, together with $\mathbb{E}|F|^2 = 1$, give,

$$2(1 - \mathbb{E}(G/F)) \le 2\mathbb{E}(\ln(|F|/\lambda)1_A) \le \begin{cases} \mathbb{E}(|F|^2 1_A)/\lambda^2 \\ 2\mathbb{E}(|F|1_A)/\lambda. \end{cases} \tag{1.4.9}$$

Here, the top and bottom lines of the split inequality represent two separate, valid upper estimates. This proves Inequality (1.4.6). Finally, apply the Cauchy–Schwarz inequality to the product $F - G = F(1 - G/F)$. As $|G/F| \le 1$, we may use (1.4.8) and (1.4.9) to obtain

$$\mathbb{E}|F - G| \le (\mathbb{E}|F|^2)^{1/2}(2(1 - \mathbb{E}G/F))^{1/2} \le \mathbb{E}|F|^2/\lambda. \qquad \square$$

1.5 Martingale Embedding and Projection

We let $(\Omega, \mathcal{F}, (\mathcal{F}_t), \mathbb{P})$ denote Wiener's filtered probability space, let (z_t) denote complex Brownian motion, and define the (almost surely finite) stopping time $\tau = \inf\{t > 0 : |z_t| > 1\}$. In this section, we construct a left inverse for the martingale embedding operator $M(f) = f(z_\tau)$ and verify basic L^p-estimates.

Let $\mathcal{G}$ denote the stopping time sigma-algebra defined by τ, thus

$$\mathcal{G} = \{A \in \mathcal{F} : A \cap \{\tau \leq t\} \in \mathcal{F}_t, t > 0\}.$$

Lemma 1.5.1 shows that a $\mathcal{G}$-measurable random variable on Ω is represented by a measurable function f defined on

$$\mathbb{T} = \{\zeta \in \mathbb{C} : |\zeta| = 1\}.$$

Representing accordingly the conditional expectation operator $F \to \mathbb{E}(F|\mathcal{G})$ provides the first step towards constructing a left inverse of $M(f) = f(z_\tau)$.

Lemma 1.5.1 *Let $1 \leq p < \infty$. Assume that $X : \Omega \to \mathbb{R}$ is $\mathcal{G}$-measurable and $\mathbb{E}|X|^p < \infty$. There exists a uniquely determined p-integrable $f : \mathbb{T} \to \mathbb{R}$ such that*

$$\mathbb{E}|X|^p = \int_{\mathbb{T}} |f|^p dm \quad \textit{and} \quad X = f(z_\tau).$$

Proof For simplicity of notation we treat the case $p = 1$ and set $Z = z_\tau$. Define on $\mathbb{T}$ a measure ν by

$$\nu(E) = \int_{Z^{-1}(E)} X d\mathbb{P}, \tag{1.5.1}$$

with $E \subseteq \mathbb{T}$ measurable. Since, by Kakutani's Theorem (equation (1.1.54)), $\mathbb{P}(\{z_\tau \in E\}) = m(E)$, and since X is integrable, ν is absolutely continuous with respect to m. By the Radon–Nikodym theorem, there exists a unique integrable $f : \mathbb{T} \to \mathbb{R}$ such that

$$\int_E f dm = \nu(E), \quad E \subseteq \mathbb{T}, \quad \text{and} \quad \mathbb{E}|X| = \int_{\mathbb{T}} |f| dm. \tag{1.5.2}$$

We next claim that $X = f(Z)$. Taking into account that both random variables are measurable with respect to $\mathcal{G} = \left\{Z^{-1}(E) : E \subseteq \mathbb{T}\right\}$ it suffices to show that

$$\int_{Z^{-1}(E)} X d\mathbb{P} = \int_{Z^{-1}(E)} f(Z) d\mathbb{P}. \tag{1.5.3}$$

To see (1.5.3) we invoke Kakutani's theorem again. Indeed, the integral form of the identities $\mathbb{P}(\{z_\tau \in E\}) = m(E)$, is just

$$\int_{Z^{-1}(E)} f(Z) d\mathbb{P} = \int_E f dm. \tag{1.5.4}$$

Hence, in combination with (1.5.2) we obtain (1.5.3). □

Theorem 1.5.2 *Let $1 \le p \le \infty$. The martingale embedding $M(f) = f(z_\tau)$ defines a contractive linear operator $M\colon L^p(\mathbb{T}) \to L^p(\Omega)$. There exists, a contraction $N_0\colon L^p(\Omega) \to L^p(\mathbb{T})$ such that*

$$N_0(M(f)) = f,$$

for $f \in L^p(\mathbb{T})$, and,

$$\int_{\mathbb{T}} (N_0 F) g dm = \mathbb{E}(F(Mg)),$$

for $F \in L^p(\Omega)$ and $g \in L^q(\mathbb{T})$, where $1/p + 1/q = 1$.

Proof By Lemma 1.5.1, for each $F \in L^p(\Omega)$ there exists a unique $f \in L^p(\mathbb{T})$ such that $\mathbb{E}(F|\mathcal{G}) = f(Z)$. Putting $N_0(F) = f$ defines a contractive linear operator

$$N_0\colon L^p(\Omega) \to L^p(\mathbb{T}),$$

for $1 \le p \le \infty$. The operator N_0 is called the Doob projection. As $M(g) = g(Z)$, for $g \in L^p(\mathbb{T})$, Kakutani's theorem implies that

$$M\colon L^p(\mathbb{T}) \to L^p(\Omega)$$

is a contraction for $1 \le p \le \infty$. Taking into account the uniqueness assertion in Lemma 1.5.1, we obtain

$$N_0(M(f)) = f, \tag{1.5.5}$$

for any $f \in L^p(\mathbb{T})$. Moreover, for $F \in L^2(\Omega)$ and $g \in L^2(\mathbb{T})$ we have

$$\int_{\mathbb{T}} (N_0 F) g dm = \mathbb{E}(F(Mg)). \tag{1.5.6}$$

By (1.5.5) and (1.5.6), the operator $MN_0\colon L^2(\Omega) \to L^2(\Omega)$ is a self-adjoint projection. Its range is $L^2(\Omega, \mathcal{G})$. □

Stochastic integral representation of M: We continue by analyzing more closely the martingale embedding operator M. Let $f \in L^1(\mathbb{T})$ and let

$$f(z) = \int_{\mathbb{T}} P(z, \zeta) f(\zeta) dm(\zeta), \qquad z \in \mathbb{D},$$

denote its harmonic extension to the unit disk. As $\partial_z \overline{\partial_z} f(z) = 0$, and since we set $Z = z_\tau$, applying Itô's formula (1.1.49) to $M(f) = f(z_\tau)$ gives

$$M(f) = \mathbb{E}M(f) + \int_0^\infty 1_{\{t \le \tau\}} (\partial_z f(z_s) dz_s + \overline{\partial_z} f(z_s) d\overline{z}_s). \tag{1.5.7}$$

1.6 Projecting Holomorphic Martingales

We return to the martingale embedding operator M and its left inverse N_0, defined in Section 1.5. Here, we investigate the operator N_0 in considerably more detail and exploit Itô's formula to study its effect on holomorphic martingales and the quadratic variation process.

We begin by defining Doob's operator N and utilizing stochastic calculus to derive its basic properties. We show that N coincides with N_0 and that M, N intertwine the Hilbert transform H with the stochastic Hilbert transform $\mathcal{H}$ as follows:

$$\mathcal{H}(Mf) = M(Hf) \quad \text{and} \quad N(\mathcal{H}\mathcal{F}) = \mathcal{H}(N\mathcal{F}),$$

where $f \in L^p(\mathbb{T})$ and $F \in L^p(\Omega)$.

Defining Doob's Projection

We let $(\Omega, (\mathcal{F}_t), \mathbb{P})$ denote the filtered Wiener space, and let (z_t) denote complex Brownian motion on Ω with $\langle z_t, \overline{z_t}\rangle = 2t$ and $\langle z_t, z_t\rangle = 0$. We employ the increasing families of stopping times,

$$\tau_r = \inf\{t > 0 : |z_t| > r\}, \quad 0 < r \leq 1, \quad \tau_1 = \tau,$$

and let $\mathcal{G}_r = \mathcal{F}_{\tau_r}$ denote the stopping time sigma-algebra generated by τ_r. Recall that by Itô's formula, for any (bounded) harmonic function $u\colon \mathbb{D} \to \mathbb{R}$, the process $(u(z_t)\colon 0 < t < \tau)$ forms a martingale with respect to $(\mathcal{F}_t)$. Moreover, $(u(z_{\tau_r})\colon 0 < r < 1)$ forms a martingale with respect to the filtration $(\mathcal{G}_r)$. Indeed, by Theorem 1.1.9,

$$u(z_{\tau_s}) = \mathbb{E}(u(z_{\tau_r})|\mathcal{G}_s) \quad \text{for} \quad 0 < s < r < 1.$$

We now turn to the construction of the martingale operator N.

Step 1: We begin by recalling the optional stopping theorem (Theorem 1.1.9). Given $F \in L^1(\Omega)$ and $F_t = \mathbb{E}(F|\mathcal{F}_t)$, we let F_{τ_r} be obtained by sampling the process $(F_t : t > 0)$ at τ_r. Theorem 1.1.9 asserts that

$$F_{\tau_r} = \mathbb{E}(F|\mathcal{G}_r),$$

for $0 < r \leq 1$, and that

$$F_{t\wedge\tau_r} = \mathbb{E}(F|\mathcal{F}_{t\wedge\tau_r}) = \mathbb{E}(\mathbb{E}(F|\mathcal{G}_r)|\mathcal{F}_t) \quad \text{for} \quad t > 0.$$

Step 2: For $z \in \mathbb{D}$ and $\zeta \in \mathbb{T}$, the Poisson kernel is given by

$$P(z,\zeta) = (1 - |z|^2)/|\zeta - z|^2.$$

For $\zeta \in \mathbb{T}$ fixed, $\partial_z \overline{\partial_z} P(z,\zeta) = 0$, $z \in \mathbb{D}$. Hence, in view of Itô's formula, the process $(P(z_{\tau_s},\zeta) : 0 < s < 1)$ forms a martingale with respect to the stopping time sigma-algebras $(\mathcal{G}_s)$. Indeed, we have

$$\mathbb{E}(P(z_{\tau_r},\zeta)|\mathcal{G}_s) = P(z_{\tau_s},\zeta) \quad \text{for} \quad r \geq s.$$

Step 3: Fix $1 \leq p \leq \infty$, $0 < s < 1$. We let $S_p^s = L^p(\Omega, \mathcal{G}_s, \mathbb{P})$. Thus,

$$F = \mathbb{E}(F|\mathcal{G}_s), \tag{1.6.1}$$

for $F \in S_p^s$. We define Doob's projection $N(F)$, for $F \in S_p^s$, by integration against Poisson kernels,

$$N(F)(\zeta) = \lim_{r\to 1} \mathbb{E}(FP(z_{\tau_r},\zeta)), \qquad \zeta \in \mathbb{T}. \tag{1.6.2}$$

For $F \in S_p^s$, the limit in (1.6.2) is well defined. Indeed, as $P(z,\zeta)$ is harmonic with respect to the variable z, on Wiener space the process $(P(z_{\tau_s},\zeta) : 0 < s \leq 1)$ forms a $(\mathcal{G}_s)$-martingale, and (1.6.1) gives that

$$\mathbb{E}(FP(z_{\tau_r},\zeta)) = \mathbb{E}(\mathbb{E}(FP(z_{\tau_s},\zeta)|\mathcal{G}_s)) = \mathbb{E}(FP(z_{\tau_s},\zeta)), \quad \text{for} \quad 1 > r \geq s,$$

for any $F \in S_p^s$. Summing up, for any $F \in S_p^s$ the limit (1.6.2) defining $N(F)$ exists in $L^p(\mathbb{T})$ and pointwise, for almost every $\zeta \in \mathbb{T}$. With the next theorem we complete the definition of N as an operator on all of $L^p(\Omega)$.

Theorem 1.6.1 *Let $1 \leq p \leq \infty$. There exists a unique extension of N to $L^p(\Omega)$ (still denoted by N), so that for any $F \in L^p(\Omega)$,*

$$\|N(F)\|_{L^p(\mathbb{T})} \leq \|\mathbb{E}(F|\mathcal{F}_\tau)\|_{L^p(\Omega)}. \tag{1.6.3}$$

Moreover, $|N(F)|^p \leq N(|F|^p)$ almost surely, and if $F \geq 0$ then $N(F) \geq 0$.

Proof Fix $F \in L^p(\Omega)$ and $0 < s < 1$. Put $G_s = \mathbb{E}(F|\mathcal{F}_{\tau_s})$. As for $0 < r < 1$ we have $\mathbb{E}P(z_{\tau_r},\zeta) = 1$, each of the random variables $P(z_{\tau_r},\zeta)$ is a probability density on Wiener space. Hence, by Hölder's inequality, followed by integration over $\zeta \in \mathbb{T}$, we obtain

$$\int_{\mathbb{T}} \mathbb{E}(|G_s|P(z_{\tau_r},\zeta))^p dm(\zeta) \leq \int_{\mathbb{T}} \mathbb{E}\left(|G_s|^p P(z_{\tau_r},\zeta)\right) dm(\zeta).$$

Next, recall that for $s \leq r < 1$, we have $N(G_s)(\zeta) = \mathbb{E}(G_s P(z_{\tau_r},\zeta))$, and that $\int_{\mathbb{T}} P(z,\zeta) dm(\zeta) = 1$. In view of the above, we proved

$$\int_{\mathbb{T}} |N(G_s)|^p dm \leq \mathbb{E}|G_s|^p. \tag{1.6.4}$$

Let Y^p denote the L^p norm closure of $\{S_p^s : 0 < s < 1\}$. By Inequality (1.6.4), the operator N defined by the limit (1.6.2) admits a norm one extension to Y^p. On the other hand, the submartingale convergence theorem (Theorem 1.1.10) implies that $\mathbb{E}(F|\mathcal{F}_\tau) \in Y^p$. Finally, putting $N(F) = N(\mathbb{E}(F|\mathcal{F}_\tau))$ defines the desired norm one extension of the limit (1.6.2) to $L^p(\Omega)$ satisfying Inequality (1.6.3).

Next, assume that $F \geq 0$. As $\mathbb{E}(F|\mathcal{F}_\tau) \geq 0$, by the limit (1.6.2),

$$N(F)(\zeta) = \lim_{r\to 1} \mathbb{E}(\mathbb{E}(F|\mathcal{F}_\tau)P(z_{\tau_r}, \zeta)) \geq 0,$$

for almost every $\zeta \in \mathbb{T}$. Moreover, for $F = 1_\Omega$ we have $N(1_\Omega) = 1_\mathbb{T}$.

Finally, we turn to proving the pointwise estimate $|N(F)|^p \leq N(|F|^p)$. Fix $\zeta \in \mathbb{T}$, put $x = |N(F)(\zeta)|$. Since $1 \leq p < \infty$, the map $y \to |y|^p$ is convex. Hence, given $x \geq 0$ there exist $a = a(x)$ and $b = b(x)$ satisfying $x^p = ax + b$, and such that for any $y \in \mathbb{R}$ we have the inequality $|y|^p \geq ay + b$. Then using that N maps nonnegative functions to nonnegative functions, we obtain

$$|N(F)(\zeta)|^p = a|N(F)(\zeta)| + b \leq (N(a|F| + b))(\zeta) \leq N(|F|^p)(\zeta). \qquad \square$$

Applying the covariance formula: Itô's formula and the optional stopping theorem (Theorem 1.1.9), applied to Poisson kernels yields

$$P(z_{\tau_r}, \zeta) = 1 + \int_0^{\tau_r} \partial P(z_t, \zeta)dz_t + \bar{\partial} P(z_t, \zeta)d\overline{z_t}. \tag{1.6.5}$$

Let $F \in L^1(\Omega)$ with Itô integral representation $F = F_0 + \int_0^\infty X_t dz_t + Y_t d\overline{z_t}$. Theorem 1.1.12 and equation (1.1.41) give

$$\mathbb{E}(FP(z_{\tau_r}, \zeta)) = F_0 + 2\mathbb{E} \int_0^{\tau_r} \partial P(z_t, \zeta)Y_t + \bar{\partial} P(z_t, \zeta)X_t dt, \tag{1.6.6}$$

for $\zeta \in \mathbb{T}$. By Theorem 1.6.1 we may pass to the limit as $r \to 1$, which gives a useful formula for Doob's projection displaying the dependence on the adapted processes (X_t) and (Y_t). This will later enable us to systematically exploit the metric and algebraic information encoded in Itô's integral representation of F.

Embedding and Projection

Next, we show that Doob's projection N is the left inverse of the martingale embedding operator $M(f) = f(z_\tau)$.

Theorem 1.6.2 *Let $1 \leq p \leq \infty$, and $F \in L^p(\Omega)$, $f \in L^p(\mathbb{T})$, and $g \in L^q(\mathbb{T})$, with $1/p + 1/q = 1$. Then,*

$$\int_\mathbb{T} N(F)g\,dm = \mathbb{E}(Fg(z_\tau)), \tag{1.6.7}$$

and

$$N(f(z_\tau)) = f. \tag{1.6.8}$$

Proof Fix $g \in L^q(\mathbb{T})$, with harmonic extension $g(z) = \int_{\mathbb{T}} P(z, \zeta)g(\zeta)dm(\zeta)$. With respect to the stopping time σ-algebras, the random variables $g(z_{\tau_s})$ form a bounded martingale in $L^p(\Omega)$. Next, we fix $F \in L^p(\Omega)$ and put $G_s = \mathbb{E}(F|\mathcal{F}_{\tau_s})$, where $0 < s < 1$. As

$$N(G_s)(\zeta) = \mathbb{E}(G_s P(z_{\tau_s}, \zeta)),$$

by averaging over $\zeta \in \mathbb{T}$, and by using that $g(z_{\tau_s})$ is a $(\mathcal{G}_s)$-martingale, we obtain the identity,

$$\int_{\mathbb{T}} N(G_s)gdm = \mathbb{E}\left(G_s \int_{\mathbb{T}} P(z_{\tau_s}, \zeta)g(\zeta)dm(\zeta)\right) = \mathbb{E}(G_s g(z_\tau)).$$

Let $G = \mathbb{E}(F|\mathcal{F}_\tau)$. The martingale convergence theorem gives $\|G - G_s\|_p \to 0$, and Theorem 1.6.1 implies that $N(G) \in L^p(\Omega)$. As z_τ is uniformly distributed over $\mathbb{T}$, and $g(z_\tau) \in L^q(\Omega)$, we obtain the limiting case of the identity above,

$$\int_{\mathbb{T}} N(G)gdm = \lim_{s\to 1} \int_{\mathbb{T}} N(G_s)gdm = \lim_{s\to 1} \mathbb{E}(G_s g(z_\tau)) = \mathbb{E}(Gg(z_\tau)).$$

As $N(G) = N(F)$ and $\mathbb{E}(Gg(z_\tau)) = \mathbb{E}(Fg(z_\tau))$ we arrive at (1.6.7). The reproducing property (1.6.8) of Doob's projection is a special case of (1.6.7). To see this, fix $f \in L^p(\mathbb{T})$, apply the first assertion of Theorem 1.6.2 to $F = f(z_\tau)$, and use the fact that z_τ is uniformly distributed over $\mathbb{T}$. This gives

$$\int_{\mathbb{T}} N(f(z_\tau))gdm = \mathbb{E}(f(z_\tau)g(z_\tau)) = \int_{\mathbb{T}} fgdm, \tag{1.6.9}$$

for any $g \in L^q(\mathbb{T})$. By duality, (1.6.9) implies $N(f(z_\tau)) = f$, as claimed. □

Remark: In view of Theorem 1.6.2, it is now easy to verify that the martingale operator N_0 defined in equation (1.5.5) coincides with N. Indeed, fix $G \in L^p(\Omega)$, and recall that $MN_0G = \mathbb{E}(G|\mathcal{F}_\tau)$. Next, recall that by definition $NG = N\mathbb{E}(G|\mathcal{F}_\tau)$. Now use $\mathrm{Id} = NM$ and collect terms,

$$MNG = MN\mathbb{E}(G|\mathcal{F}_\tau) = MN(MN_0G) = M(NM)N_0G = MN_0G.$$

As $\|N_0G - NG\|_p = \|M(N_0G - NG)\|_p = 0$ we obtained $N = N_0$. Our next application of Theorem 1.6.2 allows us to identify the Fourier coefficients of $N(F)$ directly in terms of F.

Theorem 1.6.3 *Let $F \in L^p(\Omega)$. Then the Fourier coefficients of $N(F) \in L^p(\mathbb{T})$ are given by*

$$c_n = \mathbb{E}(Fz_\tau^{-n}), \qquad n \in \mathbb{Z}. \tag{1.6.10}$$

Proof For $F \in L^p(\Omega)$, by Theorem 1.6.1 we have $N(F) \in L^p(\mathbb{T})$. Apply Theorem 1.6.2 with $g(\zeta) = (\zeta^{-n})$. This gives

$$\int_{\mathbb{T}} N(F)gdm = \mathbb{E}(Fz_\tau^{-n}). \tag{1.6.11}$$

The left-hand side of (1.6.11) is the nth Fourier coefficient of NF. The right-hand side is just c_n. □

Remark: We give a second proof of Theorem 1.6.3 using the Fourier expansion of the Poisson kernel $P(z, e^{i\alpha})$. Fix $z \in \mathbb{D}$. As, by equation (1.1.58), the Fourier series expansion of the Poisson kernels is given as $P(z, e^{i\alpha}) = \sum_{n=-\infty}^{\infty} z^{-n} e^{in\alpha}$, we have

$$\mathbb{E}\left(FP(z_{\tau_r}, e^{i\alpha})\right) = \sum_{n=-\infty}^{\infty} \mathbb{E}\left(Fz_{\tau_r}^{-n}\right) e^{in\alpha}.$$

Take L^p limits as $r \to 1$ to obtain $N(F)(e^{i\alpha}) = \sum_{n=-\infty}^{\infty} c_n e^{in\alpha}$, where $c_n = \mathbb{E}\left(Fz_\tau^{-n}\right)$. Thus we verified Theorem 1.6.3 using the Fourier series of $P(z, e^{i\alpha})$. A very interesting consequence of Theorem 1.6.3 arises when F is a holomorphic random variable.

Theorem 1.6.4 *Let $1 \le p \le \infty$. If $f \in H^p(\mathbb{T})$ then $M(f) \in H^p(\Omega)$. If $F \in H^p(\Omega)$, then $N(F) \in H^p(\mathbb{T})$ and we obtain the following commuting diagram:*

$$\begin{array}{ccc} H^p(\mathbb{T}) & \xrightarrow{\quad \text{Id} \quad} & H^p(\mathbb{T}) \\ & {\scriptstyle M} \searrow \qquad \nearrow {\scriptstyle N} & \\ & H^p(\Omega) & \end{array} \tag{1.6.12}$$

Proof Let $F \in H^p(\Omega)$ with Itô integral $F = \mathbb{E}F + \int X_s dz_s$. Theorem 1.6.1 implies $N(F) \in L^p(\mathbb{T})$. By Theorem 1.1.12 and the covariance formula (1.1.41) we obtained

$$\mathbb{E}(Fz_\tau^n) = 0, \quad n \in \mathbb{N}.$$

Finally, invoking (1.6.10) shows that $N(F) \in H^p(\mathbb{T})$. □

Intertwining $\mathcal{H}$ and H

We next relate the stochastic Hilbert transform to its analytic counterpart. We noticed in Section 1.4 that for $f \in L^2(\mathbb{T})$, the stochastic Hilbert transform $\mathcal{H}$ acts on the random variable $F = f(z_\tau)$ by

$$\mathcal{H}(f(z_\tau)) = (Hf)(z_\tau),$$

where H denotes the Hilbert transform on $L^2(\mathbb{T})$ defined by the relation

$$Hz^n = (-i)\,\mathrm{sign}(n)z^n, \qquad n \in \mathbb{Z}, \quad z \in \mathbb{T}. \tag{1.6.13}$$

Recall that on holomorphic random variables the stochastic Hilbert transform acts by multiplication with $(-i)$. Thus, for $G \in H^2(\Omega)$ with $\mathbb{E}G = 0$,

$$\mathcal{H}G = (-i)\,G, \qquad \mathcal{H}\overline{G} = (+i)\,\overline{G}. \tag{1.6.14}$$

Theorem 1.6.5 *Let $1 < p < \infty$. For any $F \in L^p(\Omega)$, $N(\mathcal{H}(F)) = H(N(F))$, and hence the following diagram commutes*

$$\begin{array}{ccccc} L^p(\mathbb{T}) & \xrightarrow{M} & L^p(\Omega) & \xrightarrow{N} & L^p(\mathbb{T}) \\ \downarrow{\scriptstyle H} & & \downarrow{\scriptstyle \mathcal{H}} & & \downarrow{\scriptstyle H} \\ L^p(\mathbb{T}) & \xrightarrow[M]{} & L^p(\Omega) & \xrightarrow[N]{} & L^p(\mathbb{T}) \end{array} \tag{1.6.15}$$

where $M(f) = f(z_\tau)$.

Proof Fix $F \in L^2(\Omega)$ and $\mathbb{E}F = 0$. Put $c_n = \mathbb{E}\,(F{z_\tau}^{n})$ and $d_n = \mathbb{E}\,(\mathcal{H}(F)z_\tau^{-n})$. Theorem 1.6.3 asserts that c_n is the nth Fourier coefficient of $N(F)$, and that d_n is the nth Fourier coefficient of $N(\mathcal{H}F)$. In view of (1.6.13), it suffices to prove that the Fourier coefficients satisfy the identity

$$d_n = (-i)\,\mathrm{sign}(n)c_n, \qquad n \in \mathbb{Z}\backslash\{0\}. \tag{1.6.16}$$

To this end we select $G_1, G_2 \in H^2(\Omega)$ such that $F = G_1 + \overline{G_2}$. Applying the stochastic Hilbert transform to F and taking into account the identities (1.6.14) we have

$$\mathbb{E}F + \mathcal{H}(F) = (-i)\,G_1 + i\,\overline{G_2}.$$

In view of the covariance formula (1.1.41) for any $n \in \mathbb{N}$, we obtain

$$\mathbb{E}\,(\mathcal{H}(F)z_\tau^n) = (+i)\,\mathbb{E}\left(\overline{G_2}z_\tau^n\right) = (+i)\,\mathbb{E}\,(Fz_\tau^n)\,. \tag{1.6.17}$$

Taking into account that $z_\tau^{-n} = \overline{z_\tau^n}$, we find similarly that, for $n \in \mathbb{N}$,

$$\mathbb{E}\,(\mathcal{H}(F)z_\tau^{-n}) = (-i)\,\mathbb{E}\,(G_1 z_\tau^{-n}) = (-i)\,\mathbb{E}\,(Fz_\tau^{-n})\,. \tag{1.6.18}$$

Comparing (1.6.17) and (1.6.18), we obtain (1.6.16). □

Summary: Let $1 < p < \infty$. Combining Theorem 1.6.2 and Theorem 1.6.5, and merging diagram (1.6.15) with equation (1.6.8), we obtain the following commutative diagram:

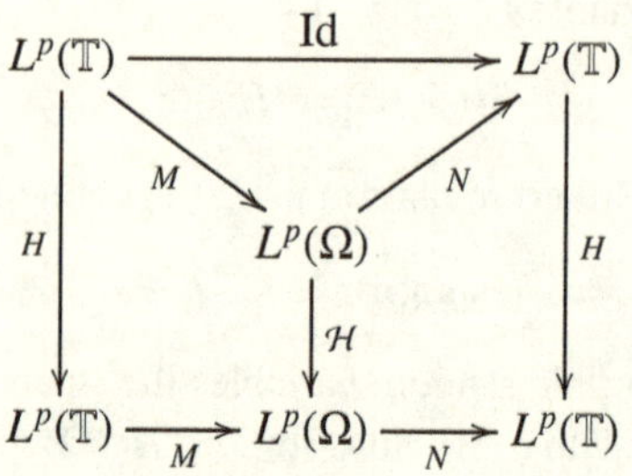

(1.6.19)

1.7 Applications to $H^p(\mathbb{T})$

In Section 1.6, we defined Doob's martingale projection N and showed that it is a contraction between L^p-spaces, and that by mapping $H^p(\Omega)$ to $H^p(\mathbb{T})$ it preserves analyticity. Moreover,

$$\mathrm{Id}_{L^p(\mathbb{T})} = NM,$$

where $Mf = f(\zeta_t)$. By exploiting these basic properties of N, we transfer to $H^p(\mathbb{T})$, the Marcinkiewicz decomposition theorem (Theorem 1.2.8) and the complex convexity theorem (Theorem 1.2.10).

We present now probabilistic proofs for two important theorems on $H^p(\mathbb{T})$. Theorem 1.7.1 was originally obtained by Jones (1983), and Theorem 1.7.2 is due to Bourgain (1983a).

Theorem 1.7.1 (Jones's decomposition theorem) *Let $1 < p < \infty$, $\lambda > 0$, and $f \in H^p_0(\mathbb{T})$. Define $h = N(f(z_\rho))$ as the Doob projection of the holomorphic random variable $f(z_\rho)$, where*

$$\rho = \inf\{0 < t < \tau : |f(z_t)| > \lambda\}.$$

Then $h \in H^\infty(\mathbb{T})$, with $|h| \leq \lambda$, and

$$\int_{\mathbb{T}} |f - h| dm \leq C_1 (C_2/\lambda)^{p-1} \int_{\mathbb{T}} |f|^p dm, \tag{1.7.1}$$

where $C_1, C_2 > 0$ are independent of $1 < p < \infty$.

Proof Put $F = f(z_\tau)$ and $G = \mathbb{E}(F|\mathcal{F}_\rho)$. As $F \in H^1(\Omega)$, the conditional expectation G is a holomorphic random variable satisfying $|G| \leq \lambda$ for which – in view of Inequalities (1.2.48) and (1.2.52) – we have

$$\mathbb{E}|F - G| \leq C_1 (C_2/\lambda)^{p-1} \mathbb{E}|F|^p. \tag{1.7.2}$$

Recall that Doob's projection is a contraction on L^∞. Hence, by Theorem 1.6.4 we have $h \in H^\infty(\mathbb{T})$ with $|h| \le \lambda$. As Doob's projection is simultaneously an L^1 contraction, we have $\int_{\mathbb{T}} |f - h| dm \le \mathbb{E}(|F - G|)$. Taking into account that $F = M(f)$, this gives $\mathbb{E}|F|^p = \int_{\mathbb{T}} |f|^p dm$, which finishes the proof of (1.7.1). □

Remarks: For many applications to interpolation theory it suffices to apply Jones's decomposition theorem (Theorem 1.7.1) with $p = 2$. In this case, Theorem 1.4.4 provides a short proof: For $h \in H^2(\mathbb{T})$ and $\lambda > 0$, there exists $h_0 \in H^\infty(\mathbb{T})$ such that $|h_0| \le \lambda$ and $\|h - h_0\|_1 \le \|h\|_2^2/\lambda$. The proof now proceeds in three simple steps.

(i) We recall here the canonical inner-outer factorization for Hardy spaces. Let $h \in H^2(\mathbb{T})$ with $\|h\|_2 = 1$ and put $r = \ln|h|$. Then $r \in L^1(\mathbb{T})$ and putting $f = \exp(r + iH(r))$ defines an element in $H^2(\mathbb{T})$ such that $|h| = |f|$. Moreover, the quotient $b = (h/f)$ defines an element in $H^\infty(\mathbb{T})$ with $h = bf$. We use inner-outer factorization to reduce the decomposition problem in $H^2(\mathbb{T})$ to the case of outer functions.

(ii) We apply the construction of Theorem 1.4.4 to the outer function f. Let $\lambda > 0$ and put $A = \{|f| > \lambda\}$. Define $x = r1_{\mathbb{T}\setminus A} + (\ln\lambda)1_A$ and form the outer function

$$f_0 = \exp(x + iH(x)).$$

We have $f_0 \in H^\infty(\mathbb{T})$ with $|f_0| \le \lambda$. Taking into account that $\|f\|_2 = 1$, Theorem 1.4.4 gives the crucial L^1 estimate,

$$\|f - f_0\|_1 \le 1/\lambda. \tag{1.7.3}$$

(iii) Finally, define $h_0 = bf_0$. Then $|h_0| \le \lambda$, and $h - h_0 = b(f - f_0)$ with $|b| = 1$. Hence, by Inequality (1.7.3), we finally conclude that $\|h - h_0\|_1 \le 1/\lambda$.

Theorem 1.7.2 *For $f \in H_0^1(\mathbb{T})$ and $z \in \mathbb{C}$, define $g = N(f(z_\rho))$ as the Doob projection of the holomorphic random variable $f(z_\rho)$, where ρ is the stopping time*

$$\rho = \inf\{t < \tau : |f(z_t)| > 12|z|\}.$$

Then $g \in H^\infty(\mathbb{T})$, $|g| \le 12|z|$, and

$$\left(|z|^2 + \kappa^2 \int_{\mathbb{T}} |g|^2 dm\right)^{1/2} + \kappa \int_{\mathbb{T}} |f - g| dm \le \int_{\mathbb{T}} |z + f| dm, \tag{1.7.4}$$

whenever $\kappa \le 1/60$.

Proof First, the composition $F = f(z_\tau)$ defines a holomorphic random variable on Wiener space. Since $F \in H^1(\Omega)$, its conditional expectation $G = \mathbb{E}(F|\mathcal{F}_\rho)$

is a bounded holomorphic random variable satisfying $|G| \leq 12|z|$. By Theorem 1.2.10, we have the complex convexity estimate,

$$\left(|z|^2 + \kappa^2 \mathbb{E}|G|^2\right)^{1/2} + \kappa \mathbb{E}|F - G| \leq \mathbb{E}|z + F|. \tag{1.7.5}$$

Next, recall that we defined g by applying Doob's projection to G. Thus by Theorem 1.6.4, we have $g \in H^\infty$ and $|g| \leq 12|z|$. By Theorem 1.6.1, Doob's projection operator is simultaneously contracting the L^2 and L^1 norms. Hence, the terms on the left-hand side of Inequality (1.7.4) are smaller than (or equal to) those on the left-hand side of Inequality (1.7.5). Finally, as $M(f) = F$, we have $\int_{\mathbb{T}} |z + f| dm = \mathbb{E}|z + F|$. Thus, we obtain that Inequality (1.7.5) implies Inequality (1.7.4). □

1.8 Projecting Square Functions

Recall that in Section 1.2.3 we introduced the space $SL^\infty(\Omega)$, which consists of those square integrable random variables F on Wiener space Ω for which the quadratic variation $\langle F \rangle$ is uniformly bounded on Ω.

Here, we begin by defining the space $SL^\infty(\mathbb{T})$, consisting of $f \in L^2(\mathbb{T})$ with uniformly bounded Littlewood–Paley square functions, and we record the observation that the martingale operator M maps $SL^\infty(\mathbb{T})$ into $SL^\infty(\Omega)$. The main results of this section assert that Doob's projection operator N maps $SL^\infty(\Omega)$ to $SL^\infty(\mathbb{T})$, and that the analytic counterpart of Theorem 1.2.9 holds true. We closely examine the effect of Doob's projection N on the quadratic variation process with the aim of proving the crucial point-wise estimates of Theorem 1.8.1.

Fix a real-valued $u \in L^2(\mathbb{T})$ and let $u(z)$, $z \in \mathbb{D}$, denote its harmonic extension to the unit disk. The Littlewood–Paley integral of u is defined as

$$g_{\mathbb{D}}^2(u)(\zeta) = \frac{4}{\pi} \int_{\mathbb{D}} |\partial u(w)|^2 \log \frac{1}{|w|} P(w, \zeta) dA(w), \tag{1.8.1}$$

where $\zeta \in \mathbb{T}$, $\partial = (\partial_x - i\partial_y)/2$ and where $P(w, \zeta)$ denotes the Poisson kernel defined by equation (1.1.55).

The Littlewood–Paley integral of u admits an interesting stochastic interpretation. As in the previous section we work with complex Brownian motion, started at $0 \in \mathbb{D}$, normalized such that $\langle z_t, \overline{z_t} \rangle = 2t$, and stopped at $\tau = \inf\{t > 0 : |z_t| > 1\}$. Itô's formula gives

$$u(z_\tau) = u(0) + \int_0^\tau \partial u(z_s) dz_s + \overline{\partial} u(z_s) d\overline{z_s}, \tag{1.8.2}$$

and hence

$$\langle u(z_\tau)\rangle = 4\int_0^\tau |\partial u(z_s)|^2 ds.$$

The identity,

$$N\left(\langle u(z_\tau)\rangle\right) = g_{\mathbb{D}}^2(u). \tag{1.8.3}$$

holds true. Indeed, for $r < 1$, the covariance formula (1.1.42) yields

$$\mathbb{E}(\langle u(z_\tau)\rangle P(z_{\tau_r},\zeta)) = 4\mathbb{E}\int_0^{\tau_r} |\partial u(z_t)|^2 P(z_t,\zeta)dt, \tag{1.8.4}$$

where $\tau_r = \inf\{t > 0 : |z_t| > r\}$. Finally, invoking the occupation time formula (1.1.57), and passing to the limit as $r \to 1$ we obtain

$$4\mathbb{E}\int_0^\tau |\partial u(z_t)|^2 P(z_t,\zeta)dt = \frac{4}{\pi}\int_{\mathbb{D}} |\partial u(w)|^2 \log\frac{1}{|w|} P(w,\zeta)dA(w).$$

The space $SL^\infty(\Omega)$, defined in Section 1.2.3, consists of those $F \in L^2(\Omega)$ for which $\langle F\rangle \in L^\infty(\Omega)$. Its norm is given by

$$\|F\|_{SL^\infty(\Omega)} = |\mathbb{E}(F)| + \left\|\langle F\rangle^{1/2}\right\|_\infty.$$

We define $SL^\infty(\mathbb{T})$ to consist of those $f \in L^2(\mathbb{T})$ for which $g_{\mathbb{D}}(f) \in L^\infty(\Omega)$, and put

$$\|f\|_{SL^\infty(\mathbb{T})} = \left|\int_{\mathbb{T}} f dm\right| + \|g_{\mathbb{D}}(f)\|_\infty.$$

In view of (1.8.3), the martingale embedding operator $Mf = f(z_\tau)$ satisfies

$$\|M\colon SL^\infty(\mathbb{T}) \to SL^\infty(\Omega)\| = 1.$$

The main result of this section implies that Doob's projection N acts as a bounded operator from $SL^\infty(\Omega)$ to $SL^\infty(\mathbb{T})$. Consider a real-valued $G \in L^2(\Omega)$ with stochastic integral representation

$$G = G_0 + \int_0^\infty X_t dz_t + \overline{X_t} d\overline{z_t}, \tag{1.8.5}$$

where (X_t) is complex valued and adapted. Recall that the martingale square function of G, defined to be the limit of its increasing quaratic variation process, is then given by

$$\langle G\rangle = 4\int_0^\infty |X_t|^2 dt.$$

Applying Doob's projection gives

$$N(\langle G\rangle)(\zeta) = \lim_{r\to 1}\mathbb{E}(\langle G\rangle P(z_{\tau_r},\zeta)). \tag{1.8.6}$$

As $(P(z_t,\zeta), t > 0)$ forms an $(\mathcal{F}_t)$-martingale, equation (1.1.42) allows us to rewrite (1.8.6) as

$$N(\langle G\rangle)(\zeta) = 4\mathbb{E}\int_0^{\tau} |X_t|^2 P(z_t,\zeta)dt. \tag{1.8.7}$$

Our main result gives pointwise estimates for the Littlewood–Paley function $g^2_{\mathbb{D}}(N(G))$ by $N(\langle G\rangle)$.

Theorem 1.8.1 *Let $G \in L^2(\Omega)$ be real valued and $u = N(G)$. Then the commutation relation between Doob's projection and the quadratic variation is given by the pointwise estimate*

$$N(\langle u(z_\tau)\rangle) \le CN(\langle G\rangle),$$

where $C > 0$ is universal constant.

The Lusin function: The Lusin function of $u \in L^2(\mathbb{T})$ is defined as

$$S^2(u)(\zeta) = \int_{\Gamma(\zeta)} |\partial u(w)|^2 dA(w), \tag{1.8.8}$$

where $\Gamma(\zeta) = \{w \in \mathbb{D} : |\zeta - w| \le 2(1-|w|)\}$ and where $u(w)$ denotes the harmonic extension of u. Since $-\log|w|P(w,\zeta) > c$ for $w \in \Gamma(\zeta)$, the Littlewood–Paley integral provides a pointwise upper bound for the Lusin area function, so that in view of (1.8.3),

$$S^2(u) \le CN(\langle u(z_\tau)\rangle).$$

In the following subsections we limit ourselves to proving that

$$S^2(u) \le CN(\langle G\rangle) \tag{1.8.9}$$

hold true, for $\zeta \in \mathbb{T}$. The inequality asserted in Theorem 1.8.1 is stronger than Inequality (1.8.9). We will next present a consequence of Theorem 1.8.1, and thereafter turn to proving Inequality (1.8.9).

Application to $H^\infty(\mathbb{T}) \cap SL^\infty(\mathbb{T})$

Theorem 1.8.1 implies that the linear operator N maps $SL^\infty(\Omega)$ boundedly into $SL^\infty(\mathbb{T})$. We use this fact now to transfer the content of Theorem 1.2.9 to $SL^\infty(\mathbb{T})$.

Theorem 1.8.2 *Let $1 < p < \infty$, $f \in H^p_0(\mathbb{T})$, and $\lambda > 0$. Put*

$$\rho = \inf\left\{0 < t < \tau : |f(z_t)| > \lambda \text{ or } \langle f(z_t)\rangle > \lambda^2\right\} \quad \text{and} \quad h = N(f(z_\rho)).$$

Then $h \in H^\infty(\mathbb{T}) \cap SL^\infty(\mathbb{T})$, satisfying

$$|h| \le \lambda,\ g_{\mathbb{D}}(h) \le C\lambda, \quad \text{and} \quad \int_{\mathbb{T}} |f-h|dm \le (C\lambda)^{1-p}\int_{\mathbb{T}} |f|^p dm, \tag{1.8.10}$$

where $C = C(p)$.

Proof Fix $f \in H_0^p(\mathbb{T})$ and $\lambda > 0$. Put $F = f(z_T)$. Then $F \in H^p(\Omega)$ with $\mathbb{E}F = 0$. Define $G = \mathbb{E}(F|\mathcal{F}_\rho)$ and note that $G = f(z_\rho)$. Applying Theorem 1.2.9 gives

$$|G| \leq \lambda, \quad \langle G\rangle^{1/2} \leq \lambda, \text{ and } \mathbb{E}|F - G| \leq C\lambda^{1-p}\mathbb{E}|F|^p.$$

As holomorphic martingales are stable under stopping we have $G \in H^\infty(\Omega)$. By Theorem 1.6.4, Doob's projection operator preserves H^∞ spaces, and hence $h = N(G)$ satisfies $h \in H^\infty(\mathbb{T})$ with $|h| \leq \lambda$. Theorem 1.8.1 asserts that Doob's projection operator preserves SL^∞ spaces so that

$$\|g_{\mathbb{D}}(h)\|_\infty \leq C\left\|\langle G\rangle^{1/2}\right\|_\infty.$$

It remains to repeat the proof of Theorem 1.7.1. We first use that Doob's projection is an L^1 contraction, hence $\int_{\mathbb{T}} |f - h| dm \leq \mathbb{E}(|F - G|)$. Next, recall that $N(F) = f$, which yields $\mathbb{E}|F|^p = \int_{\mathbb{T}} |f|^p dm$, and thereby the last remaining estimate in (1.8.10). □

1.8.1 Whitney Squares and Localization

Let $G \in L^2(\Omega)$ be real valued, with stochastic integral representation

$$G = G_0 + \int_0^\infty X_t dz_t + \overline{X_t} d\overline{z_t}, \tag{1.8.11}$$

where (X_t) is a complex-valued adapted process. Recall that Itô's formula gave equation (1.6.5), that is,

$$P(z_{T_r}, \zeta) = 1 + \int_0^{T_r} \partial_z P(z_t, \zeta) dz_t + \overline{\partial_z} P(z_t, \zeta) d\overline{z_t}. \tag{1.8.12}$$

As $N(G)(\zeta) = \lim_{r\to 1} \mathbb{E}(GP(z_{T_r}, \zeta))$, Theorem 1.1.12 and equation (1.1.41) yield the representation

$$N(G)(\zeta) = G_0 + 2\mathbb{E}\int_0^T \partial_z P(z_t, \zeta)\overline{X_t} + \overline{\partial_z} P(z_t, \zeta) X_t dt, \tag{1.8.13}$$

for $\zeta \in \mathbb{T}$. Put $u = N(G)$. We denote by $u(w)$, $w \in \mathbb{D}$, the harmonic extension of $u \in L^2(\mathbb{T})$ to the unit disk. Taking into account that $u(w)$ is given by integration with respect to the Poisson kernel, we obtain

$$u(w) = G_0 + 2\mathbb{E}\int_0^T \int_{\mathbb{T}} \left[\partial_z P(z_t, \zeta)\overline{X_t} + \overline{\partial_z} P(z_t, \zeta) X_t\right] P(w, \zeta) dm(\zeta) dt. \tag{1.8.14}$$

Thus, in view of equations (1.8.4) and (1.8.7), Theorem 1.8.1 asserts that

$$\mathbb{E}\int_0^T |\partial_z u(z_t)|^2 P(z_t, \zeta) dt \leq C\mathbb{E}\int_0^T |X_t|^2 P(z_t, \zeta) dt, \tag{1.8.15}$$

where $C > 0$ is independent of $\zeta \in \mathbb{T}$ and G. Finally, we recall that by equation (1.8.3) the term on the left-hand side is just the Littlewood–Paley integral of $u = N(G)$.

Integral Kernels

We base the proof of Inequality (1.8.15) on estimates for $\partial_w u(w)$, derived from (1.8.14). To this end, let

$$f(w,\zeta) = \partial_w P(w,\zeta),$$

where $w \in \mathbb{D}$, $\zeta \in \mathbb{T}$. For $z, w \in \mathbb{D}$, define

$$K_1(w,z) = \int_{\mathbb{T}} \partial_z P(z,\zeta) f(w,\zeta) dm(\zeta), \qquad K_2(w,z) = \int_{\mathbb{T}} \overline{\partial_z} P(z,\zeta) f(w,\zeta) dm(\zeta),$$

and put

$$K(w,z) = \left(|K_1(w,z)|^2 + |K_2(w,z)|^2\right)^{1/2}. \tag{1.8.16}$$

The following proposition summarizes the discussion above and thus is our first basic step in the proof of Theorem 1.8.1.

Proposition 1.8.3 *Let $G \in L^2(\Omega)$ with stochastic integral* (1.8.11). *Let $u(w)$ be given by* (1.8.14). *Then*

$$|\partial_w u(w)| \leq 2\sqrt{2}\mathbb{E} \int_0^{\tau} K(w,z_t)|X_t| dt, \qquad w \in \mathbb{D}, \tag{1.8.17}$$

and

$$(1 - |w|)K(w,z) \leq \frac{C}{|w-z|^2 + (1-|w|)^2}, \tag{1.8.18}$$

for $w, z \in \mathbb{D}$ satisfying $|w| \leq |z|$.

Proof Applying the complex gradient ∂_w to both sides of (1.8.14), and separating terms, we obtain

$$\partial_w u(w) = 2\mathbb{E} \int_0^{\tau} X_t K_1(w,z_t) + \overline{X_t} K_2(w,z_t) dt. \tag{1.8.19}$$

As

$$\left|X_t K_1(w,z_t) + \overline{X_t} K_2(w,z_t)\right| \leq \sqrt{2} K(w,z_t)|X_t|,$$

the identity (1.8.19) gives Inequality (1.8.17). Fix $w, z \in \mathbb{D}$ with $|w| \leq |z|$. The kernels $K_i(w,z)$, $i \in \{1,2\}$ are defined, respectively, by the densities

$$\zeta \mapsto \partial_z P(z,\zeta)\partial_w P(w,\zeta) \quad \text{and} \quad \zeta \mapsto \overline{\partial_z} P(z,\zeta)\partial_w P(w,\zeta).$$

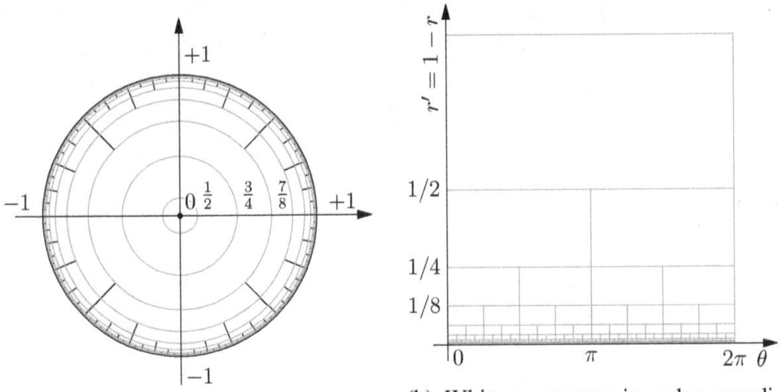

(a) Whitney squares in the disk

(b) Whitney squares in polar coordinates

Figure 1.1 Whitney squares.

Integrating by parts and exploiting their oscillatory nature gives the basic kernel estimates,

$$(1 - |w|)|K_i(w, z)| \leq \frac{C}{|w - z|^2 + (1 - |w|)^2},$$

for $i \in \{1, 2\}$, which yields Inequality (1.8.18). □

Whitney Squares

Let $n \in \mathbb{N} \cup \{0\}$ and let $m \in \mathbb{N}$ with $1 \leq m \leq 2^n$. Define the Whitney square for the unit disk by

$$W_m^n = \left\{re^{i\theta} : 2^{-n-1} \leq 1 - r < 2^{-n}, (m-1)2^{-n}2\pi \leq \theta < m2^{-n}2\pi\right\}.$$

By inspection (see Figure 1.1), any two distinct Whitney squares are disjoint and their union covers the unit disk. Moreover,

$$c \cdot \operatorname{diam}(W_m^n) \leq \operatorname{dist}(W_m^n, \mathbb{T}) \leq C \cdot \operatorname{diam}(W_m^n),$$

where $0 < c \leq C < \infty$ are independent of m, n. We let $\mathcal{W}$ denote the collection of all Whitney squares in $\mathbb{D}$. Fix $\zeta \in \mathbb{T}$ and define $C = C(\zeta)$ to be

$$C(\zeta) = \{W \in \mathcal{W} : \operatorname{dist}(W, \zeta) \leq 16 \operatorname{dist}(W)\}. \tag{1.8.20}$$

See Figure 1.2 for an illustration of $C(\zeta)$ with $\zeta = -\sqrt{-1}$. We call $C(\zeta)$ the cone of Whitney cubes connecting $0 \in \mathbb{D}$ to $\zeta \in \mathbb{T}$. In this section we systematically use the notation

$$d(V) = \operatorname{dist}(V, \mathbb{T}), \quad d(W, V) = \inf\{|w - v| : v \in V, w \in W\}. \tag{1.8.21}$$

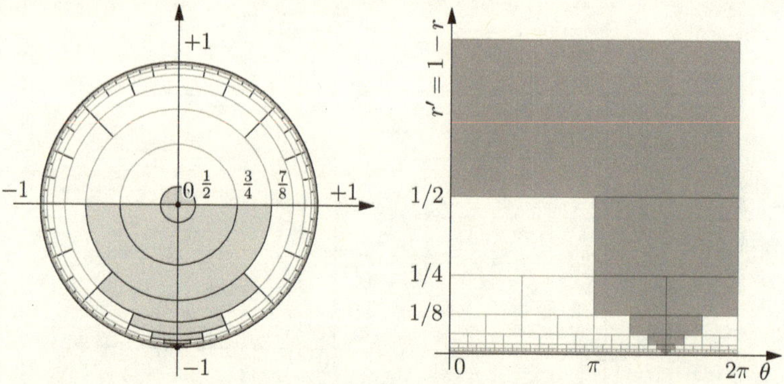

(a) Whitney squares in $C(-\sqrt{-1})$ (b) $C(-\sqrt{-1})$ in polar coordinates

Figure 1.2 Depicting $C(\zeta)$ with $\zeta = -\sqrt{-1}$.

Matrix Operators

Let $\zeta \in \mathbb{T}$. In equation (1.8.20), we defined $C = C(\zeta)$, the cone of Whitney cubes connecting $0 \in \mathbb{D}$ to $\zeta \in \mathbb{T}$; For $W \in C$, $V \in \mathcal{W}$, set

$$k(W, V) = \sup_{w \in W, v \in V} K(w, v),$$

where $K(w, v)$ is given by equation (1.8.16), and put

$$L(W, V) = d(W)d(V)k(W, V).$$

By Inequality (1.8.18), the following structural estimates hold true,

$$L(W, V) \leq \frac{C \min\{d(W), d(V)\}}{d(W, V)^2 + \max\{d(W), d(V)\}^2}, \tag{1.8.22}$$

where $C < \infty$ is independent of W and V. Note that condition (1.8.22) actually encodes two distinct estimates in a compressed form. If necessary, we may easily retrieve those by considering separately the cases

$$d(W) \leq d(V) \quad \text{and} \quad d(W) \geq d(V).$$

Next, define

$$m(W, V) = d(V, \zeta)L(W, V), \tag{1.8.23}$$

where $d(V, \zeta) = \text{dist}(V, \zeta)$.

Let $G \in L^p(\Omega)$ with stochastic integral (1.8.11), $V \in \mathcal{W}$. We define

$$a_V(G)^2 = \mathbb{E} \int_0^\tau 1_V(z_t)|X_t|^2 P(z_t, \zeta)dt. \tag{1.8.24}$$

Our next proposition shows that $m(W, V)$ is the weight by which the Whitney square $V \in \mathcal{W}$ contributes to the size of $|\partial NG(w)|$ where $w \in W \in C = C(\zeta)$.

Proposition 1.8.4 *Let $G \in L^p(\Omega)$. Then for each $W \in C$ and $w \in W$,*

$$d(W)|\partial NG(w)| \leq C \sum_{V \in \mathcal{W}} m(W, V) a_V(G), \tag{1.8.25}$$

where $(m(W, V) : W \in C, V \in \mathcal{W})$ is defined by (1.8.23).

Proof Let $W \in C$ and $w \in W$. By Proposition 1.8.3, we obtain

$$|\partial NG(w)| \leq 2\sqrt{2} \sum_{V \in \mathcal{W}} \mathbb{E} \int_0^T 1_V(z_t) K(w, z_t)|X_t| dt. \tag{1.8.26}$$

We rewrite each summand on the right-hand side of Inequality (1.8.26) as

$$\mathbb{E} \int_0^T 1_V(z_t) K(w, z_t) P^{-1/2}(z_t, \zeta) P^{1/2}(z_t, \zeta)|X_t| dt. \tag{1.8.27}$$

The Cauchy–Schwarz inequality shows that (1.8.27) is bounded as follows

$$\left(\mathbb{E} \int_0^T 1_V(z_t) K^2(w, z_t) P^{-1}(z_t, \zeta) dt\right)^{1/2} \left(\mathbb{E} \int_0^T 1_V(z_t)|X_t|^2 P(z_t, \zeta) dt\right)^{1/2}. \tag{1.8.28}$$

Note that the second factor on the right-hand side of (1.8.28) is just a_V. By the occupation time formula (1.1.57) we may rewrite and estimate the left-hand side of (1.8.28), as follows

$$\begin{aligned} &\left(\frac{1}{\pi} \int_V K(w, v)^2 P^{-1}(v, \zeta)| \log \frac{1}{|v|}| dA(v)\right)^{1/2} \\ &\qquad \leq k(W, V) \left(\frac{1}{\pi} \int_V P^{-1}(v, \zeta)| \log \frac{1}{|v|}| dA(v)\right)^{1/2}. \end{aligned} \tag{1.8.29}$$

By (1.8.28) and (1.8.29) we obtain

$$d(W)\mathbb{E} \int_0^T 1_V(z_t) K(w, z_t)|X_t| dt \leq C d(W) d(V) d(V, \zeta) k(W, V) a_V(G). \tag{1.8.30}$$

Recall that $d(W)d(V)d(V, \zeta)k(W, V) = m(W, V)$. Thus combining Inequalities (1.8.26) and (1.8.30) gives (1.8.25). □

The conditional proof of (1.8.9)**:** Let $\mathcal{L} \subseteq \mathcal{W}$. We denote by $\ell^2(\mathcal{L})$ the Hilbert space of those sequences $a = (a_W : W \in \mathcal{L})$ for which

$$\|a\|^2_{\ell^2(\mathcal{L})} = \sum_{W \in \mathcal{L}} |a_W|^2 < \infty.$$

Recall that we fixed $\zeta \in \mathbb{T}$ and defined $C = C(\zeta)$ in equation (1.8.20) Recall further, that depending on $\zeta \in \mathbb{T}$ the matrix $(m(W, V))$ is fixed in equation (1.8.23). We use it to define the linear operator M on $\ell^2(\mathcal{W})$ by

$$(Ma)_W = \sum_{V \in \mathcal{W}} m(W, V) a_V, \qquad W \in C. \tag{1.8.31}$$

In Section 1.8.2 we prove Theorem 1.8.6, asserting that M gives rise to a bounded operator between the Hilbert spaces $\ell^2(\mathcal{W})$ and $\ell^2(C)$ with norm independent of $\zeta \in \mathbb{T}$. Put

$$\Gamma = \Gamma(\zeta) = \{z \in \mathbb{D} : |\zeta - z| \leq 2(1 - |z|)\},$$

and observe that

$$\int_\Gamma |\partial_w u(w)|^2 dA(w) \leq \sum_{W \in C} \int_W |\partial_w NG(w)|^2 dA(w). \tag{1.8.32}$$

By Proposition 1.8.4, for any $W \in C(\zeta)$,

$$\int_W |\partial NG(w)|^2 dA(w) \leq C_1 \left(\sum_{V \in \mathcal{W}} m(W, V) a_V \right)^2,$$

where $a_V = a_V(G)$ is defined by equation (1.8.24). Invoking Theorem 1.8.6 gives the following estimate for the Lusin function,

$$S^2(u)(\zeta) \leq C_1 \|Ma\|^2_{\ell^2(C)} \leq C_2 \|a\|^2_{\ell^2(\mathcal{W})}, \tag{1.8.33}$$

with $C_2 < \infty$ independent of $\zeta \in \mathbb{T}$. Finally, since the Whitney cubes $\mathcal{W}$ form a disjoint cover of the unit disk, we have

$$\|a\|^2_{\ell^2(\mathcal{W})} = \mathbb{E} \int_0^T |X_t|^2 P(z_t, \zeta) dt. \tag{1.8.34}$$

Merging (1.8.32)–(1.8.34) finishes the proof.

This proof of Theorem 1.8.1 relied on the fact that the matrix M, defined by equation (1.8.23), gives rise to a bounded operator between the Hilbert spaces $\ell^2(\mathcal{W})$ and $\ell^2(C)$. The next subsection is devoted to proving this fact.

1.8.2 Schur's Lemma

In this section we exploit Schur's Lemma, and verify that the operator M, defined by equation (1.8.23), satisfies the correct ℓ^2 inequalities.

Let $A = (A(l, n))$ be a matrix with nonnegative entries. It induces a linear operator on the vector space of finitely supported real-valued sequences $x = (x(n))$,

$$(Ax)(l) = \sum_n a(l, n)x(n).$$

Schur's Lemma gives a simple and useful upper estimate for the operator norm of A on ℓ^2.

Theorem 1.8.5 (Schur's lemma) *If there exist nonnegative sequences $(p(l))$ and $(q(n))$ and constants $C_1, C_2 > 0$ such that $\sum_k a(k, n)p(k) \leq C_1 q(n)$, and $\sum_k a(l, k)q(k) \leq C_2 p(l)$, for any $l, n \in \mathbb{N}$, then*

$$\|Ax\|_{\ell^2} \leq (C_1 C_2)^{1/2} \|x\|_{\ell^2}$$

for any compactly supported real-valued sequence $x = (x(n))$.

Proof Fix $x \in \ell^2$, and put $y(l) = \sum_{k=1}^{\infty} a(l, k)x(k)$. Write

$$a(l, k)x(k) = \left(a(l, k)\frac{q(k)}{p(l)}\right)^{1/2} \left(a(l, k)\frac{p(l)}{q(k)}\right)^{1/2} x(k).$$

Take the sum over k, and apply the Cauchy–Schwarz inequality to the term on the right-hand side. This gives

$$|y(l)|^2 \leq C_2 \sum_{k=1}^{\infty} a(l, k)\frac{p(l)}{q(k)}|x(k)|^2.$$

Take the sum over l, and change the order of summation to get,

$$\begin{aligned}\sum_{l=1}^{\infty} |y(l)|^2 &\leq C_2 \sum_{k=1}^{\infty} \sum_{l=1}^{\infty} a(l, k)\frac{p(l)}{q(k)}|x(k)|^2 \\ &\leq C_1 C_2 \sum_{k=1}^{\infty} |x(k)|^2.\end{aligned}$$

□

Theorem 1.8.5 (Schur's Lemma) will now be applied to the matrix M defined in (1.8.31). Fix $\zeta \in \mathbb{T}$. Let $C = C(\zeta)$ denote the cone of Whitney squares in $\mathbb{D}$ connecting the origin to $\zeta \in \mathbb{T}$. We fix a nonnegative matrix

$$m(W, V) = d(V, \zeta)L(W, V), \tag{1.8.35}$$

where $W \in C$, $V \in \mathcal{W}$, and where $L(W, V)$ satisfies the following estimates: If $d(W) \leq d(V)$, then

$$0 \leq L(W, V) \leq \frac{Cd(W)}{d(W, V)^2 + d(V)^2}. \tag{1.8.36}$$

If, conversely, $d(W) \geq d(V)$, then

$$0 \leq L(W, V) \leq \frac{Cd(V)}{d(W, V)^2 + d(W)^2}. \tag{1.8.37}$$

Theorem 1.8.6 *If* $M = (m(W, V))$ *is given by* (1.8.35) *and if* $L = (L(W, V))$ *satisfies Inequalities* (1.8.36) *and* (1.8.37), *then*

$$\|Ma\|_{\ell^2(C)} \leq C\|a\|_{\ell^2(\mathcal{W})}, \tag{1.8.38}$$

where $C = C(C_0)$.

Proof We decompose M into the sum of three matrices M_i, $i \leq 3$, and exploit Schur's Lemma to prove that each of the summands is bounded. To this end, define three matrices $A_i = (a_i(W, V))$ with

$$a_i(W, V) \in \{0, 1\}$$

by applying the following rules. Let $W \in C$ and $V \in \mathcal{W}$.

(i) Define $a_1(W, V) = 1$ if $d(W) \leq d(V)$ and put $a_1(W, V) = 0$ otherwise.
(ii) Define $a_2(W, V) = 1$ if $d(W) > d(V)$ and $d(V, \zeta) \leq d(W)$. Otherwise, we set $a_2(W, V) = 0$.
(iii) Define $a_3(W, V) = 1$ if $d(W) > d(V)$ and $d(V, \zeta) > d(W)$. Otherwise, we put $a_3(W, V) = 0$.

Define the decomposition of M as $M = M_1 + M_2 + M_3$, where

$$m_i(W, V) = m(W, V)a_i(W, V), \qquad i \in \{1, 2, 3\}, \qquad W \in C, V \in \mathcal{W}. \tag{1.8.39}$$

In Lemmas 1.8.7–1.8.9 we exploit Schur's Lemma to prove that the matrices $M_i, i \in \{1, 2, 3\}$, satisfy the norm estimates

$$\|M_i a\|_{\ell^2(C)} \leq C\|a\|_{\ell^2(\mathcal{W})}, \tag{1.8.40}$$

where $C = C(C_0)$. As $M = M_1+M_2+M_3$, Inequality (1.8.40) implies Inequality (1.8.38). Let $W \in C$. The indicator functions a_i define a splitting of $\mathcal{W}$ as $\mathcal{W} = \mathcal{W}_1(W) \cup \mathcal{W}_2(W) \cup \mathcal{W}_3(W)$, where $\mathcal{W}_i(W) = \{V \in \mathcal{W} : a_i(W, V) = 1\}$. See Figure 1.3. Simultaneously, for $V \in \mathcal{W}$ we decompose $C = C_1(V) \cup C_2(V) \cup C_3(V)$ where $C_i(V) = \{W \in C : a_i(W, V) = 1\}$. See Figure 1.4. □

Boundedness of M_1

In order to show that $\left\|M_1 : \ell^2(\mathcal{W}) \to \ell^2(C)\right\| \leq C$ we will verify that the matrix $(m_1(W, V) : W \in C, V \in \mathcal{W})$ satisfies the hypothesis of Theorem 1.8.5.

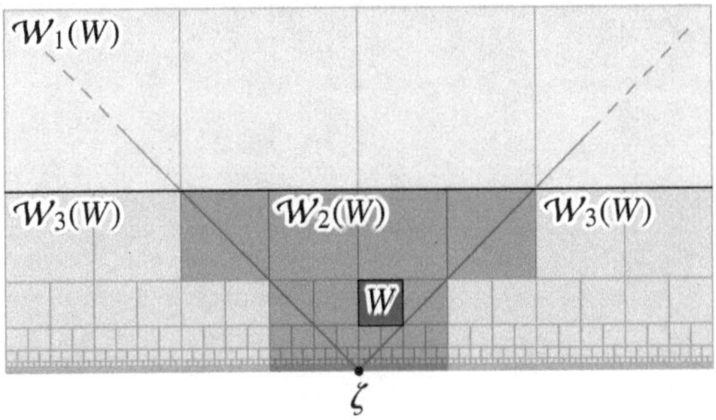

Figure 1.3 Proof of Theorem 1.8.6: $W \in C$ defines the splitting $\mathcal{W} = \mathcal{W}_1(W) \cup \mathcal{W}_2(W) \cup \mathcal{W}_3(W)$ where $\mathcal{W}_i(W) = \{V \in \mathcal{W} : a_i(W, V) = 1\}$.

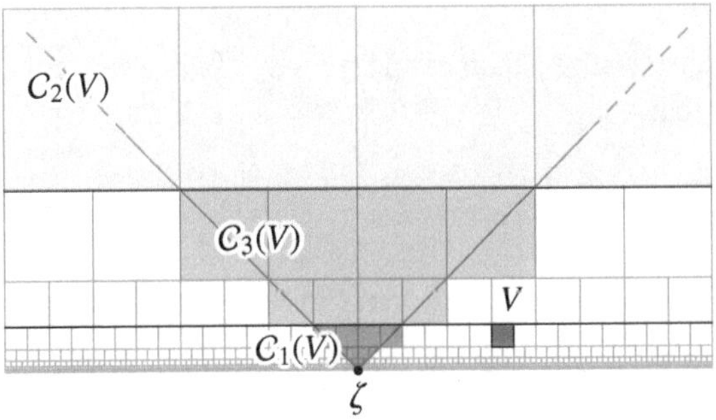

Figure 1.4 Proof of Theorem 1.8.6: $V \in \mathcal{W}$ defines the splitting $C = C_1(V) \cup C_2(V) \cup C_3(V)$ with $C_i(V) = \{W \in C : a_i(W, V) = 1\}$.

Lemma 1.8.7 *Let* $0 < \epsilon < 1$. *Let* $V \in \mathcal{W}$ *and* $q(V) = d(V)^\epsilon d(\zeta, V)^{-\epsilon}$. *Then*

$$\sup_{W \in C} \sum_{V \in \mathcal{W}} m_1(W, V) q(V) \le C, \tag{1.8.41}$$

and

$$\sup_{V \in \mathcal{W}} \sum_{W \in C} m_1(W, V) q(V)^{-1} \le C. \tag{1.8.42}$$

Proof Fix first $W \in C$ and $V \in \mathcal{W}$. Note if $a_1(W, V) \neq 0$ then $d(W) \le d(V)$ and $d(\zeta, V) \le Cd(W, V)$. Hence invoking the assumption (1.8.36) gives the estimates for the entries in the matrix M_1,

$$m_1(W, V) \le \frac{C_0 d(W)}{d(W, V) + d(V)}. \tag{1.8.43}$$

Verification of (1.8.41): Fix $W \in C$ and put $\mathcal{W}_1(W) = \{V \in \mathcal{W} : a_1(W, V) = 1\}$. See Figure 1.5. We enumerate those Whitney squares as $\mathcal{W}_1(W) = \{V_\ell^k\}$, where $k \in \mathbb{N}$, and $\ell \in \mathbb{Z}$ satisfy $k \le C|\ln d(W)|$, and $|\ell| \le C2^{-k}/d(W)$. By

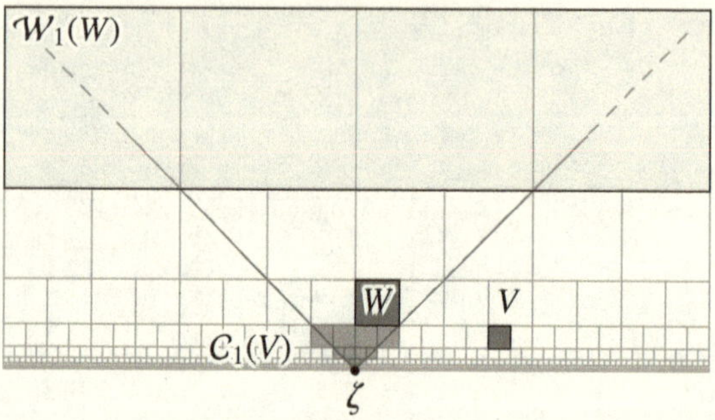

Figure 1.5 Proof of Lemma 1.8.7: For $W \in C$ and $V \in \mathcal{W}$, the figure depicts $\mathcal{W}_1(W) = \{V \in \mathcal{W} : a_1(W, V) = 1\}$ and $C_1(V) = \{W \in C : a_1(W, V) = 1\}$.

counting the squares along their levels we may easily arrange the enumeration in such a way that

$$cd(W) \leq 2^{-k} d\left(V_\ell^k\right) \leq Cd(W), \tag{1.8.44}$$

and

$$c(|\ell| + 1)d\left(V_\ell^k\right) \leq d\left(W, V_\ell^k\right) + d\left(V_\ell^k\right) \leq C(|\ell| + 1)d\left(V_\ell^k\right). \tag{1.8.45}$$

In view of the matrix estimate (1.8.43), these conditions yield

$$m_1\left(W, V_\ell^k\right) q\left(V_\ell^k\right) \leq C2^{-k}(|\ell| + 1)^{-1-\epsilon}, \tag{1.8.46}$$

whenever $V_\ell^k \in \mathcal{W}_1(W)$. Finally, we take the sum of the estimates (1.8.46) and obtain

$$\sum m_1\left(W, V_\ell^k\right) q\left(V_\ell^k\right) \leq C.$$

In view of Inequality (1.8.46), the resulting upper bound on the right-hand side may be chosen to be independent of $W \in C$. Thus we verified (1.8.41).

Verification of (1.8.42): Fix $V \in \mathcal{W}$, and define the collection $C_1(V) = \{W \in C : a_1(W, V) = 1\}$. See Figure 1.5. Choose an enumeration of those Whitney squares as $C_1(V) = \{W_i\}$, where $i \in \mathbb{N}$, satisfying $d(W_1) = d(V)$ and $d(W_i) \leq 2^{-\alpha i} d(W_1)$, where $0 < \alpha < 1$ is chosen independently of the Whitney square $V \in \mathcal{W}$. As $a_1(W_i, V) = 1$, we get

$$d(\zeta, V) + d(V) \leq C(d(W_i, V) + d(V)).$$

Recalling our choice for $q(V)$ and invoking the matrix estimate (1.8.43) yields,

$$m_1(W_i, V)q(V)^{-1} \leq C(2^{-\alpha i})^{1-\epsilon} \tag{1.8.47}$$

Finally, taking the sum of the estimates (1.8.47) completes the verification of Inequality (1.8.42) □

Boundedness of M_2

In order to show that $\left\|M_2 : \ell^2(\mathcal{W}) \to \ell^2(C)\right\| \le C$ we will verify that the matrix $(m_2(W,V) : W \in C, V \in \mathcal{W})$ satisfies the hypothesis of Theorem 1.8.5.

Lemma 1.8.8 *There exists $C > 0$ so that*

$$\sup_{W \in C} \sum_{V \in \mathcal{W}} m_2(W,V) \le C, \tag{1.8.48}$$

and

$$\sup_{V \in \mathcal{W}} \sum_{W \in C} m_2(W,V) \le C. \tag{1.8.49}$$

Proof First, fix $W \in C$ and $V \in \mathcal{W}$. Note that if $a_2(W,V) \neq 0$ then $d(V) \le d(W)$ and $d(V,\zeta) \le Cd(W)$. Hence, invoking assumption (1.8.37) gives the estimates for the entries in the matrix M_2,

$$m_2(W,V) \le \frac{C_0 d(V,\zeta) d(V)}{d(W)^2}. \tag{1.8.50}$$

Verification of (1.8.48): Fix $W \in C$. We isolate the relevant Whitney squares by forming $\mathcal{W}_2(W) = \{V \in \mathcal{W} : a_2(W,V) = 1\}$. See Figure 1.6. We enumerate those as $\mathcal{W}_2(W) = \left\{V_\ell^k\right\}$ where $k \in \mathbb{N}$ and $|\ell| \le C2^k$ such that

$$c2^{-k} d(W) \le d\left(V_\ell^k\right) \le Cd(W)2^{-k}, \tag{1.8.51}$$

and

$$c(|\ell|+1)d\left(V_\ell^k\right) \le d\left(V_\ell^k,\zeta\right) \le C(|\ell|+1)d\left(V_\ell^k\right). \tag{1.8.52}$$

Using the matrix estimate 1.8.50 together with Inequalities (1.8.51) and (1.8.52) we find

$$m_2\left(W, V_\ell^k\right) \le C2^{-2k}(|\ell|+1). \tag{1.8.53}$$

Summing the estimates (1.8.53) over the range $k \in \mathbb{N}$ and $|\ell| \le C2^k$ gives an estimate of the left-hand side in Inequality (1.8.48)by a bound independent of $W \in C$.

Verification of (1.8.49)**:** Fix $V \in \mathcal{W}$ and put $C_2(V) = \{W \in C : a_2(W,V) = 1\}$. Note that then $d(V,\zeta) \le d(W)$ for $W \in C_2(V)$. See Figure 1.6. Choose an enumeration of the relevant Whitney squares as $C_2(V) = \{W_1, W_2, \ldots, W_L\}$, so that $d(W_L) \ge 1/2$, and

$$cd(\zeta,V) \le d(W_1) \le Cd(\zeta,V).$$

Recall that $d(V,\zeta) \le d(W_i)$ and $d(V) \le d(W_i)$. Hence, by Inequality (1.8.50) we obtain that

$$m_2(W_i,V) \le Cd(V)d(W_i)^{-1}.$$

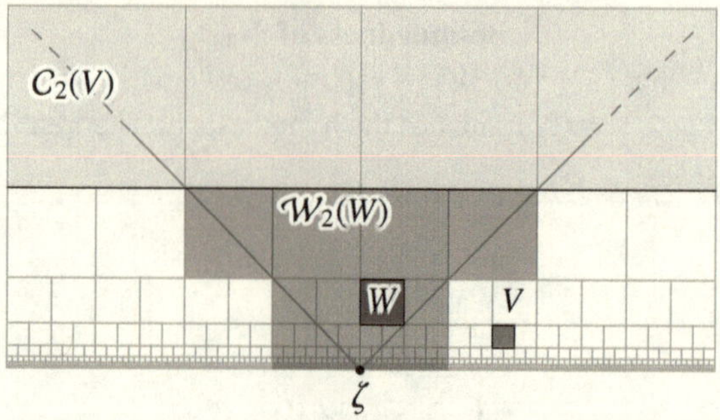

Figure 1.6 Proof of Lemma 1.8.8: For $W \in C$ and $V \in \mathcal{W}$, the figure depicts $\mathcal{W}_2(W) = \{V \in \mathcal{W} : a_2(W, V) = 1\}$ and $C_2(V) = \{W \in C : a_2(W, V) = 1\}$.

Taking the sum over $1 \le i \le L$ and invoking $\sum_{i=1}^{L} d(W_i)^{-1} \le Cd(W_1)^{-1}$ gives Inequality (1.8.49). □

Boundedness of M_3

In order to show that $\left\| M_3 : \ell^2(\mathcal{W}) \to \ell^2(C) \right\| \le C$ we will verify that the matrix $(m_3(W, V) : W \in C, V \in \mathcal{W})$ satisfies the hypothesis of Theorem 1.8.5.

Lemma 1.8.9 *Let $W \in C, V \in \mathcal{W}$, and $0 < \epsilon < 1$. If $q(V) = d(V)^{\epsilon} d(\zeta, V)^{\epsilon}$ and $p(W) = d(W)^{2\epsilon}$, then there exists $C > 0$, so that*

$$\sup_{W \in C} \sum_{V \in \mathcal{W}} m_3(W, V) \frac{q(V)}{p(W)} \le C, \tag{1.8.54}$$

and

$$\sup_{V \in \mathcal{W}} \sum_{W \in C} m_3(W, V) \frac{p(W)}{q(V)} \le C. \tag{1.8.55}$$

Proof First, fix $W \in C$ and $V \in \mathcal{W}$. If $a_3(W, V) \neq 0$ then $d(V) < d(W)$ and $d(V, \zeta) \ge d(W)$. Hence $d(V, \zeta) \ge cd(W)$. Invoking assumption (1.8.37) gives the estimates for the entries in the matrix M_3,

$$m_3(W, V) \le \frac{C_0 d(V)}{d(V, \zeta)}. \tag{1.8.56}$$

Verification of (1.8.54): Fix $W \in C$. Define the relevant Whitney squares by putting $\mathcal{W}_3(W) = \{V \in \mathcal{W} : a_3(W, V) = 1\}$. As the squares in $\mathcal{W}_3(W)$ satisfy $d(V) < d(W)$ and $d(V, \zeta) \ge cd(W)$, we may enumerate the relevant Whitney cubes as $\mathcal{W}_3(W) = \left\{V_\ell^k\right\}$ where $k \in \mathbb{N}$ and $c2^k \le |\ell| \le C2^k / d(W)$, such that

$$c2^{-k} d(W) \le d\left(V_\ell^k\right) \le Cd(W) 2^{-k}, \tag{1.8.57}$$

and

$$c(|\ell| + 1)d\left(V_\ell^k\right) \le d\left(V_\ell^k, \zeta\right) \le C(|\ell| + 1)d\left(V_\ell^k\right). \tag{1.8.58}$$

Note that by Inequality (1.8.56) we have $m_3\left(W, V_\ell^k\right) \le C(|\ell| + 1)^{-1}$. Next, we recall that $q(V) = d(V)^\epsilon d(V, \zeta)^\epsilon$ and $p(W) = d(W)^{2\epsilon}$. Invoking Inequalities (1.8.57) and (1.8.58) first gives $q\left(V_\ell^k\right)/p(W) \le C2^{-2\epsilon k}(|\ell| + 1)^\epsilon$, and then

$$m_3\left(W, V_\ell^k\right)\frac{q\left(V_\ell^k\right)}{p(W)} \le C2^{-2\epsilon k}(|\ell| + 1)^{-1+\epsilon}. \tag{1.8.59}$$

Next, observe the elementary estimate

$$\sum_{k=1}^{\infty}\sum_{|\ell|\ge 2^k} 2^{-2\epsilon k}(|\ell| + 1)^{-1+\epsilon} \le C_\epsilon,$$

where the constant $C_\epsilon < \infty$ depends only on $0 < \epsilon < 1$. Hence, summing the estimates (1.8.59) over $V_\ell^k \in \mathcal{W}_3(W)$ gives Inequality (1.8.54) where the right-hand side estimate is independent of $W \in C$.

Verification of (1.8.55): Fix $V \in \mathcal{W}$ and define $C_3(V) = \{W \in C : a_3(W, V) = 1\}$. We view $C_3(V)$ as a chain of Whitney squares connecting its largest square, say W_L, to its smallest one, say W_1. Recall that for $W \in C_3(V)$ we have $d(V) \le d(W) \le d(V, \zeta)$. Hence, we may write $C_3(V) = \{W_1, W_2, \ldots, W_L\}$ such that $d(W_{i+C}) > 4d(W_i)$,

$$cd(V) \le d(W_1) \le Cd(V), \tag{1.8.60}$$

and

$$cd(V, \zeta) \le d(W_L) \le Cd(V, \zeta). \tag{1.8.61}$$

Evaluating a geometric sum then gives $\sum_{i=1}^L d(W_i)^{2\epsilon} \le C_\epsilon d(\zeta, V)^{2\epsilon}$. By Inequality (1.8.56) for $W_i \in C_3(V)$, we get the matrix estimate

$$m_3(W_i, V)\frac{p(W_i)}{q(V)} \le \frac{C_\epsilon d(W_i)^{2\epsilon} d(V)^{1-\epsilon}}{d(\zeta, V)^{1+\epsilon}}. \tag{1.8.62}$$

Summing the estimates (1.8.62) over $1 \le i \le L$ and using the geometric sum calculated just after equation (1.8.61) we get

$$\sum_{i=1}^{L} m_3(W_i, V)\frac{p(W_i)}{q(V)} \le \frac{C_\epsilon d(V)^{1-\epsilon}}{d(\zeta, V)^{1-\epsilon}}.$$

As V is Whitney square, the term on the right-hand side is clearly bounded independent of $V \in C$. □

1.9 Notes

Section 1.1 forms a list of basic material covered by many texts on probability and martingale theory.

Section 1.1.1 reviews conditional expectations, filtered probability spaces, and discrete time martingales using Kahane (1985) and Neveu (1975).

Section 1.1.2 contains basic theorems on Brownian martingales, stochastic integrals, Itô's formula, and conformal invariance of complex Brownian motion. We used the books of Bass (1995), Durrett (1984), Petersen (1977) and Revuz and Yor (1991). See also Doob (1954) and Weisz (1994).

Section 1.1.3 introduces harmonic and subharmonic functions and their relation to Brownian motion. Basic sources are the books by Garnett (1981), Koosis (1980), and Ransford (1995).

Section 1.1.4 reviews the Poisson, Fejer, and de La Vallee Poussin kernels, the Hilbert transform, Hardy spaces $H^p(\mathbb{T})$, and the disk algebra $A(\mathbb{T})$. The books by García-Cuerva and Rubio de Francia (1985), Garnett (1981), Katznelson (1968), Koosis (1980), and Pełczyński (1977) were our basic sources. The paragraph on the trace class follows Wojtaszczyk (1991, Chapter III.G).

For Sections 1.1.5 and 1.1.6 we used Wojtaszczyk (1991).

Section 1.2 introduces the concept of holomorphic random variables, which goes back to Föllmer (1974), Getoor and Sharpe (1972) and Varopoulos (1980, 1981), who introduced the important algebra of bounded holomorphic martingales and demonstrated its significance for real, complex, and functional analysis! The spectrum of the algebra $H^\infty(\Omega)$ was investigated by Carne (1982). Arai (1986b) proves inner-outer factorization for the algebra $H^\infty(\Omega)$, and in Arai (1986a) provides quantitative estimates for the corona solutions obtained by Varopoulos's stochastic method (Varopoulos, 1980).

In Section 1.2.1 the use of Itô's formula is modeled after Durrett (1984). Theorem 1.2.3 is due to Garling (1988). The proof of Theorem 1.2.4 arose in discussions with Peter Yuditskii, who views it as a continuous version of the nonlinear telescoping lemma (Lemma 2.5.3).

Section 1.2.2 contains the complex convexity estimates of Theorem 1.2.5 obtained inependently by Bonami (1970, Chapter III, Theorem 7), Davis, Garling, and Tomczak-Jaegermann (1984) and Weissler (1980). More recently, Aleksandrov (2007) obtained an elegant analytic proof. Theorem 1.2.6 is in Müller (2012). Proposition 1.2.7 is due to Bourgain (1983a). Obtaining the best value for $\kappa > 0$ remains an open problem. We refer to Osekowski (2012) for systematic investigations of sharp martingale inequalities.

In Section 1.2.3, Theorem 1.2.8 is contained in Müller (1993). Theorem 1.2.9 is taken from Jones and Müller (2004). Theorems 1.2.10 and 1.2.11 are in Müller (2012).

In Section 1.3, Theorem 1.3.1 (the Burkholder–Davis–Gundy inequalities) crystallizes a large section of Burkholder's pioneering work in stochastic analysis. More recently, Burkholder's contributions centered on unconditional convergence of martingales with values in Banach spaces, and on the systematic search for the best constants in martingale inequalities (Burkholder [1973, 1979, 1981, 1983, 1989, 2001], Burkholder, Davis, and Gundy [1972], and Burkholder and Gundy [1970])

We followed Revuz and Yor (1991) in our presentation of the Burkholder–Davis–Gundy inequalities (Theorem 1.3.1). Theorem 1.3.2 is due to Burkholder (1979).

In Section 1.4, the stochastic Hilbert transform was introduced by Getoor and Sharpe (1972). Striking ideas by Varopoulos (1980) established the importance of the stochastic Hilbert transform to complex analysis. Our presentation incorporates work from Garling unpublished paper (Garling, D. J. H. $L^1(\Omega)/H^1(\Omega)$ is a Grothendieck space of cotype 2. pages 1–18, 1992.) which contains, inter alia, an exposition of the results of Varopoulos (1980). Theorem 1.4.4 was obtained in joint work with Peter Yuditskii (Müller and Yuditskii, 2019)

Section 1.5 reviews martingale embedding and projection operators using Petersen (1977) as basic source.

In Section 1.6, the books by Durrett (1984) and Petersen (1977) were our basic reference for defining Doob's projection, Theorem 1.6.1, and Theorem 1.6.2. Theorem 1.6.3, Theorem 1.6.4, and Theorem 1.6.5 are due to Varopoulos (1980).

In Section 1.7, Jones's decomposition theorem (Theorem 1.7.1) is due to Jones (1983). The probabilistic proof in the text follows Müller (1993). Theorem 1.7.2 is due to Bourgain (1983a). We presented the proof given in Müller (2012).

From Section 1.8, Theorems 1.8.1 and 1.8.2 were obtained in Jones and Müller (2004). A real variable proof of Theorem 1.8.2, not relying on stochastic methods, was obtained by Anisimov and Kislyakov (2004).

Section 1.8.1 organizes the proof by reduction to a repeated application to Schur's Lemma. It is related in spirit to the applications of Schur's Lemma in Coifman, Jones, and Semmes (1989) and David (1991, 1998). In the field of harmonic analysis, the space $SL^\infty(\mathbb{T})$ appeared explicitly in the work of

Y. Meyer (1968). Its probabilistic analogue $SL^\infty(\Omega)$ arose in the Cameron Martin transformation, for which we refer to the book of Durrett (1984). Also see Chang, Wilson, and Wolff (1985). An explicit connection between probabilistic square functions defining $SL^\infty(\Omega)$ and $SL^\infty(\mathbb{T})$ was presented in P.A. Meyer (1976).

For the treatment of dyadic SL^∞ we refer to Jones and Müller (2004), the recent work of Lechner (2018, 2019), and to the monograph by Müller (2005, Chapter 1). It should be noted that the Banach space properties of the $SL^\infty(\mathcal{F}_n)$ class associated to a general filtered probability space $(\Omega, (\mathcal{F}_n), \mu)$ are vastly untouched.

We emphasize the conformal invariant nature of the results presented in Section 1.8, and refer to conformal invariance of Green's functions, Poisson/Martin kernels, and complex Brownian motion. See, for example, Müller and Riegler (2020).

2

Hardy Martingales

In this chapter we introduce the space $H^p(\mathbb{T}^{\mathbb{N}})$ of p-integrable Hardy martingales. In Section 2.3, we present two important classes of Hardy martingales obtained by discretizing the holomorphic martingales

$$(f(z_t): t < \tau), \quad f \in H^1(\mathbb{T}),$$

where $(z_t: t < \tau)$ denotes complex Brownian motion, stopped at the boundary of the unit disk.

In Section 2.4, we consider examples generated by the Riesz projection on $L^p(\mathbb{T}^{\mathbb{N}})$, $1 < p < \infty$, and the orthogonal projection onto the space of homogeneous polynomials on $\mathbb{T}^{\mathbb{N}}$ of a fixed finite degree.

In Section 2.5, we prove square function estimates for L^1-bounded Hardy martingales, and obtain Davis and Garsia inequalities specifically tailored to Hardy martingales.

2.1 Bochner–Lebesque Spaces

Here, we review Bochner measurable functions, vector–valued conditional expectation operators, and martingales. We use the books by Diestel and Uhl (1977, Chapters II and IV), and Pisier (2016) as basic sources.

Bochner measurable: Fix a measurable space $(F, \mathcal{F})$ and a Banach space X. Let $\mathcal{B} = \mathcal{B}(X)$ denote the sigma-algebra generated by the open subsets of X (the Borel sigma-algebra). We say that $f: F \to X$ is a simple function if it is $\mathcal{F} - \mathcal{B}$ measurable and assumes only finitely many values. By definition, a function $f: F \to X$ is Bochner measurable if there exists a sequence of simple functions $(f_n: F \to X)$ such that

$$\lim_{n\to\infty} \|f(\omega) - f_n(\omega)\|_X = 0, \qquad \omega \in F.$$

A function $f : F \to X$ is Bochner measurable if and only if it is $\mathcal{F} - \mathcal{B}$ measurable *and* its range, $f(F)$, is contained in a *separable* linear subspace of X. We let $\mathcal{L}^0(F, \mathcal{F}, X)$ denote the linear space of all Bochner measurable functions $f : F \to X$.

Measure distance: Now let $(F, \mathcal{F}, \mu)$ be a finite measure space. Given $f, g \in \mathcal{L}^0(X) = \mathcal{L}^0(F, \mathcal{F}, X)$, we define the measure distance

$$d_\mu(f, g) = \inf\{\epsilon > 0 : \mu(\{\omega \in F : \|f(\omega) - g(\omega)\|_X > \epsilon\}) < \epsilon\}.$$

We say that a Bochner measurable function $f : F \to X$ is vanishing μ-almost everywhere, if $\mu(\{\omega \in F : \|f(\omega)\|_X \neq 0\}) = 0$. We let

$$\mathcal{N}(X) = \mathcal{N}(F, \mathcal{F}, \mu, X)$$

denote the space of Bochner measurable functions vanishing μ-almost everywhere. Note that $d_\mu(f, g) = 0$ if $f - g \in \mathcal{N}(X)$. The quotient space

$$L^0(F, \mathcal{F}, \mu, X) = \mathcal{L}^0(X)/\mathcal{N}(X),$$

equipped with the distance induced by d_μ, becomes a complete metric space. When feasible, our notation suppresses the dependence on $(F, \mathcal{F}, \mu, X)$ and we write $L^0(X) = L^0(F, \mathcal{F}, \mu, X)$.

The spaces $L^p(X)$: If $1 \leq p < \infty$, we say that $f : F \to X$ is p-Bochner integrable if it is Bochner measurable, and

$$\|f\|_p = \left(\int_F \|f(\omega)\|_X^p d\mu(\omega) \right)^{1/p} < \infty.$$

Similarly, if $p = \infty$, we say that a Bochner measurable f is p-Bochner integrable if

$$\|f\|_\infty = \mu - \text{ess sup}_{\omega \in F} \|f(\omega)\|_X < \infty.$$

We let $\mathcal{L}^p(X) = \mathcal{L}^p(F, \mathcal{F}, \mu, X)$ denote the space of p-Bochner integrable functions with values in X. Note that for any $f \in \mathcal{L}^p(X)$, we have $\|f\|_p = 0$ if f vanishes μ-almost everywhere. For $1 \leq p \leq \infty$, the quotient space

$$L^p(F, \mathcal{F}, \mu, X) = \mathcal{L}^p(X)/\mathcal{N}(X),$$

equipped with the natural quotient norm induced by $\| \cdot \|_p$, is a Banach space, called the Bochner–Lebesgue space. We write $L^p(X) = L^p(F, \mathcal{F}, \mu, X)$ whenever the underlying measure space is clear from the context we are working in.

Continuous operators: Fix (S, Σ, λ), a (second) measure space, and a Banach space Y. Let $L^0(Y) = L^0(S, \Sigma, \lambda, Y)$ be the space of equivalence classes of Bochner measurable functions $f: S \to Y$, and consider a linear operator

$$\mathcal{T}: L^1(X) \to L^0(Y).$$

The continuity of $\mathcal{T}$ with respect to the metrics $\|\cdot\|_{L^1(X)}$ and d_λ is equivalent to the following condition: For any $\epsilon > 0$, there exists $\delta > 0$ such that

$$\int_F \|f(\omega)\|_X d\mu(\omega) < \delta \quad \text{implies} \quad \lambda(\{s \in S: \|(\mathcal{T}f)(s)\|_Y > \epsilon\}) < \epsilon,$$

for any $f \in L^1(X)$. More concisely, $\mathcal{T}$ is continuous if for $\epsilon > 0$ there exists $R > 0$ such that

$$\lambda\Big(\{s \in S: \|(\mathcal{T}f)(s)\|_Y > R\|f\|_{L^1(X)}\}\Big) < \epsilon,$$

for any $f \in L^1(X)$.

The Bochner integral: Let X be a Banach space. Recall that $f: F \to X$ is a simple function if there exist $n \in \mathbb{N}$, $\mathcal{F}$-measurable sets $F_1, \ldots, F_n$ and $x_1, \ldots, x_n \in X$ such that

$$f = \sum_{i=1}^n x_i 1_{F_i}.$$

The integral of f is then defined as $\int_F f d\mu = \sum_{i=1}^n \mu(F_i)x_i$. The triangle inequality in X gives

$$\left\|\int_F f d\mu\right\|_X \leq \sum_{i=1}^n \mu(F_i)\|x_i\|_X = \|f\|_{L^1(X)}. \tag{2.1.1}$$

We next define the integral for $f \in L^1(X)$. Since the space of simple functions is dense in $L^1(X)$, we obtain a sequence of simple functions (f_n) such that $\lim_{n\to\infty} \|f - f_n\|_{L^1(X)} = 0$. By (2.1.1), the integrals $\left(\int_F f_n d\mu\right)$ form a Cauchy sequence in X, and putting

$$\int_F f d\mu = \lim_{n\to\infty} \int_F f_n d\mu$$

(uniquely and independent of the approximating sequence (f_n)) defines the Bochner integral of $f \in L^1(X)$.

Let $U \subseteq \mathbb{C}$ be an open subset, and let $u: U \to X$ be continuous. We say that u is harmonic if for any $z \in U$ there exists $r_0 = r_0(z) > 0$ such that

$$u(z) = \int_{\mathbb{T}} u(z + r\zeta) dm(\zeta), \quad \text{for} \quad r < r(z).$$

We remark that $u\colon U \to X$ is harmonic (in the above sense) if for every $x^* \in X^*$ the composition $x^* \circ u\colon U \to \mathbb{C}$ is harmonic in the ordinary sense.

Let $\mathbb{T} = \{\zeta \in \mathbb{C} : |\zeta| = 1\}$. We let m denote the unique rotation invariant probability measure defined on the sigma-algebra of Lebesgue measurable subsets of $\mathbb{T}$. (Thus m is the Haar measure on $\mathbb{T}$.) Let $1 \le p \le \infty$. We define the Hardy space $H^p(\mathbb{T}, X) \subset L^p(\mathbb{T}, X)$ to consist of those $f \in L^p(\mathbb{T}, X)$ for which

$$\int_{\mathbb{T}} f(\zeta)\zeta^n dm(\zeta) = 0, \quad n \in \mathbb{N}. \tag{2.1.2}$$

Since $H^p(\mathbb{T}, X)$ is closed in $L^p(\mathbb{T}, X)$, it is a Banach space when equipped with the $L^p(\mathbb{T}, X)$ norm.

The space $H^p(\mathbb{T}, X)$ is characterized in terms of Poisson integrals. In view of the Fourier expansion of the kernels (see equation (1.1.58)), if $f \in H^p(\mathbb{T}, X)$, then its Poisson integrals,

$$F(z) = \int_{\mathbb{T}} P(z, \zeta) f(\zeta) dm(\zeta),$$

define an analytic function $F\colon \mathbb{D} \to X$. Precisely, for every bounded linear functional $x^*\colon X \to \mathbb{C}$, the composition $x^* \circ F$ is an analytic complex-valued function in $\mathbb{D}$.

The converse implication holds true as well. Indeed, if the Poisson integrals of $f \in L^p(\mathbb{T}, X)$ give rise to an X-valued analytic function in $\mathbb{D}$, then $f \in H^p(\mathbb{T}, X)$.

Positive Operators and the Algebraic Tensor Product $L^p \otimes X$

Let $L^p = L^p(F, \mathcal{F}, \mu)$ denote the Lebesgue space of p-integrable scalar-valued functions, let X be a Banach space, and let E be a linear subspace of L^p. We denote

$$E \otimes X = \operatorname{span}\{f \otimes x\colon f \in E,\ x \in X\}.$$

We will frequently use that, for $1 \le p < \infty$, the algebraic tensor product $L^p \otimes X$ is dense in the Bochner–Lebesgue space $L^p(X)$ and, more generally, that for any dense subspace E of L^p, the algebraic tensor product $E \otimes X$ is dense in $L^p(X)$.

For a linear operator $T\colon L^p \to L^p$, we define $\mathcal{T} = T \otimes \operatorname{Id}_X$ on $E \otimes X$ by

$$T \otimes \operatorname{Id}_X(f \otimes x) = T(f) \otimes x, \quad f \in E,\ x \in X,$$

and by linear extension to $E \otimes X = \operatorname{span}\{f \otimes x\colon f \in E,\ x \in X\}$. We say that $T\colon L^p \to L^p$ is a positive operator if

$$f \ge 0 \quad \text{implies} \quad Tf \ge 0, \quad \text{for } f \in L^p.$$

Theorem 2.1.1 *Let $1 \le p < \infty$. If $T\colon L^p \to L^p$ is a positive, continuous, linear operator, then $\mathcal{T} = T \otimes \mathrm{Id}_X$ extends uniquely to a bounded linear operator on $L^p(X)$ satisfying*

$$\|\mathcal{T}\colon L^p(X) \to L^p(X)\| = \|T\colon L^p \to L^p\|.$$

Proof Let $n \in \mathbb{N}$ and let $E_j \subseteq \Omega$ such that $E_i \cap E_j = \emptyset$ for $i \neq j$ and $1 \le i, j \le n$. Put $f = \sum_{j=1}^n x_j f_j$ where $f_j = 1_{E_j}$ and $x_j \in X$. Then

$$\int_F \|f\|_X^p d\mu = \int_F \left(\sum_{j=1}^n \|x_j\|_X f_j \right)^p d\mu.$$

Since T is positive, we have the pointwise estimates $Tf_j \ge 0$. We will also show below that $|T(g)|^p \le T(|g|^p)$, for $g \in L^p(F)$. With $T \otimes \mathrm{Id}(f) = \sum_{j=1}^n x_j T(f_j)$, we now have

$$\begin{aligned}
\int_F \|(T \otimes \mathrm{Id})(f)\|_X^p d\mu &\le \int_F \left(\sum_{j=1}^n \|x_j\|_X |Tf_j| \right)^p d\mu \\
&= \int_F \left(T\left(\sum_{j=1}^n \|x_j\|_X f_j \right) \right)^p d\mu \\
&\le \|T\|_p^p \int_F \left(\sum_{j=1}^n \|x_j\|_X f_j \right)^p .
\end{aligned}$$

Now let $Z = \operatorname{span}\{\sum_{j=1}^n x_j 1_{E_j}\}$, where $x_j \in X$ and $E_j \subseteq F$ are pairwise disjoint measurable sets for $j \le n$ and $n \in \mathbb{N}$. Since Z is a dense subset in $L^p(F, X)$ the vector-valued $\mathcal{T} = T \otimes \mathrm{Id}_X$ extends uniquely to a bounded linear operator on $L^p(F, X)$, independent of the Banach space X.

Finally, we turn to proving the pointwise estimate $|T(g)|^p \le T(|g|^p)$. Fix $\omega \in F$, put $x = |T(g)(\omega)|$. Since $1 \le p < \infty$, the map $y \to |y|^p$ is convex. Hence, given $x \ge 0$, there exist $a = a(x)$ and $b = b(x)$ satisfying $x^p = ax + b$, and such that for any $y \in \mathbb{R}$ we have the inequality $|y|^p \ge ay + b$. Then using that T maps nonnegative functions to nonnegative functions we obtain

$$|T(g)(\omega)|^p = a|T(g)(\omega)| + b \le (T(a|g| + b))(\omega) \le T\left(|g|^p\right)(\omega). \qquad \square$$

Theorem 2.1.2 *Let $T\colon L^1 \to L^0$ be a positive, continuous, linear operator, that is, for $\epsilon > 0$, there exists $R > 0$ satisfying*

$$\mu(\{\omega \in F\colon |Tf(\omega)| > R\|f\|_{L^1}\}) < \epsilon, \tag{2.1.3}$$

for $f \in L^1$. *Then* $\mathcal{T} = T \otimes \mathrm{Id}_X$ *extends uniquely to a continuous linear operator* $\mathcal{T}: L^1(X) \to L^0(X)$ *such that for* $y \in L^1(X)$ *and* $\epsilon > 0$,

$$\mu\left(\{\omega \in F: \|\mathcal{T}y(\omega)\|_X > R\|y\|_{L^1(X)}\}\right) < \epsilon, \tag{2.1.4}$$

where $R > 0$ *is given by* (2.1.3).

Proof Fix $y \in L^1 \otimes X$ and a linear functional $x^* \in X^*$ with $\|x^*\| = 1$. By homogeneity of (2.1.4) we may assume that $\|y\|_{L^1(X)} = 1$. By linearity,

$$\langle(\mathcal{T}y)(\cdot), x^*\rangle = T\langle y(\cdot), x^*\rangle.$$

For, $\omega \in F$, put $g(\omega) = \|y(\omega)\|_X$. Since T is a positive operator, we have $T\langle y(\cdot), x^*\rangle \leq T(g)$. The Hahn–Banach theorem implies the pointwise estimates

$$\|\mathcal{T}(y)\|_X \leq T(g). \tag{2.1.5}$$

Clearly, the estimate (2.1.5) gives the inclusion

$$\{\|\mathcal{T}(y)\|_X > R\} \subseteq \{|T(g)| > R\}.$$

Since $g \in L^1(F, \mu)$ with $\|g\|_{L^1} = 1$, applying (2.1.3) to g gives (2.1.4). □

In this book, the most frequent applications of Theorem 2.1.1 and Theorem 2.1.2 are to conditional expectations, and to convolution operators with respect to Poisson kernels (1.1.55), Fejer kernels (1.1.63), and de La Vallee Poussin kernels (1.1.66).

Vector-Valued Martingales

Let X be a Banach space. Recall that we fixed a finite measure space $(F, \mathcal{F}, \mu)$. Let $\mathcal{G} \subseteq \mathcal{F}$ be a sub-sigma-algebra such that $F \in \mathcal{G}$. Taking the conditional expectation with respect to $\mathcal{G}$ defines a positive linear operator on L^p by putting

$$\mathbb{E}_{\mathcal{G}}(f) = \mathbb{E}(f|\mathcal{G}), \quad f \in L^p.$$

Since $\|\mathbb{E}_{\mathcal{G}}: L^p \to L^p\| = 1$, applying Theorem 2.1.1 gives

$$\left\|\mathbb{E}_{\mathcal{G}} \otimes \mathrm{Id}_X: L^p(X) \to L^p(X)\right\| = 1, \quad 1 \leq p \leq \infty. \tag{2.1.6}$$

Throughout, we let $(\mathcal{G}_k)$ be a sequence of increasing sub-sigma-algebras of $\mathcal{F}$, (that is, each $\mathcal{G}_k$ is a sigma-algebra, $\mathcal{G}_k \subseteq \mathcal{G}_{k+1} \subseteq \mathcal{F}$, and $F \in \mathcal{G}_k$). A sequence of μ-Bochner integrable functions $(f_k: F \to X)$ is a $(\mathcal{G}_k)$-martingale if

$$\mathbb{E}_{\mathcal{G}_k}(f_{k+1}) = f_k.$$

In the following paragraphs we summarize basic martingale theorems and inequalities that hold true independent of the underlying Banach spaces X. This includes Doob's inequality, the Davis decomposition, and the convergence theorem for bounded, closed martingales.

Martingale inequalities: Let X be a Banach space, and let $(f_k\colon F \to X)$ be an X-valued $(\mathcal{G}_k)$-martingale. Doob's inequalities (Theorem 1.1.4) assume the following form:

$$t\mu\left(\max_{k\leq n} \|f_k\|_X > t\right) \leq \|f_n\|_{L^1(X)}, \quad t > 0, n \in \mathbb{N}, \tag{2.1.7}$$

and

$$\left\|\max_{k\leq n} \|f_k\|_X\right\|_{L^p(\mu)} \leq \frac{p}{p-1}\|f_n\|_{L^p(X)}, \quad p > 1, n \in \mathbb{N}. \tag{2.1.8}$$

Similarly, Theorem 1.1.7 (the Davis decomposition) directly extends to vector-valued martingales. Thus $(f_k\colon F \to X)$ may be decomposed into the sum of $(\mathcal{G}_k)$-martingale sequences $(g_k\colon F \to X)$ and $(b_k\colon F \to X)$ satisfying

$$f_n = g_n + b_n, \quad n \in \mathbb{N}, \tag{2.1.9}$$

$$\|\Delta g_n\|_X \leq 8 \max_{k\leq n-1} \|f_k\|_X, \quad n \in \mathbb{N}, \tag{2.1.10}$$

and

$$\sum_{k=1}^{n} \|\Delta b_k\|_{L^1(X)} \leq 8 \left\|\max_{k\leq n} \|f_k\|_X\right\|_{L^1(\mu)}, \quad n \in \mathbb{N}. \tag{2.1.11}$$

Closed martingales: We say that a $(\mathcal{G}_k)$-martingale sequence $(f_k\colon F \to X)$ is closed if there exists $f \in L^1(X)$ such that

$$f_k = \mathbb{E}(f|\mathcal{G}_k), \qquad k \in \mathbb{N}.$$

The main theorem on closed martingales asserts that their $L^p(X)$ boundedness implies their convergence in $L^p(X)$ and almost surely. The fact that this property holds true for any Banach space X separates the class of closed vector-valued martingales from ones that are merely bounded in $L^p(X)$.

Theorem 2.1.3 *Let X be a Banach space, and $1 \leq p < \infty$. An $L^p(X)$ bounded martingale $(f_k\colon F \to X)$ is closed if and only if there exists $f \in L^p(F, X)$ such that*

$$\lim_{k\to\infty} \|f - f_k\|_{L^p(F,X)} = 0.$$

In this case, there exists $E \subset F$ with $\mu(E) = 0$ and

$$\lim_{k\to\infty} \|f_k(\omega) - f(\omega)\|_X = 0,$$

for every $\omega \in F \setminus E$.

2.2 Martingales on $\mathbb{T}^{\mathbb{N}}$

Let $\mathbb{T} = \{e^{i\theta} : \theta \in [0, 2\pi[\}$ be the torus equipped with the normalized angular measure. Let $\mathbb{T}^{\mathbb{N}} = \left\{(x_i)_{i=1}^{\infty}\right\}$ be its countable product equipped with its product Haar measure $\mathbb{P}$. We let $\mathbb{E}$ denote expectation with respect to $\mathbb{P}$.

Let $(\Omega, \mathcal{G}, \mu)$ be a measure space, let X be a Banach space, and let $(\mathcal{G}_k)$ be a sequence of increasing sub-sigma-algebras of $\mathcal{G}$ (that is, each $\mathcal{G}_k$ is a sigma-algebra, $\mathcal{G}_k \subseteq \mathcal{G}_{k+1} \subseteq \mathcal{G}$, and $\Omega \in \mathcal{G}_k$). A sequence of μ-Bochner integrable functions $(g_k \colon \Omega \to X)$ is a $(\mathcal{G}_k)$-martingale if

$$\mathbb{E}(g_{k+1}|\mathcal{G}_k) = g_k. \tag{2.2.1}$$

We say that $\Delta g_k = g_k - g_{k-1}$ is the martingale difference sequence associated to (g_k). We recall that for vector-valued martingales, Doob's inequalities (2.1.7) and (2.1.8) hold true independently of the Banach space X.

Any martingale $(g_k \colon \Omega \to X)$ admits a representation on the infinite torus product $\mathbb{T}^{\mathbb{N}}$. Specifically, there exists a sequence of $\mathbb{P}$-Bochner integrable functions $(F_k \colon \mathbb{T}^{\mathbb{N}} \to X)$ such that

- almost surely F_k depends only on $\mathbb{T}^k$, the first k coordinates of $\mathbb{T}^{\mathbb{N}}$,
- for almost every $x \in \mathbb{T}^k$, the function $z \to F_{k+1}(x, z)$ is contained in $L^1(\mathbb{T}, X)$ and

$$\int_{\mathbb{T}} F_{k+1}(x, z)dm(z) = F_k(x), \text{ and} \tag{2.2.2}$$

- we have $\mathbb{P}\{F_k \in B\} = \mu\{g_k \in B\}$ for any Borel measurable $B \subseteq X$.

Montgomery-Smith (1998) and Dellacherie and Meyer (1975, 1978, 1982) present proofs of the martingale representation theorem reviewed above. In Section 1.1.1 we introduced the canonical filtration on the infinite product space $\mathbb{T}^{\mathbb{N}}$. We denoted by $\mathcal{F}_k$ the sigma-algebra on $\mathbb{T}^{\mathbb{N}}$ generated by the cylinder sets

$$\left\{(A_1, \dots, A_k, \mathbb{T}^{\mathbb{N}})\right\},$$

where A_i, $i \leq k$, are measurable subsets of $\mathbb{T}$. Equivalently, the coordinate projections

$$p_j \colon \mathbb{T}^{\mathbb{N}} \to \mathbb{T}, \qquad j \leq k, \tag{2.2.3}$$

generate $\mathcal{F}_k$; here $p_j(z) = z_j$ and $z \in \mathbb{T}^{\mathbb{N}}$ with $z = (z_j)$.

We let $\mathbb{E}_k$ denote the conditional expectation with respect to the sigma-algebra $\mathcal{F}_k$. Explicitly, for almost every $x \in \mathbb{T}^k$, we have

$$\mathbb{E}_k(F)(x) = \int_{\mathbb{T}^{\mathbb{N}}} F(x, z)d\mathbb{P}(z), \tag{2.2.4}$$

for $F \in L^1(\mathbb{T}^{\mathbb{N}})$ and $k \in \mathbb{N}$. Conversely, conditioning $F \in L^1(\mathbb{T}^{\mathbb{N}})$ to $\mathcal{F}_k$ amounts to fixing $x \in \mathbb{T}^k$ and setting

$$f(z) = F(x, z), \quad \text{where} \quad z \in \mathbb{T}^{\mathbb{N}}. \tag{2.2.5}$$

Thus, almost surely $f \in L^1(\mathbb{T}^{\mathbb{N}})$. In the remaining chapters we choose the filtered probability space $(\mathbb{T}^{\mathbb{N}}, (\mathcal{F}_k), \mathbb{P})$ as our preferred model for discussing discrete-time martingales.

2.2.1 Hardy Martingales

Let $H^1(\mathbb{T}) \subset L^1(\mathbb{T})$ consist of those integrable functions for which the harmonic extension to the unit disk is analytic. An $(\mathcal{F}_k)$-martingale (F_k) is called a Hardy martingale if conditioned on $\mathcal{F}_{k-1}$,

$$F_k \text{ defines an element in } H^1(\mathbb{T}).$$

Fix a Hardy martingale (F_k). Conditioning to $\mathcal{F}_{k-1}$ amounts to fixing $x \in \mathbb{T}^{k-1}$ and putting $h_0 = F_{k-1}(x)$ and $h(z) = F_k(x, z)$, where $z \in \mathbb{T}$. We have $h \in H^1(\mathbb{T})$, and $h_0 = \int_{\mathbb{T}} h dm$. Hence the subharmonicity estimates (1.2.17) give

$$|h_0|^{\alpha} \le \int_{\mathbb{T}} |h|^{\alpha} dm,$$

for $\alpha > 0$. Thus we verified the important submartingale estimates,

$$|F_{k\ 1}|^{\alpha} < \mathbb{E}_{k-1}|F_k|^{\alpha}, \tag{2.2.6}$$

for $k \in \mathbb{N}$ and $\alpha > 0$. Following the proof of Garling's theorem (Theorem 1.2.3), we apply Doob's maximal inequality (1.1.6) to the nonnegative submartingale $\left(|F_k|^{1/p}\right)$, where $p > 1$. This gives

$$\mathbb{E}\left(\max_{k \le n} |F_k|\right) \le C_p^p \mathbb{E}|F_n|, \tag{2.2.7}$$

where $C_p = p/(p-1)$. Since $e = \inf\{(p/(p-1))^p : 1 < p < \infty\}$, we get

$$\mathbb{E}\left(\max_{k \le n} |F_k|\right) \le e\mathbb{E}|F_n|. \tag{2.2.8}$$

Consequently, any L^1-bounded Hardy martingale (F_n) is uniformly integrable. Hence, by Doob's theorem (Theorem 1.1.2), there exists $F \in L^1(\mathbb{T}^{\mathbb{N}})$ such that $F_n = \mathbb{E}_n(F)$, for any $n \in \mathbb{N}$. Define, for $1 \le p \le \infty$,

$$H^p(\mathbb{T}^{\mathbb{N}}) = \left\{F \in L^p(\mathbb{T}^{\mathbb{N}}) : (\mathbb{E}_k(F))_{k=1}^{\infty} \text{ forms a Hardy martingale}\right\}. \tag{2.2.9}$$

By (2.2.8), $H^p(\mathbb{T}^{\mathbb{N}})$ is a closed subspace of $L^p(\mathbb{T}^{\mathbb{N}})$, and corresponds 1:1 to the L^p bounded Hardy martingales for $1 \le p \le \infty$. We let

$$P\colon L^2(\mathbb{T}^{\mathbb{N}}) \to H^2(\mathbb{T}^{\mathbb{N}}) \tag{2.2.10}$$

denote the orthogonal projection onto $H^2(\mathbb{T}^{\mathbb{N}})$, called the Riesz projection. In Section 2.4.1, we give two separate proofs of the fact that P extends to a bounded operator on $L^p(\mathbb{T}^{\mathbb{N}})$, for $1 < p < \infty$ (with $p = 1$ and $p = \infty$ excluded).

Simple Hardy martingales: We define the class of simple Hardy martingales. Let $G \in L^1(\mathbb{T}^{\mathbb{N}})$, put $G_j = \mathbb{E}_j G$ and $\Delta G_j(z) = G_j(z) - G_{j-1}(z)$. We say that G is a simple Hardy martingale if

$$\Delta G_j(z) = f_{j-1}(z) z_j, \qquad z \in \mathbb{T}^{\mathbb{N}},$$

where f_{j-1} depends only on the first $j-1$ coordinates of $z \in \mathbb{T}^{\mathbb{N}}$. Next, we define the orthogonal projection onto the span of simple Hardy martingales. Let $G = (G_j)$ be a square integrable martingale on $\mathbb{T}^{\mathbb{N}}$. Put $g_j = \Delta G_j$ and define P_0 by putting

$$P_0 G = \sum_{j=1}^{\infty} \mathbb{E}_{j-1}(g_j \overline{p_j}) p_j, \tag{2.2.11}$$

where $p_j(z) = z_j$ with $z = (z_j)$ and $z_j \in \mathbb{T}$. Thus P_0 is the orthogonal projection onto the space of simple Hardy martingales contained in $L^2(\mathbb{T}^{\mathbb{N}})$. In Section 2.4.1 we will show that P_0 extends to a bounded operator on L^p when $1 < p < \infty$. The projection P_0 does not extend boundedly to L^1. However, P_0 is bounded on $H^1(\mathbb{T}^{\mathbb{N}})$, by Theorem 2.5.6 and the square function characterization of Theorem 2.5.4. Moreover, it is known that P_0 is of weak type 1:1. This may be deduced from Theorem 2.4.7 and the fact that uniformly bounded martingale transforms are of weak type 1:1.

Throughout the literature, simple Hardy martingales are called **analytic martingales**; see for instance, Bourgain and Davis (1986), Davis et al. (1984), Edgar (1986), and Pisier (1992a, 2016).

Steinhaus martingales: Let $G = (G_n)$ be a square integrable martingale on $\mathbb{T}^{\mathbb{N}}$. We say that G is a Steinhaus martingale if there exist scalar coefficients a_j such that $\Delta G_j(z) = a_j z_j$, where $z = (z_j)$ and $z_j \in \mathbb{T}$. Khintchine's inequality for Steinhaus martingales (see Wojtaszczyk [1991, I.B.8]) asserts that, for $0 < p < \infty$, there exist $c_p, C_p > 0$ such that for any Steinhaus martingale,

$$c_p \|G\|_{L^p(\mathbb{T}^{\mathbb{N}})} \leq \|G\|_{L^2(\mathbb{T}^{\mathbb{N}})} \leq C_p \|G\|_{L^p(\mathbb{T}^{\mathbb{N}})}. \tag{2.2.12}$$

For $p = 1$, the best constant $C_1 = 2/\sqrt{\pi}$ in (2.2.12) was determined by Sawa (1985). The best constants for C_p, $0 < p < 1$, were obtained by König (2014).

Given $G \in L^2(\mathbb{T}^{\mathbb{N}})$ the orthogonal projection onto Steinhaus martingales is given by

$$S(G) = \sum_{j=1}^{\infty} \mathbb{E}(g_j \overline{p_j}) p_j,$$

where $g_j = \Delta G_j$ and $p_j(z) = z_j$. Since p_j is an orthogonal sequence in $L^2(\mathbb{T}^{\mathbb{N}})$, Khintchine's inequality applied to $S(G)$ yields $\|S\|_p \leq C_p$ for $2 \leq p < \infty$ and, by duality, $\|S\|_p \leq C_q$ for $1 < p \leq 2$ and $1/p + 1/q = 1$. However, S is unbounded on L^1, and, moreover, S does not satisfy weak type 1:1 estimates. See Section 2.4.2. Steinhaus martingales are exactly the objects called *Steinhaus series* in the books by Kahane (1968, 1985, Section 1.5).

Let X be a complex Banach space, and let (x_k) be a sequence in X. Putting

$$F_n(z) = \sum_{k=1}^{n} z_k x_k, \quad n \in \mathbb{N}, z = (z_k) \in \mathbb{T}^{\mathbb{N}}, \tag{2.2.13}$$

defines an X-valued Steinhaus martingale. Kahane's inequality asserts that for $1 \leq p < \infty$ there exist $0 < c_p, C_p < \infty$, such that

$$c_p \left(\mathbb{E}_{(z)} \|F_n(z)\|_X^p\right)^{1/p} \leq \left(\mathbb{E}_{(z)} \|F_n(z)\|_X^2\right)^{1/2} \leq C_p \left(\mathbb{E}_{(z)} \|F_n(z)\|_X^p\right)^{1/p}, \tag{2.2.14}$$

for $n \in \mathbb{N}$ and any X-valued Steinhaus martingale (F_n). (For the proof of Kahane's inequality we refer to Kahane [1985] or Wojtaszczyk [1991].)

2.2.2 Homogeneous Polynomials

We introduce the class of martingales formed by trigonometric series of fixed finite multiplicity. Let $k \in \mathbb{N}$ and denote by $C(k)$ the collection of those subsets $S \subset \mathbb{N}$ for which $|S| = k$. If $S \in C(k)$ and $\varepsilon \in \{+1, -1\}^S$, define

$$z_S^{\varepsilon} = \prod_{j \in S} z_j^{\varepsilon_j},$$

where $z = (z_i)$ with $z_i \in \mathbb{T}$. We put $p_S^{\varepsilon}(z) = z_S^{\varepsilon}$. If $m = \max S$ and $\varepsilon_m = 1$, then p_S^{ε} is a Hardy martingale. To see this, it remains to write $p_S^{\varepsilon}(z) = \left(\prod_{j \in S \setminus \{m\}} z_j^{\varepsilon_j}\right) z_m$.

Let $\mathcal{P}_k = \operatorname{span}\left\{p_S^{\varepsilon} : S \in C(k), \varepsilon \in \{+1, -1\}^S\right\}$. Thus $\mathcal{P}_k$ is spanned by homogeneous polynomials of degree k.

We next identify the orthogonal projection onto $\mathcal{P}_k$. Let $S, S' \in C(k)$, $\varepsilon' \in \{+1, -1\}^{S'}$ and $\varepsilon \in \{+1, -1\}^S$. If $S \neq S'$ or $\varepsilon \neq \varepsilon'$ then it is easy to see that $\langle p_S^{\varepsilon}, p_{S'}^{\varepsilon'} \rangle = 0$, where the complex scalar product is the one of $L^2(\mathbb{T}^{\mathbb{N}})$. The orthogonal projection onto the L^2 closure of $\mathcal{P}_k$ is hence given by

$$S^{(k)}(f) = \sum_{S, \varepsilon} \left\langle f, p_S^{\varepsilon} \right\rangle p_S^{\varepsilon}, \tag{2.2.15}$$

where the sum is taken over $S \in C(k)$ and $\varepsilon \in \{+1,-1\}^S$. Clearly, $\mathcal{S}^{(k)}$ is a contraction on L^2. In Section 2.4.2, we will consider the entire sequence $\left(\mathcal{S}^{(k)}\right)_{k=1}^{\infty}$ as the object of interest and identify its generating function $\sum \alpha^k \mathcal{S}^{(k)}$. Thus we obtain the norm estimates $\left\|\mathcal{S}^{(k)}\right\|_p \le C_p^k$, for $1 < p < \infty$ and $k \in \mathbb{N}$.

2.2.3 Dyadic Martingales

Let $\mathcal{D}$ denote the dyadic sigma-algebra on $\mathbb{T}^{\mathbb{N}}$, generated by the independent Rademacher functions $\sigma_k \colon \mathbb{T}^{\mathbb{N}} \to \{-1, 1\}$, which are defined by

$$\sigma_k(x) = \operatorname{sign}(\cos_k(x)),$$

where $x = (x_k)$ and $\cos_k(x) = \Re x_k$. Let $\mathcal{D}_k$ denote the finite algebra of subsets in $\mathbb{T}^{\mathbb{N}}$ generated by $\sigma_1, \ldots, \sigma_k$. The conditional expectation $\mathbb{E}(\cdot|\mathcal{D}_k)$ is given as the integral operator

$$\mathbb{E}(F|\mathcal{D}_k)(w) = \int_{\mathbb{T}^{\mathbb{N}}} F(z) \prod_{m=1}^{k} (1 + \sigma_m(w)\sigma_m(z)) d\mathbb{P}(z), \qquad w \in \mathbb{T}^{\mathbb{N}}, \tag{2.2.16}$$

and similarly,

$$\begin{aligned} \mathbb{E}(F|\mathcal{D})(w) &= \lim_{n\to\infty} \int_{\mathbb{T}^{\mathbb{N}}} F(z) \prod_{k=1}^{n} (1 + \sigma_k(w)\sigma_k(z)) d\mathbb{P}(z) \\ &= \int_{\mathbb{T}^{\mathbb{N}}} F(z) \lim_{n\to\infty} \prod_{k=1}^{n} (1 + \sigma_k(w)\sigma_k(z)) d\mathbb{P}(z), \qquad w \in \mathbb{T}^{\mathbb{N}}, \end{aligned} \tag{2.2.17}$$

where, by Doob's theorem (Theorem 1.1.2), the respective limits in the first and second equation exist in $L^1(\mathbb{T}^{\mathbb{N}})$ and $\mathbb{P}$-almost everywhere.

Denoting by $\mathbb{P}_{\mathcal{D}}$ the restriction of the Haar measure $\mathbb{P}$ on $\mathbb{T}^{\mathbb{N}}$ to $\mathcal{D}$, we obtain the filtered probability space $(\mathbb{T}^{\mathbb{N}}, (\mathcal{D}_k), \mathbb{P}_{\mathcal{D}})$. Any martingale with respect to $(\mathbb{T}^{\mathbb{N}}, (\mathcal{D}_k), \mathbb{P}_{\mathcal{D}})$ is called a dyadic martingale on $\mathbb{T}^{\mathbb{N}}$.

Comparing $(\mathcal{F}_k)$ and $(\mathcal{D}_k)$: The canonical filtration $(\mathcal{F}_k)$ on $\mathbb{T}^{\mathbb{N}}$ is the one generated by the coordinate projections p_k (see the coordinate projection (2.2.3)). Thus $(\mathcal{F}_k)$ and $(\mathcal{D}_k)$ are linked by the identities

$$\mathcal{D}_k = \mathcal{F}_k \cap \mathcal{D},$$

for $k \in \mathbb{N}$. Consequently,

$$\mathbb{E}(\mathbb{E}(F|\mathcal{F}_k)|\mathcal{D}) = \mathbb{E}(F|\mathcal{D}_k), \tag{2.2.18}$$

for any integrable $F \colon \mathbb{T}^{\mathbb{N}} \to \mathbb{C}$.

Consequences of equation (2.2.18): By the conditional Minkowski inequality, for any $(\mathcal{F}_k)$-martingale $F = (F_k)$ we have the pointwise estimates

$$\left(\sum_{k=1}^{\infty} |\mathbb{E}(\Delta F_k|\mathcal{D}_k)|^2\right)^{1/2} \leq \mathbb{E}\left(\left(\sum_{k=1}^{\infty} |\Delta F_k|^2\right)^{1/2} |\mathcal{D}\right). \tag{2.2.19}$$

In summary, the Hardy space of dyadic martingales $H^1(\mathbb{T}^{\mathbb{N}}, (\mathcal{D}_k), \mathbb{P}_{\mathcal{D}})$ is linked to $H^1(\mathbb{T}^{\mathbb{N}}, (\mathcal{F}_k), \mathbb{P})$ by the following commutative diagram,

$$\begin{array}{ccc} H^1(\mathbb{T}^{\mathbb{N}}, (\mathcal{D}_k), \mathbb{P}_{\mathcal{D}}) & \xrightarrow{\quad \mathrm{Id} \quad} & H^1(\mathbb{T}^{\mathbb{N}}, (\mathcal{D}_k), \mathbb{P}_{\mathcal{D}}) \\ & {\scriptstyle J} \searrow \qquad \nearrow {\scriptstyle \mathbb{E}_{\mathcal{D}}} & \\ & H^1(\mathbb{T}^{\mathbb{N}}, (\mathcal{F}_k), \mathbb{P}) & \end{array} \tag{2.2.20}$$

where J denotes the formal inclusion map, and $\mathbb{E}_{\mathcal{D}}(F) = \mathbb{E}(F|\mathcal{D})$ is the conditional expectation.

Illustrating further the representation (2.2.18) of the dyadic filtration, we remark that

$$\mathbb{E}(F|\mathcal{D}) = \sum_{k=1}^{\infty} \mathbb{E}(\sigma_k \Delta F_k|\mathcal{D}_{k-1})\sigma_k, \tag{2.2.21}$$

where $\Delta F_k = \mathbb{E}(F|\mathcal{F}_k) - \mathbb{E}(F|\mathcal{F}_{k-1})$. By Doob's theorem (Theorem 1.1.2), for $F \in L^p(\mathbb{T}^{\mathbb{N}})$, convergence in (2.2.21) holds in L^p and almost surely. By equations (2.2.16) and (2.2.18) we have,

$$\mathbb{E}(\sigma_k \Delta F_k|\mathcal{D}_{k-1}) = \mathbb{E}(\sigma_k \mathbb{E}(F|\mathcal{F}_k)|\mathcal{D}_{k-1}) = \mathbb{E}(\sigma_k F|\mathcal{D}_{k-1}), \tag{2.2.22}$$

which may be used to simplify the right-hand side of (2.2.21). We refer to Maurey's paper for further developments along the lines delineated above (Maurey, 1980).

Khintchine's inequality: On $\mathrm{span}\{\sigma_k : k \in \mathbb{N}\}$, the norm in L^2 and the norm in L^p $(0 < p < \infty)$ are equivalent. More precisely, for $f = \sum a_k \sigma_k$ we have

$$c_p \|f\|_p \leq \left(\sum |a_k|^2\right)^{1/2} \leq C_p \|f\|_p. \tag{2.2.23}$$

Szarek (1976) determined the value $C_1 = \sqrt{2}$. (Also see Wojtaszczyk [1991, I.B.8].) By comparison, it is only true for $1 < p < \infty$ that the orthogonal projection onto the L^p-closure of $\mathrm{span}\{\sigma_k : k \in \mathbb{N}\}$ bounded on $L^p(\mathbb{T}^{\mathbb{N}}, \mathcal{F}, \mathbb{P})$. It is given by

$$R(f) = \sum_{k=1}^{\infty} \langle f, \sigma_k \rangle \sigma_k, \qquad f \in L^p(\mathbb{T}^{\mathbb{N}}, \mathcal{F}, \mathbb{P}),$$

where $\langle f, \sigma_k \rangle = \mathbb{E}(f\sigma_k)$, and satisfies $\|R\|_p \sim \sqrt{p}$ for $p \to \infty$ and $\|R\|_p = \|R\|_q$ where $1/p + 1/q = 1$. Thus,

$$\left(\sum_{k=1}^{\infty} |\langle f, \sigma_k \rangle|^2 \right)^{1/2} \leq C_p \|f\|_q, \qquad f \in L^q(\mathbb{T}^{\mathbb{N}}, \mathcal{F}, \mathbb{P}), \tag{2.2.24}$$

for $1 < p < \infty$, with $C_p \sim (q-1)^{(-1/2)}$ for $q \to 1$.

2.3 Examples

Here, we use Brownian motion to obtain important, explicit constructions of Hardy martingales. We begin by discretizing complex Brownian motion in such a way that $f \in H^1(\mathbb{T})$ is mapped to $F \in H^1(\mathbb{T}^{\mathbb{N}})$ and that ΔF_k is uniformly bounded. This gives Maurey's embedding introduced in Maurey (1980).

We present a second discretization of Brownian motion transforming $f \in H^1(\mathbb{T})$ into $F \in H^1(\mathbb{T}^{\mathbb{N}})$, such that the martingale difference sequence of F controls any given sequence of lacunary Fourier coeffients of f. Pisier (1997) exploited this embedding in his construction of a counterexample to Halmos' conjecture on polynomially bounded operators. We illustrate Pisier's construction and derive Paley's theorem on lacunary Fourier coeffients.

Maurey's Embedding

We define a representation of $H^1(\mathbb{T})$ as a linear subspace of Hardy martingales with a uniformly small martingale difference sequence. The embedding is defined by composition with a carefully designed Hardy martingale ($\varphi_k : \mathbb{T}^{\mathbb{N}} \to \mathbb{D}$), whose limit is uniformly distributed over $\mathbb{T}$.

Fix $0 < \epsilon < 1/2$ and $w \in \mathbb{T}^{\mathbb{N}}$ with $w = (w_k)$. Put $\varphi_0 \in \mathbb{D}$ and

$$\varphi_k(w) = \varphi_{k-1}(w) + \epsilon(1 - |\varphi_{k-1}(w)|)^2 w_k, \qquad k \in \mathbb{N}. \tag{2.3.1}$$

See Figure 2.1. Since $(1-|\varphi_{k-1}(w)|)^2 < 1-|\varphi_{k-1}(w)|$, it is easy to see that $|\varphi_k| \leq 1$ and $\mathbb{E}_{k-1}(\varphi_k) = \varphi_{k-1}$. Clearly, conditioned to $\mathcal{F}_{k-1}$, we have $\varphi_k(w) = a + bw_k$. Thus (φ_k) forms a simple Hardy martingale, uniformly bounded and hence convergent in L^1. Note that

$$|\varphi_k - \varphi_{k-1}| = \epsilon(1 - |\varphi_{k-1}|)^2.$$

Since (φ_k) is a Cauchy sequence, we get $(1 - |\varphi_{k-1}|) \to 0$ and $\varphi = \lim \varphi_k$ takes values in $\mathbb{T}$. Note that

$$\mathbb{P}\{\varphi \in B\} = \int_B \frac{1 - |\varphi_0|^2}{|\zeta - \varphi_0|^2} dm(\zeta), \tag{2.3.2}$$

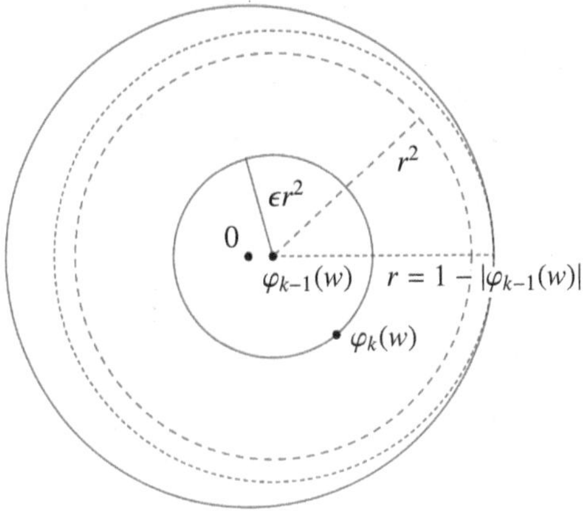

Figure 2.1 The Discretization (2.3.1). Given $w \in \mathbb{T}^{\mathbb{N}}$ and $\varphi_{k-1}(w_1,\dots,w_{k-1})$ set $\varphi_k(w_1,\dots,w_k) = \varphi_{k-1}(w_1,\dots,w_{k-1}) + \epsilon(1 - |\varphi_{k-1}(w_1,\dots,w_{k-1})|)w_k$. Conditioned on $(w_1,\dots,w_{k-1})$, the variable $\varphi_k(w_1,\dots,w_k)$ is uniformly distributed on the circle with center $\varphi_{k-1}(w_1,\dots,w_{k-1})$ and radius $\epsilon(1 - |\varphi_{k-1}(w_1,\dots,w_{k-1})|)$.

for any measurable $B \subseteq \mathbb{T}$. Indeed, starting first with $\varphi_0 = 0$, the limit function φ is invariant under rotation. Hence $\mathbb{P}\{\varphi \in B\} = m(B)$ for any measurable $B \subseteq \mathbb{T}$. Second, for $\varphi_0 \in \mathbb{D}$ fixed we observe that the iteration process (2.3.1) is invariant under Moebius transformations

$$M(\varphi_0, \zeta) = (\varphi_0 + \zeta)/(1 + \bar{z}\zeta), \qquad \zeta \in \mathbb{D},$$

and transports the uniform distribution $dm(\zeta)$ on $\mathbb{T}$ into $\frac{1-|\varphi_0|^2}{|\zeta-\varphi_0|^2}dm(\zeta)$. Figure 2.2 presents a simulation of the martingale (φ_k) and its limit φ.

We now turn to Maurey's embedding theorem.

Theorem 2.3.1 *Set $\varphi_0 = 0$ and let φ_k be defined by the recursion* (2.3.1). *For any $f \in H^1(\mathbb{T})$, the composition*

$$F_k(w) = f(\varphi_k(w)), \qquad w \in \mathbb{T}^{\mathbb{N}},$$

defines an integrable Hardy martingale $F = (F_k)$ for which

$$\mathbb{E}(|F|) = \int_{\mathbb{T}} |f| dm \quad \text{and} \quad |\Delta F_n| \leq 2\epsilon \int_{\mathbb{T}} |f| dm. \tag{2.3.3}$$

Proof It is straightforward to see that (F_n) is indeed an integrable Hardy martingale and that

$$\mathbb{E}|F| = \int_{\mathbb{T}} |f| dm.$$

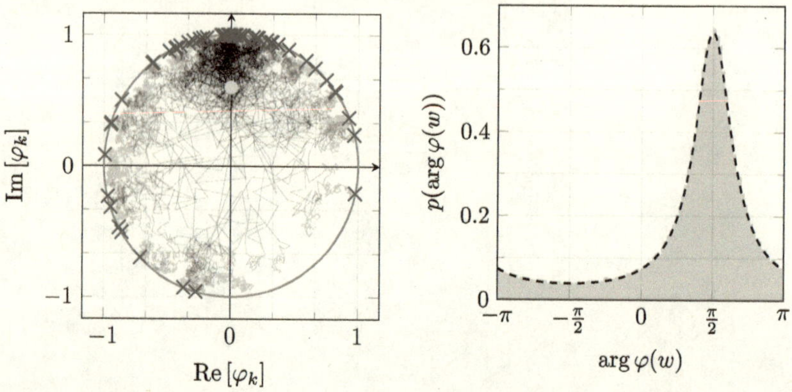

(a) Fifty samples of $(\varphi_k(w))$ and its limit $\varphi(w)$ are shown.

(b) The distribution of 10^5 samples of $\varphi(w)$ and $p(\theta) = (1 - |\varphi_0|^2)/(2\pi|e^{i\theta} - \varphi_0|^2)$ are compared.

Figure 2.2 We discretize using $\epsilon = 0.8$. The initial value is given by $\varphi_0 = 0.6i$. The pseudorandom numbers (w_k) were generated using standard NumPy 1.1.5.1. routines. The stopping criterion was $1 - |\varphi_k(w)| < 10^{-10}$.

We now turn to the pointwise estimates in (2.3.3). Fix $w \in \mathbb{T}^{\mathbb{N}}$, and $n \in \mathbb{N}$. Then

$$\Delta F_n(w) = f(\varphi_n(w)) - f(\varphi_{n-1}(w)).$$

Put next $z = \varphi_n(w)$, $u = \varphi_{n-1}(w)$ and use the Cauchy integral formula to obtain

$$f(z) - f(u) = \int_{\mathbb{T}} \left\{ \frac{\zeta}{\zeta - z} - \frac{\zeta}{\zeta - u} \right\} f(\zeta) dm(\zeta).$$

By the triangle inequality, we get

$$|f(z) - f(u)| \leq \frac{|z - u|}{(1 - |u|)(1 - |z|)} \int_{\mathbb{T}} |f| dm, \tag{2.3.4}$$

We use the defining recursion (2.3.1) to see that

$$|\varphi_n(w) - \varphi_{n-1}(w)| = \epsilon(1 - |\varphi_{n-1}(w)|)^2 \tag{2.3.5}$$

and

$$(1 - |\varphi_{n-1}(w)|) \leq (1 - \epsilon)^{-1}(1 - |\varphi_n(w)|). \tag{2.3.6}$$

Since we put $z = \varphi_n(w)$ and $u = \varphi_{n-1}(w)$, the relations (2.3.5) and (2.3.6) imply that

$$\frac{|z - u|}{(1 - |u|)(1 - |z|)} \leq 2\epsilon. \tag{2.3.7}$$

Combining the estimates (2.3.4) and (2.3.7) yields the following pointwise bounds for the martingale differences

$$|\Delta F_n| \le 2\epsilon \int_{\mathbb{T}} |f| dm. \qquad \square$$

Maurey's embedding has a reformulation in terms of complex Brownian motion (z_t) started at the origin. Define the stopping times

$$\rho_1 = \inf\{t > 0 : |z_t| > \epsilon\}, \quad \rho_k = \inf\left\{t > \rho_{k-1} : |z_t - z_{\rho_{k-1}}| > \epsilon(1 - |z_{\rho_{k-1}}|)^2\right\}.$$

Then on Wiener space, the stopped sequence (z_{ρ_k}) has the same joint distribution as the Hardy martingale (φ_k). We may therefore identify them, and, furthermore, we may identify the sequence

$$F_k = f(z_{\rho_k}), \qquad f \in H^1(\mathbb{T}),$$

with the Hardy martingales arising in Maurey's embedding.

Discretizing Brownian Motion

We introduce a second – frequently used – representation of $f \in H^1(\mathbb{T})$ as a Hardy martingale. We will design the embedding in such a way that the sequence of lacunary Fourier coefficients of f is well controlled by the induced sequence of martingale differences. The construction uses the Moebius transforms (conformal self maps of the unit disk),

$$M(z, \zeta) = (z + \zeta)/(1 + \bar{z}\zeta), \qquad z, \zeta \in \mathbb{D}.$$

Let $r_n = 1 - 2^{-n}$, $w = (w_n) \in \mathbb{T}^{\mathbb{N}}$, and define the sequence

$$h_1(w) = r_1 M(0, w_1), \quad h_n(w) = r_n M\left(\frac{h_{n-1}(w)}{r_n}, w_n\right), \quad n \ge 2.$$

Lemma 2.3.2 *The sequence (h_n) is a Hardy martingale so that $h_n : \mathbb{T}^{\mathbb{N}} \to \mathbb{D}$ is uniformly distributed on the circle $(1-2^{-n})\cdot\mathbb{T}$, and, conditioned to $h_{n-1}(w) = z$, the distribution of*

$$w_n \to h_n(w_1, \ldots, w_n)$$

coincides with that of the Poisson kernel

$$w_n \to r_n P\left(\frac{z}{r_n}, w_n\right),$$

where $P(z, \zeta) = (1 - |z|^2)/|\zeta - z|^2$.

Proof We verify first that (h_n) is a Hardy martingale.

Step 1 By construction, h_n depends only on $w_1, \ldots, w_n$, hence h_n is $\mathcal{F}_n$-measurable. The martingale property $\mathbb{E}_{n-1}(h_n) = h_{n-1}$ follows by integration,

$$\begin{aligned}\int_{\mathbb{T}} h_n(w)dm(w_n) &= r_n \int_{\mathbb{T}} M\left(\frac{h_{n-1}(w)}{r_n}, w_n\right) dm(w_n)\\ &= r_n M\left(\frac{h_{n-1}(w)}{r_n}, 0\right)\\ &= h_{n-1}(w).\end{aligned}$$

By using a power series expansion, we check that the martingale differences are analytic. Expand $M\left(\frac{h_{n-1}(w)}{r_n}, \zeta\right)$ at $\zeta = 0$ and evaluate at $\zeta = w_n$. This gives,

$$\begin{aligned}h_n(w) &= h_n(w_1, \ldots, w_{n-1}, 0) + \sum_{k=1}^{\infty} r_n \frac{M^{(k)}\left(\frac{h_{n-1}(w)}{r_n}, 0\right)}{k!} w_n^k\\ &= h_{n-1}(w) + \sum_{k=1}^{\infty} r_n \frac{M^{(k)}\left(\frac{h_{n-1}(w)}{r_n}, 0\right)}{k!} w_n^k.\end{aligned}$$

This completes the verification that (h_n) is a Hardy martingale.

Step 2 Recall the following link between Moebius transforms $M(z, \zeta) = (z + \zeta)/(1 + \bar{z}\zeta)$ and Poisson kernels:

$$P(z, \zeta) = (1 - |z|^2)/|\zeta - z|^2, \quad \zeta \in \mathbb{T}, z \in \mathbb{D}.$$

For fixed $z \in \mathbb{D}$, the Moebius transform $M(z, \cdot) \colon \mathbb{T} \to \mathbb{T}$ is a bijection such that for $A \subseteq \mathbb{T}$,

$$m(M^{-1}(z, A)) = \int_A P(z, \zeta)dm(\zeta).$$

Moreover, by inspection $M(z, 0) = z$.

Step 3 We inductively verify that $h_n \colon \mathbb{T}^{\mathbb{N}} \to \mathbb{D}$ is uniformly distributed over $(1 - 2^{-n}) \cdot \mathbb{T}$. Clearly this holds for h_1. Assume now that $h_{n-1} \colon \mathbb{T}^{\mathbb{N}} \to \mathbb{D}$ is uniformly distributed over $(1 - 2^{-n+1}) \cdot \mathbb{T}$. Fix $B \subseteq \mathbb{T}$, then calculate

$$\begin{aligned}\mathbb{P}\{w \in \mathbb{T}^{\mathbb{N}} : h_n(w) \in r_n \cdot B\} &= \mathbb{P}\left\{w \in \mathbb{T}^{\mathbb{N}} : M\left(\frac{h_{n-1}(w)}{r_n}, w_n\right) \in B\right\}\\ &= \int_{\mathbb{T}} \mathbb{P}\left\{w \in \mathbb{T}^{\mathbb{N}} : M\left(\frac{r_{n-1}\zeta}{r_n}, w_n\right) \in B\right\} dm(\zeta)\\ &= \int_{\mathbb{T}} \int_B P_{\frac{r_{n-1}\zeta}{r_n}}(w_n)dm(\zeta)dm(w_n) = m(B).\end{aligned}$$

Fix $z \in \mathbb{D}$, $|z| = r_{n-1}$. A slight modification of the above calculation shows that conditioned to the event $h_{n-1}(w) = z$, the distribution of $w_n \to h_n(w_1, \ldots, w_n)$ is given by Poisson kernel

$$w_n \to r_n P\left(\frac{z}{r_n}, w_n\right).$$

Step 4 Summing up, the sequence (h_n) forms a Hardy martingale with values in the unit disk. By the martingale convergence theorem it converges, that is,

$$h = \lim_{n\to\infty} h_n,$$

exists almost everywhere and in L^1; moreover, h is uniformly distributed over the boundary of the unit disk. □

Remarks: The independent variable $w_n \in \mathbb{T}$ can be computed from $h_n(w)$ and $h_{n-1}(w)$ by inverting the Moebius transform. Fix w_n and $\frac{h_{n-1}(w)}{r_n}$. Put, for $\zeta \in \mathbb{T}$,

$$M_n(\zeta) = M\left(\frac{h_{n-1}(w)}{r_n}, \zeta\right).$$

The inverse of $M_n(\zeta)$ is given as

$$M_n^{-1}(\zeta) = \frac{\zeta - \frac{h_{n-1}(w)}{r_n}}{1 - \frac{\bar{h}_{n-1}(w)}{r_n}\zeta}.$$

Inserting $\zeta = h_n(w)/r_n$ yields

$$M_n^{-1}(h_n(w)/r_n) = M_n^{-1}(M_n(w_n)) = w_n.$$

Consequently,

$$w_n = r_n \frac{h_n(w) - h_{n-1}(w)}{r_n^2 - \bar{h}_{n-1}(w)h_n(w)}.$$

Finally, we remark that (h_n) has the same joint distribution as the sequence of stopped complex Brownian motion (z_{τ_n}), where

$$\tau_n = \inf\{t > \tau_{n-1} : |z_t| > 1 - 2^{-n}\}, \qquad \tau_0 = 0.$$

Moreover, the independent variables w_n in $\mathbb{T}^{\mathbb{N}}$ are distributionally equivalent to the following random variables defined on Wiener space:

$$\zeta_n = r_n \frac{z_{\tau_n} - z_{\tau_{n-1}}}{r_n^2 - \bar{z}_{\tau_{n-1}} z_{\tau_n}}.$$

Theorem 2.3.3 *If $f \in H^1(\mathbb{T})$, then the composition*

$$F_n(w) = f(h_n(w)), \quad w \in \mathbb{T}^{\mathbb{N}}, \tag{2.3.8}$$

is an $L^1(\mathbb{T}^{\mathbb{N}})$ convergent Hardy martingale such that

$$\mathbb{E}|\lim F_n| = \int_{\mathbb{T}} |f| dm.$$

If $2^n \leq k \leq 2^{n+1}$, $r_n = 1 - 2^{-n}$, and

$$G_k(w) = \bar{w}_n \left[\frac{h_{n-1}(w)}{r_{n-1}} \right]^k \frac{r_n}{\left(r_n^2 - r_{n-1}^2\right) k r_{n-1}^{k-1}}, \tag{2.3.9}$$

then

$$\widehat{f}(k) = \mathbb{E}(G_k \Delta F_n) \quad \textit{and} \quad \sup_k \|G_k\|_{L^\infty(\mathbb{T}^{\mathbb{N}})} \leq C_0, \tag{2.3.10}$$

where $C_0 < \infty$ is a universal constant.

Proof Let $f \in H^1(\mathbb{T})$ define the $F_n \colon \mathbb{T}^{\mathbb{N}} \to \mathbb{C}$ by

$$F_n(w) = f(h_n(w)). \tag{2.3.11}$$

Step 1 We verify that $F_n \colon \mathbb{T}^{\mathbb{N}} \to \mathbb{C}$ forms an $L^1(\mathbb{T}^{\mathbb{N}})$-bounded Hardy martingale. Since F_n depends only on $w_1, \ldots, w_n$ it is measurable with respect to $\mathcal{F}_n$. To verify the martingale property $\mathbb{E}_{n-1} F_n = F_{n-1}$, we observe that

$$\begin{aligned} \int_{\mathbb{T}} f(h_n(w)) dm(w_n) &= f\left(r_n M\left(\frac{h_{n-1}(w)}{r_n}, 0\right)\right) \\ &= f(h_{n-1}(w)). \end{aligned}$$

The following power series expansion verifies that $F_n - F_{n-1}$ is analytic:

$$f(h_n(w)) = f(h_n(w_1, \ldots, w_{n-1}, 0)) + \sum_{k=1}^{\infty} \frac{f^{(k)}\left(r_n M\left(\frac{h_{n-1}(w)}{r_n}, 0\right)\right)}{k!} w_n^k.$$

Thus, we showed that (F_n) is a Hardy martingale. By Lemma 2.3.2, the sequence (2.3.11) is equiintegrable, hence $L^1(\mathbb{T}^{\mathbb{N}})$ convergent, and

$$\mathbb{E}|\lim F_n| = \int_{\mathbb{T}} |f| dm.$$

Step 2 Next, fix $n \in \mathbb{N}$. We prove that $\widehat{f}(k) = \mathbb{E}(G_k \Delta F_n)$ for $(2^n \le k \le 2^{n+1})$, and that G_k is uniformly bounded.

Fix $2^n \le k \le 2^{n+1}$. By the Cauchy integral formula applied to the analytic function $w_n \mapsto f(h_n(w)) - f(h_{n-1}(w))$ we obtain,

$$f'(h_{n-1}(w))\frac{r_n^2 - r_{n-1}^2}{r_n} = \int_{\mathbb{T}} \bar{w}_n(f(h_n(w)) - f(h_{n-1}(w)))dm(w_n).$$

With this formula we linked the martingale differences to the derivative of f. Apply the Cauchy integral formula again to get the Taylor coefficients from the derivative,

$$kr_{n-1}^{k-1}\hat{f}(k) = \int_{\mathbb{T}} \bar{\zeta}^{k-1} f'(r_{n-1}\zeta)dm(\zeta).$$

Since $h_{n-1}(w)/r_{n-1}$ is uniformly distributed over $\mathbb{T}$, we may replace $\zeta \in \mathbb{T}$ by $h_{n-1}(w)/r_{n-1}$ and integration over $\mathbb{T}$ by integration over $\mathbb{T}^{n-1}$. So that the right-hand side integral coincides with

$$\int_{\mathbb{T}^{n-1}} \left[\frac{h_{n-1}(w)}{r_{n-1}}\right]^k f'(h_{n-1}(w))d\mathbb{P}(w).$$

Next, insert the formula above and obtain

$$kr_{n-1}^{k-1}\hat{f}(k) = \int_{\mathbb{T}^n} \left[\frac{h_{n-1}(w)}{r_{n-1}}\right]^k \bar{w}_n \frac{r_n}{r_n^2 - r_{n-1}^2}(f(h_n(w)) - f(h_{n-1}(w)))d\mathbb{P}(w).$$

Step 3 By taking into consideration that $r_n = 1 - 2^{-n}$ and $2^n \le k \le 2^{n+1}$, it is easy to see that the weights,

$$G_k(w) = \bar{w}_n \left[\frac{h_{n-1}(w)}{r_{n-1}}\right]^k \frac{r_n}{\left(r_n^2 - r_{n-1}^2\right) kr_{n-1}^{k-1}},$$

form a uniformly bounded sequence satisfying the identity and the estimate of (2.3.10). □

The procedure of constructing Hardy martingales may be rephrased in terms of complex Brownian motion (z_t) started at the origin $0 \in \mathbb{C}$. Here, use the stopping times $\tau_0 = 0$ and

$$\tau_n = \inf\{t > \tau_{n-1} : |z_t| > 1 - 2^{-n}\}.$$

As we remarked above, on Wiener space, the sequence of stopped Brownian motion (z_{τ_n}) has the same joint distribution as the Hardy martingale (h_n) on the infinite torus product. Therefore, for $f \in H^1(\mathbb{T})$ we identify the Brownian martingale $(f(z_{\tau_n}))$ with the Hardy martingales constructed in Theorem 2.3.3.

We present Pisier's proof of Paley's theorem on lacunary Fourier coefficients of H^1 functions using the embedding of $H^1(\mathbb{T})$ defined by Theorem 2.3.3.

Corollary 2.3.4 (Paley's theorem) *Let* $f \in H^1(\mathbb{T})$ *and* $k_n \in (2^n, 2^{n+1}]$, *where* $n \in \mathbb{N}$. *Then*

$$\left(\sum_{n=1} \left|\widehat{f}(k_n)\right|^2\right)^{1/2} \leq C\|f\|_{L^1(\mathbb{T})}.$$

Proof Given $f \in H^1(\mathbb{T})$, define the Hardy martingale

$$F_n(w) = f(h_n(w)), \qquad w \in \mathbb{T}^{\mathbb{N}}.$$

In Section 2.5, we will prove Theorem 2.5.8 asserting that

$$\left(\sum_{k=1}^{n} \left(\|\Delta F_k\|_{L^1(\mathbb{T}^{\mathbb{N}})}\right)^2\right)^{1/2} \leq \sqrt{2}\|F_n\|_{L^1(\mathbb{T}^{\mathbb{N}})} \tag{2.3.12}$$

for any integrable, scalar-valued Hardy martingale. Apply Theorem 2.3.3 to the differences ΔF_n and estimate

$$\left|\widehat{f}(k_n)\right| \leq \int_{\mathbb{T}^{\mathbb{N}}} |G_{n,k_n}\Delta F_n| d\mathbb{P} \leq \|G_{n,k_n}\|_\infty \|\Delta F_n\|_{L^1}.$$

Hence, by (2.3.12),

$$\sum_{n=1} \left|\widehat{f}(k_n)\right|^2 \leq C\|F\|^2_{L^1(\mathbb{T}^{\mathbb{N}})}.$$

Since $\|F\|_1 = \|f\|_1$ this completes the proof. □

Remark: Let X be a complex Banach space. In Section 2.1, defined by Equation (2.1.2), we introduced the vector-valued Hardy space $H^1(\mathbb{T}, X) \subset L^1(\mathbb{T}, X)$. The kth Fourier coefficient of $f \in L^1(\mathbb{T}, X)$ is given by,

$$\widehat{f}(k) = \int_{\mathbb{T}} f(\zeta)\overline{\zeta}^k dm(\zeta). \tag{2.3.13}$$

Pisier's proof of Corollary 2.3.4 yields that, for any $f \in H^1(\mathbb{T}, X)$, there exists an X-valued, L^1-bounded Hardy martingale $\left(F_n : \mathbb{T}^{\mathbb{N}} \to X\right)$ such that

$$\sup_n \mathbb{E}\|F_n\|_X \leq \int_{\mathbb{T}} \|f\|_X dm, \tag{2.3.14}$$

and

$$\left\|\widehat{f}(k)\right\|_X \leq C_0 \mathbb{E}\|\Delta F_n\|_X, \qquad 2^n < k \leq 2^n, \quad n \in \mathbb{N}, \tag{2.3.15}$$

where $C_0 < \infty$ is determined by equation (2.3.10). We emphasize that the estimates (2.3.14) and (2.3.15) hold true for any complex Banach space X.

Pisier's inequality (2.3.15) provides a link between the Hardy martingale cotype condition in Section 5.3, and Banach spaces for which vector-valued versions of Paley's theorem hold true. This topic is investigated in depth by Blasco and Pełczyński (1991).

2.4 Classes of Martingales and Projections

In this section we examine more closely $H^p(\mathbb{T}^\mathbb{N})$, the space of L^p-bounded Hardy martingales, and $\overline{\operatorname{span}}_{L^p}\mathcal{P}^{(k)}$, the L^p-closure of the homogeneous polynomials of degree k.

We prove that the Riesz projection onto $H^p(\mathbb{T}^\mathbb{N})$ is bounded for $1 < p < \infty$ and, hence, that $H^p(\mathbb{T}^\mathbb{N})$ is a complemented subspace of $L^p(\mathbb{T}^\mathbb{N})$. We present two separate methods of proof due to Garling (1991) and Bourgain (1983c, 1984b, 1984c), respectively. Concerning homogeneous polynomials, we present Pisier's theorem (Theorem 2.4.15), asserting that for $1 < p$, $q < \infty$ there exists $C = C(p, q)$ such that

$$\left\| \mathcal{S}^{(k)} \otimes \mathrm{Id}_{L^q} : L^p(L^q) \to L^p(L^q) \right\| \le C^k, \quad k \in \mathbb{N}, \tag{2.4.1}$$

where $L^p(L^q) = L^p\left(\mathbb{T}^\mathbb{N}, L^q(\Omega)\right)$, and where $\mathcal{S}^{(k)}$ denotes the orthogonal projection onto the L^2-closure of $\mathcal{P}_k$ (see equation (2.2.15)). We give Pisier's proof (Pisier, 1980, 1982a, 1986), which exploits generating functions, Riesz products, and integral estimates for holomorphic semigroups of operators.

2.4.1 Hardy Martingales

In Section 2.2.1, equation (2.2.10), we introduced the Riesz projection acting on $L^2(\mathbb{T}^\mathbb{N})$. Here, we show that it extends to a bounded operator on $L^p(\mathbb{T}^\mathbb{N})$ provided that $1 < p < \infty$.

We present two methods of proof. One, due to Garling (1990), builds on his realization that the framework of abstract harmonic analysis on compact abelian groups applies directly to the Hardy martingale class, and utilizes analyticity as in the work of Yudin (1989). A second method, due to Bourgain (1983c, 1984b, 1984c), proceeds by reduction to the Hilbert transform on $L^p(\mathbb{T})$.

Yudin's Lemma and its Consequences

Garling (1991) pointed out the significance of Yudin's lemma (Lemma 2.4.1) in bounding the Riesz projection onto Hardy martingales, and more generally to L^p-estimates of martingale transfom operators. Yudin's lemma may be viewed as an L^{2n} version of Pythagoras's L^2-theorem.

Lemma 2.4.1 *Let $n \in \mathbb{N}$. If $F, G \in L^{2n}(\mathbb{T}^\mathbb{N})$ and $\mathbb{E}F^n\overline{G^n} = 0$, then*

$$\mathbb{E}\left(\left|F^{2n}\right| + \left|G^{2n}\right|\right) \le n^{2n}\mathbb{E}\left(|F + G|^{2n}\right).$$

Proof We have $w^n - z^n = (w - z)\left(\sum_{m=0}^{n-1} w^m z^{n-1-m}\right)$, for any $w, z \in \mathbb{C}$, and $n \in \mathbb{N}$. Next,

$$\left|w^m z^{n-1-m}\right| \le \max(|w|, |z|)^{n-1},$$

for $0 \le m \le n-1$. This gives the basic point-wise estimate

$$|w^n - z^n| \le n|w-z| \max\left(\left|w^{n-1}\right|, \left|z^{n-1}\right|\right). \qquad (2.4.2)$$

Note that $|F^{2n}| + |G^{2n}| = |F^n - (-G)^n|^2 + F^n(-\overline{G^n}) + \overline{F^n}(-G)^n$. Taking its expectation and using that $\mathbb{E}F^n\overline{G^n} = 0$ gives

$$\mathbb{E}\left(\left|F^{2n}\right| + \left|G^{2n}\right|\right) = \mathbb{E}\left|F^n - (-G)^n\right|^2. \qquad (2.4.3)$$

The pointwise estimate (2.4.2) in combination with Hölder's inequality for the conjugate exponents $n, n/(n-1)$ gives,

$$\begin{aligned}
\mathbb{E}\left|F^n - (-G)^n\right|^2 &\le n^2 \mathbb{E}|F+G|^2 \max\left(|F|^{n-1}, |G|^{n-1}\right)^2 \\
&\le n^2 \left(\mathbb{E}|F+G|^{2n}\right)^{1/n} \left(\mathbb{E} \max\left(\left|F^{2n}\right|, \left|G^{2n}\right|\right)\right)^{(n-1)/n} \qquad (2.4.4) \\
&\le n^2 \left(\mathbb{E}|F+G|^{2n}\right)^{1/n} \left(\mathbb{E}\left|F^{2n}\right| + \left|G^{2n}\right|\right)^{(n-1)/n}.
\end{aligned}$$

Combining (2.4.3) and (2.4.4) and canceling the term $\left(\mathbb{E}\left(\left|F^{2n}\right| + \left|G^{2n}\right|\right)\right)^{(n-1)/n}$ finishes the proof. □

The Riesz projection: Garling's first application of Yudin's lemma (Garling, 1991) is to prove that the Riesz projection on $\mathbb{T}^{\mathbb{N}}$, defined in (2.2.10), satisfies L^p estimates.

Let $\mathbb{Z}^{(\mathbb{N})}$ denote the additive group of all integer sequences $n = (n_i)$, for which only finitely many entries are nonvanishing. By means of the (finite) products,

$$z^n = \prod z_i^{n_i}, \quad z \in \mathbb{T}^{\mathbb{N}}, \qquad n \in \mathbb{Z}^{(\mathbb{N})},$$

the elements in $\mathbb{Z}^{(\mathbb{N})}$ define the characters of the infinite torus group $\mathbb{T}^{\mathbb{N}}$. Thus $\mathbb{Z}^{(\mathbb{N})}$ is the dual group of $\mathbb{T}^{\mathbb{N}}$. We define Fourier coefficients of $F \in L^p(\mathbb{T}^{\mathbb{N}})$ by integration,

$$\widehat{F}(n) = \int_{\mathbb{T}^{\mathbb{N}}} F(z)\overline{z^n} d\mathbb{P}(z), \quad n \in \mathbb{Z}^{(\mathbb{N})},$$

and put $\operatorname{spec} F = \left\{n \in \mathbb{Z}^{(\mathbb{N})} : \hat{F}(n) \neq 0\right\}$. We say that F is a trigonometric polynomial on $\mathbb{T}^{\mathbb{N}}$ if $\operatorname{spec} F$ is a finite subset of $\mathbb{Z}^{(\mathbb{N})}$.

Let $0 \in \mathbb{Z}^{(\mathbb{N})}$ be the neutral group element. For $k \in \mathbb{N}$ we put

$$\mathbb{Z}(k) = \left\{n \in \mathbb{Z}^{(\mathbb{N})} : n = (n_1, n_2, \ldots, n_k, 0)\right\}.$$

Note that $\mathbb{Z}(k)$ is related to the conditional expectation operators $\mathbb{E}_k(\cdot)$ by the identity

$$\mathbb{E}_k(F) = \sum_{n \in \mathbb{Z}(k)} \widehat{F}(n) z^n.$$

Next, set $\Lambda_0 = \{0\}$, and $\Lambda_k = \{n \in \mathbb{Z}(k) : n_k \geq 1\}$, for $k \geq 1$ and put

$$\Lambda = \bigcup \Lambda_k. \tag{2.4.5}$$

By construction, Λ is closed under addition, satisfying

$$\Lambda \cup \{-\Lambda\} = \mathbb{Z}^{(\mathbb{N})} \quad \text{and} \quad \Lambda \cap \{-\Lambda\} = \{0\}.$$

Let $1 \leq p \leq \infty$. For any L^p-bounded Hardy martingale (F_n), we obtain $\operatorname{spec} F \subseteq \Lambda$ where $F = \lim F_n$ denotes the almost sure pointwise limit. To see this, it suffices to note that for the Hardy martingale (F_k) we have

$$\mathbb{E}_k(F) = F_k = \sum_{n \in \Lambda_k} \widehat{F}(n) z^n.$$

Conversely, if $F \in L^p(\mathbb{T}^{\mathbb{N}})$ and $\operatorname{spec} F \subseteq \Lambda$, then the sequence of conditional expectations $(\mathbb{E}_k(F))$ forms an L^p-bounded Hardy martingale. In summary,

$$H^p(\mathbb{T}^{\mathbb{N}}) = \left\{F \in L^p(\mathbb{T}^{\mathbb{N}}) : \operatorname{spec} F \subseteq \Lambda\right\},$$

and the orthogonal Riesz projection, defined by (2.2.10), is given by

$$PG(z) = \sum_{n \in \Lambda} \widehat{G}(n) z^n, \quad G \in L^2(\mathbb{T}^{\mathbb{N}}).$$

Next, we show that P extends boundedly to $L^p(\mathbb{T}^{\mathbb{N}})$ for $1 < p < \infty$. We let $H_0^p(\mathbb{T}^{\mathbb{N}})$ denote the subspace of those $F \in H^p(\mathbb{T}^{\mathbb{N}})$ for which $\mathbb{E}(F) = 0$.

Theorem 2.4.2 *If $1 < p < \infty$ then the Riesz projection P extends to a bounded linear projection*

$$\left\|P : L^p(\mathbb{T}^{\mathbb{N}}) \to H^p(\mathbb{T}^{\mathbb{N}})\right\| \leq C_p,$$

where $C_p \leq Cp^2/(p-1)$, and the kernel of P is given by $\overline{H}_0^p(\mathbb{T}^{\mathbb{N}})$.

Proof Let $F \in L^\infty(\mathbb{T}^{\mathbb{N}})$. Since $PF(z) = \sum_{n \in \Lambda} \widehat{F}(n) z^n$, and Λ is closed under addition and satisfies $\Lambda \cap \{-\Lambda\} = 0$, we have

$$P(F)^n \in H^2(\mathbb{T}^{\mathbb{N}}) \quad \text{and} \quad (\mathrm{Id} - P)(F)^n \in \overline{H}_0^2(\mathbb{T}^{\mathbb{N}}),$$

for any $n \in \mathbb{N}$. Consequently,

$$\mathbb{E}\left(P(F)^n \overline{(\mathrm{Id} - P)(F)}^n\right) = 0. \tag{2.4.6}$$

By (2.4.6), Yudin's lemma (Lemma 2.4.1) may be applied. It gives

$$\mathbb{E}\left(|P(F)|^{2n} + |(F - P(F)|^{2n}\right) \leq n^{2n} \mathbb{E}\left(|F|^{2n}\right),$$

and hence the bounded extension of P to L^{2n}. Complex interpolation extends the boundedness to the range $2 \leq p < \infty$ and by duality to the remaining values $1 < p \leq 2$. □

Projection onto simple Hardy martingales: Recall that $G = (G_j)$ is a simple Hardy martingale if $\Delta G_j(z) = f_{j-1}(z)z_j$, where f_{j-1} depends only on the first $j-1$ coordinates of $z \in \mathbb{T}^{\mathbb{N}}$. Simple Hardy martingales are characterized in terms of their Fourier spectrum as follows: Put $\Sigma_j = \{n \in \mathbb{Z}(j) : n_j = 1\}$ and $\Sigma = \bigcup \Sigma_j$. Then $G \in L^2(\mathbb{T}^{\mathbb{N}})$ is a simple Hardy martingale if its Fourier spectrum is contained in Σ.

Recall that P_0, the orthogonal projection onto the span of simple Hardy martingales, is defined by equation (2.2.11). We now turn to describing its action using the Fourier expansion for the group $\mathbb{T}^{\mathbb{N}}$. To this end we first introduce a slight variation of the Riesz projection. Let $G \in L^2(\mathbb{T}^{\mathbb{N}})$. Define

$$QG(z) = \sum_{j=1}^{\infty}\sum_{m=2}^{\infty} c_{j,m}(z)z_j^m.$$

We note that $\|Q\|_p \le C\|P\|_p$ for $1 < p < \infty$, and that the difference

$$P_0 = P - Q$$

is the orthogonal projection onto the subspace of $H^2(\mathbb{T}^{\mathbb{N}})$ formed by simple Hardy martingales. Thus P_0 admits a bounded extension to L^p, for $1 < p < \infty$ with $\|P_0\|_p \le C\|P\|_p$.

Martingale transforms: Yudin's lemma yields norm estimates for martingale transforms acting on $H^p(\mathbb{T}^{\mathbb{N}})$ Garling (1991). Let $A \subseteq \mathbb{N}$. Let $F = (F_n)$ be a convergent martingale in L^p with $1 < p < \infty$. Define the martingale transform

$$T_A(F) = \sum_{k\in A} \Delta F_k, \quad F \in L^2(\mathbb{T}^{\mathbb{N}}).$$

Clearly, T_A is a martingale transform satisfying $\|T_A F\|_2 \le \|F\|_2$ and $T_A^2 = T_A$. We show that T_A extends to a bounded projection on $L^p(\mathbb{T}^{\mathbb{N}})$. This is done in two independent steps. First we extend to $H^p(\mathbb{T}^{\mathbb{N}})$, then we exploit the boundedness of the Riesz transform to pass from $H^p(\mathbb{T}^{\mathbb{N}})$ to $L^p(\mathbb{T}^{\mathbb{N}})$.

Note that if $F, G \in H^1(\mathbb{T}^{\mathbb{N}})$ such that $F \cdot G \in L^1(\mathbb{T}^{\mathbb{N}})$, then

$$\Delta(FG)_k = F_{k-1}\Delta G_k + G_{k-1}\Delta F_k + \Delta G_k \Delta F_k. \tag{2.4.7}$$

Thus the class of Hardy martingales is closed under the formation of pointwise products.

Theorem 2.4.3 *Let $1 < p < \infty$. For any $A \subseteq \mathbb{N}$, the martingale transform T_A extends boundedly to $H^p(\mathbb{T}^{\mathbb{N}})$, with*

$$\left\|T_A : H^p(\mathbb{T}^{\mathbb{N}}) \to H^p(\mathbb{T}^{\mathbb{N}})\right\| \le A_p,$$

where $A_p \le Cp^2/(p-1)$.

Proof Fix $F \in H^\infty(\mathbb{T}^\mathbb{N})$ and assume that there exists $N > 1$ so that $\Delta F_k = 0$ for $k \geq N$. Then $T_A(F)$ defines a Hardy martingale and $\Delta(T_A F)_k = 0$ for $k \notin A$.

We claim that for any $n \in \mathbb{N}$, $G = (T_A(F))^n$ may be expanded as,

$$G = \sum_{k \in A} \Delta G_k. \tag{2.4.8}$$

To see this, we proceed by induction. By the identity (2.4.7), Hardy martingales are stable under the formation of pointwise products. We apply the identity (2.4.7) to $G = (T_A F)^n$ and $T_A F$. This gives

$$\Delta(G T_A F)_k = G_{k-1}\Delta(T_A F)_k + (T_A F)_{k-1}\Delta G_k + \Delta G_k \Delta(T_A F)_k.$$

Hence $\Delta(G T_A F)_k = 0$ if $k \notin A$. This completes the inductive verification of (2.4.8).

Put $B = \mathbb{N}\backslash A$. Invoking the expansion (2.4.8) we obtain

$$(T_A F)^n \overline{(T_B F)^n} = \sum_{k \in A} \sum_{l \notin A} \Delta(T_A F)^n_k \overline{\Delta(T_B F)^n_l}.$$

Taking expectations gives $\mathbb{E}(T_A F)^n \overline{(T_B F)}^n = 0$, where $n \in \mathbb{N}$. Since $F = T_A F + T_B F$, Yudin's lemma (Lemma 2.4.1), yields that

$$\mathbb{E}\left(|T_A F|^{2n} + |T_B F|^{2n}\right) \leq n^{2n}\mathbb{E}\left(|F|^{2n}\right).$$

This gives the bounded extension of T_A to $H^{2n}(\mathbb{T}^\mathbb{N})$, with $\|T_A\|_{2n} \leq n$. By interpolation, we extend the boundedness to $2 \leq p < \infty$, and by duality to the values $1 < p \leq 2$. □

By Theorem 2.4.2, the Riesz projection on $L^p(\mathbb{T}^\mathbb{N})$ is bounded, it maps $L^p(\mathbb{T}^\mathbb{N})$ onto $H^p(\mathbb{T}^\mathbb{N})$, and annihilates $\overline{H^p_0(\mathbb{T}^\mathbb{N})}$. We use the Riesz projection to transfer the estimates of Theorem 2.4.3 from $H^p(\mathbb{T}^\mathbb{N})$ to $L^p(\mathbb{T}^\mathbb{N})$. Thus Theorem 2.4.3 yields Burkholder's L^p-estimates for martingale transform operators.

Theorem 2.4.4 *Let $1 < p < \infty$. For any $A \subseteq \mathbb{N}$, the martingale transform T_A extends boundedly to $L^p(\mathbb{T}^\mathbb{N})$, with*

$$\left\|T_A : L^p(\mathbb{T}^\mathbb{N}) \to L^p(\mathbb{T}^\mathbb{N})\right\| \leq A_p,$$

where $A_p \leq C\|P\|_p p^2/(p-1)$ and where P denotes the Riesz projection.

Proof Let $2 \leq p < \infty$. Let $F \in L^p(\mathbb{T}^\mathbb{N})$. Let P denote the Riesz projection and write $F = P(F) + (F - P(F))$. Since $P(F) \in H^p(\mathbb{T}^\mathbb{N})$ and $\overline{F - P(F)} \in H^p(\mathbb{T}^\mathbb{N})$, we may apply Theorem 2.4.3 and get

$$\|T_A P(F))\|_p \leq Cp\|P(F)\|_p$$

and

$$\|T_A(F - P(F))\|_p = \left\|T_A(\overline{F - P(F)})\right\|_p \leq Cp\|F - P(F)\|_p.$$

Since $2 \leq p < \infty$, the Riesz projection P is bounded $L^p(\mathbb{T}^{\mathbb{N}})$ with norm $\leq Cp$. By duality, we obtain the corresponding estimates for the range $1 < p \leq 2$. □

The Transfer Method

We now turn to the limiting case $p \to 1$. We will show that the Riesz projection and the projection onto simple Hardy martingales are of weak type $(1:1)$. The method of proof is by transference to the Hilbert transform on $L^1(\mathbb{T})$.

The Hilbert transform on $\mathbb{T}^{\mathbb{N}}$: We first rewrite the Riesz projection displaying its action on martingale differences. Let $G = (G_n)$ be a square integrable martingale on $\mathbb{T}^{\mathbb{N}}$. Put $g_j = \Delta G_j$ and let p_j denote the jth coordinate projection defined by $p_j(z) = z_j$, where $z \in \mathbb{T}^{\mathbb{N}}$. Conditioned on the first $j-1$ coordinates, the Fourier coefficients of g_j are given by

$$c_{j,m}(z) = (\mathbb{E}_{j-1} g_j \overline{p_j}^m)(z).$$

Hence, the Fourier series of g_j, expanded along powers of z_j, turns out to be $g_j(z) = \sum_{m=-\infty}^{\infty} c_{j,m}(z) z_j^m$. Note that $c_{j,0}(z) = \mathbb{E}_{j-1}\Delta G_j = 0$. Let $P_j g_j$ denote the projection onto the analytic part of g_j, that is, we let

$$P_j g_j(z) = \sum_{m=1}^{\infty} c_{j,m}(z) z_j^m$$

and define

$$P(G)(z) = \sum_{j=1}^{\infty} P_j g_j(z) = \sum_{j=1}^{\infty} \sum_{m=1}^{\infty} c_{j,m}(z) z_j^m.$$

Similarly, let $H_j g_j$ denote the Hilbert transform of g_j acting on the variable z_j. We next define the Hilbert transform on $L^2(\mathbb{T}^{\mathbb{N}})$ by

$$\mathbb{H}(G)(z) = \sum_{j=1}^{\infty} H_j(g_j)(z) = \sum_{j=1}^{\infty} \sum_{m=-\infty}^{\infty} (-i)\operatorname{sign}(m) c_{j,m}(z) z_j^m.$$

Since $H_j(g_j) = H_j(G_j)$ and $\mathbb{E}_{j-1} H_j(g_j) = 0$, the Hilbert transform $\mathbb{H}$ acts like a martingale transform operator, satisfying

$$\Delta(\mathbb{H}G)_j = \mathbb{H}(\Delta G_j).$$

Note that $\mathbb{H}$ is linked to the Riesz projection P by the identity

$$P = (1/2)(\mathrm{Id} - i\mathbb{H}).$$

We frequently write Fourier coefficients as $c_{j,m}(z) = \mathbb{E}_{j-1}(g_j(z)\overline{z_j}^m)$.

Bedrosian's lemma: We use notational convention as follows. Given $y \in \mathbb{T}$ and an infinite sequence of integers $M = (M_k)$, we put $y_k = y^{M(k)}$ and define $y^M \in \mathbb{T}^{\mathbb{N}}$ by $y^M = (y^{M_k})$. For $z \in \mathbb{T}^{\mathbb{N}}$ we write $z \cdot y^M = (z_k y_k)$ and put

$$G(y, z, M) = F(z \cdot y^M),$$

where $F\colon \mathbb{T}^{\mathbb{N}} \to \mathbb{C}$ is a trigonometric polynomial, say.

We prove the weak type estimates for $\mathbb{H}$ by exploiting the following connection to the Hilbert transform $H\colon L^1(\mathbb{T}) \to L^{1,\infty}(\mathbb{T})$: Depending on F there exists a sequence of integers $M = (M_k)$ such that

$$(\mathbb{H}F)(z \cdot y^M) = (HG(\cdot\,, z\,, M))(y),$$

where on the right-hand side, the Hilbert transform H acts on the first variable of G. The above assertion is usually called the transfer identity. The key to its proof is Bedrosian's lemma (Bedrosian, 1963).

Lemma 2.4.5 (Bedrosian's lemma) *Let $K, N \in \mathbb{N}$ satisfying $K > N + 1$. Let $f\colon \mathbb{T} \to \mathbb{C}$ be a trigonometric polynomial with Fourier spectrum contained in $[-N + 1, N - 1]$ and let g be a trigonometric polynomial satisfying*

$$g(y) - \sum_{k \neq 0} d_k y^{kK},$$

where $y \in \mathbb{T}$, and $d_k \in \mathbb{C}$. Then $H(fg) = fHg$.

Proof Let $f(y) = \sum_{|n|<N} c_n y^n$, where $y \in \mathbb{T}$. Expanding fg and applying the Hilbert transform gives

$$H(fg)(y) = \sum_{|n|<N} \sum_{k \neq 0} (-i)\operatorname{sign}(kK + n) c_n d_k y^{kK+n},$$

and, similarly,

$$f(y)(Hg)(y) = \sum_{|n|<N} \sum_{k \neq 0} (-i)\operatorname{sign}(kK) c_n d_k y^{kK+n}.$$

Now invoke the assumption that $K > N+1$. This gives $\operatorname{sign}(kK) = \operatorname{sign}(kK+n)$, for $k \neq 0$ and $|n| < N$. Comparing the above expansions of $H(fg)$ and $f(Hg)$ verifies our claim that $(Hfg)(y) = f(y)(Hg)(y)$. □

Bedrosian's lemma, formulated in Bedrosian (1963) for the Hilbert transform on the real line, was later rediscovered by Bourgain (1984c). The proof of the transfer identity consists of iteratively applying Bedrosian's lemma.

Proposition 2.4.6 *Let $F\colon \mathbb{T}^{\mathbb{N}} \to \mathbb{C}$ be a finite trigonometric polynomial. Then there exists an increasing sequence of integers $M = (M_k)$ such that for*

$$G(y, z, M) = F(z \cdot y^M),$$

with $z \in \mathbb{T}^{\mathbb{N}}$ and $y \in \mathbb{T}$, the following identity holds true:

$$(\mathbb{H}F)(z \cdot y^M) = (HG(\cdot, z, M))(y),$$

where H acts on the first variable of G.

Proof Let F be a trigonometric polynomial on $\mathbb{T}^{\mathbb{N}}$. Form $F_k = \mathbb{E}_k(F)$. Thus for $z \in \mathbb{T}^{\mathbb{N}}$ with $z = (z_k)$, we have $F_k(z) = F_k(z_1, \dots, z_k)$ and (F_k) forms a martingale with respect to the product filtration on $\mathbb{T}^{\mathbb{N}}$. Accordingly, we construct the sequence $M = (M_k)$ by induction over k exploiting Bedrosian's lemma and the recursion $F_k = F_{k-1} + \Delta F_{k-1}$.

Step 1 Fix $z \in \mathbb{T}^{\mathbb{N}}$ with $z = (z_k)$ and $y, w \in \mathbb{T}$. Put $M_1 = 0$. Clearly $F_2 = F_1 + \Delta F_2$, and ΔF_2 is a trigonometric polynomial in (z_1, z_2) satisfying $\mathbb{E}_1(\Delta F_2) = 0$. By Lemma 2.4.5, there exists $M_2 \in \mathbb{N}$ depending on the polynomial degree of ΔF_2 such that with $M = (M_1, M_2, 0, 0, \dots)$,

$$G_2(y, z, M) = F_2(z \cdot y^M),$$

satisfies

$$(\mathbb{H}F_2)(z \cdot w^M) = (HG_2(\cdot, z, M))(w),$$

where the Hilbert transform H acts on the first variable of G_2.

Step k Suppose that we already defined $M = (M_1, \dots, M_{k-1}, 0, \dots)$ such that

$$G_{k-1}(y, z, M) = F_{k-1}(z \cdot y^M)$$

satisfies

$$\mathbb{H}F_{k-1}(z \cdot w^M) = (HG_{k-1}(\cdot, z, M))(w),$$

where the Hilbert transform H acts on the first variable of G_{k-1}. Next we determine M_k. Since $F_k = F_{k-1} + \Delta F_{k-1}$ is a trigonometric polynomial in $z_1, \dots, z_k$ and $\mathbb{E}_{k-1}(\Delta F_{k-1}) = 0$, by Lemma 2.4.5 we may determine $M_k > M_{k-1}$ such that with $M = (M_1, \dots, M_{k-1}, M_k, 0, \dots)$ and $G_k(y, z, M) = F_k(z \cdot y^M)$, we obtain

$$\mathbb{H}F_k(z \cdot w^M) = (HG_k(\cdot, z, M))(w),$$

where now H acts on the first variable of G_k. Since F is a trigonometric polynomial, there exists $k \in \mathbb{N}$ such that $F_k = F$. Hence, after finitely many iterations we determine the desired integer vector M. □

We next obtain the weak type estimates for the Hilbert transform $\mathbb{H}$. By means of the transfer identity, this result is a consequence of inequality (1.1.67), Kolmogorov's estimates for H.

Theorem 2.4.7 *The Hilbert transform $\mathbb{H}$ extends to a bounded operator between $L^1(\mathbb{T}^{\mathbb{N}})$ and $L^{1,\infty}(\mathbb{T}^{\mathbb{N}})$. There exists $C_1 < \infty$ such that*

$$\left\| \mathbb{H}\colon L^1(\mathbb{T}^{\mathbb{N}}) \to L^{1,\infty}(\mathbb{T}^{\mathbb{N}}) \right\| \le C_1.$$

Proof We utilize inequality (1.1.16) to show that $\mathbb{H}$ is of weak type $(1:1)$. Let F be a trigonometric polynomial on $\mathbb{T}^{\mathbb{N}}$. By Proposition 2.4.6 there exists a sequence of integers $M = (M_k)$, such that

$$G(y, z, M) = F(z \cdot y^M),$$

satisfies the transfer identity

$$(\mathbb{H}F)(zw^M) = (HG(\cdot, z, M))(w),$$

where $z \in \mathbb{T}^{\mathbb{N}}$, $y \in \mathbb{T}$, and where H is the Hilbert transform acting on the first variable of G. By inequality (1.1.16) it suffices to verify that,

$$\int_{\mathbb{T}^{\mathbb{N}}} |\,\mathbb{H}F(z)|^{1/2} |\varphi(z)| d\mathbb{P}(z) \le C_1 \left(\int_{\mathbb{T}^{\mathbb{N}}} |\,F(z)| d\mathbb{P}(z) \int_{\mathbb{T}^{\mathbb{N}}} |\varphi(z)| d\mathbb{P}(z) \right)^{1/2},$$

for any measurable testing function $\varphi\colon \mathbb{T}^{\mathbb{N}} \to \mathbb{C}$ satisfying $|\varphi| \le 1$. To this end we exploit rotational invariance of the measure $\mathbb{P}$ and the transfer identity as follows:

$$\begin{aligned} &\int_{\mathbb{T}^{\mathbb{N}}} |\,\mathbb{H}F(z)|^{1/2} |\varphi(z)| d\mathbb{P}(z) \\ &\quad = \int_{\mathbb{T}^{\mathbb{N}}} \int_{\mathbb{T}} \left| \mathbb{H}F(z \cdot w^M) \right|^{1/2} \left| \varphi(z \cdot w^M) \right| dm(w) d\mathbb{P}(z) \qquad (2.4.9) \\ &\quad = \int_{\mathbb{T}^{\mathbb{N}}} \int_{\mathbb{T}} |(HG(\cdot, z, M))(w)|^{1/2} \left| \varphi(z \cdot w^M) \right| dm(w) d\mathbb{P}(z). \end{aligned}$$

Now we fix $z \in \mathbb{T}^{\mathbb{N}}$, use the estimate (1.1.67), Kolmogorov's weak type $(1:1)$ estimates for the Hilbert transform H, and invoke inequality (1.1.16). This gives

$$\begin{aligned} &\int_{\mathbb{T}} |(HG(\cdot, z, M))(w)|^{1/2} \left| \varphi(z \cdot w^M) \right| dm(w) \\ &\quad \le C_1 \left(\int_{\mathbb{T}} |\,G(y, z, M)| dm(y) \int_{\mathbb{T}} \left| \varphi(z \cdot w^M) \right| dm(w) \right)^{1/2}. \end{aligned}$$

Next, integrating over $z \in \mathbb{T}^{\mathbb{N}}$, and applying the Cauchy–Schwarz inequality to the product on the right-hand side above, shows that the integrals appearing in (2.4.9) are bounded by

$$C_1 \left(\int_{\mathbb{T}^{\mathbb{N}}} \int_{\mathbb{T}} \left| F(z \cdot y^M) \right| dm(y) d\mathbb{P}(z) \int_{\mathbb{T}^{\mathbb{N}}} \int_{\mathbb{T}} \left| \varphi(z \cdot y^M) \right| dm(y) d\mathbb{P}(z) \right)^{1/2}.$$

By rotational invariance of the Haar measure $\mathbb{P}$, this term is just the same as

$$C_1\left(\int_{\mathbb{T}^{\mathbb{N}}} |F(z)|d\mathbb{P}(z)\int_{\mathbb{T}^{\mathbb{N}}} |\varphi(z)|d\mathbb{P}(z)\right)^{1/2}. \tag{2.4.10}$$

In summary, we proved that (2.4.9) is bounded by (2.4.10) and that $\mathbb{H}$ is of weak type (1:1), as claimed. □

We further illustrate the use of the transfer method by giving a second proof of the $L^p(\mathbb{T}^{\mathbb{N}})$ estimates for $\mathbb{H}$.

Theorem 2.4.8 *If $1 < p < \infty$, then $\mathbb{H}$ extends to a bounded operator on $L^p(\mathbb{T}^{\mathbb{N}})$ with*

$$\left\|\mathbb{H}\colon L^p(\mathbb{T}^{\mathbb{N}}) \to L^p(\mathbb{T}^{\mathbb{N}})\right\| \le C_p,$$

where C_p coincides with the norm of H on $L^p(\mathbb{T})$.

Proof Let F be a trigonometric polynomial on $\mathbb{T}^{\mathbb{N}}$. By Proposition 2.4.6, there exists a sequence of integers $M = (M_k)$ such that $G(y, z, M) = F(z\cdot y^M)$ satisfies the transfer identity,

$$(\mathbb{H}F)(zy^M) = (HG(\cdot, z, M))(y),$$

for $z \in \mathbb{T}^{\mathbb{N}}, y \in \mathbb{T}$. Here, H acts on the first variable of G. By rotational invariance, for any $w \in \mathbb{T}$,

$$\int_{\mathbb{T}^{\mathbb{N}}} |\mathbb{H}F(z)|^p d\mathbb{P}(z) = \int_{\mathbb{T}^{\mathbb{N}}} \left|\mathbb{H}F(z\cdot w^M)\right|^p d\mathbb{P}(z).$$

Averaging the right-hand side over $w \in \mathbb{T}$ and applying Fubini's theorem gives

$$\int_{\mathbb{T}^{\mathbb{N}}} |\mathbb{H}F(z)|^p d\mathbb{P}(z) = \int_{\mathbb{T}^{\mathbb{N}}}\int_{\mathbb{T}} \left|\mathbb{H}F(z\cdot w^M)\right|^p dm(w)d\mathbb{P}(z).$$

Next we fix $z \in \mathbb{T}^{\mathbb{N}}$. We exploit the transfer identity and the boundedness of H on $L^p(\mathbb{T})$ to obtain

$$\begin{aligned}\int_{\mathbb{T}} \left|\mathbb{H}F(z\cdot w^M)\right|^p dm(w) &= \int_{\mathbb{T}} |(HG(\cdot, z, M))(w)|^p dm(w)\\ &\le C_p^p \int_{\mathbb{T}} |G(y, z, M)|^p dm(y).\end{aligned} \tag{2.4.11}$$

Taking into account that $G(y, z, M) = F(z\cdot y^M)$ and using rotational invariance of the measure $\mathbb{P}$ again, we find that

$$\int_{\mathbb{T}} |G(y, z, M)|^p dm(y)d\mathbb{P}(z) = \int_{\mathbb{T}^{\mathbb{N}}} |F(z)|^p d\mathbb{P}(z). \tag{2.4.12}$$

Finally, combining (2.4.11) and (2.4.12) gives,

$$\int_{\mathbb{T}^{\mathbb{N}}} |\,\mathbb{H}F(z)|^p d\mathbb{P}(z) \le C_p^p \int_{\mathbb{T}^{\mathbb{N}}} |\,F(z)|^p d\mathbb{P}(z),$$

where C_p is the norm of the Hilbert transform H on $L^p(\mathbb{T})$. □

2.4.2 Homogeneous Polynomials

In this section we consider spaces of homogeneous polynomials on $\mathbb{T}^{\mathbb{N}}$. We present Pisier's L^p bounds (Pisier, 1980, 1982a, 1986) for the associated orthogonal projections, utilizing vector-valued Riesz products and complex interpolation.

Let $k \in \mathbb{N}$ and denote by $C(k)$ the collection of those subsets $S \subset \mathbb{N}$ for which $|S| = k$. If $S \in C(k)$ and $\varepsilon \in \{+1, -1\}^S$, define $p_S^\varepsilon(z) = \prod_{j\in S} z_j^{\varepsilon_j}$ where $z \in \mathbb{T}^{\mathbb{N}}$. Let

$$\mathcal{P}_k = \operatorname{span}\left\{p_S^\varepsilon : \ S \in C(k),\ \varepsilon \in \{+1,-1\}^S\right\}.$$

We noted in Section 2.2.2 that the orthogonal projection onto the L^2 closure of $\mathcal{P}_k$ is given by

$$\mathcal{S}^{(k)}(f) = \sum_{S,\varepsilon} \left\langle f, p_S^\varepsilon\right\rangle p_S^\varepsilon, \tag{2.4.13}$$

where the sum is taken over $S \in C(k)$ and $\varepsilon \in \{+1, -1\}^S$. Clearly $\mathcal{S}^{(k)}$ is a contraction on L^2. Our next theorem asserts that $\mathcal{S}^{(k)}$ admits a bounded extension to L^p where $1 < p < \infty$, with additional information on the dependence of the norm on p and $k \in \mathbb{N}$. Observing that the generating function of the sequence $\left(\mathcal{S}^{(k)}\right)_{k=1}^\infty$ is given by a Riesz product, provides the key to the following L^p estimates. For $n \in \mathbb{N}$, $\alpha \in \mathbb{D}$, and $z = (z_j)$ with $z_j \in \mathbb{T}$ we let

$$J_\alpha(z) = \prod_{j=1}^{n} (1 + \alpha(z_j + \overline{z_j})/2).$$

We begin by showing that the integral operator,

$$(\mathcal{J}_\alpha f)(z) = \mathbb{E}_{(w)}(J_\alpha(z\cdot\overline{w})f(w)), \qquad f \in L^2(\mathbb{T}^{\mathbb{N}}),$$

defines the generating function of the sequence of operators $\mathcal{S}^{(k)}$.

Proposition 2.4.9 *If f is a trigonometric polynomial on $\mathbb{T}^{\mathbb{N}}$ and if $n \in \mathbb{N}$ appearing in the Riesz product is large enough, then for any $\alpha \in \mathbb{C}$,*

$$(\mathcal{J}_\alpha f) = \mathbb{E}(f) + \sum_{k=1}^{\infty} \left(\frac{\alpha}{2}\right)^k \left(\mathcal{S}^{(k)} f\right),$$

and $\|\mathcal{J}_\alpha f\|_2 = \|\mathcal{J}_{|\alpha|} f\|_2$.

Proof Fix $n \in \mathbb{N}$ and $k \le n$. Denote by $C(n,k)$ the collection of all subsets in $\{1, \ldots, n\}$ consisting of k different elements. Fix a sequence of scalars (b_j) then,

$$\prod_{j=1}^{n}(1+b_j) = 1 + \sum_{j=1}^{n} b_j + \sum_{k=2}^{n} \sum_{S\in C(n,k)} \prod_{j\in S} b_j. \tag{2.4.14}$$

Next, put $b_j = (z_j + \overline{z_j})/2$, where $z_j \in \mathbb{T}$. Since $\overline{z_j} = z_j^{-1}$, we have $\prod_{i\in S} b_i = \sum_{\varepsilon\in\{-1,1\}^S} 2^{-k} \prod_{j\in S}\{z_j\}^{\varepsilon_j}$ for any $S \in C(n,k)$. This gives the expansion

$$J_\alpha(z) = 1 + \sum_{k=1}^{n} \frac{\alpha^k}{2^k} \sum_{S\in C(n,k)} \sum_{\varepsilon\in\{-1,1\}^S} \prod_{j\in S} \{z_j\}^{\varepsilon_j},$$

which, by integration, yields the first assertion of the proposition. The second one follows from the first and the orthogonality relation

$$\left\| \sum_{k=1}^{\infty} \left(\frac{\alpha}{2}\right)^k \left(S^{(k)} f\right) \right\|_2^2 = \sum_{k=0}^{\infty} \left(\frac{|\alpha|}{2}\right)^{2k} \|S^{(k)} f\|_2^2 . \qquad \square$$

The next proposition asserts that the Riesz products J_x generate nice and simple operators when $x \in [-1, 1]$.

Proposition 2.4.10 *For* $1 \le p \le \infty$ *and* $x \in [-1,1]$, *the operator* $\mathcal{J}_x$ *defines a contraction on* $L^p(\mathbb{T}^\mathbb{N})$, *and* $f \ge 0$ *implies* $\mathcal{J}_x f \ge 0$. *Moreover, for any (!) Banach space* X *and* $x \in [-1,1]$, *the vector-valued convolution operator*

$$(\mathcal{J}_x f)(z) = \mathbb{E}_{(w)}(J_x(z \cdot \overline{w}) f(w)), \qquad f \in L^p(\mathbb{T}^\mathbb{N}, X),$$

is a contraction on $L^p(X)$ *for* $1 \le p \le \infty$.

Proof Let $\zeta \in \mathbb{T}$ and $x \in [-1,1]$. Then $1 + x(\zeta + \overline{\zeta})/2 \ge 0$. Hence the Riesz product J_x is nonnegative and $\mathbb{E}(|J_x|) = \mathbb{E}(J_x) = 1$. Consequently, $\mathcal{J}_x$ is a convolution operator defined by a nonnegative kernel in the unit ball of L^1. By Theorem 2.1.1, $\mathcal{J}_x$ extends to a contraction on $L^p(X)$ for any Banach space X and any $1 \le p \le \infty$. $\square$

Theorem 2.4.11 *There exists* $C < \infty$ *such that the following holds. For any* $1 < p \le 2$ *there exists* $\delta = \delta(p)$ *satisfying*

$$\|\mathcal{J}_\alpha : L^p \to L^p\| < C,$$

for any $\alpha \in \{z \in \mathbb{D} : |z| < \delta\}$, *and* $\delta(p) \sim (p-1)$ *as* $p \to 1$.

Proof Fix $1 < p \leq 2$ and let $0 < \theta < 1$ be determined by the relation $1/p = 1 - \theta/2$. Next, choose any $\alpha \in \mathbb{D}$ satisfying

$$\frac{2}{\pi}\left|\operatorname{Arg}\left(\frac{\alpha+1}{\alpha-1}\right)\right| = \theta.$$

We let $S = \{x+iy : 0 < x < 1,\ y \in \mathbb{R}\}$ denote the vertical strip $S_0 = \{iy : y \in \mathbb{R}\}$ and $S_1 = 1 + S_0$. By Caratheodory's theorem (see Theorem 2.4.16), there exists an analytic map $\varphi \colon S \to \mathbb{D}$ with continuous extension to $S \cup S_0 \cup S_1$ such that

$$\varphi(S_0) \subseteq [-1, 1], \quad \varphi(S_1) \subseteq \mathbb{T}, \quad \text{and} \quad \varphi(\theta) = \alpha.$$

Fix $f_0 \in L^1$, $f_1 \in L^2$, $g_0 \in L^\infty$, and $g_1 \in L^2$. On the strip S we define the analytic map

$$G(\beta) = \left\langle \mathcal{J}_{\varphi(\beta)}\left(f_0^{1-\beta} f_1^{\beta}\right), \left(g_0^{1-\beta} g_1^{\beta}\right)\right\rangle, \quad \beta \in S.$$

It is easy to see that G extends continuously to $S \cup S_0 \cup S_1$ and that by the three lines theorem (see Müller (2005, Section 2.3) for the presentation of the three lines theorem)

$$|G(\theta)| \leq \sup_{y \in \mathbb{R}} |G(iy)|^{1-\theta} \sup_{y \in \mathbb{R}} |G(1+iy)|^{\theta}.$$

By Proposition 2.4.9 and Proposition 2.4.10 we have,

$$\sup_{y \in \mathbb{R}} |G(iy)| \leq \|f_0\|_1 \|g_0\|_\infty, \quad \text{and} \quad \sup_{y \in \mathbb{R}} |G(1+iy)| \leq \|f_1\|_2 \|g_1\|_2.$$

Hence, if $f = f_0^{1-\beta} f_1^{\beta}$ and $g = g_0^{1-\beta} g_1^{\beta}$ then with $\alpha = \varphi(\theta)$, where φ is given by Caratheodory's theorem (Theorem 2.4.16), we have

$$|\langle \mathcal{J}_\alpha f, g\rangle| \leq (\|f_0\|_1 \|g_0\|_\infty)^{1-\theta} (\|f_1\|_2 \|g_1\|_2)^{\theta}.$$

By the maximum principle, there exists $C = C(p)$ and $\delta = \delta(p)$ such that

$$|\langle \mathcal{J}_\alpha f, g\rangle| \leq C \|f\|_p \|g\|_q,$$

for any $\alpha \in \mathbb{D}$ with $|\alpha| < \delta$. □

Theorem 2.4.12 *If $1 < p < \infty$ and $k \in \mathbb{N}$, then $\left\|S^{(k)} : L^p \to L^p\right\| \leq C^k$, where $C = C(p)$ with $C(p) \leq C_0 p^2/(p-1)$.*

Proof Let $1 < p \leq 2$, and let $\delta = \delta(p)$ be determined by Theorem 2.4.11. Let f be a trigonometric polynomial on $\mathbb{T}^{\mathbb{N}}$. Applying Cauchy's integral formula to the power series expansion for $\mathcal{J}_\alpha(f)$ obtained in Proposition 2.4.9 gives the identities

$$S^{(k)}(f) = \left(\frac{2}{\delta}\right)^k \int_0^{2\pi} \mathcal{J}_{(\delta e^{it})}(f) e^{-tik} \frac{dt}{2\pi},$$

for any $k \in \mathbb{N}$. By the triangle inequality, Theorem 2.4.11 yields the estimate $\|S^{(k)}\|_p \le C(2/\delta)^k$. Since $\delta(p) \sim (p-1)$ as $p \to 1$, we obtain that $\|S^{(k)}\|_p \le C(2/(p-1))^k$. □

Unlike the Riesz projection and the projection onto simple Hardy martingales, the projections onto Steinhaus martingales is not of weak type 1 : 1.

Theorem 2.4.13 $S^{(1)}: L^2(\mathbb{T}^{\mathbb{N}}) \to L^2(\mathbb{T}^{\mathbb{N}})$, *the orthogonal projection onto Steinhaus series, does not satisfy weak type* 1 : 1 *estimates, i.e.,*

$$\sup_{\mathbb{E}|G|=1} \sup_{\lambda>0} \lambda \mathbb{P}\left(\left\{|S^{(1)}G| > \lambda\right\}\right) = \infty.$$

Proof We give a rough sketch of the proof, which applies to several classes of Fourier multiplier operators. By Khintchine's inequality for Steinhaus martingales (2.2.12), there exist $C_1 < \infty$ and $C_{1/2} < \infty$, such that

$$\|S^{(1)}(f)\|_2 \le C_1 \|S^{(1)}(f)\|_1 \le C_{1/2} \|S^{(1)}(f)\|_{1/2}, \tag{2.4.15}$$

for any trigonometric polynomial f on $\mathbb{T}^{\mathbb{N}}$. Assume now that $S^{(1)}$ is of weak type 1: 1. By direct integration, this assumption implies that

$$\|S^{(1)}: L^1(\mathbb{T}^{\mathbb{N}}) \to L^{1/2}(\mathbb{T}^{\mathbb{N}})\| < \infty. \tag{2.4.16}$$

In view of (2.4.15) and (2.4.16), the projection $S^{(1)}: L^1(\mathbb{T}^{\mathbb{N}}) \to L^1(\mathbb{T}^{\mathbb{N}})$ is bounded, and its range is isomorphic to a Hilbert space. This is a contradiction, showing that $S^{(1)}$ is not of weak type 1:1. □

A Vector-Valued Extension of $S^{(k)}$

We let (Ω, μ) be a probability space and write $L^p(L^q) = L^p(\mathbb{T}^{\mathbb{N}}, L^q(\Omega))$ to denote the $L^q(\Omega)$-valued Bochner–Lebesgue space over the probability space $\mathbb{T}^{\mathbb{N}}$.

Let $f: \mathbb{T}^{\mathbb{N}} \to L^q(\Omega)$ be a trigonometric polynomial with coefficients in $L^q(\Omega)$. Define the vector-valued convolution integral operator

$$(\mathcal{J}_\alpha f)(z) = \mathbb{E}(J_\alpha(z \cdot \overline{w}) f(w)),$$

where, again, we have

$$J_\alpha(z) = \prod_{j=1}^{n} (1 + \alpha(z_j + \overline{z_j})/2).$$

As in the scalar case, the convolution operator $\mathcal{J}_\alpha$ is the generating function for the sequence of projections $S^{(k)}$. If $n \in \mathbb{N}$ arising in the definition of the product $J_\alpha(z)$ is larger than the polynomial degree of f, then

$$(\mathcal{J}_\alpha f) = \mathbb{E}(f) + \sum_{k=1}^{\infty} \left(\frac{\alpha}{2}\right)^k (S^{(k)} f),$$

for any $\alpha \in \mathbb{C}$. Our next result concerns the $L^p(L^q)$ boundedness of the vector-valued projections $S^{(k)}$ with explicit control of the dependence on k.

The first step consists in proving that the convolution operators $\mathcal{J}_x$, $x \in [-1,+1]$, extends to an analytic semigroup of operators

Theorem 2.4.14 *Let $1 < p, q < \infty$. There exists $\delta = \delta(p,q)$ and $C = C(p,q)$ such that for any $\alpha \in \{z \in \mathbb{D} : |z| < \delta\}$,*

$$\left\| \mathcal{J}_\alpha : L^p(L^q) \to L^p(L^q) \right\| < C.$$

Proof Recall that, for $x \in [-1,1]$, the kernel J_x is nonnegative and integrable with $\mathbb{E}(J_x) = 1$. Hence the scalar-valued convolution $\mathcal{J}_x$ is a positive contraction operator on L^p for $1 \leq p \leq \infty$. (The endpoints $p = 1$ and $p = \infty$ are included.) By Proposition 2.4.10, for any (!) Banach space X,

$$(\mathcal{J}_x f)(z) = \mathbb{E}_{(w)}(J_x(z \cdot \overline{w}) f(w)), \qquad f \in L^p(\mathbb{T}^{\mathbb{N}}, X),$$

defines a contraction on $L^p(X)$ for $1 \leq p \leq \infty$. Next, recall that for $1 < p < \infty$ there exists $\delta = \delta(p)$ and $C = C(p)$ such that for any $\alpha \in \{z \in \mathbb{D} : |z| < \delta\}$ we have $\|\mathcal{J}_\alpha : L^p \to L^p\| < C$ and that $\delta(p) = O(p-1)$ as $p \to 1$. By Fubini's theorem, the vector-valued operators $\mathcal{J}_\alpha$ satisfy $\|\mathcal{J}_\alpha : L^p(L^p) \to L^p(L^p)\| < C$, for $|\alpha| \leq \delta$.

Now fix $1 < p < \infty$. We apply complex interpolation between the mixed norm spaces $L^p\left(L^1\right)$ and $L^p\left(L^q\right)$. Let $1 < q < p$ and choose $0 < \theta < 1$ such that $1/q = \theta/p + 1 - \theta$. Choose $\alpha \in \mathbb{D}$ such that

$$\frac{2}{\pi}\left|\operatorname{Arg}\left(\frac{\alpha+1}{\alpha-1}\right)\right| = \theta.$$

Recall that we let $S = \{x + iy : 0 < x < 1, y \in \mathbb{R}\}$ denote the vertical strip, $S_0 = \{iy : y \in \mathbb{R}\}$ and $S_1 = 1 + S_0$. By Caratheodory's theorem (Theorem 2.4.16) there exists an analytic map $\varphi : S \to \mathbb{D}$ with continuous extension to $S \cup S_0 \cup S_1$ such that

$$\varphi(S_0) \subseteq [-1,1], \quad \varphi(S_1) \subseteq \mathbb{T}, \quad \text{and} \quad \varphi(\theta) = \alpha.$$

We fix $f_0 \in L^p(L^1)$, $g_0 \in L^p(L^\infty)$, $f_1 \in L^p(L^p)$, and $g_1 \in L^{p'}(L^{p'})$, where p' is the Hölder conjugate exponent given by $1/p + 1/p' = 1$. Define the analytic map

$$G(\beta) = \left\langle \mathcal{J}_{\varphi(\beta)}\left(f_0^{1-\beta} f^\beta\right), \left(g_0^{1-\beta} g_1^\beta\right)\right\rangle, \quad \beta \in S.$$

By the three lines theorem,

$$|G(\theta)| \leq \sup_{y\in\mathbb{R}} |G(iy)|^{1-\theta} \sup_{y\in\mathbb{R}} |G(1+iy)|^\theta.$$

In view of the above remarks, we have

$$\sup_{y\in\mathbb{R}} |G(iy)| \le \|f_0\|_{(p,1)}\|g_0\|_{(p',\infty)} \quad \text{and} \quad \sup_{y\in\mathbb{R}} |G(1+iy)| \le \|f_1\|_{(p,p)}\|g_1\|_{(p',p')}.$$

In summary, if $f = f_0^{1-\theta} f_1^{\theta}$ and $g = g_0^{1-\theta} g_1^{\theta}$, then for $\alpha = \varphi(\theta)$,

$$|\langle \mathcal{J}_\alpha f, g\rangle| \le \left(\|f_0\|_{(p,1)}\,\|g_0\|_{(p',\infty)}\right)^{1-\theta}\left(\|f_1\|_{(p,p)}\,\|g_1\|_{(p',p')}\right)^{\theta}.$$

By the maximal principle, there exists $C = C(p,q)$ and $\delta = \delta(p,q)$ such that

$$|\langle \mathcal{J}_\alpha f, g\rangle| \le C\|f\|_{(p,q)}\|g\|_{(p',q')},$$

for any $\alpha \in \mathbb{D}$ with $|\alpha| < \delta$. □

Theorem 2.4.15 *For any $1 < p, q < \infty$, there exists $C = C(p,q)$ such that for any probability space (Ω,μ),*

$$\left\|S^{(k)} : L^p(L^q) \to L^p(L^q)\right\| \le C^k \tag{2.4.17}$$

where $L^p(L^q) = L^p(\mathbb{T}^{\mathbb{N}}, L^q(\Omega))$.

Proof Let $1 < p < \infty$ and $1 < q < p$. Let $C > 0$ and $\delta > 0$ be determined by Theorem 2.4.14. Fix a $L^q(\Omega)$-valued trigonometric polynomial on $\mathbb{T}^{\mathbb{N}}$; call it f. By Cauchy's integral formula we have,

$$S^{(k)}(f) = \left(\frac{2}{\delta}\right)^k \int_0^{2\pi} J_{(\delta e^{it})} e^{-tik}\frac{dt}{2\pi}.$$

Applying the triangle inequality and Theorem 2.4.14 finishes the proof. □

Caratheodory's theorem was used repeatedly above. For ease of reference, we now state it formally. Let $0 < \theta < 1$ and $D_\theta = \left\{z \in \mathbb{D} : \frac{2}{\pi}\left|\operatorname{Arg}\left(\frac{z+1}{z-1}\right)\right| = \theta\right\}$. Note that D_θ is the union of two circles, each of which passes through the points $\{+1,-1\}$, and if $E_\theta \subset \mathbb{D}$ is the simply connected domain bounded by D_θ then there exists $\delta = \delta(\theta)$ such that

$$\{z \in \mathbb{D} : |z| < \delta\} \subset E_\theta,$$

with $\delta(\theta) = O(1/\theta)$ as $\theta \to 1$.

Theorem 2.4.16 (Caratheodory's theorem) *Denote $S = \{x + iy : 0 < x < 1, y \in \mathbb{R}\}$, $S_0 = \{iy : y \in \mathbb{R}\}$ and $S_1 = 1 + S_0$. For any $0 < \theta < 1$ and $\alpha \in D_\theta$, there exists an analytic map $\varphi : S \to \mathbb{D}$ with continuous extension to $S \cup S_0 \cup S_1$ such that*

$$\varphi(S_0) \subseteq [-1,1], \quad \varphi(S_1) \subseteq \mathbb{T}, \quad \text{and} \quad \varphi(\theta) = \alpha.$$

2.5 Basic L^1 Estimates

We present L^1 estimates for the maximal function of vector-valued Hardy martingales. With the Hahn–Banach theorem, we pass from the scalar-valued to the vector-valued case. We prove the Davis and Garsia inequality for a scalar-valued Hardy martingale, and L^1 estimates for its square function. The method of proof is based on nonlinear telescoping and complex convexity estimates for $H^1(\mathbb{T})$. Extensions of the Davis and Garsia inequality to vector-valued Hardy martingales will be discussed in Chapter 5.

2.5.1 Maximal Functions

In Section 2.2 we obtained that a scalar-valued Hardy martingale (F_n) satisfies the maximal estimate

$$\mathbb{E}\left(\max_{k\leq n}|F_k|\right) \leq e\mathbb{E}|F_n|, \tag{2.5.1}$$

for $n \in \mathbb{N}$. Consequently, an L^1-bounded Hardy martingale is a uniformly integrable subset of $L^1(\mathbb{T}^{\mathbb{N}})$. We now show that (2.5.1) is not limited to the scalar-valued case. Analogous estimates hold true for Hardy martingales with values in a Banach space X.

Theorem 2.5.1 *Let X be a complex Banach space and let (F_n) be an L^1-bounded X-valued Hardy martingale. Then,*

$$\mathbb{E}\left(\sup_{n\in\mathbb{N}}\|F_n\|\right) \leq e\sup_{n\in\mathbb{N}}\mathbb{E}\|F_n\|, \tag{2.5.2}$$

and (F_n) is a uniformly integrable sequence in $L^1(X)$.

Proof Let (F_n) be an X-valued Hardy martingale such that $\sup_{n\in\mathbb{N}}\mathbb{E}\|F_n\| < \infty$. We show first that $(\|F_n\|^{\alpha})$ is a submartingale. To this end, we choose F_n and condition on $\mathcal{F}_{n-1}$ by fixing $w \in \mathbb{T}^{n-1}$ and putting

$$a = F_{n-1}(w) \quad \text{and} \quad f(z) = F_n(w,z),$$

where $z \in \mathbb{T}$. Since (F_n) is a Hardy martingale, $f \in H^1(\mathbb{T}, X)$ and $\int f dm = a$.

By the Hahn–Banach theorem, there exists $y \in X^*$ with $\|y\| = 1$ and $y(a) = \|a\|$. Next, define (the scalar-valued) $h \in H^1(\mathbb{T})$ by putting $h(z) = y(f(z))$. Applying inequality (1.2.17) in the remark following Theorem 1.2.2 gives

$$\left|\int_{\mathbb{T}} h dm\right|^{\alpha} \leq \int_{\mathbb{T}} |h|^{\alpha} dm.$$

Since the integral on the left-hand side equals $y(a) = \|F_{n-1}(w)\|$, and since $|h(z)| \le \|f(z)\|$, we obtain the submartingale estimate,

$$\|F_{n-1}\|^\alpha = \left\| \int_{\mathbb{T}} f dm \right\|^\alpha \le \int_{\mathbb{T}} \|f\|^\alpha dm = \mathbb{E}_{n-1}(\|F_n\|^\alpha),$$

for any $n \in \mathbb{N}$ and $0 < \alpha \le 1$. Now we repeat the proof of Theorem 1.2.3. We fix $p > 1$, and apply Doob's L^p-inequality to the submartingale $(\|F_n\|^{\alpha/p})$. This gives

$$\mathbb{E}\left(\sup_n \|F_n\|^\alpha \right) \le C_p^p \sup_n \mathbb{E}(\|F_n\|^\alpha),$$

where $C_p = p/(p-1)$. Since $e > C_p^p$ for any $1 < p < \infty$, we get (2.5.2).

Finally, we prove that (F_n) is a uniformly integrable sequence in $L^1(X)$. Put $g = \sup_{n \in \mathbb{N}} \|F_n\|$. Then $\int_A \|F_n\| d\mathbb{P} \le \int_A g d\mathbb{P}$ for any $n \in \mathbb{N}$ and $A \in \mathbb{T}^{\mathbb{N}}$ measurable. By (2.5.2), g is integrable, hence for any $\epsilon > 0$ there exists $\delta > 0$ such that $\int_A \|F_n\| d\mathbb{P} < \epsilon$ for any $n \in \mathbb{N}$ and $A \subseteq \mathbb{T}^{\mathbb{N}}$ with $\mathbb{P}(A) < \delta$. □

Corollary 2.5.2 *Let (F_n) be an L^1-bounded, X-valued Hardy martingale. If (F_n) converges in measure, then it also converges in $L^1(X)$, and there exists $F \in L^1(X)$ such that $F_n = \mathbb{E}_n(F)$, for any $n \in \mathbb{N}$.*

Proof By Theorem 2.5.1 (F_n) is uniformly integrable, and by assumption it converges in measure. Combining these conditions, we find that (F_n) converges in the norm of $L^1(X)$. Hence there exists $F \in L^1(X)$ such that $F_n = \mathbb{E}_n(F)$, for any $n \in \mathbb{N}$. □

2.5.2 Square Functions

Here we exploit nonlinear telescoping and complex uniform convexity estimates (1.2.32), to obtain square function estimates for L^1-bounded Hardy martingales. The same method of proof, in Theorem 2.5.4, yields the Davis and Garsia inequality (2.5.13).

The basic strategy is as follows:

(i) Fix a stage k and condition the martingale to $\mathcal{F}_{k-1}$.
(ii) Apply complex uniform convexity estimates to ΔF_k.
(iii) Use nonlinear telescoping to extract the desired martingale inequality from complex uniform convexity and invoke the maximal function estimate (2.5.1).

We begin by proving the nonlinear telescoping inequality, which provides a general framework for obtaining square function estimates. The idea of the proof is related to the one applied in Theorem 1.2.4 for continuous martingales.

Lemma 2.5.3 (nonlinear telescoping lemma) *Let $n \in \mathbb{N}$. Let $M_1, \ldots, M_n$, $v_1, \ldots, v_n$, and $w_1, \ldots, w_n$ be nonnegative and integrable. If*

$$\mathbb{E}\left(M_{k-1}^2 + v_k^2\right)^{1/2} + \mathbb{E}w_k \le \mathbb{E}M_k \quad \textit{for } 1 \le k \le n, \tag{2.5.3}$$

then

$$\mathbb{E}\left(\sum_{k=1}^{n} v_k^2\right)^{1/2} + \mathbb{E}\sum_{k=1}^{n} w_k \le 2(\mathbb{E}M_n)^{1/2}\left(\mathbb{E}\max_{k\le n} M_k\right)^{1/2}. \tag{2.5.4}$$

Proof Let $0 \le \epsilon \le 1$ be defined by

$$\epsilon^2 = (\mathbb{E}M_n)\left(\mathbb{E}\max_{k\le n} M_k\right)^{-1}. \tag{2.5.5}$$

Next, choose nonnegative $s_k \in L^\infty$ so that $\sum_{k=1}^n s_k^2 \le \epsilon^2$. Clearly, we have

$$1 - s_k^2 < \left(1 - s_k^2\right)^{1/2}.$$

Multiply by $M_{k-1} > 0$, add $v_k s_k$, and use the Cauchy–Schwarz inequality. This gives

$$M_{k-1}\left(1 - s_k^2\right) + v_k s_k \le M_{k-1}\left(1 - s_k^2\right)^{1/2} + v_k s_k \le \left(M_{k-1}^2 + v_k^2\right)^{1/2}.$$

Subtracting $M_{k-1}\left(1 - s_k^2\right)$ gives

$$v_k s_k \le s_k^2 M_{k-1} + \left(M_{k-1}^2 + v_k^2\right)^{1/2} - M_{k-1}. \tag{2.5.6}$$

Integrating the pointwise estimates (2.5.6) gives

$$\mathbb{E}(v_k s_k) \le \mathbb{E}\left(s_k^2 M_{k-1}\right) + \mathbb{E}\left(M_{k-1}^2 + v_k^2\right)^{1/2} - \mathbb{E}M_{k-1}. \tag{2.5.7}$$

Next, apply the hypothesis (2.5.3) to the central term $\mathbb{E}\left(M_{k-1}^2 + v_k^2\right)^{1/2}$ appearing in the integrated estimates (2.5.7). This gives

$$\mathbb{E}(v_k s_k) \le \mathbb{E}\left(s_k^2 M_{k-1}\right) + \mathbb{E}M_k - \mathbb{E}M_{k-1} - \mathbb{E}w_k. \tag{2.5.8}$$

Taking the sum over $k \le n$ and exploiting the telescoping nature of the right-hand side of (2.5.8) yields,

$$\begin{aligned}\mathbb{E}\left(\sum_{k=1}^{n} v_k s_k\right) + \sum_{k=1}^{n} \mathbb{E}w_k &\le \mathbb{E}M_n + \mathbb{E}\left(\sum_{k=1}^{n} s_k^2 M_{k-1}\right) \\ &\le \mathbb{E}M_n + \epsilon^2 \mathbb{E}\max_{k\le n} M_{k-1}.\end{aligned} \tag{2.5.9}$$

Since Inquality (2.5.9) holds for every choice of $s_k \in L^\infty$ such that $\sum_{k=1}^n s_k^2 \le \epsilon^2$, we may take the supremum and obtain, by duality, the square function estimate

$$\epsilon\mathbb{E}\left(\sum_{k=1}^{n} v_k^2\right)^{1/2} + \sum_{k=1}^{n} \mathbb{E}w_k \le 2\mathbb{E}M_n. \tag{2.5.10}$$

It remains to divide by $0 < \epsilon \le 1$. □

Remark: If, in the hypothesis of Lemma 2.5.3, we assume just that $w_1, \ldots, w_n$ are integrable, and not necessarily nonnegative, then by estimate (2.5.10) the proof given above yields that

$$\mathbb{E}\left(\sum_{k=1}^{n} v_k^2\right)^{1/2} + \frac{1}{\epsilon}\mathbb{E}\sum_{k=1}^{n} w_k \le 2(\mathbb{E}M_n)^{1/2}\left(\mathbb{E}\max_{k\le n} M_k\right)^{1/2}, \tag{2.5.11}$$

where $\epsilon^2 = (\mathbb{E}M_n)(\mathbb{E}\max_{k\le n} M_k)^{-1}$.

Theorem 2.5.4 *There exists $C > 0$ so that for every Hardy martingale $F = (F_k)$,*

$$\mathbb{E}\left(\sum_{k=1}^{n} |\Delta F_k|^2\right)^{1/2} \le C\mathbb{E}|F_n|, \tag{2.5.12}$$

and there exists a Hardy martingale $G = (G_k)$ such that $|\Delta G_k| \le C|F_{k-1}|$, and

$$\mathbb{E}\left(\sum_{k=1}^{n} \mathbb{E}_{k-1}|\Delta G_k|^2\right)^{1/2} + \mathbb{E}\left(\sum_{k=1}^{n} |\Delta(F-G)_k|\right) \le C\mathbb{E}|F_n|. \tag{2.5.13}$$

Proof The first step of the proof reduces the problem to complex uniform convexity estimates. The second step exploits nonlinear telescoping.

Step 1 Fix a Hardy martingale $F = (F_k)$. Conditioning to $\mathcal{F}_{k-1}$ amounts to fixing $x \in \mathbb{T}^{k-1}$ and putting

$$h_0 = F_{k-1}(x), \quad h(z) = F_k(x,z),$$

where $z \in \mathbb{T}$. Since $F_{k-1} = \mathbb{E}_{k-1}(F_k)$ we have $h_0 = \int_{\mathbb{T}} h\,dm$. By inequalities (1.2.17) and (1.2.40),

$$\int_{\mathbb{T}} \left(|h_0|^2 + \kappa^2|h - h_0|^2\right)^{1/2} dm \le \int_{\mathbb{T}} |h|dm,$$

for $\kappa^2 \le 1/6$. This yields

$$\mathbb{E}\left(|F_{k-1}|^2 + \kappa^2|\Delta F_k|^2\right)^{1/2} \le \mathbb{E}|F_k|. \tag{2.5.14}$$

By Theorem 1.7.2, there exists $g \in H^\infty(\mathbb{T})$ such that $\int g\,dm = 0$, $|g| \le C|h_0|$, and

$$\left(|h_0|^2 + \kappa^2 \int_{\mathbb{T}} |g|^2 dm\right)^{1/2} + \kappa \int_{\mathbb{T}} |h - h_0 - g|dm \le \int_{\mathbb{T}} |h|dm,$$

for $\kappa \le 1/60$. Thus we obtain a Hardy martingale $G = (G_k)$ so that $|\Delta G_k| \le |F_{k-1}|$, and

$$\mathbb{E}\left(|F_{k-1}|^2 + \kappa^2\mathbb{E}_{k-1}|\Delta G_k|^2\right)^{1/2} + \kappa\mathbb{E}|\Delta(F-G)_k| \le \mathbb{E}|F_k|. \tag{2.5.15}$$

Step 2 Apply nonlinear telescoping, using Lemma 2.5.3 with $M_k = |F_k|$ and $v_k = \kappa|\Delta F_k|$. By (2.5.14), this gives the square function estimate

$$\mathbb{E}\left(\sum_{k=1}^{n} |\Delta F_k|^2\right)^{1/2} \leq 2\kappa^{-1}(\mathbb{E}|F_n|)^{1/2}\left(\mathbb{E}\max_{k\leq n}|F_k|\right)^{1/2}.$$

Invoke estimate (2.2.8) to remove the maximal function on the right-hand side. We next turn to the proof of the Davis and Garsia inequality (2.5.13). We apply Lemma 2.5.3 with the specification,

$$M_k = |F_k|, \quad v_k = \kappa\left(\mathbb{E}_{k-1}|\Delta G_k|^2\right)^{1/2}, \quad \text{and} \quad w_k = \kappa|\Delta(F-G)_k|,$$

and use (2.5.15). This gives a Hardy martingale $G = (G_k)$ with uniformly bounded increments $|\Delta G_k| \leq C|F_{k-1}|$, such that

$$\mathbb{E}\left(\sum_{k=1}^{n} \mathbb{E}_{k-1}|\Delta G_k|^2\right)^{1/2} + \mathbb{E}\left(\sum_{k=1}^{n}|\Delta(F-G)_k|\right) \leq C(\mathbb{E}|F_n|)^{1/2}\left(\mathbb{E}\max_{k\leq n}|F_k|\right)^{1/2}.$$

Invoking estimate (2.2.8) again, we finally arrive at (2.5.13). □

Remark: Specializing Theorem 2.5.4, for every Hardy martingale F there exists a Hardy martingale G such that

$$|\Delta G_k| \leq C|F_{k-1}| \quad \text{and} \quad \sum_{k=1}^{\infty} \mathbb{E}|\Delta B_k| \leq C \sup_{k\in\mathbb{N}} \mathbb{E}|F_k|, \tag{2.5.16}$$

where $B = F - G$ (and $C_0 < \infty$ is independent of F). The splitting $F = G + B$ is called the Davis decomposition of the Hardy martingale F. It is instructive to compare (2.5.16) with the estimates (1.1.11) and (1.1.12) in Theorem 1.1.7 – the Davis decomposition for general martingales: As compensation for sacrificing generality, we obtained improved estimates for the decomposing martingales G and B.

Further Applications

Following are two straightforward applications of nonlinear telescoping to the Burkholder–Gundy inequality and the Lépingle's estimate (which often serves as a converse to Burkholder–Gundy). We use below the inequalities of Minkowski and Hölder, in the following form:

$$\left(M^2 + \left(\int_{\mathbb{T}} u\,dm\right)^2\right)^{1/2} \leq \int_{\mathbb{T}} (M^2 + u^2)^{1/2} dm \leq \left(M^2 + \int_{\mathbb{T}} u^2 dm\right)^{1/2}, \tag{2.5.17}$$

where $u \geq 0$ is square integrable and $M > 0$ a scalar.

Theorem 2.5.5 *For every $(\mathcal{F}_k)$-martingale (F_k),*

$$\frac{1}{2}\mathbb{E}\left(\sum_{k=1}^{n}(\mathbb{E}_{k-1}|\Delta_k F|)^2\right)^{1/2} \le \mathbb{E}\left(\sum_{k=1}^{n}|\Delta_k F|^2\right)^{1/2} \le 2\mathbb{E}\left(\sum_{k=1}^{n}\mathbb{E}_{k-1}|\Delta_k F|^2\right)^{1/2}. \tag{2.5.18}$$

Proof We first prove the left-hand side estimate (Lépingle's inequality). Define simply $M_k = \left(\sum_{m=1}^{k}|\Delta_m F|^2\right)^{1/2}$. Conditioning to $\mathcal{F}_{k-1}$ and using Minkowski's inequality (2.5.17), we obtain

$$\mathbb{E}\left(M_{k-1}^2 + \mathbb{E}_{k-1}^2|\Delta_k F|\right)^{1/2} \le \mathbb{E}M_k.$$

Apply Lemma 2.5.3, with $v_k = \mathbb{E}_{k-1}|\Delta_k F|$ and $w_k = 0$, to get the left-hand side estimate (Lépingle's inequality).

Consider next the right-hand side estimate, due to Burkholder and Gundy. Let now M_k be the conditional square function defined by

$$M_k^2 = \sum_{m=1}^{k}\mathbb{E}_{m-1}|\Delta_m F|^2.$$

Conditioning to $\mathcal{F}_{k-1}$ and using Minkowski's inequality (2.5.17) gives

$$\mathbb{E}\left(M_{k-1}^2 + |\Delta_k F|^2\right)^{1/2} \le \mathbb{E}M_k.$$

Apply Lemma 2.5.3, with $v_k = |\Delta_k F|$ and $w_k = 0$, to obtain the right-hand side estimate (Burkholder–Gundy inequality). □

We use Lépingle's inequality to prove that the orthogonal projection onto simple Hardy martingales extends to a bounded operator on martingale H^1 spaces. Recall that by $\mathcal{F}_j$ we denote the sigma-algebra on $\mathbb{T}^{\mathbb{N}}$ generated by the coordinate projections

$$p_k\colon \mathbb{T}^{\mathbb{N}} \to \mathbb{T}, \qquad k \le j,$$

where $p_k(z) = z_k$ and $z \in \mathbb{T}^{\mathbb{N}}$ with $z = (z_k)$. Thus $(\mathbb{T}^{\mathbb{N}}, (\mathcal{F}_j), \mathbb{P})$ is a filtered probability space. We let $\mathbb{E}_j$ denote the conditional expectation with respect to $\mathcal{F}_j$. Let $G \in L^2(\mathbb{T}^{\mathbb{N}})$, put $G_j = \mathbb{E}_j G$ and $\Delta G_j = G_j - G_{j-1}$. In inequality (2.2.11) we noted that the orthogonal projection onto simple Hardy martingales is given by

$$P_0 G = \sum_{j=1}^{\infty}\mathbb{E}_{j-1}(\Delta G_j\overline{p_j})p_j.$$

As shown above, P_0 extends to a bounded operator on L^p $(1 < p < \infty)$. We now prove that P_0 is bounded on the martingale space $H^1(\mathbb{T}^{\mathbb{N}}, (\mathcal{F}_j), \mathbb{P})$.

Theorem 2.5.6 *The orthogonal projection P_0 extends boundedly to H^1, such that*

$$\|P_0 : H^1 \to H^1\| \leq 2,$$

where $H^1 = H^1(\mathbb{T}^{\mathbb{N}}, (\mathcal{F}_j), \mathbb{P})$.

Proof Let $G = (G_j)$ be a martingale with respect to $(\mathbb{T}^{\mathbb{N}}, (\mathcal{F}_j), \mathbb{P})$. Then with $H^1 = H^1(\mathbb{T}^{\mathbb{N}}, (\mathcal{F}_j), \mathbb{P})$, we have

$$\|G\|_{H^1} = \mathbb{E}\left(\sum |\Delta G_j|^2\right)^{1/2}, \tag{2.5.19}$$

where $\Delta G_j = G_j - G_{j-1}$. Let $z \in \mathbb{T}^{\mathbb{N}}$. Since $|p_j(z)| = 1$ for $z \in \mathbb{T}^{\mathbb{N}}$, we have

$$\left|\mathbb{E}_{j-1}(\Delta G_j \overline{p_j}) p_j\right| \leq \mathbb{E}_{j-1}|\Delta G_j|,$$

hence in view of Lépingle's inequality,

$$\mathbb{E}\left(\sum |\mathbb{E}_{j-1}(\Delta G_j \overline{p_j}) p_j|^2\right)^{1/2} \leq 2\mathbb{E}\left(\sum |\Delta G_j|^2\right)^{1/2}.$$

Since $\|P_0 G\|_{H^1} = \mathbb{E}(\sum |\mathbb{E}_{j-1}(\Delta G_j \overline{p_j}) p_j|^2)^{1/2}$, it remains to invoke (2.5.19). □

Remark: Recall that we let $\mathcal{P}_k$ be the span of homogeneous polynomials of degree k where $k \in \mathbb{N}$. Let $\mathcal{S}^{(k)}$ be the orthogonal projection onto the L^2 closure of $\mathcal{P}_k$, defined by equation (2.4.13). The proof of Theorem 2.5.6 may be modified to show that the projection $\mathcal{S}^{(k)}$ is bounded on the martingale space $H^1(\mathbb{T}^{\mathbb{N}}, (\mathcal{F}_j), \mathbb{P})$, with norm depending on k only.

2.5.3 Riesz Factorization and Complex Convexity

In this section, we utilize the F. Riesz factorization theorem for $H^p(\mathbb{T})$ (reviewed in Section 1.1 (Preliminaries, see equation (1.1.76)). It was invented as a tool for extrapolation, and asserts that for $0 < p < \infty$ and $h \in H^p(\mathbb{T})$ there exist $f, g \in H^{2p}(\mathbb{T})$ such that $h = f \cdot g$ and

$$\int_{\mathbb{T}} |h|^p dm = \left(\int_{\mathbb{T}} |f|^{2p} dm \int_{\mathbb{T}} |g|^{2p} dm\right)^{1/2}.$$

We will prove that for any Hardy martingale (F_k),

$$\left(\sum (\mathbb{E}|\Delta F_k|)^2\right)^{1/2} \leq \sqrt{2}\mathbb{E}|F|, \tag{2.5.20}$$

where the emphasis is on the constant $\sqrt{2}$. Proposition 2.5.7 – due to Haagerup and Pisier (1989) – is the key ingredient yielding (2.5.20) by telescoping and conditioning. See Theorem 2.5.8. Moreover, in Theorem 2.5.9 we obtain the estimate, analogous to (2.5.20), for L^p-valued Hardy martingales ($0 < p \leq 2$).

Proposition 2.5.7 (Haagerup–Pisier inequality) *Let $p \in \{2, 1\} \cup \{2^{-n} : n \in \mathbb{N}\}$. There exists $\delta(p) > 0$, such that for any $f \in H^p(\mathbb{T})$,*

$$|f_0|^2 + \delta(p)\left(\int_{\mathbb{T}} |f - f_0|^p dm\right)^{2/p} \le \left(\int_{\mathbb{T}} |f|^p dm\right)^{2/p}, \tag{2.5.21}$$

where $f_0 = \int f dm$. We have $\delta(2) = 1$, $\delta(1) = 1/2$, and $\delta(p/2) = \delta(p)/(2^{4/p-1})$.

Proof By orthogonality, for $p = 2$ we have equality in condition (2.5.21) with $\delta(2) = 1$. Now assume that condition (2.5.21) holds true for $0 < p \le 2$. We will prove that, for $h \in H^{p/2}(\mathbb{T})$,

$$|h_0|^2 + \frac{\delta(p)}{A}\left(\int_{\mathbb{T}} |h - h_0|^{p/2} dm\right)^{4/p} \le \left(\int_{\mathbb{T}} |h|^{p/2} dm\right)^{4/p}, \tag{2.5.22}$$

where $A = 2^{4/p-1}$. To this end, fix $h \in H^{p/2}(\mathbb{T})$. By the Riesz factorization theorem there exist $f, g \in H^p(\mathbb{T})$ such that

$$h = f \cdot g, \quad h_0 = f_0 \cdot g_0, \quad \int_{\mathbb{T}} |h|^{p/2} dm = \left(\int_{\mathbb{T}} |f|^p dm \int_{\mathbb{T}} |g|^p dm\right)^2, \tag{2.5.23}$$

where $f_0 = \int_{\mathbb{T}} f dm$ and $g_0 = \int_{\mathbb{T}} g dm$. Using $h = f \cdot g$ and $h_0 = f_0 \cdot g_0$ we write

$$h - h_0 = (f - f_0)g_0 + f(g - g_0).$$

By Hölder's inequality and arithmetic, the above identity yields the estimate,

$$\begin{aligned}\frac{1}{A}\left(\int_{\mathbb{T}} (L^1)|h - h_0|^{p/2} dm\right)^{4/p} &\le \left(\int_{\mathbb{T}} |f - f_0|^p dm\right)^{2/p} |g_0|^2 \\ &\quad + \left(\int_{\mathbb{T}} |f|^p dm \int_{\mathbb{T}} |g - g_0|^p dm\right)^{2/p}.\end{aligned}$$

Hence with $h_0 = f_0 \cdot g_0$, the term on the left-hand side of estimate (2.5.22) is bounded by

$$\left(|f_0|^2 + \delta(p)\left(\int_{\mathbb{T}} |f - f_0|^p dm\right)^{2/p}\right)|g_0|^2 + \delta(p)\left(\int_{\mathbb{T}} |f|^p dm \int_{\mathbb{T}} |g - g_0|^p dm\right)^{2/p}. \tag{2.5.24}$$

Invoking assumption (2.5.21) two consecutive times we obtain the following upper bound for (2.5.24):

$$\left(\int_{\mathbb{T}} |f|^p dm\right)^{2/p}\left(|g_0|^2 + \delta(p)\left(\int_{\mathbb{T}} |g - g_0|^p dm\right)^{2/p}\right) \le \left(\int_{\mathbb{T}} |f|^p dm \int_{\mathbb{T}} |g|^p dm\right)^{2/p}.$$

Invoking the factorization identity (2.5.23) finishes the proof of (2.5.22). □

Remark: If $1 < p \le 2$, then (2.5.21) holds true for any $f \in L^p(\mathbb{T})$. See, for instance, Garling (2007, Theorem 9.8.1). Hence, we may start the extrapolation above with any $1 < p \le 2$. This gives (2.5.21) for any $0 < p \le 2$. See Xu (1990) for the noncommutative extension.

Following, are two applications of Proposition 2.5.7. The first extends estimate (2.5.20) to scalar-valued L^p-bounded Hardy martingales, and the second to L^1-bounded Hardy martingales with values in L^p for $0 < p \le 2$.

Theorem 2.5.8 *Let $p \in \{2, 1\} \cup \{2^{-n} : n \in \mathbb{N}\}$. There exists $A(p) < \infty$ such that for any Hardy martingale $F = (F_k)$,*

$$\sum (\mathbb{E}|\Delta F_k|^p)^{2/p} \le A(p)(\mathbb{E}|F|^p)^{2/p}. \tag{2.5.25}$$

We have $A(2) = 1$, $A(1) = 2$, and $A(p/2) = A(p)(2^{4/p-1})$.

Proof Fix (F_k) and $k \in \mathbb{N}$. We fix $x \in \mathbb{T}^{k-1}$ and put $h_0 = F_{k-1}(x)$. For $z \in \mathbb{T}$ set $h(z) = F_k(x, z)$. Since $F_{k-1} = \mathbb{E}_{k-1}(F_k)$, we have $h_0 = \int_{\mathbb{T}} h dm$. Hence Proposition 2.5.7 gives

$$\left(|F_{k-1}|^2 + \delta(p)(\mathbb{E}_{k-1}|\Delta F_k|^p)^{2/p}\right)^{p/2} \le \mathbb{E}_{k-1}|F_k|^p. \tag{2.5.26}$$

Averaging gives

$$\mathbb{E}\left(|F_k|^2 + \delta(p)(\mathbb{E}_{k-1}|\Delta F_k|^p)^{2/p}\right)^{p/2} \le \mathbb{E}|F_k|^p.$$

Hence, by Minkowski's inequality,

$$\left((\mathbb{E}|F_{k-1}|^p)^{2/p} + \delta(p)(\mathbb{E}|\Delta F_k|^p)^{2/p}\right)^{1/2} \le (\mathbb{E}|F_k|^p)^{1/p}.$$

Squaring the above inequalities and evaluating the resulting telescoping series gives

$$\sum_{k=1}^{\infty} (\mathbb{E}|\Delta F_k|^p)^{2/p} \le A(p)(\mathbb{E}|F|^p)^{2/p},$$

where $A(p) = \delta(p)^{-1}$. This completes the proof of (2.5.25). □

We turn to our first vector valued-version of the estimate (2.5.20). Let (Ω, μ) be a finite measure space and put $L^p = L^p(\Omega, \mu)$ for $0 < p \le 2$. For L^p-valued Hardy martingales we have an analogue of Theorem 2.5.8.

Theorem 2.5.9 *Let $p \in \{2, 1\} \cup \{2^{-n} : n \in \mathbb{N}\}$. There exists $A(p) < \infty$ such that for any L^p-valued Hardy martingale $F = (F_k)$,*

$$\sum_{k=1}^{n} (\mathbb{E}\|\Delta F_k\|_{L^p})^2 \le A(p)(\mathbb{E}\|F_n\|_{L^p})^2, \tag{2.5.27}$$

for any $n \in \mathbb{N}$. We have $A(2) = 1$, $A(1) = 2$, and $A(p/2) = A(p)(2^{4/p-1})$.

Proof We verify first that

$$(\mathbb{E}\|F_{k-1}\|_{L^p})^2 + \delta(p)(\mathbb{E}\|\Delta F_k\|_{L^p})^2 \leq (\mathbb{E}\|F_k\|_{L^p})^2, \tag{2.5.28}$$

for any L^p-valued Hardy martingale (F_k). The constant $\delta(p) > 0$ is the one arising in Proposition 2.5.7. To check (2.5.28) we fix $x \in \mathbb{T}^{k-1}$, and set

$$f_0 = F_{k-1}(x), \qquad f(z) = F_k(x, z), \qquad z \in \mathbb{T}.$$

Fix $\omega \in \Omega$ and apply the Haagerup–Pisier inequality (2.5.21) of Proposition 2.5.7 to the scalar-valued H^p function $\zeta \to f(\zeta, \omega)$. This gives

$$|f_0(\omega)|^2 + \delta(p)\left(\int_{\mathbb{T}} |f(\zeta, \omega) - f_0(\omega)|^p dm(\zeta)\right)^{2/p} \leq \left(\int_{\mathbb{T}} |f(\zeta, \omega)|^p dm(\zeta)\right)^{2/p}.$$

Taking expectations with respect to $\omega \in \Omega$, and using Minkowski's inequality and Fubini's theorem gives

$$\left(\|f_0\|_{L^p}^2 + \delta(p)\left(\int_{\mathbb{T}} \|f - f_0\|_{L^p} dm\right)^2\right)^{1/2} \leq \int_{\mathbb{T}} \|f\|_{L^p} dm, \tag{2.5.29}$$

which yields (2.5.28) as follows: First rewrite (2.5.29) to obtain

$$\left(\|F_{k-1}\|_{L^p}^2 + \delta(p)(\mathbb{E}_{k-1}\|\Delta F_k\|_{L^p})^2\right)^{1/2} \leq \mathbb{E}_{k-1}\|F_k\|_{L^p}. \tag{2.5.30}$$

Taking expectations on both sides gives

$$\mathbb{E}\left(\|F_k\|_{L^p}^2 + \delta(p)(\mathbb{E}_{k-1}\|\Delta F_k\|_{L^p})^2\right)^{1/2} \leq \mathbb{E}\|F_k\|_{L^p}. \tag{2.5.31}$$

Minkowski's inequality applied to the left-hand side of (2.5.31) yields

$$\left((\mathbb{E}\|F_{k-1}\|_{L^p})^2 + \delta(p)(\mathbb{E}\|\Delta F_k\|_{L^p})^2\right)^{1/2} \leq \mathbb{E}\left(\|F_k\|_{L^p}^2 + \delta(p)(\mathbb{E}_{k-1}\|\Delta F_k\|_{L^p})^2\right)^{1/2}. \tag{2.5.32}$$

Combining (2.5.31) and (2.5.32), we obtain

$$(\mathbb{E}\|F_{k-1}\|_{L^p})^2 + \delta(p)(\mathbb{E}\|\Delta F_k\|_{L^p})^2 \leq (\mathbb{E}\|F_k\|_{L^p})^2. \tag{2.5.33}$$

Summing the telescoping series resulting from (2.5.33), we get (2.5.27) with $A(p) = \delta(p)^{-1}$. □

2.6 Notes

Section 2.1 reviews the Bochner–Lebesgue spaces and basic results on vector-valued martingales based on Diestel and Uhl (1977) and Pisier (2016).

In Section 2.2, Hardy martingales arise with the probabilistic investigation of Banach spaces of analytic functions, where they appeared in the work of Bourgain (1983a), Garling (1988, 1991), and Maurey (1980). Our earliest references to vector-valued Hardy martingales are contained in the work of

Bourgain (1984b), Edgar (1985), Garling (1988), Ghoussoub, Lindenstrauss, and Maurey (1989), Haagerup and Pisier (1989), Xu (1988). The term Hardy martingales was introduced by Garling (1988). For the class of simple Hardy martingales we refer to Davis, Garling, and Tomczak-Jaegermann (1984) and Edgar (1986). For martingales associated to homogeneous polynomials, including Steinhaus martingales, we refer to Pisier (1980, 1982a, 1986).

Bochner (1939, 1959) and Helson (1958, 1959) provide a general framework of analyticity on locally compact abelian groups in terms of positive cones in the dual group. We refer also to the survey by Asmar and Hewitt (1988).

Section 2.3 is based on the work of Maurey (1980) and Pisier (1996, Section 2) and Pisier (1997). Theorem 2.3.1 is due to Maurey (1980). Lemma 2.3.2 and its proof is taken from Xu (1998), who in turn refers to unpublished communications by Maurey. Theorem 2.3.3 and the proof of Corollary 2.3.4 (Paley's theorem on lacunary Taylor coefficients) is taken from Pisier (1996).

Section 2.4 presents results of Bedrosian (1963), Bourgain (1983c, 1984b, 1984c), Garling (1991), and Pisier (1980, 1982a, 1986). Theorem 2.4.2, Theorem 2.4.3, and Theorem 2.4.4 are due to Garling (1991). Lemma 2.4.5 is originally due to Bedrosian (1963). For Proposition 2.4.6 we use Bourgain (1983c, 1984b, 1984c) as sources. The proofs presented for Theorem 2.4.7 and Theorem 2.4.8 are suggested by Bourgain (1983c, 1984b, 1984c). Subsection 2.4.2 is devoted to presenting Pisier's theorem (Theorem 2.4.12) and Theorem 2.4.15. We use Pisier (1980, 1982a, 1986) as sources.

In Section 2.5, Theorem 2.5.1 is due to Garling (1988). The nonlinear telescoping lemma (Lemma 2.5.3) is extracted from Bourgain (1983a). Kwapien kindly pointed out (in a private communication) that the origin of the method is in Bourgain (1980b, Section 3).

The square function estimate (2.5.12) of Theorem 2.5.4 is due to Bourgain (1983a). The Davis and Garsia inequality (2.5.13) of Theorem 2.5.4 is obtained in Müller (2012). Theorem 2.5.5 is included for illustrating the flexibility of nonlinear telescoping method. It contains the inequalities of Burkholder and Gundy (1970) and Lépingle (1978) respectively. See also Bourgain (1983a) and Delbaen and Schachermayer (1995) for different proofs of Lépingle's inequality. The recent work by Passenbrunner (2020) establishes Lépingle's inequality for B-spline projection operators. For extensions of martingale inequalities in nonseparable limiting cases we refer to Jiao et al. (2017) and Jiao, Xie, and Zhou (2015).

Applications of nonlinear telescoping to the construction of a new idempotent Fourier multiplier on the Hardy space $H^1(\mathbb{T} \times \mathbb{T})$ are obtained by Rzeszut and Wojciechowski (2017).

Theorem 2.5.6 is due to Bourgain (1983a). The proof of Proposition 2.5.7 is obtained, by specialization, from Haagerup and Pisier (1989).

3

Embedding L^1 in L^1/H^1_0

This chapter presents the result of Bourgain (1983a) that L^1 is isomorphic to a subspace of L^1/H^1_0. The proof – which is brutally hard – shows the existence of a nonatomic sigma-algebra Σ on $\mathbb{T}$ such that the quotient map,

$$q_\Sigma : L^1(\Sigma) \to L^1(\mathbb{T})/H^1_0(\mathbb{T}), \qquad q_\Sigma(f) = \{f + H^1_0(\mathbb{T})\},$$

is invertible on its range. By duality, the transposed operator of q_Σ is onto, and, hence, the conditional expectation $\mathbb{E}_\Sigma : H^\infty(\mathbb{T}) \to L^\infty(\Sigma)$, defined by $\mathbb{E}_\Sigma(f) = \mathbb{E}(f|\Sigma)$, is a surjective operator. Thus for any $g \in L^\infty(\Sigma)$, there exists $h \in H^\infty(\mathbb{T})$ such that $\mathbb{E}_\Sigma(h) = g$.

3.1 Hardy Martingales and Dyadic Perturbations

The norm closed subspace $H^1(\mathbb{T}^\mathbb{N}) \subset L^1(\mathbb{T}^\mathbb{N})$ consists of the L^1-bounded Hardy martingales, identified with their L^1 limits. We let $\mathcal{D}$ denote the dyadic sigma-algebra on $\mathbb{T}^\mathbb{N}$ and $L^1_\mathcal{D}(\mathbb{T}^\mathbb{N})$ the subspace of $L^1(\mathbb{T}^\mathbb{N})$ consisting of $\mathcal{D}$ measurable functions. (See Section 2.2.)

In this section, we prove Bourgain's embedding theorem (Theorem 3.1.12), which asserts that the canonical quotient map,

$$q : L^1_\mathcal{D}(\mathbb{T}^\mathbb{N}) \to L^1(\mathbb{T}^\mathbb{N})/H^1_0(\mathbb{T}^\mathbb{N}), \qquad q(D) = \{D + H^1_0(\mathbb{T}^\mathbb{N})\},$$

is invertible on its range, where we set $H^1_0(\mathbb{T}^\mathbb{N}) = \{F \in H^1(\mathbb{T}^\mathbb{N}) : \mathbb{E}(F) = 0\}$. This amounts to showing that there exists $A_0 < \infty$ such that

$$\|D\|_{L^1(\mathbb{T}^\mathbb{N})} \le A_0 \inf\left\{\|D - F\|_{L^1(\mathbb{T}^\mathbb{N})} : F \in H^1(\mathbb{T}^\mathbb{N}),\ \mathbb{E}(F) = 0\right\}, \qquad (3.1.1)$$

for each $D \in L^1_\mathcal{D}(\mathbb{T}^\mathbb{N})$. It takes Sections 3.1.1 through 3.1.5 to establish the single estimate (3.1.1). In Section 3.1.1 and 3.1.2 we develop the complex analytic components of its proof. Section 3.1.3 contains a suitably adapted version of the Garnett–Jones inequality. Section 3.1.4 presents the initial sine-cosine decomposition of Hardy martingales and provides the crucial connection between the L^1-embedding problem and the Garnett–Jones theorem (Theorem 3.1.10). We complete the proof of the embedding estimates (3.1.1) in Section 3.1.5.

3.1.1 Davis Decomposition

We recall that we let $\mathcal{D}$ denote the dyadic sigma-algebra on $\mathbb{T}^{\mathbb{N}}$. Let $\sigma\colon \mathbb{T} \to \{-1, +1\}$ be defined by

$$\sigma(\zeta) = \operatorname{sign}(\Re\zeta), \qquad \zeta \in \mathbb{T}.$$

Given $k \in \mathbb{N}$, let $\mathcal{D}_k$ denote the finite algebra of subsets in $\mathbb{T}^{\mathbb{N}}$ generated by $\sigma_1, \ldots, \sigma_k$, where we defined $\sigma_k\colon \mathbb{T}^{\mathbb{N}} \to \{-1, 1\}$, by

$$\sigma_k(x) = \sigma(\zeta_k),$$

with $x = (\zeta_k)$. Denoting by $\mathbb{P}_{\mathcal{D}}$ the restriction of the Haar measure $\mathbb{P}$ on $\mathbb{T}^{\mathbb{N}}$ to $\mathcal{D}$, we obtain the filtered probability space $(\mathbb{T}^{\mathbb{N}}, (\mathcal{D}_k), \mathbb{P}_{\mathcal{D}})$.

The main result in this section is Theorem 3.1.3, which determines the extent to which the Davis and Garsia inequality in Theorem 2.5.4 is stable under dyadic perturbation. The first step is provided by the decomposition theorem for dyadic perturbations of Hardy martingales.

Theorem 3.1.1 *For every Hardy martingale $F = (F_k)$ and dyadic martingale $D = (D_k)$ there exists a Hardy martingale $G = (G_k)$ so that*

$$|\Delta G_k| \le C_0 |F_{k-1} - D_{k-1}| \tag{3.1.2}$$

and

$$\mathbb{E}|F_{k-1} - D_{k-1}| + \frac{1}{8}\mathbb{E}|\Delta B_k| \le \mathbb{E}|F_k - D_k|, \tag{3.1.3}$$

where $B = F - G$ and $C_0 \le 24$.

Proof Fix $k \le n$. We condition to $\mathcal{F}_{k-1}$. Thus we fix $x \in \mathbb{T}^{k-1}$ to define $z, b \in \mathbb{C}$, and $h \in H_0^1(\mathbb{T})$ by putting

$$z = F_{k-1}(x) - D_{k-1}(x), \quad h(y) = \Delta F_k(x, y), \quad b\sigma(y) = \Delta D_k(x, y), \qquad y \in \mathbb{T},$$

where, we recall, $\sigma(z) = \operatorname{sign}(\Re z)$. Lemma 3.1.2 applied to z, h, and b asserts that there exists $g \in H_0^\infty(\mathbb{T})$ such that defining

$$\Delta G_k(x, y) = g(y) \quad \text{and} \quad \Delta B_k(x, y) = h(y) - g(y)$$

yields a decomposition of F into the sum of Hardy martingales G and B satisfying estimates (3.1.2) and (3.1.3), with $C_0 \le 24$. □

Remark: The Davis decomposition for Hardy martingales follows from Theorem 3.1.1 by specialization. Indeed, given a Hardy martingale F and applying Theorem 3.1.1 with $D = 0$ yields a Hardy martingale G such that

$$|\Delta G_k| \leq C_0|F_{k-1}| \quad \text{and} \quad \sum_{k=1}^{\infty} \mathbb{E}|\Delta B_k| \leq 8 \sup_{k\in\mathbb{N}} \mathbb{E}|F_k|, \tag{3.1.4}$$

where $B = F - G$ and $C_0 \leq 24$.

Lemma 3.1.2 *For $h \in H^1_0(\mathbb{T})$ and $z \in \mathbb{C}$ there exists $g \in H^\infty_0(\mathbb{T})$ such that $|g| \leq C_0|z|$, and*

$$|z| + \frac{1}{8}\int_{\mathbb{T}} |h(\zeta) - g(\zeta)|dm(\zeta) \leq \int_{\mathbb{T}} |z + h(\zeta) - b\sigma(\zeta)|dm(\zeta), \tag{3.1.5}$$

for any $b \in \mathbb{C}$, where $\sigma(\zeta) = \operatorname{sign}(\Re\zeta)$. The constant satisfies $C_0 \leq 24$.

Proof The proof consists of probabilistic and complex analytic components. We let (z_t) denote complex Brownian motion defined on Wiener's filtered probability space Ω, and we set

$$\tau = \inf\{t > 0 : |z_t| > 1\}.$$

Recall that in Section 1.6, we introduced Doob's contractive projection $N: L^p(\Omega) \to L^p(\mathbb{T})$, $1 \leq p \leq \infty$. By Theorem 1.6.4, it maps holomorphic random variables into the Hardy space $H^p(\mathbb{T})$. Thus $N: H^p(\Omega) \to H^p(\mathbb{T})$ is a contraction for $1 \leq p \leq \infty$.

Fix $h \in H^1_0(\mathbb{T})$ and $z \in \mathbb{C}$. We first define $g \in H^1_0(\mathbb{T})$ and thereafter, by integrating against properly chosen testing functions, we get lower estimates for

$$\int_{\mathbb{T}} |z + h - b\sigma|dm.$$

Next, put $\rho = \inf\{t < \tau : |h(z_t)| > C_0|z|\}$ and $A = \{\rho < \tau\}$. See Figure 3.1. Then for $g = N(h(z_\rho))$, we have

$$\mathbb{E}(|h(z_\tau)1_A|) \geq \frac{1}{2}\int_{\mathbb{T}} |h - g|dm, \tag{3.1.6}$$

where $g \in H^\infty_0(\mathbb{T})$ and $\|g\|_\infty \leq C_0|z|$. We will distinguish between the cases $|b| \leq 8|z|$ and $|b| \geq 8|z|$. In the first case we exploit outer functions and Havin's lemma. The second case uses simple exponentials instead.

Case $|b| \leq 8|z|$: Define $p_0(\zeta) = N(1_A)(\zeta) + N(1_A)(\overline{\zeta})$ and put $p = p_0/4$. Hence $p(\zeta) = p\left(\overline{\zeta}\right)$ and $0 \leq p \leq 1/2$. By the covariance formula, $\int pdm = \mathbb{P}(A)/2$, and

$$\int |h|pdm = (1/2)\mathbb{E}(|h(z_\tau 1_A|).$$

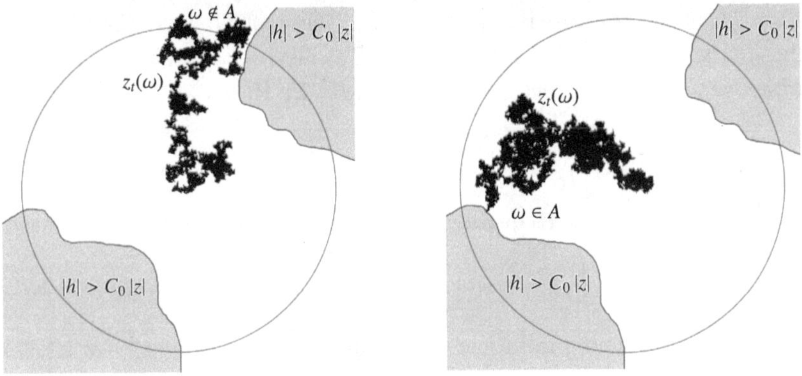

Figure 3.1 The set $A = \{\rho < \tau\}$. The left-hand side depicts a Brownian path $(z_t(\omega))_t$ for which $\rho(\omega) = \infty$ and $\omega \notin A$. The Brownian path on the right-hand side gives $\rho(\omega) < \infty$ and $\omega \in A$.

Combining this with $|h(z_\rho)1_A| = C_0|z|1_A$ and using that $|b| \le 8|z|$, we obtain

$$\int_{\mathbb{T}} |z + h - b\sigma| p dm \ge \left(\frac{1}{2} - \frac{9}{2C_0}\right) \mathbb{E}(|h(z_\tau)1_A|). \tag{3.1.7}$$

Let $q = \exp(\ln(1-p) + iH\ln(1-p))$, where H denotes the Hilbert transform for $\mathbb{T}$. Clearly, $q \in H^\infty(\mathbb{T})$ is an outer function and $\int hqdm = 0$ since $h \in H^1(\mathbb{T})$. Put $q_2 = \Im q$ and $q_1 = \Re q$. Then by inspection, $q_2(\zeta) = q_2(\overline{\zeta})$, $\int q_2 dm = 0$, and $\int q_2\sigma dm = 0$. Consequently,

$$\int_{\mathbb{T}} (z + h - b\sigma) q dm = \int_{\mathbb{T}} (z - b\sigma) q_1 dm.$$

Below, we will verify the crucial estimate

$$\int_{\mathbb{T}} q_1 dm > 1 - C_1 \mathbb{P}(A) \tag{3.1.8}$$

with $C_1 = 3$. Since $|b| \le 8|z|$ we now get,

$$\int_{\mathbb{T}} |z + h - b\sigma| \cdot |q| dm \ge \left| \int_{\mathbb{T}} (z + h - b\sigma) q dm \right| \ge |z| - 9C_1|z|\mathbb{P}(A). \tag{3.1.9}$$

Since $p + |q| = 1$, adding estimates (3.1.7) and (3.1.9) gives

$$\int_{\mathbb{T}} |z + h - b\sigma| dm \ge |z| + \left(\frac{1}{2} - \frac{9}{2C_0} - \frac{9C_1}{C_0}\right) \mathbb{E}(|h(z_\tau)1_A|). \tag{3.1.10}$$

Finally, choose $C_0 \ge 8C_1$ so that $(1/2 - (9/2 + 9C_1)/C_0) > 1/4$ and combine the inequalities (3.1.10) and (3.1.6) to obtain (3.1.5).

Case $|b| > 8|z|$: This case is straightforward. The testing functions involved are the simple exponentials. As $\sigma(\zeta) = \operatorname{sign}(\Re\zeta)$, we have $\int \sigma(\zeta)\zeta dm(\zeta) = 2/\pi$. Moreover, $\int (z + h(\zeta))\zeta dm(\zeta) = 0$ since $h \in H^1(\mathbb{T})$. Hence

$$\int_{\mathbb{T}} |z + h(\zeta) - b\sigma(\zeta)| dm(\zeta) \geq 2|b|/\pi. \tag{3.1.11}$$

Next, using the triangle inequality and invoking $|b| > 8|z|$ gives the hypothesis

$$\int_{\mathbb{T}} |z + h - b\sigma| dm \geq \int_{\mathbb{T}} |h| dm - 9|b|/8. \tag{3.1.12}$$

Let $\alpha = 1/4$. Forming the convex combination $(1 - \alpha)(3.1.11) + \alpha(3.1.12)$, using that $|z| \leq |b|/8$, and that $\int |h - g| dm \leq 2 \int |h| dm$, gives (3.1.5).

Verification of (3.1.8): Recall that $q_1 = (1 - p)\cos(H(\ln(1 - p))$. Inserting the pointwise estimate $\cos(x) \geq 1 - x^2/2$ into the defining formula for q_1 gives the lower bound

$$q_1 \geq (1 - p) - \frac{1}{2}(H(\ln(1 - p)))^2. \tag{3.1.13}$$

We thus reduced the lower L^1 estimate for q_1 to an upper L^2 estimate for the Hilbert transform. Since $0 \leq p \leq 1/2$, we have $(\ln(1 - p))^2 \leq 2p$, and hence,

$$\int_{\mathbb{T}} (H(\ln(1 - p)))^2 dm \leq 2 \int_{\mathbb{T}} (\ln(1 - p))^2 dm \leq 4 \int_{\mathbb{T}} p dm. \tag{3.1.14}$$

Combining estimates (3.1.13) and (3.1.14) gives (3.1.8). □

We next prove the Davis and Garsia inequality for dyadic perturbations of the Hardy martingale. In its proof we exploit Theorem 3.1.1 and the nonlinear telescoping lemma (Lemma 2.5.3).

Theorem 3.1.3 *Let F be an integrable Hardy martingale and $D_k = \mathbb{E}(F_k|\mathcal{D})$. Then there exist Hardy martingales G and B with $F = G + B$, and an adapted unimodular sequence $W = (w_k)$ such that*

$$y_k = \Im(w_{k-1}\Delta(G_k - D_k))$$

satisfies

$$\mathbb{E}\left(\sum_{k=1}^{n} \mathbb{E}_{k-1}|y_k|^2\right)^{1/2} \leq C(\mathbb{E}|F_n - D_n|)^{1/2}\left(\mathbb{E}\max_{1\leq k\leq n} |F_k - D_k|\right)^{1/2}, \tag{3.1.15}$$

and

$$\sum_{k=1}^{n} \mathbb{E}|\Delta B_k| \leq C\mathbb{E}|F_n - D_n|. \tag{3.1.16}$$

Proof We apply Theorem 3.1.1 to the Hardy martingale $F = (F_k)_{k=1}^n$ and its dyadic projection $D = (D_k)_{k=1}^n$, where $D_k = \mathbb{E}(F_k|\mathcal{D})$. We thus obtain Hardy martingales G and B with $F = G + B$, satisfying estimates (3.1.2) and (3.1.3), respectively.

To check (3.1.16), it suffices to evaluate the sum of the telescoping series defined by (3.1.3). Next, use the estimate (3.1.2), fix $k \le n$, and write

$$z = F_{k-1} - D_{k-1}, \quad w = \bar{z}/|z|, \quad g = \Delta G_k, \quad \text{and} \quad b = D_k.$$

Apply Lemma 3.1.4 below with the above parameters, and put

$$w_{k-1} = w \quad \text{and} \quad y_k = \Im(w_{k-1}\Delta(G_k - D_k)).$$

This yields

$$\mathbb{E}\left(|F_{k-1} - D_{k-1}|^2 + \alpha^2\mathbb{E}_{k-1}|y_k|^2\right)^{1/2} \le \mathbb{E}|F_k - D_k| + \mathbb{E}|\Delta B_k|.$$

Applying the nonlinear telescoping estimate (2.5.11) and taking into account (3.1.16), we arrive at (3.1.15). □

Lemma 3.1.4 *Let $C \ge 1$. Let $b, z \in \mathbb{C}$, and $w = \bar{z}/|z|$. If $g \in L_0^\infty(\mathbb{T})$ with $|g| \le C|z|$ and $y = \Im(w \cdot (g - b\sigma))$, then*

$$\left(|z|^2 + \alpha^2 \int_{\mathbb{T}} y^2\right)^{1/2} \le \int_{\mathbb{T}} |z + g - b\sigma| dm, \tag{3.1.17}$$

where $\alpha = \alpha(C)$.

Proof Put $A = 4C$. Given $b \in \mathbb{C}$, we distinguish between the cases $|b| \le A|z|$ and $|b| \ge A|z|$.

Case $|b| \le A|z|$: By scaling and rotation we may assume that $z = w = 1$. Since $\int g - b\sigma dm = 0$, for any $0 < \tau < 1$ we have

$$\int_{\mathbb{T}} |1 + g - b\sigma| dm \ge \int_{\mathbb{T}} |1 + \tau(g - b\sigma)| dm. \tag{3.1.18}$$

Write $g - b\sigma = u + iy$ where u, y are real valued. We have $|u + iv| \le (C + A)|z|$, by the assumption of the present case. Choose $\tau < (2A + 2C)^{-1}$. Then rewrite

$$|1 + \tau(g - b\sigma)| = ((1 + \tau u)^2 + \tau^2 y^2)^{1/2}. \tag{3.1.19}$$

Put $\alpha^2 = (\tau^{-1} + A)^{-2}/6$. Hence, by arithmetic (3.1.19) gives

$$|1 + \tau(g - b\sigma)| = (1 + \tau u)\left(1 + \frac{\tau^2 y^2}{(1 + \tau u)^2}\right)^{1/2} \ge 1 + \tau u + \alpha^2 y^2. \tag{3.1.20}$$

Since $g - b\sigma$ has vanishing mean, we have $\int u dm = 0$. Taking the expectation in (3.1.20) and using estimate (3.1.18) gives

$$\int_{\mathbb{T}} |1 + g - b\sigma| dm \geq 1 + \alpha^2 \int_{\mathbb{T}} y^2 dm. \tag{3.1.21}$$

Finally, since the right-hand side is ≥ 1 we may replace it by its square root, preserving the inequality (3.1.21).

Case $|b| \geq A|z|$: In this case we exploit that b majorizes the terms around it. Applying a rotation, we may assume that $w = 1$. Define $q = -\sigma b/|b|$. Since $|q| = 1$,

$$\int_{\mathbb{T}} |z + g - b\sigma| dm \geq \int_{\mathbb{T}} (z + g - b\sigma)\overline{q} dm \geq |b| - \int_{\mathbb{T}} |g| dm.$$

Since $A = 4C$, $|g| \leq C|z|$ and $|b| \geq A|z|$, and taking $\alpha = (1 + C/A)^{-1}$ gives

$$|b| \geq \alpha \left(\int |g - b\sigma|^2 dm \right)^{1/2} \quad \text{and} \quad |b| - \int |g| dm \geq |z| + |b|A^{-1}.$$

Summing up,

$$\int_{\mathbb{T}} |z + g - b\sigma| dm \geq |z| + \alpha \left(\int_{\mathbb{T}} |g - b\sigma|^2 dm \right)^{1/2},$$

where $\alpha = \alpha(C, A)$. Thus we verified the estimate (3.1.17) in the case $|b| \geq A|z|$. □

Summary and Outlook

The assertion of Theorem 3.1.3 may be written more concisely in terms of the martingale norms $\mathcal{P} = \mathcal{P}(\mathbb{T}^{\mathbb{N}}, (\mathcal{F}_k), \mathbb{P})$ and $H^1 = H^1(\mathbb{T}^{\mathbb{N}}, (\mathcal{F}_k), \mathbb{P})$ introduced in Section 1.1.

Recall that the conditional square function of a martingale $X = (X_k)$, defined by $\left(\sum \mathbb{E}_{k-1}|\Delta X_k|^2\right)^{1/2}$ gives rise to the $\mathcal{P}$-norm of X,

$$\|X\|_{\mathcal{P}} = \left\| \left(|X_0|^2 + \sum \mathbb{E}_{k-1} |\Delta X_k|^2 \right)^{1/2} \right\|_1, \tag{3.1.22}$$

where $\Delta X_k = X_k - X_{k-1}$ and $X_0 = \mathbb{E}X$. Compare that to the martingale H^1-norm defined in terms of the martingale square function,

$$\|X\|_{H^1} = \left\| \left(|X_0|^2 + \sum |\Delta X_k|^2 \right)^{1/2} \right\|_1. \tag{3.1.23}$$

Theorem 2.5.5 (the Burkholder–Gundy inequality) asserts that $\|X\|_{H^1} \leq 2\|X\|_{\mathcal{P}}$. Moreover, by Theorem 1.1.6 (Davis's theorem) we have

$$\mathbb{E} \sup_{k<\infty} |X_k| \leq C_1 \|X\|_{H^1}, \tag{3.1.24}$$

where $C_1 < \infty$ is independent of $X = (X_k)$. For any bounded $(\mathcal{F}_k)$-adapted sequence $W = (w_k)$, we define a martingale transform operator acting on $(\mathcal{F}_k)$-martingales $X = (X_k)$ by

$$T_W(X) = \Im\left[\sum w_{k-1}\Delta X_k\right]. \tag{3.1.25}$$

Clearly, T_W defines a bounded operator on $\mathcal{P}$ for any bounded weight W.

The assertions of Theorem 3.1.3 may be expressed in terms of the martingale spaces H^1, $\mathcal{P}$, and the martingale transform T_W. Indeed, if F is a Hardy martingale, and $F = G + B$, together with $W = (w_k)$ is determined by Theorem 3.1.3, then the estimate (3.1.15) just states that

$$\|T_W(G - D)\|_{\mathcal{P}} \le C\|F - D\|_{H^1}^{1/2}\|F - D\|_1^{1/2}, \tag{3.1.26}$$

where $D = \mathbb{E}(F|\mathcal{D})$.

Consequences of (3.1.15) and (3.1.16):

(i) First, $T_W(G) = T_W(G - D) + T_W(D)$. By estimates (2.2.19) and (2.5.12), $\|T_W(D)\|_{\mathcal{P}} \le \|D\|_{H^1} \le C\|F\|_1$, and by estimates (3.1.15) and (2.2.8), $\|T_W(G - D)\|_{\mathcal{P}} \le C\|F\|_1$. Thus (3.1.15) implies the estimate

$$\|T_W(G)\|_{\mathcal{P}} \le C\|F\|_1. \tag{3.1.27}$$

(ii) Since G is a Hardy martingale, $\mathbb{E}_{k-1}|\Delta G_k|^2 = 2\mathbb{E}_{k-1}|\Im(w_{k-1} \cdot \Delta G_k)|^2$, and hence, $\|G\|_{\mathcal{P}} = \sqrt{2}\|T_W(G)\|_{\mathcal{P}}$. (See also the related identity (3.1.34).)

(iii) Since $\mathbb{E}(F|\mathcal{D}) - \mathbb{E}(G|\mathcal{D}) = \mathbb{E}(F - G|\mathcal{D})$, the triangle inequality in L^1 and (3.1.16) give

$$\|\mathbb{E}(F|\mathcal{D}) - \mathbb{E}(G|\mathcal{D})\|_1 \le \sum_{k=1}^{\infty} \|\Delta_k(F - G)\|_1 \le C\|F - \mathbb{E}(F|\mathcal{D})\|_1. \tag{3.1.28}$$

Since the dyadic filtration $(\mathcal{D}_k)$ is interwined to $(\mathcal{F}_k)$ by $\mathcal{D}_k = \mathcal{F}_k \cap \mathcal{D}$, we have $\mathbb{E}_{k-1}|\Delta_k(\mathbb{E}(F|\mathcal{D}) - \mathbb{E}(G|\mathcal{D}))|^2 = |\mathbb{E}(\Delta_k(F - G)|\mathcal{D})|^2$ and hence,

$$\|\mathbb{E}(F|\mathcal{D}) - \mathbb{E}(G|\mathcal{D})\|_{\mathcal{P}} \le \sum_{k=1}^{\infty} \|\Delta_k(F - G)\|_1 \le C\|F - \mathbb{E}(F|\mathcal{D})\|_1. \tag{3.1.29}$$

(iv) Next, write $G - \mathbb{E}(G|\mathcal{D}) = F - \mathbb{E}(F|\mathcal{D}) + (G - F) - \mathbb{E}(G - F|\mathcal{D})$. Then the triangle inequality in L^1 followed by the estimates (3.1.16) and (3.1.28) gives

$$\begin{aligned}\|G - \mathbb{E}(G|\mathcal{D})\|_1 &\le \|F - \mathbb{E}(F|\mathcal{D})\|_1 + \sum_{k=1}^{\infty} \|\Delta_k(F - G)\|_1 + C\|F - \mathbb{E}(F|\mathcal{D})\|_1 \\ &\le (1 + 2C)\|F - \mathbb{E}(F|\mathcal{D})\|_1.\end{aligned} \tag{3.1.30}$$

Outlook: Sections 3.1.2– 3.1.4 are devoted to proving the martingale estimate (3.1.31): For any adapted sequence $W = (w_k)$ satisfying $|w_k| = 1$ and any Hardy martingale G,

$$\|G\|_{L^1} \le C\left(\|G - D\|_{L^1}^{1/2}\|G\|_{\mathcal{P}}^{1/2} + \|T_W(G - D)\|_{\mathcal{P}}^{1/4}\|G\|_{\mathcal{P}}^{3/4}\right), \tag{3.1.31}$$

where $D = \mathbb{E}(G|\mathcal{D})$. In Section 3.1.5 we will deduce the estimate (3.1.1) using estimates (3.1.31), (3.1.27), (3.1.26), and (3.1.16).

3.1.2 Cosine Martingales

We begin by associating to any Hardy martingale G its cosine part U, defined by (3.1.33), and we observe in (3.1.36) that the predictable martingale norm of U coincides with that of $T_W(G)$. Theorem 3.1.8 is the main result of this section; it determines the extent to which the estimate (3.1.36) is stable under dyadic perturbations.

Let $h \in H^2(\mathbb{T})$, with $\int_{\mathbb{T}} h = 0$, and put $u(z) = (h(z) + h(\overline{z}))/2$ to denote the even part of h. Then by Fourier expansion and Plancherel's formula, we have the elementary identity,

$$\frac{1}{2}\int_{\mathbb{T}} |h|^2 dm = \int_{\mathbb{T}} |u|^2 dm = \int_{\mathbb{T}} |\Im h|^2 dm. \tag{3.1.32}$$

In this section we need to investigate how this identity transforms under perturbation by $\sigma(e^{i\theta}) = \text{sign}(\cos(\theta))$, and we will record the consequences for Hardy martingales. The resulting expressions will be obtained by elementary methods.

To begin, we let $G = (G_k)$ be a Hardy martingale and let $U = (U_k)$ be its cosine part given by

$$U_k(x, y) = [G_k(x, y) + G_k(x, \overline{y})] / 2, \tag{3.1.33}$$

where $x \in \mathbb{T}^{k-1}$ and $y \in \mathbb{T}$. Since the Hilbert transform, applied to functions with vanishing mean, preserves the L^2 norm, we have

$$\mathbb{E}_{k-1}|\Delta G_k|^2 = 2\mathbb{E}_{k-1}|\Delta U_k|^2 = 2\mathbb{E}_{k-1}|\Im w_{k-1}\Delta G_k|^2, \tag{3.1.34}$$

for each adapted sequence $W = (w_k)$ with $|w_k| = 1$, and, consequently,

$$\left\|\left(\sum \mathbb{E}_{k-1}|\Delta U_k|^2\right)^{1/2}\right\|_1 = \left\|\left(\sum \mathbb{E}_{k-1}|\Im w_{k-1}\Delta G_k|^2\right)^{1/2}\right\|_1. \tag{3.1.35}$$

Using the $\mathcal{P}$-norm, defined in equation (3.1.22), of U and $T_W(G)$, we may rewrite the expression (3.1.35) as follows,

$$\|U\|_{\mathcal{P}} = \|T_W(G)\|_{\mathcal{P}}. \tag{3.1.36}$$

The principal result of the present section is Theorem 3.1.8, which asserts that (3.1.36) is stable under dyadic perturbations as follows: For any adapted sequence $W = (w_k)$ satisfying $|w_k| = 1$,

$$\|U - \mathbb{E}(U|\mathcal{D})\|_{\mathcal{P}} \leq C\|T_W(G - \mathbb{E}(G|\mathcal{D}))\|_{\mathcal{P}}^{1/2}\|G\|_{\mathcal{P}}^{1/2}, \tag{3.1.37}$$

where

$$T_W(G - \mathbb{E}(G|\mathcal{D})) = \Im\left[\sum w_{k-1}\Delta(G - \mathbb{E}(G|\mathcal{D}))_k\right].$$

By means of Proposition 3.1.5 and Proposition 3.1.7 we determine the effect of dyadic perturbation for each martingale difference separately. In the proof of Theorem 3.1.8, we merge their content and derive the perturbation estimate (3.1.37).

Recall that we put $\sigma(\zeta) = \text{sign}(\Re\zeta)$, $\zeta \in \mathbb{T}$. Subsequently, given an integrable $u\colon \mathbb{T} \to \mathbb{C}$, we denote

$$\langle u, \sigma\rangle = \int_{\mathbb{T}} u(\zeta)\sigma(\zeta)dm(\zeta).$$

Proposition 3.1.5 *Let $h \in H_0^2(\mathbb{T})$ and $u(z) = (h(z) + h(\bar{z}))/2$ be the even part of h. Then for $w, b \in \mathbb{C}$, with $|w| = 1$,*

$$\Im^2(w\cdot(\langle u,\sigma\rangle - b)) + \Re^2(w\cdot\langle u,\sigma\rangle) + \int_{\mathbb{T}} |u - \langle u,\sigma\rangle\sigma|^2 dm = \int_{\mathbb{T}} \Im^2(w\cdot(h - b\sigma))dm.$$

Proof Let $L_G^2(\mathbb{T})$ consisting of those $f \in L^2(\mathbb{T})$ for which $f(y) = f(\bar{y})$, for almost every $y \in \mathbb{T}$. We let $w_0 = 1_{\mathbb{T}}$, $w_1 = \sigma$, and choose $\{w_k : k \geq 2\}$ such that $\{w_k : k \geq 0\}$ is an orthonormal basis for $L_G^2(\mathbb{T})$. Let H be the Hilbert transfom. Then $\{w_k, Hw_k : k \geq 0\}$ forms an orthonormal basis in $L^2(\mathbb{T})$. Moreover, in the Hardy space $H^2(\mathbb{T})$ the analytic system

$$\{(w_k + iHw_k) : k \geq 0\}$$

is orthogonal, complete, and $\|w_k + iHw_k\|_2 = \sqrt{2}$ for $k \geq 1$. Fix $h \in H_0^2(\mathbb{T})$ and $w, b \in \mathbb{C}$, with $|w| = 1$. Replacing h by wh and b by wb shows that it suffices to prove the proposition with $w = 1$. Since $\int_{\mathbb{T}} udm = 0$, we have

$$u = \sum_{n=1}^{\infty} c_n w_n.$$

Apply the Hilbert transform and rearrange terms to get

$$h - b\sigma = (c_1 - b)\sigma + ic_1 H\sigma + \sum_{n=2}^{\infty} c_n(w_n + iHw_n). \tag{3.1.38}$$

Then, taking imaginary parts gives

$$\Im(h - b\sigma) = \Im(c_1 - b)\sigma + (\Re c_1)H\sigma + \sum_{n=2}^{\infty} \Im c_n w_n + \Re c_n H w_n. \tag{3.1.39}$$

By orthogonality, the identity (3.1.39) yields

$$\int_{\mathbb{T}} \Im^2(h - b\sigma)dm = \Im^2(c_1 - b) + \Re^2 c_1 + \sum_{n=2}^{\infty} |c_n|^2. \tag{3.1.40}$$

On the other hand, since $\int_{\mathbb{T}} u dm = 0$, $c_1 = \langle u, \sigma\rangle$, and $w_1 = \sigma$, we get

$$\int_{\mathbb{T}} |u - \langle u, \sigma\rangle\sigma|^2 dm = \sum_{n=2}^{\infty} |c_n|^2. \tag{3.1.41}$$

Comparing equations (3.1.40) and (3.1.41) completes the proof. □

The proof of Proposition 3.1.7 below uses two elementary inequalities that we present first.

Lemma 3.1.6 *Let $\mu, b \in \mathbb{C}$ and*

$$|\mu| + \frac{|\mu - b|^2}{|\mu| + |b|} = a. \tag{3.1.42}$$

Then for any $w \in \mathbb{T}$,

$$(a - |b|)^2 \leq 4\left(\Im^2(w \cdot (\mu - b)) + \Re^2(w \cdot \mu)\right), \tag{3.1.43}$$

and

$$|\mu - b|^2 \leq 2(a^2 - |\mu|^2). \tag{3.1.44}$$

Proof By rotational invariance it suffices to prove estimate (3.1.43) for $w = 1$. Let $\mu = m_1 + im_2$ and $b = b_1 + ib_2$. By definition (3.1.42), we have

$$a - |b| = \frac{|\mu|^2 - |b|^2 + |\mu - b|^2}{|\mu| + |b|}.$$

Expand and regroup the terms in the numerator to obtain

$$|\mu|^2 - |b|^2 + |\mu - b|^2 = 2m_1(m_1 - b_1) + 2m_2(m_2 - b_2). \tag{3.1.45}$$

By the Cauchy–Schwarz inequality, the right-hand side (3.1.45) is bounded by

$$2\left(m_1^2 + (m_2 - b_2)^2\right)^{1/2}\left(m_2^2 + (m_1 - b_1)^2\right)^{1/2}.$$

Note that $m_1 = \Re\mu$ and $m_2 - b_2 = \Im(\mu - b)$. It remains to observe that

$$\left(m_2^2 + (m_1 - b_1)^2\right)^{1/2} \leq |\mu| + |b|,$$

or, equivalently,

$$m_1^2 + m_2^2 - 2m_1b_2 + b_1^2 \leq |\mu|^2 + 2|\mu||b| + |b|^2,$$

which is obviously true.

Next we turn to verifying the estimate (3.1.44). We have $a^2 - |\mu|^2 = (a + |\mu|)(a - |\mu|)$, and hence,

$$a^2 - |\mu|^2 = \left[2|\mu| + \frac{|\mu - b|^2}{|\mu| + |b|}\right]\frac{|\mu - b|^2}{|\mu| + |b|}. \tag{3.1.46}$$

In view of (3.1.46), we get the estimate (3.1.44) by showing that

$$2|\mu|^2 + 2|\mu||b| + |\mu - b|^2 \geq \frac{1}{2}(|\mu| + |b|)^2. \tag{3.1.47}$$

The left-hand side of (3.1.47) is larger than $|\mu|^2 + |b|^2$ while the right-hand side is smaller $|\mu|^2 + |b|^2$. □

Next, we merge the inequalities of Lemma 3.1.6 and the identity obtained in Proposition 3.1.5.

Proposition 3.1.7 *There exists $C_0 > 0$ so that the following holds. Let w, $b \in \mathbb{C}$, with $|w| = 1$, $h \in H_0^2(\mathbb{T})$. Let u be the even part of h and put*

$$|\langle u, \sigma\rangle| + \frac{|\langle u, \sigma\rangle - b|^2}{|b| + |\langle u, \sigma\rangle|} = a.$$

Then,

$$\int_{\mathbb{T}} |u - b\sigma|^2 dm(y) \leq 8\left(a^2 - |\langle u, \sigma\rangle|^2\right) + \int_{\mathbb{T}} |u - \langle u, \sigma\rangle\sigma|^2 dm(y), \tag{3.1.48}$$

and

$$(a - |b|)^2 + \int_{\mathbb{T}} |u - \langle u, \sigma\rangle\sigma|^2 dm(y) \leq 8\int_{\mathbb{T}} \Im^2(w \cdot (h - b\sigma))dm(y). \tag{3.1.49}$$

Proof Put

$$J^2 = \int_{\mathbb{T}} \Im^2(w \cdot (h - b\sigma))dm(y). \tag{3.1.50}$$

The proof exploits the basic identities (3.1.52) and (3.1.51) for the integral J^2 and $\int_{\mathbb{T}} |u - b\sigma|^2 dm(y)$ and intertwines those using arithmetic as in the estimates (3.1.42)–(3.1.44).

Use the straightforward identity,

$$\int_{\mathbb{T}} |u - b\sigma|^2 dm(y) = |\langle u, \sigma\rangle - b|^2 + \int_{\mathbb{T}} |u - \langle u, \sigma\rangle\sigma|^2 dm(y). \tag{3.1.51}$$

Applying the estimate (3.1.44) gives $|\langle u,\sigma\rangle - b|^2 \le 8(a^2 - |\langle u,\sigma\rangle|^2)$. Hence, the identity (3.1.51) gives

$$\int_{\mathbb{T}} |u - b\sigma|^2 dm(y) \le 8(a^2 - |\langle u,\sigma\rangle|^2) + \int_{\mathbb{T}} |u - \langle u,\sigma\rangle\sigma|^2 dm(y).$$

Proposition 3.1.5 asserts that

$$\Im^2(w \cdot (\langle u,\sigma\rangle - b)) + \Re^2(w \cdot \langle u,\sigma\rangle) + \int_{\mathbb{T}} |u - \langle u,\sigma\rangle\sigma|^2 dm(y) = J^2. \quad (3.1.52)$$

Apply the estimate (3.1.43) with $\mu = \langle u,\sigma\rangle$ to the left-hand side in (3.1.52). This gives

$$(a - |b|)^2 + \int_{\mathbb{T}} |u - \langle u,\sigma\rangle\sigma|^2 dm(y) \le 8J^2. \qquad \square$$

Finally, we translate the inequalities (3.1.48) and (3.1.49) into a single martingale estimate.

Theorem 3.1.8 *Let $G = (G_k)_{k=1}^n$ be a Hardy martingale and let $U = (U_k)_{k=1}^n$ be its cosine martingale given by equation (3.1.33). Then, for any adapted sequence $W = (w_k)$ satisfying $|w_k| = 1$, we have*

$$\|U - \mathbb{E}(U|\mathcal{D})\|_{\mathcal{P}} \le C\|T_W(G - \mathbb{E}(G|\mathcal{D}))\|_{\mathcal{P}}^{1/2}\|G\|_{\mathcal{P}}^{1/2}, \quad (3.1.53)$$

where T_W is the martingale transform operator

$$T_W(G - \mathbb{E}(G|\mathcal{D})) = \Im\left[\sum w_{k-1}\Delta(G_k - \mathbb{E}(G_k|\mathcal{D}))\right].$$

Proof Let $\{g_k\}$ be the martingale difference sequence of the Hardy martingale $G = (G_k)$, and let $\{u_k\}$ be the martingale difference sequence of the associated cosine martingale $U = (U_k)$. Set

$$b_k = \mathbb{E}_{\mathcal{D}}\mathbb{E}_{k-1}(u_k\sigma_k). \quad (3.1.54)$$

By equations (2.2.21) and (2.2.18) we get $\mathbb{E}_{\mathcal{D}}(u_k) = b_k\sigma_k$. The conditional Minkowski's inequality (2.2.19) gives

$$\mathbb{E}\mathbb{E}_{\mathcal{D}}\left(\sum_{k=1}^{\infty} |\mathbb{E}_{k-1}(u_k\sigma_k)|^2\right)^{1/2} \ge \mathbb{E}\left(\sum_{k=1}^{\infty} |\mathbb{E}_{\mathcal{D}}\mathbb{E}_{k-1}(u_k\sigma_k)|^2\right)^{1/2}. \quad (3.1.55)$$

Step 1 Let $Y^2 = \sum_{k=1}^{\infty} |\mathbb{E}_{k-1}(u_k\sigma_k)|^2$ and $Z^2 = \sum_{k=1}^{\infty} |b_k|^2$. We restate the estimate (3.1.55) as

$$\mathbb{E}(Y) \ge \mathbb{E}(Z). \quad (3.1.56)$$

Step 2 Since $\mathbb{E}_{\mathcal{D}}(g_k) = \mathbb{E}_{\mathcal{D}}(u_k)$, the square of the conditioned square functions of $T_W(G - \mathbb{E}_{\mathcal{D}}G)$ coincides with

$$\sum \mathbb{E}_{k-1}|\Im(w_{k-1} \cdot (g_k - b_k\sigma_k))|^2. \quad (3.1.57)$$

Step 3 The sequence $\{u_k - b_k\sigma_k\}$ is the martingale difference sequence of $U - \mathbb{E}_{\mathcal{D}}(U)$. The square of its conditioned square functions is hence given by

$$\sum \mathbb{E}_{k-1}|u_k - b_k\sigma_k|^2. \tag{3.1.58}$$

Following the pattern of the estimate (3.1.42), define

$$a_k = |\mathbb{E}_{k-1}(u_k\sigma_k)| + \frac{|\mathbb{E}_{k-1}(u_k\sigma_k) - b_k|^2}{|\mathbb{E}_{k-1}(u_k\sigma_k)| + |b_k|},$$

and

$$v_k = u_k - \mathbb{E}_{k-1}(u_k\sigma_k)\sigma_k, \qquad r_k^2 = \mathbb{E}_{k-1}|v_k|^2.$$

By the inequality (3.1.48),

$$\mathbb{E}_{k-1}|u_k - b_k\sigma_k|^2 \leq 8\left(a_k^2 + r_k^2 - |\mathbb{E}_{k-1}(u_k\sigma_k)|^2\right). \tag{3.1.59}$$

Step 4 With $X^2 = \sum_{k=1}^{\infty} a_k^2 + r_k^2$, we have $X \geq Y$ and

$$\|U - \mathbb{E}_{\mathcal{D}}(U)\|_{\mathcal{P}} \leq \sqrt{8}\mathbb{E}(X^2 - Y^2)^{1/2} \leq \sqrt{8}(\mathbb{E}(X-Y))^{1/2}(\mathbb{E}(X+Y))^{1/2}. \tag{3.1.60}$$

We claim that for the factor $\mathbb{E}(X + Y)$ in the inequality (3.1.60), there holds the upper bound

$$\mathbb{E}(X + Y) \leq C\|U\|_{\mathcal{P}} = 2C\|G\|_{\mathcal{P}}. \tag{3.1.61}$$

Indeed, by elementary arithmetic we obtain

$$a_k^2 \leq 2|\mathbb{E}_{k-1}(u_k\sigma_k)|^2 + 2|\mathbb{E}_{k-1}(u_k\sigma_k) - b_k|^2 \leq 6\mathbb{E}_{k-1}|u_k|^2 + 4|b_k|^2,$$

and

$$r_k^2 \leq 2\mathbb{E}_{k-1}|u_k|^2 + 2|\mathbb{E}_{k-1}(u_k\sigma_k)|^2 \leq 4\mathbb{E}_{k-1}|u_k|^2,$$

which gives the pointwise estimate

$$X \leq \sqrt{10}\left(\sum \mathbb{E}_{k-1}|u_k|^2\right)^{1/2} + 2\left(\sum |b_k|^2\right)^{1/2}. \tag{3.1.62}$$

In view of the estimate (3.1.55), we have

$$\mathbb{E}\left(\sum |\mathbb{E}_{\mathcal{D}}\mathbb{E}_{k-1}(\sigma_k u_k)|^2\right)^{1/2} \leq \mathbb{E}\left(\sum |\mathbb{E}_{k-1}(\sigma_k u_k)|^2\right)^{1/2} \leq \mathbb{E}\left(\sum \mathbb{E}_{k-1}|u_k|^2\right)^{1/2}.$$

Since $b_k = \mathbb{E}_{\mathcal{D}}\mathbb{E}_{k-1}(u_k\sigma_k)$, the above estimate and the pointwise inequalities (3.1.62) give

$$\mathbb{E}(X) \leq (\sqrt{10} + 2)\mathbb{E}\left(\sum \mathbb{E}_{k-1}|u_k|^2\right)^{1/2}. \tag{3.1.63}$$

Since $Y \leq X$, the inequality (3.1.63) gives the estimate in (3.1.61) with $C = 2(\sqrt{10} + 2)$. Recall that by equation (3.1.34) we have $\|U\|_{\mathcal{P}} = 2\|G\|_{\mathcal{P}}$.

Step 5 We turn to estimating $\mathbb{E}(X - Y)$. First, we obtain $\mathbb{E}(X - Y) \leq \mathbb{E}(X - Z)$ by recalling the estimate (3.1.56). Second, applying the triangle inequality to the square functions gives

$$X - Z \leq \left(\sum_{k=1}^{\infty}(a_k - |b_k|)^2 + r_k^2\right)^{1/2}.$$

By inequality (3.1.49),

$$(a_k - |b_k|)^2 + r_k^2 \leq 8\mathbb{E}_{k-1}|\Im(w_{k-1} \cdot (g_k - b_k\sigma_k))|^2. \tag{3.1.64}$$

Thus, in view of the expression (3.1.57),

$$\mathbb{E}(X - Y) \leq \mathbb{E}(X - Z) \leq C\|T_W(G - \mathbb{E}_{\mathcal{D}}G)\|_{\mathcal{P}}. \tag{3.1.65}$$

Applying the estimates (3.1.61) and (3.1.65) to the factors in the right-hand side of (3.1.60) finishes the proof. □

3.1.3 Rademacher Coefficients

In this section we review the Garnett–Jones inequality (Garnett and Jones, 1978, 1982) for Rademacher coefficients of dyadic martingales. Recall that $(\mathbb{T}^{\mathbb{N}}, (\mathcal{D}_k), \mathbb{P}_{\mathcal{D}})$ denotes the dyadic filtration on $\mathbb{T}^{\mathbb{N}}$ defined in Section 2.2. For a given dyadic martingale $D = (D_k)$, with $d_k = \Delta D_k$, we define its kth Rademacher coefficient by

$$\langle d_k, \sigma_k\rangle = \mathbb{E}(d_k\sigma_k),$$

where the kth Rademacher function $\sigma_k\colon \mathbb{T}^{\mathbb{N}} \to \{-1, 1\}$ is defined by $\sigma_k(x) = \operatorname{sign}(\cos_k(x))$, with $\cos_k(x) = \Re x_k$ and $x = (x_k)$. The theorem of Garnett–Jones provides a powerful estimate for Rademacher coefficients of a dyadic martingales.

Theorem 3.1.9 *There exists $C < \infty$ such that for any dyadic martingale (D_k) and $n \in \mathbb{N}$,*

$$\sum_{k=1}^{n}|\langle d_k, \sigma_k\rangle|^2 \leq C\|D_n\|_1 \left\|\left(\sum_{k=1}^{n}|d_k|^2\right)^{1/2}\right\|_1, \tag{3.1.66}$$

where $d_k = \Delta D_k$.

We apply the Garnett–Jones inequality to families of dyadic martingales constructed as follows. Denote, as always, the product filtration on $\mathbb{T}^{\mathbb{N}}$ by $(\mathcal{F}_k)$. Let (V_k) be an $(\mathcal{F}_k)$-martingale with $v_k = \Delta V_k$. For $w, x \in \mathbb{T}^{\mathbb{N}}$, put

$$D_n(w, x) = \sum_{k=1}^{n} d_k(w, x),$$

where

$$d_k(w,x) = v_k\left(x_1^{\sigma_1(w)},\dots,x_{k-1}^{\sigma_{k-1}(w)},x_k\right)\sigma_k(w). \tag{3.1.67}$$

Thus for fixed x, the sequence $(D_n(\cdot,x))$ forms a dyadic martingale on $\mathbb{T}^{\mathbb{N}}$ and

$$A(v_k)(x) = \mathbb{E}_{(w)}(d_k(w,x)\sigma_k(w)) \tag{3.1.68}$$

is its kth Rademacher coefficient. We next restate the Garnett–Jones inequality for x-parametrized dyadic martingales.

Proposition 3.1.10 *Let (V_k) be an $(\mathcal{F}_k)$-martingale on $\mathbb{T}^{\mathbb{N}}$ satisfying*

$$\Delta V_k(z,y) = -\Delta V_k(z,\overline{y}), \tag{3.1.69}$$

for $z \in \mathbb{T}^{k-1}$ and $y \in \mathbb{T}$. Then,

$$\left\|\left(\sum_{k=1}^n |A(v_k)|^2\right)^{1/2}\right\|_1 \le C\|V_n\|_1^{1/2}\left\|\left(\sum_{k=1}^n |v_k|^2\right)^{1/2}\right\|_1^{1/2}, \tag{3.1.70}$$

where $v_k = \Delta V_k$ and $A(v_k)$ is defined by (3.1.68).

Proof Given (V_k) and $v_k = \Delta V_k$, define $d_k(w,x)$ by (3.1.67). By the Garnett–Jones inequality, for each $x \in \mathbb{T}^{\mathbb{N}}$, fixed,

$$\sum_{k=1}^n |A(v_k)(x)|^2 \le C\mathbb{E}_{(w)}|D_n(w,x)|\mathbb{E}_{(w)}\left(\sum_{k=1}^n |d_k(x,w)|^2\right)^{1/2}. \tag{3.1.71}$$

Take the square root on both sides of (3.1.71) and integrate with respect to the remaining variable x. Then, apply the Cauchy–Schwarz inequality to see that the left-hand side of (3.1.70) is bounded by a multiple of

$$(\mathbb{E}_{(x)}\mathbb{E}_{(w)}|D(w,x)|)^{1/2}\cdot\left(\mathbb{E}_{(x)}\mathbb{E}_{(w)}\left(\sum |d_k(x,w)|^2\right)^{1/2}\right)^{1/2}. \tag{3.1.72}$$

Since (V_k) satisfies (3.1.69), we may rewrite (3.1.68) as

$$d_k(w,x) = v_k\left(x_1^{\sigma_1(w)},\dots,x_{k-1}^{\sigma_{k-1}(w)},x_k^{\sigma_k(w)}\right). \tag{3.1.73}$$

Changing the order of integration in the expression (3.1.72) and using identity (3.1.73) completes the proof. □

In the sections below, we use the Garnett–Jones inequality in combination Lépingle's inequality (2.5.18), which asserts that, for any martingale $X = (X_k)$,

$$\left\|\left(|X_0|^2 + \sum(\mathbb{E}_{k-1}|\Delta X_k|)^2\right)^{1/2}\right\|_1 \le 2\left\|\left(|X_0|^2 + \sum|\Delta X_k|^2\right)^{1/2}\right\|_1. \tag{3.1.74}$$

where $\Delta X_k = X_k - X_{k-1}$.

3.1.4 Sine-Cosine Decomposition

Let $f \in H^1(\mathbb{T})$. Recall that the even part of f is $u(y) = (f(y) + f(\bar{y}))/2$, and its odd part $v = f - u$. Note that,

$$\int_{\mathbb{T}} u(y)\Re(y)dm(y) = (-i)\int_{\mathbb{T}} v(y)\Im(y)dm(y). \tag{3.1.75}$$

Let $G = (G_k)$ be a Hardy martingale. Recall that in equation (3.1.33) we defined its cosine part by

$$U_k(x,y) = \left[G_k(x,y) + G_k(x,\bar{y})\right]/2,$$

where $x \in \mathbb{T}^{k-1}$ and $y \in \mathbb{T}$. Now set $V_k = G_k - U_k$, and call $V = (V_k)$ the sine part of G.

For $w \in \mathbb{T}^{\mathbb{N}}$ with $w = (w_k)$, we set $\cos_k(w) = \Re w_k$ and $\sin_k(w) = \Im w_k$. In view of (3.1.75), the sine part V_k is linked to U_k by the following important identity:

$$\mathbb{E}_{k-1}(U_k \cos_k) = (-i)\mathbb{E}_{k-1}(V_k \sin_k). \tag{3.1.76}$$

Moreover, we have $\mathbb{E}(V_k|\mathcal{D}) = 0$. We put $\alpha = 2/\pi$ and define the martingale $Z = (Z_k)$ by

$$\Delta Z_k = (-i)\alpha\left[\mathbb{E}_{k-1}(V_k \sin_k)\right]\sigma_k. \tag{3.1.77}$$

We recall that $H^1 = H^1\left(\mathbb{T}^{\mathbb{N}}, (\mathcal{F}_j), \mathbb{P}\right)$ and that the martingale H^1 norm of $X = (X_k)$ is given by

$$\|X\|_{H^1} = \left\|\left(|X_0|^2 + \sum |\Delta X_k|^2\right)^{1/2}\right\|_1,$$

where $\Delta X_k = X_k - X_{k-1}$. See equation (3.1.23).

We next exploit the martingale inequalities of Garnett–Jones (3.1.66) and Lépingle (3.1.74)

Proposition 3.1.11 *Let $G = (G_k)$ be a Hardy martingale, let $V = (V_k)$ be its sine part, and let $Z = (Z_k)$ be given by* (3.1.77). *Then*

$$\|\mathbb{E}(G|\mathcal{D})\|_{H^1} \le C\|V\|_{L^1}^{1/2}\|V\|_{H^1}^{1/2} + C\|Z - \mathbb{E}(G|\mathcal{D})\|_{H^1} \tag{3.1.78}$$

and

$$\|Z - \mathbb{E}(G|\mathcal{D})\|_{H^1} \le 2\alpha\|U - \mathbb{E}(G|\mathcal{D})\|_{H^1}, \tag{3.1.79}$$

where $U = (U_k)$ denotes the cosine martingale asociated to G.

Proof Let Z be given by (3.1.77). Put $z_k = \Delta Z_k$. Use (3.1.68) to define the parametrized Rademacher coefficient $q_k(x) = A(z_k)(x)$, $x \in \mathbb{T}^{\mathbb{N}}$, and let $Q = (Q_k)$ be the martingale with difference sequance $\Delta Q_k = q_k$. Next, write

$$\mathbb{E}(G|\mathcal{D}) = Q + \mathbb{E}(G|\mathcal{D}) - Q.$$

Estimating Q: Let $v_k = \Delta V_k$. We first observe that $|A(z_k)(x)| \le \mathbb{E}_{k-1}|A(v_k)(x)|$, for $x \in \mathbb{T}^{\mathbb{N}}$. Hence the inequalities of Lépingle (3.1.74) and Garnett–Jones (3.1.70) give

$$\left\| \left(\sum_{k=1}^{n} |q_k|^2 \right)^{1/2} \right\|_1 \le 2 \left\| \left(\sum_{k=1}^{n} |A(v_k)|^2 \right)^{1/2} \right\|_1 \le C\|V\|_{L^1}^{1/2}\|V\|_{H^1}^{1/2}.$$

Estimating $\mathbb{E}(G|\mathcal{D}) - Q$: Let $D_k = \mathbb{E}(G_k|\mathcal{D})$ and put $d_k = \Delta D_k$. Since

$$d_k(x_1, \ldots, x_k) = d_k\left(x_1^{\pm 1}, \ldots, x_k^{\pm 1}\right),$$

for $x \in \mathbb{T}^{\mathbb{N}}$, we have the identity $A(d_k)(x) = d_k(x)$. Recalling that $q_k(x) = A(z_k)(x)$, this gives

$$q_k(x) - d_k(x) = A(z_k - d_k)(x).$$

Hence invoking Minkowski's inequality gives,

$$\left\| \left(\sum_{k=1}^{n} |q_k - d_k|^2 \right)^{1/2} \right\|_1 = \left\| \left(\sum_{k=1}^{n} |A(z_k - d_k)|^2 \right)^{1/2} \right\|_1 \le \left\| \left(\sum_{k=1}^{n} |z_k - d_k|^2 \right)^{1/2} \right\|_1 .$$

This completes the verification of (3.1.78).

Estimating $\mathbb{E}(G|\mathcal{D}) - Z$: We now turn to proving the estimate (3.1.79). Invoking the identity (3.1.76) we obtain

$$z_k = (-i)\alpha\mathbb{E}_{k-1}(V_k \sin_k)\sigma_k = \alpha\mathbb{E}_{k-1}(U_k \cos_k)\sigma_k = \alpha\mathbb{E}_{k-1}(u_k \cos_k)\sigma_k.$$

Since $d_k = m_{k-1}\sigma_k$, we have $m_{k-1} = \alpha\mathbb{E}_{k-1}(d_k \cos_k)$ and, hence,

$$|z_k - d_k| = |\alpha\mathbb{E}_{k-1}(u_k \cos_k) - m_{k-1}| \le \alpha\mathbb{E}_{k-1}|(u_k - d_k)\cos_k|.$$

Since $|\mathbb{E}_{k-1}((u_k - d_k)\cos_k)| \le \mathbb{E}_{k-1}|u_k - d_k|$, Lépingle's inequality (2.5.18) gives

$$\left\| \left(\sum_{k=1}^{n} |z_k - d_k|^2 \right)^{1/2} \right\|_1 \le \alpha \left\| \left(\sum_{k=1}^{n} (\mathbb{E}_{k-1}|u_k - d_k|)^2 \right)^{1/2} \right\|_1 \le 2\alpha \left\| \left(\sum_{k=1}^{n} |u_k - d_k| \right)^{1/2} \right\|_1 ,$$

which finishes the proof of the estimate (3.1.79). □

Summary of Sections 3.1.2 through 3.1.4

The martingale inequality (3.1.80) summarizes Sections 3.1.2 through 3.1.4. For any Hardy martingale G and any adapted sequence $W = (w_k)$ satisfying $|w_k| = 1$,

$$\|G\|_{L^1} \le C\left(\|G - D\|_{L^1}^{1/2}\|G\|_{\mathcal{P}}^{1/2} + \|T_W(G - D)\|_{\mathcal{P}}^{1/4}\|G\|_{\mathcal{P}}^{3/4}\right), \tag{3.1.80}$$

where $D = \mathbb{E}(G|\mathcal{D})$. Indeed, Proposition 3.1.11 and Theorem 2.5.5 give

$$\|D\|_{L^1} \leq C\left(\|V\|_{L^1}^{1/2}\|V\|_{\mathcal{P}}^{1/2} + \|U - D\|_{\mathcal{P}}\right), \tag{3.1.81}$$

where $D = \mathbb{E}(G|\mathcal{D}) = \mathbb{E}(U|\mathcal{D})$, and where U, V form the sine-cosine decomposition of G. By the triangle inequality, we clearly have

$$\|V\|_{L^1} \leq \|G - D\|_{L^1} + \|U - D\|_{L^1}.$$

Since $\|V\|_{\mathcal{P}} \leq C\|G\|_{\mathcal{P}}$ and $\|U - D\|_{\mathcal{P}} \leq C\|G\|_{\mathcal{P}}$, the estimate (3.1.81) gives

$$\|D\|_{L^1} \leq C\left(\|G - D\|_{L^1}^{1/2} + \|U - D\|_{\mathcal{P}}^{1/2}\right)\|G\|_{\mathcal{P}}^{1/2}. \tag{3.1.82}$$

By Theorem 3.1.8, applied to $U - D$ on the right-hand side of (3.1.82), for any adapted sequence $W = (w_k)$ satisfying $|w_k| = 1$,

$$\|D\|_{L^1} \leq C\left(\|G - D\|_{L^1}^{1/2}\|G\|_{\mathcal{P}}^{1/2} + \|T_W(G - D)\|_{\mathcal{P}}^{1/4}\|G\|_{\mathcal{P}}^{3/4}\right). \tag{3.1.83}$$

It remains to replace D by G in the left-hand side of (3.1.83), and this is easily done. If $\|G\|_{L^1} \leq 2\|D\|_{L^1}$, then (3.1.80) follows directly. If, however, $\|G\|_{L^1} \geq 2\|D\|_{L^1}$, then $2\|G - D\|_{L^1} \geq \|G\|_{L^1}$ and (3.1.80) holds for trivial reasons.

3.1.5 Separation in $L^1(\mathbb{T}^{\mathbb{N}})$

We let $\mathcal{D}$ denote the dyadic sigma-algebra on $\mathbb{T}^{\mathbb{N}}$ and $L^1_{\mathcal{D}}(\mathbb{T}^{\mathbb{N}})$ the subspace of $L^1(\mathbb{T}^{\mathbb{N}})$ consisting of $\mathcal{D}$ measurable functions. In this section we finally show that the canonical quotient map

$$q\colon L^1_{\mathcal{D}}(\mathbb{T}^{\mathbb{N}}) \to L^1(\mathbb{T}^{\mathbb{N}})/H^1_0(\mathbb{T}^{\mathbb{N}}), \qquad q(D) = \left\{D + H^1_0(\mathbb{T}^{\mathbb{N}})\right\},$$

is invertible on its range. Thus, the quotient space $L^1(\mathbb{T}^{\mathbb{N}})/H^1_0(\mathbb{T}^{\mathbb{N}})$ contains a subspace isomorphic to L^1.

Theorem 3.1.12 (Martingale separation theorem) *There exist constants $A_0, C < \infty$ such that for any $F \in H^1(\mathbb{T}^{\mathbb{N}})$ with $\mathbb{E}(F) = 0$,*

$$\|F\|_{L^1(\mathbb{T}^{\mathbb{N}})} \leq C\|F - \mathbb{E}(F|\mathcal{D})\|_{L^1(\mathbb{T}^{\mathbb{N}})}, \tag{3.1.84}$$

and, for any $D \in L^1_{\mathcal{D}}$,

$$\|D\|_{L^1(\mathbb{T}^{\mathbb{N}})} \leq A_0 \inf\left\{\|D - F\|_{L^1(\mathbb{T}^{\mathbb{N}})}\right\}, \tag{3.1.85}$$

where the infimum is taken over $F \in H^1(\mathbb{T}^{\mathbb{N}})$ satisfying $\mathbb{E}(F) = 0$.

Proof Fix a Hardy martingale F, and apply Theorem 3.1.3 to obtain the decomposition $F = G + B$, where G, B are Hardy martingales satisfying equations (3.1.15), (3.1.16), and (3.1.27). Separately, since G is a Hardy martingale, the estimate (3.1.80) holds true.

Step 1 Use (3.1.16) directly to get

$$\|B\|_{L^1} \le \sum_{k=1}^{\infty} \mathbb{E}|\Delta B_k| \le C_1 \|F - \mathbb{E}(F|\mathcal{D})\|_{L^1}, \tag{3.1.86}$$

and invoke inequalities (3.1.30) and (3.1.29) asserting that

$$\|G - \mathbb{E}(G|\mathcal{D})\|_{L^1} \le C\|F - \mathbb{E}(F|\mathcal{D})\|_{L^1}, \tag{3.1.87}$$

and

$$\|\mathbb{E}(F|\mathcal{D}) - \mathbb{E}(G|\mathcal{D})\|_{\mathcal{P}} \le C\|F - \mathbb{E}(F|\mathcal{D})\|_{L^1}. \tag{3.1.88}$$

Step 2 By (3.1.80) we have

$$\|G\|_{L^1} \le C\left(\|T(G - \mathbb{E}(G|\mathcal{D}))\|_{\mathcal{P}}^{1/4}\|G\|_{\mathcal{P}}^{3/4} + \|G - \mathbb{E}(G|\mathcal{D})\|_{L^1}^{1/2}\|G\|_{\mathcal{P}}^{1/2}\right), \tag{3.1.89}$$

where $T = T_W$ and $W = (w_k)$ is determined by Theorem 3.1.3 such that the estimate (3.1.15) holds true.

Step 3 Theorem 3.1.3 controls the terms on the right-hand side of (3.1.89). Indeed, (3.1.15) and (3.1.88) give

$$\|T(G - \mathbb{E}(G|\mathcal{D}))\|_{\mathcal{P}} \le C\|F - \mathbb{E}(F|\mathcal{D})\|_{L^1}^{1/2}\|F\|_{L^1}^{1/2}. \tag{3.1.90}$$

Recall that by the estimate (3.1.27), $\|G\|_{\mathcal{P}} \le C\|F\|_{L^1}$. Hence, using (3.1.90) and (3.1.87), the right-hand side of (3.1.89) gives

$$\|G\|_{L^1} \le C\|F - \mathbb{E}(F|\mathcal{D})\|_{L^1}^{1/8}\|F\|_{L^1}^{7/8}. \tag{3.1.91}$$

Step 4 Since $F = G + B$, (3.1.86) and (3.1.91) give

$$\|F\|_{L^1} \le C\|F - \mathbb{E}(F|\mathcal{D})\|_{L^1}^{\alpha}\|F\|_{L^1}^{1-\alpha}, \quad \text{with} \quad \alpha = 1/8.$$

Canceling the factor $\|F\|_{L^1}^{1-\alpha}$ on both sides, we obtain (3.1.84).

Step 5 It remains to show that (3.1.84) yields (3.1.85) by direct arithmetic. Fix $D \in L^1_{\mathcal{D}}$ and define

$$L_\alpha = \inf\left\{\|D - F\|_1 : \|F\|_1 \ge \alpha\|D\|_1,\ F \in H^1_0(\mathbb{T}^{\mathbb{N}})\right\},$$

where $0 < \alpha < 1$. Then by the triangle inequality,

$$\|q(D)\|_{L^1(\mathbb{T}^{\mathbb{N}})/H^1_0(\mathbb{T}^{\mathbb{N}})} \ge \min\{(1-\alpha)\|D\|_{L^1}, L_\alpha\}.$$

Fix $F \in H^1_0(\mathbb{T}^{\mathbb{N}})$, with $\|F\|_1 \ge \alpha\|D\|_1$, and use that D is $\mathcal{D}$ measurable. This gives

$$\|D - F\|_{L^1(\mathbb{T}^{\mathbb{N}})} \ge \|D - \mathbb{E}(F|\mathcal{D})\|_{L^1(\mathbb{T}^{\mathbb{N}})} \ge \|F - \mathbb{E}(F|\mathcal{D})\|_{L^1(\mathbb{T}^{\mathbb{N}})} - \|D - F\|_{L^1(\mathbb{T}^{\mathbb{N}})}.$$

Since, by (3.1.84), $||F - \mathbb{E}(F|\mathcal{D})||_1 \geq C^{-1}||F||_1$, we find

$$2||D - F||_1 \geq C^{-1}||F||_1 \geq \alpha C^{-1}||D||_1.$$

Solving $1 - \alpha = \alpha C^{-1}/2$ for α gives $||q(D)||_{L^1/H^1_0} \geq A_0^{-1}||D||_1$, where $A_0^{-1} = 1/(2C + 1)$. □

3.2 The Quotient Space $L^1(\mathbb{T})/H_0^1(\mathbb{T})$

In this section we prove that L^1 is isomorphic to a subspace of $L^1(\mathbb{T})/H_0^1(\mathbb{T})$ by exploiting the martingale embedding of Section 3.1. To this end we further develop the transfer method, shifting the embedding problem from $\mathbb{T}$ to $\mathbb{T}^{\mathbb{N}}$ where the solution was given by Theorem 3.1.12.

3.2.1 The Meyer–Bonami Transfer Operator

Here, we discuss the Meyer–Bonami transfer operator (Bonami [1970]; Meyer [1968]), summarizing its main properties in Theorem 3.2.4. Proposition 3.2.5 presents its action on products of almost independent functions.

The Basic Building Block

Bernstein's estimate and the Poincare inequality are used to obtain the L^1-version of Meyer's basic estimate (3.2.1), transferring $L^1(\mathbb{T})$-estimates, for polynomials on $\mathbb{T}$, to $L^1(\mathbb{T} \times \mathbb{T})$. Our presentation relies on Meyer (1968, Chapter IV. A. 2) and the work of Kazaniecki and Wojciechowski (2014).

Proposition 3.2.1 *Let $k, n \in \mathbb{N}$, let $f, g\colon \mathbb{T} \to \mathbb{C}$ be trigonometric polynomials of degree at most k, and form*

$$h_1(z) = f(z) + z^n g(z), \qquad h_2(y, z) = f(z) + yg(z), \qquad y, z \in \mathbb{T}.$$

There exists $C > 0$ (independent of f and g) such that for $\epsilon > 0$ and $n > Ck/\epsilon$,

$$(1 - \epsilon) \int_{\mathbb{T}} |h_1(z)| dm(z) \leq \int_{\mathbb{T}} \int_{\mathbb{T}} |h_2(y, z)| dm(y) dm(z) \leq (1 + \epsilon) \int_{\mathbb{T}} |h_1(z)| dm(z). \tag{3.2.1}$$

Proof The proof of Proposition 3.2.1 is based on Lemma 3.2.2 (Poincare's inequality) and Lemma 3.2.3 (Bernstein's estimate). In particular, it uses a basic fact (verified in Step 1) on the approximation of trigonometric polynomials by functions that are constant on dyadic intervals.

Step 1 Let $k \in \mathbb{N}$, let $f: \mathbb{T} \to \mathbb{C}$ be a trigonometric polynomial of degree at most k. For $n, j \in \mathbb{N}$ put $I_j = \{e^{i\theta}: 2\pi(j-1)n^{-1} \leq \theta < 2\pi jn^{-1}\}$, and form the partition $\mathcal{E}_n = \{I_j: 1 \leq j \leq n\}$ to define the conditional expectation operator,

$$E_n(f)(z) = \sum_{I\in\mathcal{E}_n} \frac{1}{m(I)} \int_I f dm 1_I(z), \quad z \in \mathbb{T}.$$

We claim that

$$\int_{\mathbb{T}} |f - E_n(f)| dm \leq C_E \frac{k}{n} \int_{\mathbb{T}} |f| dm, \tag{3.2.2}$$

where $C_E = 2\pi C_P$, and where $C_P < \infty$, is the constant appearing in Poincare's inequality, Lemma 3.2.2. We verify (3.2.2) by invoking Lemma 3.2.2 and Lemma 3.2.3 and taking into account that f is a polynomial of degree $\leq k$,

$$\begin{aligned}
\int_{\mathbb{T}} |f - E_n(f)| dm &= \sum_{I\in\mathcal{E}_n} \int_I \left| f(\zeta) - \int_I f(w) \frac{dm(w)}{m(I)} \right| dm(\zeta) \\
&\leq C_P \sum_{I\in\mathcal{E}_n} \operatorname{diam}(I) \int_I \left| f'(\zeta) \right| dm(\zeta) \\
&\leq 2\pi C_P \frac{k}{n} \sum_{I\in\mathcal{E}_n} \int_I |f(\zeta)|\, dm(\zeta).
\end{aligned}$$

Step 2 Let $g: \mathbb{T} \to \mathbb{C}$ be a second trigonometric polynomial of degree at most k, form

$$\widetilde{h_1}(z) = E_n f(z) + z^n E_n g(z), \qquad \widetilde{h_2}(y,z) = E_n f(z) + y E_n g(z), \qquad y, z \in \mathbb{T},$$

and set

$$D_n = \|f - E_n f\|_{L^1(\mathbb{T})} + \|g - E_n g\|_{L^1(\mathbb{T})}.$$

By the triangle inequality, we obtain

$$\left\|h_1 - \widetilde{h_1}\right\|_{L^1(\mathbb{T})} \leq D_n \quad \text{and} \quad \left\|h_2 - \widetilde{h_2}\right\|_{L^1(\mathbb{T}\times\mathbb{T})} \leq D_n. \tag{3.2.3}$$

We claim that we have the lower estimate,

$$\|h_2\|_{L^1(\mathbb{T}\times\mathbb{T})} \geq (1/2)\left(\|f\|_{L^1(\mathbb{T})} + \|g\|_{L^1(\mathbb{T})}\right), \tag{3.2.4}$$

and the equality,

$$\left\|\widetilde{h_1}\right\|_{L^1(\mathbb{T})} = \left\|\widetilde{h_2}\right\|_{L^1(\mathbb{T}\times\mathbb{T})}. \tag{3.2.5}$$

Assuming that expressions (3.2.4) and (3.2.5) hold true, we finish the proof as follows. By (3.2.5) we may rewrite

$$\begin{aligned}
&\left| \|h_1\|_{L^1(\mathbb{T}\times\mathbb{T})} - \|h_2\|_{L^1(\mathbb{T}\times\mathbb{T})} \right| \\
&\quad = \left| \|h_1\|_{L^1(\mathbb{T}\times\mathbb{T})} - \left\|\widetilde{h_1}\right\|_{L^1(\mathbb{T}\times\mathbb{T})} - \|h_2\|_{L^1(\mathbb{T}\times\mathbb{T})} + \left\|\widetilde{h_2}\right\|_{L^1(\mathbb{T}\times\mathbb{T})} \right|
\end{aligned} \tag{3.2.6}$$

Now we estimate the right-hand side of (3.2.6) by applying the triangle inequality and inequalities (3.2.3), (3.2.2), and (3.2.4), as follows:

$$\begin{aligned}\left\|h_1 - \widetilde{h_1}\right\|_{L^1(\mathbb{T}\times\mathbb{T})} + \left\|h_2 - \widetilde{h_2}\right\|_{L^1(\mathbb{T}\times\mathbb{T})} &\leq 2\left(\|f - E_n f\|_{L^1(\mathbb{T})} + \|g - E_n g\|_{L^1(\mathbb{T})}\right)\\ &\leq 2C_E\frac{k}{n}\left(\|f\|_{L^1(\mathbb{T})} + \|g\|_{L^1(\mathbb{T})}\right)\\ &\leq 4C_E\frac{k}{n}\|h_2\|_{L^1(\mathbb{T}\times\mathbb{T})}.\end{aligned}$$

Step 3 It remains to verify expressions (3.2.4) and (3.2.5). To this end recall that $\mathcal{E}_n = \left\{I_j \colon 1 \leq j \leq n\right\}$ where $I_j = \{e^{i\theta} \colon 2\pi(j-1)n^{-1} \leq \theta < 2\pi jn^{-1}\}$. Given $I = I_j$ we set $w_I = 2\pi(j-1)n^{-1}$. Now the following observations give the equality (3.2.5):

$$\begin{aligned}\int_{\mathbb{T}} |E_n f(z) + z^n E_n g(z)| dm(z) &= \sum_{I\in\mathcal{E}_n} \int_I |E_n f(w_I) + z^n E_n g(w_I)| dm(z)\\ &= \sum_{I\in\mathcal{E}_n} m(I) \int_{\mathbb{T}} |E_n f(w_I) + y E_n g(w_I)| dm(y)\\ &= \int_{\mathbb{T}}\int_{\mathbb{T}} |E_n f(w) + y E_n g(w)| dm(y) dm(w).\end{aligned}$$

Finally, we turn to (3.2.4). We have

$$\|g\|_{L^1(\mathbb{T})} = \int_{\mathbb{T}} \left| \int_{\mathbb{T}} \bar{y}(f(z) + yg(z)) dm(y) \right| dm(z) \leq \|h_2\|_{L^1(\mathbb{T}\times\mathbb{T})},$$

and, similarly, $\|f\|_{L^1(\mathbb{T})} \leq \|h_2\|_{L^1(\mathbb{T}\times\mathbb{T})}$. □

We next turn to proving Lemma 3.2.2 and Lemma 3.2.3, as used above. For the proof of the L^1-Poincare inequality, we use Acosta and Durán (2004) as reference.

Lemma 3.2.2 (Poincare's inequality) *Let $I \subseteq \mathbb{T}$ be connected and let $f \colon I \to \mathbb{C}$ be differentiable, then*

$$\int_I \left| f(\zeta) - \int_I f(w) \frac{dm(w)}{m(I)} \right| dm(\zeta) \leq C_P \operatorname{diam}(I) \int_I |f'(\zeta)| dm(\zeta).$$

Proof Fix $\alpha > 0$. Assume that $\varphi \colon [0, \alpha] \to \mathbb{R}$ is continuously differentiable and satisfies $\int_0^\alpha \varphi(x) dx = 0$. We claim that

$$\alpha\varphi(y) = \int_0^y \varphi'(x) x dx - \int_y^\alpha \varphi'(x)(\alpha - x) dx, \qquad 0 < y < \alpha. \tag{3.2.7}$$

Indeed, since $\int_0^\alpha \varphi(x) dx = 0$, integration by parts gives

$$\int_0^\alpha \varphi'(x) x dx = \varphi(\alpha)\alpha - \varphi(0) \cdot 0 - \int_0^\alpha \varphi(x) dx = \varphi(\alpha)\alpha.$$

Hence, by arithmetic, for the right-hand side term of (3.2.7) we have

$$\int_0^y \varphi'(x)xdx - \int_y^\alpha \varphi'(x)(\alpha - x)dx = \int_0^\alpha \varphi'(x)xdx + \varphi(y)\alpha - \varphi(\alpha)\alpha = \varphi(y)\alpha.$$

Thus, by (3.2.7) we obtain the pointwise estimate

$$\alpha|\varphi(y)| \leq \int_0^y |\varphi'(x)|xdx + \int_y^\alpha |\varphi'(x)|(\alpha - x)dx, \qquad 0 < y < \alpha. \tag{3.2.8}$$

Integrating this estimate over $0 < y < \alpha$, and invoking Fubini's theorem we obtain

$$\begin{aligned}\alpha \int_0^\alpha |\varphi(y)|dy &\leq \int_0^\alpha \int_0^y |\varphi'(x)|xdxdy + \int_0^\alpha \int_y^\alpha |\varphi'(x)|(\alpha - x)dxdy \\ &= \int_0^\alpha |\varphi'(x)|x(\alpha - x)dx + \int_0^\alpha |\varphi'(x)|(\alpha - x)xdx.\end{aligned} \tag{3.2.9}$$

Since $x(\alpha - x) \leq \alpha^2/4$, for $0 < x < \alpha$, we obtain from (3.2.9), the following L^1 Poincare inequality for functions $\varphi\colon [0, \alpha] \to \mathbb{R}$, satisfying $\int_0^\alpha \varphi dx = 0$,

$$\int_0^\alpha |\varphi(y)|dy \leq \frac{\alpha}{2} \int_0^\alpha |\varphi'(x)|dx. \tag{3.2.10}$$

Finally, we remark that the constant $\alpha/2$ in the estimate (3.2.10) is the best possible. □

We refer to the book by Zygmund (1988, Volume II, Chapter X, Theorem 3.16) for Bernstein's inequality.

Lemma 3.2.3 (Bernstein's inequality) *Let $k \in \mathbb{N}$, and let $f\colon \mathbb{T} \to \mathbb{C}$ be a trigonometric polynomial of degree at most k. Then,*

$$\int_\mathbb{T} |f'(\zeta)|dm(\zeta) \leq k \int_\mathbb{T} |f(\zeta)|dm(\zeta). \tag{3.2.11}$$

Proof Let $\zeta \in \mathbb{T}$, $c_\nu \in \mathbb{R}$, $|\nu| \leq k$, and $f(\zeta) = \sum_{\nu=-k}^k c_\nu \zeta^\nu$. Then $f'(\zeta) = \sum_{\nu=-k}^k c_\nu \nu \zeta^{\nu-1}$. Now, for $m \in \{1, \dots 2k\}$ set $\beta_m = k^{-1}(m - 1/2)\pi$ and $\alpha_m = k^{-1} \sin^{-2}(\beta_m/2)$. We refer to Zygmund (1988, Volume II, p. 10) for the identities

$$f'(\zeta) = \sum_{m=1}^{2k} f\left(\zeta e^{i\beta_m}\right) \alpha_m (-1)^{m+1} \quad \text{and} \quad \sum_{m=1}^{2k} \alpha_m = k. \tag{3.2.12}$$

Integrating these identities with respect to the Haar measure on $\mathbb{T}$ yields (3.2.11). □

The Meyer–Bonami Embedding

We define a class of subspaces in $L^1(\mathbb{T})$ and in $H^1(\mathbb{T})$ by imposing restrictions on their Fourier spectrum and we show that each of these subspaces generates canonical martingale classes on the infinite torus product. Given $E \subseteq \mathbb{Z}$, define $L^1_E(\mathbb{T}) \subseteq L^1(\mathbb{T})$ by imposing $f \in L^1_E(\mathbb{T})$ if $\hat{f}(k) = 0$ for $k \notin E$, where $\hat{f}(k) = \int_{\mathbb{T}} f(\zeta)\overline{\zeta^k}dm(\zeta)$. Accordingly, we set

$$H^1_E(\mathbb{T}) = L^1_E(\mathbb{T}) \cap H^1(\mathbb{T}).$$

Below (in Theorem 3.2.4 and Proposition 3.2.5), we will construct a class of integer sets E and linear operators mapping $H^1_E(\mathbb{T})$ into $H^1(\mathbb{T}^{\mathbb{N}})$, and simultaneously Riesz products on $\mathbb{T}$ into Riesz products on $\mathbb{T}^{\mathbb{N}}$.

We emphasize that $H^1(\mathbb{T}^{\mathbb{N}}) \subset L^1(\mathbb{T}^{\mathbb{N}})$ denotes the linear space consisting of the L^1 limits of L^1-bounded Hardy martingales. We first review the beginning paragraphs of Section 2.4.1 where we characterized $H^1(\mathbb{T}^{\mathbb{N}})$ in terms of the Fourier spectrum. Recall that $\mathbb{Z}^{(\mathbb{N})}$ denotes the additive group of integer sequences $n = (n_i)$, with only finitely many entries nonvanishing and that

$$\mathbb{Z}(k) = \left\{n \in \mathbb{Z}^{(\mathbb{N})} : n = (n_1, n_2, \dots, n_k, 0)\right\}.$$

Put $\Lambda_0 = \{0\}$,

$$\Lambda_k = \{n \in \mathbb{Z}(k) : n_k \geq 1\}, \quad \text{and} \quad \Lambda = \bigcup \Lambda_k. \tag{3.2.13}$$

By means of the products $z^n = \prod z_i^{n_i}$, where $z \in \mathbb{T}^{\mathbb{N}}$ and $n \in \mathbb{Z}^{(\mathbb{N})}$, we define

$$\widehat{F}(n) = \int_{\mathbb{T}^{\mathbb{N}}} F(z)\overline{z^n}d\mathbb{P}(z), \quad n \in \mathbb{Z}^{(\mathbb{N})},$$

and set $\operatorname{spec} F = \left\{n \in \mathbb{Z}^{(\mathbb{N})} : \widehat{F}(n) \neq 0\right\}$ for $F \in L^1(\mathbb{T}^{\mathbb{N}})$. In Section 2.4.1 we obtained the identity

$$H^1(\mathbb{T}^{\mathbb{N}}) = \left\{F \in L^1(\mathbb{T}^{\mathbb{N}}) : \operatorname{spec} F \subseteq \Lambda\right\}.$$

That is to say, if (F_n) is an L^1-bounded Hardy martingale and $F = \lim F_n$ its L^1 limit, then $\operatorname{spec} F \subseteq \Lambda$; and, conversely, if $F \in L^1(\mathbb{T}^{\mathbb{N}})$ and $\operatorname{spec} F \subseteq \Lambda$, then the sequence of conditional expectations $(\mathbb{E}_k(F))$ forms an L^1-bounded Hardy martingale.

Fix a sequence (a_m) of strictly increasing positive integers. We recursively determine the integer vector ν as

$$\nu = (n_m) \text{ where } n_1 = 1 \text{ and } n_m = 4^m \max_{1 \leq i \leq m-1} a_i n_i, \tag{3.2.14}$$

put $A = \left\{\mu \in \mathbb{Z}^{(\mathbb{N})} : |\mu_i| \leq a_i\right\}$ and $E = \{\mu \cdot \nu \in \mathbb{Z} : \mu \in A\}$. We define the operator T on L^1_E by putting

$$(Tz^{\mu\cdot\nu})(w) = w^{\mu}, \quad w \in \mathbb{T}^{\mathbb{N}}, \quad \mu \in A, \tag{3.2.15}$$

where $w = (w_i)$, $\mu = (\mu_i)$, and $w^{\mu} = \prod w_i^{\mu_i}$. Since the trigonometric polynomials in L_E^1 admit a unique expansion in terms of the monomials $\{z^{\mu\cdot\nu} : \mu \in A\}$, the operator T has a well-defined linear extension to the polynomials with spectrum in $E \subseteq \mathbb{Z}$. Separately, these polynomials are a dense subset of L_E^1. We next prove that T extends to a bounded operator on L_E^1, invertible on its range.

Theorem 3.2.4 *The operator T admits a bounded extension*

$$T : L_E^1(\mathbb{T}) \to L^1(\mathbb{T}^{\mathbb{N}}),$$

which is invertible on its range. It preserves analyticity , $TH_E^1 \subseteq H^1(\mathbb{T}^{\mathbb{N}})$, and martingale structure

$$T(\operatorname{span}\{z^m : m \in D_k, z \in \mathbb{T}\}) \subseteq \operatorname{span}\left\{w^{\mu} : \mu \in \mathbb{Z}(k), w \in \mathbb{T}^{\mathbb{N}}\right\},$$

where $D_k = \{\mu \cdot \nu : \mu \in \mathbb{Z}(k)\}$.

Proof We start with an observation due to Meyer (1968). Let f, g be trigonometric polynomials with spectrum in $[-k, k]$. Let $h(x, y) = f(x) + yg(x)$, where $x, y \in \mathbb{T}$. By Proposition 3.2.1, for every $\epsilon > 0$ and $n > Ck/\epsilon$ we obtain

$$(1 - \epsilon) \int_{\mathbb{T}\times\mathbb{T}} |h| dm \otimes dm \le \int_{\mathbb{T}} |f(x) + x^n g(x)| dm(x) \le (1 + \epsilon) \int_{\mathbb{T}\times\mathbb{T}} |h| dm \otimes dm.$$

Iteratively applied, Meyer's observation gives the metric assertion of the theorem. For $f \in L_E^1$,

$$c\|f\|_{L^1(\mathbb{T})} \le \|T(f)\|_{L^1} \le C\|f\|_{L^1(\mathbb{T})}. \tag{3.2.16}$$

Hence T extends to a bounded operator on L_E^1 invertible on its range.

We next exhibit the martingale structure of $T\left(L_E^1\right)$. Fix $f \in L_E^1$, $w \in \mathbb{T}^{\mathbb{N}}$, and $k, m \in \mathbb{N}$ to form

$$b_m(k, w) = \sum_{\mu \in A \cap \mathbb{Z}(m-1)} \widehat{f}(\mu \cdot \nu + kn_m) w^{\mu}.$$

Put $M(m) = [-a_m, a_m] \backslash \{0\}$ and

$$d_m(w) = \sum_{k \in M(m)} b_m(k, w) w_m^k.$$

Thus, d_m is a martingale difference sequence with respect to the canonical filtration on the infinite torus product, and, moreover,

$$Tf = \sum_{m=1}^{\infty} d_m. \tag{3.2.17}$$

Finally, we show that T preserves analyticity. Recall first that the integer vector $\nu = (n_m)$ is defined in expression (3.2.14), and appeared in the definition of the operator T. Next, observe that if $k \in \mathbb{Z}$, $\mu \in A \cap \mathbb{Z}(m-1)$, and $\mu \cdot \nu + kn_m > 0$, then $k \geq 1$. Thus, for fixed $x \in \mathbb{T}^{n-1}$ and $f \in L_E^1 \cap H_0^1(\mathbb{T})$, $d_m(x, y)$ is analytic with respect to the variable $y \in \mathbb{T}$. Hence (3.2.17) defines a Hardy martingale. □

Almost Independent Functions

Preparing for the construction of Bourgain's embedding operator, we show that the Meyer–Bonami embedding T respects product structures and almost independent functions. In particular, it maps Riesz products on $\mathbb{T}$ to Riesz products on $\mathbb{T}^{\mathbb{N}}$.

We specify the term "almost-independence" as follows: Let (a_j) and (n_j) be integer sequences such that the expression (3.2.14) holds true. Let

$$M(j) = [-a_j, a_j] \setminus \{0\}, \tag{3.2.18}$$

and let s_j be a trigonometric polynomial whose frequency support is contained in $M(j)$. That is, we demand that

$$s_j(\zeta) = \sum_{m \in M(j)} \widehat{\sigma_j(m)} \zeta^m.$$

Then we say that $s_j(\zeta^{n_j})$ forms an almost independent sequence of polynomials.

Proposition 3.2.5 *Let $s_j(\zeta^{n_j})$ be an almost independent sequence of polynomials. Let h_j be given by $h_j(\zeta) = s_j(\zeta^{n_j})$. Then for any $n \in \mathbb{N}$,*

$$T \prod_{j=1}^{n} (1 + h_j) = \prod_{j=1}^{n} T(1 + h_j).$$

Proof Let $w \in \mathbb{T}^{\mathbb{N}}$. Since $T(1 + h_j)(w) = (1 + s_j(w_j))$, the claimed identity is just

$$T \left(\prod_{j=1}^{n} (1 + h_j) \right) (w) = \prod_{j=1}^{n} (1 + s_j(w_j)). \tag{3.2.19}$$

Let $C(n, k)$ be the collection of all subsets in $\{1, \ldots, n\}$ consisting of k elements. Expanding the product yields

$$\prod_{j=1}^{n} (1 + h_j) = \sum_{k=1}^{n} \sum_{S \subseteq C(n,k)} \prod_{j \in S} h_j. \tag{3.2.20}$$

Now fix $S \subseteq C(n, k)$. Then $S = \{t_1, \ldots t_k\}$ and we assume the elements in S are indexed by their size. Since $h_j(\zeta) = s_j(\zeta^{n_j})$, we obtain

$$\prod_{j\in S} h_j(\zeta) = \sum_{m_1\in M(1)} \cdots \sum_{m_k\in M(k)} \prod_{l=1}^{k} \widehat{s_{t_l}}(m_l)\zeta^{n_{t_l}m_l}, \tag{3.2.21}$$

where $\widehat{s_{t_l}}(m_l)$ is the m_lth Fourier coefficient of the polynomial s_{t_l}. Separating coefficients and variables, we write

$$\prod_{l=1}^{k} \widehat{s_{t_l}}(m_l)\zeta^{n_l m_l} = \prod_{l=1}^{k} \widehat{s_{t_l}}(m_l) \prod_{l=1}^{k} \zeta^{n_l m_l}. \tag{3.2.22}$$

Apply the transfer operator T to (functions of) the variable ζ. This gives

$$T\left(\prod_{l=1}^{k} \zeta^{n_l m_l}\right)(w) = \prod_{l=1}^{k} w_l^{m_l}, \quad w = (w_l) \in \mathbb{T}^{\mathbb{N}}. \tag{3.2.23}$$

By (3.2.21), (3.2.22), and (3.2.23) we obtain that

$$T\left(\prod_{j\in S} h_j\right)(w) = \prod_{j\in S} s_j(w_j). \tag{3.2.24}$$

Inserting (3.2.24) into the product expansion formula (3.2.20), we obtain the identity (3.2.19). □

In the course of the above proof we showed that the Fourier coefficients of $g = \prod_{j=1}^{n}(1 + h_j)$ satisfy the relation $\widehat{g}(0) = 1$ and $\widehat{g}(m) = 0$ unless there exists $S \subseteq \{1, \ldots, n\}$ and $b_k \in M(k)$ such that $m = \sum_{k\in S} b_k n_k$, in which case

$$\widehat{g}(m) = \prod_{k\in S} \widehat{s_k}(b_k). \tag{3.2.25}$$

3.2.2 The Bourgain Transfer Operator

Recall that $\mathcal{D}$ denotes the dyadic sigma-algebra on $\mathbb{T}^{\mathbb{N}}$. Let $L^1_{\mathcal{D}}(\mathbb{T}^{\mathbb{N}})$ be the subspace of $L^1(\mathbb{T}^{\mathbb{N}})$ consisting of $\mathcal{D}$-measurable functions. Accordingly, $\mathbb{E}_{\mathcal{D}}\colon L^1(\mathbb{T}^{\mathbb{N}}) \to L^1_{\mathcal{D}}(\mathbb{T}^{\mathbb{N}})$ denotes the conditional expectation operator.

The next theorem summarizes Bourgain's analysis of the Meyer–Bonami operator T. It yields a nonatomic sigma-algebra Σ on $\mathbb{T}$ such that the identity on $L^1(\mathbb{T}, \Sigma)$ factors through the transfer operator T.

Theorem 3.2.6 *There exists a diffuse sigma-algebra Σ on $\mathbb{T}$, and a bounded linear operator $J\colon L^1(\mathbb{T}) \to L^1(\mathbb{T}^{\mathbb{N}})$ satisfying the following conditions. The restriction of J to $L^1(\Sigma)$ is invertible on its range, that is, there exists $A_1 < \infty$ such that*

$$\|h\|_{L^1(\mathbb{T})} \le A_1 \|Jh\|_{L^1}, \qquad h \in L^1(\Sigma). \tag{3.2.26}$$

The operator J preserves analyticity as follows,

$$J(H^1(\mathbb{T})) \subseteq H^1(\mathbb{T}^{\mathbb{N}}). \tag{3.2.27}$$

On the closed subspace $J\left(L^1(\Sigma)\right)$, the conditional expectation $\mathbb{E}_{\mathcal{D}}$ is a small perturbation of the identity,

$$\|(Jh) - \mathbb{E}_{\mathcal{D}}(Jh)\|_{L^1} \leq (4A_0)^{-1}\|\mathbb{E}_{\mathcal{D}}(Jh)\|_{L^1}, \qquad h \in L^1(\Sigma), \tag{3.2.28}$$

where the constant $A_0 < \infty$ satisfies inequality (3.1.85).

Proof The proof of Theorem 3.2.6 utilizes the transfer operator of Meyer–Bonami and exploits its properties recorded in Theorem 3.2.4 and Proposition 3.2.5. We first introduce the operator J using almost independent copies of Fejer kernels. In the second part, we define Σ, the nonatomic sigma-algebra, and show that the restriction of J to $L^1(\Sigma)$ is a small perturbation of the identity. In the final part we verify that the elements of the closed space $JL^1(\Sigma)$ satisfy the identity (3.2.28).

Part 1: Defining J and Σ

Let $a \in \mathbb{N}$. We let F_a denote the Fejer kernel defined by,

$$F_a(\zeta) = \sum_{m=-a}^{a} \widehat{F_a}(m)\zeta^m,$$

where $\widehat{F_a}(m) = 1 - |m|/(1 + a)$. Recall that we denote $\sigma(\zeta) = \operatorname{sign}\Re(\zeta)$. Let $A_0 < \infty$ be the constant arising in inequality (3.1.85). Chose $\epsilon = \epsilon(A_0) > 0$ sufficiently small. By Fejer's theorem we may choose an increasing sequence of integers (a_j) such that

$$\|F_{a_j} * \sigma - \sigma\|_{L^1(\mathbb{T})} \leq 2^{-j}\epsilon. \tag{3.2.29}$$

Next, we take a sequence of integers (n_j) satisfying the recursion (3.2.14). With this choice we form the highly periodized Fejer kernels $k_j(\zeta) = F_{a_j}(\zeta^{n_j})$ and put

$$K(\zeta) = \prod_{j=1}^{n} k_j(\zeta). \tag{3.2.30}$$

Let T be the Meyer–Bonami operator defined in inequality (3.2.15). For $g \in L^1(\mathbb{T})$ put $Jg = T(K * g)$. Clearly, the linear operator

$$J : L^1(\mathbb{T}) \to L^1(\mathbb{T}^{\mathbb{N}})$$

is bounded. By Theorem 3.2.4, J preserves analyticity and martingale structure. Indeed, we have the expression (3.2.27) and

$$J(\operatorname{span}\{z^m : m \in D_k\}) \subset \operatorname{span}\{w^\mu : \mu \in \mathbb{Z}(k)\},$$

where $D_k = \{\mu \cdot \nu : \mu \in \mathbb{Z}(k)\}$ and $\mathbb{Z}(k) = \{n \in \mathbb{Z}^{(\mathbb{N})} : n = (n_1, n_2, \ldots, n_k, 0)\}$.

Next, we define the sigma-algebra for which the conclusion of Bourgain's transfer theorem holds true. Let (a_j) satisfy (3.2.29), and let (n_j) be given by the recursion (3.2.14). We define Σ to be the sigma-algebra on $\mathbb{T}$ generated by the sequence of Rademacher functions $(\sigma(\zeta^{n_j})\colon j \in \mathbb{N})$, where $\zeta \in \mathbb{T}$.

Part 2: Comparing $K * \mathbb{E}_\Sigma$ and $\mathbb{E}_\Sigma$

We now turn to the estimates for J claimed in Theorem 3.2.6. The first step in this direction consists in proving that on $L^1(\Sigma)$ the convolution by K is a small perturbation of the identity.

To this end we exhibit the multiplicative structure of $\widehat{K}$. By the recursion (3.2.14) and (3.2.25) the Fourier coefficients of the product K are the product of the Fourier coefficients of its factors. Specifically, for any $S \subseteq \{1, \ldots, n\}$,

$$\widehat{K}\left(\sum_{k \in S} b_k n_k\right) = \prod_{k \in S} \widehat{F_{a_k}}(b_k), \qquad b_k \in M(k), \tag{3.2.31}$$

where $M(k) = [-a_k, a_k]\backslash\{0\}$ is defined by equation (3.2.18). Otherwise, we have $\widehat{K}(m) = 0$. The following identity will be repeatedly applied in the course of proving Theorem 3.2.6.

Proposition 3.2.7 *Let $\sigma \in L^1(\mathbb{T})$, with $\int \sigma dm = 0$. Then $h_j(\zeta) = \sigma(\zeta^{n_j})$ satisfies*

$$K * \left(\prod_{j=1}^{n} (1 + h_j)\right) = \prod_{j=1}^{n} (1 + k_j * h_j). \tag{3.2.32}$$

Proof Expanding the products appearing on both sides of (3.2.32) shows that the identity (3.2.32) follows from the claim that for any $S \subseteq \{1, \ldots, n\}$,

$$K * \left(\prod_{j \in S} h_j\right) = \prod_{j \in S} k_j * h_j. \tag{3.2.33}$$

Thus, it suffices to verify (3.2.33) for $S = \{1, \ldots, k\}$. The left-hand side product coincides with

$$\sum_{m_1 \in M(1)} \cdots \sum_{m_k \in M(k)} \widehat{K}\left(\sum_{l \in S} m_l n_l\right) \prod_{l=1}^{k} \widehat{\sigma}(m_l) \zeta^{n_l m_l}. \tag{3.2.34}$$

By equation (3.2.31), the term (3.2.34) coincides with

$$\sum_{m_1 \in M(1)} \cdots \sum_{m_k \in M(k)} \prod_{l=1}^{k} \widehat{F_{a_l}}(m_l)\widehat{\sigma}(m_l)\zeta^{n_l m_l}. \tag{3.2.35}$$

A direct expansion of the product on the right-hand side of (3.2.33) leads to

$$\sum_{m_1 \in M(1)} \cdots \sum_{m_k \in M(k)} \prod_{l=1}^{k} \widehat{F_{a_l} * \sigma}(m_l)\zeta^{n_l m_l},$$

which coincides with expression (3.2.35). □

Proposition 3.2.8 *If* (a_j) *and* (n_j) *satisfy the inequalities* (3.2.29) *and* (3.2.14) *and if* K *is defined by* (3.2.30), *then*

$$\|K * h - h\|_{L^1(\mathbb{T})} \le \epsilon \|h\|_{L^1(\mathbb{T})}, \quad h \in L^1(\Sigma).$$

Proof Let $\mathbb{E}_\Sigma$ be the conditional expectation operator acting on $L^1(\mathbb{T})$ onto $L^1(\Sigma)$. We interpret $\mathbb{E}_\Sigma$ and $K * \mathbb{E}_\Sigma$ as integral operators. The first part of the proof consists in identifying the respective integral kernels. Then, by estimating the difference between the explicit kernels we will be able to prove that

$$\|K * (\mathbb{E}_\Sigma g) - \mathbb{E}_\Sigma g\|_{L^1(\mathbb{T})} \le \epsilon \|g\|_{L^1(\mathbb{T})}, \qquad g \in L^1(\mathbb{T}).$$

Step 1 We begin with the representation of $\mathbb{E}_\Sigma$. Let $g \in L^1(\mathbb{T})$ and $h = \mathbb{E}_\Sigma g$. Note that h is given by integration against the kernel

$$B(z, \zeta) = \prod_{k=1}^{\infty} (1 + \sigma(z^{n_k})\sigma(\zeta^{n_k})).$$

Thus,

$$h(z) = \int_{\mathbb{T}} B(z, \zeta) g(\zeta) dm, \tag{3.2.36}$$

where $z, \zeta \in \mathbb{T}$. Next, we turn to the kernel of $K * \mathbb{E}_\Sigma$. Put $s_k = F_{a_k} * \sigma$, fix again $z, \zeta \in \mathbb{T}$, and define

$$A(z, \zeta) = \prod_{k=1}^{\infty} (1 + s_k(z^{n_k})\sigma(\zeta^{n_k})).$$

Proposition 3.2.7 asserts that convolving with respect to z yields $K *_z B(z, \zeta) = A(z, \zeta)$. This gives the integral representation

$$K * h(z) = \int_{\mathbb{T}} A(z, \zeta) g(\zeta) dm(\zeta). \tag{3.2.37}$$

Step 2 Next we prove that the kernels A, B are close in the following sense:

$$\sup_{\zeta\in\mathbb{T}} \int_{\mathbb{T}} |A(z,\zeta) - B(z,\zeta)| dm(z) \leq \epsilon. \tag{3.2.38}$$

In view of (3.2.36) and (3.2.37) the estimate (3.2.38) finishes the proof of the proposition. To this end we fix $\zeta \in \mathbb{T}$ and put $\tau_k = \sigma(\zeta^{n_k})$. Let $A_0 = B_0 = 1$, and for $j \in \mathbb{N}$ put

$$A_j(z) = \prod_{k=1}^{j} (1 + s_k(z^{n_k})\tau_k), \qquad B_j(z) = \prod_{k=1}^{j} (1 + \sigma(z^{n_k})\tau_k).$$

Rewrite the difference $A_j(z) - B_j(z)$ of the kernels as follows:

$$A_{j-1}(z)\tau_j(s_j(z^{n_j}) - \sigma(z^{n_j})) + (A_{j-1}(z) - B_{j-1}(z))\tau_j(1 + \sigma(z^{n_j})).$$

Next, take absolute values and exploit that (z^{n_k}) is an almost independent sequence. Since $A_{j-1} \geq 0$,

$$\int_{\mathbb{T}} A_{j-1}(z) dm(z) = 1 \quad \text{and} \quad \int_{\mathbb{T}} (1 \pm \sigma(z)) dm(z) = 1,$$

and by invoking the estimate (3.2.29) it is easy to see that $\int_{\mathbb{T}} |A_j(z) - B_j(z)| dm(z)$ is bounded by

$$(1+\delta_j) \int_{\mathbb{T}} |A_{j-1}(z) - B_{j-1}(z)| dm(z) + (1+\delta_j) \int_{\mathbb{T}} |s_j(z) - \sigma(z)| dm(z),$$

where $\delta_j \leq 2^{-j}$. Iterating the last estimate proves (3.2.38). □

Remark: We infer from Proposition 3.2.8 that the restriction to $L^1(\Sigma)$ of the transfer operator J is a small perturbation of the identity. Hence the restriction is invertible on its range and the estimate (3.2.26) holds true. This then implies that $JL^1(\Sigma)$ is a closed subspace of $L^1(\mathbb{T}^{\mathbb{N}})$.

Part 3: Almost Dyadic Martingales on $\mathbb{T}^{\mathbb{N}}$

The last part in the proof of Theorem 3.2.6 consists in proving that the restriction of $\mathbb{E}_{\mathcal{D}}$ to $JL^1(\Sigma)$ is a small perturbation of the identity. The next proposition shows that (3.2.28) holds true.

Proposition 3.2.9 *Let $g \in L^1(\mathbb{T})$. If the integer sequences (a_j) and (n_j) satisfy expressions* (3.2.29) *and* (3.2.14)*, and $h = \mathbb{E}_\Sigma g$, then*

$$\|\mathbb{E}_{\mathcal{D}} J(h) - J(h)\|_{L^1} \leq \epsilon \|h\|_{L^1}. \tag{3.2.39}$$

Proof We fix $g \in L^1(\mathbb{T})$ and put $h = \mathbb{E}_\Sigma g$. We isolate the integral representation of $\mathbb{E}_{\mathcal{D}} Jh$ and compare it to the one for Jh. This will yield the estimate (3.2.39). We again use the sequence $s_k = F_{a_k} * \sigma$ approximating $\sigma(\zeta) = \operatorname{sign}(\cos(\zeta))$.

Step 1 We begin with $J\mathbb{E}_\Sigma$. First, $X(w,\zeta) = \prod_{k=1}^{\infty}(1 + s_k(w_k)\sigma(\zeta^{n_k}))$, where $w \in \mathbb{T}^{\mathbb{N}}$ and $\zeta \in \mathbb{T}$, is the kernel of $J\mathbb{E}_\Sigma$. Indeed, by Proposition 3.2.7, we have

$$Jh(w) = \int_{\mathbb{T}} X(w,\zeta)g(\zeta)dm(\zeta). \tag{3.2.40}$$

By equation (2.2.17), $\mathbb{E}_{\mathcal{D}}$ is an integral operator with kernel $\prod_{k=1}^{\infty}(1 + \sigma_k(w)\sigma_k(z))$. Taking the conditional expectation of equation (3.2.40) and performing a routine calculation shows that, for $w \in \mathbb{T}^{\mathbb{N}}$,

$$\mathbb{E}_{\mathcal{D}}(Jh)(w) = \int_{\mathbb{T}} Y(w,\zeta)g(\zeta)dm(\zeta), \tag{3.2.41}$$

where $Y(w,\zeta) = \prod_{k=1}^{\infty} 1 + \gamma_k\sigma(w_k)\sigma(\zeta^{n_k})dm(z)$ and $\gamma_k = \int_{\mathbb{T}} \sigma(z)s_k(z)dm(z)$.

Step 2 In view of (3.2.40) and (3.2.41), the inequality (3.2.39) follows from estimating the difference between the kernels X and Y. We claim that

$$\sup_{\zeta\in\mathbb{T}} \mathbb{E}_w(|X(w,\zeta) - Y(w,\zeta)|) \leq \epsilon. \tag{3.2.42}$$

To see this, we fix $\zeta \in \mathbb{T}$ and put $\tau_k = \sigma(\zeta^{n_k})$. Define the partial products $X_0 = Y_0 = 1$, and

$$X_j(w) = \prod_{k=1}^{j}(1 + s_k(w_k)\tau_k), \qquad Y_j(w) = \prod_{k=1}^{j}(1 + \gamma_k\sigma(w_k)\tau_k),$$

where $j \in \mathbb{N}$ and $w \in \mathbb{T}^{\mathbb{N}}$. Rewrite the difference $X_j(w) - Y_j(w)$ as

$$X_{j-1}(w)\tau_j(s_j(w_j) - \sigma(w_j)) + (X_{j-1}(w) - Y_{j-1}(w))\tau_j(1 + \gamma_j\sigma(w_j)).$$

The second summand in the above term coincides with

$$(X_{j-1}(w) - Y_{j-1}(w))\tau_j(1 + \sigma(w_j)) - (X_{j-1}(w) - Y_{j-1}(w))\tau_j(1 - \gamma_j)\sigma(w_j)).$$

Hence,

$$\mathbb{E}|X_j - Y_j| \leq (1 + \epsilon_j)\mathbb{E}|X_{j-1} - Y_{j-1}| + 2\epsilon_j\mathbb{E}|X_j|,$$

where we put $\int_{\mathbb{T}} |s_j - \sigma|dm = \epsilon_j$. Since $X_j > 0$, we have $\mathbb{E}|X_j| = 1$. Hence iterating the above estimate down to $j = 1$ gives $\mathbb{E}|X_j - Y_j| \leq C\sum_{k=1}^{j} \epsilon_k$. By our initial choice of $\epsilon_k = \epsilon 2^{-k}$, this gives the estimate (3.2.42). □

3.2.3 Separation in $L^1(\mathbb{T})$

Finally, we show that $L^1(\mathbb{T})/H_0^1(\mathbb{T})$ contains a copy of L^1. The realization of the embedding is the natural one. Let q denote the quotient mapping

$$q(g) = \left\{g + H_0^1(\mathbb{T})\right\}, \qquad g \in L(\mathbb{T}),$$

associating to $g \in L^1(\mathbb{T})$ its equivalent class in $L^1(\mathbb{T})/H_0^1(\mathbb{T})$. We will prove that, in $\mathbb{T}$, there exists a nonatomic sigma-algebra Σ, such that the restricted quotient map

$$q_\Sigma : L^1(\Sigma) \to L^1(\mathbb{T})/H_0^1(\mathbb{T}), \qquad q_\Sigma(h) = \left\{h + H_0^1(\mathbb{T})\right\}, \tag{3.2.43}$$

is invertible on its range. Specifically, we will show that, for each $h \in L^1(\Sigma)$,

$$\|h\|_{L^1(\mathbb{T})} \le A \inf\left\{\|h - f\|_{L^1(\mathbb{T})} : f \in H_0^1(\mathbb{T})\right\}. \tag{3.2.44}$$

The conditional expectation operator $\mathbb{E}_\Sigma : H^\infty(\mathbb{T}) \to L^\infty(\Sigma)$ is just the transpose of the operator of q_Σ. Indeed, for any $f \in H^\infty(\mathbb{T})$ and $h \in L^1(\Sigma)$ we have,

$$\langle h, \mathbb{E}(f|\Sigma)\rangle = \langle h, f\rangle = \langle q_\Sigma(h), f\rangle = \langle h, q_\Sigma^*(f)\rangle.$$

The estimate (3.2.44) asserts that the range of q_Σ is closed in $L^1(\mathbb{T})/H_0^1(\mathbb{T})$. In view of the Hahn–Banach theorem, its transpose is surjective.

Theorem 3.2.10 *There exists a non atomic sigma-algebra Σ and $A < \infty$ such that for each $h \in L^1(\Sigma)$,*

$$\|h\|_{L^1(\mathbb{T})} \le A \inf\left\{\|h - f\|_{L^1(\mathbb{T})}\right\},$$

where the infimum is extended over $f \in H_0^1(\mathbb{T})$. By duality, the conditional expectation operator

$$\mathbb{E}_\Sigma : H^\infty(\mathbb{T}) \to L^\infty(\Sigma)$$

is surjective.

Proof Let Σ denote the diffuse sigma-algebra in $\mathbb{T}$, determined in the course of proving Theorem 3.2.6. Let $h \in L^1(\Sigma)$ and $f \in H_0^1(\mathbb{T})$. Put $H = Jh$, $F = Jf$, and $D = \mathbb{E}_\mathcal{D} H$. By (3.2.27), F is a Hardy martingale and $\mathbb{E}(F) = 0$. Since D is a dyadic martingale on $\mathbb{T}^\mathbb{N}$, Theorem 3.1.12 gives

$$\|D\|_{L^1} \le A_0\|D - F\|_{L^1}. \tag{3.2.45}$$

Write $D - F = (D - H) + (H - F)$. Since $D = \mathbb{E}_\mathcal{D} H$, combining the estimate (3.2.28) with the expression (3.2.45) gives

$$\|D\|_{L^1} \le A_0\|H - F\|_{L^1} + \frac{1}{4}\|D\|_{L^1}. \tag{3.2.46}$$

Since $H - F = J(h - f)$, we get

$$\|D\|_{L^1} \leq 2A_0\|J\| \cdot \|h - f\|_{L^1(\mathbb{T})}. \tag{3.2.47}$$

By the estimate (3.2.26), the restriction of J to $L^1(\Sigma)$ is an invertible operator. Hence, in view of inequality (3.2.28) we have

$$\|h\|_{L^1(\mathbb{T})} \leq 2A_1\|D\|_{L^1}. \tag{3.2.48}$$

Combining (3.2.47) and (3.2.48) we finally obtain the separation estimate

$$\|h\|_{L^1(\mathbb{T})} \leq A\|h - f\|_{L^1(\mathbb{T})},$$

for arbitrary $h \in L^1(\Sigma)$ and $f \in H^1_0(\mathbb{T})$. □

Remark: Summing up the estimates of the proof in compressed form, we have

$$\|J\| \cdot \|q_\Sigma(h)\|_{L^1(\mathbb{T}^\mathbb{N})/H^1_0(\mathbb{T}^\mathbb{N})} \geq \|q(H)\|_{L^1(\mathbb{T}^\mathbb{N})/H^1_0(\mathbb{T}^\mathbb{N})}$$

and

$$\begin{aligned}\|q(H)\|_{L^1(\mathbb{T}^\mathbb{N})/H^1_0(\mathbb{T}^\mathbb{N})} &\geq \|q(D)\|_{L^1(\mathbb{T}^\mathbb{N})/H^1_0(\mathbb{T}^\mathbb{N})} - \|D - H\|_{L^1(\mathbb{T}^\mathbb{N})} \\ &\geq \frac{\|D\|_{L^1(\mathbb{T}^\mathbb{N})}}{2A_0} \geq \frac{\|h\|_{L^1(\mathbb{T})}}{4A_0A_1}\end{aligned}$$

□

Concluding remarks and open problems:

(i) We use the Hahn–Banach theorem and duality to conclude that there exists a nonatomic sigma-algebra on $\mathbb{T}$ such that the conditional expectation operator $\mathbb{E}_\Sigma : H^\infty(\mathbb{T}) \to L^\infty(\Sigma)$ is surjective. Thus, for any $g \in L^\infty(\Sigma)$ there exists $h \in H^\infty(\mathbb{T})$ that solves the equation

$$\mathbb{E}_\Sigma(h) = g. \tag{3.2.49}$$

During the Winter School in Harmonic Analysis at Okemo Mountain (1999), Peter W. Jones set forth a constructive, complex analytic method for solving equation (3.2.49). Jones's explicit approach avoids martingale techniques and the Hahn–Banach theorem.

(ii) Let S be a compact metric space and let $M(S)$ denote the space of signed Radon measures on S equipped with the total variation norm of measures. Let $C(S)$ denote the space of continuous functions on S equipped with the supremum norm. By the Riesz representation theorem, the bilinear form, defined on $M(S) \times C(S)$ by

$$(\mu, f) \to \int_S f d\mu,$$

identifies $M(S)$ as the dual space of $C(S)$. We say that a subspace $Z \subseteq M(S)$ is weak-$*$ closed if it is closed with respect to the $\sigma(M(S), C(S))$ topology. Clearly, for a probability measure $\mu \in M(S)$, any subspace of $L^1(S,\mu)$ is a subspace of $M(S)$.

Let μ denote a nonatomic probability measure on S. Let X be a norm closed subspace of $L^1(S,\mu)$, and assume that X is a weak-$*$ closed subspace of $M(S)$. Talagrand (1990) conjectured that under these assumptions, $L^1(S,\mu)/X$ contains a subspace isomorphic to $L^1(S,\mu)$.

3.3 Notes

Chapter 3 presents the result of Bourgain (1983a) that L^1 embeds into L^1/H^1 (Theorem 3.2.10).

Section 3.1 contains the proof of the martingale separation theorem (Theorem 3.1.12), the first main theorem in Bourgain (1983a). Theorems 3.1.1 and 3.1.3 are taken from Müller (1993, 2012, 2014). Theorem 3.1.8 uses Bourgain (1983a) as source. Propositions 3.1.10 and 3.1.11 are Bourgain's interpretation of the Garnett–Jones Theorem (Garnett, 1981; Garnett and Jones, 1978, 1982).

Section 3.2 presents the transfer of the martingale separation theorem (Theorem 3.1.12) to the unit disk. Theorem 3.2.4 goes back to Meyer (1968) and Bonami (1970). Our presentation incorporates the paper by Kazaniecki and Wojciechowski (2014). The source for Theorem 3.2.6 is Bourgain (1983a). Theorem 3.2.10 is the eponymous theorem of Bourgain (1983a).

4

Embedding L^1 in X or L^1/X

This chapter presents those L^1 subspaces X – other than $H^1(\mathbb{T})$ – for which it is known that L^1 embeds into X or L^1/X. Thereafter, we introduce refined techniques of L^1 embedding and show the existence of a normalized weakly null sequence in L^1 that doesn't contain an unconditional basic sequence. Finally, we prove that there exists an L^1 subspace X, such that neither X nor L^1/X contain a copy of L^1.

Section 4.1 contains fundamental representation theorems for continuous linear operators acting on L^0 and L^1. We present Kwapień's characterization of operators on L^0 and the analogous theorem by Kalton for operators on L^1. In Section 4.2 we separately consider complemented subspaces of L^1, reflexive subspaces of L^1, and subspaces of L^1 isomorphic to ℓ^1. We prove that for each of these spaces, L^1 embeds into X or into L^1/X.

We present the Johnson–Maurey–Schechtman construction of a weakly-null sequence F_{JMS} in L^1, such that for any (!) infinite subsequence (φ_{n_k}) of F_{JMS}, the space $\overline{\operatorname{span}}\{\varphi_{n_k}\}$ contains a copy of L^1. In Section 4.3 we exploit random measures and prove Rosenthal's L^1-theorem. In Section 4.4 we study a very peculiar L^1-approximation problem and its combinatorial solution due to Talagrand. In Section 4.5 we define Talagrand's space $X \subset L^1$, prove that neither X nor L^1/X contain a copy of L^1, and determine two Banach spaces, T_1 and T_2, such that L^1 embeds into $T_1 \oplus T_2$, whereas neither T_1 nor T_2 contain a copy of L^1.

4.1 Random Measures Representing Operators

We fix a finite measure space $(S, \mathcal{F}, \lambda)$. In Section 1.1 we introduced $\mathcal{L}^0 = \mathcal{L}^0(S, \mathcal{F})$, the vector space of real-valued $\mathcal{F}$-measurable functions on S. We let

$\mathcal{N} = \mathcal{N}(S, \mathcal{F}, \lambda)$ consist of those $f \in \mathcal{L}^0$ for which $\lambda(\{f \neq 0\}) = 0$, and formed the quotient space

$$L^0 = L^0(S, \mathcal{F}, \lambda) = \mathcal{L}^0/\mathcal{N},$$

equipped with the measure distance d_λ. Recall that for $f, g \in \mathcal{L}^0$, the measure distance is given by

$$d_\lambda(f, g) = \inf\{\epsilon > 0 : \lambda(\{|f - g| > \epsilon\}) < \epsilon\}, \tag{4.1.1}$$

and $d_\lambda(f, g) = 0$ if $f - g \in \mathcal{N}$. Note that for (f_n) and g in L^0 the convergence $d_\lambda(f_n, g) \to 0$ is equivalent to the convergence in λ-measure of (f_n) to g. As observed in Section 1.1, a linear operator

$$T\colon L^0(\Omega, \mathcal{B}, \mu) \to L^0(S, \mathcal{F}, \lambda), \tag{4.1.2}$$

is continuous with respect to the measure distances d_μ and d_λ if and only if for any $\epsilon > 0$ there exists $\delta > 0$ such that, for $f \in L^0(\Omega, \mathcal{B}, \mu)$,

$$\mu(\{|f| > \delta\}) < \delta \quad \text{implies} \quad \lambda(\{|Tf| > \epsilon\}) < \epsilon.$$

Random Measures

We set $\Omega = [0, 1]$, let $\mathcal{B}$ denote the Borel sigma-algebra on $[0, 1]$, and denote by μ the Lebesgue measure on $[0, 1]$. We say that

$$\mathcal{T} = \{\mu_x\colon \mathcal{B} \to \mathbb{R},\ x \in S\}, \tag{4.1.3}$$

is a random measure from $(S, \mathcal{F})$ to $(\Omega, \mathcal{B})$, if each μ_x is a finite measure, and if $x \to \mu_x(A)$ is $\mathcal{F}$-measurable for all $A \in \mathcal{B}$. The random measure (4.1.3) defines a linear operator

$$\widetilde{T}\colon \text{span}\{1_A\colon A \in \mathcal{B}\} \to \mathcal{L}^0(S, \mathcal{F})$$

by

$$\left(\widetilde{T}1_A\right)(x) = \mu_x(A) = \int_\Omega 1_A d\mu_x, \tag{4.1.4}$$

for $x \in S$, $A \in \mathcal{B}$, and linear extension. We assume that $\mathcal{T} = \{\mu_x : x \in S\}$ satisfies the following condition:

$$\text{If } \mu(N) = 0 \text{ then } \mu_x(N) = 0, \text{ for } \lambda\text{-almost every } x \in S. \tag{4.1.5}$$

Condition (4.1.5) allows us to obtain a well-defined extension of equation (4.1.4) acting on equivalence classes of measurable functions. If

$$[f]_1 = \{f + \mathcal{N}(\Omega, \mathcal{B}, \mu)\}, \quad f \in \mathcal{L}^0(\Omega, \mathcal{B}),$$

and

$$[g]_2 = \{g + \mathcal{N}(S, \mathcal{F}, \lambda)\}, \quad g \in \mathcal{L}^0(S, \mathcal{F}),$$

then condition (4.1.5) translates into the following identity between equivalence classes,

$$T([f]_1) = \left[\widetilde{T}(f)\right]_2, \quad f \in \operatorname{span}\{1_A : A \in \mathcal{B}\}.$$

Summing up, by means of equation (4.1.4), any random measure $\mathcal{T} = \{\mu_x : x \in S\}$ satisfying condition (4.1.5) gives rise to a well-defined linear operator acting on the quotient spaces,

$$T : \operatorname{span}\{1_A : A \in \mathcal{B}\}/\mathcal{N}(\Omega, \mathcal{B}, \mu) \to L^0(S, \mathcal{F}, \lambda).$$

In Section 4.1.1 we present Kwapień's theorem (Theorem 4.1.3), which shows that any measure-continuous linear operator

$$T : L^0(\Omega, \mathcal{B}, \mu) \to L^0(S, \mathcal{F}, \lambda)$$

is uniquely represented by a random measure $\mathcal{T} = \{\mu_x : \mathcal{B} \to \mathbb{R}, x \in S\}$ satisfying condition (4.1.5) and

$$(T1_A)(x) = \mu_x(A),$$

for $x \in S$ and $A \in \mathcal{B}$, and provides the intrinsic description of the measures $\mu_x \in \mathcal{T}$.

In Section 4.1.2 we develop the idea in the context of operators on L^1 spaces. We show Kalton's theorem (Theorem 4.1.15), which asserts that any bounded linear operator

$$T : L^1(\Omega, \mathcal{B}, \mu) \to L^1(S, \mathcal{F}, \lambda)$$

is represented by a random measure $\mathcal{T} = \{\mu_x : \mathcal{B} \to \mathbb{R}, x \in S\}$, and characterize the measures $\mu_x \in \mathcal{T}$ in the spirit of Kwapień's original conditions.

4.1.1 Kwapień's Representation

We begin by fixing notation. We put $\Omega = [0, 1]$ and let $\mathcal{B}$ denote the Borel sigma-algebra on $[0, 1]$. For $N \in \mathbb{N}$, μ_N denotes the Lebesgue measure on $[0, 1]^N$. Thus $\mu = \mu_1$ is Lebesgue measure on $\Omega = [0, 1]$. Separately, we fix a finite, complete measure space $(S, \mathcal{F}, \lambda)$.

Overview of Section 4.1.1

Kwapień's Theorem 4.1.3 asserts that for every continuous linear operator

$$T\colon L^0(\Omega,\mathcal{B},\mu)\to L^0(S,\mathcal{F},\lambda)$$

there exists a unique a random measure $\mathcal{T}=\{\mu_x\colon\mathcal{B}\to\mathbb{R},\ x\in S\}$ satisfying condition (4.1.5) such that

$$Tf(x)=\int_\Omega f d\mu_x,\quad f\in L^0(\Omega,\mu),$$

for λ-almost every $x\in S$, and, moreover, $\mu_x\in\mathcal{T}$ is purely atomic with finite support. Specifically, for λ-almost every $x\in S$, there exist $a_n(x)\in\mathbb{R}$, and $\sigma(x)\in\Omega$ such that

$$\mu_x=\sum_{n=1}^{A(x)}a_n(x)\delta_{\sigma_n(x)},$$

where $\delta_{\sigma_m(x)}$ denotes the Dirac measure at $\delta_{\sigma_m(x)}$. In the course of the proof, we determine the functions a_n, σ_n and $A\colon S\to\mathbb{N}\cup\{\infty\}$ constructively, and show that they are Borel measurable and satisfy the conditions

(i) $\sigma_n(x)\neq\sigma_m(x),\ 1\le m\neq n<A(x)$,
(ii) $\lambda(\{A=\infty\})=0$.

The random measure $\mathcal{T}=\{\mu_x : x\in S\}$, exhibited by Theorem 4.1.3, is called the Kwapień representation of T. Conversely, if a random measure satisfies condition (4.1.5) and consists of finitely supported, purely atomic measures, λ-almost surely, then equation (4.1.4) gives rise to a continuous linear operator equation (4.1.2). (Theorem 4.1.5.)

Dyadic Intervals

Recall that $I\subseteq\Omega=[0,1[$ is a dyadic interval if there exists $n\in\mathbb{N}\cup\{0\}$ and $1\le i\le 2^n$ such that $I=[(i-1)/2^n,i/2^n[$. We let $\mathcal{I}$ denote the collection of all dyadic intervals contained in the unit interval $[0,1[$. For $I\in\mathcal{I}$, there exist uniquely determined $I_1,I_2\in\mathcal{I}$ such that $I_1\cap I_2=\emptyset$ and $I_1\cup I_2=I$. In that case, $|I_1|=|I_2|=|I|/2$. If $\inf I_1=\inf I$ we say that I_1 is the left half of I, and

$$h_I=1_{I_1}-1_{I_2},\tag{4.1.6}$$

is called the L^∞-normalized Haar function, supported on I. Finally, we denote $\mathcal{I}_n=\{I\in\mathcal{I}\colon|I|=2^{-n}\}$. Several layers of dyadic intervals are depicted in Figure 4.1.

Lemma 4.1.1 *Let $T\colon L^0(\Omega,\mathcal{B},\mu)\to L^0(S,\mathcal{F},\lambda)$ be a continuous linear operator. Let*

$$E_I=\{(T1_I)\neq 0\},\qquad I\in\mathcal{I}.$$

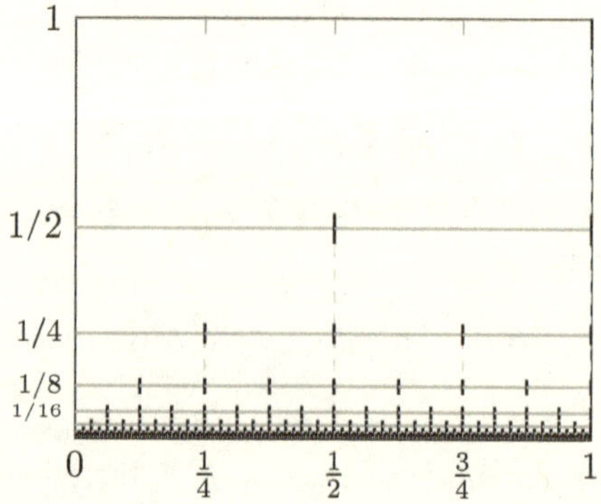

Figure 4.1 Dyadic intervals $I \subseteq [0, 1)$ represented in the unit square as $I \times |I|$.

Then, for $\epsilon > 0$, there exists $\delta > 0$ such that for any $\mathcal{K} \subset \mathcal{I}$, consisting of pairwise disjoint dyadic intervals,

$$\mu\left(\bigcup_{I \in \mathcal{K}} I\right) < \delta \quad \textit{implies} \quad \lambda\left(\bigcup_{I \in \mathcal{K}} E_I\right) < \epsilon.$$

Proof Fix $\epsilon > 0$, and choose $\delta > 0$ such that $\lambda(\{|Tf| > \epsilon\}) < \epsilon$ for $f \in L^0(\Omega)$ with $\mu(\{|f| > \delta\}) < \delta$.

Step 1 (Defining testing functions) Let $\mathcal{K} \subset \mathcal{I}$ consist of pairwise disjoint dyadic intervals satisfying $\mu(\bigcup \mathcal{K}) < \delta$. Without loss of generality, we may assume that $\mathcal{K}$ is of finite cardinality, with $\operatorname{card}(\mathcal{K}) < N$ say. Fix $y_I \in \mathbb{R}$, $I \in \mathcal{K}$, put $y = (y_I)$, and form

$$f_y = \sum_{I \in \mathcal{K}} y_I 1_I.$$

Note that $\mu(\{|f_y| > \delta\}) < \delta$, independent of $y \in \mathbb{R}^N$. Defining the testing functions $w_I = T1_I$ gives,

$$\lambda\left(\left\{\left|\sum_{I \in \mathcal{K}} y_I w_I\right| > \epsilon\right\}\right) < \epsilon, \tag{4.1.7}$$

for any choice of $(y_I) \in \mathbb{R}^N$. By rescaling, and by exploiting that Inequality (4.1.7) holds true for any choice of vectors $(y_I) \in \mathbb{R}^N$ we obtain,

$$\lambda\left(\left\{\sum_{I \in \mathcal{K}} y_I w_I \neq 0\right\}\right) < \epsilon. \tag{4.1.8}$$

Step 2 (Randomization) We show that there exist $y = (y_I) \in [0,1]^N$ such that

$$\lambda\left(\left\{\sum_{I\in\mathcal{K}} y_I w_I \neq 0\right\}\right) = \lambda\left(\bigcup_{I\in\mathcal{K}} E_I\right). \tag{4.1.9}$$

To this end, we fix $w \in \mathbb{R}^N$ with $w \neq 0$, and form $L = \{y \in \mathbb{R}^N : \langle w, y\rangle = 0\}$. Since $\dim L = N - 1$, we have

$$\mu_N\left(L \cap [0,1]^N\right) = 0, \tag{4.1.10}$$

where μ_N denotes the Lebesgue measure on $[0,1]^N$. Given $y = (y_I) \in [0,1]^N$, define

$$p_I(y) = y_I, \qquad I \in \mathcal{K}.$$

Fix $x \in \bigcup\{E_I : I \in \mathcal{K}\}$ and let $w = (w_I(x))$. Then $w \neq 0$ and, hence, (4.1.10) gives

$$\mu_N\left(\left\{y \in [0,1]^N : \sum_{I\in\mathcal{K}} p_I(y)w_I(x) \neq 0\right\}\right) = 1. \tag{4.1.11}$$

Conversely, if $x \notin \bigcup\{E_I : I \in \mathcal{K}\}$ then $w_I(x) = 0$ for $I \in \mathcal{K}$, and hence

$$\mu_N\left(\left\{y \in [0,1]^N : \sum_{I\in\mathcal{K}} p_I(y)w_I(x) \neq 0\right\}\right) = 0. \tag{4.1.12}$$

Integrating – with respect to λ – the equations (4.1.11) and (4.1.12) over $x \in S$ we obtain,

$$\int_S \mu_N\left(\left\{y \in [0,1]^N : \sum_{I\in\mathcal{K}} p_I(y)w_I(x) \neq 0\right\}\right) d\lambda(x) = \lambda\left(\bigcup_{I\in\mathcal{K}} E_I\right).$$

Fubini's theorem gives

$$\int_{[0,1]^N} \lambda\left(\left\{x \in S : \sum_{I\in\mathcal{K}} p_I(y)w_I(x) \neq 0\right\}\right) d\mu_N(y) = \lambda\left(\bigcup_{I\in\mathcal{K}} E_I\right).$$

Finally, applying the mean value theorem to the continuous functions $(p_I : I \in \mathcal{K})$, shows that there exists $y \in [0,1]^N$ such that equation (4.1.9) holds true. Combining the assertions (4.1.8) and (4.1.9) finishes the proof. □

The Key to Kwapień's Representation Theorem

Proposition 4.1.2 introduces a random selection method and decoupling estimates. We first shorten the notation by putting $L^0(\Omega) = L^0(\Omega, \mathcal{B}, \mu)$ and $L^0(S) = L^0(S, \mathcal{F}, \lambda)$, and fix a continuous linear operator

$$T : L^0(\Omega) \to L^0(S).$$

We utilize the continuity of T by studying its behavior on the testing functions (4.1.13). Recall that $\mathcal{I}$ denotes the collection of dyadic intervals contained in the unit interval. We put

$$E_I = \{T1_I \neq 0\} \quad \text{for} \quad I \in \mathcal{I}.$$

Given $n \in \mathbb{N}$, we let $\mathcal{I}_n = \{I \in \mathcal{I} : |I| = 2^{-n}\}$, define $w_n \colon S \to \{0, \dots, 2^n\}$ by

$$w_n(T) = \sum_{I \in \mathcal{I}_n} 1_{E_I}, \tag{4.1.13}$$

and put

$$A(T) = \sup_{n \in \mathbb{N}} w_n(T). \tag{4.1.14}$$

The sequence $(w_n(T))$ is nondecreasing, pointwise. To verify the monotonicity of $(w_n(T))$ note that if $I \in \mathcal{I}_n$ and $I_1, I_2 \in \mathcal{I}_{n+1}$ satisfy $I = I_1 \cup I_2$, then

$$E_I \subseteq E_{I_1} \cup E_{I_2}.$$

Hence, $w_n(T)(x) \leq w_{n+1}(T)(x)$, for $x \in S$.

Proposition 4.1.2 *If $T \colon L^0(\Omega) \to L^0(S)$ is a linear and continuous operator, then $A(T) \colon S \to \mathbb{N} \cup \{\infty\}$ defined by* (4.1.14) *satisfies*

$$\lambda(\{A(T) = \infty\}) = 0.$$

Proof Fix $\epsilon > 0$. Set $C_0 = \lambda(S)$ and put $\epsilon_0 = \epsilon/(4C_0)$. By Lemma 4.1.1, there exists $\delta = \delta(\epsilon)$ such that

$$\mu\left(\bigcup_{I \in \mathcal{K}} I\right) < \delta \quad \text{implies} \quad \lambda\left(\bigcup_{I \in \mathcal{K}} E_I\right) < C_0 \epsilon_0, \tag{4.1.15}$$

for any collection of pairwise disjoint dyadic intervals $\mathcal{K}$. Now fix $n \in \mathbb{N}$ such that $2(2^n\delta)^{-1} \leq \epsilon_0$. Let $M > (8/\delta)$, and put $S_M = \{x \in S : w_n(T)(x) > M\}$. Inequality (4.1.23), in the third step of the proof, gives $\lambda(S_M) < \epsilon$. Since the sequence $(w_n(T))$ is nondecreasing, pointwise, this shows that Proposition 4.1.2 holds true.

Step 1 (Random selectors) Here $n \in \mathbb{N}$ is fixed and $\delta > 0$ is determined by (4.1.15). Recall that we let μ denote the Lebesque measure on $[0, 1]$. Choose independent random variables $(Y_I \colon [0, 1] \to \{0, 1\} : I \in \mathcal{I}_n)$ satisfying

$$\mu(\{Y_I = 1\}) = \delta/2.$$

For $x \in S_M$, define the collection of pairwise disjoint dyadic intervals $\mathcal{K}(x) = \{I \in \mathcal{I}_n : x \in E_I\}$ and set

$$R_0(x) = \{y \in [0, 1] : Y_I(y) 1_{E_I}(x) = 0, \quad \text{for any } I \in \mathcal{I}_n\}.$$

If $y \in R_0(x)$ and $I \in \mathcal{K}(x)$ then $Y_I(y) = 0$. Since $\mathrm{card}(\mathcal{K}(x)) \geq M$, we obtain

$$\mu(R_0(x)) \leq (1 - (\delta/2))^M.$$

Next, let

$$E = \left\{ (y, x) \in [0, 1] \times S : \sum_{I \in \mathcal{I}_n} Y_I(y) 1_{E_I}(x) \geq 1 \right\}. \tag{4.1.16}$$

Note that if $x \in S$ and $y \in [0, 1] \setminus R_0(x)$ then $(y, x) \in E$. Testing this observation on $S_M \subseteq S$ gives,

$$\mu \otimes \lambda(E) \geq \inf_{x \in S_M} \mu([0, 1] \setminus R_0(x)) \lambda(S_M) \geq (1 - (1 - (\delta/2))^M) \lambda(S_M). \tag{4.1.17}$$

Using that $M > (8/\delta)$, we obtain $(1 - (1 - (\delta/2))^M) \geq 1/2$ and hence Inequality (4.1.17) gives

$$\mu \otimes \lambda(E) \geq (1/2) \lambda(S_M).$$

Step 2 (Decoupling estimate) Let $N = 2^n$ and form

$$R_1 = \left\{ y \in [0, 1] : \sum_{I \in \mathcal{I}_n} Y_I(y) < \delta N \right\}, \quad R_2 = [0, 1] \setminus R_1. \tag{4.1.18}$$

Recall that $E \subseteq [0, 1] \times S$ was defined in (4.1.16). Next, we will prove the crucial decoupling estimate

$$\mu \otimes \lambda(E) \leq \sup_{\mathcal{K}} \lambda \left(\bigcup_{I \in \mathcal{K}} E_I \right) + C_0 \mu(R_2), \tag{4.1.19}$$

where the supremum is taken over $\mathcal{K} \subset \mathcal{I}_n$ with $\mathrm{card}(\mathcal{K}) \leq N\delta$. To this end, first consider $y \in R_1$, and let

$$\mathcal{K}(y) = \{ I \in \mathcal{I}_n : Y_I(y) = 1 \}.$$

We have $\mathrm{card}(\mathcal{K}(y)) < \delta N$ and

$$\sum_{I \in \mathcal{I}_n} Y_I(y) 1_{E_I}(x) = \sum_{I \in \mathcal{K}(y)} 1_{E_I}(x), \tag{4.1.20}$$

for $x \in S$. Integrating the identity (4.1.20) over $x \in S$ gives

$$\lambda \left(\left\{ x \in S : \sum_{I \in \mathcal{I}_n} Y_I(y) 1_{E_I}(x) \geq 1 \right\} \right) = \lambda \left(\bigcup_{I \in \mathcal{K}(y)} E_I \right),$$

and, since $\mu(\bigcup \mathcal{K}(y)) < \delta$ we obtain

$$\mu \otimes \lambda \left(\left\{ (y, x) \in R_1 \times S : \sum_{I \in \mathcal{I}_n} Y_I(y) 1_{E_I}(x) \geq 1 \right\} \right) \leq \sup_{\mathcal{K}} \lambda \left(\bigcup_{I \in \mathcal{K}} E_I \right), \tag{4.1.21}$$

where the supremum is taken over $\mathcal{K} \subset \mathcal{I}_n$ with $\mu(\bigcup \mathcal{K}) < \delta$. Next, we simply use that

$$\left\{(y, x) \in R_2 \times S : \sum_{I \in \mathcal{I}_n} Y_I(y) 1_{E_I}(x) \geq 1\right\} \subseteq R_2 \times \bigcup_{I \in \mathcal{I}_n} E_I,$$

and hence

$$\mu \otimes \lambda\left(\left\{(y, x) \in R_2 \times S : \sum_{I \in \mathcal{I}_n} Y_I(y) 1_{E_I}(x) \geq 1\right\}\right) \leq C_0 \mu(R_2). \tag{4.1.22}$$

Since $[0, 1] = R_1 \cup R_2$, adding the estimates (4.1.21) and (4.1.22) gives (4.1.19).

Step 3 (Summing up) By our initial choice of $\delta > 0$, combining estimate (4.1.19) with (4.1.17) we obtain

$$(1 - (1 - (\delta/2))^M)\lambda(\{w_n(T) > M\}) \leq C_0(\epsilon_0 + \mu(R_2)),$$

where R_2 is defined by (4.1.18). Recall that $(Y_I : [0, 1] \to \{0, 1\} : I \in \mathcal{I}_n)$ are independent random variables satisfying

$$\int_0^1 (Y_I - \delta/2) d\mu = 0.$$

Hence $\int_0^1 (\sum_{I \in \mathcal{I}_n} (Y_I - \delta/2))^2 d\mu = \sum_{I \in \mathcal{I}_n} \int_0^1 (Y_I - \delta/2)^2 d\mu$, and Chebyshev's inequality gives

$$\mu(R_2) = \mu\left(\left\{\sum_{I \in \mathcal{I}_n} (Y_I - \delta/2) \geq N\delta/2\right\}\right) \leq \frac{\sum_{I \in \mathcal{I}_n} \int_0^1 (Y_I - \delta/2)^2 d\mu}{(N\delta/2)^2} \leq \frac{2}{N\delta}.$$

Consequently,

$$(1 - (1 - (\delta/2))^M)\lambda(\{w_n(T) > M\}) \leq C_0(\epsilon_0 + 2(N\delta)^{-1}). \tag{4.1.23}$$

Since $M > (8/\delta)$, $2^n = N$ and $2(N\delta)^{-1} < \epsilon_0$, by simple arithmetic, Inequality (4.1.23) yields

$$\lambda(\{w_n(T) > M\}) \leq 4C_0\epsilon_0.$$

Invoking that the sequence $(w_n(T))$ is nondecreasing, we have $\lambda(\{w_n(T) > M\}) \leq \epsilon$ for all $n \in \mathbb{N}$, and

$$\lambda\left(\left\{\sup_n w_n(T) = \infty\right\}\right) = 0. \qquad \square$$

Kwapień's Representation

We are now prepared to state Kwapień's representation theorem.

Theorem 4.1.3 (Kwapien's representation theorem) *Let $T: L^0(\Omega, \mathcal{B}, \mu) \to L^0(S, \mathcal{F}, \lambda)$ be a linear operator. Then T is $d_\mu - d_\lambda$ continuous if and only if there exists a random measure*

$$\mathcal{T} = \{\mu_x : \mathcal{B} \to \mathbb{R},\ x \in S\}$$

satisfying condition (4.1.5) *such that for λ-almost every $x \in S$, $\mu_x \in \mathcal{T}$ is purely atomic with finite support, and*

$$Tf(x) = \int_\Omega f d\mu_x, \tag{4.1.24}$$

for $f \in L^0(\Omega, \mathcal{B}, \mu)$, and λ-almost every $x \in S$. Moreover, if a random measure $\{\nu_x : \mathcal{B} \to \mathbb{R},\ x \in S\}$ represents T by (4.1.24)*, then*

$$\lambda(\{x \in S : \nu_x \neq \mu_x\}) = 0. \tag{4.1.25}$$

Remarks: It is instructive to make explicit the condition that a random measure $\{\mu_x : \mathcal{B} \to \mathbb{R} : x \in S\}$ consists of finitely supported, purely atomic measures, for λ-almost every $x \in S$.

Let $A: S \to \mathbb{N} \cup \{\infty\}$ be measurable. Let $(a_m : S \to \mathbb{R})$ and $(\sigma_m : S \to \Omega)$ be sequences of measurable functions, and assume that for $x \in S$ the equation

$$\mu_x(B) = \sum_{m=1}^{A(x)} a_m(x)\delta_{\sigma_m(x)}(B), \qquad B \in \mathcal{B}, \tag{4.1.26}$$

defines a Borel measure on Ω.

(i) The condition that the measures $\{\mu_x : x \in S\}$ are purely atomic and finitely supported, for λ-almost every $x \in S$, is equivalent to

$$\lambda(\{A = \infty\}) = 0.$$

(ii) Separately, the condition that

$$\lambda\left(\sigma_m^{-1}(N) \cap \{a_m \neq 0\}\right) = 0, \qquad m \in \mathbb{N}, \tag{4.1.27}$$

for any $N \subset \Omega$ satisfying $\mu(N) = 0$, is equivalent to $\{\mu_x : x \in S\}$ satisfying condition (4.1.5).

(iii) We remark that condition (4.1.27) implies that for $m \in \mathbb{N}$ and $\epsilon > 0$ there exists $\delta > 0$ such that for any $D \subset \Omega$ with $\mu(D) < \delta$ we have

$$\lambda\left(\sigma_m^{-1}(D) \cap \{a_m \neq 0\}\right) < \epsilon.$$

Indeed, if not, there would be $m_0 \in \mathbb{N}$, $\epsilon_0 > 0$, and a sequence $D_n \subset \Omega$ with $\mu(D_n) < 2^{-n}$ such that

$$\lambda\left(\sigma_{m_0}^{-1}(D_n) \cap \{a_{m_0} \neq 0\}\right) > \epsilon_0,$$

for $n \in \mathbb{N}$. Taking $N = \bigcap_{k=1}^{\infty} \bigcup_{n=k}^{\infty} D_n$ gives $N \subset \Omega$ with $\mu(N) = 0$ for which condition (4.1.27) fails.

Kwapień's proof of Theorem 4.1.3 is constructive and explicit (Kwapień, 1973). It exhibits the representing of measures of an operator by analyzing its response to the selected family of testing functions investigated in Proposition 4.1.2. The following result gives the main implication in Theorem 4.1.3.

Theorem 4.1.4 *For every continuous linear operator $T: L^0(\Omega, \mathcal{B}, \mu) \to L^0(S, \mathcal{F}, \lambda)$, there exist measurable functions*

$$(A: S \to \mathbb{N} \cup \{\infty\}), \quad (a_m: S \to \mathbb{R}), \quad \textit{and} \quad (\sigma_m: S \to \Omega),$$

satisfying $A(x) < \infty$ and $\sigma_n(x) \neq \sigma_m(x)$, for $1 \leq m \neq n < A(x)$, as well as

$$Tf(x) = \sum_{m=1}^{A(x)} a_m(x) f(\sigma_m(x)), \qquad f \in L^0(\Omega, \mathcal{B}, \mu), \tag{4.1.28}$$

for λ-almost every $x \in S$. Moreover, if $N \in \mathcal{B}$ and $m \in \mathbb{N}$ then

$$\mu(N) = 0 \quad \textit{implies} \quad \lambda\left(\sigma_m^{-1}(N) \cap \{a_m \neq 0\}\right) = 0. \tag{4.1.29}$$

Proof We put $w_I = T1_I$ and $E_I = \{w_I \neq 0\}$ for $I \in \mathcal{I}$. Let $n \in \mathbb{N}$ and $\mathcal{I}_n = \{I \in \mathcal{I} : |I| = 2^{-n}\}$. We consider the intervals in $\mathcal{I}_n$ as ordered from left to right, and for $I, J \in \mathcal{I}_n$ we put

$$J < I \quad \text{if} \quad \sup J < \sup I. \tag{4.1.30}$$

For $I \in \mathcal{I}_n$ and $I_1, I_2 \in \mathcal{I}_{n+1}$ satisfying $I = I_1 \cup I_2$, we have

$$E_I \subseteq E_{I_1} \cup E_{I_2}. \tag{4.1.31}$$

The proof of Theorem 4.1.4 is organized as follows.

(i) In Step 1 we define A, σ_m, and a_m.
(ii) In Step 2 we show that $\sigma_m(x) \neq \sigma_k(x)$, $1 \leq m \neq k \leq A(x)$.
(iii) In Step 3 we prove that A, σ_m, and a_m satisfy (4.1.28).
(iv) In Step 4 we verify that condition (4.1.29) holds true.

Step 1a First we define $A\colon S \to \mathbb{N} \cup \{\infty\}$. We let $w_n = \sum \{1_{E_I} : I \in \mathcal{I}_n\}$ and put

$$A = \sup_{n \in \mathbb{N}} w_n.$$

By Proposition 4.1.2, $\lambda(\{A = \infty\}) = 0$. We set $S_1 = S \backslash \{A = \infty\}$. For a given $x \in S_1$, there exists $n(x) \in \mathbb{N}$ such that $A(x) = w_n(x)$ for any $n \geq n(x)$. Moreover, we have $E_I = E_{I_1} \cup E_{I_2}$, and hence,

$$1_{E_I}(x) = 1_{E_{I_1}}(x) + 1_{E_{I_2}}(x), \tag{4.1.32}$$

for $n \geq n(x)$, $I \in \mathcal{I}_n$ and $I_1, I_2 \in \mathcal{I}_{n+1}$, satisfying $I = I_1 \cup I_2$.

Step 1b Second, we define the sequence $(\sigma_m\colon S \to \Omega)$. To this end, fix $n \in \mathbb{N}$, $x \in S_1$, and $1 \leq m \leq w_n(x)$. Define the dyadic interval

$$I(n, x, m) = \min \left\{ I \in \mathcal{I}_n : \sum_{J \leq I} 1_{E_J}(x) \geq m \right\}, \tag{4.1.33}$$

where the minimum is taken with respect to the order (4.1.30), and let $\sigma_m^n(x)$ be the right endpoint of the dyadic interval $I(n, x, m)$. From expression (4.1.31), we infer that $\sigma_m^{n+1}(x) \leq \sigma_m^n(x)$, and that the limit

$$\sigma_m(x) = \lim_{n \to \infty} \sigma_m^n(x), \tag{4.1.34}$$

is well defined. Figure 4.2 depicts the limiting process that defines $\sigma_m(x)$ in equation (4.1.34).

Step 1c Now we turn to defining $(a_m\colon S \to \mathbb{R})$. We begin by fixing $n \in \mathbb{N}$, $x \in S_1$, and $1 \leq m \leq w_n(x)$. Let $K = I(n, x, m)$ be given by definition (4.1.33), and put

$$S(K) = \{z \in S_1 : I(n, z, m) = K \text{ and } m \leq w_n(z)\}.$$

We define

$$a_m^n(z) = 1_{S(K)}(z)(T 1_K)(z), \qquad z \in S_1.$$

Recall that for any $x \in S_1$, we chose $n(x) \in \mathbb{N}$ such that $A(x) = w_n(x)$ for any $n \geq n(x)$. Retracing the steps involved in the definition of $a_m^n(x)$, we obtain $a_m^{n+1}(x) = a_m^n(x)$ for any $n \geq n(x)$. Hence the limit

$$a_m(x) = \lim_{n \to \infty} a_m^n(x)$$

is well defined.

Having completed the definition of A, σ_m, and a_m, we record a useful observation for later reference. Fix $x \in S_1$ and $m \in \mathbb{N}$ satisfying $1 \leq m \leq A(x)$.

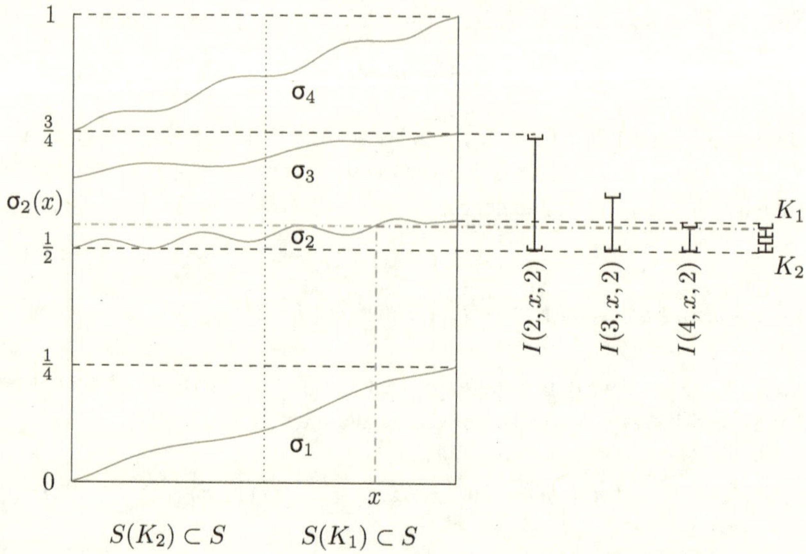

Figure 4.2 The construction of σ_2: The intervals $I(n, x, 2)$ are linearly ordered by inclusion and determine $\sigma_2(x) = \bigcap_{n=1}^{\infty} \overline{I(n, x, 2)}$.

Fix $n \geq n(x)$, $k \in \mathbb{N} \cup \{0\}$, and let $K_k = I(n + k, x, m)$ (which is well defined since $w_{n+k}(x) = A(x)$). Then, invoking (4.1.32) we obtain

$$a_m(x) = (T1_{K_k})(x). \tag{4.1.35}$$

Step 2 (Proving $\sigma_m(x) \neq \sigma_k(x)$, $1 \leq m \neq k < A(x)$) Recall that in Step 1 we verified $\lambda(\{A = \infty\}) = 0$ and put $S_1 = S \setminus \{A = \infty\}$.

Let $x \in S_1$ and $1 \leq m < A(x)$. We claim that there exists $i(x) \in \mathbb{N}$ such that, for $i \geq i(x)$,

$$\sigma_m(x) \in I(i, x, m), \tag{4.1.36}$$

and

$$|\sigma_m(x) - \sigma_{m+1}(x)| \geq 2^{-i}. \tag{4.1.37}$$

To this end we examine the matrix $(\sigma_m^n(x), 1 \leq m < A(x), n > n(x))$. First, if $m + 1 \leq A(x)$ then $\sigma_m^n(x) \leq \sigma_{m+1}^n(x)$, and

$$I(n, x, m) \cap I(n, x, m + 1) = \emptyset. \tag{4.1.38}$$

Next, since $n \geq n(x)$, for any for $k \in \mathbb{N}$ we have,

$$\sigma_m^{n+k}(x) \in \overline{I(n, x, m)} = I(n, x, m) \cup \{\sigma_m^n(x)\}.$$

Moreover, a moments reflection reveals that there exists $k(n, x, m) \in \mathbb{N}$ such that

$$\sigma_m^{n+k}(x) \in I(n, x, m),$$

for $k > k(n, x, m)$. Since $\sigma_m^{n+k}(x) \leq \sigma_m^n(x)$, and since dyadic intervals are left closed, we obtain

$$\sigma_m(x) \in I(n, x, m). \tag{4.1.39}$$

Replacing m by $m + 1$ in expression (4.1.39) and invoking (4.1.38) shows that $\sigma_m(x)$ and $\sigma_{m+1}(x)$ are contained in separate dyadic intervals. Taking into account that $1 \leq m < A(x) < \infty$, we obtain $i(x) \in \mathbb{N}$ such that for $i \geq i(x)$ the separation estimate (4.1.37) holds true.

Step 3 (Identifying T) Let $f \in \operatorname{span}\{1_I : I \in \mathcal{I}\}$ and fix $x \in S_1$. Let $i(x) > n(x) \in \mathbb{N}$ be determined such that (4.1.36) and (4.1.37) hold true for $i \geq i(x)$. Clearly, there exist $i \geq i(x)$ and coefficients $(v(J) : J \in \mathcal{I}_i)$ such that

$$f = \sum_{J \in \mathcal{I}_i} v(J) 1_J.$$

Now let

$$\mathcal{S}(i, x) = \{I(i, x, m) : 1 \leq m \leq A(x)\}.$$

If $I \in \mathcal{I}_i \backslash \mathcal{S}(i, x)$, then $x \notin E_I$ and $(T 1_I)(x) = 0$. Let $1 \leq m \leq A(x)$ and let $J_m \in \mathcal{I}_i$ be defined by $\sigma_m(x) \in J_m$. In view of equation (4.1.35) we have

$$v(J_m) = f(\sigma_m(x)) \quad \text{and} \quad (T 1_{J_m})(x) = a_m(x).$$

Summing up, we verified the identity,

$$\begin{aligned}(Tf)(x) &= \sum_{J \in \mathcal{I}_i} v(J)(T 1_J)(x) = \sum_{m=1}^{A(x)} v(J_m)(T 1_{J_m})(x) \\ &= \sum_{m=1}^{A(x)} v(J_m) a_m(x) = \sum_{m=1}^{A(x)} f(\sigma_m(x)) a_m(x).\end{aligned} \tag{4.1.40}$$

Since $\lambda(S \backslash S_1) = 0$ and $\operatorname{span}\{1_I : I \in \mathcal{I}\}$ is dense in $L^0(\Omega, \mathcal{B}, \mu)$, we finished the proof of (4.1.28).

Step 4 (Verifying condition (4.1.29)) Fix $\epsilon > 0$ and choose $\delta > 0$ such that the conclusion of Lemma 4.1.1 holds true. Fix $N \subset \Omega$ with $\mu(N) = 0$. Choose an open set $U \supset N$ satisfying $\mu(U) < \delta$. Fix $x \in S$ with $A(x) < \infty$. Fix $1 \leq m \leq A(x)$ such that $\sigma_m(x) \in U$ and $a_m(x) \neq 0$. Let $i(x)$ be determined as in the previous steps of the proof. Let $i \geq i(x)$ and choose $K \in \mathcal{I}_i$ such that $\sigma_m(x) \in K$. Note that $\sigma_m(x) \in K$ implies $x \in E_K$. Consequently, with $\mathcal{I}_i(U) = \{K \in \mathcal{I}_i, K \subseteq U\}$ and

$$B_i = \bigcup\{E_K : K \in \mathcal{I}_i(U)\},$$

we have

$$\sigma_m^{-1}(N) \cap \{1 \leq m \leq A\} \cap \{a_m \neq 0\} \subseteq \bigcup_{i=1}^{\infty} B_i.$$

Since $\mu(\bigcup \mathcal{I}_i(U)) < \delta$, Lemma 4.1.1 implies that $\lambda(B_i) < \epsilon$. Taking into account that $B_{i+1} \supseteq B_i$, we have $\lambda(\bigcup B_i) < \epsilon$, for arbitrary $\epsilon > 0$. Hence,

$$\lambda\left(\sigma_m^{-1}(N) \cap \{1 \leq m \leq A\} \cap \{a_m \neq 0\}\right) = 0.$$

Finally, if $m > A(x)$ then $a_m(x) = 0$, and hence, $\lambda(\sigma_m^{-1}(N) \cap \{a_m \neq 0\}) = 0$, for any $m \in \mathbb{N}$ and $N \subset \Omega$ with $\mu(N) = 0$. □

We complete the proof of Theorem 4.1.3 by showing that any random measure consisting of finitely supported purely atomic measures gives rise to a continuous linear operator between L^0 spaces.

Theorem 4.1.5 *Let $A\colon S \to \mathbb{N} \cup \{\infty\}$ be measurable with $\lambda(\{A = \infty\}) = 0$. Let $(a_m\colon S \to \mathbb{R})$ and $(\sigma_m\colon S \to \Omega)$ be sequences of measurable functions such that*

$$\lambda\left(\sigma_m^{-1}(N) \cap \{a_m \neq 0\}\right) = 0, \tag{4.1.41}$$

for any $N \subset \Omega$ with $\mu(N) = 0$ and $m \in \mathbb{N}$. Let $f\colon \Omega \to \mathbb{R}$ be $\mathcal{B}$-measurable and put

$$Tf(x) = \sum_{m=1}^{A(x)} a_m(x) f(\sigma_m(x)), \qquad x \in S. \tag{4.1.42}$$

Then the linear operator $T\colon L^0(\Omega, \mathcal{B}, \mu) \to L^0(S, \mathcal{F}, \lambda)$ defined by (4.1.42) is continous.

Proof First, recall that by (4.1.41) the formula (4.1.42) gives rise to a well-defined linear operator acting between equivalence classes of measurable functions. We turn to proving continuity of T. Fix $\epsilon > 0$, $k = k(\epsilon)$, and $i = i(k, \epsilon)$ such that

$$\lambda(\{A > k\}) < \epsilon/4 \quad \text{and} \quad \lambda(\{a_m > i\}) < \epsilon/(4k),$$

for $m \leq k$. Define $G \subseteq S$ by

$$G = \{A \leq k\} \cap \bigcap_{m=1}^{k} \{a_m \leq i\}.$$

Our choice of k and i gives $\mu(S\backslash G) \leq \epsilon/2$. Let $\epsilon_1 = \epsilon/(4k)$. Recall that there exists $0 < \delta < \epsilon_1/i$ such that

$$\lambda\left(\sigma_m^{-1}(D) \cap \{a_m \neq 0\}\right) < \epsilon_1, \tag{4.1.43}$$

for $m \leq k$ and $D \subset \Omega$ with $\mu(D) < \delta$. Pick $f \in L^0(\Omega)$ satisfying $\mu(\{|f| > \delta\}) < \delta$, and form

$$R = \bigcap_{m=1}^{k} S\backslash E_m, \quad \text{where} \quad E_m = \sigma_m^{-1}(\{|f| > \delta\} \cap \{a_m \neq 0\}).$$

Apply (4.1.43) with $D = \{|f| > \delta\}$ to obtain $\mu(S\backslash R) < \epsilon/2$. For $x \in G \cap R$ we have

$$A(x) \leq k \quad \text{and} \quad |a_m(x) f(\sigma_m(x))| \leq \delta i,$$

for $m \leq k$. Since $ik\delta < \epsilon$, we obtain $|Tf(x)| \leq \epsilon$, for $x \in G \cap R$, and hence,

$$\{|Tf| > \epsilon\} \subseteq S\backslash(G \cap R).$$

Since $\lambda(S\backslash(G \cap R)) \leq \lambda(S\backslash G) + \lambda(S\backslash R) < \epsilon$, we obtain $\lambda(\{|Tf| > \epsilon\}) < \epsilon$. □

Applications

We now illustrate the power of Theorem 4.1.3 and Theorem 4.1.4 by presenting two of their numerous consequences.

(i) For any continuous linear operator T, either $T(L^0(\Omega))$ or $(\mathrm{Id} - T)(L^0(\Omega))$ contain a *complemented* copy of $L^0(\Omega)$. See Corollary 4.1.6. The surprising element of Corollary 4.1.6 comes from noting that conditional expectation operators are typically *unbounded* on $L^0(\Omega)$.

(ii) Perhaps the most unexpected consequence of Theorem 4.1.3 asserts that *any nonvanishing* continuous linear operator on $L^0(\Omega)$ fixes a copy of $L^0(\Omega)$. See Corollary 4.1.7.

Let E be a Borel measurable subset of Ω with $\mu(E) > 0$. Let $\mathcal{B}_E$ be the sigma-algebra of Borel subsets contained in E and let $\mu_{|E}$ be the restriction of the Lebesgue measure to E. Then we denote

$$L^0(E) = L^0(E, \mathcal{B}_E, \mu_{|E}).$$

Corollary 4.1.6 *For any continuous linear operator $T\colon L^0(\Omega) \to L^0(\Omega)$ there exists a Borel subset $E \subseteq \Omega$ satisfying $\mu(E) > 0$ such that either*

$$T(L^0(E)) \quad \textit{or} \quad (\mathrm{Id} - T)(L^0(E)),$$

contain a complemented (!) subspace isomorphic to $L^0(E)$.

Proof Let $\mathcal{T} = \{\mu_x : x \in \Omega\}$ be the random measure providing the Kwapień representation of T. Put

$$\nu_x = \delta_x - \mu_x, \quad x \in \Omega.$$

The random measure $\{\nu_x : x \in \Omega\}$ forms the Kwapień representation of $\mathrm{Id} - T$. We next consider

$$C_1 = \{x \in \Omega : \mu_x(\{x\}) \geq 1/2\} \quad \text{and} \quad C_2 = \{x \in \Omega : \nu_x(\{x\}) \geq 1/2\}.$$

Since $1 = \mu_x(\{x\}) + \nu_x(\{x\})$ for $x \in \Omega$, either $\mu(C_1) > 0$ or $\mu(C_2) > 0$. Without loss of generality, we assume that $\mu(C_1) > 0$. Recall that μ_x is given explicitly by

$$\mu_x = \sum_{n=1}^{A(x)} a_n(x)\delta_{\sigma_n(x)}, \qquad x \in \Omega,$$

where $\sigma_{n_1}(x) \neq \sigma_{n_2}(x)$, for $1 \leq n_1 \neq n_2 < A(x)$, and $A(x) < \infty$ for μ-almost every $x \in \Omega$. Consequently, for any $x \in C_1$ with $A(x) < \infty$ there exists $n \in \mathbb{N}$ such that $\sigma_n(x) = x$, and hence

$$\mu_x(\{x\}) = a_n(x)\delta_{\sigma_n(x)}(\{x\}). \tag{4.1.44}$$

Now fix $m \in \mathbb{N}$, and consider

$$E_m = \{x \in C_1 : A(x) \geq m,\ \sigma_m(x) = x,\ a_m(x) \geq 1/2\}. \tag{4.1.45}$$

Since $\mu(C_1) > 0$, there exists $m_0 \in \mathbb{N}$ such that $\mu(E_{m_0}) > 0$. We put $a = a_{m_0}$ and $E = E_{m_0}$. By (4.1.44) we observe the following identities (between atomic measures),

$$\mu_x = a(x)\delta_x, \qquad x \in E.$$

Thus we showed that $(Tf)(x) = a(x)f(x)$ for $f \in L^0(E)$. Since (4.1.45) gives $a(x) \geq 1/2$ for $x \in E$, the linear operator

$$Vf(x) = (1/a(x))f(x), \quad x \in E, \quad f \in L^0(E),$$

is continuous on $L^0(E)$. Moreover, $VTf(x) = TVf(x) = f(x)$ for $x \in E$ and $f \in L^0(E)$, and hence V inverts T on $T(L^0(E))$. □

Write $L^0(\Omega) = L^0(\Omega, \mathcal{B}, \mu)$, let X be a linear subspace of $L^0(\Omega)$, and assume that $T\colon L^0(\Omega) \to L^0(\Omega)$ is a continuous linear operator. We denote by

$$T_{|X}\colon X \to T(X)$$

the restriction of T defined as $T_{|X}(f) = T(f)$, for $f \in X$. We say that $T_{|X}$ is an isomorphism if it is bijective and if $T_{|X}^{-1}\colon T(X) \to X$ is $d_\mu - d_\mu$ continuous. In this case we say that T fixes a copy of X.

Corollary 4.1.7 *Let $T: L^0(\Omega, \mathcal{B}, \mu) \to L^0(\Omega, \mathcal{B}, \mu)$ be a continuous linear operator satisfying $T \neq 0$. There exists a Borel measurable $E \subseteq \Omega$ with $\mu(E) > 0$ such that $T_{|L^0(E)}$ is an isomorphism.*

Proof Let $\mathcal{T} = \{\mu_x : x \in \Omega\}$ be the Kwapień random measure of T. It consists of finitely supported, purely atomic measures

$$\mu_x = \sum_{n=1}^{A(x)} a_n(x)\delta_{\sigma_n(x)}, \qquad x \in \Omega,$$

where $a_n: \Omega \to \mathbb{R}$, $\sigma_n: \Omega \to \Omega$ are Borel measurable, $A(x) < \infty$ for μ-almost every $x \in \Omega$, and where $\sigma_{n_1}(x) \neq \sigma_{n_2}(x)$, for $1 \leq n_1 \neq n_2 < A(x)$.

Step 1 Since $T \neq 0$, there exists a Borel subset $C \subseteq \Omega$ with $\mu(C) > 0$ and $l \in \mathbb{N}$ such that $|a_l(x)| > 0$ for $x \in C$. We put $a = a_l$, $\sigma = \sigma_l$, and define the continuous linear operator,

$$T_0 f(x) = 1_C(x)a(x)f(\sigma(x)),$$

for $x \in \Omega$ and $f \in L^0(\Omega)$. Note that the Kwapień random measure of T_0 is given by

$$\nu_x = 1_C(x)a(x)\delta_{\sigma(x)}, \quad x \in \Omega.$$

Step 2 Since $\sigma_{n_1}(x) \neq \sigma_{n_2}(x)$, for $1 \leq n_1 \neq n_2 < A(x)$, and $A(x) < \infty$ for μ-almost every $x \in \Omega$, there exist $k \in \mathbb{N}$, $\alpha > 0$, and a Borel subset $C_1 \subseteq C$ satisfying $\mu(C_1) > 0$ such that

$$A(x) \leq k \quad \text{and} \quad |\sigma_{n_1}(x) - \sigma_{n_2}(x)| > \alpha, \tag{4.1.46}$$

for $1 \leq n_1 \neq n_2 \leq k$ and $x \in C_1$. Next, fix $m \in \mathbb{N}$ such that $2^{-m} < \alpha/2$, and define

$$T_m f = \sum_{I \in \mathcal{I}_m} 1_{E(I)} T(f 1_I),$$

where $E(I) = C \cap \sigma^{-1}(I)$. The Kwapień random measure of T_m is then given by

$$\mu_{m,x} = \sum_{n=1}^{A(x)} \sum_{I \in \mathcal{I}_m} 1_{E(I)}(x)a_n(x)\delta_{\sigma_n(x)\cap I}. \tag{4.1.47}$$

In the next step of the proof we show that $\mu_{m,x} = \nu_x$ for $x \in C_1$.

Step 3 Using (4.1.46) and $2^{-m} < \alpha/2$, we observe that for $x \in C_1$ and $I \in \mathcal{I}_m$ there exists at most one $n \leq A(x)$ such that $\sigma_n(x) \in I$. In that case we set $I(n) = I$. Consequently, we are able to simplify the sum on the right-hand side of equation (4.1.47) and obtain

$$\mu_{m,x} = \sum_{n=1}^{A(x)} 1_{E(I(n))}(x) a_n(x) \delta_{\sigma_n(x) \cap I(n)}.$$

Next, observe that if $x \in C_1$ and $1_{E(I(n))}(x) \neq 0$, then $x \in E(I(n))$ and $\sigma(x) \in I(n)$. Since in Step 1 we set $\sigma = \sigma_l$, the assumption that $1_{E(I(n))}(x) \neq 0$ implies that $n = l$, and consequently,

$$\mu_{m,x} = a_l(x)\delta_{\sigma_l(x)} = \nu_x, \qquad x \in C_1.$$

Recall that $|a(x)| > 0$ for $x \in C_1$, and that, by the continuity of T, the condition $\mu(C_1) > 0$ implies that $G_1 = \sigma(C_1)$ satisfies $\mu(G_1) > 0$. Summing up, we obtained that

$$T_{m|L^0(G_1)} = T_{0|L^0(G_1)} \neq 0.$$

Step 4 By Lusin's theorem there exist $\beta > 0$ and a closed subset C_2 of C_1 with $\mu(C_2) > 0$ such that the restriction $\sigma_{|C_2}$ is a continuous mapping, and, moreover, $|a(x)| > \beta$ for $x \in C_2$. Note that $G_2 = \sigma(C_2)$ is a closed subset of Ω and that $\mu(G_2) > 0$. Let $\mathcal{B}_{G_2}$ denote the sigma-algebra consisting of the Borel subsets of G_2, and define

$$\nu(A) = \mu(\sigma^{-1}(A)), \qquad A \in \mathcal{B}_{G_2}.$$

We claim that the measure ν is absolutely continuous with respect to μ. To this end fix $\epsilon > 0$, small enough for $\epsilon_1 = \epsilon/\beta$ to satisfy $\epsilon_1 < 1$. Since $T_0 \colon L^0(\Omega) \to L^0(\Omega)$ is continuous, there exists $\delta > 0$ such that

$$\mu(\{x \in \Omega : |(T_0 1_A)(x)| > \epsilon\} < \epsilon,$$

for any $A \in \mathcal{B}$ satisfying $\mu(A) < \delta$. We next show that $\nu(A \cap G_2) < \epsilon$ for $A \in \mathcal{B}$ with $\mu(A) < \delta$. Indeed, since $|a(x)| > \beta$ on C_2, we get

$$\begin{aligned} \nu(A \cap G_2) &= \mu(\{1_{\sigma^{-1}(A \cap G_2)} > \epsilon_1\}) \\ &\leq \mu\left(\left\{x \in C_2 : |a(x)| 1_{\sigma^{-1}(A)}(x) > \beta\epsilon_1\right\}\right) \\ &\leq \mu(\{|T_0 1_A| > \epsilon\}) < \epsilon. \end{aligned}$$

Step 5 As ν is absolutely continuous with respect to μ, the Radon-Nikodym theorem yields the existence of a μ-integrable $u \colon G_2 \to \mathbb{R}_0^+$ such that

$$\nu(A) = \int_A u(y)d\mu(y), \qquad A \in \mathcal{B}_{G_2}.$$

Since $\nu(G_2) = \mu(C_2) > 0$, there exists a Borel subset $G_3 \subseteq G_2$ with $\mu(G_3) > 0$ and $\beta_1 > 0$ such that $u > \beta_1$ on G_3. We next verify that with $C_3 = \sigma^{-1}(G_3)$, the measure estimate

$$\beta_1\mu(\{y \in G_3 : |f(y)| > \delta/\beta\}) \leq \mu(\{x \in C_3 : |(T_0 f)(x)| > \delta\}), \tag{4.1.48}$$

holds true for $\delta > 0$ and $f \in L^0(G_3)$. Indeed, since $(T_0 f)(x) = a(x)f(\sigma(x))$ and $|a(x)| > \beta$ on C_3, we have

$$\begin{aligned}\mu(\{x \in C_3 : |(T_0 f)(x)| > \delta\}) &\geq \mu(\{x \in C_3 : |f(\sigma(x))| > \delta/\beta\})\\ &= \nu(\{y \in G_3 : |f(y)| > \delta/\beta\})\\ &= \int_{\{y\in G_3:|f(y)|>\delta/\beta\}} u(y)d\mu(y)\\ &\geq \beta_1\mu(\{y \in G_3 : |f(y)| > \delta/\beta\}).\end{aligned}$$

We turn to proving that $T_{0|L^0(G_3)}$ has a continuous inverse. To this end we fix $\epsilon > 0$ and choose $\delta > 0$ such that $\delta/\beta < \epsilon$ and $\delta/\beta_1 < \epsilon$. Then, by (4.1.48), for any $f \in L^0(G_3)$ satisfying

$$\mu(\{x \in C_3 : |(T_0 f)(x)| > \delta\}) < \delta,$$

we obtain that

$$\mu(\{y \in G_3 : |f(y)| > \epsilon\}) \leq \beta_1^{-1}\mu(\{x \in C_3 : |(T_0 f)(x)| > \delta\}) \leq \beta_1^{-1}\delta < \epsilon.$$

Step 6 Summarizing the above proof, we showed that $T_{0|L^0(G_3)}$ is bijective, has a continuous inverse, and that

$$T_{0|L^0(G_3)} = T_{m|L^0(G_3)}.$$

Taking into account that the sets $\{E(I) : I \in \mathcal{I}_m\}$ are pairwise disjoint, this shows that there exists $I \in \mathcal{I}_m$ satisfying $\mu(G_3 \cap I) > 0$ such that $T_{|L^0(G_3\cap I)}$ is bijective and has a continuous inverse. It suffices to put $E = G_3 \cap I$. □

Remark: Let $G = G_3$ and T_0 be determined in the above proof of Theorem 4.1.7. We obtained $T_0 f(x) = a(x)f(\sigma(x))$ with $|a(x)| \geq \beta$ for $x \in G$. Hence,

$$T_0(L^0(G)) \text{ is contained in } L^0(\sigma^{-1}(G)).$$

Moreover, $g \in L^0(\sigma^{-1}(G))$ is contained in $T(L^0(G))$ if and only if g is measurable with respect to the sigma-algebra $\{\sigma^{-1}(A) : A \in \mathcal{B}_G\}$. Since G is a Borel subset with $\mu(G_3) > 0$, the assumption that $\sigma\colon \sigma^{-1}(G) \to G$ is bijective implies that

$$\mathcal{B}_{\sigma^{-1}(G)} = \{\sigma^{-1}(A) : A \in \mathcal{B}_G\},$$

and that $T(L^0(G)) = L^0(\sigma^{-1}(G))$. Summing up, if $\infty > |a(x)| \geq \beta$ for $x \in G$, and if $\sigma\colon \sigma^{-1}(G) \to G$ is bijective, then $T(L^0(G))$ is complemented in $L^0(\Omega)$ and the projection is given by the multiplication operator,

$$P_{\sigma^{-1}(G)}(f) = 1_{\sigma^{-1}(G)}f, \quad f \in L^0(G).$$

Kwapień's conjecture famously links Banach subspaces of L^0 to those of L^1. Specifically Kwapień (1970) asked if every Banach space E isomorphic to a subspace of L^0 is actually isomorphic to a subspace of L^1. The strongest result in its support is due to Kalton (1985), who showed that the conclusion of Kwapień's conjecture holds true if the Banach space E is isomorphic to $(\sum E)_1$.

4.1.2 Kalton's Representation

We begin by briefly describing the content of Kalton's representation theorem. We let Ω be a compact metric space, $\mathcal{B}$ be the Borel sigma-algebra generated by the open subsets of Ω, and we fix a probability measure $\mu\colon \mathcal{B} \to [0,1]$. Separately, we fix a finite measure space $(S, \mathcal{F}, \lambda)$. Kalton's theorem (Theorem 4.1.15) asserts that every linear operator

$$T\colon L^1(\Omega, \mathcal{B}, \mu) \to L^1(S, \mathcal{F}, \lambda) \tag{4.1.49}$$

is represented by a random measure $\mathcal{T}$ from $(S, \mathcal{F})$ to $(\Omega, \mathcal{B})$ and provides structural information on the measures in $\mathcal{T}$.

Overview of Section 4.1.2

Theorem 4.1.9 shows that for every operator (4.1.49) there exists a unique random measure $\mathcal{T} = \{\mu_x\colon \mathcal{B} \to \mathbb{R},\ x \in S\}$ satisfying condition (4.1.5) such that

$$Tf(x) = \int_\Omega f d\mu_x,$$

for λ-almost every $x \in S$ and $f \in L^1(\Omega, \mu)$. The representing random measure $\mathcal{T}$ is obtained by applying Theorem 4.1.8 (the disintegration theorem) to the product measure

$$\beta(A \times B) = \int_B (T1_A)(x)d\lambda(x),$$

where $A \in \mathcal{B}$ and $B \in \mathcal{F}$.

For a fixed $x \in S$ and $\mu_x \in \mathcal{T}$, there exists a sequence of distinct points $\sigma_n \in \Omega$ with $\sigma_n = \sigma_n(x)$, a sequence of scalar coefficients $a_n \in \mathbb{C}$ with $a_n = a_n(x)$, and with decreasing modulus such that

$$\mu_x = \sum_{n=1}^{\infty} a_n(x)\delta_{\sigma_n(x)} + \nu_x,$$

where the measure ν_x is a nonatomic measure on Ω, and where $\delta_{\sigma_n(x)}$ denotes the Dirac measure supported in the point $\sigma_n(x) \in \Omega$. To emphasize, we have

(i) $|a_{n+1}(x)| \leq |a_n(x)|$,
(ii) $\sigma_n(x) \neq \sigma_m(x)$ for $m \neq m$,
(iii) $\nu_x(\{t\}) = 0$ for every $t \in \Omega$.

The main result in this section is Kalton's theorem (Theorem 4.1.15). It provides a detailed study of the *dependence on* $x \in S$ in the weight $a_n(x)$ and the Dirac-singularity $\sigma_n(x)$. We build the analysis on the Kuratowski and Ryll-Nardzewski selection theorem (Theorem 4.1.10), and will show that $\sigma_n \colon S \to \Omega$ and $a_n \colon S \to \mathbb{C}$ are measurable mappings satisfying

$$\sum_{n=1}^{\infty} \int_{\sigma_n^{-1}(A)} |a_n(x)| d\lambda(x) \leq \|T\| \mu(A), \quad \text{for} \quad A \in \mathcal{B}. \tag{4.1.50}$$

Put $T_n f(x) = a_n(x) f(\sigma_n(x))$. We call $\{T_n\}$ the sequence of weighted composition operators associated to the atomic parts of the representing measures $\mathcal{T}$. The estimate (4.1.50) yields

$$\sum_{n=1}^{\infty} \|T_n f\|_1 \leq \|T\| \|f\|_1, \tag{4.1.51}$$

for $f \in L^1(\Omega)$. Thus, decomposing the representing measures $\mathcal{T} = \{\mu_x : x \in S\}$ into their respective atomic and diffuse parts gives a splitting of the operator as $T = T^{(a)} + T^{(d)}$, where

$$T^{(a)} f = \sum_{n=1}^{\infty} T_n f$$

and where $(T^{(d)} f)(x) = \int_\Omega f d\nu_x$, with ν_x being the diffuse part of μ_x. To emphasize, the measurability assertions of $\sigma_n \colon S \to \Omega$ and $a_n \colon S \to \mathbb{C}$ in Theorem 4.1.15 are derived from the Kuratowski and Ryll-Nardzewski selection theorem.

Representing Measures

Recall that we fixed Ω to be a compact metric. By the Riesz representation theorem, the space of signed Radon measures $\mathcal{M}(\Omega)$, is the dual of $C(\Omega)$, the

Banach space of continuous functions on Ω with the supremum norm, and integration gives the bilinear form facilitating the duality,

$$\langle \mu, f \rangle = \int_\Omega f d\mu,$$

where $\mu \in \mathcal{M}(\Omega)$ and $f \in C(\Omega)$. We recall as well that any collection of probability measures on Ω forms a weak-* compact subset of $\mathcal{M}(\Omega)$.

Our first aim in this section is Theorem 4.1.9, which exhibits a random measure representing linear operators on L^1. We base its proof on the disintegration theorem for product measures (Theorem 4.1.8).

Theorem 4.1.8 (Disintegration theorem) *Let $(S, \mathcal{F})$ be a measurable space. Let β be a probability measure on the product space $(\Omega \times S, \mathcal{B} \times \mathcal{F})$. Let λ be the probability measure on $(S, \mathcal{F})$ given by*

$$\lambda(B) = \beta(\Omega \times B), \quad B \in \mathcal{F}.$$

Then there exists a family of probability measures $\{\gamma_s : \mathcal{B} \to [0,1],\ s \in S\}$ such that for all $A \in \mathcal{B}$, the map $s \to \gamma_s(A)$ is λ-integrable, and

$$\beta(A \times B) = \int_B \gamma_s(A) d\lambda(s), \tag{4.1.52}$$

for $B \in \mathcal{F}$.

The first step in the proof of Kalton's representation theorem exploits disintegration of measure (Theorem 4.1.8). Let $\mathcal{B}$ denote the Borel sigma-algebra on Ω. We fix now a probability space $(\Omega, \mathcal{B}, \mu)$ together with a second finite measure space $(S, \mathcal{F}, \lambda)$.

Theorem 4.1.9 *Let $T : L^1(\Omega, \mathcal{B}, \mu) \to L^1(S, \mathcal{F}, \lambda)$ be a bounded linear operator. There exists a uniquely defined family of finite measures $\mathcal{T} = \{\mu_x : \mathcal{B} \to \mathbb{R},\ x \in S\}$ such that for any $f \in L^1(\Omega, \mu)$ and λ-almost every $x \in S$, we have*

$$Tf(x) = \int_\Omega f d\mu_x, \tag{4.1.53}$$

where the right-hand side is an absolutely converging Lebesgue integral. Moreover, for every $A \in \mathcal{B}$ the map $x \to \mu_x(A)$ is λ-integrable and satisfies

$$\int_S |\mu_x(A)| d\lambda(x) \leq \|T\| \mu(A), \tag{4.1.54}$$

for all $A \in \mathcal{B}$. (In particular, condition (4.1.5) *holds true for $\mathcal{T}$.)*

Proof It suffices to prove the theorem in the case when T is a positive operator operator. We may therefore assume that $Tf \geq 0$ if $f > 0$, and we normalize by $\|T\| = 1$. Given $A \in \mathcal{B}$ and $B \in \mathcal{F}$ we put

$$\beta(A \times B) = \int_B (T1_A)(x) d\lambda(x). \tag{4.1.55}$$

It is routine to verify that β uniquely extends to a probability measure on the product space $(\Omega \times S, \mathcal{B} \times \mathcal{F})$. We still denote the extension by β. We put $w = (T1_\Omega)$ and let λ_1 be the probability measure on $(S, \mathcal{F})$ defined by

$$\lambda_1(B) = \beta(\Omega \times B) = \int_B w d\lambda.$$

Clearly we have $d\lambda_1 = w d\lambda$. Applying Theorem 4.1.8 to the product measure β, gives a family of probability measures $\{\gamma_x : \mathcal{B} \to [0,1],\ x \in S\}$, such that the map $x \to \gamma_x(A)$ is λ_1-integrable, and

$$\beta(A \times B) = \int_B \gamma_x(A) d\lambda_1(x) = \int_B \gamma_x(A) w(x) d\lambda(x), \tag{4.1.56}$$

for all $A \in \mathcal{B}$ and $B \in \mathcal{F}$. Moreover, by (4.1.55), $\int_S \gamma_x(A) d\lambda_1(x) \leq \mu(A)$.

Finally, we draw the consequences of equation (4.1.56). Since measurable step functions are dense in L^1, combining (4.1.56) and (4.1.55) gives

$$(T1_A)(x) = w(x)\gamma_x(A),$$

for λ-almost every $x \in S$ and all $A \in \mathcal{B}$. Defining $\mu_x(A) = w(x)\gamma_x(A)$ for all $A \in \mathcal{B}$ and again invoking the L^1-density of measurable step functions we obtain the representation (4.1.53). □

Measurable Selection Theorems

Recall that we let Ω be a compact metric space, $\mathcal{B}$ be the Borel sigma-algebra generated by the open subsets of Ω, and that $\mu : \mathcal{B} \to [0,1]$ is a regular probability measure. An atom in $(\Omega, \mathcal{B}, \mu)$, is a $\mathcal{B}$-measurable set A with $\mu(A) \neq 0$ such that $\mu(B) = 0$ or $\mu(B) = \mu(A)$, for every $B \in \mathcal{B}, B \subseteq A$. We say that the measure space $(\Omega, \mathcal{B}, \mu)$ is diffuse, or nonatomic, if it does not contain any atom. Conversely, it is called atomic if Ω may be written as a partition of countably many measurable elements that are either atoms or null sets. A classical result of Sierpinski asserts that for compact Ω and regular measure μ, the measure space $(\Omega, \mathcal{B}, \mu)$ is diffuse if and only if $\mu(\{t\}) = 0$ for $t \in \Omega$. See Li and Queffélec (2018, Vol 1, p 369). The Jordan decomposition theorem for finite measures asserts that μ may be decomposed as

$$\mu = \mu_a + \mu_d,$$

such that $(\Omega, \mathcal{B}, \mu_a)$ is atomic and $(\Omega, \mathcal{B}, \mu_d)$ is diffuse.

Theorem 4.1.9 represents a bounded operator $T\colon L^1(\Omega,\mathcal{B},\mu)\to L^1(S,\mathcal{F},\lambda)$ by a random measure $\mathcal{T}=\{\mu_x : x\in S\}$. In Theorem 4.1.15 we will split each representing measure $\mu_x\in\mathcal{T}$ as

$$\mu_x = \nu_x + (\mu_x - \nu_x),$$

where ν_x is diffuse, $(\mu_x-\nu_x)$ is atomic, and, most importantly, the mapping $x\to\nu_x(A)$ is suitably $\mathcal{F}$-measurable for any $A\in\mathcal{B}$. The first step in this direction consists in proving Theorem 4.1.13.

We begin with a cornerstone of real analysis, the Kuratowski and Ryll-Nardzewski selection theorem (1965). Recall that a topological space Y is called a Polish space if it is homeomorphic to a separable, complete metric space. We denote the Borel sigma-algebra on Y by $\mathcal{B}_Y$.

Theorem 4.1.10 (Kuratowski and Ryll–Nardzewski selection theorem) *Let $(A,\mathcal{A})$ be a measure space, let Y be a Polish topological space, and let $\mathcal{F}_Y$ denote the collection of closed subsets in Y. Assume that $F\colon A\to\mathcal{F}_Y$ satisfies*

$$F(a)\neq\emptyset, \tag{4.1.57}$$

for $a\in A$, and

$$\{a\in A : F(a)\cap V\neq\emptyset\}\in\mathcal{A}, \tag{4.1.58}$$

for any open subset $V\subseteq Y$. Then there exists an $\mathcal{A}-\mathcal{B}_Y$ measurable $h\colon A\to Y$ such that

$$h(a)\in F(a), \tag{4.1.59}$$

for $a\in A$.

Proof By assumption, the topology on Y is metrizable, by a metric d say, such that (Y,d) is a separable and complete metric space. We may assume that $\operatorname{diam}(Y)<1$. We claim that there exists a sequence of $\mathcal{A}-\mathcal{B}_Y$ measurable functions $h_n\colon A\to Y$ such that

$$\operatorname{dist}(h_n(a),F(a))<2^{-n} \tag{4.1.60}$$

and

$$d(h_n(a),h_{n-1}(a))<2\cdot 2^{-n}, \tag{4.1.61}$$

for $a\in A$ and $n\in\mathbb{N}\cup\{0\}$. The pointwise limit $h(a)=\lim_{n\to\infty}h_n(a)$ exists and satisfies the conclusion of Theorem 4.1.10.

Step 1 Pick any $y_0 \in Y$. Define $h_0 : A \to Y$ by $h_0(a) = y_0$ for $a \in A$. Assume that we already defined $h_1, \ldots, h_{n-1}$ satisfying the estimates (4.1.60) and (4.1.61). We proceed to define h_n. Let (y_j) be a dense sequence in Y, put $B_j^n = \{y \in Y : d(y_j, y) < 2^{-n}\}$, and define

$$C_j^n = \left\{a \in A : F(a) \cap B_j^n \neq \emptyset\right\} \qquad \text{and} \qquad D_j^n = h_{n-1}^{-1}\{B_j^n\}. \tag{4.1.62}$$

By assumption (4.1.58) we have $C_j^n \in \mathcal{A}$. Since we assume that h_{n-1} is $\mathcal{A}-\mathcal{B}_Y$-measurable, we have $D_j^n \in \mathcal{A}$. Now consider $A_j^n = C_j^n \cap D_j^n$. We claim that

$$A = \bigcup_{j=1}^{\infty} A_j^n. \tag{4.1.63}$$

We will prove that any $a \in A$ is contained in the union on the right-hand side. Evaluating (4.1.60) for h_{n-1} shows that there exists $y \in F(a)$ such that

$$d(h_{n-1}(a), y) < 2^{-n+1}.$$

Since (y_j) is dense in Y, there exists $y_k \in \{y_j\}$ close enough to y such that the following two conditions hold true,

$$d(y_k, y) < 2 \cdot 2^{-n} - d(h_{n-1}(a), y) \qquad \text{and} \qquad d(y_k, y) < 2^{-n}. \tag{4.1.64}$$

Using the triangle inequality and the first estimate in (4.1.64) we obtain,

$$d(h_{n-1}(a), y_k) \le d(y_k, y) + d(h_{n-1}(a), y) \le 2 \cdot 2^{-n}. \tag{4.1.65}$$

Since $y \in F(a)$, the second estimate in (4.1.64) gives

$$\operatorname{dist}(y_k, F(a)) \le d(y_k, y) < 2^{-n}. \tag{4.1.66}$$

Comparing (4.1.65) and (4.1.66) with the defining conditions for C_k^n and D_k^n shows that $a \in A_k^n$ and so the claim (4.1.63) holds true.

By claim (4.1.63), we may uniquely determine $k(a) \in \mathbb{N}$ such that $a \in A_{k(a)}^n \setminus \bigcup_{j<k(a)} A_j^n$, and, finally, define

$$h_n(a) = y_{k(a)}. \tag{4.1.67}$$

Step 2 It remains to verify that h_n is $\mathcal{A} - \mathcal{B}_Y$-measurable and satisfies conditions (4.1.60) and (4.1.61). We begin with measurability. Let V be a Borel subset of Y. Then, by definition of h_n, we have

$$h_n^{-1}\{V\} = \bigcup \left\{A_k^n \setminus \bigcup_{j<k} A_j^n : y_k \in Y\right\}. \tag{4.1.68}$$

Invoking that $A_j^n \in \mathcal{A}$ for $j \in \mathbb{N}$, we infer that $h_n^{-1}\{V\} \in \mathcal{A}$. Next, unwind the condition $a \in A_{k(a)}^n$. Since then $a \in C_{k(a)}^n$, by definition (4.1.62) we have

$$\operatorname{dist}(y_{k(a)}, F(a)) < 2^{-n}. \tag{4.1.69}$$

On the other hand, since $a \in D_{k(a)}^n$, again (4.1.62) gives

$$d(y_{k(a)}, h_{n-1}(a)) < 2 \cdot 2^{-n}. \tag{4.1.70}$$

Taking into account that $h_n(a) = y_{k(a)}$, we finally obtain (4.1.60) and (4.1.61) by appealing to (4.1.69) and (4.1.70). □

Souslin sets: Let A be metric a space. We say that $S \subseteq A$ is called a Souslin set if there exists a Polish space Y and a continuous surjection $g\colon Y \to S$. In short, any continuous image of a complete, separable, metrizable topological space is called a Souslin set. We recall well-known basic facts on Souslin sets.

(i) If (S_n) is a sequence of Souslin sets in A, then $\bigcup S_n$ and $\bigcap S_n$ are Souslin sets.
(ii) If A is a Souslin set, and $\mathcal{B}_A$ denotes its Borel sigma-algebra (generated by the open sets in A), then each $S \in \mathcal{B}_A$ is a Souslin set.

We now turn to proving Theorem 4.1.12, which asserts that every Souslin set may be approximated, from above and below, by Borel sets. We begin with monotone convergence for outer measures.

Lemma 4.1.11 *Fix a measure space $(A, \mathcal{B})$. Let $\mu\colon \mathcal{B} \to [0, \infty)$ be a measure and define the outer measure $\mu^*: 2^A \to [0, \infty)$ by*

$$\mu^*(D) = \inf\{\mu(B) : B \in \mathcal{B},\ B \supseteq D\},$$

where $D \subseteq A$. Then, given $D \subseteq A$ there exists $C \in \mathcal{B}$ such that $D \subseteq C$ and

$$\mu^*(D) = \mu(C). \tag{4.1.71}$$

If a a sequence of sets (D_n) satisfies $A \supseteq D_{n+1} \supseteq D_n$, for $n \in \mathbb{N}$, then

$$\lim_{n\to\infty} \mu^*(D_n) = \mu^*\left(\bigcup_{n=1}^{\infty} D_n\right). \tag{4.1.72}$$

Proof To verify (4.1.71), first choose $B_n \in \mathcal{B}$, $B_n \supseteq D$ such that

$$\mu(B_n) \leq \mu^*(D) + 1/n.$$

Put $C = \bigcap B_n$. Then $C \in \mathcal{B}$ and $\mu(C) \leq \mu^*(D)$. Since, moreover, $C \supseteq D$, the converse inequality $\mu(C) \geq \mu^*(D)$ is clear, by definition.

Now we turn to proving equation (4.1.72). We assume that $\mu^*(D_n) < \infty$ for $n \in \mathbb{N}$, since otherwise there is nothing to prove. By equation (4.1.71), there exist $B_n \in \mathcal{B}$ such that $B_n \supseteq D_n$ and $\mu^*(D_n) = \mu(B_n)$. For $k \leq n$, we get $D_k \subseteq B_k \cap B_n \subseteq B_k$, and

$$\mu^*(D_k) = \mu(B_k \cap B_n) \leq \mu(B_k) = \mu^*(D_k).$$

Hence,

$$\mu(B_k \cap B_n) = \mu(B_k),$$

for $k \leq n$. Since

$$\mu(B_k) = \mu(B_k \cap B_n) - \mu(B_k \backslash B_n),$$

and $\mu(B_k) < \infty$ we have $\mu(B_k \backslash B_n) = 0$. Hence, $\mu\left(\bigcup_{k=1}^n B_k\right) = \mu(B_n)$ and by passing to the limit gives

$$\mu\left(\bigcup_{k=1}^{\infty} B_k\right) = \lim_{n\to\infty} \mu\left(\bigcup_{k=1}^{n} B_k\right) = \lim_{n\to\infty} \mu(B_n).$$

Summing up, we obtain

$$\lim_{n\to\infty} \mu(B_n) = \lim_{n\to\infty} \mu^*(D_n) \leq \mu^*\left(\bigcup_{k=1}^{\infty} D_k\right) \leq \mu\left(\bigcup_{k=1}^{\infty} B_k\right) = \lim_{n\to\infty} \mu(B_n).$$

Since the left- and right-hand sides are identical, we must have equality signs throughout the equation above. This proves equation (4.1.72). □

Universally measurable: Let A be a metric space. Let $\mathcal{B}_A$ denote the Borel sigma-algebra on A. Let $\nu\colon \mathcal{B}_A \to [0, \infty)$ be a measure. We say that $D \subset A$ is ν-measurable if there exist $C_1, C_2 \in \mathcal{B}_A$ such that $C_1 \subseteq D \subseteq C_2$ and $\nu(C_2 \backslash C_1) = 0$. Finally, we recall that $D \subseteq A$ is called universally measurable if it is ν-measurable, for every measure $\nu\colon \mathcal{B}_A \to [0, \infty)$. The collection of all universally measurable subsets forms a sigma-algebra on A denoted by $\mathcal{U}_A$.

Theorem 4.1.12 *Let A be a metric space. Let S be a Souslin subset of A. Then $S \in \mathcal{U}_A$.*

Proof Let Y be a Polish space. Let $g\colon Y \to S$ be a continuous surjection. Put $\mathcal{B} = \mathcal{B}_A$ and fix a measure $\mu\colon \mathcal{B} \to [0, \infty)$. We will prove that there exist $C_1, C_2 \in \mathcal{B}$ such that $C_1 \subseteq S \subseteq C_2$ and $\mu(C_2 \backslash C_1) = 0$.

In Lemma 4.1.11 we associated to μ the outer measure μ^*. By (4.1.71) there exists $C_2 \in \mathcal{B}$ such that $S \subseteq C_2$ and $\mu^*(S) = \mu(C_2)$. The main part of the proof consists in showing that there exists $C_1 \in \mathcal{B}$ such that $C_1 \subseteq S$ and

$$\mu^*(S) = \mu(C_1).$$

We turn to the multistep construction of the set C_1.

Step 1 Fix $\epsilon > 0$ and set $\mathcal{Y}_0 = \{Y\}$. Since Y is a Polish space, there exists a sequence of closed subsets $Y_m \subseteq Y$ such that $Y = \bigcup Y_m$ and $\mathrm{diam}(Y_m) < 2^{-1}$. By equation (4.1.72), the monotone convergence of outer measures, there exists $n(1) \in \mathbb{N}$ such that

$$A_1 = g\left(\bigcup_{m=1}^{n(1)} Y_m\right)$$

satisfies $\mu^*(A_1) > \mu^*(S) - \epsilon/2$. Define $\mathcal{Y}_1 = \{Y_m : m \leq n(1)\}$.

Step 2 We claim that, for each $k \in \mathbb{N}$, there exists a finite collection of closed subsets of Y, denoted $\mathcal{Y}_k$, such that the following conditions hold true:

(i) Each $T \in \mathcal{Y}_k$ satisfies $\mathrm{diam}(T) < 2^{-k}$.

(ii) If $A_k = g(\bigcup \mathcal{Y}_k)$, then

$$\mu^*(A_k) > \mu^*(S) - \sum_{j=1}^{k} \epsilon 2^{-j}. \tag{4.1.73}$$

(iii) For any $T \in \mathcal{Y}_k$ there exists $T_1 \in \mathcal{Y}_{k-1}$ such that $T \subset T_1$. Consequently, $\bigcup \mathcal{Y}_k \subset \bigcup \mathcal{Y}_{k-1}$ and $A_k \subseteq A_{k-1}$.

We assume that $\mathcal{Y}_1, \ldots, \mathcal{Y}_{k-1}$ are constructed. We proceed to define $\mathcal{Y}_k$. Fix $T \in \mathcal{Y}_{k-1}$. There exist closed sets $Y(T, m) \subset T$ such that $\mathrm{diam}(Y(T, m)) < 2^{-k}$ and $T = \bigcup_{m=1}^{\infty} Y(T, m)$. Since $\mathcal{Y}_{k-1}$ has finite cardinality, by Lemma 4.1.11 we may determine $n(k) \in \mathbb{N}$ such that

$$\mu^*(A_k) > \mu^*(S) - \sum_{j=1}^{k} \epsilon 2^{-j},$$

where $A_k = \bigcup_{m=1}^{n(k)} \bigcup_{T \in \mathcal{Y}_{k-1}} Y(T, m)$. It remains to define

$$\mathcal{Y}_k = \{Y(T, m) : T \in \mathcal{Y}_{k-1}, m \leq n(k)\},$$

and the induction continues.

Step 3 Since $\bigcup \mathcal{Y}_k \subset \bigcup \mathcal{Y}_{k-1}$ we have $A_k \subseteq A_{k-1}$. Let $\overline{A}_k$ denote the closure of A_k in the metric space A. We have $\overline{A}_k \in \mathcal{B}$. Now form

$$C^{(\epsilon)} = \bigcap_{k=1}^{\infty} \overline{A}_k. \tag{4.1.74}$$

Clearly $C^{(\epsilon)} \in \mathcal{B}$ and, by Inequality (4.1.73) and by monotone convergence of the measure μ,

$$\mu(C^{(\epsilon)}) \geq \mu^*(S) - \sum_{j=1}^{\infty} \epsilon 2^{-j}. \tag{4.1.75}$$

Step 4 We claim that $C^{(\epsilon)} \subseteq S$. To this end we fix $x \in C^{(\epsilon)}$ and consider

$$\mathcal{D}_k = \left\{T \in \mathcal{Y}_k : B(x, 2^{-k}) \cap g(T) \neq \emptyset\right\}.$$

Clearly, we have $\mathcal{D}_k \neq \emptyset$ for $k \in \mathbb{N}$. Fix $k_1, k_2 \in \mathbb{N}$ with $k_1 < k_2$ and observe the following. If $T_1 \in \mathcal{Y}_{k_1}$, $T_2 \in \mathcal{D}_{k_2}$, and $T_1 \supseteq T_2$, then $T_1 \in \mathcal{D}_{k_1}$.

Put $k(1) = 1$. By the above observation, there exists $T_1 \in \mathcal{D}_{k(1)}$ and an infinite set of natural numbers N_1 such that

$$\{T \in \mathcal{D}_k : T \subseteq T_1\} \neq \emptyset,$$

for $k \in N_1$. We claim that there exists a sequence of infinite subsets (N_n) of $\mathbb{N}$ and a sequence of subsets (T_n) of Y such that $N_n \subseteq N_{n-1}$, $T_n \subseteq T_{n-1}$,

$$T_n \in \mathcal{D}_{k(n)}, \tag{4.1.76}$$

where $k(n) = \inf\{k \in N_{n-1} : k \geq k(n-1)\}$ and $\{T \in \mathcal{D}_k : T \subseteq T_n\} \neq \emptyset$, for $k \in N_n$. The construction is done inductively, where the argument needed for the general step is analogous to the one we presented in the first step. By (4.1.76) for $t_n \in T_n$, we have

$$d(x, g(t_n)) < 2^{-n}.$$

Recall that T_n is a closed subset of Y, that $\operatorname{diam}(T_n) \leq 2^{-n}$, and that $T_n \subseteq T_{n-1}$. Hence, there exists $t \in \bigcap_{n=1}^{\infty} T_n$. The continuity of g gives $d(x, g(t)) = 0$, hence $x = g(t)$ and $x \in g(Y) = S$.

Step 5 Next, fix $m \in \mathbb{N}$. Form $C^{(\epsilon)} \subseteq S$ with $\epsilon = 1/m$ according to equation (4.1.74) We have $C^{(1/m)} \in \mathcal{B}$, and by (4.1.75),

$$\mu\left(C^{(1/m)}\right) \geq \mu^*(S) - 1/m.$$

Finally, define $C_1 = \bigcup_{m=1}^{\infty} C^{(1/m)}$. Clearly, $C_1 \in \mathcal{B}$, $C_1 \subseteq S$, and

$$\mu(C_1) \geq \mu\left(C^{(1/m)}\right) > \mu^*(S) - 1/m,$$

for $m \in \mathbb{N}$. Hence $\mu(C_1) = \mu^*(S)$. Above we determined that $C_2 \in \mathcal{B}$ such that $S \subseteq C_2$ and $\mu(C_2) = \mu^*(S)$. Hence we obtained $C_1 \subseteq S \subseteq C_2$ and

$$\mu(C_2 \backslash C_1) = \mu(C_2) - \mu(C_1) = 0.$$

□

Theorem 4.1.13 *Let Z be a Souslin space. For every metric space A and every continuous surjection $u\colon Z \to A$ there exists a $\mathcal{U}_A - \mathcal{B}_Z$-measurable $f\colon A \to Z$,* *such that* $u(f(a)) = a$, *for* $a \in A$. *Thus* $\mathrm{id}_A = u \circ f$ *and the following diagram commutes:*

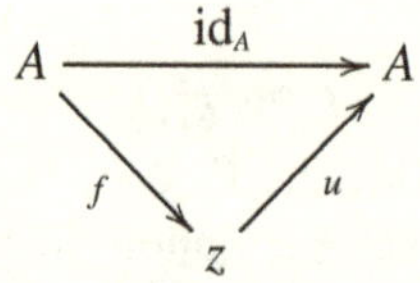

(4.1.77)

Proof Since Z is a Souslin space, there exists a Polish space Y and a continuous surjection $v\colon Y \to Z$. Thus $u \circ v\colon Y \to A$ is continuous and surjective. Put $g = u \circ v$ and define,

$$F(a) = g^{-1}\{a\},$$

for $a \in A$. Since g is a continuous surjection, $F(a)$ is a nonempty closed subset of Y. Let V be an open subset of Y. Since an open subset of a Polish space is also Polish, $g(V)$ is a Souslin subset of A, hence $g(V) \in \mathcal{U}_A$, by Theorem 4.1.12.

We apply Theorem 4.1.10 with $\mathcal{A} = \mathcal{U}_A$ to obtain a $\mathcal{U}_A - \mathcal{B}_Y$-measurable $h\colon A \to Y$, such that $g \circ h = \mathrm{id}_A$, where $\mathrm{id}_A(a) = a$ for $a \in A$. Thus the following diagram commutes:

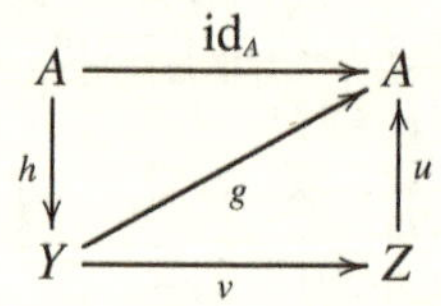

(4.1.78)

Put $f = v \circ h$. Since v is continuous, $f\colon A \to Z$ is $\mathcal{U}_A - \mathcal{B}_Z$-measurable, and, moreover,

$$u(f(a)) = g(h(a)) = a,$$

for $a \in A$. □

Kalton's Characterisation of Representing Measures

Let Ω be a compact metric space, and $M = \mathcal{M}(\Omega)$ be the space of signed Radon measures on Ω. We denote by $M_c \subset M$ the subspace consisting of diffuse measures on Ω. Thus $\nu \in M_c$ if $\nu(\{t\}) = 0$ for every $t \in \Omega$. Let $\mathcal{B}_M$ denote the sigma-algebra generated by the weak-* open subsets of M.

Lemma 4.1.14 *M_c is $\mathcal{B}_M$-measurable, i.e., $M_c \in \mathcal{B}_M$.*

Proof First fix an open set $U \subseteq \Omega$. Let $\nu \in M$ and let (ν_α) be a net in M such that $\nu_\alpha \to \nu$ in the weak-* topology in M. Let $|\nu_\alpha|$ and $|\nu|$ denote the total variation of those measures. Then,

$$\liminf |\nu_\alpha|(U) \leq |\nu|(U).$$

Hence, $\{\nu \in M : |\nu|(U) \leq \epsilon\}$ is a weak-* closed subset of M, for any $\epsilon > 0$ and open sets $U \subseteq \Omega$.

Since Ω is compact, for $n \in \mathbb{N}$ there exists an integer $m(n)$ and open sets $U_1^n, \ldots, U_{m(n)}^n$ such that $\operatorname{diam}(U_m^n) < 1/n$ for $1 \leq m \leq m(n)$, and $\Omega = \bigcup_{m=1}^{m(n)} U_m^n$. Hence, $\nu \in M_c$ if and only if

$$\lim_{n\to\infty} \sup_{1\leq m\leq m(n)} |\nu| (U_m^n) = 0.$$

Hence M_c is determined by countable operations (unions and intersections) on weak-* closed subsets. □

We are now able to complete the proof of Kalton's representation theorem. Recall that $M = \mathcal{M}(\Omega)$, and that $\mathcal{U}_M$ denotes the sigma-algebra of universally measurable subsets of M. Let $\mu \in M$ be a fixed probability measure and $(S, \mathcal{F}, \lambda)$ a finite complete measure space.

Theorem 4.1.15 (Kalton's theorem) *Let $T\colon L^1(\Omega, \mathcal{B}, \mu) \to L^1(S, \mathcal{F}, \lambda)$ be an operator with representing measures $\mathcal{T} = \{\mu_x\colon \mathcal{B} \to \mathbb{R},\ x \in S\}$ given by Theorem 4.1.9. There exists a sequence of $\mathcal{F}$ measurable weights $a_n\colon S \to \mathbb{R}$ and $\mathcal{F} \to \mathcal{B}$ measurable mappings $\sigma_n\colon S \to \Omega$ satisfying*

$$|a_{n+1}(x)| \leq |a_n(x)|, \qquad \sigma_n(x) \neq \sigma_m(x) \text{ for } m \neq m, \tag{4.1.79}$$

and

$$\mu_x = \sum a_n(x)\delta_{\sigma_n(x)} + \nu_x,$$

for λ-almost every $x \in S$, where the series of Dirac measures converges in the weak-$*$ topology of M, and where $\{\nu_x : x \in S\}$ is a family of diffuse Borel measures on Ω such that $x \to \nu_x(A)$ is λ-integrable for all Borel subsets A of Ω.

Proof Motivated by the restricting conditions (4.1.79), we define sets

$$F = \left\{(w_n) \in \Omega^{\mathbb{N}} : w_n \neq w_m \text{ for } m \neq m\right\} \quad \text{and} \quad E = \left\{(s_n) \in \ell^1 : |s_{n+1}| \leq |s_n|\right\}.$$

We let F carry the product topology of $\Omega^{\mathbb{N}}$, and E the norm induced by ℓ^1.

Step 1 Next, define the continuous map $d\colon M \times E \times F \to M$, by putting

$$d(\nu, s, w) = \nu - \sum s_n \delta_{w_n},$$

and let $Z = \{z \in M \times E \times F : d(z) \in M_c\}$. Z is a Borel set in $M \times E \times F$ and hence forms a Souslin space. A moments reflection reveals that the mapping $u\colon Z \to M$ defined by $u(\nu, s, w) = \nu$ is continuous and surjective. By Theorem 4.1.13, there exists a $\mathcal{U}_M - \mathcal{B}_Z$-measurable $f\colon M \to Z$ such that $u(f(\nu)) = \nu$ for all $\nu \in M$. Since $f(\nu) \in Z$, we have

$$f(\nu) = (\nu, (s_n(\nu)), (w_n(\nu))), \tag{4.1.80}$$

where $(s_n(\nu)) \in E$, $(w_n(\nu)) \in F$, and $d(f(\nu)) \in M_c$. Equation (4.1.80) gives rise to $\mathcal{U}_M$-measurable functions $s_n\colon M \to \mathbb{R}$ and to $\mathcal{U}_M - \mathcal{B}$ measurable functions $w_n\colon M \to \Omega$.

Since the composition $d \circ f\colon M \to M_c$ is $\mathcal{U}_M - \mathcal{B}_M$-measurable, we showed that splitting off the atomic part, determines a $\mathcal{U}_M - \mathcal{B}_M$-measurable operation.

Step 2 We apply the above splitting operation to $\mathcal{T} = \{\mu_x : x \in S\}$. Equation (4.1.80), defines weights $a_n\colon S \to \mathbb{R}$ by $a_n(x) = s_n(\mu_x)$, and mappings $\sigma_n\colon S \to \Omega$ by $\sigma_n(x) = w_n(\mu_x)$.

We first prove that σ_n is a $\mathcal{F} - \mathcal{B}$-measurable mapping. To this end fix $B \subseteq \mathcal{B}$. As w_n is a $\mathcal{U}_M - \mathcal{B}$-measurable mapping, the set $w_n^{-1}(B)$ is universally measurable. Exploiting this fact, we define a finite Borel measure $\gamma\colon \mathcal{B}_M \to [0, \infty)$ by

$$\gamma(C) = \lambda\{x \in S : \mu_x \in C\}.$$

As $w_n^{-1}(B)$ is universally measurable, it is γ-measurable as well. Hence, there exist $C_1, C_2 \in \mathcal{B}_M$, such that

$$C_1 \subseteq w_n^{-1}(B) \subseteq C_2$$

and $\gamma(C_2 \backslash C_1) = 0$. Since $\sigma_n^{-1}(B) = \{x \in S : \mu_x \in w_n^{-1}(B)\}$, we get

$$\{x \in S : \mu_x \in C_1\} \subseteq \sigma_n^{-1}(B) \subseteq \{x \in S : \mu_x \in C_2\}$$

and $\lambda\{x \in S : \mu_x \in C_2 \backslash C_1\} = \gamma(C_2 \backslash C_1) = 0$. Recalling that the measure space $(S, \mathcal{F}, \lambda)$ is complete, we finally verify that $\sigma_n^{-1}(B) \in \mathcal{F}$, proving that σ_n is $\mathcal{F} - \mathcal{B}$-measurable.

A straightforward adaptation of the argument in Step 2 shows that $a_n\colon S \to \mathbb{R}$ is $\mathcal{F}$-measurable. $\square$

4.2 L^1 Embedding

Chapter 3 was devoted to proving that L^1 embeds into L^1/H^1. We recall the following basic facts about $H^1(\mathbb{T})$:

(i) By Paley's theorem (Corollary 2.3.4), $H^1(\mathbb{T})$ contains a complemented copy of ℓ^2, and hence is not isomorphic to a complemented subspace in L^1.
(ii) The dual of $H^1(\mathbb{T})$ is not separable, which prevents it from being reflexive.
(iii) $H^1(\mathbb{T})$ is not isomorphic to ℓ^1, again by Paley's theorem.

In this section we prove that L^1 embeds into X or L^1/X, for the following classes of $X \subseteq L^1$:

(i) Complemented subspaces (Section 4.2.1);
(ii) Reflexive subspaces (Section 4.2.2);
(iii) Copies of ℓ^1 (Section 4.2.3).

By the Dunford–Pettis theorem, a complemented subspace of L^1 is not reflexive. Hence the first and second classes of subspaces are disjoint. The third class is not contained in the first one. Indeed, Bourgain (1980a, 1981) constructed an example of a noncomplemented subspace of L^1 that is isomorphic to ℓ^1. Finally, as we noted above, $H^1(\mathbb{T})$ is not contained in any of the three classes we consider in this section.

In Section 4.2.4 we present a result of Johnson, Maurey, and Schechtman (2007), which asserts that there exists an L^1-normalized, weakly null sequence (φ_i), such that L^1 embeds into

$$Z[(i_k)] = \overline{\text{span}}^{L^1}\{\varphi_{i_k} : k \in \mathbb{N}\},$$

for any infinite sequence (i_k). That is, no matter how lacunary the subsequence (i_k) may be, the space $Z[(i_k)]$ contains a subsequence equivalent to the entire Haar system in L^1.

4.2.1 Enflo's Subspace Theorem

Kalton's theorem (Theorem 4.1.15) allows us to study operators on L^1 by analyzing their representing measures. Thus we obtain Enflo's subspace theorem, which asserts that either X or $L^1(\Omega)/X$ contains a copy of L^1, for every complemented subspace $X \subseteq L^1$.

Theorem 4.2.1 *Let $\mathcal{T} = \{\mu_x : x \in \Omega\}$ be the Kalton representation of $T\colon L^1(\Omega) \to L^1(\Omega)$, let $E = \{x \in \Omega : \mu_x(\{x\}) \geq 1/2\}$, and let* $0 < \epsilon < 1/2$.

If $\mu(E) > 0$, then there exists $A \subset \Omega$ with $\mu(A) > 0$ such that for any $B \subseteq A$, with $\mu(B) > 0$, the restriction

$$1_B T : L^1(B) \to L^1(B)$$

defined by $(1_B T f)(x) = 1_B(x)(Tf)(x)$ is a surjective operator and

$$\int_B |Tf| d\mu \geq \left(\frac{1}{2} - \epsilon\right) \int_B |f| d\mu,$$

for any $f \in L^1(B)$.

Proof This proof follows the steps outlined by Kalton (1979) and Capon (1980).

Step 1 Denote $E = \{x \in \Omega : \mu_x(\{x\}) \geq 1/2\}$. Let $\mu_x = \sum_{n=1}^{\infty} a_n(x)\delta_{\sigma_n(x)} + \nu_x$ be the decomposition of $\mu_x \in \mathcal{T}$ into its atomic respectively diffuse components. Since $\nu_x(\{x\}) = 0$ and since $\sigma_n(x) \neq \sigma_m(x)$ for $n \neq m$, there exists $n = n(x)$ such that the evaluation of μ_x on the singleton set $\{x\}$ reduces to the identity,

$$\mu_x(\{x\}) = a_n(x)\delta_{\sigma_n(x)}(\{x\}).$$

Hence $E = \bigcup E_n$, where $E_n = \{x \in \Omega : \sigma_n(x) = x,\ a_n(x) > 1/2\}$. Since $\mu(E) > 0$, there exists $n_0 \in \mathbb{N}$ such that $\mu(E_{n_0}) > 0$. Setting $F = E_{n_0}$ and $a(x) = a_{n_0}(x)$, we obtain the identity (between measures on Ω)

$$\mu_x = a(x)\delta_x + \nu_x, \qquad x \in F,$$

where $a(x) \geq 1/2$ and $\nu_x(\{x\}) = 0$. Define $T_0 f(x) = 1_F(x)a(x)f(x)$. It is easy to see that its Kalton representation is given by the measures $\mu_{0,x} = 1_F(x)a(x)\delta_x$, where $x \in \Omega$.

Step 2 Since Ω is a compact metric space for $n \in \mathbb{N}$, there exists a finite set $L(n) \subset \mathbb{N}$ and a measurable partition $\{B(k,n) : k \in L(n)\}$ of Ω such that $\operatorname{diam}(B(k,n)) < 1/n$. We associate to $\{B(k,n)\}$ the operator

$$T_n f(x) = \sum_{k \in L(n)} 1_{B(k,n)}(x) T(1_{B(k,n)} f)(x).$$

We will show that for any $\epsilon, \delta > 0$, there exists $N \in \mathbb{N}$ such that for any $n > N$ there exists a measurable set $A_0 \subseteq F$ such that $\mu(A_0) > (1 - \delta)\mu(F)$, and

$$\int_B |(T_0 - T_n)f| d\mu \leq \epsilon \int_B |f| d\mu, \tag{4.2.1}$$

for any $B \subseteq A_0$ and $f \in L^1(A_0)$.

We fix the Kalton representation $\mathcal{T}_n = \{\mu_{n,x} : x \in \Omega\}$ of the operator T_n. It acts on functions by the identity $\langle f, \mu_{n,x}\rangle = \sum_{k\in L(n)} 1_{B(k,n)}(x)\langle 1_{B(k,n)}f, \mu_x\rangle$. We now claim that

$$\lim_{n\to\infty} \int_\Omega |\mu_{n,x} - \mu_{0,x}|(\Omega)d\mu(x) = 0. \tag{4.2.2}$$

Indeed, if $x \notin F$ then $\mu_{n,x} = \mu_{0,x} = 0$. Let us consider then $x \in F$ and determine uniquely $k(n) \in L(n)$ such that $\mu_{n,x} = 1_{B(k(n),n)}(x)\mu_x$. Since $\mu_{0,x} = a(x)\delta_x$, we have

$$|\mu_{n,x} - \mu_{0,x}|(\Omega) = |\nu_x|(B(k(n), n)),$$

for $x \in F$. Using that ν_x is a diffuse measure we get $\lim_{n\to\infty} |\nu_x|(B(k(n), n)) = 0$, proving that $\lim_{n\to\infty} |\mu_{n,x} - \mu_{0,x}|(\Omega) = 0$ holds true. Invoking Lebesgue's theorem on dominated convergence gives (4.2.2).

Step 3 The signed measures $\{\mu_{0,x} - \mu_{n,x}\}$ represent the operator $S_n = T_0 - T_n$, and the measure ρ_n, given by

$$\rho_n(E) = \int_\Omega |\mu_{n,x} - \mu_{0,x}|(E)d\mu(x),$$

satisfies $0 \le \rho_n(E) \le \|S_n\|\mu(E)$. Hence ρ_n is absolutely continuous with repect to μ. Let $g_n \ge 0$ be its Radon–Nikodym density. Thus $\rho_n(E) = \int_E g_n d\mu$, where $E \subseteq \Omega$. By (4.2.2) we have $\rho_n(\Omega) \to 0$. Hence for any $\epsilon, \delta > 0$ fixed, there exists $N \in \mathbb{N}$ such that for any $n > N$, the measure of $A_0 = F \cap \{g_n < \epsilon\}$ is large in the sense that

$$\mu(A_0) > \mu(F) - \mu\{g_n < \epsilon\} > (1 - \delta)\mu(F).$$

On the other hand, by choice of A_0, for any measurable $B \subseteq A_0$ we have

$$\rho_n(B) = \int_B g_n d\mu \le \epsilon\mu(B).$$

In summary, we verified that

$$\sup_{B\subseteq A_0} \int_\Omega |\mu_{n,x} - \mu_{0,x}|(B)\frac{d\mu(x)}{\mu(B)} \le \epsilon.$$

In view of Inequality (4.1.54), the last estimate for the representing measures yields (4.2.1).

Step 4 Let $A_0 = F \cap \{g_n < \epsilon\}$, as considered in the previous step. Since $\{B(k, n) : k \in L(n)\}$ is a measurable partition of Ω, there exists $k \in L(n)$ such that $A = A_0 \cap B_{k,n}$ satisfies $\mu(A) > 0$. We now verify that A is a set satisfying the requirements of the theorem. First, note that if $B \subseteq A$, $\mu(B) > 0$, $f \in L^1(B)$, and $x \in B$, then $(T_n f)(x) = (Tf)(x)$. Consequently, in view of the operator estimates obtained in step 3 we have $\|1_B(Tf - T_0 f)\|_1 \leq \epsilon\|f\|_1$ and, hence,

$$\|1_B(Tf)\|_1 \geq \|T_0 f\|_1 - \|1_B(Tf - T_0 f)\|_1 \geq \|T_0 f\|_1 - \epsilon\|f\|_1,$$

for $f \in L^1(B)$. On the other hand, $f \in L^1(B)$ implies $\|T_0 f\|_1 \geq (1/2)\|f\|_1$.

Step 5 Here, we show, that for any $B \subseteq A$, with $\mu(B) > 0$, the restriction $1_B T\colon L^1(B) \to L^1(B)$ is a surjective operator. Again we will compare $1_B T$ to the multiplier T_0. For $x \in B$ define the multiplier inverse to T_0 by $Vf(x) = f(x)/a(x)$. Since $a(x) > 1/2$, the identity $T_0 V f = f$ and the estimate $\|V(f)\|_1 \leq 2\|f\|_1$ hold true, for $f \in L^1(B)$. Moreover, on $L^1(B)$ the operators $(1_B T - T_0)V$ and $(1_B TV - \mathrm{Id})$ coincide. Hence,

$$\left\|(1_B TV - \mathrm{Id})\colon L^1(B) \to L^1(B)\right\| \leq 2\epsilon.$$

Since $\epsilon < 1/2$, a perturbation argument shows that the operator $1_B TV$ is surjective and invoking that V is the multiplier given by the bounded function $(1/a(x))$, this proves that $1_B T$ maps $L^1(B)$ onto itself. □

Remark: Let T be a linear operator on L^1 satisfying the hypothesis of Theorem 4.2.1, and assume that for $A \subseteq \Omega$ the theorem's conclusion holds true. Fix $B \subseteq A$ with $\mu(B) > 0$. Then the multiplier $L_B\colon T(L^1(B)) \to L^1(B)$ given by $L_B(Tf) = 1_B \cdot (Tf)$ is well defined and bijective; hence, by Banach's isomorphism theorem, its inverse, denoted R_B, is also bounded. The operator

$$P\colon L^1(\Omega) \to T(L^1(B)),$$

where $Pf = R_B(1_B f)$ is bounded, surjective, and idempotent. In summary, the subspace $T(L^1(B))$ is complemented in $L^1(\Omega)$ and isomorphic to $L^1(B)$.

Theorem 4.2.2 (Enflo's theorem) *If X is a complemented subspace of $L^1(\Omega)$, then X or $L^1(\Omega)/X$ contain a complemented subspace isomorphic to $L^1(\Omega)$.*

Proof Let X be a complemented subspace of $L^1(\Omega)$ and let $P\colon L^1(\Omega) \to L^1(\Omega)$ be a projection onto X. That is, $P^2 = P$ and $P(L^1(\Omega)) = X$. Let $\mathcal{P} = \{\mu_x : x \in \Omega\}$ and $\mathcal{R} = \{\rho_x : x \in \Omega\}$ denote the Kalton representation of P and $R = \mathrm{Id} - P$, respectively. Now define $M, N \subseteq \Omega$ by

$$M = \{x \in \Omega : \mu_x\{x\} \geq 1/2\} \quad \text{and} \quad N = \{x \in \Omega : \rho_x\{x\} \geq 1/2\}.$$

Since $P + R = \mathrm{Id}$, we have $\mu_x + \rho_x = \delta_x$ for $x \in \Omega$ and $\mu(M) + \mu(N) > 0$. Now, if $\mu(M) > 0$ then by Theorem 4.2.1, P fixes a complemented copy of $L^1(\Omega)$, hence X contains a complemented copy of $L^1(\Omega)$. If $\mu(N) > 0$, then, by Theorem 4.2.1, R fixes a complemented copy of $L^1(\Omega)$, and $Y = R(L^1(\Omega))$

contains a complemented copy of $L^1(\Omega)$. Since X is complemented in $L^1(\Omega)$, its complement, Y, identifies with the quotient space $L^1(\Omega)/X$. □

We record a straightforward consequence of Enflo's Theorem.

Corollary 4.2.3 *Let Y_1, Y_2 be Banach spaces such that L^1 is isomorphic to $Y_1 \oplus Y_2$. Then either Y_1 or Y_2 contains a complemented copy of L^1.*

Proof There exists a linear subspace $X \subseteq L^1$ such that X is isomorphic to Y_1 and L^1/X is isomorphic to Y_2, and hence $L^1 = X \oplus L^1/X$. By Enflo's theorem (Theorem 4.2.2) either Y_1 or Y_2 contains a complemented copy of L^1. □

Remarks: We recall that Pełczyński's decomposition method provides the connection between projections on the one side, and isomorphisms between Banach spaces on the other side. We refer to Müller (2005, Section 2.2) for a concise presentation. In our present context, Pełczyński's decomposition method allows us to reinterpret Corollary 4.2.3 as follows: If L^1 is isomorphic to $Y_1 \oplus Y_2$, then either Y_1 or Y_2, are isomorphic to L^1. (Spaces sharing this property are called primary spaces.)

Compare this to Talagrand's theorem (Theorem 4.5.12), which asserts that there exist two Banach spaces Y_1 and Y_2 such that L^1 is isomorphic to a *subspace* of $Y_1 \oplus Y_2$ and neither Y_1 nor Y_2 contain a copy of L^1.

4.2.2 Reflexive Subspaces

Let $(\Omega, \mathcal{L}, m)$ be a measure space whose nonatomic part has positive measure. We assume that there exists $\Omega_0 \in \mathcal{L}$ with $m(\Omega_0) > 0$ and $m(\{x\}) = 0$ for $x \in \Omega_0$. Put $L^1 = L^1(\Omega, \mathcal{L}, m)$. Under this assumption, L^1 contains an infinite dimensional reflexive subspace R. Clearly, L^1 does not embed into one of its reflexive subspaces. By proving that L^1/R contains a copy of L^1, we show that the three space problem for L^1 embedding holds true for every reflexive subspace $R \subset L^1$. The proof uses the Dunford–Pettis theorem (Theorem 1.1.22), which asserts that weakly compact sets in L^1 are uniformly integrable.

Theorem 4.2.4 *If $R \subset L^1$ is a reflexive subspace then the quotient space L^1/R contains a copy of L^1.*

Proof Put $B_R = \{f \in R : \|f\|_1 \leq 1\}$. Since R is a reflexive subspace, Theorem 1.1.23 implies that B_R is weakly compact in L^1.

By the Dunford–Pettis theorem (Theorem 1.1.22), B_R is uniformly integrable. Hence there exists $t_0 > 0$ such that

$$\sup_{f \in B_R} \int_{\{|f| > t_0\}} |f| dm < \frac{1}{8}.$$

Next, set $\delta = t_0^{-1}/8$, and choose $A \in \mathcal{L}$ satisfying $m(A) \leq \delta$. Then for any $f \in B_R$,

$$\int_A |f|dm = \int_{\{|f|\leq t_0\}\cap A} |f|dm + \int_{\{|f|>t_0\}} |f|dm < \frac{1}{4}.$$

Thus, by rescaling, for any $A \in \mathcal{L}$ with $m(A) \leq \delta$ we have

$$\int_A |g|dm \leq \frac{1}{4}\|g\|_1, \qquad g \in R. \tag{4.2.3}$$

Now fix any $A \subset \Omega$ with $0 < m(A) \leq \delta$, and put $F = L^1(A, dm)$. Clearly, F is isomorphic to L^1. Let $q\colon L^1 \to L^1/R$ be the canonical quotient map defined by $q(f) = \{f + R\}$. We show next that the quotient map, restricted to F, is an embedding. We claim that

$$\|f\|_1 \geq \|q(f)\|_{L^1/R} \geq \frac{1}{2}\|f\|_1, \qquad f \in F. \tag{4.2.4}$$

The right-hand side of Inequality (4.2.4) is obvious. We turn to proving the left-hand side. To this end, fix $f \in F$ and note that, by the triangle inequality,

$$\inf\{\|f - g\|_1 : g \in R\} = \inf\{\|f - g\|_1 : \|g\|_1 \leq 2\|f\|_1, g \in R\}. \tag{4.2.5}$$

Now let $g \in R$ with $\|g\|_1 \leq 2\|f\|_1$. Since f is supported in A, we have with equations (4.2.3) and (4.2.5),

$$\begin{aligned} \|f - g\|_1 &\geq \|1_A(f - g)\|_1 \\ &\geq \|f\|_1 - \|1_A g\|_1 \\ &\geq \|f\|_1 - \frac{1}{4}\|g\|_1 \geq \frac{1}{2}\|f\|_1. \end{aligned} \tag{4.2.6}$$

In summary, by claim (4.2.4) the space F is isomorphic to a subspace of L^1/R. □

4.2.3 Copies of ℓ^1

Let $(\Omega, \mathcal{L}, m)$ be a measure space whose nonatomic part has positive measure, and let $L^1 = L^1(\Omega, \mathcal{L}, m)$. The dichotomy of Kadec and Pełczyński (1961/62) states that a subspace of L^1 is either reflexive, or contains a complemented copy of ℓ^1. Having treated the reflexive case in Section 4.2.2 we now turn to subspaces $X \subseteq L^1$ isomorphic to ℓ^1. Since the nonatomic part of $(\Omega, \mathcal{L}, m)$ has positive measure, $L^1(\Omega, \mathcal{L}, m)$ is not isomorphic to a subspace of ℓ^1. We prove in this section that L^1/X contains a copy of L^1, for every $X \subset L^1$ isomorphic to ℓ^1.

We simplify our notation and assume from the beginning that $m(\{\omega\}) = 0$ for $\omega \in \Omega$. We denote $L_0^1 = \{f \in L^1 : \int f dm = 0\}$. Let $X \subset L_0^1$, and let

$$d(\ell^1, X) = \inf\left\{\left\|T : X \to \ell^1\right\| \cdot \left\|T^{-1} : \ell^1 \to X\right\|\right\},$$

where the infimum is taken over all bijective linear operators $T: X \to \ell^1$, and let $q: L^1 \to L^1/X$ defined by

$$q(f) = \{f + X\}.$$

Using that $d(\ell^1, X) < \infty$, we will construct a nonatomic sigma-algebra $\Sigma \subseteq \mathcal{L}$ such that the restricted quotient map $q: L^1(\Omega, \Sigma) \to L^1/X$ is invertible on its range. Theorem 4.2.5 is the crucial step of the proof.

Theorem 4.2.5 *Let $X \subseteq L^1_0(\Omega)$ and $K = d(\ell^1, X)$. For any $\epsilon > 0$, there exists a nonatomic sigma-algebra $\Sigma \subseteq \mathcal{L}$ so that*

$$\left\|x - \mathbb{E}(x|\Sigma)\right\|_{L^1} \geq (1 - \epsilon)K^{-1}\|x\|_{L^1},$$

for $x \in X$. Moreover, if $c_1 = (1 - \epsilon)K^{-1}$ and $c_2 = c_1/(2 + c_1)$, then

$$\inf\{\|f - x\|_1 : x \in X\} \geq c_2\|f\|_1,$$

for $f \in L^1(\Omega, \Sigma, m)$.

With the following lemma we start the proof of Theorem 4.2.5. It is the only specialized tool by which the proof exploits the hypothesis that X is isomorphic to ℓ^1.

Lemma 4.2.6 *Let (g_n) be a sequence in ℓ^1 satisfying $\|g_n\|_{\ell^1} = 1$, $n \in \mathbb{N}$. For any $\epsilon > 0$, there exists a subsequence $(n(k))$ such that*

$$\left\|\sum \alpha_k g_{n(k)}\right\|_{\ell^1} > 1 - \epsilon, \tag{4.2.7}$$

whenever $\alpha_k > 0$ and $\sum \alpha_k = 1$.

Proof Let $g_n \in \ell^1$, with $g_n = (g_{n,p})_{p=1}^{\infty}$ and $\|g_n\|_{\ell^1} = 1$. Fix $\epsilon > 0$. By combinatorial means we select a subsequence $(n(k))$ and signs $b_p \in \{+1, -1, 0\}$, such that

$$\sum_{p=1}^{\infty} b_p g_{n(k),p} \geq 1 - \epsilon, \qquad k \in \mathbb{N}. \tag{4.2.8}$$

Once this is done the verification of Inequality (4.2.7) is straightforward.

Step 1 By a Cantor diagonal sequence argument we may find a subsequence of integers $(\nu(k))$ and (a_p) so that $\lim_{k\to\infty} g_{\nu(k),p} = a_p$ and $\sum |a_p| \leq 1$. We now relabel the sequence $(g_{\nu(k)})$ by putting $g_k = g_{\nu(k)}$.

Next, we determine a subsequence $(n(k))$ of $(\nu(k))$ and a sequence of signs, $b_p \in \{+1, -1, 0\}$ so that (4.2.8) holds true. First, select $m(0) \in \mathbb{N}$ and $n(0) \in \mathbb{N}$ so that

$$\sum_{p=m(0)+1}^{\infty} |a_p| \le \epsilon/3 \quad \text{and} \quad \sum_{p=0}^{m(0)} |g_{n(0),p} - a_p| \le \epsilon/3.$$

Continuing inductively, we obtain two sequences $m(k) > m(k-1)$ and $n(k) > n(k-1)$ so that

$$\sum_{p=m(k)+1}^{\infty} |g_{n(k-1),p}| \le \epsilon/3 \quad \text{and} \quad \sum_{p=0}^{m(k)} |g_{n(k),p} - a_p| \le 4^{-k}\epsilon/3. \tag{4.2.9}$$

Now define $b_p \in \{+1, -1, 0\}$ by putting $b_p = \operatorname{sign} a_p$, for $p \le m(0)$, and

$$b_p = \operatorname{sign} g_{n(k),p}, \qquad m(k) + 1 \le p \le m(k+1). \tag{4.2.10}$$

Step 2 With the above choice of $(n(k))$ and b_p, we now verify the estimate (4.2.8). By (4.2.9) we have

$$\sum_{p=1}^{\infty} b_p g_{n(k),p} \ge \sum_{p=1}^{m(k+1)} b_p g_{n(k),p} - \frac{\epsilon}{3}. \tag{4.2.11}$$

By (4.2.10), the definition of b_p yields

$$\sum_{p=1}^{m(k+1)} b_p g_{n(k),p} = \sum_{p=1}^{m(k)} b_p g_{n(k),p} + \sum_{p=m(k)+1}^{m(k+1)} |g_{n(k),p}|. \tag{4.2.12}$$

For $1 \le j \le k-1$ and $m(j) + 1 \le p \le m(j+1)$, write

$$b_p g_{n(k),p} = b_p g_{n(j),p} + b_p(g_{n(k),p} - g_{n(j),p}),$$
$$|g_{n(j),p}| = |g_{n(k),p}| - (|g_{n(k),p}| - |g_{n(j),p}|).$$

The triangle inequality together with (4.2.9) gives

$$\begin{aligned} \sum_{p=1}^{m(k)} b_p g_{n(k),p} &= \sum_{j=1}^{k-1} \sum_{m(j)+1}^{m(j+1)} b_p g_{n(k),p} \\ &\ge \sum_{j=1}^{k-1} \sum_{m(j)+1}^{m(j+1)} b_p g_{n(j),p} - |g_{n(k),p} - g_{n(j),p}| \\ &\ge \sum_{p=1}^{m(k)} |g_{n(k),p}| - 2\sum_{p=1}^{m(k)} |a_p - g_{n(k),p}| \\ &\ge \sum_{p=1}^{m(k)} |g_{n(k),p}| - \frac{2\epsilon}{3}. \end{aligned} \tag{4.2.13}$$

Finally, combining Inequalities (4.2.13) with (4.2.11) and (4.2.12) we arrive at Inequality (4.2.8).

Step 3 We next fix $\alpha_k > 0$ and $\sum \alpha_k = 1$. Then using (4.2.8) we obtain that

$$\begin{aligned}\left\|\sum_{k=1}^{\infty} \alpha_k g_{n(k)}\right\|_{\ell^1} &= \sum_{p=1}^{\infty}\left|\sum_{k=1}^{\infty} \alpha_k g_{n(k),p}\right| \\ &\geq \sum_{p=1}^{\infty} b_p \sum_{k=1}^{\infty} \alpha_k g_{n(k),p} \\ &= \sum_{k=1}^{\infty} \alpha_k \sum_{p=1}^{\infty} b_p g_{n(k),p} \\ &\geq (1-\epsilon) \sum_{k=1}^{\infty} \alpha_k. \end{aligned}$$ □

We continue with the proof of Theorem 4.2.5. Let $(\sigma_n : \Omega \to \{-1, +1\})$ be a sequence of $\mathcal{L}$-measurable, independent functions satisfying $\int \sigma_n dm = 0$. Put $\sigma_0 = 1$. (The routine construction (σ_n) uses that $m(\{\omega\}) = 0$ for $\omega \in \Omega$.) Let $N \subseteq \mathbb{N}$ and define $\Sigma(N) \subseteq \mathcal{L}$ to be the sigma-algebra generated by $\{\sigma_n : n \in N\}$.

Lemma 4.2.7 *Let $T : \ell^1 \to X$ be an isomorphism. Assume that there exists a set of integers $M = \{0, m(1), \ldots, m(n)\}$ and $0 < \delta_n < 1$, such that*

$$\left\|x - \mathbb{E}(x|\Sigma(M))\right\|_{L^1} \geq \frac{1-\delta_n}{\left\|T^{-1}\right\|}\left\|T^{-1}x\right\|_{\ell^1},$$

for $x \in X$. Then, for any $1 > \delta_{n+1} > \delta_n$, there exists $m(n+1) > m(n)$ such that with $N = \{m(n+1)\} \cup M$ we have

$$\left\|x - \mathbb{E}(x|\Sigma(N))\right\|_{L^1} \geq \frac{1-\delta_{n+1}}{\left\|T^{-1}\right\|}\left\|T^{-1}x\right\|_{\ell^1}, \qquad x \in X.$$

Proof Let $\delta > \delta_n$, and set $\epsilon = (\delta - \delta_n)/100$. Assume that the conclusuion of Lemma 4.2.7 fails. Then for any $p > m(n)$, there exists $x_p \in X$ so that

$$\left\|x_p - \mathbb{E}(x_p|\Sigma(M \cup \{p\}))\right\|_{L^1} \leq \frac{1-\delta}{\left\|T^{-1}\right\|}\left\|T^{-1}x_p\right\|_{\ell^1}. \tag{4.2.14}$$

If $f_p = x_p/\left\|T^{-1}x_p\right\|_{\ell^1}$ .and $g_p = T^{-1}f_p$, then $\|g_p\|_{\ell^1} = 1$. Using Lemma 4.2.6 and passing to a subsequence, we may assume that for any choice of $\alpha_p > 0$, with $\sum \alpha_p = 1$, we have

$$\left\|\sum \alpha_p g_p\right\|_{\ell^1} > 1 - \epsilon. \tag{4.2.15}$$

Taking conditional expectations and forming $h_p = \mathbb{E}(f_p|\Sigma(M \cup \{p\}))$, we may rewrite Inequality (4.2.14) as

$$\|f_p - h_p\|_{L^1} \leq (1-\delta)\left\|T^{-1}\right\|^{-1}. \tag{4.2.16}$$

Independent of p, the atoms of $\Sigma(M \cup \{p\})$ have measure equal to 2^{-n-1}. Since $\|h\|_1 \leq \|T\|$, the sequence $\{h_p\}$ is uniformly integrable. Hence, by passing to a subsequence we obtain $h \in L^1(\Omega)$ such that

$$h_p \longrightarrow h \quad \text{in} \quad \sigma(L^1, L^\infty).$$

Note that the weak limit h is measurable with respect to $\Sigma(M)$. By Mazur's theorem there exist $\alpha_p > 0$, with $\sum \alpha_p = 1$, such that the sums $\sum_{p=1}^m \alpha_p h_p$ converge to h in the L^1-norm. Let $f = \sum \alpha_p f_p$ and $g = \sum \alpha_p g_p$. Then $f \in X$, $Tg = f$, and (4.2.15) yields $\|g\|_{\ell^1} > 1 - \epsilon$. Moreover,

$$h = \mathbb{E}(h|\Sigma(M)) = \mathbb{E}(f|\Sigma(M)). \tag{4.2.17}$$

Invoking Inequality (4.2.16) gives,

$$\|f - h\|_{L^1} \leq \sum \alpha_p \|f_p - h_p\|_{L^1} \leq (1 - \delta)\left\|T^{-1}\right\|^{-1}. \tag{4.2.18}$$

Since $g = T^{-1}f$, the lower estimate $\|g\|_{\ell^1} > 1 - \epsilon$ gives $1 - \epsilon \leq \left\|T^{-1}f\right\|_{\ell^1}$. Combining this with (4.2.17) and (4.2.18), we obtain

$$\|f - \mathbb{E}(f|\Sigma(M))\|_{L^1} \leq \frac{(1 - \delta)\left\|T^{-1}f\right\|_{\ell^1}}{(1 - \epsilon)\left\|T^{-1}\right\|}. \tag{4.2.19}$$

By the choices made at the beginning of the proof we have $(1 - \delta)/(1 - \epsilon) < 1 - \delta_n$, hence Inequality (4.2.19) finally gives

$$\|f - \mathbb{E}(f|\Sigma(M))\|_{L^1} < \frac{1 - \delta_n}{\left\|T^{-1}\right\|}\left\|T^{-1}f\right\|_{\ell^1}.$$

Since $f \in X$, this contradicts the hypothesis. □

Proof of Theorem 4.2.5: Let $\epsilon > 0$ and $\tau > \epsilon/2$. Choose $T : \ell^1 \to X$ so that

$$\|T\|\left\|T^{-1}\right\| \leq (1 + \tau)d(\ell^1, X).$$

Put $M_0 = \{0\}$, then $\Sigma(M_0) = \{\emptyset, \Omega\}$. Since $X \subset L^1_0(\Omega)$, we have $\mathbb{E}(x|\Sigma(M_0)) = 0$, for $x \in X$, and hence,

$$\|x - \mathbb{E}(x|\Sigma(M_0))\|_{L^1} \geq \left\|T^{-1}x\right\|_{\ell^1} / \left\|T^{-1}\right\|,$$

for $x \in X$. Next, we Lemma 4.2.7 repeatedly, and obtain an infinite, increasing chain of integer sets $M_n \subseteq \mathbb{N}$ such that $\Sigma = \sigma(\bigcup \Sigma(M_n))$ satisfies the conclusion of the theorem.

Now we turn to proving the second part of the theorem. Fix $f \in L^1(\Omega, \Sigma)$ and define $L_\alpha = \inf\{\|f - x\|_1 : \|x\|_1 \geq \alpha\|f\|_1, x \in X\}$, where $0 < \alpha < 1$. Then by the triangle inequality,

$$\|q(f)\|_{L^1/X} \geq \min\{(1 - \alpha)\|f\|_{L^1}, L_\alpha\}. \tag{4.2.20}$$

Fix $x \in X$, with $\|x\|_1 \geq \alpha\|f\|_1$, and use that f is Σ-measurable. This gives

$$\|f - x\|_{L^1} \geq \|f - \mathbb{E}(x|\Sigma)\|_{L^1} \geq \|x - \mathbb{E}(x|\Sigma)\|_{L^1} - \|f - x\|_{L^1}. \tag{4.2.21}$$

Invoking $\|x - \mathbb{E}(x|\Sigma)\|_1 \geq c_1\|x\|_1$ in Inequality (4.2.21) yields $2\|f - x\|_1 \geq \alpha c_1\|f\|_1$. Solving $1 - \alpha = \alpha c_1/2$ for α, and using (4.2.20) gives $\|q(f)\|_{L^1/X} \geq c_2\|f\|_1$, where $c_2 = c_1/(2 + c_1)$.

Reviewing the second part of the above proof shows that the following remark holds true.

Remark: Let $X \subseteq L^1_0(\Omega)$, $c_1 > 0$, and let Σ be a sub-sigma-algebra of $\mathcal{L}$ such that

$$\|x - \mathbb{E}(x|\Sigma)\|_{L^1} \geq c_1\|x\|_{L^1}, \tag{4.2.22}$$

for any $x \in X$. Then, with $c_2 = c_1/(2 + c_1)$, the quotient map $q\colon L^1 \to L^1/X$ satisfies

$$\|q(f)\|_{L^1/X} \geq c_2\|f\|_{L^1},$$

for $f \in L^1(\Omega, \Sigma, m)$.

4.2.4 The Johnson–Maurey–Schechtman Embedding

In this section we present a refined method, due to Johnson, Maurey, and Schechtman (2007), for constructing subspaces of L^1 isomorphic to L^1. To explain its motivation and background we review (some of) the history of the unconditional subsequence property (USP). Recall that a sequence (y_i) in a Banach space X is K-unconditional if

$$\left\|\sum_{i=1}^{m} \varepsilon_i a_i y_i\right\|_Y \leq K \left\|\sum_{i=1}^{m} a_i y_i\right\|_Y,$$

for any $a_i \in \mathbb{R}$, any choice of signs $\varepsilon_i \in \{+1, -1\}$, and $m \in \mathbb{N}$. We say that a Banach space X satisfies the unconditional subsequence property if every normalized weakly null sequence (x_i) in X contains an unconditional subsequence. Bessaga and Pełczyński (1958a) conjectured that the USP holds true for every Banach space. Later, Maurey and Rosenthal (1977) constructed a specific Banach space for which the USP fails, and queried whether L^1 is already a counterexample to the conjecture of Bessaga and Pełczyński, or not. This brings us to the work of Johnson, Maurey, and Schechtman (2007) who determined a normalized weakly null sequence (φ_i) in L^1 such that for any choice of $n_k \uparrow \infty$,

$$L^1 \text{ embeds into } \overline{\operatorname{span}}^{L^1}\{\varphi_{n_k} : k \in \mathbb{N}\}.$$

Recall the notion of block representability. Let (x_i) and (y_i) be sequences in Banach spaces X and Y, and let $K \geq 1$. We say that (x_i) is K-block represented in (y_i) if there exist finite subsets $M_i \subset \mathbb{N}$ satisfying $\max M_i < \min M_{i+1}$, and scalar coefficients $\{c_j : j \in \mathbb{N}\}$ such that

$$z_i = \sum_{j \in M_i} c_j y_j, \tag{4.2.23}$$

$i \in \mathbb{N}$, satisfies

$$\frac{1}{K} \left\| \sum a_i z_i \right\|_Y \leq \left\| \sum a_i x_i \right\|_X \leq K \left\| \sum a_i z_i \right\|_Y, \tag{4.2.24}$$

for any finitely supported scalar sequence $\{a_i\}$. Theorem 4.2.11 states that there exists a normalized weakly null sequence $F_{JMS} = (\varphi_i : i \in \mathbb{N})$ in L^1 such that for *any infinite* subset $\mathbb{M} \subseteq \mathbb{N}$ and $\epsilon > 0$, the Haar system is $(1 + \epsilon)$-block represented in $(\varphi_i : i \in \mathbb{M})$.

The section ends with the result of Pełczyński (1961), asserting that $L^1[0,1[$ does not embed into a Banach space with an unconditional basis.

Sums of Independent Random Variables

Fix $\epsilon > 0$, $S \in \mathbb{N}$, and define $\varphi : [0,1] \to [-1,1]$, by

$$\varphi(s) = \begin{cases} 1 & \text{if } \; 0 \leq s < \epsilon/(2S); \\ 0 & \text{if } \; \epsilon/(2S) \leq s < 1 - \epsilon/(2S); \\ -1 & \text{if } \; 1 - \epsilon/(2S) \leq s < 1. \end{cases} \tag{4.2.25}$$

See Figure 4.3. Let $\varphi_{([0,1],i)}$ denote a sequence of independent copies of φ. Clearly, we have

$$\int_0^1 \varphi_{([0,1],i)} dt = 0 \quad \text{and} \quad \int_0^1 \left|\varphi_{([0,1],i)}\right| dt = \frac{\epsilon}{S}, \tag{4.2.26}$$

for $1 \leq i \leq S$.

Let $\mathcal{U}$ be the algebra of those subsets of the unit interval that can be written as finite union of dyadic intervals. Let $\epsilon > 0$, $H \in \mathcal{U}$, and $S = S(H) \in \mathbb{N}$. Choose $\mathcal{U}$ measurable, independent random variabes $\varphi_{(H,i)} : H \to \{1, 0, -1\}$ such that

$$\int \varphi_{(H,i)} dt = 0 \quad \text{and} \quad \int \left|\varphi_{(H,i)}\right| dt = \frac{\epsilon |H|}{S}, \tag{4.2.27}$$

for $1 \leq i \leq S$. We may consider $\varphi_{(H,i)}$ as defined on $[0,1[$ and supported on H. Next, set

$$g_H = \sum_{i=1}^{S} \left|\varphi_{(H,i)}\right| \quad \text{and} \quad d_H = \sum_{i=1}^{S} \varphi_{(H,i)}. \tag{4.2.28}$$

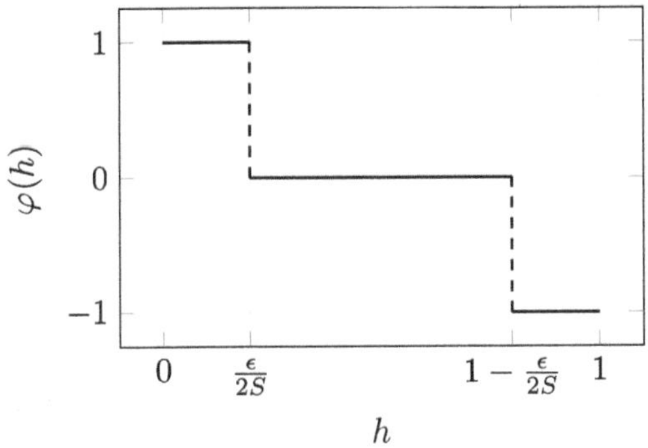

Figure 4.3 The graph of φ defined in (4.2.25). Independent copies of φ yield the random variables $\varphi_{([0,1],i)}$ satisfying condition (4.2.26).

Figure 4.4 describes g_H and d_H by their distribution functions.

Lemma 4.2.8 *Fix $H \in \mathcal{U}$ and $S = S(H)$. If $A = \{g_H = 1\} \cap \{d_H = 1\}$ and $B = \{g_H = 1\} \cap \{d_H = -1\}$, then*

$$|A| = |B| = \frac{|\{g_H = 1\}|}{2} \geq \frac{\epsilon}{2}|H|\left(1 - \epsilon|H|\frac{S-1}{S}\right) \tag{4.2.29}$$

and

$$\|(1_A - 1_B) - d_H\|_{L^1[0,1]} \leq \epsilon^2|H|^2\frac{S-1}{S}. \tag{4.2.30}$$

Proof First, note that $\{|\varphi_{(H,i)}| : 1 \leq i < S\}$ are identically distributed and independent with values in $\{1, 0\}$. Hence $g_H(x) = 1$ if there exists $1 \leq i_0 \leq S$ such that $|\varphi_{(H,i_0)}(x)| = 1$ and $|\varphi_{(H,i)}(x)| = 0$ for $1 \leq i \neq i_0 \leq S$. Hence,

$$|\{g_H = 1\}| = \epsilon|H|\left(1 - \frac{\epsilon|H|}{S}\right)^{S-1}.$$

By Bernoulli's inequality we have

$$\left(1 - \frac{\epsilon|H|}{S}\right)^{S-1} \geq \left(1 - \epsilon|H|\frac{S-1}{S}\right).$$

By symmetry we have $|A| = |B| = |\{g_H = 1\}|/2$. Thus we completed the verification of Inequality (4.2.29). We next turn to Inequality (4.2.30). First, note that $d_H 1_{\{g_H=1\}} = 1_A - 1_B$ and that $|d_H| \leq g_H$. Hence,

$$|(1_A - 1_B) - d_H| = \left|d_H(1 - 1_{\{g_H=1\}})\right| \leq g_H\left(1 - 1_{\{g_H=1\}}\right). \tag{4.2.31}$$

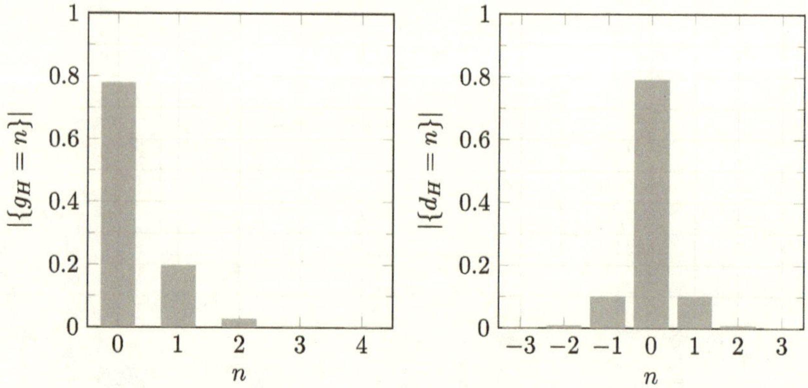

(a) The distribution of $g_{[0,1[}$ (b) The distribution of $d_{[0,1[}$

Figure 4.4 Illustrating Lemma 4.2.8 using 10^6 Monte Carlo Samples for $S = 100$, $\epsilon = 0.25$ gives $|A| = 0.098087$, $|B| = 0.097616$.

Note that condition (4.2.27) gives $\int g_H dt = \epsilon|H|$ and that $1_{\{g_H=1\}} = 1_{\{g_H=1\}} g_H$. Invoking Inequalities (4.2.31) and (4.2.29) we obtain

$$\begin{aligned} \int_H |(1_A - 1_B) - d_H| dt &\le \int_H g_H dt - \int_{\{g_H=1\}} g_H dt \\ &\le \epsilon|H|\left(1 - \left(1 - \epsilon|H|\frac{S-1}{S}\right)\right) \\ &\le \epsilon^2|H|^2 \frac{S-1}{S}. \end{aligned} \tag{4.2.32}$$

□

An iteration yields the first proto-Haar function. Fix $m \in \mathbb{N}$, put $M = m2^m$, and $\epsilon = 2^{-m}$. Given $H \in \mathcal{U}$ and $S = S(H)$, we let $\varphi_{(H,i)} : H \to \{-1, 0, 1\}$, $1 \le i \le S$, be independent, $\mathcal{U}$-measurable, and satisfying the constraints (4.2.27).

Proposition 4.2.9 *For each $H \in \mathcal{U}$, there exist*

(i) *disjoint, $\mathcal{U}$-measurable sets $A, B \subseteq H$, satisfying $|A| = |B|$ and*

$$|H \backslash (A \cup B)| < 2|H|e^{-m}; \tag{4.2.33}$$

(ii) *disjoint, $\mathcal{U}$-measurable sets $H_j \subseteq H$ $(1 \le j \le M)$ such that*

$$\left\| (1_A - 1_B) - \sum_{j=1}^{M} d_{H_j} \right\|_{L^1[0,1]} \le |H|^2 m 2^{-m}, \tag{4.2.34}$$

where d_{H_j} is defined by (4.2.28).

Proof To begin the first step of the construction, we put $H_1 = H$, and form $g_1 = g_{H_1}$ and $d_1 = d_{H_1}$ according to the constraints (4.2.27). Set $T_1 = \{g_1 = 1\}$,

$$A_1 = T_1 \cap \{d_1 = 1\}, \quad \text{and} \quad B_1 = T_1 \cap \{d_1 = -1\}.$$

Next, put $H_2 = H_1 \backslash T_1$. Assume that H_j together with A_{j-1}, B_{j-1}, are defined for $j < M$. Form $g_j = g_{H_j}$ and $d_j = d_{H_j}$ according to the constraints (4.2.27). Put $T_j = \{g_j = 1\}$,

$$A_j = T_j \cap \{d_j = 1\}, \quad \text{and} \quad B_j = T_j \cap \{d_j = -1\}.$$

Then $|A_j| = |B_j| = |T_j|/2$ and by Lemma 4.2.8,

$$|T_j| \geq |H_j|\epsilon(1-\epsilon) \tag{4.2.35}$$

and

$$\left\|(1_{A_j} - 1_{B_j}) - d_j\right\|_{L^1} \leq \epsilon^2|H|^2. \tag{4.2.36}$$

Finally, we put $H_{j+1} = H_j \backslash T_j$. The inductive construction ends with $j = M$. Now define $A = \bigcup_{j=1}^M A_j$ and $B = \bigcup_{j=1}^M B_j$. The sets A, B are disjoint and satisfy

$$|A| = |B| \quad \text{and} \quad A \cup B \subseteq H.$$

Iterating (4.2.35) gives

$$|H \backslash (A \cup B)| \leq |H|[1 - \epsilon(1-\epsilon)]^M.$$

Putting $d = \sum_{j=1}^M d_j$ and summing (4.2.36) gives

$$\|(1_A - 1_B) - d\|_{L^1} \leq M\epsilon^2|H|^2.$$

Taking into account that $\epsilon = 2^{-m}$ and $M = m2^m$, finishes the proof. □

Defining F_{JMS}: Recall that we let $\mathcal{U}$ denote the algebra of those subsets in the unit interval that may be written as finite union of dyadic intervals. Thus $\mathcal{U}$ is a countable collection of subsets in $[0, 1[$. We fix a sequence of sets $(E(n))_{n=1}^\infty$ such that $E(n) \in \mathcal{U}$ and

$$\{n \in \mathbb{N} : E(n) = H\}$$

is infinite, for any $H \in \mathcal{U}$. Let $(k_n)_{n=1}^\infty$ be an increasing sequence of powers of two and put $a_n = k_n^{3/4}$. We impose the condition that k_n increases fast enough such that

$$\sum_{n=1}^\infty \frac{a_n}{k_n} = \sum_{n=1}^\infty k_n^{-1/4} \leq 1. \tag{4.2.37}$$

Next, for fixed $n \in \mathbb{N}$, we choose a sequence of $\mathcal{U}$-measurable functions $\varphi_{(n,i)}: E(n) \to \{1, 0, -1\}$ such that

$$\int \varphi_{(n,i)} dt = 0 \quad \text{and} \quad \int |\varphi_{(n,i)}| dt = \frac{|E(n)|}{k_n}, \tag{4.2.38}$$

for $i \in \mathbb{N}$. We consider $\varphi_{(n,i)}$ as defined on $[0, 1[$ and supported on $E(n)$. Note that by independence and (4.2.38), there exists $1 \leq C(n) < \infty$ such that

$$\left\| \sum_{i=1}^{\infty} b_i \varphi_{(n,i)} \right\|_1 \leq C(n) \left(\sum_{i=1}^{\infty} |b_i|^2 \right)^{1/2}, \tag{4.2.39}$$

for any square summable sequence (b_i). Without loss of generality, we may take $(C(n))$ to be an increasing sequence of scalars.

Next, we impose a second growth condition on the sequence $a_n = k_n^{3/4}$. We demand that

$$a_N^{-1/3} C(N-1) \sum_{n=1}^{N-1} a_n + a_N^{1/4} \sum_{n=N+1}^{\infty} a_n^{-1/4} \leq 2^{-N-1}, \tag{4.2.40}$$

for $N \geq 2$. We let (k_n) be any increasing sequence of powers of two satisfying (4.2.37) and (4.2.40), where $a_n = k_n^{3/4}$. Next, define

$$\varphi_i = \sum_{n=1}^{\infty} a_n \varphi_{(n,i)}, \quad i \in \mathbb{N}. \tag{4.2.41}$$

By (4.2.37) the series on the right-hand side is absolutely convergent in L^1 and $\|\varphi_i\|_1 \leq 1$. By (4.2.39) and (4.2.37) it follows that $\varphi_i \longrightarrow 0$ weakly in L^1. We put

$$F_{JMS} = (\varphi_i : i \in \mathbb{N}). \tag{4.2.42}$$

Theorem 4.2.10 *Let $\mathbb{K} \subseteq \mathbb{N}$ be infinite, let $\tau > 0$, and $H \in \mathcal{U}$. Then there exist $f \in \operatorname{span}\{\varphi_i \in F_{JMS} : i \in \mathbb{K}\}$, disjoint $A, B \in \mathcal{U}$ such that $A \cup B = H$,*

$$|A| = |B| = (1/2)|H|, \quad \textit{and} \quad \|(1_A - 1_B) - f\|_1 \leq 2\tau|H|. \tag{4.2.43}$$

Proof Let $m \in \mathbb{N}$ such that $m2^{-m} < \tau/2$. Put $M = m2^{-m}$ and $\epsilon = 2^{-m}$.

Part 1 Set $H(1) = H$. Select an integer $n(1) > m + 1$ such that $E(n(1)) = H(1)$ and

$$2^{-n(1)} \leq 2^{-m-1} |H(1)|^2.$$

Note that $k_{n(1)}$ is an integer multiple of 2^m. Put $S(1) = 2^{-m} k_{n(1)}$, select $\sigma(1) \subset \mathbb{K}$ with cardinality $S(1)$, and define

$$g_1 = \sum_{i \in \sigma(1)} |\varphi_{(n(1),i)}| \quad \text{and} \quad d_1 = \sum_{i \in \sigma(1)} \varphi_{(n(1),i)}.$$

Put $T_1 = \{g_1 = 1\}$,

$$A_1 = T_1 \cap \{d_1 = 1\}, \quad \text{and} \quad B_1 = T_1 \cap \{d_1 = -1\}. \tag{4.2.44}$$

Fix $1 < j < M$. Assume we are given $n(j-1) \in \mathbb{N}$, $\sigma(j-1) \subset \mathbb{K}$ together with $H_{j-1} \in \mathcal{U}$ satisfying $H_{j-1} \subseteq H$, and disjoint sets $A_{j-1} \in \mathcal{U}$ and $B_{j-1} \in \mathcal{U}$ such that

$$A_{j-1} \cup B_{j-1} \subseteq H_{j-1}.$$

Set $H_j = H_{j-1} \backslash (A_{j-1} \cup B_{j-1})$. Choose $n(j) > n(j-1)$ such that

$$2^{-n(j)} \leq 2^{-m-j}|H(1)|^2 \quad \text{and} \quad E(n(j)) = H_j. \tag{4.2.45}$$

Since $2^{-m}k_{n(j)} \in \mathbb{N}$, we may choose $\sigma(j) \subset \mathbb{K}$ with cardinality $S(j) = 2^{-m}k_{n(j)}$ and $\max \sigma(j-1) < \min \sigma(j)$. Define

$$g_j = \sum_{i \in \sigma(j)} |\varphi_{(n(j),i)}| \quad \text{and} \quad d_j = \sum_{i \in \sigma(j)} \varphi_{(n(j),i)}.$$

Put $T_j = \{g_j = 1\}$,

$$A_j = T_j \cap \{d_j = 1\}, \quad \text{and} \quad B_j = T_j \cap \{d_j = -1\}. \tag{4.2.46}$$

We stop at $j = M$. Now, set $\widetilde{A} = \bigcup_{j=1}^M A_j$ and $\widetilde{B} = \bigcup_{j=1}^M B_j$. By construction, $\widetilde{A}, \widetilde{B}$ are disjoint subsets of H. Let $d = \sum_{j=1}^M d_j$. In view of Proposition 4.2.9 we have

$$\left|\widetilde{A}\right| = \left|\widetilde{B}\right| \geq (1/2)|H|(1-\tau) \quad \text{and} \quad \left\|(1_{\widetilde{A}} - 1_{\widetilde{B}}) - d\right\|_1 \leq (1/2)|H|^2\tau. \tag{4.2.47}$$

Part 2 We now link the above construction to $\operatorname{span}\{\varphi_i \in F_{JMS} : i \in \mathbb{K}\}$. Fix $1 \leq j \leq M$, and put

$$h_j = a_{n(j)}{}^{-1} \sum_{i \in \sigma(j)} \varphi_i.$$

We prove next that $\|h_j - d_j\|_1 \leq |H|^2 2^{-m-j-1}$. Comparing the definitions of (d_j) and (φ_i) shows that

$$h_j - d_j = \sum_{n=1}^{n(j)-1} \frac{a_n}{a_{n(j)}} \sum_{i \in \sigma(j)} \varphi_{(n,i)} + \sum_{n=n(j)+1}^{\infty} \frac{a_n}{a_{n(j)}} \sum_{i \in \sigma(j)} \varphi_{(n,i)}.$$

By conditions (4.2.38) and (4.2.39),

$$\|h_j - d_j\|_1 \leq \frac{S(j)^{1/2}}{a_{n(j)}} \sum_{n=1}^{n(j)-1} C(n)a_n + \frac{S(j)}{a_{n(j)}} \sum_{n=n(j)+1}^{\infty} \frac{a_n|E(n)|}{k_n}.$$

Recall that $a_n = k_n^{3/4}$. Since $S(j) \le k_{n(j)}$ and $|E(n)| \le 1$, the right-hand side of the term above is bounded by

$$C(n(j)-1)a_{n(j)}^{-1/3} \sum_{n=1}^{n(j)-1} a_n + a_{n(j)}^{1/4} \sum_{n=n(j)+1}^{\infty} a_n^{-1/4} \le 2^{-n(j)-1}.$$

Hence, in view of condition (4.2.45) we arrive at $\|h_j - d_j\| \le |H|2^{-m-j-1}$.

Part 3 Recall that in the first part of the proof we put $d = \sum_{j=1}^{M} d_j$. Finally, let $f = \sum_{j=1}^{M} h_j$. Since $\sigma(j) \subset \mathbb{K}$ is of finite cardinality, we have $f \in \operatorname{span}\{\varphi_i \in F_{JMS} : i \in \mathbb{K}\}$. Summing the estimates obtained in Part 2 of the proof gives

$$\|d - f\|_1 \le 2^{-m}|H|.$$

Recall that according to Part 1 of the proof, $\widetilde{A}, \widetilde{B} \in \mathcal{U}$ are disjoint subsets of H satisfying condition (4.2.47). Hence $D = H \setminus (\widetilde{A} \cup \widetilde{B})$ satisfies

$$|D| \le \tau|H|.$$

Since $D \in \mathcal{U}$ there exist $D_1, D_2 \in \mathcal{U}$ such that $D_1 \cap D_2 = \emptyset$, $D = D_1 \cap D_2$, and $|D_1| = |D_2| = |D|/2$. Finally, we set

$$A = \widetilde{A} \cup D_1 \quad \text{and} \quad B = \widetilde{B} \cup D_2.$$

Thus $A, B \in \mathcal{U}$ satisfy $A \cap B = \emptyset$, $A \cup B = H$, and $|A| = |B| = (1/2)|H|$. Invoking (4.2.47) shows that

$$\|(1_A - 1_B) - f\|_1 \le \left\|(1_A - 1_B) - (1_{\widetilde{A}} - 1_{\widetilde{B}})\right\|_1 + \left\|(1_{\widetilde{A}} - 1_{\widetilde{B}}) - d\right\|_1 + \|d - f\|_1 \le 2\tau|H|.$$

□

Generating a Dyadic Tree

We turn to proving Theorem 4.2.11, the main result of this section. We obtain it by induction from Theorem 4.2.10 by asserting the following: If $\mathbb{M} \subset \mathbb{N}$ is infinite, then for any measurable $H \subseteq [0,1]$ and $\epsilon > 0$ there exists $g \in \operatorname{span}\{\varphi_i : i \in \mathbb{M}\}$ and disjoint measurable sets $A, B \subseteq H$ satisfying

$$|A| = |B| = |H|/2 \quad \text{and} \quad \|(1_A - 1_B) - g\|_1 < \epsilon.$$

Dyadic trees: Let $\mathcal{I}$ denote the collection of dyadic intervals contained in the unit interval $[0,1]$. Dyadic intervals are ordered, linearly, as follows: Given $I, J \in \mathcal{I}$ put

$$J \prec I \text{ if } \quad \begin{cases} |I| = |J| \text{ and } \sup J < \sup I; \\ |J| > |I|. \end{cases}$$

Let $\mathcal{E} = \{E_I \subseteq [0,1] : I \in \mathcal{I}\}$ be a collection of measurable sets such that $|E_I| = |I|$ for $I \in \mathcal{I}$. We say that $\mathcal{E}$ is a dyadic tree if

$$E_I = E_{I_1} \cup E_{I_2} \quad \text{and} \quad E_{I_1} \cap E_{I_2} = \emptyset,$$

for every $I, I_1, I_2 \in \mathcal{I}$ satisfying $I_1 \cap I_2 = \emptyset$, and $I = I_1 \cup I_2$. Set

$$d_I = 1_{E_{I_1}} - 1_{E_{I_2}}.$$

We call $\{d_I : I \in \mathcal{I}\}$ the L^∞-normalized Haar system associated to the dyadic tree $\mathcal{E}$.

Assuming that $\mathbb{M} \subseteq \mathbb{N}$ is infinite, we show that there exists a dyadic tree (E_I) such that its associated Haar system is very well approximated by a (suitably chosen) block basis of $(\varphi_i : i \in \mathbb{M})$. Thus we obtain that the Haar system is $(1+\epsilon)$ represented in $(\varphi_i : i \in \mathbb{M})$ for any $\epsilon > 0$. In Section 1.1.6 we showed that the Haar system is not unconditional in L^1. Hence $(\varphi_i : i \in \mathbb{M})$ fails to be an unconditional basis in the L^1-closure of $\operatorname{span}\{\varphi_i : i \in \mathbb{M}\}$.

Theorem 4.2.11 *For any infinite subset $\mathbb{M} \subseteq \mathbb{N}$ and $\tau > 0$, there exist finite subsets $M_I \subseteq \mathbb{M}$ and $g_I \in \operatorname{span}\{\varphi_i : i \in M_I\}$ satisfying*

$$\max M_J < \min M_I,$$

for $I, J \in \mathcal{I}$ with $J < I$, and a dyadic tree (E_I) with associated Haar system (d_I) such that

$$\|d_I - g_I\|_1 \leq \tau |I|^2,$$

for $I \in \mathcal{I}$.

Proof Given $M \subseteq \mathbb{N}$ in the proof below we write $F(M) = \operatorname{span}\{\varphi_i : i \in M\}$. Fix $\tau > 0$ and $\mathbb{M} \subseteq \mathbb{N}$. Set $E_{[0,1]} = [0,1]$. Put $I = [0,1]$, $I_1 = [0,1/2]$, and $I_2 = [1/2, 0]$.

Step 1 (Defining $d_{[0,1]}, M_{[0,1]}$) Apply Theorem 4.2.10 with $\mathbb{K} = \mathbb{M}$ and $H = E_I$. This yields disjoint sets $E_{I_1}, E_{I_2} \in \mathcal{U}$ such that

$$E_{I_1} \cup E_{I_2} = E_I, \quad |E_{I_1}| = |E_{I_2}| = 1/2,$$

and a finite subset $M_I \subset \mathbb{M}$ together with $g_I \in F(M_I)$ such that $\|d_I - g_I\|_1 \leq \tau$, where $d_I = 1_{E_{I_1}} - 1_{E_{I_2}}$.

Step 2 (Defining d_I, M_I) Fix $I \in \mathcal{I}$. Assume that $E_J \in \mathcal{U}$, d_J, and M_J are defined for $J < I$, and that $E_I \in \mathcal{U}$ already exists. We proceed to define d_I and M_I. Choose $m_0 \in \mathbb{N}$ such that

$$m_0 \geq \max M_J,$$

for $J \in \mathcal{I}$ with $J < I$. Let I_1, I_2 be (the uniquely defined) disjoint dyadic intervals satisfying $I = I_1 \cup I_2$. Apply Theorem 4.2.10 with $\mathbb{K} = \mathbb{M} \cap [m_0 + 1, \infty[$ and $H = E_I$. This gives disjoint sets $E_{I_1}, E_{I_2} \in \mathcal{U}$ satisfying

$$E_{I_1} \cup E_{I_2} = E_I, \quad \text{with} \quad |E_{I_1}| = |E_{I_2}| = |E_I|/2,$$

a finite set $M_I \subset \mathbb{M} \cap [m_0 + 1, \infty[$, and $g_I \in F(M_I)$ such that

$$\|d_I - g_I\|_1 \leq \tau |I|^2,$$

where $d_I = 1_{E_{I_1}} - 1_{E_{I_2}}$. This completes the induction step. □

In Section 1.1.6 we determined directly the fact that the Haar system is not unconditional in L^1. Deeper is the following result of Pełczyński (1961), which asserts that L^1 is not isomorphic to a subspace of a Banach space with an unconditional basis.

Theorem 4.2.12 (Pełczyński's theorem) *For any Banach space X and linear injection $T\colon L^1 \to X$,*

$$\|T\| \cdot \left\|T^{-1}|_{T(L^1)}\right\| < \infty \quad \textit{implies} \quad \mathrm{ubc}(X) = \infty. \tag{4.2.48}$$

Proof Let $r_n\colon [0,1] \to \{-1, 1\}$ denote the sequence of independent Rademacher functions. For $x \in L^1$,

$$x r_n \to 0 \quad \text{in} \quad \sigma(L^1, L^\infty). \tag{4.2.49}$$

Since $1 + r_n \geq 0$, we have $|x + r_n x| = |x| + |x| r_n$, pointwise, and hence by (4.2.49)

$$\int_0^1 |x + r_n x| dt = \int_0^1 |x| dt + \int_0^1 r_n |x| dt \to \int_0^1 |x| dt. \tag{4.2.50}$$

Let X be a Banach space with an unconditional Schauder basis $\{x_k\}$. Assume there exists a linear injection $T\colon L^1 \to X$ for which the implication of the theorem fails to hold true. Choose $N \in \mathbb{N}$ such that

$$\sqrt{N} > 2^4 \|T\| \cdot \left\|T^{-1}|_{T(L^1)}\right\| \cdot \mathrm{ubc}(X). \tag{4.2.51}$$

We now turn to showing that this assumption leads to a contradiction. Fix $x \in L^1$ with $\|x\|_1 = 1$ and $\epsilon > 0$.

(i) Put $y_1 = x$ and choose $n(1) > 1$ such that

$$(1 - \epsilon 2^{-1})\|y_1\|_1 \leq \|y_1 + r_{n(1)} y_1\|_1 \leq (1 + \epsilon 2^{-1})\|y_1\|_1.$$

(ii) Given $y_1, \ldots, y_{k-1} \in L^1$ and $n(1) < \cdots < n(k-1)$, put $y_k = r_{n(k-1)} \sum_{i=1}^{k-1} y_i$ and choose $n(k) > n(k-1)$ such that with $w_k = \sum_{i=1}^{k} y_i$,

$$(1 - \epsilon 2^{-k})\|w_k\|_1 \leq \|w_k + r_{n(k)} w_k\|_1 \leq (1 + \epsilon 2^{-k})\|w_k\|_1.$$

Stopping the procedure at $k = N$, gives $y_1, \ldots, y_N \in L^1$ satisfying

$$\left\| \sum_{i=1}^{k} y_i \right\|_1 \leq 2 \quad \text{and} \quad \|y_k\|_1 \geq 1/2, \quad \text{for} \quad 1 \leq k \leq N. \tag{4.2.52}$$

In view of condition (4.2.49), the indices $n(1) < \cdots < n(N)$ may be chosen recursively, such that, in X, there exists a block basis $\{z_k\}$ of the unconditional basis $\{x_k\}$, satisfying

$$\|Ty_k - z_k\|_X \leq \epsilon \|T\| 2^{-k}, \quad 1 \leq k \leq N. \tag{4.2.53}$$

Thus, for any choice of $\varepsilon_k \in \{-1, +1\}$ we have

$$\begin{aligned} \left\| \sum_{k=1}^{N} \varepsilon_k T y_k \right\|_X &\leq \sum_{k=1}^{N} \|Ty_k - z_k\|_X + \left\| \sum_{k=1}^{N} \varepsilon_k z_k \right\|_X \\ &\leq \sum_{k=1}^{N} \|Ty_k - z_k\|_X + \mathrm{ubc}(X) \left\| \sum_{k=1}^{N} z_k \right\|_X \\ &\leq 2 \sum_{k=1}^{N} \|Ty_k - z_k\|_X + \mathrm{ubc}(X) \left\| \sum_{k=1}^{N} T y_k \right\|_X \\ &\leq 4\|T\| \cdot \mathrm{ubc}(X). \end{aligned} \tag{4.2.54}$$

Khintchine's inequality (2.2.24) and Inequality (4.2.54) give

$$\begin{aligned} \left(\sum_{k=1}^{N} \|y_k\|_1^2 \right)^{1/2} &\leq C_1 Av_{\{ve_k\}} \left\| \sum_{k=1}^{N} \varepsilon_k y_k \right\|_1 \\ &\leq C_1 \left\| T^{-1}|_{T(L^1)} \right\| Av_{\{\varepsilon_k\}} \left\| \sum_{k=1}^{N} \varepsilon_k T y_k \right\|_1 \\ &\leq C_1 4 \|T\| \cdot \left\| T^{-1}|_{T(L^1)} \right\| \mathrm{ubc}(X), \end{aligned} \tag{4.2.55}$$

where $C_1 = \sqrt{2}$. Merging estimates (4.2.54) with (4.2.55) and invoking (4.2.52) yields

$$\sqrt{N} \leq 2^4 \|T\| \cdot \left\| T^{-1}|_{T(L^1)} \right\| \cdot \mathrm{ubc}(X), \tag{4.2.56}$$

contradicting our choice of N in definition (4.2.51). □

4.3 Rosenthal's L^1-Theorem

We recall the following convention. Given three Banach spaces E_1, E_2, E_3 and a bounded linear operator $A \colon E_1 \to E_2$, we say that A fixes a copy of E_3 if there exists a linear subspace $F \subseteq E_1$ satisfying the following conditions:

(i) F is isomorphic to E_3.
(ii) There exists $c > 0$, such that $\|A(x)\|_{E_2} \geq c\|x\|_{E_1}$ for $x \in F$.

This section presents a close analysis of operators fixing a copy of $L^1(\Omega, \mathcal{L}, \mu)$. For simplicity, we restrict ourselves to the most important case where $(\Omega, \mathcal{L}, \mu)$ denotes the Lebesgue probability space on the unit interval $[0, 1]$. Subsequently, we set $L^1 = L^1(\Omega, \mathcal{L}, \mu)$.

In Chapter 3 and Section 4.2, we showed that L^1 embeds into L^1/X when either $X = H^1$, X is reflexive, or X is isomorphic to ℓ^1. Moreover, if X is a complemented subspace of L^1 then either X or L^1/X contains a copy of L^1. The remaining Sections 4.3– 4.5 of Chapter 4 are devoted to proving Talagrand's result (Talagrand, 1990), that there exists a subspace $X \subset L^1$ such that neither X nor L^1/X contains a copy of L^1.

The basic framework for Talagrand's construction is provided by Rosenthal's L^1-theorem (Theorem 4.3.1) which is the main result of the present section. Rosenthal's investigation (Rosenthal [1984, 1983]) of operators on L^1 starts with Kalton's representation theorem (Theorem 4.1.15) and proceeds by a thorough study of the associated representing random measures.

Theorem 4.3.1 *Let X be a Banach space and let $S\colon L^1 \to X$ be a bounded linear operator fixing a copy of L^1. Then there exists a nonatomic sigma-algebra $\Sigma \subseteq \mathcal{L}$ and $B \in \Sigma$ with $\mu(B) > 0$ and $c > 0$ such that*

$$\|Sf\|_X \geq c\|f\|_1,$$

for any $f \in L^1(B, \Sigma)$ satisfying $\int f = 0$.

We next specialize Theorem 4.3.1 by letting S be the canonical quotient map $q\colon L^1 \to L^1/Y$, defined by $q(f) = f + Y$, and X be a quotient space L^1/Y.

Corollary 4.3.2 *If $Y \subseteq L^1$ is a closed subspace, and if the quotient map $q\colon L^1 \to L^1/Y$ fixes a copy of L^1, then there exists a nonatomic sigma-algebra $\Sigma \subseteq \mathcal{L}$ and $B \in \Sigma$ with $\mu(B) > 0$ and $c > 0$ such that*

$$\inf\{\|f - y\|_{L^1} : y \in Y\} > c\|f\|_1,$$

for any $f \in L^1(B, \Sigma)$ satisfying $\int f d\mu = 0$.

In the case when $Y \subseteq L^1$ is a dual Banach space, Corollary 4.3.2 may be strengthened considerably. This is because of the following theorem due to Ghoussoub and Rosenthal on the factorization of operators $V\colon L^1 \to L^1/Y$, where $Y \subset L^1$ is a dual Banach space.

Recall that $I \subseteq [0, 1]$ is a dyadic interval if there exists $n \in \mathbb{N} \cup \{0\}$ and $k \in \{1, \ldots, 2^n\}$ such that $I = \{t : 2^{-n}(k - 1) \leq t < 2^{-n}k\}$. We let $\mathcal{I}_n$ denote the collection of all dyadic intervals $I \subseteq [0, 1[$ satisfying $|I| = 2^{-n}$.

Theorem 4.3.3 (Ghoussoub–Rosenthal theorem) *Let $Y \subset L^1$ be a dual Banach space and let $V : L^1 \to L^1/Y$ be an operator. There exists an operator $J : L^1 \to L^1$ such that*

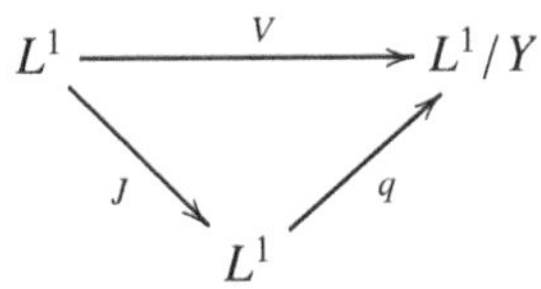

(4.3.1)

where $q : L^1 \to L^1/Y$ is the canonical quotient map given by $q(f) = f + Y$.

Proof Let $V : L^1 \to L^1/Y$ be a bounded linear operator and let $q : L^1 \to L^1/Y$ be the canonical quotient map. Let $I \in \mathcal{I}$, and choose $b_I \in L^1$, such that

$$\|b_I\|_1 \le 2\|V\| \quad \text{and} \quad q(b_I) = V(1_I/|I|).$$

Let $\mathcal{I}_n(I) = \{K \subseteq I : K \in \mathcal{I}_n\}$, where $n \in \mathbb{N}$ satisfies $2^{-n} \le |I|$, and put

$$w_{I,n} = \frac{1}{|I|} \sum_{K \in \mathcal{I}_n(I)} |K| b_K. \tag{4.3.2}$$

We have $q(w_{I,n}) = a_I$. Clearly, if $I = I_1 \cup I_2$ with $I_1, I_2 \in \mathcal{I}$, then $|I_1| = |I_2| = |I|/2$, and if, moreover, $2^{-n} \le |I_1|$, then

$$w_{I,n} = (w_{I_1,n} + w_{I_2,n})/2. \tag{4.3.3}$$

By definition, the set of differences $\{w_{I,n} - w_{I,1} : n \in \mathbb{N}\}$ is uniformly bounded, and, since $q(w_{I,n} - w_{I,1}) = 0$, we have $w_{I,n} - w_{I,1} \in Y$. Since $Y \subset L^1$ is a dual space, the unit ball of Y – equipped with the weak-* topology – is compact and metrizable. By a Cantor diagonal sequence argument we may determine $k_m \uparrow \infty$ such that $\{w_{I,k_m} : m \in \mathbb{N}\}$ is weak-* convergent. Let $w_I \in L^1$ denote its limit. By (4.3.3) and (4.3.2) we get the relations,

$$w_I = (w_{I_1} + w_{I_2})/2, \quad q(w_I) = a_I, \quad \|w_I\|_1 \le 4\|V\|,$$

which permit us to define (consistently) an operator J by linear extension of the map $J(1_I/|I|) = w_I$. We first observe the factoring identity,

$$qJ(1_I/|I|) = q(w_I) = a_I = V(1_I/|I|).$$

To see that J has a bounded extension, we fix a collection of pairwise disjoint dyadic intervals $\mathcal{E}$, coefficients $x_I \in \mathbb{R}$. Then, using that $J(1_I) = |I| w_I$ and $\|w_I\|_1 \le 4\|V\|$ we get,

$$\left\| J\left(\sum_{I \in \mathcal{E}} x_I 1_I\right)\right\|_1 \le 4\|V\| \sum_{I \in \mathcal{E}} |x_I| \cdot |I|. \tag{4.3.4}$$

Since the intervals in $\mathcal{E}$ are chosen to be pairewise disjoint, the estimates (4.3.4) yield that J admits a unique extension to a bounded linear operator on L^1 satisfying $qJ = V$. □

Recall that for $X = H^1$, X reflexive, or X isomorphic to ℓ^1, the quotient space L^1/X contains a copy of L^1. While varying significantly in depth and complexity, the respective proofs had one aspect in common: In each case we determined a nonatomic sigma-algebra Σ such that the embedding of L^1 is provided by the restricted quotient map

$$q_\Sigma : L^1(\Sigma) \to L^1/X, \qquad q_\Sigma(f) = \{f + X\}.$$

The following result (Theorem 4.3.4) explains, to some extent, the common structure in Theorem 3.2.10, Theorem 4.2.4, and Theorem 4.2.5.

Theorem 4.3.4 *Let $Y \subseteq L^1$ be a dual Banach space. If L^1/Y contains a copy of L^1 then there exists $B \subseteq \Omega$ with $\mu(B) > 0$, a nonatomic sigma-algebra $\Sigma \subset \mathcal{L}$ on B, and $c > 0$ such that*

$$\inf\{\|f - y\|_{L^1} : y \in Y\} > c\|f\|_{L^1},$$

for any $f \in L^1(B, \Sigma)$ and $\int f d\mu = 0$.

Proof Since L^1/Y contains a copy of L^1 there exists a bounded linear operator $V : L^1 \to L^1/Y$ and a constant $c > 0$ such that

$$\|V(f)\|_{L^1/Y} \geq c\|f\|_1, \tag{4.3.5}$$

for any $f \in L^1$. Let $q : L^1 \to L^1/Y$ be the canonical quotient map defined by $q(f) = f + Y$. Invoking the Ghoussoub–Rosenthal theorem (Theorem 4.3.3) we obtain a bounded linear operator $J : L^1 \to L^1$ such that

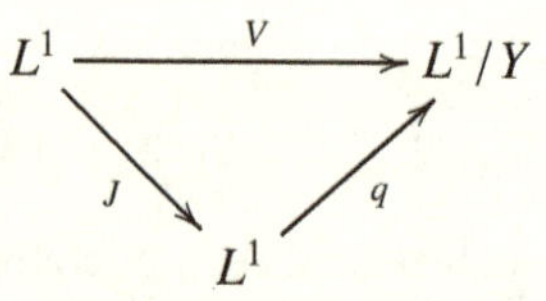

(4.3.6)

Since the above diagram commutes we have $V = qJ$ and hence, by (4.3.5),

$$\|J(f)\|_{L^1} \geq \|q(J(f))\|_{L^1/Y} = \|V(f)\|_{L^1/Y} \geq c\|f\|_{L^1},$$

for $f \in L^1$. Thus the operator J fixes a copy of L^1, and $J(L^1)$ is a closed subspace of L^1 isomorphic to L^1. Moreover, restricted to $J(L^1)$, the canonical projection q is an isomorphism. Indeed, if $g \in J(L^1)$ there exists $f \in L^1$ such that $Jf = g$, $\|f\|_1 \geq c\|g\|_1$, and

$$\|q(g)\|_{L^1/Y} = \|V(f)\|_{L^1/Y} \geq c\|f\|_{L^1} \geq c\|g\|_1/\|J\|.$$

In summary, $q: L^1 \to L^1/Y$ fixes a copy of L^1. Applying Corollary 4.3.2 finishes the proof. $\square$

Rosenthal (1984) proves Theorem 4.3.1 by showing that there are no operators $T: L^1 \to L^1$, but weighted composition operators $T_0(f)(x) = a(x)f(\sigma(x))$. The passage from T to T_0 is made possible by replacing L^1 with $L^1(A, \mathcal{A}, \mu_{|\mathcal{A}})$, where $(A, \mathcal{A}, \mu_{|\mathcal{A}})$ will be a carefully selected, nonatomic measure space contained in $(\Omega, \mathcal{L}, \mu)$.

Theorem 4.3.5 (Rosenthal's theorem) *For any bounded linear operator $T: L^1 \to L^1$ and $\epsilon > 0$ there exists a diffuse measure space $(A, \mathcal{A}) \subseteq (\Omega, \mathcal{L})$ with $\mu(A) > 0$, and a weighted composition operator T_0 given by*

$$(T_0 f)(x) = a(x)f(\sigma(x))$$

such that

$$\left\|(T - T_0): L_0^1(A, \mathcal{A}) \to L^1\right\| < \epsilon,$$

where $L_0^1(A, \mathcal{A}) = \left\{f \in L^1(A, \mathcal{A}) : \int f = 0\right\}$.

Theorem 4.3.5 forms the crucial step in the proof of Theorem 4.3.1, as well as in the proof of Theorem 4.3.6 on sign-preserving operators between L^1 spaces. Recall that a bounded linear operator $T: L^1 \to Y$ is sign preserving if there exists $\delta > 0$ and $E \subseteq [0, 1]$ with $|E| > 0$ such that

$$\|Td\|_Y \geq \delta\|d\|_1,$$

for any measurable $d: [0, 1] \to \{-1, 0, +1\}$ satisfying $|d| = 1_E$ and $\int_0^1 d(t)dt = 0$. In the special case where the operator T acts between L^1 spaces, Rosenthal (1984) obtained the following remarkable result.

Theorem 4.3.6 *Any sign preserving operator $T: L^1 \to L^1$ fixes a complemented copy of L^1.*

The proof of Theorem 4.3.6 begins by applying Kalton's representation theorem to $T: L^1 \to L^1$. Let $T = T^{(a)} + T^{(d)}$ be the splitting of T into its atomic and diffuse parts. By Theorem 4.3.5, if $T: L^1 \to L^1$ is sign preserving, then $T^{(a)} \neq 0$ and, moreover, restricting T to a – suitably chosen – diffuse measure space, we may assume that T is just a weighted composition operator $T_0(f)(x) = a(x)f(\sigma(x))$. Finally, for T_0 the conclusion of Theorem 4.3.6 is easily verified. We present the details of the proof at the end of Section 4.3.2.

The roadmap to Theorem 4.3.5: In the following subsections we present, in detail, the proofs of Theorem 4.3.5 and Theorem 4.3.1. We start by applying Kalton's representation theorem (Theorem 4.1.15) to $T\colon L^1 \to L^1$, obtaining a uniquely defined random measure $\mathcal{T} = \{\mu_x : x \in \Omega\}$, such that

$$Tf(x) = \int_\Omega f d\mu_x,$$

almost surely, for $x \in \Omega$. Moreover, we obtain measurable $\sigma_n\colon \Omega \to \Omega$ and weights $a_n\colon \Omega \to \mathbb{C}$ satisfying $|a_{n+1}(x)| \le |a_n(x)|$ and $\sigma_n(x) \ne \sigma_m(x)$ for $1 \le n \ne n < \infty$, such that

$$Tf(x) = \sum_{k=1}^{\infty} a_n(x) f(\sigma_n(x)) + Rf(x), \tag{4.3.7}$$

where $Rf(x) = \int_\Omega f d\nu_x$, and $\{\nu_x : x \in \Omega\}$ is a random measure consisting only of nonatomic measures on Ω. The partial sums of the weighted composition operators

$$T_n f(x) = a_n(x) f(\sigma_n(x))$$

satisfy the pointwise estimates (4.1.51), that is, $\sum \|T_n f\|_1 \le \|T\| \|f\|_1$. We put $T^{(d)} = R$, $T^{(a)} = T - T^{(d)}$, and we call $T^{(a)}$ the atomic part of the Kalton decompositon and $T^{(d)}$ its diffuse part. Proving Theorem 4.3.10 and Theorem 4.3.9 are the main objectives in the following subsections. Thereafter, we will show how those results combined yield the proof of Rosenthal's theorem (Theorem 4.3.5).

4.3.1 Diffuse Random Measures

We consider an operator $R\colon L^1(\Omega) \to L^1(\Omega)$ whose Kalton representation consists of diffuse measures only. Thus $Rf(x) = \int_\Omega f d\nu_x$, where

$$\nu_x(\{t\}) = 0,$$

for every $t \in \Omega$, and almost every $x \in \Omega$.

For simplicity of notation we let $\Omega = [0, 1]$. Recall that we let $\mathcal{I}$ denote the collection of dyadic intervals contained in the unit interval, and that we put $\mathcal{I}_n = \{I \in \mathcal{I} : |I| = 2^{-n}\}$. Enflo's maximal function of R is defined by

$$\lambda_R(x) = \lim_{n\to\infty} \max_{I \in \mathcal{I}_n} |R|(1_I)(x),$$

where $|R|$ denotes the *linear* operator given by $|R|(f)(x) = \int_{[0,1]} f d|\nu_x|$.

Proposition 4.3.7 *If $R\colon L^1 \to L^1$ is a bounded operator whose Kalton representation $\mathcal{T} = \{\nu_x : x \in \Omega\}$ consists of nonatomic measures only, then $\lambda_R(x) = 0$.*

Proof Given a dyadic interval I we put $b_I = |R|(1_I)$ and $\sigma(I) = \inf\{t : t \in I\}$. Thus $\sigma(I)$ denotes the left endpoint of I. Define the measures

$$\rho_x^n = \sum_{I \in \mathcal{I}_n} b_I(x)\delta_{\sigma(I)}.$$

These measures approximate, at scale 2^{-n}, the representing measures of $|R|$. Indeed, by the martingale convergence theorem, the sequence $\{\rho_x^n : n \in \mathbb{N}\}$ is w^*-convergent to $|\nu_x|$.

Given $n \in \mathbb{N}$, $x \in [0,1]$, let $I(x,n) \in \mathcal{I}_n$ satisfy $\max_{I \in \mathcal{I}_n} b_I(x) = b_{I(x,n)}(x)$ and denote $\sigma(x,n) = \inf\{t : t \in I(x,n)\}$. Thus we have the obvious relation betwen measures,

$$\rho_x^n \geq b_{I(x,n)}(x)\delta_{\sigma(x,n)}. \tag{4.3.8}$$

By passing to a subsequence, we may assume that $\sigma(x,n)$ is a convergent sequence, and let $\sigma(x) = \lim_{n\to\infty} \sigma(x,n)$. Moreover, the pointwise limit

$$\lim_{n\to\infty} b_{I(x,n)}(x) = \lambda_R(x)$$

exists, and also the w^* limit $\lim_{n\to\infty} \rho_x^n = |\nu_x|$. Thus passing to the limit in the relation (4.3.8) we get $|\nu_x| \geq \lambda_R(x)\delta_{\sigma(x)}$. Since we assumed that $|\nu_x|$ is a nonatomic measure, the last inequality implies that $\lambda_R(x)$ vanishes. Indeed, inserting the singleton set $\{\sigma(x)\}$ we find,

$$0 = |\nu_x|\{\sigma(x)\} \geq \lambda_R(x)\delta_{\sigma(x)}\{\sigma(x)\} = \lambda_R(x).$$

Since, by definition, $\lambda_R(x) \geq 0$, we obtain $\lambda_R(x) = 0$ as claimed. □

Proposition 4.3.8 *Let $R\colon L^1 \to L^1$ be a bounded operator whose Kalton representation $\mathcal{T} = \{\nu_x : x \in \Omega\}$ consists of nonatomic measures only. Then there exist pairwise disjoint collections $\{\mathcal{G}_I : I \in \mathcal{I}\}$ consisting of pairwise disjoint dyadic intervals, and signs $\{\varepsilon_K : K \in \mathcal{G}_I\}$ such that the block basis*

$$g_I = \sum_{K \in \mathcal{G}_I} \varepsilon_K h_K, \qquad I \in \mathcal{I},$$

satisfies the following conditions:

(i) $\|Rg_I\|_1 \leq \epsilon|I|^3$, *for* $I \in \mathcal{I}$.
(ii) *If* $I \in \mathcal{I}$ *and* $I = I_0 \cup I_1$ *with* $I_0, I_1 \in \mathcal{I}$, *then*

$$\operatorname{supp} g_{I_0} = \{g_I = 1\} \quad \textit{and} \quad \operatorname{supp} g_{I_1} = \{g_I = -1\}.$$

(iii) $|\operatorname{supp} g_I| = |I|$ *for* $I \in \mathcal{I}$.

Proof Let $\{r_K : K \in \mathcal{I}\}$ be an enumeration of the independent, $\{\pm\}$-valued Rademacher system. The proof proceeds by induction.

Step 1 Start with $I = [0, 1]$. Let $n \in \mathbb{N}$ and $\mathcal{I}_n = \{K \in \mathcal{I} : |K| = 2^{-n}\}$. Put

$$a_t = \sum_{K \in \mathcal{I}_n} r_K(t) h_K$$

and $b_t = R(a_t)$. By the inequalities of Hölder and Cauchy–Schwarz, we have the estimate

$$\int_0^1 \|b_t\|_1 dt \le \left(\|\max_{K \in \mathcal{I}_n} |Rh_K|\|_1\right)^{1/2} \left(\sum_{K \in \mathcal{I}_n} \|Rh_K\|_1\right)^{1/2}.$$

Since $\limsup_{n \to \infty} \max_{K \in \mathcal{I}_n} |Rh_K(x)| \le \lambda_R(x)$, by Proposition 4.3.7 there exists $n = n([0, 1])$ such that $\|\max_{K \in \mathcal{I}_n} |Rh_K|\|_1 < \epsilon$. Choose $t_0 \in [0, 1]$ such that $\|b_{t_0}\|_1 < \epsilon$, define signs $\varepsilon_K = r_K(t_0)$, $K \in \mathcal{I}_n$, and finally put $\mathcal{G}_{[0,1]} = \mathcal{I}_n$.

Step 2 At stage $i \in \mathbb{N}$ of the construction we assume that $\mathcal{G}_K$ for $|K| > 2^{-i}$ have been determined. Now fix $I \in \mathcal{I}$ with $|I| = 2^{-i}$ and let I_p denote the dyadic predecessor of I, that is, $|I_p| = 2|I|$ and $I_p \supseteq I$. If I is the right half of I_p set

$$\mathcal{L}_I(n) = \{K \in \mathcal{I}_n : K \subset \{g_{I_p} = 1\}\},$$

and if, conversely, I is the left-hand side of I_p set

$$\mathcal{L}_I(n) = \{K \in \mathcal{I}_n : K \subset \{g_{I_p} = -1\}\}.$$

Accordingly, for $n \ge i$ we form $a_{I,t} = \sum_{K \in \mathcal{L}_I(n)} r_K(t) h_K$ and put $b_{I,t} = R(a_{I,t})$. By the inequalities of Hölder and Cauchy-Schwarz, we have the estimate

$$\int_0^1 \|b_{I,t}\|_1 dt \le \left(\|\max_{K \in \mathcal{L}_I(n)} |Rh_K|\|_1\right)^{1/2} \left(\sum_{K \in \mathcal{L}_I(n)} \|Rh_K\|_1\right)^{1/2}.$$

Again, by Proposition 4.3.7 for $\epsilon(I) = \epsilon^2 |I|^4 / \|R\|$ there exists $n = n(I)$ large enough such that

$$\left\| \max_{K \in \mathcal{L}_I(n)} |Rh_K| \right\|_1 \le \epsilon(I).$$

Choose $t_0 \in [0, 1]$ such that $\|b_{I,t_0}\|_1 \le \epsilon(I)$. Set $\varepsilon_K = r_K(t_0)$ for $K \in \mathcal{L}_I(n)$ and $\mathcal{G}_I = \mathcal{L}_I(n)$. □

Remark: Put $G_I = \operatorname{supp} g_I$. By the definition of g_I, we have $G_I = G_{I_0} \cup G_{I_1}$, $G_{I_0} \cap G_{I_1} = \emptyset$, and $|G_{I_0}| = |G_{I_1}| = |G_I|/2$.

Theorem 4.3.9 *Let $R\colon L^1 \to L^1$ be a bounded operator whose Kalton representation $\mathcal{T} = \{\nu_x : x \in \Omega\}$ consists of nonatomic measures only. Then for any $\epsilon > 0$ there exists a nonatomic sigma-algebra $\mathcal{A}$ on Ω such that*

$$\|Rf\|_1 < \epsilon\|f\|_1, \tag{4.3.9}$$

for any $\mathcal{A}$ measurable $f \in L_0^1(\Omega)$.

Proof Apply Proposition 4.3.8 to $R\colon L^1 \to L^1$. Let $g_I = \sum_{K \in \mathcal{G}_I} \epsilon_K h_K$ be the resulting block basis satisfying $\|Rg_I\|_1 \le \epsilon|I|^3$, for all $I \in \mathcal{I}$. Define $\mathcal{A}$ to be the sigma-algebra generated by $\{g_I : I \in \mathcal{I}\}$. By Proposition 4.3.8, and the remark thereafter, $f_I = g_I/\|g_I\|_2^2$, with $I \in \mathcal{I}$, forms a martingale difference sequence spanning a dense subspace in $L_0^1(\Omega, \mathcal{A})$. Hence, each $f \in L_0^1(\Omega, \mathcal{A})$ admits a norm-convergent expansion as

$$f = \sum_{I \in \mathcal{I}} \langle f, f_I \rangle f_I.$$

Since $\|Rf_I\|_1 < \epsilon|I|^3$, we obtain (4.3.9) by applying the triangle inequality. □

4.3.2 Atomic Random Measures

Here, we verify Rosenthal's theorem (Theorem 4.3.5) for operators $T\colon L^1 \to L^1$ whose Kalton representation consists of atomic measures only.

Theorem 4.3.10 *Let $T\colon L^1 \to L^1$ be a bounded linear operator and $T = T^{(a)} + T^{(d)}$ its Kalton decomposition. For any $\epsilon > 0$, there exists $S \subseteq \Omega$ with $\mu(S) > 0$ and a weighted composition operator, $(T_0 f)(x) = a(x) f(\tau(x))$, such that*

$$\left\|(T^{(a)} - T_0)f\right\|_1 < \epsilon\|f\|_1,$$

for any $f \in L^1(S)$.

Proof Fix $T\colon L^1 \to L^1$ bounded and linear. Let $\{T_n\}$ denote the sequence of weighted composition operators associated to the atomic parts of its representing measures. Thus $T^{(a)} = \sum_{n=1}^{\infty} T_n$. By the estimates (4.1.51) we may apply Lemma 4.3.11 to the sequence $\{T_n\}$. Hence for $\epsilon > 0$, there exists $m \in \mathbb{N}$ and $A \subseteq \Omega$ such that $\mu(A) > 1 - \epsilon$ and

$$\left\|\sum_{n=m+1}^{\infty} T_n f\right\|_1 \le \epsilon\|f\|_1, \tag{4.3.10}$$

for $f \in L^1(A)$. It therefore suffices to construct a weighted composition operator T_0 and $S \subseteq A$ with $\mu(S) > 0$ such T_0 and $\sum_{n=1}^m T_n$ are close as operators on $L^1(S)$. The operator T_0 will be determined in Proposition 4.3.15 below. □

Submartingale convergence: Let (f_k) be a sequence of nonnegative random variables satisfying $\mathbb{E}(f_k) \leq C$ for $k \in \mathbb{N}$. We assume that (f_k) is adapted to a filtered probability space $(\Omega, (\Sigma_k))$ and that $f_k \leq \mathbb{E}_k(f_{k+m})$, for $m, k \in \mathbb{N}$. The submartingale convergence theorem (Theorem 1.1.3) asserts that (f_k) converges almost surely with $f = \lim f_k$ satisfying $\mathbb{E}(f) \leq C$. If, moreover, (f_k) is uniformly integrable then

$$f_k \leq \mathbb{E}(f|\Sigma_k),$$

for any $k \in \mathbb{N}$. We draw consequences of the submartingale convergence theorem to operators on L^1 taking values in $(\sum L^1)_1$. Recall that

$$\left(\sum_1^\infty L^1\right)_1 = \left\{(f_i) : f_i \in L^1 \quad \text{and} \quad \sum_{i=1}^\infty \|f_i\|_{L^1} < \infty\right\}$$

equipped with the norm

$$\|(f_i)\| = \sum_{i=1}^\infty \|f_i\|_{L^1}.$$

Lemma 4.3.11 *Let $M < \infty$. Let $T_n : L^1 \to L^1$ be a sequence of operators such that*

$$\sum \|T_n(f)\|_1 \leq M\|f\|_1,$$

for any $f \in L^1$. Then there exist weights $h_n : [0,1] \to [0, M]$ satisfying

$$\|T_n f\| \leq \|f h_n\|_1, \tag{4.3.11}$$

for all $f \in L^1$ and $\sum h_n \leq M$. Moreover, for $\epsilon > 0$ there exists $m \in \mathbb{N}$ such that $A = \{\sum_{n>m} h_n \leq \epsilon\}$ satisfies $|A| > 1 - \epsilon$ and

$$\sum_{n=m+1}^\infty \|T_n(f 1_A)\|_1 \leq \epsilon \|f\|, \tag{4.3.12}$$

for any $f \in L^1$.

Proof Fix $k, n \in \mathbb{N}$ and let $\mathcal{I}_k$ denote the collection of all dyadic intervals I with $|I| = 2^{-k}$. Define

$$h_{n,k} = \sum_{I \in \mathcal{I}_k} \|T_n(1_I/|I|)\|_1 1_I.$$

For $I \in \mathcal{I}_k$ fixed and $x \in I$, we clearly have $h_{n,k}(x) \leq M$. We next verify that for n fixed the sequence $(h_{n,k})_{k=1}^{\infty}$ is a nonnegative submartingale and is uniformly bounded. To this end, let $I \in \mathcal{I}_k$, $m \in \mathbb{N}$, and put $\mathcal{I}_{k+m}(I) = \{K \in \mathcal{I}_{k+m} : K \subseteq I\}$. Then,

$$\begin{aligned} \int_I h_{n,k+m} d\mu &\geq \left\| T_n \left(\sum_{K \in \mathcal{I}_{k+m}(I)} 1_K \right) \right\|_1 \\ &= \|T_n 1_I\|_1 . \end{aligned} \tag{4.3.13}$$

Let Σ_k be the finite algebra generated by $\mathcal{I}_k$. The estimate (4.3.13) translates to $\mathbb{E}(h_{n,k+m}|\Sigma_k) \geq h_{n,k}$. Summing up the first part of the proof, we showed that $(h_{n,k})_{k=1}^{\infty}$ is a nonnegative, uniformly integrable submartingale. By the submartingale convergence theorem (Theorem 1.1.3), $(h_{n,k})_{k=1}^{\infty}$ converges almost surely such that $h_n = \lim h_{n,k}$ satisfies $h_{n,k} \leq \mathbb{E}(h_n|\Sigma_k)$.

For $I \in \mathcal{I}_k$ fixed we thus proved $\|T_n 1_I\|_1 \leq \int 1_I h_n d\mu$. Taking the sum over $I \in \mathcal{I}_k$ finally gives

$$\left\| T_n \left(\sum_{I \in \mathcal{I}_k} b_I 1_I \right) \right\|_1 \leq \int_0^1 \left(\sum_{I \in \mathcal{I}_k} b_I 1_I \right) h_n d\mu. \tag{4.3.14}$$

Since step functions are dense in L^1, Inequality (4.3.14) gives (4.3.11), which in turn gives $\sum_{n=1}^{\infty} h_n \leq M$. □

We restate the content of Lemma 4.3.11 in terms of a linear operator

$$T : L^1 \to \left(\sum_1^{\infty} L^1 \right)_1 .$$

Given $\epsilon > 0$, Lemma 4.3.11 asserts that there exist operators J and En, a subset $A \subseteq [0, 1]$ with $|A| > 1 - \epsilon$ and $m \in \mathbb{N}$ such that the following diagram is commutative

$$\begin{array}{ccccc} L^1(A) & \xrightarrow{\;i\;} & L^1([0,1]) & \xrightarrow{\;T\;} & \left(\sum_1^{\infty} L^1\right)_1 \\ & {\scriptstyle J}\searrow & & \nearrow{\scriptstyle En} & \\ & & \left(\sum_1^{m} L^1\right)_1 & & \end{array} \tag{4.3.15}$$

where i denotes the inclusion operator $i(f) = f$.

Measure exhaustion: We say that a linear operator $T : L^1 \to L^1$ is nonnegative if $f \geq 0$ implies $T(f) \geq 0$. The next Lemma may be viewed as the Hahn–Jordan decomposition for nonnegative operators on L^1. Its proof displays a typical use of Zorn's Lemma.

Lemma 4.3.12 *Let $T: L^1(\Omega) \to L^1(\Omega)$ be a bounded, nonnegative linear operator. Let $\epsilon > 0$ and assume that $E \subseteq \Omega$ is measurable, $\mu(E) > 0$, and*

$$\|T1_E\|_1 \leq \epsilon\|1_E\|_1. \tag{4.3.16}$$

Then there exists $S \subseteq E$ with $\mu(S) > 0$ such that

$$\|T1_G\|_1 \leq \epsilon\|1_G\|_1,$$

for any $G \subseteq S$ with $\mu(G) > 0$.

Proof Assume to the contrary that, for any $S \subseteq E$ with $\mu(S) > 0$, there exists $G \subseteq S$ with $\mu(G) > 0$ such that $\|T1_G\|_1 > \epsilon\|1_G\|_1$. In two separate parts we show that this assumption leads to a contradiction.

Part 1 By an exhaustion of measure argument, as presented in the second part of the proof, there exists $N \subseteq \mathbb{N}$ and pairwise disjoint sets $G_i \subseteq E$ satisfying $\mu(E\setminus\bigcup_{i\in N} G_i) = 0$ and $\|T1_{G_i}\|_1 > \epsilon\|1_{G_i}\|_1$, for any $i \in N$. Since then,

$$\|T1_E\|_1 = \left\|T\sum_{i\in N} 1_{G_i}\right\|_1 = \sum_{i\in N}\|T1_{G_i}\|_1 > \sum_{i\in N}\epsilon\|1_{G_i}\|_1 = \epsilon\|1_E\|_1,$$

we obtain a lower estimate for $\|T1_E\|_1$, contradicting (4.3.16).

Part 2 We next turn to the measure exhaustion argument based on Zorn's Lemma. We say that a collection of measurable sets $\mathcal{E}$ satisfies property L if the following three conditions are satisfied:

1) $G \subseteq E$ with $\mu(G) > 0$ for any $G \in \mathcal{E}$,
2) $G_1 \cap G_2 = \emptyset$ for any $G_1, G_2 \in \mathcal{E}$ with $G_1 \neq G_2$,
3) $\|T1_G\|_1 > \epsilon\|1_G\|_1$, for any $G \in \mathcal{E}$.

We first remark that the first two conditions of property L imply that $\mathcal{E}$ is necessarily of countable cardinality. Next, put

$$\mathcal{L} = \{\mathcal{E} : \mathcal{E} \text{ has Property } L\}.$$

We define a partial ordering on $\mathcal{L}$ by decreeing that $\mathcal{E}_1 \in \mathcal{L}$ is smaller than $\mathcal{E}_2 \in \mathcal{L}$ if $\mathcal{E}_1 \subseteq \mathcal{E}_2$.

We claim that there exists a maximal element in $\mathcal{L}$, and that any maximal element $\mathcal{M} \in \mathcal{L}$ satisfies

$$\mu\left(E\setminus\bigcup_{G\in\mathcal{M}} G\right) = 0.$$

We use Zorn's Lemma to prove the existence of a maximal element in $\mathcal{L}$. Let $\{\mathcal{E}_i : i \in I\}$ be a totally ordered subset of $\mathcal{L}$. Put $\mathcal{E} = \bigcup_{i\in I}\mathcal{E}_i$. It is easy to check that $\mathcal{E} \in \mathcal{L}$. Indeed, if $E_1, E_2 \in \mathcal{E}$ with $E_1 \neq E_2$, let $\alpha, \beta \in I$ such that $E_1 \in \mathcal{E}_\alpha$ and $E_2 \in \mathcal{E}_\beta$. Since $\{\mathcal{E}_i : i \in I\}$ is totally ordered we have $\mathcal{E}_\alpha \subseteq \mathcal{E}_\beta$ or $\mathcal{E}_\beta \subseteq \mathcal{E}_\alpha$. Assuming that $\mathcal{E}_\alpha \subseteq \mathcal{E}_\beta$, we have $E_1, E_2 \in \mathcal{E}_\beta$ and hence $E_1 \cap E_2 = \emptyset$. Thus we verified that $\mathcal{E}$ satisfies the second condition of property L. The first and third conditions are trivially satisfied for $\mathcal{E}$ since each of the collections $\mathcal{E}_i$ with $i \in I$ satisfy property L. Thus we checked that every totally ordered subset of $\mathcal{L}$ admitts an upper bound in $\mathcal{L}$. By Zorn's lemma, there exists a maximal element in $\mathcal{L}$. Call it $\mathcal{M}$ and put

$$S = E \backslash \bigcup_{G \in \mathcal{M}} G.$$

If $\mu(S) > 0$, then, by assumption, there would exist $G \subseteq S$ such that $\mu(G) > 0$ and $\|T1_G\|_1 > \epsilon \|1_G\|_1$. Hence,

$$\mathcal{M}_0 = \{G\} \cup \mathcal{M}$$

satisfies property L as well as $\mathcal{M} \subseteq \mathcal{M}_0$ and $\mathcal{M} \neq \mathcal{M}_0$. Clearly, this contradicts the maximality of $\mathcal{M}$, and we conclude that $\mu(S) = 0$. □

We now turn to Proposition 4.3.15, addressing the problem of approximating the finite sum of the weighted composition operators T_n by a single such operator. We prepare the proof with Lemma 4.3.13 and Lemma 4.3.14.

Lemma 4.3.13 *Let $m \in \mathbb{N}$, $\epsilon > 0$, $E \subseteq \Omega$ with $\mu(E) > 0$. Let $\sigma_n : \Omega \to \Omega$ be measurable and assume $\sigma_k(x) \neq \sigma_n(x)$ for $1 \leq k \neq n \leq m$. Then there exists a measurable $S \subseteq E$ with $\mu(S) > 0$ such that*

$$\mu\left(\sigma_n^{-1}(S) \cap \sigma_k^{-1}(S)\right) < \epsilon\mu(S), \tag{4.3.17}$$

for $1 \leq k \neq n \leq m$.

Proof Recall that Ω is a compact metric space. Let d denote the metric on Ω. Fix $\epsilon > 0$, $1 \leq k \neq n \leq m$ and put $f = \sigma_n$ and $g = \sigma_k$. By assumption we have $f(x) \neq g(x)$. Define, for $i \in \mathbb{N}$,

$$E_i = \{x \in \Omega : d(f(x), g(x)) > 1/i\}.$$

Clearly, the sets E_i are increasing and $\mu(\Omega \backslash \bigcup E_i) = 0$. Hence there exists $i = i(\epsilon, E, \Omega)$ such that $\mu(\Omega \backslash E_i) < \epsilon\mu(E)$.

Let $\{F_j\}$ be a countable, measurable partition of E such that $\operatorname{diam}(F_j) < 1/i$ and $\mu(F_j) > 0$. Set $F = \bigcup f^{-1}(F_j) \cap g^{-1}(F_j)$, and note that $F \cap E_i = \emptyset$. We claim that there exists $j \in \mathbb{N}$ such that

$$\mu\left(f^{-1}(F_j) \cap g^{-1}(F_j)\right) \leq \epsilon\mu(F_j). \tag{4.3.18}$$

If this were not the case then the set F would satisfy a lower measure estimate,

$$\mu(F) > \epsilon \sum \mu(F_j) = \epsilon\mu(E).$$

On the other hand, we have $\mu(F \cap (\Omega\backslash E_i)) < \epsilon\mu(E)$. Hence, by arithmetic, $\mu(F \cap E_i) > 0$, contradicting the fact that $F \cap E_i = \emptyset$.

Let F_j satisfy $\mu(F_j) > 0$ and (4.3.18). Now we replace E by F_j and select a different pair of $k \neq n$ and repeat the above argument. After $m(m-1)/2$ such iterations we arrive at the set S for which (4.3.18) holds true. □

Recall that the weighted composition operators $T_n f = a_n(x) f(\sigma_n(x))$ were introduced in definition (4.3.7), and that $m \in \mathbb{N}$ was defined such that condition (4.3.10) holds true.

Lemma 4.3.14 *Let $\eta > 0$ and $E \subseteq \Omega$ with $\mu(E) > 0$. There exists a measurable $S \subseteq E$ with $\mu(S) > 0$, such that for $1 \leq k \neq n \leq m$, the sets $E_k = \sigma_k^{-1}(S)$, and $E_n = \sigma_n^{-1}(S)$ satisfy*

$$\int_{E_k \cap E_n} |T_n f| d\mu \leq \eta \int_S |f| d\mu, \qquad f \in L^1(S).$$

Proof Fix $1 \leq k \neq n \leq m$. Let a_n be the weight of the composition operator T_n, $\lambda > 0$ and put $b_n = a_n 1_{\{|a_n|>\lambda\}}$ and $g_n = a_n - b_n$. Since a_n is integrable, we may select $\lambda > 0$ such that

$$\int_\Omega |b_n| d\mu \leq \eta\mu(E)/2.$$

Write $T_n = G_n + B_n$ where $G_n(f)(x) = g_n(x) f(\sigma_n(x))$ and $B_n = T_n - G_n$. We estimate the operators B_n and G_n separately.

We begin with the linear operator $|B_n|$. By choice of b_n we have

$$\||B_n|(1_E)\|_1 \leq \int_\Omega |b_n| d\mu \leq \eta\mu(E)/2.$$

Applying Lemma 4.3.12, there exists $S_1 \subseteq E$ with $\mu(S_1) > 0$ such that

$$\||B_n|(f)\|_1 \leq \eta/2\|f\|_1, \quad \text{for} \quad f \in L^1(S_1).$$

Next, we turn to estimating the linear operator $|G_n|$. Apply Lemma 4.3.13 to the set S_1 using $\epsilon = \eta/(2\lambda)$. This gives $S_2 \subseteq S_1$ measurable, such that for $A_n = \sigma_n^{-1}(S_2)$ and $A_k = \sigma_k^{-1}(S_2)$, we obtain

$$\int_{A_k \cap A_n} |g_n(x)| 1_{S_2}(\sigma_n(x)) d\mu \leq \lambda\mu(A_k \cap A_n) \leq \eta\mu(S_2)/2.$$

To facilitate a direct application of Lemma 4.3.12 we rewrite the above estimate as

$$\|1_{A_k}(1_{A_n}|G_n|)(1_{S_2})\|_1 \leq \eta\mu(S_2)/2.$$

Now, applying Lemma 4.3.12 to the operator $1_{A_k}(1_{A_n}|G_n|)$ we obtain $S_3 \subseteq S_2$, with $\mu(S_3) > 0$, such that

$$\big\|1_{A_k}(1_{A_n}|G_n|)(f)\|_1 \le \eta\big\|f\|_1/2, \quad \text{for} \quad f \in L^1(S_3).$$

Thus we obtain that

$$\int_{A_k \cap A_n} |T_n f| d\mu \le \eta \int_{S_3} |f| d\mu, \quad \text{for} \quad f \in L^1(S_3).$$

Finally, we replace E by S_3 and pick a different pair of $k \neq n$. Repeating this procedure $m(m-1)/2$ times, we arrive at the set $S \subseteq E$ with the required properties. □

Proposition 4.3.15 *For any $\epsilon > 0$ there exists $S \subseteq \Omega$ with $\mu(S) > 0$ and a weighted composition operator, $(T_0 f)(x) = a(x) f(\tau(x))$, such that*

$$\left\| \left(\sum_{n=1}^{m} T_n f \right) - T_0 f \right\|_1 < \epsilon \|f\|_1,$$

for any $f \in L^1(S)$.

Proof Let $m \in \mathbb{N}$, $\epsilon > 0$, and put $\eta = \epsilon/(m(m-1))$. Apply Lemma 4.3.14 to the weighted composition operators $T_1, \ldots, T_m$, with $\eta > 0$ fixed above. Accordingly, we choose $S \subseteq \Omega$ with $\mu(S) > 0$ such that

$$\int_{E_k \cap E_n} |T_n f| d\mu \le \eta \int_S |f| d\mu, \quad 1 \le k \neq n \le m, \qquad f \in L^1(S),$$

where $E_j = \sigma_j^{-1}(S)$, and $1 \le j \le m$. Now fix $1 \le n \le m$, let $Z_n = \bigcup_{k \neq n} E_n \cap E_k$, form $D_n = E_n \backslash Z_n$, and put

$$(V_n f)(x) = 1_{D_n}(x)(T_n f)(x).$$

Note the pointwise relations,

$$|(V_n - T_n)(f)(x)| = |1_{Z_n}(x)(T_n f)(x)| \le \sum_{k \neq n} 1_{E_k \cap E_n}(x) |T_n f(x)|. \tag{4.3.19}$$

Next we integrate, over $x \in S$, the estimates (4.3.19). Invoking Lemma 4.3.14, this gives

$$\|(V_n - T_n)(f)\|_1 \le (m-1)\eta \|f\|_1, \quad f \in L^1(S). \tag{4.3.20}$$

We are now in the position to define the approximating weighted composition operator: By construction, $D_n \cap D_k = \emptyset$ for $1 \le n \neq k \le m$. For $x \in D_n$, put $a(x) = a_n(x)$, $\tau(x) = \sigma_n(x)$, and form

$$(T_0 f)(x) = a(x) f(\tau(x)).$$

Thus we have $T_0 = \sum_{n=1}^{m} V_n$. The estimates (4.3.20) directly yield

$$\left\| \sum_{n=1}^{m} (T_n - T_0)(f) \right\|_1 \leq \sum_{n=1}^{m} \|(T_n - T_0)(f)\|_1$$
$$\leq \eta m(m-1)\|f\|_1,$$

for any $f \in L^1(S)$. By our choice of $\eta > 0$, this completes the proof. □

Putting It Together

We show next how Theorem 4.3.10 and Theorem 4.3.9 combined give the proof of Theorem 4.3.5. Thereafter, we verify that Theorem 4.3.5 implies Theorem 4.3.1 and Theorem 4.3.6, with almost identical arguments.

Proof of Theorem 4.3.5: We apply Kalton's representation theorem to the operator $T\colon L^1 \to L^1$. Let $\mathcal{T} = \{\mu_x : x \in \Omega\}$ denote the collection of representing measures, and let

$$\mu_x = \sum_{n=1}^{\infty} a_n(x)\delta_{\sigma_n(x)} + \nu_x,$$

denote the splitting into atomic and diffuse components. This gives the corresponding decomposition of the operator as $T = T^{(a)} + T^{(d)}$ with $T^{(d)} f(x) = \int f d\nu_x$. In Theorem 4.3.10, we proved that there exists a weighted composition operator T_0 and $A \subset \Omega$ with $\mu(A) > 0$ such that

$$\left\|(T^{(a)} - T_0)f\right\|_1 \leq \epsilon\|f\|_1,$$

for $f \in L^1(A)$. Next, by Theorem 4.3.9 there exists a diffuse sigma-algebra $\mathcal{A}$ on A with $(A, \mathcal{A}) \subseteq (\Omega, \mathcal{B})$ such that $T^{(d)}$ almost annihilates $L^1(A, \mathcal{A})$ in the sense that

$$\left\|T^{(d)} f\right\|_1 \leq \epsilon\|f\|_1, \quad f \in L^1(A, \mathcal{A}).$$

Proof of Theorem 4.3.1: Let $S\colon L^1 \to X$ fix a copy of L^1. Let $c > 0$, and let $G \subseteq L^1$ be a linear subspace isomorphic to L^1 such that $\|Sf\|_X \geq c\|f\|_1$ for any $f \in G$. Let $T\colon L^1 \to L^1$ be an isomorphism onto its range such that $T(L^1) = G$. Next fix $\epsilon > 0$. Apply Theorem 4.3.5 to obtain a weighted composition operator

$$(T_0 f)(x) = a(x) f(\tau(x)), \tag{4.3.21}$$

a measurable set $A \subseteq \Omega$ with $\mu(A) > 0$, and a diffuse sigma-algebra $\mathcal{A}$ on A such that

$$\|(T - T_0)(f)\|_1 < \epsilon\|f\|_1, \tag{4.3.22}$$

for any $f \in L^1(A, \mathcal{A})$ with $\int f d\mu = 0$. Fix $0 < \epsilon << \|T\|$. Select $A_1 \subset A$ such that $\mu(A_1) > 0$ and

$$1 \leq \sup_{x \in A_1} |(\mathbb{E}_{\mathcal{A}} a)(x)| / \inf_{x \in A_1} |(\mathbb{E}_{\mathcal{A}} a)(x)| \leq (1 + \epsilon).$$

Without loss of generality, $a(x) > 0$ for $x \in A_1$. Define then $\Sigma_0 = \{\tau^{-1} E : E \subseteq A_1, E \in \mathcal{A}\}$. We claim that Σ_0 is a diffuse sigma-algebra. To see this we note that by Inequality (4.3.22) the operator T_0 is bounded, and that

$$\left\|1_{\tau^{-1}(E)}\right\|_1 \leq C_0 \|T_0(1_E)\|_1 \leq C_1 \|1_E\|_1,$$

implying that $\mu\left(\tau^{-1}(E)\right) \leq C_3 \mu(E)$ for any $E \in \mathcal{A}$. Summing up, by operator (4.3.21) and Inequality (4.3.22) we showed that $\|Sy\|_X \geq c_0 \|y\|_1$, for any $y \in L^1(\Sigma_0)$ with $\int y = 0$.

Proof of Theorem 4.3.6: The following is a minor adjustment of the proof of Theorem 4.3.1. Fix a sign-preserving operator $T \colon L^1 \to L^1$. Let $T = T^{(a)} + T^{(d)}$ be its decomposition into atomic and diffuse parts obtained by Kalton's representation. By Theorem 4.3.9, $T^{(a)} \neq 0$. Fix $0 < \epsilon \ll \left\|T^{(a)}\right\|$. Invoking Theorem 4.3.5 yields a weighted composition operator

$$(T_0 f)(x) = a(x) f(\tau(x)), \tag{4.3.23}$$

a measurable set $A \subseteq \Omega$ with $\mu(A) > 0$ and a diffuse sigma-algebra $\mathcal{A}$ on A such that

$$\|(T - T_0)(f)\|_1 < \epsilon \|f\|_1, \tag{4.3.24}$$

for $f \in L^1(A, \mathcal{A})$ and $T_0 \neq 0$. Select $A_1 \subset A$ such that $\mu(A_1) > 0$ and

$$1 \leq \sup_{x \in A_1} |(\mathbb{E}_{\mathcal{A}} a)(x)| / \inf_{x \in A_1} |(\mathbb{E}_{\mathcal{A}} a)(x)| \leq (1 + \epsilon).$$

Without loss of generality, $a(x) > 0$ for $x \in A_1$. Define then $\Sigma_0 = \{\tau^{-1} E : E \subseteq A_1, E \in \mathcal{A}\}$. As above, Σ_0 is a diffuse sigma-algebra. Hence by expressions (4.3.23) and (4.3.24), $\|Ty\|_X \geq c_0 \|y\|_1$, for any $y \in L^1_0(\Sigma_0)$.

4.4 Approximation in L^1

In Sections 3.1.5, 3.2.3, 4.2.2, and 4.2.3 we established L^1-*separation estimates* and thereby obtained an important list of Banach spaces X for which L^1/X contains a copy of L^1. In Section 4.5 we will present Talagrand's construction of $X \subset L^1$ such that neither X nor L^1/X contain a copy of L^1.

In the present section we treat the L^1-*approximation problem* that forms the core of Talagrand's work (Talagrand, 1990). We show that there exists a sequence of countable sets $F_n \subset L^1[0,1]$ with $\|g\|_1 \geq 1/3$ for $g \in F_n$, such that the following two conditions are satisfied:

(1) For any $\epsilon > 0$, there exists $n_0 \in \mathbb{N}$ such that for any $n \geq n_0$ and any $g \in \text{conv}(F_n)$,

$$|\{|g| > \epsilon\}| < \epsilon. \tag{4.4.1}$$

(2) For any diffuse sigma-algebra Σ_0 in $[0,1]$, any $A \in \Sigma_0$ with $|A| > 0$, any $n \in \mathbb{N}$ with $n \geq \left|\log_2 |A|\right|$, and any $\epsilon > 0$, there exist

$$f \in L^1(A, \Sigma_0) \quad \text{and} \quad g \in \text{conv}(F_n)$$

such that

$$\int_A f dt = 0, \quad \|f\|_1 \geq 1/4 \quad \text{and} \quad \|f - g\|_1 < \epsilon. \tag{4.4.2}$$

One immediately notices strong repelling forces between the two conditions stated above. Those forces were overcome by Talagrand (1990) using great combinatorial power displayed in the following subsections.

Defining $\mathbf{F_n}$: Let I_j, J_j, $0 \leq j \leq k$, be dyadic intervals. We say that pairs of dyadic intervals $(I_0, J_0), \ldots, (I_j, J_j), \ldots, (I_k, J_k)$ form the admissible cone C, written as

$$C = \{(I_0, J_0), \ldots, (I_j, J_j), \ldots, (I_k, J_k)\}, \tag{4.4.3}$$

if $J_j \subsetneq I_j$ and $I_{j+1} \subsetneq J_j$. See Figure 4.5. We call J_k the tip of the cone C and I_0 its base, and we write $t(C) = J_k$ and $b(C) = I_0$ to denote the tip and, respectively, the base of C. The cone C generates a blocking of Haar functions as follows,

$$f(C) = \frac{1}{2(k+1)} \sum_{j=0}^{k} f_{(I_j,J_j)}, \tag{4.4.4}$$

where $f_{(I,J)} = 1_I/|I| - 1_J/|J|$. Finally, for fixed n, we let F_n be the collection of all those $f(C)$ that are generated by admissible cones C with cardinality $n+1$. Thus,

$$F_n = \{f(C) : |C| = n+1\}, \tag{4.4.5}$$

where $|C|$ denotes the cardinality of C.

Overview: The verification of the properties (4.4.1) and (4.4.2) is carried out in two separate subsections below. Our main ingredients are:

(i) a combinatorially fortified version of the upcrossing lemma for non negative martingales, and
(ii) Lyaponov's convexity theorem on the range of vector measures.

It takes Section 4.4.1, to prove that F_n, defined by (4.4.5), satisfies of property (4.4.1). See Theorem 4.4.6. In Section 4.4.2, Theorem 4.4.8, we verify that property (4.4.2) holds true.

4.4.1 Dubins's Lemma and its Consequences

Here we prove that F_n satisfies property (4.4.1). Our starting point is the upcrossing lemma of Dubins. Recall that Theorem 1.1.3 (the martingale convergence theorem) asserts that an L^1-bounded martingale is convergent almost everywhere. In its proof, the upcrossing lemma counts and estimates the oscillations of an L^1-bounded martingale. In this section we build the analysis on Dubins's upcrossing lemma, asserting that for nonnegative martingales the number of upcrossings decays exponentially fast.

Recall that $\mathcal{I}$ denotes the collection of dyadic intervals contained in $[0, 1]$. Fix $b \in L^1[0, 1]$ and write

$$b_K = \int_K b dt/|K|, \quad \text{for} \quad K \in \mathcal{I}.$$

Fix $n \in \mathbb{N}$ and $I_0, I \in \mathcal{I}$ satisfying $I_0 \supset I$ and $|I_0|/|I| = 2^n$. Let $(K(m) : 0 \le m \le n)$ denote the enumeration of

$$\{K \in \mathcal{I} : I_0 \supseteq K \supseteq I\},$$

uniquely determined by the conditions $K(0) = I_0, K(n) = I$, and $|K(m-1)| = 2|K(m)|$, for $1 \le m \le n$. See Figure 4.5. Putting $a_m = b_{K(m)}$ defines the vector $(a_m : 0 \le m \le n)$.

Next, fix $x, y \in \mathbb{R}$ with $0 < x < y < \infty$ and $k \in \mathbb{N}$ with $k < n$. We say that $(b_K : I_0 \supseteq K \supseteq I)$ performs *at least* k upcrossings of the interval (x, y) if there exist strictly increasing integers $1 \le m(0) < \cdots < m(2k-1) \le n$ such that

$$a_{m(2j-2)} < x < y < a_{m(2j-1)},$$

where $1 \le j \le k$. See Figure 4.6. Finally, we let $\mathcal{B}_k = \mathcal{B}_k(b, x, y)$ denote the collection of all dyadic intervals I for which $(b_K : [0, 1] \supseteq K \supseteq I)$ performs at least k upcrossings of the interval (x, y). Dubins's upcrossing lemma provides estimates of the set covered by the collection $\mathcal{B}_k$. We refer to Neveu (1975) for its proof presented below.

Lemma 4.4.1 *Let $b \in L^1[0, 1]$ with $b \ge 0$. Let $x, y \in \mathbb{R}$ such that $0 \le x < y < \infty$. Then for any $k \in \mathbb{N}$, the collection $\mathcal{B}_k = \mathcal{B}_k(b, x, y)$ satisfies the measure estimate*

$$\left|\bigcup \mathcal{B}_k\right| \le (x/y)^k. \tag{4.4.6}$$

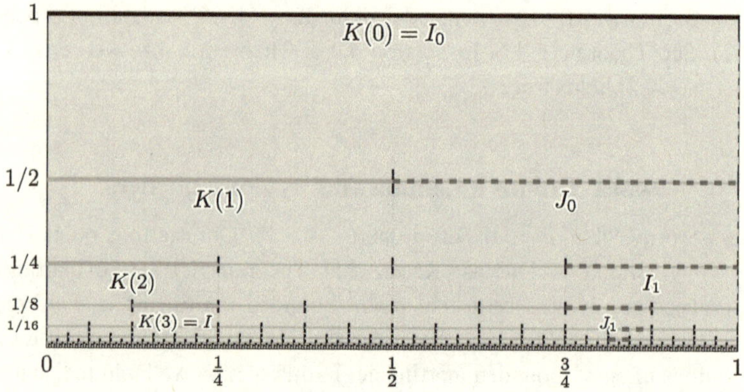

Figure 4.5 Here, $C = \{(I_0, J_0), (I_1, J_1)\}$ forms an admissible cone. The intervals $K(0) = J_0, K(1), K(2), K(3) = I$ give the enumeration of $\{K \in \mathcal{I} : I \subseteq K \subseteq J_0\}$.

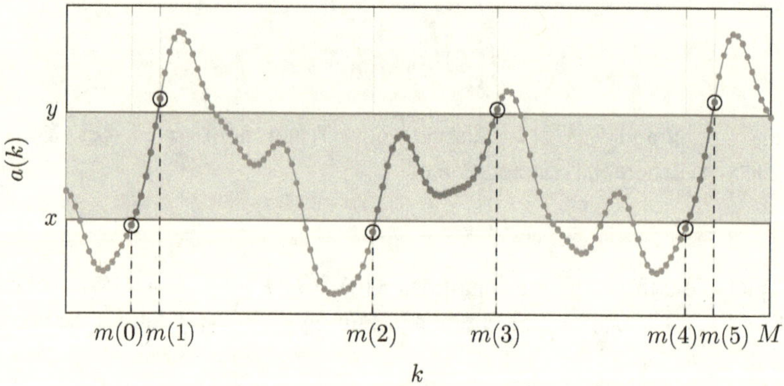

Figure 4.6 The sequence $(a(k) : 1 \le k \le M)$ performs at least three upcrossings of the interval (x, y).

Proof We obtain the measure estimate (4.4.6) by iterating the observation contained in Step 1 below. The inductive proof employs a stopping argument run on dyadic intervals. Throughout, we freely use the fact that given any collection of dyadic intervals, its maximal intervals form a collection of pairwise disjoint dyadic intervals.

Step 1 Fix $b \in L^1[0, 1]$ and nonnegative. Let $x, y \in \mathbb{R}$ with $0 < x < y < \infty$. Next, fix $I \in \mathcal{I}$ and assume $b_I < x$. Consider $\mathcal{B}(I) = \{J \in \mathcal{I} : J \subseteq I, b_J > y\}$, and let $\mathcal{Y}(I)$ denote the collection of maximal intervals in $\mathcal{B}(I)$. Since $\mathcal{Y}(I)$ consists of pairwise disjoint dyadic intervals, we have

$$\sum_{J\in\mathcal{Y}(I)} |J| < \frac{x}{y}|I|. \tag{4.4.7}$$

Indeed, set $Y = \bigcup \mathcal{Y}(I)$, and use that $b \geq 0$, to obtain

$$x|I| > \int_I b dt \geq \int_Y b dt = \sum_{J\in\mathcal{Y}(I)} \int_J b dt > y \sum_{J\in\mathcal{Y}(I)} |J|.$$

Dividing by $y > 0$ gives (4.4.7).

Step 2 Fix $k \in \mathbb{N}$, put $\mathcal{B}_0 = \mathcal{I}$, and let $\mathcal{B}_i = \mathcal{B}_i(b, x, y)$, for $1 \leq i \leq k$. Clearly, the collections $\mathcal{B}_i$ are decreasing in i. We now assume that $\mathcal{B}_k \neq \emptyset$, since otherwise there is nothing to prove.

Put $\mathcal{Y}_0 = \{[0, 1]\}$. We claim that there exist disjoint collections of dyadic intervals, $\mathcal{Y}_1, \ldots, \mathcal{Y}_k$, such that, for any $1 \leq i \leq k$, the following conditions are satisfied:

(i) The collection $\mathcal{Y}_i$ consists of pairwise disjoint dyadic intervals, such that

$$\sum_{J\in\mathcal{Y}_i} |J| < \frac{x}{y} \sum_{J\in\mathcal{Y}_{i-1}} |J|.$$

(Recall that $\mathcal{Y}_0 = \{[0, 1]\}$.)

(ii) If $J \in \mathcal{Y}_i$ there exists $J_0 \in \mathcal{Y}_{i-1}$ such that $J \subset J_0$.

(iii) If $J \in \mathcal{Y}_i$ then $(b_K : [0, 1] \supseteq K \supseteq J)$ performs exactly i upcrossings of (x, y).

(iv) If $J \in \mathcal{Y}_i$, and if J_0 is the dyadic interval satisfying $J \subset J_0$ and $2|J| = |J_0|$, then $(b_K : [0, 1] \supseteq K \supseteq J_0)$ performs exactly $i - 1$ upcrossings of (x, y).

(v) The collection $\bigcup_{J\in\mathcal{Y}_i}\{I \in \mathcal{I} : I \subseteq J\}$ coincides with $\mathcal{B}_i$.

We prove the existence of $\mathcal{Y}_1, \ldots, \mathcal{Y}_k$ inductively. The construction in the first case is almost the same as in the general case, but not identical.

We turn to the definition of $\mathcal{Y}_1$. Defining the auxiliary collection $\mathcal{Z}_1$ we considser two cases: If $\int_0^1 b dt \geq x$, we put $\mathcal{Z}_1 = \{[0, 1]\}$, otherwise we consider $\mathcal{A} = \{I \in \mathcal{I} : b_I < x\}$, and let $\mathcal{Z}_1$ be the collection of maximal intervals in $\mathcal{A}$. Clearly, $\mathcal{Z}_1$ consists of pairwise dyadic intervals and, by assumption, is nonempty.

For any $I \in \mathcal{Z}_1$, consider $\mathcal{B}(I) = \{J \in \mathcal{I} : J \subseteq I, b_J > y\}$, let $\mathcal{Y}(I)$ denote the collection of maximal intervals in $\mathcal{B}(I)$, and, taking the union over $I \in \mathcal{Z}_1$, we define

$$\mathcal{Y}_1 = \bigcup_{I\in\mathcal{Z}_1} \mathcal{Y}(I).$$

By construction, $\mathcal{Y}_1$ consists of pairwise disjoint dyadic intervals. If $J \in \mathcal{Y}_1$, then $(b_K : J \subseteq K)$ performs exactly one upcrossing of (x, y). Hence, by assumption $\mathcal{Y}_1 \neq \emptyset$. If $J \in \mathcal{Y}_1$, and if J_0 is the dyadic interval satisfying $J \subset J_0$ and $2|J| = |J_0|$, then $(b_K : J_0 \subseteq K)$ performs zero upcrossings of (x, y). Next, if J_0 is a dyadic interval such that $J_0 \cap J = \emptyset$, for any $J \in \mathcal{Y}_1$, then again $(b_K : J_0 \subseteq K)$ performs zero upcrossings of (x, y). Thus, it follows that,

$$\bigcup_{J \in \mathcal{Y}_1} \{I \in \mathcal{I} : I \subseteq J\} = \mathcal{B}_1.$$

It remains to check the measure estimate. Using (4.4.7), we obtain

$$\sum_{J \in \mathcal{Y}_1} |J| < \frac{x}{y} \sum_{I \in \mathcal{Z}_1} |I| \leq \frac{x}{y}.$$

We now assume that $j \leq k - 1$ and that nonempty collections of dyadic intervals $\mathcal{Y}_1, \ldots, \mathcal{Y}_j$ have been determined such that the above conditions (i)–(v) hold true for $1 \leq i \leq j$.

We next construct $\mathcal{Y}_{j+1}$. For $J \in \mathcal{Y}_j$ let $\mathcal{A}(J) = \{I \in \mathcal{I} : b_I < x, I \subset J\}$, and let $\mathcal{Z}(J)$ be the collection of maximal intervals in $\mathcal{A}(J)$. Taking the union over $J \in \mathcal{Y}_j$ gives

$$\mathcal{Z}_{j+1} = \bigcup_{J \in \mathcal{Y}_j} \mathcal{Z}(J).$$

Since $j + 1 \leq k$ we have $\mathcal{Z}_{j+1} \neq \emptyset$. For any $I \in \mathcal{Z}_{j+1}$ consider $\mathcal{B}(I) = \{J \in \mathcal{I} : J \subseteq I, b_J > y\}$, let $\mathcal{Y}(I)$ denote the collection of maximal intervals in $\mathcal{B}(I)$, and take the union over I to define

$$\mathcal{Y}_{j+1} = \bigcup_{I \in \mathcal{Z}_{j+1}} \mathcal{Y}(I).$$

By construction, $\mathcal{Y}_{j+1}$ consists of pairwise disjoint dyadic intervals and $\mathcal{Y}_{j+1} \cap \mathcal{Y}_m = \emptyset$, for any $m \leq j$. See Figure 4.7.

If $J \in \mathcal{Y}_{j+1}$, there exists a unique $J_1 \in \mathcal{Y}_j$ such that $J \subset J_1$. Moreover, $(b_K : J \subseteq K \subseteq J_1)$ performs exactly one upcrossing of (x, y). If J_0 is the dyadic interval satisying $J_0 \supset J$ and $|J_0| = 2|J|$, then $(b_K : J_0 \subseteq K \subseteq J_1)$ performs zero upcrossings of (x, y). If $J_1 \in \mathcal{Y}_j$ and if J_0 is a dyadic interval such that $J_0 \cap J = \emptyset$, for any $J \in \mathcal{Y}_{j+1}$, then again $(b_K : J_0 \subseteq K \subseteq J_1)$ performs zero upcrossings of (x, y). Taking into account the induction hypothesis, we conclude that

$$\bigcup_{J \in \mathcal{Y}_{j+1}} \{I \in \mathcal{I} : I \subseteq J\} = \mathcal{B}_{j+1}.$$

See Figure 4.7. It remains to check the measure estimate. Using estimate (4.4.7) we finally verify that

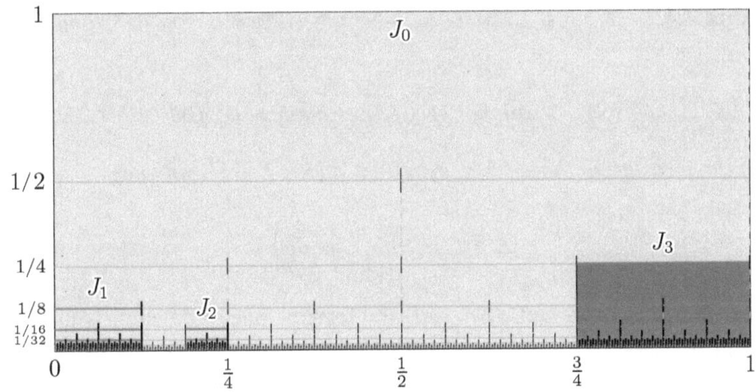

Figure 4.7 For $J_0 \in \mathcal{Y}_i$, the intervals J_1, J_2, J_3 depict $\{I \in \mathcal{Y}_{j+1} : I \subset J_0\}$. The light shaded region represents $\{I \in \mathcal{B}_j : I \subsetneq J_0\}$. The dark shaded regions represent $\{I \in \mathcal{B}_{j+1} : I \subsetneq J_0\}$.

$$\sum_{J \in \mathcal{Y}_{j+1}} |J| < \frac{x}{y} \sum_{I \subset \mathcal{Z}_{j+1}} |I| \leq \frac{x}{y} \sum_{J \in \mathcal{Y}_j} |J|.$$

We carry out the induction until $\mathcal{Y}_k$ is reached. By recursion, we have that $\sum_{J \in \mathcal{Y}_k} |J| < (x/y)^k$, and evaluating the inductive conditions at the final step, we infer that the collection of maximal intervals in $\mathcal{B}_k$ coincides with $\mathcal{Y}_k$. □

Exceptional sets: Let $0 \leq j \leq k$. Recall definition (4.4.3), which says the pairs of dyadic intervals (I_j, J_j) form an admissible cone

$$C = \{(I_0, J_0), \ldots, (I_j, J_j), \ldots, (I_k, J_k)\},$$

if $J_j \subsetneq I_j$ and $I_{j+1} \subsetneq J_j$. Recall that J_k is the tip of the cone C, denoted $t(C)$. Below, the cardinality of C is denoted $|C|$. Dubins's upcrossing lemma (Lemma 4.4.1) may be restated conveniently in terms of admissible cones as follows: Let $g \in L^\infty[0, 1]$ with $\|g\|_\infty \leq 1$. Fix $0 < \beta < \alpha < 1$. We say that an admissible cone C satisfies the (g, α, β) estimates if

$$\int_J g dt > \alpha |J|, \quad \int_I g dt < \beta |I|, \quad \forall (I, J) \in C.$$

Exploiting Lemma 4.4.1, we obtain measure estimates for exceptional sets for g defined in terms of admissible cones satisfying (g, α, β) estimates. We encode the information concisely in Lemma 4.4.3 below. Proceeding slowly, we begin with an auxiliary result.

Lemma 4.4.2 *Fix $g \in L^\infty[0,1]$ with $\|g\|_\infty \leq 1$, $0 < \beta < \alpha < 1$, and $k \in \mathbb{N}$. Define*

$$\mathcal{B} = \{t(C) : C \text{ satisfies } (g, \alpha, \beta) \text{ estimates and } |C| = k+1\}.$$

Then the measure of the set $A(g,k,\alpha,\beta) = \bigcup\{K : K \in \mathcal{B}\}$ satisfies

$$|A(g,k,\alpha,\beta)| \leq \left(\frac{1-\alpha}{1-\beta}\right)^k.$$

Proof Put $x = 1-\alpha$ and $y = 1-\beta$, and let C be one of the admissible cones defining $\mathcal{B}$. Set $b = 1-g$. Then $0 \leq b \leq 2$, and for any $(I,J) \in C$, the averages of b over J respectively I satisfy

$$b_I < x < y < b_J.$$

Recall that $t(C)$ denotes the tip of the cone C. Now the sequence of averages $\{b_K : K \supseteq t(C)\}$ performs at least k upcrossings of the interval (x,y), hence Lemma 4.4.1 gives $\left|\bigcup_{K\in\mathcal{B}} K\right| \leq (x/y)^k$. □

Fix $g \in L^\infty[0,1]$, with $\|g\|_\infty \leq 1$, $r \in \mathbb{N}$, and $-2^{r+2} \leq s \leq 2^{r+2}-1$. We next specify the set of parameters α, β, and k arising in Dubins's upcrossing lemma. Let

$$k = 2^{r+2}3(r+1), \quad \alpha = (s+1)2^{-r-2}, \quad \text{and} \quad \beta = s2^{-r-2}.$$

In view of Lemma 4.4.2, the sets

$$A_s = A\left(g, k, (s+1)2^{-r-2}, s2^{-r-2}\right), \tag{4.4.8}$$

satisfy the measure estimate $|A_s| \leq 2^{-3r-3}$. We take the union over $-2^{r+2} \leq s \leq 2^{r+2}-1$, and define

$$A = \bigcup\left\{A_s : -2^{r+2} \leq s \leq 2^{r+2}-1\right\}. \tag{4.4.9}$$

Since $|A_s| \leq 2^{-3r-3}$, the exceptional set $A = A(g)$ satisfies the measure estimates

$$|A| \leq 2^{-2r}.$$

The complement of the exceptional set, $[0,1]\backslash A$, will remain in our focus throughout the current section. We start the analysis of $[0,1]\backslash A$, by restating the pigeon hole principle of elementary combinatorics.

Lemma 4.4.3 *Let $g \in L^\infty[0,1]$, with $\|g\|_\infty \leq 1$, $r \in \mathbb{N}$, and*

$$k = 2^{r+2}3(r+1).$$

Let $A = A(g) \subset [0,1]$ *be defined by* (4.4.9). *Let* C *be an admissible cone satisfying*

$$|C| \geq k2^{r+3} \quad \textit{and} \quad t(C) \subseteq [0,1] \backslash A.$$

Then there exists a pair of dyadic intervals $(I,J) \in C$ *such that*

$$\langle g, f_{(I,J)} \rangle < 2^{-r-1}, \tag{4.4.10}$$

where $f_{(I,J)} = 1_I/|I| - 1_J/|J|$, *and where* $\langle g, f_{(I,J)} \rangle = \int_0^1 g(t) f_{(I,J)}(t)dt$.

Proof Assume to the contrary that there exists an admissible cone C satisfying $|C| \geq k2^{r+3}$ and

$$\langle g, f_{(I,J)} \rangle \geq 2^{-r-1}, \tag{4.4.11}$$

for $(I,J) \in C$. To obtain a contradiction, we now verify that (4.4.11) implies that $t(C) \subseteq A$. By (4.4.11), for any $(I,J) \in C$ there exists $s = s(I,J)$ with $-2^{r+2} \leq s < 2^{r+2}$, so that

$$\int_I g dt \geq |I|(s+1)2^{-r-2} \quad \text{and} \quad \int_J g dt < |J| s 2^{-r-2}.$$

By the pigeonhole principle, there exists $s_0 \in (-2^{r+2}, 2^{r+2})$ for which the cardinality of the pairs $\{(I,J) \in C : s_0 = s(I,J)\}$ exceeds k, where $k = 2^{r+3}3(r+1)$. Recall we put

$$A_{s_0} = A\left(g, k, (s_0+1)2^{-r-2}, s_0 2^{-r-2}\right).$$

Thus we proved that $t(C) \subseteq A_{s_0}$ and, since $A_{s_0} \subseteq A$, we have a contradiction. □

A stopping time argument: Let $C = \{(I_0,J_0), \ldots, (I_j,J_j), \ldots, (I_k,J_k)\}$ be an admissible cone, defined by (4.4.3), and let $f(C)$ be given by equation (4.4.4). Choose $p \leq k+1$ and let $f(C,p)$ denote the pth partial sum of $f(C)$. Thus,

$$f(C,p) = \frac{1}{2(k+1)} \sum_{j=0}^{p} f_{(I_j,J_j)}, \tag{4.4.12}$$

where $f_{(I,J)} = 1_I/|I| - 1_J/|J|$. To emphasize, the scaling factor in the definition of $f(C,p)$ remains $2(k+1)$.

Our aim is now to find a short sum of the form (4.4.12) that approximates $f(C)$ in measure. The resulting combinatorial difficulties will be dealt with by stopping time arguments presented in Lemma 4.4.4 and Proposition 4.4.5 below. Given $f \in L^1[0,1]$ and $g \in L^\infty[0,1]$, we denote $\langle g, f \rangle = \int_0^1 g f dt$.

Lemma 4.4.4 *Let $g \in L^\infty$ with $\|g\|_\infty \le 1$, $r \in \mathbb{N}$, and*

$$k = 2^{r+2}3(r+1).$$

Let $A = A(g)$ be defined by (4.4.9) and let C be an admissible cone satisfying

$$|C| \ge (2^{r+3}k)(2^{r+1}). \tag{4.4.13}$$

Then, for every $q^{(0)} \le |C|$ there exists $q^{(1)} \le q^{(0)}$ such that

$$h_0 = f(C, q^{(0)}) \quad \text{and} \quad h_1 = f(C, q^{(1)})$$

satisfy

$$\|h_0 - h_1\|_1 \le (q^{(0)} - q^{(1)})|C|^{-1}, \quad \{h_0 \ne h_1\} \subseteq A, \quad \text{and} \quad \langle g, h_1 \rangle \le 2^{-r-1}.$$

Proof We are given the cone $C = \{(I_0, J_0), \dots, (I_j, J_j), \dots, (I_k, J_k)\}$. We divide C into a bad part $\mathcal{B}$ and good part $\mathcal{G}$ by putting

$$\mathcal{B} = \left\{(I, J) \in C : \langle g, f_{(I,J)} \rangle > 2^{-r-1}\right\} \quad \text{and} \quad \mathcal{G} = C \setminus \mathcal{B}.$$

We now proceed by treating the cases $|\mathcal{B}| > k2^{r+3}$ and $|\mathcal{B}| \le k2^{r+3}$ separately.

Case 1 Let $|\mathcal{B}| > k2^{r+3}$. In this case we use a stopping time procedure to define $q^{(1)}$. Given $q \le k$, we denote $C(q) = \left\{(I_0, J_0), \dots, (I_j, J_j), \dots, (I_q, J_q)\right\}$. Put

$$q^{(1)} = \max\left\{q < q^{(0)} : |\mathcal{B} \cap C(q)| < k2^{r+3}\right\}.$$

Necessarily, we have $J_{q^{(1)}+1} \subseteq A$ since otherwise $\mathcal{B}$ would be a cone, contradicting the conclusion of Lemma 4.4.3. Hence $\{h_0 = h_1\} \supseteq [0,1] \setminus A$.

It remains to estimate the scalar product $\langle g, h_1 \rangle$. Put $\mathcal{B}_1 = \mathcal{B} \cap C(q^{(1)})$, and estimate

$$\begin{aligned} 2|C|\langle g, h_1 \rangle &\le \sum_{(I,J) \in \mathcal{B}_1} \left|\langle g, f_{(I,J)} \rangle\right| + \sum_{(I,J) \in \mathcal{G}} \left|\langle g, f_{(I,J)} \rangle\right| \\ &\le |\mathcal{B}_1| + 2^{-r-1}|C|. \end{aligned}$$

By the stopping time condition we have $|\mathcal{B}_1| \le k2^{r+3}$ and, comparing this with the hypothesis (4.4.13), gives $|\mathcal{B}_1| + 2^{-r-1}|C| \le 2^{-r}|C|$. Collecting terms we obtain $\langle g, h_1 \rangle \le 2^{-r-1}$.

Case 2 Assume now that $|\mathcal{B}| \le k2^{r+3}$. In this case, we simply put $q^{(1)} = q^{(0)}$ and estimate the scalar product as follows:

$$\begin{aligned} 2|C|\langle g, h_1 \rangle &= \sum_{(I,J) \in \mathcal{B}} \langle g, f_{(I,J)} \rangle + \sum_{(I,J) \in \mathcal{G}} \langle g, f_{(I,J)} \rangle \\ &\le 2|\mathcal{B}| + 2^{-r-1}|C| \\ &\le 2^{-r}|C|. \end{aligned}$$

□

We now introduce a second stopping time argument to the proof of property (4.4.1), which represents one of the core ideas in Talagrand's approach.

Proposition 4.4.5 *Let $r, L \in \mathbb{N}$. Let $C_1, \ldots, C_L$ be admissible cones of equal cardinality satisfying*

$$|C_l| \geq 2^{3r+1} 3(r+1).$$

Let $q_l^{(0)} \leq |C_l|$ and $h_{0,l} = f\left(C_l, q_l^{(0)}\right)$. If $c_l \geq 0$, with $\sum c_l \leq 1$, then for

$$f_0 = \sum_{l=1}^{L} c_l h_{0,l}, \tag{4.4.14}$$

there exists $f \in L^1$ such that $|\{f \neq f_0\}| \leq 2^{-r}$ and $\|f\|_1 \leq 2^{-r}$.

Proof For notational convenience we prepare the proof by abbreviating $d_l = c_l/|C_l|$. In view of $q_l^{(0)} \leq |C_l|$, we have

$$\sum d_l q_l^{(0)} \leq 1. \tag{4.4.15}$$

The proof proceeds by induction. The inductive step of the argument simultaneously defines a stopping criterion. If satisfied, it causes the induction to stop at this stage. If not, another induction step is executed. Let f_0 be given by the summation (4.4.14). If $\|f_0\|_1 \leq 2^{-r}$ it suffices to take $f = f_0$ and thereby finish the proof.

Step 1 Now assume that $\|f_0\|_1 > 2^{-r}$. Set $g_0 = \operatorname{sign} f_0$. Let $A(g_0)$ be the exceptional set defined by (4.4.9). We denote $A_1 = A(g_0)$ to indicate that we are in Step 1 of the induction argument. We have $A_1 \subseteq [0,1]$ and $|A_1| \leq 2^{-2r}$. Apply Lemma 4.4.4 to g_0. Thus for $l \leq L$, we obtain indices $0 < q_l^{(1)} \leq q_l^{(0)}$, such that $h_{1,l} = f\left(C_l, q_l^{(1)}\right)$ satisfies

$$\{h_{1,l} \neq h_{0,l}\} \subseteq A_1 \quad \text{and} \quad \langle g_0, h_{1,l} \rangle \leq 2^{-r-1}.$$

Consequently, for

$$f_1 = \sum_{l=1}^{L} c_l h_{1,l}, \tag{4.4.16}$$

we have $\{f_1 \neq f_0\} \subseteq A_1$. Now, if $\|f_1\|_1 \leq 2^{-r}$ we put $f = f_1$ and stop the proof at this stage. If $\|f_1\|_1 > 2^{-r}$ we continue by analyzing the sequence $\left\{q_l^{(1)}\right\}$ and observe the following: Since $c_l \geq 0$ with $\sum c_l \leq 1$, we have

$$\langle g_0, f_1 \rangle \leq 2^{-r-1}. \tag{4.4.17}$$

As $g_0 = \text{sign} f_0$, we get $\|f_0\|_1 = \langle g_0, f_0 \rangle$. Taking into account (4.4.17) and comparing the definitions (4.4.16) and (4.4.14) gives

$$\begin{aligned} \|f_0\|_1 - 2^{-r-1} &\leq \langle g_0, f_0 - f_1 \rangle \\ &\leq \sum_{l=1}^{L} c_l \|h_{1,l} - h_{0,l}\|_1 \\ &\leq \sum_{l=1}^{L} d_l \left(q_l^{(0)} - q_l^{(1)}\right). \end{aligned} \tag{4.4.18}$$

Since $\|f_0\|_1 > 2^{-r}$ and $q_l^{(1)} \leq q_l^{(0)}$, we obtain with (4.4.18) and (4.4.15) that

$$2^{-r-1} + \sum_{l=1}^{L} d_l q_l^{(1)} \leq 1.$$

We assumed that $\|f_1\|_1 > 2^{-r}$ and continue the proof by repeating the first step with f_0 replaced by f_1.

After step m**:** Now assume that we were able to iterate the above procedure m times. Specifically, we assume that we obtained indices $0 < q_l^{(m)} \leq \cdots \leq q_l^{(1)} \leq q_l^{(0)}$, and sets $A_i \subseteq [0,1]$ with $|A_i| \leq 2^{-2r}$, such that with $h_{i,l} = f\left(C_l, q_l^{(i)}\right)$ we have

$$\{h_{i-1,l} \neq h_{i,l}\} \subseteq A_i \quad \text{and} \quad \langle g_0, h_{i,l} \rangle \leq 2^{-r-1}, \quad \text{for} \quad 1 \leq i \leq m.$$

Fix $1 \leq i \leq m$. For

$$f_i = \sum_{l=1}^{L} c_l h_{i,l},$$

we obtain the inclusion $\{f_{i-1} \neq f_i\} \subseteq A_i$, the lower estimate $\|f_i\|_1 > 2^{-r}$, and by the calculation (4.4.18) – with 0 replaced by $i-1$ and 1 replaced by i – we have

$$i \cdot 2^{-r-1} + \sum_{l=1}^{L} d_l q_l^{(i)} \leq 1. \tag{4.4.19}$$

Now specialize this expression to $i = m$ to obtain

$$m \cdot 2^{-r-1} \leq 1. \tag{4.4.20}$$

Backtracking our steps, (4.4.20) holds true if we have the lower bounds

$$\|f_1\|_1 > 2^{-r}, \ldots, \|f_i\|_1 > 2^{-r}, \ldots, \|f_m\|_1 > 2^{-r}.$$

Clearly this has to stop before m exceeds 2^{r+1}. Hence our stopping criterion, $\|f_m\|_1 \le 2^{-r}$, has to be satisfied for some $m \le 2^{r+1}$. Since

$$\{f_m \neq f_0\} \subseteq A_1 \cup A_2 \cup \cdots \cup A_m,$$

and $|A_i| \le 2^{-2r}$ $(1 \le i \le m \le 2^{r+1})$, we obtain $|\{f_m \neq f_0\}| \le 2^{-r+1}$, as claimed. □

Recall that we denoted $F_n = \{f(C) : |C| = n+1\}$, where C and $f(C)$ are defined in (4.4.3) and (4.4.4) respectively. In view of Proposition 4.4.5 we arrive at our first central result, asserting that the sequence (F_n) satisfies condition (4.4.1).

Theorem 4.4.6 *For every $\epsilon > 0$ there exists $n_0 \in \mathbb{N}$ such that for $n \ge n_0$ any*

$$g \in \operatorname{conv}(F_n) \quad \textit{satisfies} \quad |\{|g| > \epsilon\}| \le \epsilon.$$

Proof Let $\epsilon > 0$. Select $r \in \mathbb{N}$ such that $2^{-r+1} < \epsilon^2$. Let $n \ge 2^{3r}3(r+1)$ and then take any $g \in \operatorname{conv}\{f(C) : |C| = n\}$. By Proposition 4.4.5 applied to g, there exists $f \in L^1$ such that $|\{g \neq f\}| \le 2^{-r}$ and $\|f\|_1 \le 2^{-r}$. Hence,

$$|\{|g| > \epsilon\}| \le |\{g \neq f\}| + |\{|f| > \epsilon\}| \le 2^{-r} + \|f\|_1/\epsilon < \epsilon. \qquad \square$$

4.4.2 Liapunov's Theorem

Having completed the proof of Theorem 4.4.6 we next show that the sequence (F_n) satisfies the competing conditions (4.4.2). The argument is based on Liapunov's theorem.

Theorem 4.4.7 *Let μ_i $(1 \le i \le n)$ denote a nonatomic measure defined on a sigma-algebra Σ. Then $\{(\mu_1(A), \ldots, \mu_n(A)) : A \in \Sigma\}$ forms a compact and convex subset of $\mathbb{R}^n$.*

General admissible cones: Fix $n \in \mathbb{N}$. Here, we isolate the elements in $\operatorname{conv}(F_n)$ by which we prove the approximation (4.4.2).

Let Σ_j denote the (purely atomic) algebra generated by dyadic intervals of length $\ge 2^{-j}$. Now let A_k, B_k be measurable subsets of the unit interval. We assume that $B_k \subset A_k \subset B_{k+1} \subset A_{k+1}$, for $0 \le k \le n-1$, and put

$$\mathcal{B} = \{(A_0, B_0), \ldots, (A_n, B_n)\}. \tag{4.4.21}$$

We call $\mathcal{B}$ a general admissible cone if there exist integers $i(k)$, $j(k)$ satisfying

$$i(0) < j(0) < \cdots i(k) < j(k) < \cdots i(n) < j(n),$$

such that $A_k \in \Sigma_{i(k)}$, $B_k \in \Sigma_{j(k)}$, and

$$\mathbb{E}(1_{B_k}|\Sigma_{i(k)}) = \frac{1}{2}1_{A_k}, \qquad \mathbb{E}(1_{A_{k+1}}|\Sigma_{j(k)}) = \frac{1}{2}1_{B_k}. \tag{4.4.22}$$

We say that A_0 is the base of the general cone $\mathcal{B}$ and denote $b(\mathcal{B}) = A_0$. Next, define the Haar packet generated by $\mathcal{B}$,

$$g(\mathcal{B}) = \frac{1}{2(n+1)}\sum_{j=0}^{n} g(A_j, B_j), \tag{4.4.23}$$

where $g(A, B) = 1_A/|A| - 1_B/|B|$. A moment's reflection reveals that $g(\mathcal{B}) \in \operatorname{conv}(F_n)$ whenever $\mathcal{B}$ is a general admissible cone.

Theorem 4.4.8 *Let Σ_0 be a diffuse sigma-algebra in $[0,1]$, let $A \in \Sigma_0$ with $|A| > 0$, and let $n \in \mathbb{N}$ satisfy $|A| > 2^{-n}$. Then for any $\epsilon > 0$, there exist $f \in L^1(A, \Sigma_0)$ and a general cone $\mathcal{B}$ such that*

$$\int_0^1 f dt = 0, \quad \|f\|_1 \geq 1/4, \quad \|f - g(\mathcal{B})\|_1 < \epsilon, \quad \text{and} \quad |\mathcal{B}| = n, \tag{4.4.24}$$

where $|\mathcal{B}|$ denotes the cardinality of $\mathcal{B}$.

Proof Let $n \in \mathbb{N}$ and $A \in \Sigma_0$ such that $|A| > 2^{-n}$. We show that for any $\epsilon > 0$ there exists a general cone,

$$\mathcal{B} = \{(A_0, B_0), \ldots, (A_n, B_n)\},$$

and a chain of Σ_0 measurable sets $K_0 \supset L_0 \supset \cdots \supset K_n \supset L_n$ satisfying $K_0 = A$, and

$$|A_k \Delta K_k| + |B_k \Delta L_k| \leq \epsilon 4^k, \qquad 0 \leq k \leq n. \tag{4.4.25}$$

Once this is done we put

$$f = \frac{1}{2n+2}\sum_{k=0}^{n}\{1_{K_k}/|K_k| - 1_{L_k}/|L_k|\},$$

and observe that $f \in L^1(A, \Sigma_0)$ and $\|f - g(\mathcal{B})\|_1 \leq \epsilon 4^n \|f\|_1$. We now turn to the inductive construction of the chain satisfying (4.4.25).

Step 1 We first recall that Σ_i denotes the algebra of sets generated by the dyadic intervals of length equal to 2^{-i} ($i \in \mathbb{N}$). Put $A = K_0$, and select $i(1) \in \mathbb{N}$ and $A_0 \in \Sigma_{i(1)}$ so that $|A_0 \Delta K_0| < \epsilon$. Suppose now that $(A_0, B_0), \ldots, (A_{k-1}, B_{k-1})$, A_k and $K_0, L_0, \ldots, K_{k-1}, L_{k-1}, K_k$ are defined and satisfy the conclusion.

Step 2 We first recall that $\mathcal{I}_i$ denotes the collection of dyadic intervals of length equal to 2^{-i} ($i \in \mathbb{N}$).

We first determine B_k, L_k. Form the vector measure

$$\lambda_k \colon (K_k, \Sigma_0 \cap K_k) \to \mathbb{R}^{\mathrm{N}}, \quad A \mapsto (|A|, |A \cap I|, I \in \mathcal{I}_{i(k)}) .$$

Clearly the range of λ_k is finite dimensional, and the measure λ_k itself is absolutely continuous with respect to Lebesgue measure on K_k. Applying Liapunov's theorem on the range of vector measures yields $L_k \subset K_k$ so that $L_k \in \Sigma_0$ and $\lambda_k(L_k) = \lambda_k(K_k)/2$. Thus we have,

$$|L_k| = |K_k|/2, \quad |L_k \cap I| = |K_k \cap I|/2, \qquad I \in \mathcal{I}_{i(k)}.$$

Lemma 4.4.9 below states that there exists an index $j(k) > i(k)$ and $B_k \in \Sigma_{i(k)}$ such that $B_k \subset A_k$ and

$$|B_k \Delta L_k| < 4|A_k \Delta K_k|, \qquad |B_k \cap I| = |A_k \cap I|/2, \qquad I \in \mathcal{I}_{i(k)}.$$

Next we determine A_{k+1} and K_{k+1}. This is done by repeating the above argument with K_k replaced by L_k and with A_k replaced by B_k. Thus we finish the induction step. □

Lemma 4.4.9 *Let $A, A_0 \in \Sigma_0$, with $A_0 \subseteq A$. Assume there exists $k \in \mathbb{N}$ such that $|A_0 \cap I| = |A \cap I|/2$, for any $I \in \mathcal{I}_k$. Then for any $\epsilon > 0$ and $B \in \Sigma_k$ satisfying*

$$|A \Delta B| < \epsilon,$$

there exists $m \in \mathbb{N}$ and $B_0 \subseteq B$ with $B_0 \in \Sigma_m$ such that $|B_0 \cap I| = |B \cap I|/2$, for any $I \in \mathcal{I}_k$. Moreover,

$$|A_0 \Delta B_0| < 4\epsilon.$$

Proof Since $B \in \Sigma_k$ for any $I \in \mathcal{I}_k$, either $I \subseteq B$ or $I \cap B = \emptyset$. Fix $I \in \mathcal{I}_k$ and let $A_1(I) = A_0 \cap I$. Enlarging $A_1(I)$, we select $A_2(I) \subseteq I$ such that $A_1(I) \subseteq A_2(I)$ and $|A_2(I) \cap I| = |I|/2$. Next, approximating $A_2(I)$ determines $m \in \mathbb{N}$ large enough and $B_1(I) \in \Sigma_m$ such that $B_1(I) \subseteq I$ and

$$|B_1(I) \Delta A_2(I)| \leq \epsilon |I|.$$

Finally, by a simple counting argument we find $B_0(I) \in \Sigma_m \cap I$, such that

$$|B_0(I) \Delta B_1(I)| \leq \epsilon |I|$$

and $|B_0(I) \cap I| = |I|/2$. Taking the union over $I \in \mathcal{I}_k$ with $I \subseteq B$, define $B_0 = \bigcup B_0(I)$ and estimate the measure of the symmetric difference,

$$|A_0 \Delta B_0| \leq \sum |A_1(I) \Delta I| + |I \Delta A_2(I)| + |A_2(I) \Delta B_1(I)| + |B_1(I) \Delta B_0(I)| \leq 4\epsilon. \ \square$$

Central Estimates of Talagrand's Construction

Let $\mathcal{B}$ be a general admissible cone defined by (4.4.22), and let $g(\mathcal{B})$ be the associated the Haar packet defined by (4.4.23). Fix $n \in \mathbb{N}$, and define

$$D_n = \{g(\mathcal{B}) : |\mathcal{B}| = n + 1, \quad \text{and} \quad |b(\mathcal{B})| \geq 2^{-n}\}, \tag{4.4.26}$$

where $|\mathcal{B}|$ denotes the cardinality of $\mathcal{B}$, where $|b(\mathcal{B})|$ denotes the measure of its base $b(\mathcal{B}) \subseteq [0, 1]$, and where $\mathcal{B}$ ranges over all general cones with the above specifications. It is easy to see that $g \in \text{conv}D_n$ implies that

$$\|g\|_\infty \leq 2^{3n}, \qquad \|g\|_1 \geq 1/3. \tag{4.4.27}$$

Moreover, by Theorem 4.4.6, for any $\epsilon > 0$ there exists $n \in \mathbb{N}$ such that any $g \in \text{conv}D_n$ satisfies the measure estimate

$$|\{|g| > \epsilon\}| < \epsilon.$$

Repeatedly applying Theorem 4.4.6, we obtain a rapidly increasing sequence of integers $(q(n))_{n=1}^\infty$ such that

$$\left|\{|g| > 2^{-2n}\}\right| < 2^{-n-1} \times 2^{-3q(n-1)-2n-2} \tag{4.4.28}$$

for any $g \in \text{conv}D_{q(n)}$. We refer to (4.4.27) and (4.4.28) as the *central estimates of Talagrand's construction.*

Separately, for ease of reference, we fix $(B_n)_{n=1}^\infty$, a sequence of pairwise disjoint, measurable subsets of the unit interval, satisfying

$$|B_n| = 2^{-3q(n-1)-2n-4}. \tag{4.4.29}$$

The Link to Rosenthal's L^1 Theorem

Recall that D_n is defined by (4.4.26). Subsequently, we use the abbreviation

$$C_n = D_{q(n)}, \quad n \in \mathbb{N}. \tag{4.4.30}$$

Let $(f_{n,k})_{k=1}^\infty$ be an enumeration of $C_n = D_{q(n)}$.

Write $L^1 = L^1([0, 1])$, put $Y_0 = L^1$, and define the Banach space $(Y_n, \|\cdot\|_{Y_n})$ by

$$Y_n = \overline{\text{span}}^{L^1}\{f_{n,k} : k \in \mathbb{N}\} \quad \text{and} \quad \|g_n\|_{Y_n} = 2^{-n}\|g_n\|_{L^1([0,1])}, \tag{4.4.31}$$

for $g_n \in Y_n$. Let $Y = (\sum_{n=0}^n Y_n)_1$. Thus $g \in Y$ if $g = (g_n)_{n=0}^\infty$, $g_n \in Y_n$, and

$$\|g\|_Y = \sum_{n=0}^\infty 2^{-n}\|g_n\|_{L^1} < \infty.$$

The central estimates (4.4.27) and (4.4.28) imply that for any $g \in Y$, with $g = (g_n)$, the partial sums $\sum_{n=0}^p g_n$, converge in measure as $p \to \infty$. Hence the summing operator is well defined for sequences $g \in Y$. We set

$$T(g) = \sum_{n=0}^{\infty} g_n, \tag{4.4.32}$$

where the series on the right-hand side converges in measure. Recall that $L^0 = L^0([0,1])$ denotes the space of all measurable functions on $[0,1]$ equipped with the topology induced by the measure distance

$$d(f,g) = \inf\{\epsilon > 0 : |\{|f-g| > \epsilon\}| < \epsilon\}.$$

Thus T maps Y continuously into L^0. Define $E = TY$, and endow it with the quotient norm

$$\|f\|_E = \inf\{\|g\|_Y : f = T(g)\}. \tag{4.4.33}$$

The space E is canonically identified with $Y/\mathrm{Ker\,T}$, the space of equivalence classes $[g] = g + \mathrm{Ker\,T}$ equipped with the norm

$$\|[g]\|_E = \inf\{\|g-h\|_Y : h \in \mathrm{Ker\,T}\}.$$

We observe at once the following useful estimates

$$\|f\|_E \le \inf_{n\in\mathbb{N}} \inf_{g_n\in Y_n} \|f-g_n\|_1 + 2^{-n}\|g_n\|_1 \quad \text{and} \quad \|f\|_E < \|f\|_1, \tag{4.4.34}$$

for $f \in L^1$. By the second estimate in (4.4.34), the inclusion operator

$$i : L^1 \to E, \tag{4.4.35}$$

defined by $i(f) = f$ is a contraction. Theorem 4.4.10 states that i does not fix a copy of L^1. In its proof we merge Theorem 4.4.8 with Rosenthal's L^1 theorem.

Theorem 4.4.10 *The formal identity operator $i: L^1 \to E$, defined by $i(f) = f$ does not fix a copy of L^1.*

Proof Recall that we let $([0,1], \mathcal{L}, dt)$ denote the Lebesgue measure space. By Corollary 4.3.2, it suffices to show that for any nonatomic sigma-algebra $\Sigma_0 \subseteq \mathcal{L}$, any $A \in \Sigma_0$, with $|A| > 0$ and any $\eta > 0$ there exists $f \in L^1(A, \Sigma_0)$ such that

$$\int_A f dt = 0 \text{ and } \|f\|_E < \eta\|f\|_{L^1}.$$

Determine $n \in \mathbb{N}$ so that $|A| > 2^{-n}$. By Theorem 4.4.8, for any $\epsilon > 0$ there exists $f \in L^1(A, \Sigma_0)$ and $g \in D_n$ so that

$$\int_A f dt = 0, \quad \text{and} \quad \|f-g\|_{L^1} \le \epsilon\|f\|_{L^1}.$$

It remains to appeal to the first estimate in (4.4.34) to conclude the proof. □

4.5 Talagrand's Examples

In this section we present Talagrand's counterexample to the three-space conjecture for L^1 embeddings. We systematically exploit the L^1 approximation results obtained in Section 4.4 and construct a subspace $X \subset L^1$ such that neither L^1/X nor X contain a copy of L^1.

Defining Talagrand's space X: Let $C_n = D_{q(n)}$ be the countable set defined by (4.4.30). Let $(f_{n,k})_{k=1}^{\infty}$ be an enumeration of the functions in C_n. Let (B_n) be the sequence of disjoint sets fixed right above condition (4.4.29), and let $\varphi\colon \mathbb{N} \times \mathbb{N} \to \mathbb{N}$ be an arbitrary injective map, fixed throughout the section. Put $A(n, k) = B_{\varphi(n,k)}$ and let

$$g_{n,k} = f_{n,k} + 2^{-n} 1_{A(n,k)}/|A(n, k)|, \qquad n, k \in \mathbb{N}. \tag{4.5.1}$$

Talagrand's space $X \subset L^1$ and its canonical building blocks X_n are defined by

$$X = \overline{\operatorname{span}}^{L^1}\{g_{n,k} : n, k \in \mathbb{N}\} \quad \text{and} \quad X_n = \overline{\operatorname{span}}^{L^1}\{g_{n,k} : k \in \mathbb{N}\}. \tag{4.5.2}$$

Clearly, $X \subset L^1$ is the closure of $\operatorname{span}\{X_n : n \in \mathbb{N}\}$ in L^1.

Theorem 4.5.1 *Neither Talagrand's space X nor L^1/X contain a copy of L^1.*

The two assertions of Theorem 4.5.1 are proven in Section 4.5.1 and Section 4.5.2, respectively.

4.5.1 Subspaces

In this section we prove that L^1 is not isomorphic to a subspace of Talagrand's space X, defined by (4.5.2). This is done in two basic steps. In Lemma 4.5.2 we show that X_n is isomorphic to ℓ^1. In Theorem 4.5.3 we prove that X is isomorphic to the ℓ^1 sum of the X_n.

Lemma 4.5.2 *For each $n \in \mathbb{N}$, the space X_n is isomorphic to ℓ^1, with Banach Mazur distance $d(X_n, \ell^1)$ bounded by 2^n. Specifically, there exists $C < \infty$ such that*

$$\sum_{n,k=1}^{\infty} 2^{-n}|a_{n,k}| \leq C \left\| \sum_{n,k=1}^{\infty} a_{n,k} g_{n,k} \right\|_1, \tag{4.5.3}$$

for any finitely supported sequence of scalars $a_{n,k}$.

Proof Let $(g_{n,k})$ be defined by (4.5.1). Fix a finitely supported sequence of scalars $(a_{n,k})$ and put $h = \sum a_{n,k} g_{n,k}$. Exploiting that $(g_{n,k})$ is almost disjointly supported, we will be able to show that

$$\int_0^1 |h|dt \geq c \sum_{n,k=1}^{\infty} 2^{-n}|a_{n,k}|. \tag{4.5.4}$$

To this end, we put

$$u_n = \sum_{k=1}^{\infty} a_{n,k} f_{n,k},$$

and let $b_n = \sum_{k=1}^{\infty} |a_{n,k}|$. The central estimates (4.4.27), (4.4.29), and (4.4.28) imply that $\|u_n\|_1 \leq 2b_n$ and $\|u_n\|_\infty \leq 2b_n 2^{3q(n)}$. Moreover, if $\ell \geq n$ we obtain

$$\int_{B_\ell} |u_n|dt \leq 2b_n 2^{q(n)}|B_\ell| \leq 2b_n 2^{-\ell-n-4}. \tag{4.5.5}$$

In the converse case, when $\ell \leq n$ we put

$$A_n = \left\{|u_n| > b_n 2^{-2n}\right\}, \quad A'_\ell = \bigcup_{n \geq \ell} A_n, \quad B'_\ell = B_\ell \backslash A'_\ell,$$

and estimate

$$\int_{B'_\ell} |u_n|dt \leq b_n 2^{-2n}|B'_\ell| \leq b_n 2^{-\ell-n-3}. \tag{4.5.6}$$

Put $u = \sum_{n=1}^{\infty} u_n$. Taking the sum over n, and using (4.5.6) for $\ell \leq n$ and (4.5.5) for $\ell \geq n$ gives

$$\int_{B'_\ell} |u|dt \leq 2^{-\ell-3} \sum_{n=1}^{\infty} 2^{-n} b_n. \tag{4.5.7}$$

Next, we put

$$f = \sum a_{n,k} 2^{-n} 1_{A(n,k)}/|A(n,k)|,$$

where the sum is taken over $k, n \in \mathbb{N}$ and where $(A(n,k))$ is the sequence of pairwise disjoint sets defined in (4.5.1). Writing $\ell = \varphi(r,s)$ gives $B_\ell = A(r,s)$. Using also that $\left|B'_\ell\right| \geq |B_\ell|/2$, we obtain the lower estimate,

$$\int_{B'_\ell} |f|\, dt \geq 2^{-r-1}|a_{r,s}|. \tag{4.5.8}$$

Since $g_{n,k} = f_{n,k} + 2^{-n} 1_{A(n,k)}/|A(n,k)|$, we have $h = u + f$. Combine the upper estimate (4.5.7) with the lower estimate (4.5.8) to obtain, for fixed $\ell = \varphi(r,s)$, the crucial lower estimate

$$\int_{B'_\ell} |h|dt \geq 2^{-r-1}|a_{r,s}| - 2^{-\ell-3} \sum_{n=1}^{\infty} 2^{-n} b_n.$$

Since (B'_ℓ) is a sequence of pairwise disjoint sets, we may take the sum over $\ell = \varphi(r, s)$ and obtain estimate (4.5.4). Finally, note that there exist $c > 0$ and $C < \infty$ such that

$$c \leq \|g_{n,k}\|_{L^1} \leq C, \tag{4.5.9}$$

for $n, k \in \mathbb{N}$. Hence for $n \in \mathbb{N}$ fixed, condition (4.5.3) shows that X_n is isomorphic to ℓ^1, and that the Banach Mazur distance of X_n to ℓ^1 is bounded by 2^n. □

By Lemma 4.5.2 we know that X_n is small compared to L^1 in the sense that it is isomorphic to ℓ^1, for $n \in \mathbb{N}$. Now, we extend this information to the closure of $\operatorname{span}\{X_n : n \in \mathbb{N}\}$ in L^1.

Theorem 4.5.3 *The space X is isomorphic to $(\sum X_n)_1$.*

Proof The proof again exploits the central estimates (4.4.27) and (4.4.28). We take advantage of the fact that functions spanned by C_{n+1} are essentially supported on very small sets, where as a scale we use the size of those sets at which a typical function in C_n is essentially constant.

Step 1 Let $h = \sum h_n$ be a finite sum of $h_n \in X_n$ and let $h_n = \sum_{k=1}^{\infty} a_{n,k} g_{n,k}$, where only finitely many of the coefficients $a_{n,k} \in \mathbb{R}$ are nonvanishing. We will show that

$$\sum_{n=1}^{\infty} \|h_n\|_1 \leq C\|h\|_1, \tag{4.5.10}$$

where $C < \infty$ is independent of $h_n \in X_n$. Denote $M = \|h\|_1$, and put $u_n = \sum_{k=1}^{\infty} a_{n,k} f_{n,k}$ and $u = \sum u_n$. Applying Lemma 4.5.2 to h gives

$$\sum_{n,k=1}^{\infty} |a_{n,k}| 2^{-n} \leq MC. \tag{4.5.11}$$

We have $\|u_n\|_1 \leq \sum_{k=1}^{\infty} |a_{n,k}|$, hence (4.5.11) gives $\|u_n\|_1 \leq MC2^n$. Moreover, with $\|g_{n,k} - f_{n,k}\|_1 = 2^{-n}$ we get $\|h_n - u_n\|_1 \leq \sum_{k=1}^{\infty} 2^{-n}|a_{n,k}|$, so that (4.5.11) gives $\|u\|_1 \leq \|h\|_1 + CM$ and

$$\sum_{n=1}^{\infty} \|h_n\|_1 \leq \sum_{n=1}^{\infty} \|u_n\|_1 + MC.$$

Thus we reduced (4.5.10) to proving that there exists $\overline{C} < \infty$ such that

$$\sum_{n=1}^{\infty} \|u_n\|_1 \leq \|u\|_1 + M\overline{C}. \tag{4.5.12}$$

Step 2 We next verify the estimate (4.5.12). Define the sets

$$A_n = \left\{|u_n| > \|u_n\|_1 2^{-2n}\right\}, \quad A'_n = \bigcup_{j>n} A_j, \quad \text{and} \quad L_n = A_n \setminus A'_n.$$

Clearly, we have

$$\sup_{t\in[0,1]\setminus A_n} |u_n(t)| \le 2^{-2n}\|u_n\|_1 \le MC2^{-n}. \tag{4.5.13}$$

Since $u_n/\|u_n\|_1 \in \operatorname{conv} C_n$, the central distributional estimates (4.4.27), (4.4.29), and (4.4.28) yield

$$|A_n| \le 2^{-n-1}|B_n| \le 2^{-3q(n-1)-3n-3}, \qquad |A'_n| \le 2^{-3q(n-1)-3n-2},$$

and

$$\|u_n\|_\infty \le 2^{3q(n)}\|u_n\|_1 \le 2^{3q(n)+n}MC.$$

First, we fix $i \le n$. By combining the measure estimate for A'_n with the uniform estimates for u_i, $i \le n$, we get

$$\int_{A'_n} |u_i|dt \le MC2^{3q(i)+i+2}\left|A'_n\right| \le MC2^{-2n}, \tag{4.5.14}$$

If, moreover, $i < n$ then (4.5.14) yields that $\int_{A_n} |u_i|dt \le MC2^{-2n}$.

Next, fix $i > n$. We get $\sup_{t\subset L_n} |u_i(t)| \le 2^{-2i}\|u_i\|_1$, and in view of the estimate (4.5.11) we have $\|u_i\|_1 \le 2^i MC$. Consequently,

$$\int_{L_n} |u_i|dt \le 2^{-i}MC. \tag{4.5.15}$$

Taking the sum over $i \ne n$, separating the cases $i > n$, where we use (4.5.15), and $i < n$, where we use (4.5.14), gives

$$\int_{L_n}\left|\sum_{i\ne n} u_i\right| dt \le 2^{-n}MC.$$

It follows that

$$\int_{L_n} |u_n|dt \le \int_{L_n} |u|dt + \int_{L_n}\left|\sum_{i\ne n} u_i\right|dt \le \int_{L_n} |u|dt + MC2^{-n}. \tag{4.5.16}$$

Since the sets $\{L_n\}$ are pairwise disjoint, summing over n in Inequality (4.5.16) gives

$$\sum_{n=1}^{\infty}\int_{L_n} |u_n|dt \le \int_0^1 |u|dt + MC. \tag{4.5.17}$$

Decomposing $[0,1] = ([0,1]\backslash A_n) \cup L_n \cup A'_n$, and invoking (4.5.13) and (4.5.14) gives

$$\|u_n\|_1 \leq \int_{L_n} |u_n| dt + MC2^{-n}.$$

Summing over n and taking into account condition (4.5.17) shows that (4.5.12) holds true. □

By Lemma 4.5.2, the space X_n is isomorphic to ℓ^1. By Theorem 4.5.3 X is isomorphic to $(\sum X_n)_1$. Finally, we show that L^1 is not isomorphic to a subspace of X.

Theorem 4.5.4 *The space* $X = \overline{\text{span}}^{L^1}\{X_n : n \in \mathbb{N}\}$ *does not contain a copy of* L^1.

Proof By Theorem 4.5.3, the space X is isomorphic to $(\sum X_n)_1$. Let V be a bounded operator on L^1 with values in $(\sum X_n)_1$. For $N \in \mathbb{N}$, we define

$$X^N = \left(\sum_{n=1}^{N} X_n\right)_1$$

and regard it as a subset of $\left(\sum_{n=1}^{\infty} X_n\right)_1$. Let $\epsilon > 0$. Lemma 4.3.11 asserts that there exists an operator W, a subset $A \subseteq [0,1]$ with $|A| > 1 - \epsilon$, and $N \in \mathbb{N}$ such that the following diagramm is commutative

$$\begin{array}{ccccc} L^1(A) & \xrightarrow{\;i\;} & L^1([0,1]) & \xrightarrow{\;V\;} & (\sum_1^\infty X_n)_1 \\ & {\scriptstyle W}\searrow & & \nearrow{\scriptstyle I} & \\ & & X^N & & \end{array} \tag{4.5.18}$$

where i and, respectively, I denote the formal inclusion operators. Moreover,

$$\|W(x) - V(x)\|_{(\sum X_n)_1} \leq \epsilon \|x\|_1,$$

for each $x \in L^1(A)$. On the other hand, if V is an embedding operator, and $\epsilon > 0$ is taken small enough, there exist $c > 0$ and $C < \infty$, such that

$$c\|x\|_1 \leq \|W(x)\|_{(\sum X_n)_1} \leq C\|x\|_1,$$

for $x \in L^1(A)$. Equivalently, $W(L^1(A))$ contains a copy of L^1. By Lemma 4.5.2, each of the spaces X_n is isomorphic to ℓ^1, and hence the same is true for X^N. Since we have $W(L^1(A)) \subseteq X^N$, we obtain a contradiction. □

4.5.2 Quotients

In this section we prove that L^1/X does not contain a subspace isomorphic to L^1, thereby completing the proof of Talagrand's theorem (Theorem 4.5.1). Moreover, we show that there exist two Banach spaces T_1 and T_2 such that L^1 is isomorphic to a *subspace* of $T_1 \oplus T_2$, and neither T_1 nor T_2 contain a copy of L^1. These examples, due to Talagrand, show the sharpness of Enflo's results presented in Section 4.2.1.

We rely heavily on the spaces Y, E, and X introduced in Section 4.4 and Section 4.5.1.

Review of Y, E and X: Recall that in the closing paragraphs of Section 4.4.2, for $n \in \mathbb{N}$, we set

$$Y_n = \overline{\operatorname{span}}^{L^1}\{f_{n,k} : k \in \mathbb{N}\},$$

where $(f_{n,k})_{k=1}^{\infty}$ is an enumeration of the countable set C_n defined in (4.4.30). Using $Y_0 = L^1$, we defined Y to consist of those sequences $g = (g_n)_{n=0}^{\infty}$ for which $g_n \in Y_n$ and

$$\|g\|_Y = \sum_{n=0}^{\infty} 2^{-n}\|g_n\|_{L^1} < \infty. \tag{4.5.19}$$

We noted that the summing operator $T(g_n) = \sum g_n$ is well defined on Y. Setting $E = TY$, we defined E as a space of measurable functions, endowed with the quotient norm

$$\|f\|_E = \inf\{\|g\|_Y : f = T(g)\}.$$

Thus E is a quotient of Y. Finally, in definition (4.5.1) we introduced $X = \overline{\operatorname{span}}^{L^1}\{g_{n,k} : n, k \in \mathbb{N}\}$, where

$$g_{n,k} = f_{n,k} + 2^{-n}1_{A(n,k)}/|A(n,k)|$$

is defined by (4.5.2). We let Z denote the closure of X in E, that is $Z = \overline{X}^E$.

Overview of the proof that L^1 doesn't embed into L^1/X: We briefly describe the approach to Theorem 4.5.11, which shifts the analysis from L^1/X to E/Z. We recall that E itself is the closure of L^1 in E or, equivalently, $E = \overline{L^1}^E$. Hence obviously $\overline{L^1}^E/\overline{X}^E = E/Z$. In this section, the main theorems assert the following:

(i) The spaces L^1/X and E/Z are isomorphic (Theorem 4.5.5).
(ii) L^1 is not isomorphic to a subspace of E/Z (Theorem 4.5.9).

The proof of Theorem 4.5.9 proceeds by two stages of reduction to Theorem 4.4.10:

(i) First, if L^1 was isomorphic to a subspace of E/Z, then L^1 is isomorphic to a subspace of E.
(ii) Second, if L^1 was isomorphic to a subspace of E, then the formal embedding $i\colon L^1 \to E$ fixes a copy of L^1, which contradicts Theorem 4.4.10.

Theorem 4.5.5 *The space L^1/X is isomorphic to E/Z. There exists $0 < C < \infty$ so that for $x \in L^1[0,1]$,*

$$C^{-1}\|x\|_{L^1/X} \le \|x\|_{E/Z} \le C\|x\|_{L^1/X}.$$

Proof Let $x \in L^1[0,1]$. Proving $\|x\|_{L^1/X} \le C\|x\|_{E/Z}$ amounts to showing that for any $y_1 \in X$ there exists $y_2 \in X$ such that

$$\|x - y_2\|_{L^1} \le C\|x - y_1\|_E. \tag{4.5.20}$$

To this end we resolve the quotient norm $\|x - y_1\|_E$ and select $g = (g_n)_{n=0}^{\infty} \in Y$ such that $x - y_1 = g_0 + \sum_{n=1}^{\infty} g_n$ and

$$\|g\|_Y \le 2\|x - y_1\|_E.$$

Since $(g_n)_{n=0}^{\infty} \in Y$, the central distributional estimates (4.4.27), (4.4.29), (4.4.28), and (4.5.19) imply that the L^0-limit $\sum g_n$ exists and is integrable. By (4.5.10), this gives

$$\sum_{n=0}^{\infty} \|g_n\|_1 \le C\left\|\sum_{n=0}^{\infty} g_n\right\|_1 < \infty.$$

Select $N \in \mathbb{N}$ large enough so that $\sum_{n=N}^{\infty} \|g_n\|_1 < \|g_0\|_1/2$. Comparing definitions (4.4.31) and (4.5.2) shows that there exist $b_n \in L^1$ with $\|b_n\|_1 \le 1$, and $x_n \in X$ such that

$$g_n = b_n\|g_n\|_1 2^{-n} + x_n.$$

This gives the splitting

$$\sum_{n=0}^{\infty} g_n = g_0 + \widetilde{g_0} + \left(\sum_{n=1}^{N} b_n\|g_n\|_1 2^{-n}\right) + \widetilde{x}, \tag{4.5.21}$$

where $\widetilde{x} = \sum_{n=1}^{N} x_n$ and $\widetilde{g_0} = \sum_{n=N+1}^{\infty} g_n$. Clearly, $\widetilde{x} \in X$, and $\|\widetilde{g_0}\|_1 \le \|g_0\|_1/2$. Hence,

$$\|(x - y_1) - \widetilde{x}\|_{L^1} \le 2\|g\|_Y.$$

Now set $y_2 = y_1 - \widetilde{x}$ to obtain (4.5.20). Proving that $\|x\|_{E/Z} \le C\|x\|_{L^1/X}$ is routine: It amounts to showing that for $y_1 \in X$ there exists $y_2 \in X$ such that

$$\|x - y_2\|_E \le C\|x - y_1\|_{L^1}.$$

Here, however, it suffices to take $y_2 = y_1$, and to use that obviously $\|w\|_E \leq \|w\|_{L^1}$ for any $w \in L^1$. □

We prove next that *any* bounded linear operator $V : L^1 \to Y/\ker T$ factors canonically through Y, using the ideas of the Ghoussoub and Rosenthal (1983), introduced earlier in the proof of Theorem 4.3.3. We make the argument work in the present context by analyzing, in Lemma 4.5.7 below, the space $\operatorname{Ker} T = \{g \in E : T(g) = 0\}$, where T is the summing operator defined by (4.4.32).

Theorem 4.5.6 *Every bounded linear operator $V : L^1 \to E$ factors boundedly through the summing operator $T : Y \to E$, i.e., there exists a bounded operator $J : L^1 \to Y$ so that*

$$V = TJ.$$

Proof We denote by $\mathcal{I}$ the collection of dyadic intervals contained in the unit interval, and let $\mathcal{I}_k = \{I \in \mathcal{I} : |I| = 2^{-k}\}$ for $k \in \mathbb{N}$.

Fix a bounded linear operator $V : L^1 \to E$. Define $a_I = V(1_I/|I|)$ for $I \in \mathcal{I}$. Then $\|a_I\|_E \leq \|V\|$. Since E is a quotient of Y, there exists $b_I \in Y$, with $\|b_I\|_Y \leq 2\|V\|$ and $T(b_I) = a_I$.

Fix $k \in \mathbb{N}$, $I \in \mathcal{I}$ with $2^{-k} \leq |I|$, and form $\mathcal{I}_k(I) = \{K \subseteq I : K \in \mathcal{I}_k\}$. Then

$$w_{I,k} = \frac{1}{|I|} \sum_{K \in \mathcal{I}_k(I)} |K| b_K \tag{4.5.22}$$

satisfies $T(w_{I,k}) = a_I$. Clearly if $I = I_1 \cup I_2$ with $I_1, I_2 \in \mathcal{I}$, then $|I_1| = |I_2| = |I|/2$ and if, moreover, $2^{-k} \leq |I_1|$, then

$$w_{I,k} = (w_{I_1,k} + w_{I_2,k})/2. \tag{4.5.23}$$

The sequences $\{w_{I,k} : k \in \mathbb{N}\}$ satisfy the hypothesis of Lemma 4.5.7. Let $k_m \uparrow \infty$ be the subsequence, and $w_I \in Y$ be the limit of the sequence $\{w_{I,k_m} : m \in \mathbb{N}\}$ obtained by means of Lemma 4.5.7. By definition (4.5.23) the limits $w_I \in Y$ satisfy

$$w_I = (w_{I_1} + w_{I_2})/2, \quad T(w_I) = a_I, \quad \|w_I\|_Y \leq 4\|V\|. \tag{4.5.24}$$

Next, define the map $J(1_I/|I|) = w_I$. By (4.5.24), the linear extension of J to $\operatorname{span}\{1_I : I \in \mathcal{I}\}$ is well defined and satisfies

$$TJ(1_I/|I|) = T(w_I) = a_I = V(1_I/|I|). \tag{4.5.25}$$

Finally, we show that J has a bounded extension to L^1. To this end we fix a collection of pairwise disjoint dyadic intervals $\mathcal{E}$, and coefficients $x_I \in \mathbb{R}, I \in \mathcal{E}$. By (4.5.25), and the triangle inequality,

$$\left\| J\left(\sum_{I \in \mathcal{E}} x_I 1_I\right)\right\|_Y \leq \|V\| \cdot \left\|\sum_{I \in \mathcal{E}} |x_I| \cdot 1_I\right\|_1 . \tag{4.5.26}$$

Since step functions are dense in L^1, the estimates (4.5.26) imply that the linear extension of the map J extends boundedly and uniquely to all of L^1. □

We now provide the required information on the factoring space $\{T(g) = 0\}$. Lemma 4.5.7 is used as a substitute for the compactness hypothesis in the original Ghoussoub–Rosenthal theorem (Theorem 4.3.3).

Lemma 4.5.7 *Let $z \in E$ and $g^{(k)} \in Y$ satisfy*

$$Tg^{(k)} = z, \qquad \left\|g^{(k)}\right\|_Y \leq 2\|z\|_E, \quad k \in \mathbb{N}.$$

Then there exists a sequence $k_m \uparrow \infty$ and $g \in Y$ with $g = (g_0, g_1, \dots)$ and $\|g\|_Y \leq 4\|z\|_E$ such that

$$g_n^{(k_m)} \longrightarrow g_n, \qquad \text{in the weak topology} \quad \sigma(L^1, L^\infty),$$

for $n \geq 1$. Moreover, $z - \sum_{n=1}^{p} g_n \longrightarrow g_0$ in measure, as $p \to \infty$, and hence $Tg = z$.

Proof We move to the front a consequence of Egorov's theorem on almost uniform convergence. Let (h_r) be a sequence in L^1 with $\|h_r\|_1 \leq A$. Let $g\colon [0,1] \to \mathbb{R}$ be measurable. If (h_r) converges to g in measure, i.e., if

$$\lim_{r\to\infty} |\{|h_r - g| > \epsilon\}| = 0,$$

for any $\epsilon > 0$, then $\|g\|_1 \leq A$.

Step 1 (Selecting the subsequence (k_n)) Assume $\|z\|_E < 1$. Then $\left\|g_n^{(k)}\right\|_1 \leq 2^{n+1}$ for any $n \in \mathbb{N} \cup \{0\}$. Next, define the set $M_n = \{f \in Y_n : \|f\|_1 \leq 2^{n+1}\}$. Clearly M_n is convex and closed in the L^1-norm. Hence, by Mazur's theorem, M_n is closed with respect to the weak topology $\sigma(L^1, L^\infty)$. By estimate (4.4.27), the set M_n is bounded in L^∞, and is hence uniformly integrable and relatively compact for the $\sigma(L^1, L^\infty)$ topology. By a Cantor diagonal sequence argument, there exists a sequence $k_m \uparrow \infty$ and there exist $g_n \in M_n$ such that

$$g_n^{(k_m)} \longrightarrow g_n \qquad \text{in the weak topology} \quad \sigma(L^1, L^\infty).$$

Since $g_n \in M_n$, we have $\|g_n\|_1 \leq 2^{n+1}$. Hence by the central estimates (4.4.27) and (4.4.28), the partial sums $\sum_{n=1}^{p} g_n$ converge in measure. Set

$$g_0 = z - \sum_{n=1}^{\infty} g_n. \tag{4.5.27}$$

Step 2 (Bounding g_0) It remains to prove that $g_0 \in L^1$. To this end we fix $r \in \mathbb{N}$ and, using Mazur's theorem, we choose $\alpha_m = \alpha_m(r) \geq 0$, with $\sum \alpha_m \leq 1$ such that

$$\max_{n\leq r} \left\| g_n - \sum_{m=1}^{\infty} \alpha_m g_n^{(k_m)} \right\|_1 \leq 2^{-r}.$$

Put $h_r = \sum_{m=1}^{\infty} \alpha_m g_0^{(k_m)}$. We have $\|h_r\|_1 \leq 2$. Invoking the hypothesis that $Tg^{(k)} = z$ gives

$$g_0^{(k_m)} = z - \sum_{n=1}^{\infty} g_n^{(k_m)}. \tag{4.5.28}$$

Multiplying this by α_m, taking the sum over m, and substracting the resulting series from the series (4.5.27) gives the expansion

$$h_r - g_0 = \sum_{1\leq n\leq r} \left(g_n - \sum_{m=1}^{\infty} \alpha_m g_n^{(k_m)} \right) + \sum_{n>r} g_n - \sum_{n>r} \sum_{m=1}^{\infty} \alpha_m g_n^{(k_m)}. \tag{4.5.29}$$

Exploiting (4.5.29) we deduce that $g_0 - h_r \longrightarrow 0$ in measure as $r \to \infty$. To see this we treat the three series separately. First, the assertion that the partial sums $\sum_{n=1}^{r} g_n$ converge in measure yields that $\sum_{n>r} g_n$ converges to 0, in measure, as $r \to \infty$. Second, we have

$$\left\| \sum_{n\leq r} \left(g_n - \sum_{m=1}^{\infty} \alpha_m g_n^{(k_m)} \right) \right\|_1 \leq r 2^{-r},$$

and, third, the choice of Y_n and Talagrand's central estimates (4.4.27) and (4.4.28) yield $\lim_{r\to\infty} \left| \left\{ \left| \sum_{n>r} \sum_{m=1}^{\infty} \alpha_m g_n^{(k_m)} \right| > \epsilon \right\} \right| = 0$. Putting it together, we obtain that

$$\lim_{r\to\infty} |\{|g_0 - h_r| > \epsilon\}| = 0.$$

Recall that $\|h_r\|_1 \leq 2$ and that, by Egorov's theorem, the norm-closed unit ball of L^1 is closed in L^0. This gives $\|g_0\|_1 \leq 4$. □

Theorem 4.5.8 *The space $(E, \|\cdot\|_E)$ does not contain a copy of L^1.*

Proof We prove the following implication: If $(E, \|\cdot\|_E)$ contains a copy of L^1, that is, if there exists a bounded linear operator $V\colon L^1 \to E$ and $\delta > 0$ satisfying

$$\|Vf\|_E > \delta \|f\|_{L^1}, \qquad f \in L^1,$$

then the inclusion operator $i\colon L^1 \to E$, $i(f) = f$, fixes a copy of L^1. The latter condition, however, contradicts Theorem 4.4.10.

Step 1 By Theorem 4.5.6, there exists a bounded linear operator $J\colon L^1 \to Y$, such that the following diagram commutes:

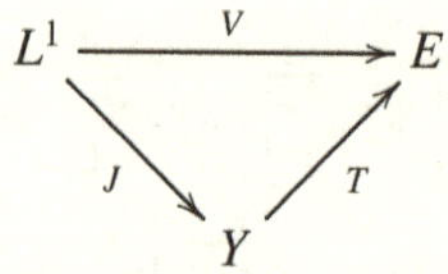

(4.5.30)

Exploiting that $V = TJ$ gives the following chain of inequalities,

$$\|f\|_{L^1} \geq c_1\|J(f)\|_Y \geq c_1\|TJ(f)\|_E = c_1\|V(f)\|_E \geq (c_1\delta)\|f\|_{L^1}, \tag{4.5.31}$$

for $f \in L^1$. We interpret the chain (4.5.31) by saying that both $J(L^1) \subseteq Y$ and $TJ(L^1) \subseteq E$ are isomorphic to L^1 and that $T\colon Y \to E$ acts as an isomorphism between $J(L^1)$ and $TJ(L^1)$. Thus we showed that T fixes a copy of L^1.

Step 2 By Lemma 4.3.11, for any $\epsilon > 0$ there exists a measurable set $A \subseteq [0, 1]$ with $|A| > 1/2$, $N \in \mathbb{N}$, and a bounded linear operator $J_1\colon L^1 \to Y$, such that

$$\|J_1(f) - J(f)\|_Y \leq \epsilon\|J(f)\|_Y, \tag{4.5.32}$$

for $f \in L^1(A)$, and the inclusion $J_1(g) \in \left(\sum_{n=0}^N Y_n\right)_1$ holds for every $g \in L^1$. If we take $\epsilon > 0$ to be small enough, then by (4.5.32), for $f \in L^1(A)$,

$$\|f\|_{L^1} \geq c_1\|J_1 f\|_Y \geq c_2\|TJ_1(f)\|_E \geq c_3\|f\|_{L^1}. \tag{4.5.33}$$

Since J_1 ranges in $\left(\sum_{n=0}^N Y_n\right)_1$, the space $TJ_1(L^1(A))$ is a subspace of L^1, which by (4.5.33) is isomorphic to its ambient space L^1. In summary, letting $j\colon L^1(A) \to L^1$ denote the inclusion map, $j(f) = f$, we showed that the following diagram is commutative

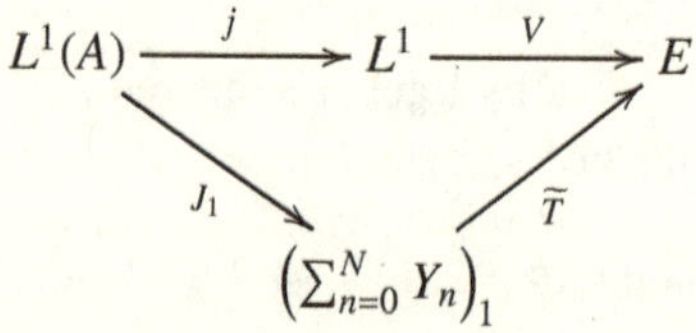

(4.5.34)

where $\widetilde{T}$ is a small perturbation of the summing operator T.

Step 3 We claim next that the formal identity $i\colon L^1 \to E$ given by $i(f) = f$ is invertible on $TJ_1(L^1(A))$. Let $P_n\colon Y \to Y_n$ be the canonical coordinate projection and put $J_{1n} = P_n J_1$. By the Inequalities (4.5.31) and (4.5.33), for $f \in L^1(A)$ and $\epsilon > 0$ small enough,

$$\begin{aligned}\|TJ_1(f)\|_E &\geq \|TJ(f)\|_E - \|TJ(f) - TJ_1(f)\|_E \\ &\geq c_4\|J(f)\|_Y - \epsilon\|J(f)\|_Y \\ &\geq c_5\|J_1(f)\|_Y.\end{aligned}$$

Since $\sum_{n=0}^{N} J_{1n}(f) = TJ_1(f)$ and $\left\|J_1(f)\right\|_Y \geq 2^{-N}\left\|\sum_{n=0}^{N} J_{1n}(f)\right\|_{L^1}$, we obtain that

$$\|TJ_1(f)\|_E = \left\|\sum_{n=0}^{N} J_{1n}(f)\right\|_E \geq c_3 2^{-N}\left\|\sum_{n=0}^{N} J_{1n}(f)\right\|_{L^1} \geq c_3 2^{-N}\|TJ_1(f)\|_E,$$

for $f \in L^1(A)$. Since, by Step 2, $TJ_1\left(L^1(A)\right)$ is isomorphic to L^1, we showed that the inclusion map $i\colon L^1 \to E$ fixes a copy of L^1. Clearly, this contradicts Theorem 4.4.10. □

Recall that we denoted $Z = \overline{\text{span}}^E\{g_{n,k},\ n,k \in \mathbb{N}\}$. In Lemma 4.5.10 we will verify that Z is isomorphic to ℓ^1. In particular, $Z \subset E$ is a dual space. This will allow us to invoke the Ghoussoub–Rosenthal theorem (Theorem 4.3.3), to operators ranging in E/Z.

Theorem 4.5.9 *The quotient space E/Z does not contain a copy of L^1.*

Proof We prove that if E/Z contains a copy of L^1, then E also contains a copy of L^1, which contradicts Theorem 4.5.8. Let $V\colon L^1 \to E/Z$ be a bounded linear operator, for which there exists $\delta > 0$ so that

$$\|V(x)\|_{E/Z} > \delta\|x\|_1,$$

for $x \in L^1$. Let $q\colon E \to E/Z$ denote the quotient map, defined by $q(f) = f + Z$, for $f \in E$. By Lemma 4.5.10, the space Z is isomorphic to ℓ^1 and therefore a dual space. Hence Theorem 4.3.3 applies to the quotient E/Z. Thus we obtain a bounded linear operator $J\colon L^1 \to E$ satisfying $V = qJ$. Hence,

$$\|x\|_1 \geq c_1\|J(x)\|_E \geq c_1\|qJ(x)\|_{E/Z} = c_1\|V(x)\|_{E/Z} > c_2\|x\|_1, \qquad x \in L^1.$$

In other words, $J\colon L^1 \to E$ is an embedding and E contains a copy of L^1, contradicting Theorem 4.5.8. □

The argument above used that $Z = \overline{\text{span}}^E\{g_{n,k},\ n,k \in \mathbb{N}\}$ is isomorphic to ℓ^1. We show next that more is true, namely that in E, the sequence $(2^n g_{n,k})$ is equivalent to the unit vector basis of ℓ^1.

Lemma 4.5.10 *There exists $C < \infty$ such that*

$$\sum_{n,k=1}^{\infty} 2^{-n}|a_{n,k}| \leq C\left\|\sum_{n,k=1}^{\infty} a_{n,k}g_{n,k}\right\|_E \leq C\sum_{n,k=1}^{\infty} 2^{-n}|a_{n,k}|, \tag{4.5.35}$$

for any finitely supported sequence of scalars $a_{n,k}$. In E, the sequence $(2^n g_{n,k})$ is equivalent to the unit vector basis of ℓ^1.

Proof Let $(a_{n,k})$ be a finitely supported sequence of scalars and put $h = \sum_{n,k=1}^{\infty} a_{n,k} g_{n,k}$. Assuming the normalization $\|h\|_E < 1$, we will show that

$$\sum_{n,k=1}^{\infty} |a_{n,k}| 2^{-n} \le C. \tag{4.5.36}$$

By definition of the norm in E, there exist $g_n \in Y_n$, $g_0 \in Y_0 = L^1$ such that

$$\sum_{n,k=1}^{\infty} a_{n,k} g_{n,k} = \sum_{n=0}^{\infty} g_n, \quad \text{and} \quad \sum_{n=0}^{\infty} 2^{-n} \|g_n\|_1 < 1. \tag{4.5.37}$$

Recall the definition of C_n in (4.4.28). Since $g_n/\|g_n\|_1$ is in the closed absolutely convex hull of C_n, there exist $b_{n,k} \in [-1, 1]$ such that

$$g_n/\|g_n\|_1 = \sum_{k=1}^{\infty} b_{n,k} f_{n,k} \quad \text{and} \quad \sum_{k=1}^{\infty} |b_{n,k}| \le 1. \tag{4.5.38}$$

We replace next $f_{n,k}$ by $g_{n,k}$ and then use the series (4.5.4) for expansions with respect to the $g_{n,k}$ system. To this end, we define $f'_n = -\sum_{k=1}^{\infty} 2^{-n} b_{n,k}/|A_{n,k}| 1_{A_{n,k}}$ so that

$$g_n/\|g_n\|_1 = f'_n + \sum_{k=1}^{\infty} b_{n,k} g_{n,k}.$$

Thus,

$$\sum_{n=0}^{\infty} g_n = \sum_{n=0}^{\infty} f'_n \|g_n\|_1 + \sum_{n,k=1}^{\infty} b_{n,k} \|g_n\|_1 g_{n,k}. \tag{4.5.39}$$

Comparing the series (4.5.39) with the expansion in (4.5.37) gives that

$$\sum_{n,k=1}^{\infty} (a_{n,k} - b_{n,k} \|g_n\|_1) g_{n,k} = f' - g_0,$$

where $f', g_0 \in L^1$, with $\|g_0\|_1 \le 1$, and $\|f'\|_1 \le 1$. By Lemma 4.5.2 we have the L^1 estimate

$$\sum_{n,k=1}^{\infty} |a_{n,k} - b_{n,k} \|g_n\|_1| 2^{-n} \le C \|f' - g_0\|_1 \le 2C. \tag{4.5.40}$$

Now observe that, by (4.5.38) and (4.5.37), we know already that,

$$\sum_{n,k=1}^{\infty} |b_{n,k}| \|g_n\|_1 2^{-n} \le \sum_{n=1}^{\infty} \|g_n\|_1 2^{-n} \le 1.$$

Thus (4.5.40) yields condition (4.5.36), proving the left-hand side estimate of (4.5.35). The right-hand side estimate follows from the triangle inequality, and the fact that $\|g_{n,k}\|_E \leq C2^{-n}$. □

Finally, we summarize the above results and obtain the main theorem of this section.

Theorem 4.5.11 *The space L^1/X does not contain a copy of L^1.*

Proof Combine Theorem 4.5.9 and Theorem 4.5.5. □

The next result, due to Talagrand, should be compared and contrasted with Enflo's theorem asserting that if L^1 is isomorphic to $Y_1 \oplus Y_2$ then either Y_1 or Y_2 are isomorphic to L^1. (See Corollary 4.2.3.)

Theorem 4.5.12 *There exist two Banach spaces T_1 and T_2 such that L^1 is isomorphic to a subspace of $T_1 \oplus T_2$ and neither T_1 nor T_2 contain a copy of L^1.*

Proof As above, we let X_n denote the spaces defined by (4.5.2). Consider the L^1-subspaces,

$$X^{(1)} = \overline{\operatorname{span}}^{L^1}\{X_{2n} : n \in \mathbb{N}\} \quad \text{and} \quad X^{(2)} = \overline{\operatorname{span}}^{L^1}\{X_{2n+1} : n \in \mathbb{N}\},$$

and form the respective quotient spaces,

$$T_1 = L^1/X^{(1)} \quad \text{and} \quad T_2 = L^1/X^{(2)}.$$

Define $q_i \colon L^1 \to T_i$ by $q_i(f) = f + X^{(i)}$ and put $j(f) = (q_1(f), q_2(f))$. Clearly,

$$j \colon L^1 \to T_1 \oplus T_2$$

is bounded with $\|j\| \leq 2$. By condition (4.5.41) below, the operator j is invertible on its range $j(L^1)$. In other words, L^1 is isomorphic to a subspace of $T_1 \oplus T_2$. The proof of Theorem 4.5.11 shows that neither $T_1 = L^1/X^{(1)}$ nor $T_2 = L^1/X^{(2)}$ contain a copy of L^1. We turn to proving that there exists $c > 0$ such that

$$\|q_1(f)\|_{T_1} + \|q_2(f)\|_{T_2} \geq c\|f\|_1, \tag{4.5.41}$$

for any $f \in L^1$. Let $C < \infty$ be the constant arising from condition (4.5.10). Fix $\epsilon > 0$ and let $f \in L^1$. Resolving the norm of the quotient spaces T_1 and, respectively, T_2 we obtain $x_1 \in X^{(1)}$ and $x_2 \in X^{(2)}$ such that $\|f - x_i\|_1 < \|q_i(f)\|_{T_i} + \epsilon$ for $i \in \{1, 2\}$. Applying the triangle inequality gives,

$$\|x_1 - x_2\|_1 < \|q_1(f)\|_{T_1} + \|q_2(f)\|_{T_2} + 2\epsilon.$$

Now apply (4.5.10) to $h = x_1 - x_2$. This gives $\|x_1\|_1 \leq C\|x_1 - x_2\|_1$ and hence,

$$\|f\|_1 \leq \|f - x_1\|_1 + \|x_1\|_1 < (C + 1)\|q_1(f)\|_{T_1} + C\|q_2(f)\|_{T_2} + (2C + 1)\epsilon.$$

Choosing $\epsilon > 0$ small enough gives (4.5.41) with $c = 1/(C + 1)$. □

4.6 Notes

Section 4.1 is devoted to the representation theorems of Kwapień (1973) and Kalton (1978).

Section 4.1.1 presents Kwapień's pioneering work on the intrinsic representation of continuous operators on L^0 (Theorem 4.1.3, Theorem 4.1.4, and Theorem 4.1.5). The proof of Proposition 4.1.2 contains the fruitful ideas of random selectors and decoupling estimates. It forms the core of Kwapień's work, published in 1973(!).

Corollaries 4.1.6 and 4.1.7 are important consequences of Kwapień's representation theorem. For their proof, our source was Kalton (1978).

Section 4.1.2 contains Kalton's work on the representation of L^1 operators. Theorem 4.1.15, the main result, is due to Kalton (1978).

We reviewed basic material on representing measures using Edgar (1975), Fakhoury (1979), Kalton (1978), and Weis (1984). Our presentation of measurable selection theorems is based on Stegall (1985), and the work of Kuratowski and Ryll-Nardzewski (1965) for Theorem 4.1.10, Diestel (1975) for Souslin sets and universally measurable sets, and Kalton (1978) for Theorem 4.1.13 and Theorem 4.1.15. For background we refer to Bogachev (2007).

Kwapień's conjecture (Kalton, 1985; Kwapień, 1970) provides a very interesting connection between the representations in Section 4.1.1 and Section 4.1.2.

Section 4.2 treats important embeddings of L^1. Subsection 4.2.1 is based on the papers by Kalton (1978, 1979, 1980). The proof of Enflo's theorem (Theorem 4.2.2) is taken in part from Capon (1980). Corollary 4.2.3 is due to Enflo, and solves a problem posed by Lindenstrauss and Pełczyński (1971). Maurey (1975a) contains Enflo's original proof providing the reduction to multiplication operators. In joint work, Lechner, Motakis, Müller, and Schlumprecht (2021, 2020) obtained a novel approach based on Haar multipliers in L^1. Recent, intrinsic characterizations of Haar multipliers in L^1 are due to Semenov and Uksusov (2012) and Wark (2015, 2017). The proof of Theorem 4.2.4 is from a private communication by L. Weis. Theorem 4.2.5 is due to Talagrand (1990). Bourgain (1980a, 1981) contains an example of a noncomplemented copy of ℓ^1 in L^1.

Subsection 4.2.4 presents the paper by Johnson, Maurey, and Schechtman (2007). For extensions, see Odell and Zheng (2010). The remarks concerning the history of the problem also incorporate Bessaga and Pełczyński (1958a), Maurey and Rosenthal (1977), and Maurey and Schechtman (1979).

Section 4.3 is based on Rosenthal (1984). It incorporates Rosenthal (1983) and Enflo and Starbird (1979). The quotient lifting argument of Theorem 4.3.4

uses Ghoussoub and Rosenthal (1983). The submartingale argument in the proof of Lemma 4.3.11 is due to Talagrand (1990). For more recent developments, see Dosev et al. (2013), Godefroy et al. (2000), Liu (1998), and Maslyuchenko et al. (2009).

The results in Sections 4.4 and 4.5 are due to Talagrand (1990). The replacement strategy for the Haar packets (4.4.12), detailed in Lemma 4.4.4 and Proposition 4.4.5, corresponds *in spirit* to the use of Carleson's "allowed pairs" and polynomials (Carleson, 1966, Sections 7,8; Hunt, 1968 Section 6,7). The proof of Dubin's lemma was taken from the book by Neveu (1975). The constructions in Section 4.5 reflect the seminal ideas of Kislyakov's lemma (Lemma 7.1.1).

The book *Narrow operators on function spaces and vector lattices* by Popov and Randrianantoanina (2013), presents the main theorems discussed in Sections 4.3 through 4.5.

5

Isomorphic Invariants

The remaining chapters of the book present vector-valued Hardy martingales with applications to the geometry and classification of complex Banach spaces. We will investigate – in considerable depth and detail – important isomorphic invariants derived from the study of Hardy martingales ranging in Banach spaces. Their convergence, unconditional convergence, and martingale inequalities tied to the notions of complex uniform convexity are recurrent topics in the chapters to come.

Section 5.1 is devoted to proving the dimension conjecture for polydisk algebras. We present Bourgain's investigation of multi-indexed Hardy martingales, resulting in deep decay-estimates for their Steinhaus coefficients. Thus we exhibit the sequence of isomorphic invariants by which Bourgain proved that the Banach spaces $A(\mathbb{T}^m)$ and $A(\mathbb{T}^n)$ are isomorphic if and only if $m = n$.

In Section 5.2 we discuss Banach spaces with the analytic Radon–Nikodym property (aRNP), and give equivalent conditions in terms of bounded vector-valued analytic function, integrable Hardy martingales, representable operators, and factorization through L^1/H^1. This line of development culminates in the real variable characterization of the aRNP by the plurisubharmonic variational principle.

In Section 5.3 we introduce quantitative measures of convergence for vector-valued Hardy martingales, including Hardy martingale cotype, H^1-trees in Banach spaces, and the concept of H^1-uniform convexity. We present three theorems due to Xu: the dichotomy theorem, the subsequence theorem, and the renorming theorem.

In Section 5.4 we investigate the unconditional convergence of vector-valued Hardy martingales and Banach spaces that are unconditional for Hardy martingale differences (the aUMD). We prove Kwapień's probabilistic characterization of Banach spaces containing c_0 and present three theorems by Garling on Hardy martingales ranging in Banach spaces with aUMD.

5.1 Isomorphic Classification of Polydisk Algebras

The polydisk algebra $A(\mathbb{T}^d)$, as introduced in Section 1.1.4, is defined to be the norm closure of the analytic polynomial span $\{z^n : n \in \mathbb{N}_0^d, z \in \mathbb{T}^d\}$ in $C(\mathbb{T}^d)$. In this section we present Bourgain's isomorphic classification of the polydisk algebras (Bourgain, 1984b). The proof exploits Pisier's L^1/H^1-valued Riesz products.

Theorem 5.1.1 *If $m \neq n$ then the polydisk algebras $A(\mathbb{T}^m)$ and $A(\mathbb{T}^n)$ are not isomorphic as Banach spaces. Moreover, if $m < n$ then the dual space $A^*(\mathbb{T}^n)$ is not isomorphic to a subspace of $A^*(\mathbb{T}^m)$.*

Open problems: Thorough discussions of the dimension conjecture beyond the class of polydisk algebras were given by Pełczyński (1977, Chapter 11) and Bourgain (1984b). We selected two of the most outstanding conjectures in the area:

(i) For $1 < n \neq d < \infty$, the ball algebra $A(\mathbb{B}_d)$ is not isomorphic to $A(\mathbb{B}_n)$. The method in Ryll and Wojtaszczyk (1983) and Wojtaszczyk (1984) provides a very promising point of entry.

(ii) The spaces $H^\infty(\mathbb{T}^2)$ and $H^\infty(\mathbb{T}^3)$ are not isomorphic. More generally, for $1 < n \neq d < \infty$ the space $H^\infty(\mathbb{T}^n)$ is not isomorphic to $H^\infty(\mathbb{T}^d)$.

The Isomorphic Invariant

We introduce multiindexed Hardy martingales, and use their Steinhaus coefficients to define isomorphic invariants by which we obtain the isomorphic classification the polydisk algebras.

Let $\Omega = \mathbb{T}^{\mathbb{N}}$ and let $\mathbb{P}$ denote the Haar probability measure on $\mathbb{T}^{\mathbb{N}}$. We let X be a complex Banach space, and fix $d \in \mathbb{N}$. For $K \in \mathbb{N}^d$, assume that $F_K : \Omega^d \to X$ is integrable. We then say that the sequence $(F_K\colon K \in \mathbb{N}^d)$ is an X-valued Hardy martingale of type (d) if, for $z \in \Omega$, $1 \leq m \leq d$, and any choice of

$$K_1 \in \mathbb{N}^{m-1}, K_2 \in \mathbb{N}^{d-m} \quad \text{and} \quad w_1 \in \Omega^{m-1}, w_2 \in \Omega^{d-m},$$

the sequence

$$F_k(z) = F_{(K_1,k,K_2)}(w_1, z, w_2), \quad k \in \mathbb{N},$$

defines an X-valued Hardy martingale satisfying

$$\int_\Omega F_k(z)d\mathbb{P}(z) = 0.$$

In short, $F = (F_K\colon K \in \mathbb{N}^d)$ is a Hardy martingale with respect to the components in Ω^d separately.

Steinhaus coefficients: Let $F = (F_K\colon K \in \mathbb{N}^d)$ be a Hardy martingale of type (d). Fix $K = (k_1, \ldots, k_d) \in \mathbb{N}^d$. For $v = (v^{(1)}, \ldots, v^{(d)}) \in \Omega^d$ with $v^{(i)} = \left(v_k^{(i)}\right)$, set $v_K = \prod_{i=1}^d v_{k_i}^{(i)}$. We define the Steinhaus coefficient $x_K \in X$ by

$$x_K = \int_{\Omega^d} F_K(v)\overline{v_K} d\mathbb{P}_d(v), \tag{5.1.1}$$

where $\mathbb{P}_d$ denotes the Haar probability measure on Ω^d.

Set

$$S(d, F, X) = \inf\left\{\|x_K\|_X\colon K \in \mathbb{N}^d\right\}, \tag{5.1.2}$$

and define

$$S(d, X) = \sup\{S(d, F, X)\colon F\},$$

where the supremum is taken over all X-valued Hardy martingales of type (d), $F = (F_K\colon K \in \mathbb{N}^d)$, satisfying,

$$\sup_{K\in\mathbb{N}^d} \int_{\Omega^d} \|F_K(v)\|_X d\mathbb{P}_d(v) \leq 1. \tag{5.1.3}$$

Proposition 5.1.2 *Let X, Y be Banach spaces. If there exists $d \in \mathbb{N}$ such that*

$$S(d, X) = 0 \quad \text{and} \quad S(d, Y) > 0,$$

then Y is not isomorphic to a subspace of X.

Proof We assume to the contrary, that there exists a linear injection $i\colon Y \to X$ and $0 < c \leq C < \infty$, such that

$$c\|y\|_Y \leq \|i(y)\|_X \leq C\|y\|_Y, \qquad y \in Y.$$

Since $S(d, Y) > 0$, there exists $\eta > 0$ and an L^1-bounded Y-valued Hardy martingale (F_K) of type (d), such that its Steinhaus sequence (y_K) satisfies

$$\|y_K\|_Y > \eta, \tag{5.1.4}$$

for any $K \in \mathbb{N}^d$. The embedding $i\colon Y \to X$ allows us to define an L^1-bounded, X-valued Hardy martingale (G_K) of type (d) by putting

$$G_K(u) = i(F_K(u)),$$

for $K \in \mathbb{N}^d$ and $u \in \Omega^d$.

Let (x_K) be the Steinhaus sequence of (G_K). Since $S(d, X) = 0$, for any $\epsilon > 0$ there exists $K \in \mathbb{N}^d$ such that $\|x_K\|_X < \epsilon$. Conversely, we have $x_K = i(y_K)$, and hence $c\|y_K\|_Y \leq \|x_K\|_X$ for any $K \in \mathbb{N}^d$. In view of condition (5.1.4), this gives a contradiction. □

Theorem 5.1.3 *Let $m, n, d \in \mathbb{N}$. If $m \leq n + 1$ and $m < d \leq n$, then*

$$S(d, A^*(\mathbb{T}^m)) = 0 \quad \text{and} \quad S(d, A^*(\mathbb{T}^n)) > 0. \tag{5.1.5}$$

Consequently, $A^*(\mathbb{T}^n)$ is not isomorphic to a subspace of $A^*(\mathbb{T}^m)$.

Proof Put $Y = A^*(\mathbb{T}^d)$. Below, in Theorem 5.1.17, we will prove that there exists an L^1-bounded, Y-valued Hardy martingale of type (d) with Steinhaus sequence (y_L), such that $\|y_L\|_Y \geq 1$, for any $L \in \mathbb{N}^d$. Hence

$$S\left(d, A^*(\mathbb{T}^d)\right) > 0.$$

Since $A(\mathbb{T}^d) \subseteq A(\mathbb{T}^n)$ for $n \geq d$, by the Hahn–Banach theorem, each functional on $A(\mathbb{T}^d)$ extends with equal norm to a functional in $A(\mathbb{T}^n)$. Consequently,

$$S(d, A^*(\mathbb{T}^n)) \geq S\left(d, A^*(\mathbb{T}^d)\right) > 0, \quad \text{for} \quad n \geq d. \tag{5.1.6}$$

See Figure 5.1. Put $X = A^*(\mathbb{T}^{d-1})$. By Theorem 5.1.18, any X-valued, type (d) Hardy martingale, F, satisfies

$$S\left(d, F, A^*(\mathbb{T}^{d-1})\right) = 0.$$

Hence, again by the Hahn–Banach theorem, we obtain

$$S(d, A^*(\mathbb{T}^m)) \leq S\left(d, A^*(\mathbb{T}^{d-1})\right) = 0, \quad \text{for} \quad m \leq d - 1. \tag{5.1.7}$$

See Figure 5.1. Combining Inequalities (5.1.6) and (5.1.7) yields Equation (5.1.5). □

Remarks:

(i) Combining Theorem 5.1.3 and Proposition 5.1.2, we obtain the isomorphic classification of the polydisk algebras as stated in Theorem 5.1.1. The main ingredients, Theorem 5.1.18 and Theorem 5.1.17, are proven in Section 5.1.3.

(ii) In Section 5.1.2 we treat the bidisk algebra $A(\mathbb{T}\times\mathbb{T})$ in considerable detail. We will show that

$$A^*(\mathbb{T} \times \mathbb{T}) \text{ is not finitely represented in } A^*(\mathbb{T}).$$

The isomorphic invariant, separating the finite-dimensional subspaces of $A^*(\mathbb{T})$ from those of $A^*(\mathbb{T} \times \mathbb{T})$, is given in terms of Hardy martingales of type (2).

(iii) In Section 5.1.3 we employ Hardy martingales of type $(d + 1)$ and show that $S\left(d + 1, A^*(\mathbb{T}^d)\right) = 0$ and $S\left(d + 1, A^*(\mathbb{T}^{d+1})\right) > 0$. Consequently,

$$A^*(\mathbb{T}^{d+1}) \text{ does not embed into } A^*(\mathbb{T}^d).$$

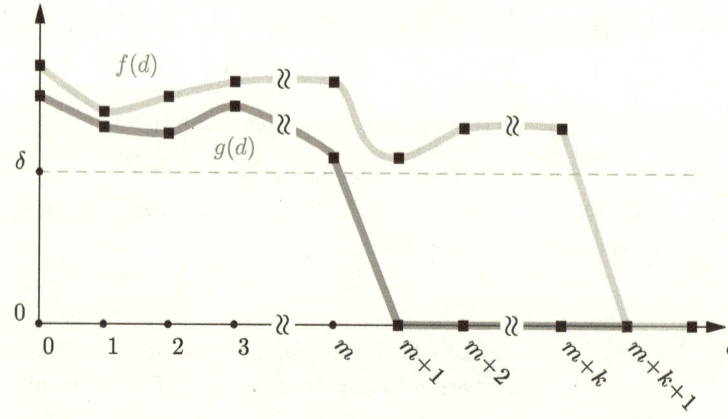

Figure 5.1 The isomorphic invariants $S(d, A^*(\mathbb{T}^n))$: With $g(d) = S(d, A^*(\mathbb{T}^m)$ and $f(d) = S(d, A^*(\mathbb{T}^{m+k})$, $g(d)$ is bounded by $f(d)$. Theorem 5.1.3 asserts that $g(d) = 0$ for $d \geq m+1$, $f(d) > 0$ for $d \leq m+k$, and that $A^*(\mathbb{T}^m)$ does not embed into $A^*(\mathbb{T}^{m+k})$ for $k \geq 1$.

5.1.1 Riesz Products and Rudin–Shapiro Martingales

Vector-valued Riesz products and Rudin–Shapiro martingales play an important role in the proof of Theorem 5.1.1. To prepare for the work in Section 5.1.2 and Section 5.1.3, we first provide a brief discussion of those subjects.

Riesz Products

Pisier (see Davis et al. [1984]) introduced L^1/H_0^1-valued Hardy martingales formed by Riesz products. Theorem 5.1.4, from Pisier's work, asserts that

$$S\left(1, L^1/H_0^1\right) \geq 1/2.$$

On the other hand, the Haagerup–Pisier theorem (Theorem 2.5.9) implies that

$$S(1, L^1) = 0.$$

Thus, by Proposition 5.1.2, Pisier's analysis of vector-valued Riesz products yields the well-known result of Pełczyński (1977, Proposition 4.1), that L^1/H_0^1 is not isomorphic to a subspace of L^1. Pełczyński's proof is based on the fact that there exists a noncompact absolutely summing operator on the disk algebra. However, more can be said using martingales: Applying Theorem 2.5.9 for $X = L^1$, we obtain

$$\sum_{k=1}^{m} \|x_k\|_X^2 \leq C(\mathbb{E}\|F_m\|_X)^2, \quad m \in \mathbb{N},$$

for any X-valued, Hardy martingale $F = (F_k)$ with Steinhaus coefficients (x_k). This gives the decay estimates

$$\inf\{\|x_k\|_X : 1 \le k \le m\} \le C^{1/2} m^{-1/2} \sup_{k\in\mathbb{N}} \mathbb{E}\|F_k\|_X, \qquad m \in \mathbb{N}.$$

Consequently, by Theorem 5.1.4 L^1/H_0^1 is not finitely represented in L^1.

In the following sections, (Theorems 5.1.12 and 5.1.17) we utilize suitable tensor products of the martingale exhibited by Theorem 5.1.4. This it plays a crucial role in Bourgain's isomorphic classification of the polydisk algebras.

Theorem 5.1.4 (Pisier's theorem) *There exists an L^1/H_0^1-valued Hardy martingale $F = (F_n)$ so that*

$$\Delta F_n(w) = f_n(w_1, \ldots, w_{n-1}) w_n,$$

$$\|\mathbb{E} f_{n-1}\|_{L^1/H_0^1} \ge \frac{1}{2},$$

and

$$\|F_n(w)\|_{L^1/H_0^1} \le 1, \quad w \in \mathbb{T}^{\mathbb{N}}.$$

Proof For $w = (m_n) \in \mathbb{T}^{\mathbb{N}}$ and $t \in [-\pi, \pi]$, define the Riesz product

$$G_{k-1}(w,t) = \prod_{m=1}^{k-1} \left\{1 + \frac{1}{2}\left\{e^{-i3^m t} w_m - e^{+i3^m t}\bar{w}_m\right\}\right\}, \tag{5.1.8}$$

and the martingale

$$F_n(w,t) = \frac{1}{2} \sum_{k=1}^{n} G_{k-1}(w,t) e^{\ i3^k t} w_k,$$

We first verify that $G_n(w,t)$ and $F_n(w,t)$ differ by an element of H_0^1. Hence in the quotient space they coincide. To this end, expand the Riesz product to obtain the recursion

$$G_n(w,t) = G_{n-1}(w,t) + \frac{1}{2} G_{n-1}(w,t) e^{-i3^n t} w_n + \frac{1}{2} G_{n-1}(w,t) e^{i3^n t} \bar{w}_n.$$

Observe that the third term on the right-hand side,

$$t \to \frac{1}{2} G_{n-1}(w,t) e^{i3^n t} \bar{w}_n,$$

defines an element in H_0^1. Hence,

$$G_n(w,t) = G_{n-1}(w,t) + \frac{1}{2} G_{n-1}(w,t) e^{-i3^n t} w_n, \quad \text{in} \quad L^1/H_0^1.$$

Iterating the above with $G_n(w,t)$ replaced by $G_{n-1}(w,t)$ gives the identity

$$G_n(w,t) = \frac{1}{2}\sum_{k=1}^{n} G_{k-1}(w,t)e^{-i3^k t}w_k \quad \text{in} \quad L^1/H_0^1.$$

In other words, the right- and left-hand sides differ by an element of H_0^1, or equivalently, the $G_n(w,t)$ defines a lifting of $F_n(w)$. This yields the pointwise upper estimate of the martingale

$$\|F_n(w)\|_{L^1/H_0^1} \le \int_{-\pi}^{\pi} |G_n(w,t)|dt \le 1.$$

Next, we turn to the lower estimates for the L^1/H_0^1-valued martingale differences,

$$\begin{aligned}\left\|\mathbb{E}_w\left(G_{n-1}(w,t)e^{-i3^n t}\right)\right\|_{L^1/H_0^1} &= \inf_{h\in H_0^1}\int_{-\pi}^{\pi}\left|\mathbb{E}_w\left(G_{n-1}(w,t)e^{-i3^n t}\right) - h(t)\right| dt \\ &= \inf_{h\in H_0^1}\int_{-\pi}^{\pi}\left|\mathbb{E}_w\left(G_{n-1}(w,t)\right) - e^{i3^n t}h(t)\right| dt \\ &\ge \inf_{h\in H_0^1}\int_{-\pi}^{\pi}\left(\mathbb{E}_w(G_{n-1}(w,t)) - e^{i3^n t}h(t)\right) dt. \\ &= \mathbb{E}_w\int_{-\pi}^{\pi} G_{n-1}(w,t)dt = 1.\end{aligned}$$

□

The following theorem highlights the isolated nature of the Riesz product $F_n(w)$ in Theorem 5.1.4. We exhibit a large class of its martingale transforms, denoted $F_n(w,z)$, for which $\|\Delta F_n(w)\| = \|\Delta F_n(w,z)\|$ and

$$C\mathbb{E}\|F_n(w,z)\|_{L^1/H_0^1} \ge \sqrt{n}, \qquad n\in\mathbb{N}. \tag{5.1.9}$$

Theorem 5.1.5 *Let X be a Banach space of cotype 2 and let $(F_k)_{k=1}^n$ be an X-valued Hardy martingale with*

$$\Delta F_k(w) = f_k(w_1,\dots,w_{k-1})w_k \quad \text{and} \quad x_k = \mathbb{E}(f_k), \qquad k\le n. \tag{5.1.10}$$

Then

$$\left(\sum_{k=1}^{n}\|x_k\|_X^2\right)^{1/2} \le C\mathbb{E}_{(w,z)}\left\|\sum_{k=1}^{n} z_{n-k}\Delta F_k\right\|_X,$$

where the integration is taken over the product space $\mathbb{T}^{\mathbb{N}}\times\mathbb{T}^{\mathbb{N}}$, and $C = C(X)$ depends only on the cotype 2 constant of the Banach space X.

Proof Fix $z\in\mathbb{T}^{\mathbb{N}}$ and put $F_n(w,z) = \sum_{k=1}^{n} z_{n-k}\Delta F_k$. The key to the proof is the Riesz product

$$P(w,z) = \prod_{k=1}^{n}\left\{1 + \frac{1}{2}\{w_k z_{n-k} + \overline{w_k z_{n-k}}\}\right\}.$$

Take the convolution of the X-valued F_n and the $\mathbb{R}$-valued P to define

$$G(w,z) = \mathbb{E}_{(s,t)}\left(F_n(s,t)P(ws^{-1}, zt^{-1})\right),$$

where the integration is taken over $\mathbb{T}^{\mathbb{N}} \times \mathbb{T}^{\mathbb{N}}$. Clearly, we have

$$\mathbb{E}_{(w,z)}\|G(w,z)\| \le \mathbb{E}_{(w,z)}\|F_n(w,z)\|. \tag{5.1.11}$$

Expanding the Riesz product and taking into account that ΔF_k is given by condition (5.1.10) we obtain

$$G(w,z) = \frac{1}{2}\sum_{k=1}^{n} x_l w_k z_{n-k}. \tag{5.1.12}$$

We justify the identity (5.1.12) in Proposition 5.1.9. Since X is a Banach space of cotype 2, we obtain for each fixed $z \in \mathbb{T}^{\mathbb{N}}$,

$$\left(\sum_{k=1}^{n} \|x_k\|^2\right)^{1/2} \le C\mathbb{E}_w \left\|\sum_{k=1}^{n} \{x_k z_{n-k}\} w_k\right\|. \tag{5.1.13}$$

Stringing together the estimates (5.1.11) and (5.1.13) completes the proof. □

In Section 7.5 we prove Bourgain's theorem that L^1/H^1 is of cotype 2 (Bourgain, 1984d). Hence Theorem 5.1.5 yields the estimates (5.1.9).

Rudin–Shapiro Martingales

Rudin–Shapiro martingales are obtained from Rudin–Shapiro polynomials by following the rules of martingale transfer (from $\mathbb{T}$ to $\mathbb{T}^{\mathbb{N}}$). We use them to obtain a uniformly bounded, scalar-valued Hardy martingale with prescribed Steinhaus coefficients in ℓ^2.

Theorem 5.1.6 *For any sequence $(c_n) \in \ell^2$, there exists a uniformly bounded Hardy martingale $F = (F_n)$ so that*

$$\sup_{n\in\mathbb{N}} |F_n| \le C\left(\sum_{n=1}^{\infty} |c_n|^2\right)^{1/2},$$

and for any $n \in \mathbb{N}$,

$$\mathbb{E}_w(\overline{w_n}\Delta_n F(w)) = c_n.$$

Proof Fix a complex sequence $(c_n) \in \ell^2$, with $\sum_{m=1}^{\infty} |c_m|^2 \le 1$. Fix $w \in \mathbb{T}^{\mathbb{N}}$ with $w = (w_n)$. Define $F_1 = G_1 = 1$ and, for $m \ge 1$,

$$F_{m+1}(w) = F_m(w) + \overline{G_m(w)} c_{m+1} w_{m+1},$$
$$G_{m+1}(w) = G_m(w) - \overline{F_m(w)} c_{m+1} w_{m+1}.$$

By construction, in $\mathbb{C}^2$ the vectors (F_m, G_m) and $\left(\overline{G_m}, -\overline{F_m}\right)$ are orthogonal. Hence we get the characteristic one step relation in the Rudin–Shapiro construction,

$$|F_{m+1}(w)|^2 + |G_{m+1}(w)|^2 = (1 + |c_{m+1}|^2)\left(|F_m(w)|^2 + |G_m(w)|^2\right).$$

Iterating, and thereby accumulating the factors $(1 + |c_{m+1}|^2)$ we obtain the uniform estimate

$$|F_n|^2 \leq C \sum_{m=1}^{\infty} |c_m|^2.$$

By definition, the sequence (G_n) is adapted to the product filtration on $\mathbb{T}^{\mathbb{N}}$ and, for (F_n), we find by resolving the recursive definition,

$$F_{n+1}(w) = \sum_{m=1}^{n} \overline{G_m(w)} c_{m+1} w_{m+1}.$$

Thus $F = (F_n)$ is a uniformly bounded, simple Hardy martingale, and

$$\mathbb{E}_w(\overline{w_n}(F_n(w) - F_{n-1}(w)) = c_n \mathbb{E}\left(\overline{G_{n-1}}\right) = c_n. \qquad \square$$

Rudin–Shapiro martingales give rise to uniformly bounded testing functions in the disk algebra. This yields lower estimates for some integrable Hardy martingales with values in L^1/H_0^1. Further applications in a similar spirit appear in the work of Pełczyński and Wojciechowski (1992a,b).

As an illustration, we verify in a very special case the inequalities needed for the cotype 2 property of the quotient space L^1/H_0^1. In the general case this important property is proved in Section 7.5.

Theorem 5.1.7 *Let $h_1, \ldots, h_n \in A(\mathbb{T})$ so that $\|h_m\|_A \leq 2$ and*

$$h_m(t) = \sum_{n=M_m}^{N_m} \hat{h}_m(k) e^{ikt}, \quad M_m > 4N_{m-1},\ t \in \mathbb{T}.$$

If $x_1, \ldots, x_n \in L^1/H_0^1$ satisfy $|\langle x_m, h_m \rangle| \geq \|x_m\|_{L^1/H_0^1}$ then

$$\mathbb{E}_w \left\| \sum_{m=1}^{n} w_m x_m \right\|_{L^1/H_0^1} \geq c \left(\sum_{m=1}^{n} \|x_m\|_{L^1/H_0^1}^2 \right)^{1/2}.$$

Proof By homogeneity we may assume that $\sum_{k=1}^{n} \|x_k\|_{L^1/H_0^1}^2 = 1$. Put $c_m = \|x_m\|$, for $m \leq n$. By Theorem 5.1.6 we may choose a Rudin–Shapiro martingale $F_m = F_m(w,t)$ satisfying

$$\mathbb{E}_w(\overline{w_m}(F_m(w,t) - F_{m-1}(w,t))) = c_m h_m(t), \qquad t \in \mathbb{T},$$

and $|F_m(w,t)| \le C$. Now consider the L^1/H^1_0-valued Hardy martingale

$$G(w) = \sum_{m=1}^{n} x_m w_m, \qquad w = (w_m) \in \mathbb{T}^{\mathbb{N}},$$

and test it against the Rudin–Shapiro martingales,

$$\begin{aligned} c\mathbb{E}_w \|G(w)\|_{L^1/H^1_0} &\ge \mathbb{E}_w \langle F_n(w,t), G(w)\rangle \\ &= \sum_{k=1}^{n} \langle \mathbb{E}_w(\overline{w_k}(F_k(w,t) - F_{k-1}(w,t))), x_k\rangle \\ &= \sum_{k=1}^{n} \langle c_k h_k, x_k\rangle \\ &\ge \sum_{k=1}^{n} c_k \cdot \|x_k\| = 1. \end{aligned}$$

□

5.1.2 The Bidisk Algebra

Here, we prove that the bidisk algebra $A(\mathbb{T}\times\mathbb{T})$ is not isomorphic to $A(\mathbb{T})$. We obtain the nonisomorphism by showing that $A^*(\mathbb{T}\times\mathbb{T})$ does not embed into $A^*(\mathbb{T})$. To this end, we investigate Hardy martingales of type (2) in considerable detail. Moreover, we prove a martingale lifting theorem showing that $A^*(\mathbb{T}\times\mathbb{T})$ is not finitely represented in $A^*(\mathbb{T})$.

Hardy Martingales of Type (2)

Multiindexed Hardy martingales are introduced in Section 5.1.1 to define an isomorphic invariant. Here, we discuss the special case of type (2) Hardy martingales.

Put $\Omega = \mathbb{T}^{\mathbb{N}}$. Let X be a quasi-Banach space, let $k, j \in \mathbb{N}$ and let $F_{k,j} \in L^1(\Omega\times\Omega, X)$ satisfy

$$F_{k,j}(w,z) = F_{k,j}(w_1,\ldots,w_k,z_1,\ldots,z_j), \quad k,j\in\mathbb{N}.$$

We say that $F = (F_{k,j})$ is an X-valued bi-Hardy martingale if for almost every $w \in \mathbb{T}^{\mathbb{N}}$ and $k \in \mathbb{N}$ fixed, the sequence

$$(F_{k,j}(w,\cdot))_{j=1}^{\infty} \tag{5.1.14}$$

is an X-valued Hardy martingale with respect to the second variable, and simultaneously for almost every $z \in \mathbb{T}^{\mathbb{N}}$ and $j \in \mathbb{N}$ fixed,

$$(F_{k,j}(\cdot,z))_{k=1}^{\infty} \tag{5.1.15}$$

is an X-valued Hardy martingale with respect to the first variable.

Martingale differences: A Hardy martingale of type (2) may be represented by an array of its martingale differences,

$$F_{k,j} = \sum_{m=1}^{k} \sum_{n=1}^{j} d_{m,n}, \tag{5.1.16}$$

where

$$d_{m,n}(w,z) = d_{m,n}(w_1, \ldots, w_m, z_1, \ldots, z_n),$$

and where for fixed $w' \in \mathbb{T}^{m-1}$ and $z' \in \mathbb{T}^{n-1}$, each of the mappings

$$z_n \to d_{m,n}(w', w_m, z', z_n) \quad \text{and} \quad w_m \to d_{m,n}(w', w_m, z', z_n) \tag{5.1.17}$$

defines an element in $H_0^1(\mathbb{T}, X)$.

Theorem 5.1.8 *Let* $0 < p \leq 2$ *and* $L^p = L^p(\Omega \times \Omega)$. *Let* $F = (F_{k,j})$ *be an* L^p*-bounded, scalar-valued Hardy martingale of type* (2)*, represented by Equation* (5.1.16)*. Then,*

$$\sum_{m=1}^{k} \sum_{n=1}^{j} \|d_{m,n}\|_{L^p}^2 \leq C(p) \|F_{k,j}\|_{L^p}^2, \qquad (k,j) \in \mathbb{N} \times \mathbb{N}.$$

Proof Fix $(k,j) \in \mathbb{N} \times \mathbb{N}$. It suffices to show that there exists $c(p) > 0$ (independent of F or (k,j)) such that

$$\|F_{k-1,j}\|_{L^p}^2 + c(p) \sum_{n=1}^{j} \|d_{k,n}\|_{L^p}^2 \leq \|F_{k,j}\|_{L^p}^2. \tag{5.1.18}$$

Once condition (5.1.18) is established, we repeat with k and j interchanged and exploit the resulting telescoping series. The proof of condition (5.1.18) begins by exploiting the representation (5.1.16). For fixed $(w,z) \in \Omega \times \Omega$, we put

$$g(w,z) = \sum_{n=1}^{j} d_{k,n}(w,z)$$

and obtain

$$F_{k,j}(w,z) = F_{k-1,j}(w,z) + g(w,z). \tag{5.1.19}$$

We keep $z \in \Omega$ fixed, and take the expectation of $|F_{k,j}(w,z)|$ with respect to $w \in \Omega$. By the complex convexity estimate (2.5.21), followed by Minkowski's inequality, we obtain for $z \in \Omega$,

$$\begin{aligned} &\left(\mathbb{E}_{(w)} |F_{k,j}(w,z)|^p\right)^{2/p} \\ &\geq \left(\mathbb{E}_{(w)} |F_{k-1,j}(w,z)|^p\right)^{2/p} + \delta(p) \left(\mathbb{E}_{(w)} |g(w,z)|^p\right)^{2/p}. \end{aligned} \tag{5.1.20}$$

Next, by Theorem 2.5.9,

$$\left(\mathbb{E}_{(w,z)} |g(w,z)|^p\right)^{2/p} \geq \delta(p) \sum_{n=1}^{j} \left(\mathbb{E}_{(w,z)} \left|d_{k,n}(w,z)\right|^p\right)^{2/p}. \tag{5.1.21}$$

Finally, we raise the estimate (5.1.20) to the power $p/2$, take the expectation with respect to $z \in \Omega$, use Minkowski's inequality, and invoke estimate (5.1.21). This gives

$$\begin{aligned} &\left(\mathbb{E}_{(w,z)} \left|F_{k,j}(w,z)\right|^p\right)^{2/p} \\ &\quad \geq \left(\mathbb{E}_{(w,z)} \left|F_{k-1,j}(w,z)\right|^p\right)^{2/p} + \delta(p)^2 \sum_{n=1}^{j} \left(\mathbb{E}_{(w,z)} \left|d_{k,n}(w,z)\right|^p\right)^{2/p}. \end{aligned} \tag{5.1.22}$$

Putting $c(p) = \delta(p)^2$, where $\delta(p)$ is the constant in condition (2.5.21), finishes the proof. □

Steinhaus coefficients: For convenience, below we use the terms "bi-Hardy martingale," "Hardy martingale of type (2)," and "biparameter Hardy martingale" interchangeably. For $k, j \in \mathbb{N}$, the Steinhaus coefficients of a bi-Hardy martingale $(F_{k,j})$ are given by

$$x_{k,j} = \mathbb{E}(F_{k,j}(w,z)\overline{w_k z_j}),$$

where the expectation is over $\Omega \times \Omega$.

Special cases:

(i) We illustrate the concept by describing an elementary procedure to generate bi-Hardy martingales with values in $A^*(\mathbb{T} \times \mathbb{T})$. Take a Hardy martingale (F_n) with values in L^1/H_0^1. For $k, j \in \mathbb{N}$ we let

$$F_{k,j}(w,z) = F_k(w_1, \ldots, w_k) \otimes F_j(z_1, \ldots, z_j),$$

where $w, z \in \Omega$ with $w = (w_k)$ and $z = (z_j)$. Thus $(F_{k,j})$ forms a bi-Hardy martingale with values in $A^*(\mathbb{T} \times \mathbb{T})$ and we have

$$\mathbb{E}(F_{k,j}(w,z)\overline{w_k z_j}) = \mathbb{E}(F_k(w)\,\overline{w_k}) \otimes \mathbb{E}(F_j(z)\,\overline{z_j}).$$

The Steinhaus coefficients of (F_n) are defined as $y_n = \mathbb{E}(F_n(z)\,\overline{z_n})$ where the expectation is over Ω. Thus the Steinhaus coefficients of $(F_{k,j})$ and (F_n) are related by $x_{k,j} = y_k \otimes y_j$.

(ii) We say that $F = (F_{k,j})$ is a simple bi-Hardy martingale if the defining Hardy martingales (5.1.14) and (5.1.15) are simple ones. In this case there exist X-valued functions $y_{m,n}$ satisfying

$$y_{m,n}(w,z) = y_{m,n}(w_1, \ldots, w_{m-1}, z_1, \ldots, z_{n-1}), \tag{5.1.23}$$

where $w, z \in \Omega$, and

$$F_{k,j}(w,z) = \sum_{m=1}^{k}\sum_{n=1}^{j} y_{m,n}(w,z)w_m z_n, \qquad (k,j) \in \mathbb{N}\times\mathbb{N}. \tag{5.1.24}$$

In view of equation (5.1.24), the Steinhaus coefficients of $F = (F_{k,j})$ are given by

$$x_{k,j} = \mathbb{E}_{(w,z)}(F_{k,j}(w,z)\overline{w_k z_j}) = \mathbb{E}_{(w,z)}(y_{k,j}(w,z)),$$

where $(k,j) \in \mathbb{N}\times\mathbb{N}$, and where the expectation is over $(w,z) \in \Omega\times\Omega$.

Riesz Products as Hardy Martingales of Type (2).

We start by giving a detailed description of the bi-Hardy martingales contained in the expansion of the Riesz product used in the proof of Theorem 5.1.5.

Proposition 5.1.9 *Let X be a complex Banach space. Fix $n \in \mathbb{N}$ and let $P(w,z)$ denote the Riesz product,*

$$P(w,z) = \prod_{i=1}^{n}\left\{1 + \frac{1}{2}\{w_i z_{n-i} + \overline{w_i z_{n-i}}\}\right\}, \tag{5.1.25}$$

where $w, z \in \Omega$ with $w = (w_k)$ and $z = (z_k)$. Let $(F_{k,j})$ be an X-valued Hardy martingale of type (2). Then the convolution of $F = F_{n,n}$ and P is a simple Hardy martingale of type (2), that is,

$$\mathbb{E}\left(F(\sigma,\tau)P(w\sigma^{-1}, z\tau^{-1})\right) = \sum_{k,j=1}^{n} y_{k,j}(w,z)w_k z_j,$$

where $y_{k,j}$ satisfies condition (5.1.23). Moreover, $y_{k,n-k}$ is a Steinhaus coefficient of $(F_{k,j})$ and is given by

$$y_{k,n-k} = \frac{1}{2}\mathbb{E}(F(w,z)\overline{w_k z_{n-k}}),$$

where $1 \le k < n$. In particular, $y_{k,n-k}$ does not depend on $w, z \in \Omega$.

Before we turn to the proof of Proposition 5.1.9 we need two lemmata analyzing in detail the Riesz product and its expansion.

Lemma 5.1.10 *Fix $n \in \mathbb{N}$ and $k \le n$. Denote by $C(n,k)$ the collection of all subsets in $\{1,\ldots,n\}$ consisting of k different elements. The Riesz product defined by (5.1.25) satisfies the identity*

$$P(w,z) = 1 + \sum_{k=1}^{n}\sum_{S\in C(n,k)}\sum_{\varepsilon\in\{-1,1\}^S} 2^{-k}\prod_{i\in S}\{w_i z_{n-i}\}^{\varepsilon_i}, \tag{5.1.26}$$

where $w, z \in \Omega$ with $w = (w_k)$ and $z = (z_k)$.

Proof To see this we fix a sequence of scalars (b_i) and expand,

$$\prod_{i=1}^{n}(1+b_i) = 1 + \sum_{k=1}^{n} \sum_{S\in C(n,k)} \prod_{i\in S} b_i. \tag{5.1.27}$$

Next we specialize, choose $w, z \in \mathbb{T}^n$, and put $b_i = \{w_i z_{n-i} + \overline{w_i z_{n-i}}\}/2$ where $i \leq n$. Then for $S \in C(n,k)$ we have,

$$\prod_{i\in S} b_i = \sum_{\varepsilon\in\{-1,1\}^S} 2^{-k} \prod_{i\in S} \{w_i z_{n-i}\}^{\varepsilon_i}. \tag{5.1.28}$$

Combine (5.1.28) and (5.1.27) to obtain formula (5.1.26) for the Riesz product $P(w,z)$. □

Lemma 5.1.11 *Let $k \geq 2$, $S \in C(n,k)$, and $\varepsilon \in \{-1,1\}^S$. Then the product*

$$\prod_{i\in S} \{w_i z_{n-i}\}^{\varepsilon_i}, \tag{5.1.29}$$

defines a Hardy martingale of type (2) *if and only if $\varepsilon = (1,\delta,1)$ where $\delta \in \{-1,1\}^{k-2}$.*

Proof Fix $k \geq 2$ and $S \in C(n,k)$. In this case we say that the product (5.1.29) is of multiplicity $2k$. To verify the claim remove from S the largest and the smallest entry,

$$l = \max S, \quad s = \min S, \quad S' = S\setminus\{l,s\}.$$

Then w_l is the largest nonvanishing component of w arising in the above product and, similarly, z_{n-s} is the largest nonvanishing component of z. Hence, the product (5.1.29) is a Hardy martingale with respect to both w and z if and only if it is of the form

$$w_s z_{n-s} \left\{ \prod_{i\in S'} \{w_i z_{n-i}\}^{\varepsilon_i} \right\} z_{n-l} w_l, \tag{5.1.30}$$

or, equivalently, if $\varepsilon = (1,\delta,1)$ where $\delta \in \{-1,1\}^{k-2}$. □

Proof of Proposition 5.1.9: Restate the assertion of the Lemma 5.1.11 by saying that if the product (5.1.29) is of multiplicity 4 or higher, then it is a Hardy martingale with respect to both w and z precisely when it can be written as

$$\left\{ w_s \prod_{i\in S'} \{w_i z_{n-i}\}^{\varepsilon_i} z_{n-l} \right\} z_{n-s} w_l, \tag{5.1.31}$$

where

$$s < \min S' \leq \max S' < l, \quad S' \in C(n, k-2), \quad \varepsilon \in \{-1, 1\}^{S'}. \tag{5.1.32}$$

If the product (5.1.29) is of multiplicity 2, then by conditions (5.1.32) it is a Hardy martingale with respect to both w and z precisely when it is of the form $w_i z_{n-i}$, where $1 \leq i < n$. Finally, we remark that in the products (5.1.31) the dependence on the components of w and z are of the form

$$y(w_s, \dots, w_{l-1}, z_{n-l-1}, \dots, z_{n-s-1}) z_{n-s} w_l, \tag{5.1.33}$$

where $s \leq n-1$ and $l \geq s+1$.

Decay Estimates for Steinhaus Coefficients

We exhibit in Theorem 5.1.12 an $A^*(\mathbb{T} \times \mathbb{T})$-valued, normalized bi-Hardy martingale for which the Steinhaus coefficients satisfy a lower bound. By contrast, in Theorem 5.1.13, we will show that the Steinhaus coefficients of an $A^*(\mathbb{T})$-valued bi-Hardy martingale satisfy

$$\liminf_{k+j \to \infty} \|x_{k,j}\|_{A^*(\mathbb{T})} = 0.$$

As we know already, by Proposition 5.1.2, this distinction gives rise to an isomorphic invariant, separating subspaces of $A^*(\mathbb{T})$ from $A^*(\mathbb{T} \times \mathbb{T})$.

Theorem 5.1.12 *Let $Y = A^*(\mathbb{T} \times \mathbb{T})$. There exists an L^1-bounded, Y-valued Hardy martingale of type* (2) *with Steinhaus sequence $(x_{k,j})$, such that $\|x_{k,j}\|_Y \geq 1$, for any $k, j \in \mathbb{N}$. Consequently,*

$$S(2, Y) > 0.$$

Proof Let $\Omega = \mathbb{T}^{\mathbb{N}}$. Put $X = A^*(\mathbb{T})$ and $Y = A^*(\mathbb{T} \times \mathbb{T})$. Recall that by Pisier's theorem (Theorem 5.1.4), there exists an L^1-bounded, L^1/H_0^1-valued Hardy martingale (F_k) such that $x_k = \mathbb{E}(F_k(v)\overline{v_k})$ satisfies $\|x_k\|_{L^1/H_0^1} = 1$ and $\Delta F_k(v) = f_k(v_1, \dots, v_{k-1}) v_k$.

Let $F_{k,j} = F_k \otimes F_j$ be its tensor product defined on $\Omega \times \Omega$. Thus, if $u = (v, z)$ with $v, z \in \Omega$ then $F_{k,j}(u) = F_k(v) \otimes F_j(z)$. Hence, clearly $(F_{k,j})$ is a Hardy martingale of type (2) with values in Y. For its Steinhaus coefficients we have the identities,

$$x_{k,j} = \mathbb{E}(F_{k,j}(u)\overline{u_{k,j}}) = \mathbb{E}(F_k(u)\overline{u_k}) \otimes \mathbb{E}(F_j(u)\overline{u_j}) = x_k \otimes x_j.$$

Now, for any pair of normalized testing functions $y_k, y_j \in A(\mathbb{T})$ we get

$$\|x_{k,j}\|_Y \geq \langle x_k, y_k \rangle \langle x_j, y_j \rangle.$$

By duality this gives $\|x_{k,j}\|_Y = 1$. □

We review the de La Vallee Poussin convolution operators. Fix $s \in \mathbb{N}$. For $k \in \mathbb{Z}$ put

$$q(k) = \begin{cases} 0, & |k| \geq 2s, \\ \text{linear}, & s < |k| \leq 2s, \\ 1, & |k| \leq s, \end{cases}$$

and form $V_s \in L^1(\mathbb{T})$ by prescribing its Fourier coefficients as $\widehat{V_s}(k) = q(k)$, for $k \in \mathbb{N}$. It is well known that $\|V_s\|_1 \leq 4$. Hence, V_s defines convolution operator on L^1 as follows,

$$(V_s f)(\zeta) = \int_{\mathbb{T}} f(w) V_s(\zeta \overline{w}) dm(\zeta), \tag{5.1.34}$$

where dm denotes the normalized Haar measure on $\mathbb{T}$.

In the setting of Theorem 5.1.5 we singled out one of the basic ideas contributing to the proof of Theorem 5.1.13 below.

Theorem 5.1.13 *Let $X = A^*(\mathbb{T})$ and let $(x_{k,j})$ be the Steinhaus sequence of an L^1-bounded X-valued Hardy martingale of type* (2). *For any $\epsilon > 0$ and any choice of $k_0, j_0 \in \mathbb{N}$, there exists $k > k_0$ and $j > j_0$ such that $\|x_{k,j}\|_X < \epsilon$. Consequently,*

$$S(2, X) = 0.$$

Proof Let $(F_{k,j})$ be an L^1-bounded X-valued Hardy martingale of type (2) and let $(x_{k,j})$ be its Steinhaus sequence. Let $\Omega = \mathbb{T}^{\mathbb{N}}$, fix $v \in \Omega$, $j \in \mathbb{N}$, and form

$$G_k^{(j)}(v) = \mathbb{E}_{(z)}(F_{k,j}(v, z)\overline{z_j}),$$

where the expectation is over $z \in \Omega$. Thus for j fixed $\left(G_k^{(j)}\right)$, is an L^1 bounded X-valued Hardy martingale such that $x_{k,j} = \mathbb{E}\left(G_k^{(j)}(v)\overline{v_k}\right)$, for any $k \in \mathbb{N}$.

Step 1 (Reduction) Fix $s \in \mathbb{N}$, and let V_s denote the convolution operator given by the de La Vallee Poussin kernel. For s, j fixed the sequence $f_k = V_s G_k^{(j)}$ is then an L^1-bounded Hardy martingale with values in $V_s(X) = \{V_s x \colon x \in X\}$. Since the latter is finite dimensional, the estimate (2.5.25) of Theorem 2.5.8 with $p = 1$ applies to the Hardy martingale (f_k). Hence there exists $C = C(s)$ such that

$$\sum_k (\mathbb{E}\|\Delta f_k\|_X)^2 \leq C \sup_k (\mathbb{E}\|f_k\|)^2.$$

Clearly, the above inequality translates into estimates for the Steinhaus coefficients of f_k and implies that

$$\sum_k \|V_s(x_{k,j})\|_X^2 < \infty,$$

for $s, j \in \mathbb{N}$. Summing up the first part of the proof, we showed that or any $s \in \mathbb{N}$, $j \in \mathbb{N}$ and $\epsilon > 0$ there exists $k \geq j$ such that

$$\left\| V_s(x_{k,j}) \right\|_X < \epsilon. \tag{5.1.35}$$

Step 2 (Iteration) Fix $n \in \mathbb{N}$ and $\epsilon_j > 0$ for $1 \leq j \leq n$. We prove next that there exist increasing integers k_j, and $y_j \in A(\mathbb{T})$ with $\|y_j\| \leq 1$ such that:

(i) If $L_j = (k_j, n - j)$ then (up to a factor 4, say) the norm of the Steinhaus coefficient $x_j = x_{L_j}$ is achieved by integration against $y_j \in A(\mathbb{T})$, satisfying

$$2\langle x_j, y_j \rangle > \|x_j\|_X - \epsilon_j.$$

(ii) There exist $s_{2j} < s_{2j+1}$ such that $\operatorname{spec}(y_j) \subseteq [s_{2j}, s_{2j+1}]$ for any $1 \leq j \leq n$, and

$$4s_{2j-1} \leq s_{2j} \leq s_{2j+1}/4.$$

We first select $\epsilon_j > 0$ where $1 \leq j \leq n$. Set $s_1 = 1$ and put $s_2 = 4s_1$. By estimate (5.1.35) there exists $k_1 > n$ such that, with $L_1 = (k_1, n-1)$ and $x_1 = x_{L_1}$, we have $\|V_{s_2}(x_1)\|_X < \epsilon_1/2$. Choose an analytic polynomial $p_1 \in A(\mathbb{T})$ with $\|p_1\| = 1$ such that $2\langle x_1, p_1 \rangle > \|x_1\|_X$. Select $s_3 \geq 4s_2$ such that $\operatorname{spec}(p_1) \subseteq [0, s_3]$ and put $y_1 = p_1 - V_{s_2}(p_1)$. Then by duality,

$$2\langle x_1, y_1 \rangle > 2\langle x_1, p_1 \rangle - 2\langle V_{s_2}(x_1), p_1 \rangle \geq \|x_1\| - \epsilon_1.$$

At stage $j - 1$ we are given $k_{j-1} \geq k_{j-2}$, the Steinhaus coeffiecient $x_{j-1} = x_{L_{j-1}}$, where $L_{j-1} = (k_{j-1}, n - (j - 1))$, and an analytic polynomial y_{j-1} normalized such that $\|y_{j-1}\|_\infty = 1$ and $\operatorname{spec}(y_{j-1}) \subseteq [s_{2j-2}, s_{2j-1}]$, where $4s_{2j-2} \leq s_{2j-1}$ and

$$2\langle x_{j-1}, y_{j-1} \rangle > \|x_{j-1}\|_X - \epsilon_{j-1}.$$

Next, we put $s_{2j} = 4s_{2j-1}$ and select $k_j \geq k_{j-1}$ and $L_j = (k_j, n - j)$ such that $x_j = x_{L_j}$ satisfies $\|V_{s_{2j}}(x_j)\|_X < \epsilon_j/2$. Choose an analytic polynomial p_j with $\|p_j\| \leq 1$ such that $2\langle x_j, p_j \rangle > \|x_j\|_X$ and select $s_{2j+1} \geq 4s_{2j}$ such that $\operatorname{spec}(p_j) \subseteq [0, s_{2j+1}]$. Finally, we put $y_j = p_j - V_{s_{2j}}(p_j)$. Then by duality,

$$2\langle x_j, y_j \rangle > \|x_j\| - \epsilon_j.$$

See Figure 5.2.

In the remaining steps of the proof we will show that for the Steinhaus coefficients $x_j = x_{L_j}$ selected above, the cotype estimates satisfied are,

$$\sum_{j=1}^{n} \|x_j\|_X^2 \leq C(\mathbb{E}\|F\|_X)^2, \tag{5.1.36}$$

where $F = F_{k,k}$ for $k \geq k_n$ and $C < \infty$ independent of n. To achieve estimate (5.1.36) we exploit Proposition 5.1.9 and Theorem 5.1.7.

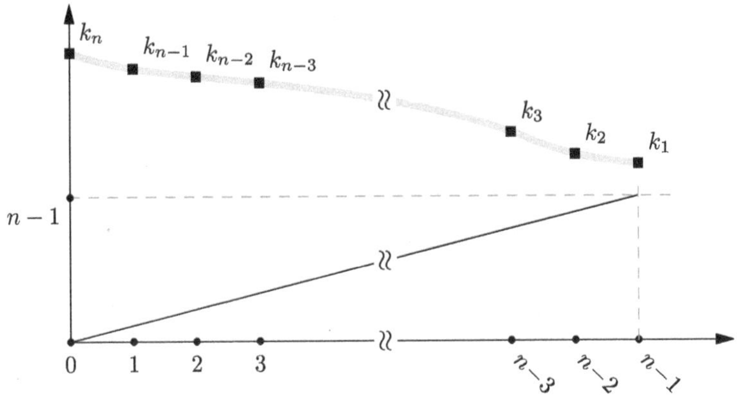

Figure 5.2 Proof of Theorem 5.1.13: The pairs $L_j = (k_j, n-j)$ for $1 \le j \le n$.

Step 3 (Riesz products) Let k_j be as determined above, and form the Riesz product

$$P(v,z) = \prod_{j=1}^{n} \left\{ 1 + \frac{1}{2}\{v_{k_j} z_{n-j} + \overline{v_{k_j} z_{n-j}}\} \right\}.$$

Fix $k = k_n$ and take the convolution of $F = F_{k,k}$ with P. Thus,

$$R(v,z) = \mathbb{E}\left(F(\sigma,\tau)P(v\sigma^{-1}, z\tau^{-1})\right)$$

gives a simple X-valued Hardy martingale of type (2). We have that R depends only on the variables v_{k_j} and z. We put $w = (w_1, \ldots, w_n)$, where $w_j = v_{k_j}$, and write $G(w,z) = R(v,z)$. Then $\mathbb{E}\|G\| \le \mathbb{E}\|F\|$, and, by Proposition 5.1.9, G gives rise to a simple X-valued Hardy martingale of type (2) such that

$$G(w,z) = \sum g_{k,j}(w,z) w_k z_j,$$

and $g_{j,n-j} = x_j/2$.

Step 4 (Rudin–Shapiro martingales) Fix $t \in \mathbb{T}, z \in \Omega$, with $z = (z_k)$. Then form the sequence $c_j = y_j(t)\|x_j\| z_{n-j}$. We fix $w \in \Omega$ with $w = (w_k)$ and inductively define the Rudin–Shapiro martingales, starting with with $U_0 = Y_0 = 1$,

$$U_j(w) = U_{j-1}(w) + \overline{Y_{j-1}(w)} c_j w_j,$$
$$Y_j(w) = Y_{j-1}(w) + \overline{U_{j-1}(w)} c_j w_j.$$

Taking the expectation in the first line gives $\mathbb{E}_{(w)}(\overline{w_j}(U_j(w) - U_{j-1}(w))) = c_j$, and by summing the telescoping series of the recursion for U_j we get

$$U_n(w) = \sum_{j=1}^{n} \overline{Y_{j-1}(w)} c_j w_j.$$

Put $U = U_n$. By orthogonality, the Rudin–Shapiro martingale U satisfies the uniform upper bound $|U| \leq \prod (1 + |c_j|^2)$. In the previous step we ensured that the spectra of $y_j \in A(\mathbb{T})$ are well separated. Hence, for fixed $w, z \in \Omega$ and $j \leq n$, the dependence on $t \in \mathbb{T}$ in $\overline{Y_{j-1}(w, z, t)} y_j(t)$ defines an element in $A(\mathbb{T})$. The same holds true for their sum $U(w, z, t)$. Retracing the steps of the recursive definition of $U = U(w, z, t)$, we have $U(w, z) \in A(\mathbb{T})$ and, in view of Theorem 5.1.7, we obtain,

$$\begin{aligned}
&\mathbb{E}_{(z)}\mathbb{E}_{(w)}\langle U(w, z), G(w, z)\rangle \\
&\quad = \sum_{j=1}^{n} \mathbb{E}_{(z)}\mathbb{E}_{(w)}(\overline{w_j z_{n-j}}\langle U_j(w) - U_{j-1}(w), x_j\rangle) \\
&\quad = \sum_{j=1}^{n} \mathbb{E}_{(z)}\overline{z_{n-j}}\mathbb{E}_{(w)}(\overline{w_j}\langle U_j(w) - U_{j-1}(w), x_j\rangle) \\
&\quad = \sum_{j=1}^{n} \langle y_j, x_j\rangle \|x_j\|.
\end{aligned}$$

In view of $\langle y_j, x_j\rangle \geq (\|x_j\| - \epsilon_j)/2$, and the uniform upper estimates for U we have thereby verified (5.1.36).

Step 5 (Summary) For $n \in \mathbb{N}$, there exist increasing integers k_j with $1 \leq j \leq n$ and $k_1 \geq n$ such that $\sum_{j=1}^{n-1} \|x_{k_j, n-j}\|^2 \leq C(\mathbb{E}\|F\|)^2$. Hence, for any choice of $k_0, j_0 \in \mathbb{N}$ there exists $k \geq k_0$ and $j \geq j_0$ such that $\|x_{k,j}\| < \epsilon$. This finishes the proof. □

Theorem 5.1.14 *The dual space $A^*(\mathbb{T} \times \mathbb{T})$ is not isomorphic to a subspace of $A^*(\mathbb{T})$, and hence the bidisk algebra $A(\mathbb{T} \times \mathbb{T})$ is not isomorphic to the disk algebra $A(\mathbb{T})$.*

Proof Let $Y = A^*(\mathbb{T} \times \mathbb{T})$ and $X = A^*(\mathbb{T})$. By Theorem 5.1.12 and Theorem 5.1.13,

$$S(2, Y) > 0 \quad \text{and} \quad S(2, X) = 0.$$

Hence appealing to Proposition 5.1.2 concludes the proof. □

Our next result shows that Theorem 5.1.13 may be strengthened considerably. To evaluate the difference between the two theorems, it suffices to compare the summary in Step 5 above with the assertion of Theorem 5.1.15. The lifting theorem for $A^*(\mathbb{T})$-valued bi-Hardy martingales is the main new ingredient responsible for the improvement expressed in estimate (5.1.37) below.

Theorem 5.1.15 *Let $(F_{k,j})$ be an L^1-bounded Hardy martingale of type (2) with values in L^1/H^1, and let $(x_{k,j})$ be the sequence of its Steinhaus coefficients. There exists $C < \infty$ such that, for any $n \in \mathbb{N}$,*

$$\sum_{k=1}^{n} \|x_{k,n-k}\|^2_{L^1/H^1_0} \leq C\left(\mathbb{E}\|F_{n,n}\|_{L^1/H^1_0}\right)^2. \tag{5.1.37}$$

Proof Fix $n \in \mathbb{N}$ and put $F = F_{n,n}$. Let $w, z \in \Omega$, and let $P(w,z)$ denote the Riesz product defined in (5.1.25). Take the convolution of F and P to define

$$Y(w,z) = \mathbb{E}_{(s,t)}\left(F(s,t)P(ws^{-1}, zt^{-1})\right), \tag{5.1.38}$$

where the integration is taken over $s, t \in \Omega$. Clearly, we have $\mathbb{E}\|Y\| \leq \mathbb{E}\|F\|$. By Lemma 5.1.10, and in view of (5.1.33), we have a Hardy martingale of type (2) such that,

$$Y(w,z) = \sum_{k,j=1}^{n} y_{k,j}(w,z)w_k z_j, \tag{5.1.39}$$

where $y_{k,j}(w,z)$ depends only on $(w_1, \ldots, w_{k-1})$ and $(z_1, \ldots, z_{j-1})$. Moreover, we have $y_{k,n-k}(w,z) = x_{k,n-k}/2$, for $k < n$. In particular, $y_{k,n-k}(w,z)$ does not depend on (w,z).

Let $0 < \alpha < 1$. Apply Theorem 5.1.16, the biparameter Hardy martingale lifting theorem, to Y. This yields a simple Hardy martingale G of type (2) with values in L^α such that,

$$G(w,z) = \sum g_{k,j}(w,z)w_k z_j.$$

where, for $w, z \in \Omega$ fixed, $g_{k,j}(w,z)$ is a lifting of $y_{k,j}(w,z)$ for any $k, j \leq n$ and satisfies

$$\int_{\mathbb{T}} \{\mathbb{E}|G(w,z,t)|^\alpha\}^{1/\alpha}\, dt \leq C_\alpha \mathbb{E}\|Y(w,z)\|, \tag{5.1.40}$$

where the expectation is taken over $w, z \in \Omega$. Since $g_{k,n-k}(w,z)$ is a lifting of $y_{k,n-k}(w,z) = x_{k,n-k}$, it follows that $g_{k,n-k}(w,z)$ does not depend on w, z. Putting $g_{k,n-k}(w,z) = g_{k,n-k}$, we have $q(g_{k,n-k}) = x_{k,n-k}$ where q is the quotient map onto L^1/H_0^1 and $\|x_{k,n-k}\|_{L^1/H_0^1} \leq \|g_{k,n-k}\|_{L^1}$. We now claim that the following estimate holds true

$$\left(\sum_{k=1}^{n} \|g_{k,n-k}\|_{L^1}^2\right)^{1/2} \leq C_\alpha \mathbb{E}\|Y\|, \tag{5.1.41}$$

where the expectation on the right-hand side is over $w, z, \in \Omega$. Turning to the proof of estimate (5.1.41), we first use Theorem 5.1.8 to obtain that for each $t \in \mathbb{T}$ fixed,

$$\left(\sum_{k,j=1}^{n} \{\mathbb{E}_{(w,z)}|g_{k,j}(w,z,t)|^\alpha\}^{2/\alpha}\right)^{1/2} \leq C\,\{\mathbb{E}_{(w,z)}|G(w,z,t)|^\alpha\}^{1/\alpha}.$$

By keeping only the terms $g_{k,n-k}(t)$ on the left-hand side and deleting the others, an integration over $t \in \mathbb{T}$ gives,

$$\int_{\mathbb{T}} \left(\sum_{k=1}^{n} |g_{k,n-k}(t)|^2\right)^{1/2} dt \leq C_\alpha \int_{\mathbb{T}} \{\mathbb{E}_{(w,z)}|G(w,z,t)|^\alpha\}^{1/\alpha}\, dt.$$

By Minkowski's inequality, the left-hand side of this estimate is larger than the left-hand side of estimate (5.1.41). Applying estimate (5.1.40) to its right-hand side gives estimate (5.1.41). We finish the proof of condition (5.1.37) by recalling that $\|x_{k,n-k}\|_{L^1/H_0^1} \leq \|g_{k,n-k}\|_{L^1}$ and that $\mathbb{E}\|Y\| \leq \mathbb{E}\|F\|$, as shown in the previous steps. □

Remark: Let $X = A^*(\mathbb{T})$. Let $F = (F_{k,j})$ be an L^1-bounded, X-valued Hardy martingale of typle (2). Theorem 5.1.13 asserts that its Steinhaus coefficients $\{x_{k,j}\}$ satisfy

$$\inf_{m\in\mathbb{N}} \inf\{\|x_{k,j}\|_X : 1 \leq k, j \leq m\} = 0. \tag{5.1.42}$$

Theorem 5.1.15 yields a striking quantitative improvement of condition (5.1.42). Indeed, letting $m \in \mathbb{N}$,

$$M = 2m,$$

and applying Theorem 5.1.15 successively with the respective specifications $n = M, n = M - 1, \ldots n = 1$, we obtain

$$\sum_{k,j=1}^{m} \|x_{k,j}\|_X^2 \leq Cm(\mathbb{E}\|F_{M,M}\|_X)^2. \tag{5.1.43}$$

By inspection, condition (5.1.43) causes an explicit decay rate for Steinhaus coefficients, resulting in

$$\inf\{\|x_{k,j}\|_X : 1 \leq k, j \leq m\} \leq C_0 m^{-1/2}, \qquad m \in \mathbb{N}, \tag{5.1.44}$$

where $C_0 = C^{1/2} \sup_{m\in\mathbb{N}} \mathbb{E}(\|F_{m,m}\|_X)$.

Lifting Theorem for Biparameter Hardy Martingales

The lifting theorem for biparameter Hardy martingales played a crucial role in the proof of Theorem 5.1.15. Its proof is based on vector-valued estimates for the Hilbert transform given by Theorem 6.1.3.

Put $\Omega = \mathbb{T}^{\mathbb{N}}$, $\Sigma = \Omega \times \Omega$, and $\sigma = \mathbb{P} \otimes \mathbb{P}$. Recall that $F = (F_{k,j})$ is a simple bi-Hardy martingale (with values in a complex Banach space X) if there exist previsible, X-valued functions $(y_{m,n})$ satisfying

$$y_{m,n}(w, z) = y_{m,n}(w_1, \ldots, w_{m-1}, z_1, \ldots, z_{n-1}) \tag{5.1.45}$$

for $(w, z) \in \Sigma$, such that

$$F_{k,j}(w, z) = \sum_{m=1}^{k} \sum_{n=1}^{j} y_{m,n}(w, z) w_m z_n,$$

for $k, j \in \mathbb{N}$.

We now use the notational convention dictated by the measure spaces involved,

$$L^p(L^\alpha) = L^p(\mathbb{T}, L^\alpha(\Sigma, \sigma)).$$

Theorem 5.1.16 *For $0 < \alpha < 1$ and $1 \leq p < \infty$, there exists $C = C(\alpha, p)$ so that the following holds. If $F = (F_{k,j})$ is a simple bi-Hardy martingale with values in L^1/H_0^1, and $A = (A_{k,j})$ is a bimartingale lifting of F with values in $L^1(\mathbb{T})$, then there exists a simple bi-Hardy martingale $G = (G_{k,j})$ with values in $L^1(\mathbb{T})$ satisfying*

$$q(G_{k,j}) = q(A_{k,j}), \qquad j, k \in \mathbb{N}, \tag{5.1.46}$$

where $q\colon L^1(\mathbb{T}) \to L^1(\mathbb{T})/H_0^1(\mathbb{T})$ is given by $q(f) = \{f + H_0^1(\mathbb{T})\}$, and

$$\|\widetilde{G}\|_{L^p(L^\alpha)} \leq C \|\widetilde{A}\|_{L^p(L^1)}, \tag{5.1.47}$$

where

$$\widetilde{G}(t, w, z) = G(w, z, t), \qquad \widetilde{A}(t, w, z) = A(w, z, t), \qquad t \in \mathbb{T}, \quad (w, z) \in \Sigma.$$

Proof For notational convenience we give the proof only in the case where $p = 1$. Let R denote the orthogonal projection onto $\overline{H_0^2(\mathbb{T})}$ with kernel $H^2(\mathbb{T})$. By Theorem 6.1.3 for $0 < \alpha < 1$, there exists $C = C(2, \alpha)$ with

$$\|\mathcal{R} : L^2(L^1) \to L^2(L^\alpha)\| \leq C,$$

where $J_\Sigma\colon L^1(\Sigma, \sigma) \to L^\alpha(\Sigma, \sigma)$ is the inclusion map, and $\mathcal{R} = R \otimes J_\Sigma$.

Step 1 Let $A = (A_{k,j})$, be an integrable bimartingale lifting of F with values in $L^1(\mathbb{T})$. Hence for almost every $t \in \mathbb{T}$,

$$\widetilde{A}(t) \colon \Sigma \to \mathbb{C}, \quad (w,z) \to \widetilde{A}(t,w,z),$$

is integrable with respect to (Σ, σ). Writing $L^1 = L^1(\Sigma)$, we put $\varphi(t) = \|\widetilde{A}(t)\|_{L^1}$ and let $g \in H^1(\mathbb{T})$ denote the outer function

$$g = \exp(\ln(\varphi + \epsilon) + iH(\ln(\varphi + \epsilon))),$$

where H denotes the Hilbert transform on $\mathbb{T}$. Hence $|g| = \varphi + \epsilon$. Define

$$G(w,z,t) = g^{1/2}(t)\mathcal{R}\left(\widetilde{A}g^{-1/2}\right)(t,w,z), \quad t \in \mathbb{T}, \quad (w,z) \in \Sigma.$$

In step 2 we first show that G is a lifting of F and is a simple bi-Hardy martingale. In Step 3 we show that

$$\|\widetilde{G}\|_{L^1(L^\alpha)} \le C_\alpha \|\widetilde{A}\|_{L^1(L^1)}.$$

Step 2 We show that G is a simple bi-Hardy martingale and verify the lifting properties (5.1.46). Since F is a simple bi-Hardy martingale, we have

$$F(w,z) = \sum_{k,j=1}^{\infty} f_{k,j}(w,z) w_k z_j, \qquad (w,z) \in \Sigma,$$

where $f_{k,j} \colon \Sigma \to L^1/H_0^1$ satisfies the constraints (5.1.45). Choose $\varphi_{k,j} \colon \Sigma \to L^1$ such that $q(\varphi_{k,j}(w,z)) = f_{k,j}(w,z)$ and put

$$\widetilde{\varphi}_{k,j}(t) \colon \Sigma \to \mathbb{C}, \quad (w,z) \to \varphi_{k,j}(w,z,t).$$

We assume, as we may, that $\widetilde{\varphi}_{k,j}$ depends smoothly on $t \in \mathbb{T}$ and define the $L^1(\mathbb{T})$-valued

$$g_{k,j} = g^{1/2}\mathcal{R}\left(\widetilde{\varphi}_{k,j} g^{-1/2}\right).$$

Since $f_{k,j} = q(\varphi_{k,j}) = q(g_{k,j})$, we obtain the identity

$$G(w,z,t) = \sum_{k,j=1}^{\infty} g_{k,j}(t,w,z) w_k z_j,$$

and verified the initial lifting conditions (5.1.46). In summary, G is a simple bi-Hardy martingale and a lifting of F.

Step 3 It remains to prove the norm estimates (5.1.47). Put

$$\mathcal{R}\left(\widetilde{A}g^{-1/2}\right)(t) \colon \Sigma \to \mathbb{C}, \quad (w,z) \to \mathcal{R}\left(\widetilde{A}g^{-1/2}\right)(t,w,z).$$

Next, observe that with $L^\alpha = L^\alpha(\Sigma)$, and for fixed $t \in \mathbb{T}$,

$$\|\widetilde{G(t)}\|_{L^\alpha} = |g(t)|^{1/2} \|\mathcal{R}(\widetilde{A}g^{-1/2})(t)\|_{L^\alpha}.$$

Apply the Cauchy–Schwarz inequality to the right-hand side of the equation to obtain

$$\int_{\mathbb{T}} \|\widetilde{G}(t)\|_{L^\alpha} dm \le \left(\int_{\mathbb{T}} |g(t)| dm\right)^{1/2} \left(\int_{\mathbb{T}} \left\|\mathcal{R}\left(\widetilde{A}g^{-1/2}\right)(t)\right\|_{L^\alpha}^2 dm\right)^{1/2}.$$

Theorem 6.1.3 gives an upper bound for the right-hand side factor,

$$\begin{aligned}\int_{\mathbb{T}} \left\|\mathcal{R}\left(\widetilde{A}g^{-1/2}\right)(t)\right\|_{L^\alpha}^2 dm &\le C_a \int_{\mathbb{T}} |g(t)|^{-1} \|\widetilde{A}(t)\|_{L^1}^2 dm \\ &= C_a \int_{\mathbb{T}} \|\widetilde{A}(t)\|_{L^1} dm.\end{aligned}$$

Recall that $|g(t)| = \|\widetilde{A}(t)\|_{L^1} + \epsilon$. Letting $\epsilon \to 0$, it is easy to see that the estimates above imply that

$$\int_{\mathbb{T}} \|\widetilde{G}(t)\|_{L^\alpha} dm \le C_\alpha \int_{\mathbb{T}} \|\widetilde{A}(t)\|_{L^1} dm. \qquad \square$$

The proof of Theorem 5.1.16 with its dependence on estimates for the vector-valued Hilbert transform is called *Bourgain's quotient-lifting method for L^1/H_0^1*. In Chapter 7 we will encounter it again, in

(i) Bourgain Davis's theorems on Hardy martingales with values in quotients of L^1 by reflexive spaces (Section 7.1.2).
(ii) Bourgain's theorem on cotype 2 property of the Banach space of L^1/H^1 (Section 7.5.1).
(iii) Kislyakov's embedding theorem (Section 7.5.2).

The versions of Bourgain's quotient-lifting method for L^1/H_0^1 presented in Chapter 7 are certainly better known than the one used in Theorem 5.1.16 for L^1/H_0^1-valued biparameter Hardy martingales.

5.1.3 Polydisk Algebras

In this section we obtain the isomorphic classification of the polydisk algebras, as stated in Theorem 5.1.1. We will show that $A^*(\mathbb{T}^k)$ does not embed into $A^*(\mathbb{T}^d)$ for $k > d$.

Theorem 5.1.3 and Proposition 5.1.2 form the framework of the proof. Correspondingly, we will compare integrable, $A^*(\mathbb{T}^d)$-valued, Hardy martingales of type $(d+1)$ to those ranging in $A^*(\mathbb{T}^{d+1})$.

Steinhaus Coefficients

The next theorem exploits Pisier's theorem (Theorem 5.1.4) to determine a normalized $A^*(\mathbb{T}^{d+1})$-valued, Hardy martingale of type $(d+1)$ with Steinhaus coefficients bounded below in norm.

We first review basic examples in the dual of $A(\mathbb{T}^d)$. Recall that, by the F. and M. Riesz theorem (Theorem 1.1.21), $L^1/H_0^1 \subset A^*(\mathbb{T})$. Let $f_j \in L^1/H_0^1$, let $s \in \mathbb{T}^d$ with $s = (s_j)$, and define

$$f(s) = f_1(s_1) \cdots f_d(s_d).$$

Then

$$\|f\|_{A^*(\mathbb{T}^d)} = \|f_1\|_{L^1/H_0^1} \cdots \|f_d\|_{L^1/H_0^1}.$$

Theorem 5.1.17 *Let $Y = A^*(\mathbb{T}^{d+1})$. There exists an L^1-bounded, Y-valued Hardy martingale of type $(d+1)$ with Steinhaus sequence (y_L) such that $\|y_L\|_Y \geq 1$, for any $L \in \mathbb{N}^{d+1}$. Consequently,*

$$S(d+1, Y) > 0.$$

Proof The proof proceeds by induction over the dimension d. The case $d = 1$ holds true by Theorem 5.1.12. We now assume that it holds for $d \in \mathbb{N}$. Put $X = A^*(\mathbb{T}^d)$. By assumption, we may pick an L^1-bounded Hardy martingale of type (d) with values in X, such that its Steinhaus sequence (x_K) satisfies $\|x_K\|_X \geq 1$ for any $K \in \mathbb{N}^d$. By Pisier's theorem (Theorem 5.1.4), there exists a uniformly bounded, L^1/H_0^1-valued Hardy martingale (G_j) such that $g_j = \mathbb{E}(G_j(z)\overline{z_j})$ satisfies $\|g_j\|_{L^1/H_0^1} = 1$. We define an L^1-bounded, Y-valued Hardy martingale of type $(d+1)$ by putting

$$F_{K,j}(v, z) = F_K(v) \otimes G_j(z),$$

where $v \in \Omega^d, z \in \Omega$, and where $K \in \mathbb{N}^d$ and $j \in \mathbb{N}$. Its Steinhaus coefficients are then given by

$$y_{K,j} = \mathbb{E}(F_{K,j}(v, z)\overline{v_K z_j}),$$

where integration on the left-hand side is over $(v, z) \in \Omega^{d+1}$ and $(K, j) \in \mathbb{N}^{d+1}$. Hence, $y_{K,j} = x_K \otimes g_j$ and $\|y_{K,j}\|_Y \geq 1$. □

We now introduce d-dimensional versions of the de La Vallee Poussin convolution operators. Fix $s \in \mathbb{N}$. For $k \in \mathbb{Z}$, define

$$r(k) = \begin{cases} 1, & |k| \geq 2s, \\ \text{linear}, & s < |k| \leq 2s, \\ 0, & |k| \leq s. \end{cases}$$

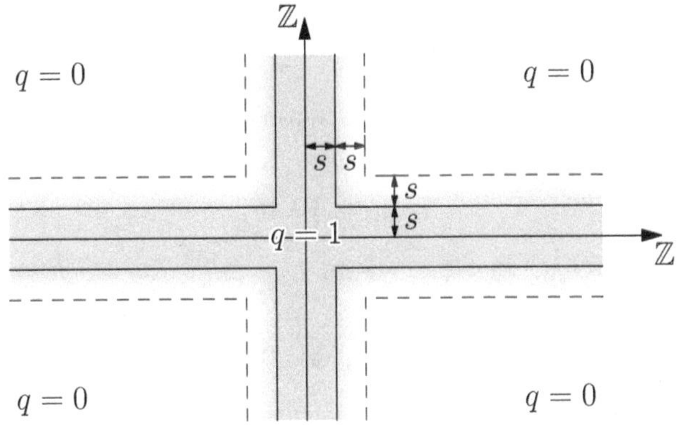

Figure 5.3 The two-dimensional de La Vallee Poussin kernels V_s: The gray scale picture of $\widehat{V_s}(k_1, k_2) = q(k_1, k_2)$, for $(k_1, k_2) \in \mathbb{N}^2$.

Given $K \in \mathbb{N}^d$ with $K = (k_1, \ldots, k_d)$, we set $q(K) = 1 - \prod_{i=1}^{d} r(k_i)$. Finally, we define $V_s \in L^1(\mathbb{T}^d)$ by prescribing its Fourier coefficients as $\widehat{V_s}(K) = q(K)$, for $K \in \mathbb{N}^d$. See Figure 5.3. By comparison with the de La Vallee Poussin kernels it is easily checked that $\|V_s\|_1 \leq 4^d$. We continue to denote V_s as the convolution operator

$$(V_s f)(\zeta) = \int_{\mathbb{T}^d} f(w) V_s(\zeta w^{-1}) dm_d(\zeta), \tag{5.1.48}$$

where $\zeta \in \mathbb{T}^d$ and m_d denotes the normalized Haar measure on $\mathbb{T}^d$. We briefly review the basic properties of V_s as a Fourier multiplier on $\mathbb{T}^d$. If $\varphi\colon \mathbb{T}^d \to \mathbb{C}$ is a trigonometric polynomial with Fourier spectrum contained in a strip of width $2s$ around the coordinate axis of $\mathbb{Z}^d$ then $V_s\varphi = \varphi$. Conversely, if the Fourier spectrum of φ is contained in

$$(\mathbb{Z}\backslash[-2s, 2s])^d,$$

then we get $V_s\varphi = 0$.

The convolution operator (5.1.48) extends to a bounded operator on $A(\mathbb{T}^d)$ with norm independent of s and, by duality and transposition, V_s is well defined on $X = A^*(\mathbb{T}^d)$ again with norm independent of s, and $\lim_{s\to\infty} V_s(x) = x$, for any $x \in X$. Finally, we remark that the range of $V_s\colon X \to X$ is isomorphic to $A^*(\mathbb{T}^{d-1})$. Here, however, the Banach–Mazur distance between $A^*(\mathbb{T}^{d-1})$ and $V_s(X)$ does not remain bounded as $s \to \infty$.

We now show that for every integrable, $A^*(\mathbb{T}^d)$-valued Hardy martingale of type $(d+1)$, there exists a subsequence of its Steinhaus coefficients that tends to zero in norm. The argument presented below is modeled on the proof of Theorem 5.1.13.

Theorem 5.1.18 *Let $X = A^*(\mathbb{T}^d)$ and let (x_L) be the Steinhaus sequence of an L^1-bounded, X-valued Hardy martingale of type $(d+1)$. Then for any $l_0 \in \mathbb{N}$,*

$$\inf\{\|x_L\|_X\} = 0, \tag{5.1.49}$$

where the infimum is extended over all $L \subset \mathbb{N}^{d+1}$ for which $\min L \geq l_0$. Consequently,

$$S(d+1, X) = 0.$$

Proof We use induction over the dimension d. By Theorem 5.1.13 the conclusion of Theorem 5.1.18 holds true for $d = 1$. We next assume that it holds true for $d-1$. We let (F_L) be an L^1-bounded, X-valued Hardy martingale of type $(d+1)$. Let (x_L) be its Steinhaus sequence, correspondingly indexed by $L \subseteq \mathbb{N}^{d+1}$.

Step 1 (Reduction) We fix $v \in \Omega^d$, $K \in \mathbb{N}^d$, and $j \in \mathbb{N}$. Then define

$$G_K^{(j)}(v) = \mathbb{E}_{(z)}(F_{K,j}(v,z)\overline{z_j}),$$

where the expectation is over $z \in \Omega$. Thus for j fixed, $\left(G_K^{(j)}\right)$ is an L^1-bounded, X-valued Hardy martingale of type (d). Its Steinhaus sequence is related to (x_L) by the identity

$$x_{(K,j)} = \mathbb{E}\left(G_K^{(j)}(v)\overline{v_K}\right), \tag{5.1.50}$$

where the expectation is over $v \in \Omega^d$, and where $K \in \mathbb{N}^d$.

Next, fix $s \in \mathbb{N}$ and let $V_s \colon X \to X$ be the Fourier multiplier (5.1.48). For $s, j \in \mathbb{N}$ fixed, the sequence

$$f_K(v) = V_s\left(G_K^{(j)}(v)\right)$$

defines an L^1-bounded Hardy martingale of type (d). Note that f_K takes values in $V_s(X)$. Since the latter space is isomorphic to $A^*(\mathbb{T}^{d-1})$, the induction hypothesis applies to f_K, and hence by condition (5.1.49), for each $k_0 \in \mathbb{N}$ and $\epsilon > 0$ there exists $K \in \mathbb{N}^d$ such that $\min K \geq k_0$ and

$$\left\|\mathbb{E}(f_K(v)\overline{v_K})\right\|_X < \epsilon.$$

In view of the identity (5.1.50) we may restate this in terms of the Steinhaus coefficients x_L of F_L as follows: For each $s, j \in \mathbb{N}$ fixed,

$$\inf\left\{\left\|V_s(x_{(K,j)})\right\|_X\right\} = 0, \tag{5.1.51}$$

where the infimum is taken over those $K \in \mathbb{N}^d$ for which $\min K \geq k_0$.

Step 2 (Iterating condition (5.1.51)) We fix $n \in \mathbb{N}$ and $\epsilon_j > 0$, where $1 \leq j \leq n$. We claim that there exist $K_j \in \mathbb{N}^d$ with $\max K_j < \min K_{j+1}$, and testing functions $y_j \in A(\mathbb{T}^d)$ with $\|y_j\| \leq 1$, so that the following conditions hold:

(i) If $L_j = (K_j, n - j)$, then up to a factor 4, the X-norm of the Steinhaus coefficient $x_j = x_{L_j}$ is achieved by testing against $y_j \in A(\mathbb{T}^d)$. Thus,

$$2\langle x_j, y_j\rangle \geq \|x_j\|_X - \epsilon_j. \tag{5.1.52}$$

(ii) There exist integers $s_{2j} < s_{2j+1}$ such that the Fourier spectrum of y_j is contained in the cube $[s_{2j}, s_{2j+1}]^d$, for any $1 \leq j < n$, and, moreover,

$$4s_{2j-1} \leq s_{2j} \leq s_{2j+1}/4.$$

Choose $\epsilon_j > 0$ sufficiently small for $1 \leq j \leq n$. We start with $s_1 = 1$ and put $s_2 = 4s_1$. There exists $K_1 \subseteq \mathbb{N}^d$ and $y_1 \in A(\mathbb{T}^d)$ and therefter we select $s_3 > 4s_2$. By condition (5.1.51), we may determine $K_1 \subseteq \mathbb{N}^d$ with $\min K_1 \geq n$ such that with $L_1 = (K_1, n-1)$ and $x_1 = x_{L_1}$ we have

$$\|V_{s_2}(x_1)\|_X < \epsilon_1/2. \tag{5.1.53}$$

Now choose a polynomial $p_1 \in A(\mathbb{T}^d)$ with $\|p_1\| \leq 1$ such that

$$2\langle x_1, p_1\rangle \geq \|x_1\|_X. \tag{5.1.54}$$

Since p_1 is an analytic polynomial, there exists $s_3 > 4s_2$ so that $\operatorname{spec}(p_1) \subseteq [0, s_3]^d$. Put $y_1 = (\mathrm{Id} - V_{s_2})(p_1)$. Then $\operatorname{spec}(y_1) \subseteq [s_2, s_3]^d$, and with estimates (5.1.53) and (5.1.54) we have

$$2\langle x_1, y_1\rangle = 2\langle x_1, p_1\rangle - |\langle V_{s_2}(x_1), p_1\rangle| \geq \|x_1\|_X - \epsilon_1.$$

At stage $j-1$ we are given $K_{j-1} \in \mathbb{N}^d$ with $\min K_{j-1} \geq \max K_{j-2}$, a Steinhaus coefficient $x_{j-1} = x_{L_{j-1}}$, where $L_{j-1} = (K_{j-1}, n-(j-1))$, and $y_{j-1} \in A(\mathbb{T}^d)$ with $\|y_{j-1}\| \leq 1$, such that

$$2\langle x_{j-1}, y_{j-1}\rangle > \|x_{j-1}\|_X - \epsilon_{j-1},$$

and $\operatorname{spec}(y_{j-1}) \subseteq [s_{2j-2}, s_{2j-1}]^d$ where $4s_{2j-2} \leq s_{2j-1}$. Put $k_{j-1} = \max K_{j-1}$ and $s_{2j} = 4s_{2j-1}$. By condition (5.1.51) we may determine $K_j \subseteq \mathbb{N}^d$ with $\min K_j \geq k_{j-1}$ so that, with $L_j = (K_j, n-j)$, and $x_j = x_{L_j}$, we have

$$\|V_{s_{2j}}(x_j)\|_X < \epsilon_j/2. \tag{5.1.55}$$

Now choose a polynomial $p_j \in A(\mathbb{T}^d)$ with $\|p_j\| \leq 1$ such that

$$2\langle x_j, p_j\rangle \geq \|x_j\|_X. \tag{5.1.56}$$

We select next $s_{2j+1} > 4s_{2j}$ so that $\operatorname{spec}(p_j) \subseteq [0, s_{2j+1}]^d$, and put $y_j = (\mathrm{Id} - V_{s_{2j}})(p_j)$. Then $\operatorname{spec}(y_j) \subseteq [s_{2j}, s_{2j+1}]^d$ and with estimates (5.1.55) and (5.1.56) it follows that $2\langle x_j, y_j\rangle \geq \|x_j\|_X - \epsilon_j$. We carry on until we reach $j = n$.

Step 3 (Riesz products) For $K \in \mathbb{N}^d$, with $K = (k_1, \dots, k_d)$, and $v \in \Omega^d$, with $v = (v^{(1)}, \dots, v^{(d)})$, let $v_K = \prod_{i=1}^d v^{(i)}_{k_i}$ where $v^{(i)} = \left(v^{(i)}_k\right)$. Fix $K_j \in \mathbb{N}^d$ as constructed in the previous step, and form $v_{K_j} \in \mathbb{T}$. For $z \in \Omega$ with $z = (z_k)$, define the Riesz product

$$P(v, z) = \prod_{j=1}^{n} \left\{1 + \frac{1}{2}\{v_{K_j} z_{n-j} + \overline{v_{K_j} z_{n-j}}\}\right\}.$$

For $L \in \mathbb{N}^{d+1}$, let

$$Z_L(v, z) = \mathbb{E}_{(\sigma,\tau)} \left(F_L(\sigma, \tau) P(v\sigma^{-1}, z\tau^{-1})\right),$$

where $\sigma \in \Omega^d, \tau \in \Omega$. The sequence (Z_L) defines an X-valued Hardy martingale of type $(d + 1)$ such that $\mathbb{E}\|Z_L\| \leq \mathbb{E}\|F_L\|$. Let $w_j = v_{K_j}$. Then $Z_L(v, z)$ is contained in

$$\operatorname{span}\left\{\prod_{j\in S} \{w_j z_{n-j}\}^{\varepsilon_j} : S \in C(n, k), \varepsilon = (1, \delta, 1), \delta \in \{-1, 1\}^{k-2}, k \leq n\right\}.$$

Hence, putting $G_{j,k}(w, z) = Z_{(K_j,k)}(v, z)$ where $w = (w_1, \dots, w_n)$ and $w_j = v_{K_j}$ defines a simple, X-valued Hardy martingale of type (2) such that

$$G_{n,n}(w, z) = \sum_{j,k=1}^{n} g_{j,k}(w, z) w_j z_k,$$

and $g_{j,n-j}(w, z) = x_{L_j}/2$ where $L_j = (K_j, n - j)$. Finally, put $G = G_{n,n}$. Since G is obtained from $F = F_{(K_n,n)}$ by convolution and a change of variables we have $\mathbb{E}\|G\| \leq \mathbb{E}\|F\|$.

Step 4 (Rudin–Shapiro martingales) We next claim that $x_j = x_{L_j}$ satisfies

$$\left(\sum_{j=1}^{n} \|x_j\|^2\right)^{1/2} \leq C\mathbb{E}(\|G\|). \tag{5.1.57}$$

The verification uses Rudin–Shapiro martingales and exploits that the Steinhaus coefficients $x_j = x_{L_j}$ may be normed by testing functions with well separated Fourier spectra. Fix $t \in \mathbb{T}^d, z \in \Omega$, with $z = (z_k)$. Then form the sequence

$$c_j = y_j(t)\|x_j\|z_{n-j}, \qquad 1 \leq m \leq n.$$

Next, fix $w \in \Omega$ with $w = (w_k)$ and form the Rudin–Shapiro martingales, $U_0 = 1$, $Y_0 = 1$, and

$$U_j(w) = U_{j-1}(w) + \overline{Y_{j-1}(w)} c_j w_j,$$
$$Y_j(w) = Y_{j-1}(w) + \overline{U_{j-1}(w)} c_j w_j.$$

Evaluating the first line of the recursion gives $\mathbb{E}_{(w)}(\overline{w_j}(U_j(w) - U_{j-1}(w))) = c_j$ and

$$U_n(w) = \sum_{j=1}^{n} \overline{Y_{j-1}(w)} c_j w_j.$$

Put $U = U_n$. By orthogonality, we have $|U|^2 \leq \prod_{j=1}^{n} (1 + |c_j|^2)$. Note that the coefficients c_j contains the factor z_{n-j}. Hence U depends not only on w but also on z. Now recall that the spectra of the testing functions $y_j \in A(\mathbb{T})$ are well separated. Hence, for fixed $w, z \in \Omega$ and $j \leq n$, the dependence on $t \in \mathbb{T}^d$, in $\overline{Y_{j-1}(w, z, t)} y_j(t)$ defines an element in $A(\mathbb{T}^d)$. The same holds true for their sum $U(w, z, t)$. Conversely, as a function on $w, z \in \Omega$, by the Rudin–Shapiro recursion, $U(w, z) \in A(\mathbb{T}^d)$ defines a Hardy martingale of type (2). Hence,

$$\mathbb{E}_{(z)} \mathbb{E}_{(w)} \langle U(w, z), G(w, z) \rangle = \sum_{j=1}^{n} \langle y_j, x_j \rangle \|x_j\|.$$

Now, taking into account that by estimate (5.1.52) we have $2\langle y_j, x_j \rangle \geq \|x_j\| - \epsilon_j$ and that $|U|^2 < C \sum_{j=1}^{n} \|x_j\|^2$, we finally obtain condition (5.1.57).

Step 5 (Summing up) For $n \in \mathbb{N}$, there exist $K_j \in \mathbb{N}^d$, with $\min K_1 \geq n$, and $\max K_{j-1} \leq \min K_j$, where $1 < j < n$, such that

$$\sum_{j=1}^{n-1} \left\| x_{(K_j, n-j)} \right\|^2 \leq C(\mathbb{E}\|G\|)^2.$$

Since G is obtained from F by convolution, we have $\mathbb{E}\|G\| \leq C\mathbb{E}\|F\|$. In summary, for any choice of $l_0 \in \mathbb{N}$ and $\epsilon > 0$ there exists $L \in \mathbb{N}^{d+1}$ with $\min L \geq l_0$ such that $\|x_L\|_X < \epsilon$. This finishes the proof of condition (5.1.49). □

Theorem 5.1.19 *For any $d \in \mathbb{N}$, the dual space $A^*(\mathbb{T}^{d+1})$ is not isomorphic to a subspace of $A^*(\mathbb{T}^d)$.*

Proof Let $Y = A^*(\mathbb{T}^{d+1})$ and $X = A^*(\mathbb{T}^d)$. By Theorem 5.1.17 and Theorem 5.1.18,

$$S(d+1, Y) > 0 \quad \text{and} \quad S(d+1, X) = 0.$$

Hence, appealing to Proposition 5.1.2 concludes the proof. □

5.2 Hardy Martingale Convergence aRNP

In this section we consider a complex Banach space X and ask if, for every bounded analytic function $f\colon \mathbb{D} \to X$, the radial limit

$$\lim_{r\to 1} f(r\zeta)$$

exists, for almost every $\zeta \in \mathbb{T}$. In that case, the Banach space X satisfies the analytic Radon–Nikodym property (aRNP). We note that the aRNP holds for L^1 (theorem of Bukhvalov and Danilevich [1982]) but not for c_0 and L^1/H^1.

In this section, we show that the aRNP is equivalent to the condition that every L^1-bounded, X-valued Hardy martingale is convergent almost surely. Separately, we obtain powerful sufficient conditions implying the aRNP (holomorphic injections, complex uniform convexity), and derive real variable characterizations of the aRNP in terms of plurisubharmonic functions (the plurisubharmonic variational principle, the existence of small plurisubharmonic slices).

5.2.1 Bounded Analytic Functions

Let X be a complex Banach space. We denote by $H^\infty(\mathbb{D}, X)$ the space of uniformly bounded, X-valued analytic functions on $\mathbb{D}$. Equipped with the norm

$$\|f\|_{H^\infty(\mathbb{D},X)} = \sup_{z\in\mathbb{D}} \|f(z)\|_X,$$

$H^\infty(\mathbb{D}, X)$ is a Banach space. Recall that for $z \in \mathbb{D}$, $\zeta \in \mathbb{T}$ we denote

$$P(z,\zeta) = (1 - |z|^2)/(|z - \zeta|^2).$$

We define $H^\infty(\mathbb{T}, X)$ to consist of those $f \in H^\infty(\mathbb{D}, X)$ for which there exists $g \in L^\infty(\mathbb{T}, X)$ such that

$$f(z) = \int_{\mathbb{T}} P(z,\zeta) g(\zeta) dm(\zeta), \qquad z \in \mathbb{D}.$$

We say that X satisfies the *aRNP* if

$$H^\infty(\mathbb{T}, X) = H^\infty(\mathbb{D}, X).$$

We give two separate characterizations of the aRNP in terms of (1) radial convergence of bounded analytic functions $f\colon \mathbb{D} \to X$, and (2) integral representations of linear operators $S\colon L^1 \to X$ that factor through the space L^1/H^1_0.

Let $0 < p < \infty$. We define $H^p(\mathbb{D}, X)$ to consist of those analytic functions $f\colon \mathbb{D} \to X$ that satisfy

$$\|f\|_{H^p(\mathbb{D},X)} = \sup_{0<r<1} \left(\int_{\mathbb{T}} \|f(r\zeta)\|_X^p dm(\zeta) \right)^{1/p} < \infty.$$

Let $H^p(\mathbb{T}, X)$ denote the subspace of those $f \in H^p(\mathbb{D}, X)$ for which there exists $g \in L^p(\mathbb{T}, X)$ such that

$$f(z) = \int_{\mathbb{T}} P(z, \zeta) g(\zeta) dm(\zeta), \qquad z \in \mathbb{D}.$$

If X is a Banach space and $1 \le p < \infty$, then $H^p(\mathbb{D}, X)$ is a Banach space as well; if $0 < p \le 1$ then

$$\|f + g\|_{H^p(\mathbb{D},X)}^p \le \|f\|_{H^p(\mathbb{D},X)}^p + \|f\|_{H^p(\mathbb{D},X)}^p,$$

for $f, g \in H^p(\mathbb{D}, X)$, and $H^p(\mathbb{D}, X)$ is a quasi-Banach space. The scale of vector-valued $H^p(\mathbb{D}, X)$ spaces is linked to $H^\infty(\mathbb{D}, X)$ by multiplication with scalar-valued analytic functions.

Theorem 5.2.1 *Let $0 < p < \infty$. For any $f \in H^p(\mathbb{D}, X)$ there exists $h \in H^\infty(\mathbb{D})$ and $g \in H^\infty(\mathbb{D}, X)$ such that $f = g/h$ and $0 < |h(z)| \le 1$, for any $z \in \mathbb{D}$.*

Proof Let $f \in H^p(\mathbb{D}, X)$. Then $\ln^+ \|f(z)\|$ is subharmonic and

$$\sup_{0<r<1} \int_{\mathbb{T}} \ln^+ \|f(r\zeta)\| dm(\zeta) < \infty.$$

Hence, there exists $u(z)$ harmonic in $\mathbb{D}$ such that $0 \le \ln^+ \|f(z)\| \le u(z)$. We then have $\ln \|f(z)\| \le u(z)$ and hence,

$$\|f(z)\| \le \exp(u(z)),$$

for any $z \in \mathbb{D}$. Let v denote the harmonic conjugate to u and put $\varphi(z) = u(z) + iv(z)$. Define $h(z) = \exp(-\varphi(z))$ and $g(z) = \exp(-\varphi(z)) f(z)$. Since φ is analytic in the unit disk, it remains to check that g, h satisfy the conclusions of the theorem.

As $|h(z)| = \exp(-u(z))$ and $u(z) \ge 0$, we have $|h(z)| \le 1$, and hence $h \in H^\infty(\mathbb{D})$. Next, $|u(z)| < \infty$ gives $|h(z)| \ne 0$. Finally,

$$\|g(z)\| = |h(z)| \|f(z)\| \le \exp(-u(z)) \|f(z)\| \le 1.$$

Hence $g \in H^\infty(\mathbb{D}, X)$. □

Remark: Bounded scalar-valued, analytic functions are uniquely determined by their radial limits: For each $h \in H^\infty(\mathbb{T})$, the radial limits $s(\zeta) = \lim_{r\to 1} h(r\zeta)$ exist almost everywhere. Now suppose that there exists a measurable set $E \subseteq \mathbb{T}$

with $m(E) > 0$ such that $s(\zeta) = 0$ for $\zeta \in E$. Then, necessarily, $h(z) = 0$ for any $z \in \mathbb{D}$. (Koosis [1980, Theorem III. B].)

We next show that the aRNP is equivalent to the existence of radial limits for $H^p(\mathbb{D}, X)$, $0 < p < \infty$.

Theorem 5.2.2 *Let X be a complex Banach space satisfying,*

$$H^\infty(\mathbb{D}, X) = H^\infty(\mathbb{T}, X).$$

Then for $0 < p < \infty$ *and* $f \in H^p(\mathbb{D}, X)$*, there exists* $M_f \subseteq \mathbb{T}$ *with* $m(\mathbb{T}\backslash M_f) = 0$ *such that* $\lim_{r\to 1} f(r\zeta)$ *exists for any* $\zeta \in M_f$.

Proof Take $f \in H^p(\mathbb{D}, X)$. In view of Theorem 5.2.1, we may write $f = g/h$ where $g \in H^\infty(\mathbb{D}, X)$ and $h \in H^\infty(\mathbb{T})$ such that $h(z) \neq 0$ for any $z \in \mathbb{D}$. Then $s(\zeta) = \lim_{r\to 1} h(r\zeta)$ exists almost everywhere with $s(\zeta) \neq 0$ in a set of full measure. By assumption, g has radial limits almost everywhere, hence the same is true for the quotient $f = g/h$. □

Poisson Integrals

Our first main result in this section is Theorem 5.2.5, which asserts that, for any complex Banach space X and $f \in H^\infty(\mathbb{D}, X)$, the following conditions are equivalent:

(1) $f \in H^\infty(\mathbb{T}, X)$;
(2) There exists $M_f \subseteq \mathbb{T}$ such that $m(\mathbb{T}\backslash M_f) = 0$ and $\lim_{r\to 1} f(r\zeta)$ exists in X, for any $\zeta \in M_f$.

Thus we obtain a vector-valued version of Fatou's theorem. The second main result is Bu's theorem (Theorem 5.2.6)), which strengthens the above equivalence as follows: If X fails the aRNP then there exists $f \in H^\infty(\mathbb{D}, X)$ such that $f(r\zeta)$, $(0 < r < 1)$ diverges, for almost all $\zeta \in \mathbb{T}$.

We turn to Fatou's theorem. The family of Poisson kernels $P(z, w) = (1 - |z|^2)/(|z - w|^2)$ gives rise to positive contractions on $L^1(\mathbb{T})$. If $0 < r < 1$ and $\zeta \in \mathbb{T}$, we define

$$P_r f(\zeta) = \int_{\mathbb{T}} P(r\zeta, w) f(w) dm(w),$$

where $\zeta \in \mathbb{T}$. Minkowski's inequality gives $\|P_r f\|_1 \leq \|f\|_1$, and by inspection, $P_r f \geq 0$ for $f \geq 0$.

Fatou's theorem describes the limits $P_r f$ as $r \to 1$.

Theorem 5.2.3 (Fatou' theorem) *Let* $f \in L^1(\mathbb{T})$*. Then there exists* $M_f \subseteq \mathbb{T}$ *with* $m(\mathbb{T}\backslash M_f) = 0$ *such that* $\lim_{r\to 1} P_r f(\zeta)$ *exists for* $\zeta \in M_f$.

Let X be a complex Banach space. On $L^1(\mathbb{T})\otimes X$ we introduce the vector-valued extension of P_r by defining,

$$\mathcal{P}_r\left(\sum_{j=1}^{n} x_j f_j\right) = \sum_{i=1}^{n} x_j P_r(f_j),$$

where $x_j \in X$ and $f_j \in L^1(\mathbb{T})$. Since $L^1(\mathbb{T}) \otimes X$ is dense in $L^1(\mathbb{T}, X)$, Theorem 2.1.1 shows that $\mathcal{P}_r$ uniquely defines an operator on $L^1(\mathbb{T}, X)$ satisfying

$$\|\mathcal{P}_r : L^1(\mathbb{T}, X) \to L^1(\mathbb{T}, X)\| \leq 1.$$

In other words, independent of the Banach space X, the metric properties of $\mathcal{P}_r$ are the same as for P_r itself. We now show that Fatou's theorem (Theorem 5.2.3) (on the almost everywhere convergence of P_r) carries over to vector-valued Poisson integrals.

Theorem 5.2.4 *Let X be a Banach space. For any $f \in L^1(\mathbb{T}, X)$ there exists $M_f \subseteq \mathbb{T}$ with $m(\mathbb{T}\setminus M_f) = 0$ such that the radial limit $\lim_{r\to 1} \mathcal{P}_r(f)(\zeta)$ exists in X, for any $\zeta \in M_f$.*

Proof In the first step we show that Fatou's theorem implies that the radial maximal function $T(f) = \sup_{r\to 1} |P_r(f)|$ is a continuous sublinear operator

$$T : L^1(\mathbb{T}) \to L^0(\mathbb{T}),$$

where the metric on the vector space $L^0(\mathbb{T})$ is given by the measure distance

$$d_m(f, g) = \inf\{\epsilon > 0 \colon m\{|f - g| > \epsilon\} < \epsilon\}.$$

In the second step we exploit that each P_r is a positive contraction on $L^1(\mathbb{T})$, and show that for any Banach space X,

$$\mathcal{T} : L^1(\mathbb{T}, X) \to L^0(\mathbb{T}), \quad \mathcal{T}(f) = \sup_{r\to 1} \|\mathcal{P}_r(f)\|_X,$$

is a continous sublinear operator. In the remaining steps we prove that this property of $\mathcal{T}$ (expressed in condition (5.2.3)) implies the existence of $\lim_{r\to 1} \mathcal{P}_r(f)(\zeta)$ in X, for almost every $\zeta \in \mathbb{T}$.

Step 1 We first note that by Fatou's theorem, for $\epsilon > 0$ there exists $R_\epsilon > 0$ such that

$$m\left(\left\{\sup_{0<r<1} |P_r(f)| > R_\epsilon \|f\|_1\right\}\right) < \epsilon. \tag{5.2.1}$$

Suppose, to the contrary, that there exists $\epsilon_0 > 0$ and a sequence $f_n \in L^1(\mathbb{T})$ with $f_n > 0$ and $\|f_n\|_1 \leq 1$, such that $\Omega_n = \{\sup_{0<r<1} P_r(f_n) > n2^n\}$ satisfies $m(\Omega_n) > \epsilon_0$. Now consider

$$f = \sum 2^{-n} f_n,$$

with $\|f\|_1 \leq 1$. Since P_r is a positive operator, we have $P_r(f) \geq P_r(2^{-n} f_n)$, yielding $\{\sup_{0<r<1} P_r(f) > n\} \supseteq \Omega_n$, and

$$m\left(\left\{\sup_{0<r<1} P_r(f) = \infty\right\}\right) \geq \inf_{n\in\mathbb{N}} m(\Omega_n). \tag{5.2.2}$$

Since $f \in L^1(\mathbb{T})$ and $m(\Omega_n) > \epsilon_0$, the measure estimate (5.2.2) contradicts Fatou's theorem.

Step 2 The estimate (5.2.1) carries over to the vector-valued setting as follows: For $f \in L^1(\mathbb{T}, X)$ and $\epsilon > 0$ we have

$$m\left(\left\{\sup_{0<r<1} \|\mathcal{P}_r(f)\|_X > R_\epsilon \|f\|_{L^1(X)}\right\}\right) < \epsilon, \tag{5.2.3}$$

where $R_\epsilon > 0$ is the same as in the scalar estimate (5.2.1). To see this we fix $f \in L^1(\mathbb{T}, X)$ satisfying $\|f\|_1 = 1$ and a linear functional $x^* \in X^*$ with $\|x^*\| = 1$. By linearity, $\langle \mathcal{P}_r f, x^*\rangle = P_r\langle f(\cdot), x^*\rangle$. For $\zeta \in \mathbb{T}$, put $g(\zeta) = \|f(\zeta)\|_X$. Since P_r is a positive operator, we have $P_r\langle f(\cdot), x^*\rangle \leq P_r(g)$. Invoking the Hahn–Banach theorem gives the pointwise bounds

$$\|\mathcal{P}_r(f)\|_X \leq P_r(g). \tag{5.2.4}$$

Since $g \in L^1(\mathbb{T})$ with $\|g\|_{L^1(\mathbb{T})} = 1$, the estimate (5.2.4) gives the inclusion

$$\left\{\sup_{0<r<1} \|\mathcal{P}_r(f)\|_X > R_\epsilon\right\} \subseteq \left\{\sup_{0<r<1} |P_r(g)| > R_\epsilon\right\}.$$

Applying estimate (5.2.1) to g gives condition (5.2.3).

Step 3 Let $\eta > 0$, $h \in L^1(\mathbb{T}, X)$, and

$$A^\eta(h) = \bigcap_{0<t<1} \bigcup_{t<r,s<1} \{\zeta \in \mathbb{T} : \|\mathcal{P}_r(h)(\zeta) - \mathcal{P}_s(h)(\zeta)\|_X > \eta\}.$$

If g is a finite linear combination of the terms $x_i \otimes g_i$ with $x_i \in X$ and $g_i \in L^1$, then $m(A^\eta(g)) = 0$.

Let $\epsilon > 0$, let R_ϵ be determined by condition (5.2.3), and let $\delta = \eta/R_\epsilon$. Given $f \in L^1(\mathbb{T}, X)$, we choose $g \in L^1(\mathbb{T}) \otimes X$ such that $\|f - g\|_1 \leq \delta$. Applying condition (5.2.3) to $f - g$ shows that the set $m(A^\eta(f - g)) \leq \epsilon$. In summary, $m(A^\eta(f)) = 0$, for $\eta > 0$ and $f \in L^1(\mathbb{T}, X)$.

Step 4 Define $M_f = \mathbb{T}\setminus\bigcup A^{1/n}(f)$, where the union is taken over $n \in \mathbb{N}$. Now $m(A^{1/n}(f)) = 0$ as we showed above, hence $m(\mathbb{T}\setminus M_f) = 0$. Finally, for $\zeta \in M_f$, and any choice of $0 < r_k < 1$ with $\lim r_k = 1$, the sequence $\mathcal{P}_{r_k}f(\zeta)$ is a Cauchy sequence in X. □

Theorem 5.2.5 *Let X be a complex Banach space, and let $f \in H^\infty(\mathbb{D}, X)$. Then the following conditions are equivalent:*

(i) *There exists $M_f \subseteq \mathbb{T}$ with $m(\mathbb{T}\setminus M_f) = 0$ such that $\lim_{r\to 1} f(r\zeta)$ exists for any $\zeta \in M_f$.*
(ii) *There exists $g \in L^\infty(\mathbb{T}, X)$ such that $f(z) = \int_{\mathbb{T}} P(z,\zeta)g(\zeta)dm(\zeta)$, for any $z \in \mathbb{D}$.*

Proof Let $f \in H^\infty(\mathbb{D}, X)$. In view of Theorem 5.2.3 it suffices to show that (i) implies (ii). Assume that the radial limits of f exist almost everywhere. Let $M_f \subseteq \mathbb{T}$ with $m(\mathbb{T}\setminus M_f) = 0$ be given by condition (i), and define

$$g(\zeta) = \lim_{r\to 1} f(r\zeta), \tag{5.2.5}$$

for any $\zeta \in M_f$. Clearly $g \in L^\infty(\mathbb{T}, X)$. We now show that f is the Poisson integral of g. To this end we fix $r, s \in (0, 1)$ with $(r + 1)/2 < s$. Then,

$$f(r\zeta) = \int_{\mathbb{T}} P\left(\frac{r\zeta}{s}, w\right) f(sw)dm(w). \tag{5.2.6}$$

Note that on the right-hand side we are taking the integral over terms that are uniformly bounded when $(r + 1)/2 < s < 1$. Hence by Lebesgue's dominated convergence theorem, we have

$$\lim_{s\to 1} \int_{\mathbb{T}} P\left(\frac{r\zeta}{s}, w\right) f(sw)dm(w) = \int_{\mathbb{T}} \lim_{s\to 1} P\left(\frac{r\zeta}{s}, w\right) f(sw)dm(w).$$

Thus in view of (5.2.5) and (5.2.6) we have shown that f is the Poisson integral of g. □

Bu's Theorem

Let X be a complex Banach space. We defined that X satisfies the aRNP if

$$H^\infty(\mathbb{D}, X) = H^\infty(\mathbb{T}, X).$$

By Theorem 5.2.5, a Banach space X satisfies the aRNP, if and only if for any $f \in H^\infty(\mathbb{D}, X)$, the radial limit $\lim_{r\to 1} f(r\zeta)$ exists for almost every $\zeta \in \mathbb{T}$. Given $f \in H^\infty(\mathbb{D}, X)$ and $0 \le r < 1$, we define $f_r \in H^\infty(\mathbb{T}, X)$ by putting

$$f_r(\zeta) = f(r\zeta), \qquad \zeta \in \mathbb{T}.$$

Thus, by Theorem 5.2.5, if a Banach space X fails to have the aRNP, then there exists $f \in H^\infty(\mathbb{D}, X)$ and a sequence $r_n \in (0,1)$ with $\lim r_n = 1$ such that the sequence $f_{r_n} : \mathbb{T} \to X$ diverges on a subset of positive measure.

This section presents the Theorem of Bu (1988), which asserts that, if a Banach space X fails the aRNP then there exists a single element $g \in H^\infty(\mathbb{D}, X)$ for which $g_r : \mathbb{T} \to X$ is *divergent* almost everywhere. This is obtained by spreading and distributing the singularities of X-valued bounded analytic functions.

Theorem 5.2.6 (Bu's theorem) *Let X be a Banach space that fails the analytic Radon–Nikodym property (aRNP). Then there exist $g \in H^\infty(\mathbb{D}, X)$, a sequence $r_n \in (0,1)$ with $\lim r_n = 1$, and $\eta > 0$ such that*

$$\limsup_{k,n\to\infty} \|g_{r_n}(\zeta) - g_{r_k}(\zeta)\|_X > \eta,$$

for almost every $\zeta \in \mathbb{T}$.

Proof Assume that X does not satisfy the aRNP. Choose $f \in H^\infty(\mathbb{D}, X)$ with $\|f\| \leq 1$ such that f does not have radial limits almost everywhere. Let $r_n \in (0,1)$ with $\lim r_n = 1$ and $\eta > 0$ such that

$$A = \left\{\zeta \in \mathbb{T} : \limsup_{k,n\to\infty} \|f_{r_n}(\zeta) - f_{r_k}(\zeta)\|_X > 2\eta\right\} \tag{5.2.7}$$

satisfies $m(A) > 0$. By Lemma 5.2.7, there exists a sequence $z_i \in \mathbb{T}$ such that $m(\mathbb{T}\setminus \bigcup z_i A) = 0$, where $zA = \{z\zeta : \zeta \in A\}$. Putting $B_1 = z_1 A$ and $B_i = z_i A \setminus \bigcup_{j=1}^{i-1} B_j$ yields a sequence of pairwise disjoint, measurable sets satisfying $\bigcup B_i = \bigcup z_i A$.

Step 1 (Construction of g) Choose $\epsilon_j > 0$ with $\sum \epsilon_j < \eta/8$, and let $\alpha_j(\zeta) = (1-\epsilon_j)1_{B_j}(\zeta) + \epsilon_j$. Thus $\ln \alpha_j \in L^1(\mathbb{T})$, and we may form the outer function $a_j \in H^\infty(\mathbb{T})$ satisfying $|a_j| = \alpha_j$. That is, $a_j = \exp(\ln \alpha_j + iH(\ln \alpha_j))$ where H denotes the Hilbert transform. Next put

$$h_j(z) = a_j(z) f(z \cdot \overline{z_j}),$$

where $z \in \mathbb{D}$ and let $g_i(z) = \sum_{j=1}^{i-1} h_j(z)$. By construction, $g_i, h_j \in H^\infty(\mathbb{D}, X)$. We claim that $\|g_i(z)\|_X \leq 1 + \eta/8$ and that for any $z \in \mathbb{D}$ the sequence $\{g_i(z)\}$ is convergent in X. Thus, we claim that the limit $g(z) = \lim_{i\to\infty} g_i(z)$ is well defined for any $z \in \mathbb{D}$, and $g \in H^\infty(\mathbb{D}, X)$.

First, since $\{B_i\}$ is a sequence of pairwise disjoint, measurable sets and since $|a_j(\zeta)| = (1-\epsilon_j)1_{B_j}(\zeta) + \epsilon_j$, the triangle inequality gives $\|g_i(z)\|_X \leq 1 + \sum \epsilon_j$. Second, if $n < k$ we have

$$\limsup_{r\to 1} \|g_n(r\zeta) - g_k(r\zeta)\|_X \le \sum_{j=n}^{\infty} \epsilon_j,$$

whenever $\zeta \in \bigcup_{j=1}^{n} B_j$. Since $\sum_{j=n}^{\infty} \epsilon_j \to 0$ and $m\left(\mathbb{T} \setminus \bigcup_{j=1}^{n} B_j\right) \to 0$ as $n \to \infty$, convolving with Poisson kernels shows that $\{g_i(z)\}$ is a Cauchy sequence for any $z \in \mathbb{D}$; hence it is convergent in X. In summary, the limit $g(z) = \lim_{i\to\infty} g_i(z)$ is well defined and $g \in H^\infty(\mathbb{D}, X)$.

Step 2 (Radial divergence almost everywhere) We next prove that g satisfies the conclusion of the theorem. To this end, we fix $j_0 \in \mathbb{N}$, $\zeta \in B_{j_0}$ and write

$$(a_{j_0} g)(r\zeta) = a_{j_0}(r\zeta)^2 f_r(\zeta \cdot \overline{z_{j_0}}) + R_{j_0}(r\zeta), \tag{5.2.8}$$

where $R_{j_0}(r\zeta) = \sum_{j\neq j_0} (h_j a_{j_0})(r\zeta)$. By comparing $|a_{j_0}|$ and $|a_j|$ for $j \neq j_0$, we get the pointwise estimate $\|R_{j_0}(r\zeta)\| \le \eta/8$. This allows us to regard R_{j_0} as a negligible error term. Since the scalar-valued outer function a_{j_0} is uniformly bounded, it has radial limits almost everywhere. Hence for almost every $\zeta \in B_{j_0}$ and k, n large enough, we have

$$\|g_{r_n}(\zeta) - g_{r_k}(\zeta)\|_X \ge \|(a_{j_0} g)_{r_n}(\zeta) - (a_{j_0} g)_{r_k}(\zeta)\|_X - \eta/8. \tag{5.2.9}$$

In view of (5.2.8) and the pointwise estimate for R_{j_0}, the right-hand side of estimate (5.2.9) is larger than

$$\left\|a_{j_0}(r_n\zeta)^2 f_{r_n}(\zeta \cdot \overline{z_{j_0}}) \quad a_{j_0}(r_k\zeta)^2 f_{r_k}(\zeta \cdot \overline{z_{j_0}})\right\|_X - 3\eta/8. \tag{5.2.10}$$

Again invoking that a_{j_0} has radial limits almost everywhere, the term in (5.2.10) is larger than

$$\|f_{r_n}(\zeta \cdot \overline{z_{j_0}}) - f_{r_k}(\zeta \cdot \overline{z_{j_0}})\|_X - \eta/2,$$

whenever k, n are large enough. Finally, since $B_{j_0} \subseteq z_{j_0} A$ for any $\zeta \in B_{j_0}$, we have $\zeta \cdot \overline{z_{j_0}} \in A$, and we infer from equation (5.2.7) that $\|f_{r_n}(\zeta \cdot \overline{z_{j_0}}) - f_{r_k}(\zeta \cdot \overline{z_{j_0}})\|_X > 2\eta$.

Summing up, for almost every $\zeta \in \bigcup B_j$ and $N \in \mathbb{N}$ there exist $k, n > N$ such that $\|g_{r_n}(\zeta) - g_{r_k}(\zeta)\|_X \ge \eta$. Since $m(\mathbb{T} \setminus \bigcup B_j) = 0$, this completes the proof □

Lemma 5.2.7 *Given any measurable $A \subseteq \mathbb{T}$ with $m(A) > 0$, there exists a sequence $z_i \in \mathbb{T}$ such that $m(\mathbb{T} \setminus \bigcup z_i A) = 0$, where $zA = \{z\zeta : \zeta \in A\}$.*

Proof The first part of the proof consists in showing that for any $B \subset \mathbb{T}$ with $0 < m(B) < 1$ and any $A \subseteq \mathbb{T}$ with $m(A) > 0$, there exists $z_0 \in \mathbb{T}$ such that $m(B \cup z_0 A) > m(B)$.

To this end we put $C = \mathbb{T} \setminus B$. By assumption we have $m(C) > 0$. We now prove the existence of $z_0 \in \mathbb{T}$ such that $m(C \cap z_0 A) > 0$. Note that for any $z \in \mathbb{T}$

we have $m(C \cap zA) = \int 1_C 1_{(zA)} dm$. Taking the expectation over $z \in \mathbb{T}$ and using Fubini's theorem, we obtain that

$$\begin{aligned}\int_{\mathbb{T}} m(C \cap zA)dm(z) &= \int_{\mathbb{T}} \int_{\mathbb{T}} 1_C(w) 1_{zA}(w) dm(w) dm(z) \\ &= \int_{\mathbb{T}} 1_C(w) \int_{\mathbb{T}} 1_A(w\bar{z}) dm(z) dm(w) \\ &= m(C)m(A).\end{aligned}$$

Since $m(C)m(A) > 0$, the above shows that there exists $z_0 \in \mathbb{T}$ such that $m(C \cap z_0 A) > 0$.

The second part of the proof exploits a measure-theoretic exhaustion argument. We put $\alpha = \sup \left\{ m(\bigcup z_i A) : (z_i) \in \mathbb{T}^{\mathbb{N}} \right\}$. Resolving the supremum defining α gives a sequence $z^{(k)} \in \mathbb{T}^{\mathbb{N}}$ with $z^{(k)} = \left(z_i^k\right)$ and $z_i^k \in \mathbb{T}$ such that

$$\alpha = \lim_{k \to \infty} m\left(\bigcup_{i \in \mathbb{N}} z_i^k A\right).$$

Next we simply put $B = \bigcup_{i,k \in \mathbb{N}} z_i^k A$. Since $\alpha \geq m(B) \geq m\left(\bigcup_{i \in \mathbb{N}} z_i^k A\right)$, we have $m(B) = \alpha$. If $m(B) = 1$, the desired sequence is $\left\{z_i^k : i, k \in \mathbb{N}\right\}$. If, however, $m(B) < 1$ then by the first part of the proof, there exists $z_0 \in \mathbb{T}$ such that $m(B \cup z_0 A) > m(B)$, and this contradicts the definition of α. □

An important **open problem,** related to Bu's theorem (Theorem 5.2.6), asks the following: Is it true, that for any Banach space X failing the aRNP, there exists a bounded analytic function $f : \mathbb{D} \to X$ such that $f(\zeta r)$ diverges for *any* $\zeta \in \mathbb{T}$ as $r \to 1$. Very strong support for the conjecture is provided by the theorem of Bu and Khaoulani (1990), who proved that if a Banach space X is isomorphic to $X \times X$ and fails the aRNP, then there exist $F \in H^\infty(\mathbb{D}, X \times X)$ and an increasing sequence (r_n) with $\lim r_n = 1$ such that,

for $z, z' \in \mathbb{D}$ satisfying $|z| \leq r_n < r_{n+1} \leq |z'|$ we have $\|F(z) - F(z')\|_{X \times X} \geq 1$.

5.2.2 Representable Operators and Factorization

Let $T : L^1(\mathbb{T}) \to X$ be bounded and linear. We say that T is a representable operator if there exists $h \in L^\infty(\mathbb{T}, X)$ such that

$$Tf = \int_{\mathbb{T}} fh dm,$$

for any $f \in L^1(\mathbb{T})$. Our next theorem relates representable operators and (the existence of) radial limits of bounded, vector-valued, analytic functions in the unit disk.

Theorem 5.2.8 *Let X be a complex Banach space, let $T\colon L^1(\mathbb{T}) \to X$ be a bounded, linear operator, and set $p_z(\zeta) = P(z, \zeta)$, where $z \in \mathbb{D}$ and $\zeta \in \mathbb{T}$. The following conditions are equivalent:*

(i) *The map $f_T(z) = T(p_z)$ defines an element in $H^\infty(\mathbb{D}, X)$.*
(ii) *For any $h \in H_0^1(\mathbb{T})$ we have $T(h) = 0$.*

Moreover, T is a representable operator if and only if $\lim_{r\to 1} f_T(r\zeta)$ exists for almost every $\zeta \in \mathbb{T}$.

Proof First, assume that $T(h) = 0$ for $h \in H_0^1(\mathbb{T})$. Recall that the trigonometric series of the Poisson kernel is given as $p_z(\zeta) = \sum_{n=-\infty}^{\infty} z^{-n}\zeta^n$, where the series converges in $L^1(\mathbb{T})$. By assumption, $T(\zeta^n) = 0$ for any $n \in \mathbb{N}$. Hence,

$$T(p_z) = \sum_{n=0}^{\infty} z^n c_n, \tag{5.2.11}$$

where $c_n = T(\zeta^{-n})$. Since clearly $\|c_n\| \le \|T\|$, the power series on the right-hand side converges for any $z \in \mathbb{D}$, and since $\|T(p_z)\| \le \|T\|$, the series on the right-hand side of (5.2.11) is uniformly bounded. Hence for $f_T(z) = T(p_z)$, we proved that $f_T \in H^\infty(\mathbb{D}, X)$.

Now set $f = f_T$. We assume that $f \in H^\infty(\mathbb{D}, X)$ with Taylor series

$$f(z) = \sum_{n=0}^{\infty} z^n c_n,$$

where $c_n \in X$. Below we define a bounded operator $R_f\colon L^1(\mathbb{T}) \to X$ such that $R_f(p_z) = T(p_z)$ and $R_f(h) = 0$ for any $h \in H_0^1(\mathbb{T})$. Since the span of $\{p_z : z \in \mathbb{D}\}$ is dense in the unit ball of $L^1(\mathbb{T})$, this shows that $T = R_f$ and that, consequently, T annihilates $H_0^1(\mathbb{T})$ as well. Now let $g \in L^1(\mathbb{T})$ with $g(\zeta) = \sum a_k\zeta^k$, where only finitely many of the coefficients are nonvanishing. Then define

$$R_f g = \sum_{k=-\infty}^{\infty} a_k c_k, \tag{5.2.12}$$

where $c_k \in X$ appears in the Taylor expansion of f. Since $a_{-k} = \int g(\zeta)\zeta^k dm$, we obtain

$$R_f g = \lim_{r\to 1} \int_{\mathbb{T}} g(\zeta) f(r\zeta) dm(\zeta),$$

and hence $\|R_f g\|_X \le \|g\|_1 \|f\|_\infty$. The series expansion of p_z gives $R_f(p_z) = f(z)$, and, by definition (5.2.12), we immediately have that $R_f h = 0$ for $h \in H_0^1$.

We next prove that the radial limits of $f = f_T$ yield a representation of T and vice versa. First, recall that T is representable if there exists $h \in L^\infty(\mathbb{T}, X)$ such that $Tg = \int ghdm$, for any $g \in L^1(\mathbb{T})$. Since the span of $\{p_z : z \in \mathbb{D}\}$ is dense in the unit ball of $L^1(\mathbb{T})$ and since T is bounded, this is equivalent to saying that $T(p_z) = \int p_z h dm$, for any $z \in \mathbb{D}$.

Finally, put $f(z) = T(p_z)$. By Theorem 5.2.5, the existence of radial limits $\lim_{r \to 1} f(r\zeta)$ almost everywhere is equivalent to the existence of $h \in L^\infty(\mathbb{T}, X)$ satisfying $f(z) = \int p_z h dm$, for any $z \in \mathbb{D}$. □

Remarks: Theorem 5.2.8 (and its proof) may be paraphrased by saying that $H^\infty(\mathbb{D}, X)$ is isometrically isomorphic to $L(L^1/H_0^1, X)$, the Banach space of all bounded linear operators $S\colon L^1/H_0^1 \to X$.

(i) As shown in the course of proving Theorem 5.2.8, for $f \in H^\infty(\mathbb{D}, X)$ the operator R_f defined by (5.2.12) satisfies $R_f(h) = 0$ for all $h \in H_0^1(\mathbb{T})$ and that $\|f\|_\infty = \|R_f\|$.

(ii) Let Z denote the Banach space of operators $T\colon L^1 \to X$ satisfying $T\left(H_0^1\right) = 0$. By Theorem 5.2.8, the equation $f_T(z) = T(p_z)$ establishes an isometric isomorphism between Z and $H^\infty(\mathbb{D}, X)$.

(iii) Let $q\colon L^1 \to L^1/H_0^1$ denote the canonical quotient map given by $q(g) = g + H_0^1$. Then $S \to S \circ q$ provides an isometric isomorphism between $L\left(L^1/H_0^1, X\right)$ and Z.

Theorem 5.2.8 combined with the remarks above yields the theorem of Dowling (1985) and Hensgen (1986), which characterizes the aRNP in terms of representable operators.

Theorem 5.2.9 *A Banach space X satisfies the analytic Radon–Nikodym property if and only if, for each bounded linear operator $S\colon L^1/H_0^1 \to X$ there exists $h \in L^\infty(\mathbb{T}, X)$, representing $Sq\colon L^1(\mathbb{T}) \to X$ as*

$$(Sq)f = \int_{\mathbb{T}} fhdm,$$

for $f \in L^1(\mathbb{T})$.

We continue in Theorem 5.2.11 the characterization of the aRNP in terms of operators. Its proof is based on the factorization theorems of Ghoussoub and Rosenthal (1983) and Lewis and Stegall (1973). We first recall the Lewis–Stegall theorem.

Theorem 5.2.10 (Lewis–Stegall theorem) *A bounded linear operator $T\colon L^1 \to X$ is representable if and only if it admits a factorisation $T = LS$ where $S\colon L^1 \to \ell^1$ and $L\colon \ell^1 \to X$ are bounded linear operators.*

The proof of the Lewis–Stegall theorem is presented in the book by Diestel and Uhl (1977) and in the recent monograph by Pisier (2016).

We next turn to the theorem of Dowling and Edgar (1988), which characterizes the aRNP by a factorization of operators.

Theorem 5.2.11 *A complex Banach space X has the aRNP if and only if every operator $T: L^1 \to X$ that factors through L^1/H_0^1 is representable.*

Proof Assume that X has the aRNP, and that the bounded linear operator $T: L^1 \to X$ admits the factorization $T = AV$ where $V: L^1 \to L^1/H_0^1$, and $A: L^1/H_0^1 \to X$ are bounded and linear. We now show that T is representable.

Recall the F. and M. Riesz theorem, which asserts that H_0^1 is a dual space contained in L^1. (See Koosis [1980, Chapter II].) By the Ghoussoub–Rosenthal theorem (Theorem 4.3.3), there exists a bounded linear operator $J: L^1 \to L^1$ such that $V = qJ$ where $q: L^1 \to L^1/H_0^1$ is the canonical quotient map. Thus

$$T = AqJ.$$

Since X has the aRNP, the operator Aq is representable by Theorem 5.2.8. The Lewis–Stegall theorem asserts that Aq, and hence T, factors through ℓ^1. Again, by the Lewis–Stegall theorem, T is a representable operator. □

We next present Pisier's approach to constructing Banach spaces that fail the aRNP. The method yields natural function spaces obtained by interpolation and factorization of operators. Here our source is Ghoussoub et al. (1989, Section V). Consider the bounded, decreasing sawtooth function

$$\varphi(\zeta) = \begin{cases} 0 & \text{if } \zeta = e^{i0}, \\ \frac{1}{2} - \frac{t}{2\pi} & \text{if } \zeta - e^{it} \text{ and } (0 < t < 2\pi). \end{cases} \tag{5.2.13}$$

By direct calculation the Fourier series expansion of $\varphi: \mathbb{T} \to [-1/2, 1/2]$ is

$$\varphi(\zeta) = \frac{1}{2\pi i} \sum_{n \in \mathbb{Z}\setminus\{0\}} \frac{\zeta^n}{n}, \qquad \zeta \in \mathbb{T}. \tag{5.2.14}$$

Define the convolution operator E (together with its representation as a Fourier multipler) by

$$E(f)(\zeta) = \int_{\mathbb{T}} f(w)\varphi(\zeta \cdot \overline{w}) dm(w) = \frac{1}{2\pi i} \sum_{n \in \mathbb{Z}\setminus\{0\}} \frac{\widehat{f(n)}}{n} \zeta^n, \tag{5.2.15}$$

where $\zeta \in \mathbb{T}$, $f \in L^1(\mathbb{T})$. Thus $E: L^1(\mathbb{T}) \to C(\mathbb{T})$ is bounded, linear, and satisfies

$$E\left(H_0^1(\mathbb{T})\right) \subset A_0(\mathbb{T}) \quad \text{and} \quad E\left(\overline{H_0^1(\mathbb{T})}\right) \subset \overline{A_0(\mathbb{T})}. \tag{5.2.16}$$

In view of condition (5.2.16), E admits a well-defined extension to a bounded linear operator (still denoted E) between the corresponding quotient spaces,

$$E\colon L^1(\mathbb{T})/H_0^1(\mathbb{T}) \to C(\mathbb{T})/A_0(\mathbb{T}).$$

Theorem 5.2.12 *Let $q\colon L^1(\mathbb{T}) \to L^1(\mathbb{T})/H_0^1(\mathbb{T})$ denote the canonical quotient map. The operator*

$$Eq\colon L^1(\mathbb{T}) \to C(\mathbb{T})/A_0(\mathbb{T})$$

is not representable by a bounded, Bochner measurable $h\colon \mathbb{T} \to C(\mathbb{T})/A_0(\mathbb{T})$.

Proof Define $F\colon \mathbb{T} \to L^\infty(\mathbb{T})$ by $F(\zeta)(w) = \varphi(\zeta\overline{w})$. Given $g \in L^1(\mathbb{T})/H_0^1(\mathbb{T})$, $f \in H^1(\mathbb{T})$, and employing bilinear pairings $\langle\, ,\,\rangle$, we rewrite the effect of the kernel operator E as

$$\begin{aligned}\langle Eg, f\rangle &= \int_{\mathbb{T}} g(\zeta) \int_{\mathbb{T}} F(\zeta)(w) f(\zeta)\, dm(w)\, dm(\zeta)\\ &= \int_{\mathbb{T}} g(\zeta)\langle F(\zeta)(\cdot), f(\cdot)\rangle dm(\zeta).\end{aligned} \tag{5.2.17}$$

Now assume that Eq is a representable operator, i.e., that there exists a bounded Bochner measurable function $h\colon \mathbb{T} \to C(\mathbb{T})/A_0(\mathbb{T})$ such that

$$(Eq)g = \int_{\mathbb{T}} ghdm, \tag{5.2.18}$$

for $g \in L^1(\mathbb{T})$. Combining equations (5.2.18) and (5.2.17) gives

$$\int_{\mathbb{T}} g(\zeta)\langle F(\zeta)(\cdot), f(\cdot)\rangle dm(\zeta) = \int_{\mathbb{T}} g(\zeta)\langle h(\zeta)(\cdot), f(\cdot)\rangle dm(\zeta), \tag{5.2.19}$$

for $g \in L^1(\mathbb{T})$ and $f \in H^1(\mathbb{T})$.

Let $q_2\colon L^\infty(\mathbb{T}) \to L^\infty(\mathbb{T})/H_0^\infty(\mathbb{T})$ denote the natural projection operator, and define

$$G\colon \mathbb{T} \to L^\infty(\mathbb{T})/H_0^\infty(\mathbb{T})$$

by $G(\zeta) = q_2F(\zeta), \zeta \in \mathbb{T}$. Since in (5.2.19), $g \in L^1(\mathbb{T})$ and $f \in H^1(\mathbb{T})$ are arbitrary, (5.2.19) implies that, for almost every $\zeta \in \mathbb{T}$,

$$G(\zeta) = h(\zeta) \quad \text{as elements in } L^\infty(\mathbb{T})/H_0^\infty(\mathbb{T})\,. \tag{5.2.20}$$

Recall that h, being Bochner measurable, is separably valued. On the other hand, we claim that there exists $c_1 > 0$ such that for $\zeta \neq \zeta_1$,

$$\|G(\zeta) - G(\zeta_1)\|_{L^\infty(\mathbb{T})/H_0^\infty(\mathbb{T})} \geq c_1. \tag{5.2.21}$$

To see that condition (5.2.21) holds true, we first invoke the duality relation between the spaces $L^\infty(\mathbb{T})/H_0^\infty(\mathbb{T})$ and $H_0^1(\mathbb{T})$, yielding

$$\begin{aligned}&\|G(\zeta) - G(\zeta_1)\|_{L^\infty(\mathbb{T})/H_0^\infty(\mathbb{T})}\\&= \sup\left\{\int_{\mathbb{T}} [\varphi(\zeta\cdot\overline{w}) - \varphi(\zeta_1\cdot\overline{w})]f(w)dm(w) : f\in H^1(\mathbb{T}), \|f\|_1 \le 1\right\}.\end{aligned} \tag{5.2.22}$$

Put $z_0 = \zeta\cdot\zeta_1$ and set $\psi(\overline{w}) = \varphi(\overline{w}) - \varphi(z_0\cdot\overline{w})$. The right-hand side of (5.2.22) equals

$$\sup\left\{\int_{\mathbb{T}} \psi(\overline{w})f(w)dm(w) : f\in H^1(\mathbb{T}), \|f\|_1 \le 1\right\}. \tag{5.2.23}$$

Since $\zeta \neq \zeta_1$, we have $|z_0 - 1| = \delta_0 > 0$, and $w \to \psi(\overline{w})$ has a jump discontinuity at e^{i0},

$$\lim_{t\to 0,\, t>0} \psi(e^{it}) = \frac{1}{2}, \quad \text{and} \quad \lim_{t\to 0,\, t<0} \psi(e^{it}) = -\frac{1}{2}. \tag{5.2.24}$$

Otherwise, in $\{w\in\mathbb{T} : |w-1| < \delta_0/2\}\setminus\{e^{i0}\}$ the function ψ is continuous. Next, for $0 < \delta < \delta_0$ define $a_\delta \colon \mathbb{T} \to \mathbb{R}$ by

$$a_\delta(e^{it}) = \begin{cases} -\delta^{-1} & \text{if} \quad t\in[-\delta, 0],\\ \delta^{-1} & \text{if} \quad t\in[\delta, 0],\\ 0 & \text{else.}\end{cases} \tag{5.2.25}$$

For each δ, $a_\delta/2$ satisfies the conditions (1.1.68). Let H denote the Hilbert transform. We have $a_\delta + iH(a_\delta) \in H^1(\mathbb{T})$ and, by estimate (1.1.69), there exists $A_1 < \infty$ such that, for any $0 < \delta < 1$,

$$\int_{\mathbb{T}} |a_\delta + iH(a_\delta)|dm \le A_1. \tag{5.2.26}$$

Since ψ, a_δ are real valued, invoking condition (5.2.24) and definition (5.2.25) yields that there exists $c > 0$ such that for $0 < \delta < \delta_0/2$,

$$\left|\int_{\mathbb{T}} \psi(a_\delta + iH(a_\delta))dm\right| \ge \left|\int_{\mathbb{T}} \psi a_\delta dm\right| \ge c. \tag{5.2.27}$$

Combining the estimates (5.2.27) and (5.2.26) yields the lower bound (5.2.21) with $c_1 = c/A_1$. □

Theorem 5.2.13 *If X is a Banach space and if there exist bounded linear operators*

$$A\colon L^1(\mathbb{T})/H_0^1(\mathbb{T}) \to X \quad \textit{and} \quad B\colon X \to C(\mathbb{T})/A_0(\mathbb{T}),$$

such that the following diagram commutes,

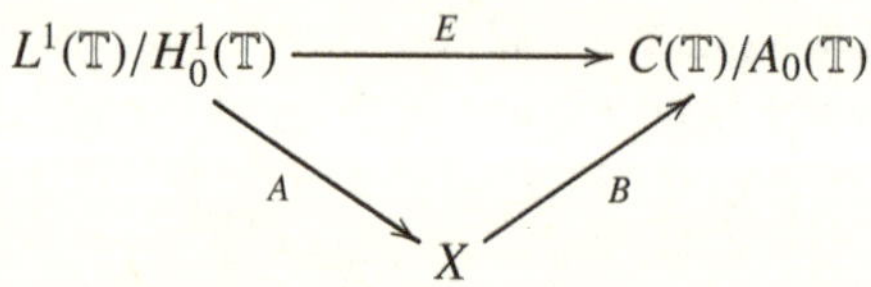

then X fails to satisfy the aRNP.

Proof Assume to the contrary that X satisfies the aRNP. Let $q\colon L^1(\mathbb{T}) \to L^1(\mathbb{T})/H^1_0(\mathbb{T})$ denote the canonical quotient map. By Theorem 5.2.9, the operator $Aq\colon L^1(\mathbb{T}) \to X$ is representable by a Bochner measurable $g\colon \mathbb{T} \to X$. Set

$$h(\zeta) = B \circ g(\zeta), \quad \zeta \in \mathbb{T}.$$

Clearly, $h\colon \mathbb{T} \to X$ is bounded, Bochner measurable, and provides a representation of $Eq = BAq$. Obviously, this contradicts Theorem 5.2.12. □

We now review the construction of the Davis–Figiel–Johnson–Pełczyński space of the operator $E\colon L^1/H^1_0 \to C/A_0$ (Davis et al., 1974). Also see Wojtaszczyk (1991, Chapter II). Define bounded, convex, and symmetric subsets of $C(\mathbb{T})/A_0(\mathbb{T})$,

$$V = \{y \in C/A_0\colon \|y\|_{C/A_0} < 1\} \quad \text{and} \quad W = \{Ew : \|w\|_{L^1/H^1_0} < 1\}.$$

Fix $n \in \mathbb{N}$ and let

$$U_n = 2^n W + 2^{-n} V.$$

Clearly we have the inclusions $2^{-n}V \subset U_n = (2^n\|E\| + 2^{-n})V$. Define a new norm on C/A_0 by

$$\|x\|_n = \inf\{\lambda > 0 : x \in \lambda U_n\}, \quad x \in C/A_0.$$

We observe the two-sided inequalities, $(2^n\|E\|+2^{-n})\|y\|_{C/A_0} \leq \|y\|_n \leq 2^n\|y\|_{C/A_0}$, and put

$$\|y\|_{Z_2} = \left(\sum_{n=1}^{\infty} \|y\|_n^2\right)^{1/2}, \quad y \in C/A_0.$$

Finally, define the Davis–Figiel–Johnson–Pełczyński space of E by

$$Z_2 = \{y \in C/A_0 : \|y\|_{Z_2} < \infty\}.$$

Defining $A\colon L^1/H^1_0 \to Z_2$ by $A(w) = E(w)$ and $B\colon Z_2 \to C/A_0$, by $B(y) = y$ gives bounded linear operators such that

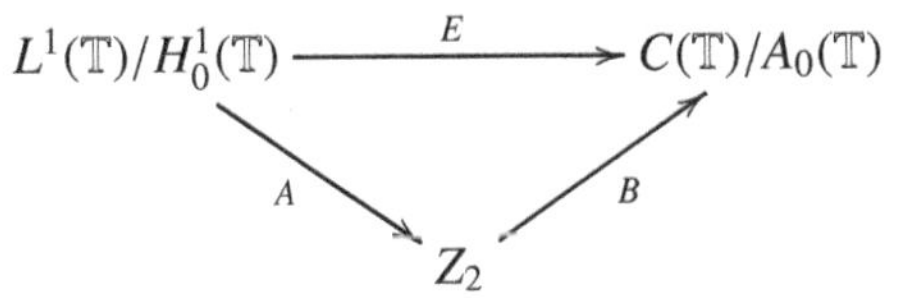

$$\|A\| \cdot \|B\| \leq 2\|E\|.$$

By Theorem 5.2.13, Z_2 is a Banach space for which the aRNP fails to hold. In contrast, Ghoussoub et al. (1989, Section V) show the following representation of $Z_2^{**}\backslash Z_2$: There exists a sequence of weak-* compact subsets $K_n \subset Z_2^{**}$ such that

$$\text{ball}\,(Z_2^{**})\,\backslash Z_2 = \bigcup K_n, \tag{5.2.28}$$

where ball(Z_2^{**}) denotes the unit ball of Z_2^{**}. Theorem 7.4.6 in Section 7.4.1 on holomorphic separation indicates that condition (5.2.28) is quite close to the aRNP.

Sections 5.2.4 and 5.2.5 review the topological and real variable motives related to the aRNP.

5.2.3 Transference

In this section we connect radial convergence of vector-valued analytic functions and the almost sure convergence of vector-valued Hardy martingales. Theorem 5.2.15 shows that a Banach space X satisfies the aRNP if and only if every L^1-bounded, X-valued Hardy martingale converges almost surely.

We begin by presenting a remarkable link between Brownian martingale convergence, radial limits, and Theorem 5.2.8 on representing operators. Let Ω denote Wiener's filtered probability space, and $\{z_t : t \in \mathbb{R}^+\}$ be complex Brownian motion started at $0 \in \mathbb{C}$. We fix the sequence $r_n = 1 - 2^{-n}$ and define the stopping times $\tau_0 = 0$ and

$$\tau_n = \inf\{t > \tau_{n-1} : |z_t| > r_n\}, \qquad \tau = \inf\{t > 0 : |z_t| > 1\}.$$

We let $\mathcal{F}_n$ and $\mathcal{F}$ denote the stopping times sigma-algebras associated to τ_n and τ, respectively. Since $\tau_n \uparrow \tau$, the sigma-algebra $\mathcal{F}$ is generated by the sequence of increasing sigma-algebras $\mathcal{F}_n$. By Ito's formula (1.1.49), stated after Theorem 1.1.14, for any bounded harmonic $f : \mathbb{D} \to X$, the sequence $f_n = f(z_{\tau_n})$ forms a martingale with respect to the filtration $(\mathcal{F}_n)$. If, moreover, f is analytic, then, by Theorem 2.3.3, (f_n) may be represented as a *Hardy martingale*. This makes it possible to link radial limits for $f : \mathbb{D} \to X$ and Brownian martingale convergence for $(f(z_{\tau_n}))$.

Theorem 5.2.14 *Let X be a complex Banach space and $f \in H^\infty(\mathbb{D}, X)$. Then the following conditions are equivalent:*

(i) *The radial limits of f exist almost everywhere.*
(ii) *The martingale $f_n = f(z_{\tau_n})$ converges almost surely.*

Proof Assume that $f_n = f(z_{\tau_n})$ converges almost surely on Wiener space. Let $G = \lim f_n$. Since the sequence $\{f_n\}$ is uniformly bounded we have $G \in L^\infty(\Omega, X)$. Consider the linear operator $T: L^1(\mathbb{T}) \to X$ given by

$$Tg = \int_{\mathbb{T}} NG \cdot g dm,$$

where $N: L^\infty(\Omega, X) \to L^\infty(\mathbb{T}, X)$ denotes Doob's projection operator. Since N is a contraction between L^∞ spaces, T is bounded, linear, and represented by NG. By Theorem 5.2.8 the map $f_0(w) = T(p_w)$ is bounded and admits radial limits almost everywhere. Hence we need to show that $f_0(w) = f(w)$ for any $w \in \mathbb{D}$. To see this, note that by the covariance formula, Doob's projection satisfies the identity

$$\int_{\mathbb{T}} g(NG) dm = \mathbb{E}(g(z_\tau)G),$$

where the expectation on the right-hand side is taken over Wiener space. Then with Lebesgue's convergence theorem we obtain

$$\mathbb{E}(p_w(z_\tau)G) = \mathbb{E}\left(p_w(z_\tau) \lim_{n\to\infty} f_n\right) = \lim_{n\to\infty} \mathbb{E}(p_w(z_\tau) f(z_{\tau_n})).$$

On the other hand, by the strong Markov property of Brownian motion and by the harmonicity of f we have

$$\mathbb{E}\left(p_w(z_\tau) f\left(z_{\tau_n}\right)\right) = \int_{\mathbb{T}} p_w(\zeta) f(r_n\zeta) dm(\zeta) = f(r_n w).$$

Passing to the limit as $n \to \infty$, we finally obtain

$$\lim_{n\to\infty} \mathbb{E}\left(p_w(z_\tau) f\left(z_{\tau_n}\right)\right) = f(w).$$

We have already observed that $\mathbb{E}(p_w(z_\tau)G) = T(p_w)$. Hence $f_0(w) = f(w)$. The same calculation also shows that $T(\zeta^n) = 0$ for any $n \in \mathbb{N}$. Next assume that $g(\zeta) = \lim_{r\to 1} f(r\zeta)$ exists almost everywhere. Then by (the conditional form of) Lebesgue's convergence theorem,

$$\mathbb{E}(g(z_\tau)|\mathcal{F}_n) = \lim_{r\to 1} \mathbb{E}(f(rz_\tau)|\mathcal{F}_n).$$

By Ito's theorem (Theorem 1.1.14) and Doob's sampling theorem (Theorem 1.1.9) we have

$$\mathbb{E}(f(rz_\tau)|\mathcal{F}_n) = f(rz_{\tau_n}),$$

and hence the limit on the right-hand side of the above equation is just $f_n = f(z_{\tau_n})$. Thus we have shown that

$$\mathbb{E}(g(z_\tau)|\mathcal{F}_n) = f_n,$$

and that (f_n) is a closed martingale. Hence (f_n) converges almost surely and in $L^1(\Omega, X)$. □

Theorem 5.2.15 *Let X be a complex Banach space. Then the following conditions on the Banach space X are equivalent:*

(i) *Every L^1-bounded, X-valued Hardy martingale converges almost surely.*
(ii) *For $h \in H^1(\mathbb{D}, X)$, radial limits $\lim_{r\to 1} h(r\zeta)$ exist almost everywhere. (The Banach space X has the aRNP.)*

Proof We prove that condition (i) implies (ii). Assume that every L^1-bounded, X-valued Hardy martingale converges almost surely. By Theorem 5.2.2 it suffices to show that for every $f \in H^\infty(\mathbb{D}, X)$, radial limits $\lim_{r\to 1} f(r\zeta)$ exist almost everywhere.

Fix $f \in H^\infty(\mathbb{D}, X)$. Put $f_n = f(z_{\tau_n})$. By Theorem 2.3.3 and the remarks thereafter, there exists an L^∞-bounded, X-valued Hardy martingale (F_n) distributionally equivalent to the Brownian martingale (f_n). By assumption on X, the Hardy martingale (F_n) converges almost surely on $\mathbb{T}^\mathbb{N}$. Theorem 2.3.3 asserts that (f_n) converges on Wiener space and by Theorem 5.2.14 we infer that $\lim_{r\to 1} f(r\zeta)$ exists almost everywhere in $\mathbb{T}$.

We next prove that condition (ii) implies (i). We assume that each $f \in H^1(\mathbb{D}, X)$ has radial limits almost everywhere. Let (F_n) be an L^1-bounded, X-valued Hardy martingale. We use a transfer method to show that (F_n) converges almost surely. We first treat the special case where each F_n is a trigonometric polynomial. Once this is done the general case follows by a Borel–Cantelli argument.

To prepare for the transfer we introduce convenient notation. Given $y \in \mathbb{T}$ and an integer sequence $M = (M(k))$ we let $y_k = y^{M(k)}$ and define $y^M \in \mathbb{T}^\mathbb{N}$ by $y^M = (y_k)$. Next, write

$$G_k(y, z, M) = F_k(z \cdot y^M),$$

where $z \in \mathbb{T}^\mathbb{N}$ with $z = (z_k)$ and $z \cdot y^M = (z_k y_k)$. By rotation invariance of the Haar measure on $\mathbb{T}^\mathbb{N}$, the joint distribution of the sequences F_k and $G_k(y, \cdot, M)$ are identical for any M and $y \in \mathbb{T}$. In particular,

$$\mathbb{P}\left\{z \in \mathbb{T}^\mathbb{N} : \|G_k(y, z, M)\| > t\right\} = \mathbb{P}\left\{z \in \mathbb{T}^\mathbb{N} : \|F_k(z)\| > t\right\}, \tag{5.2.29}$$

for $M \in \mathbb{N}^\mathbb{N}$, $y \in \mathbb{T}$, and $t > 0$. We now define a nonnegative submartingale on $\mathbb{T}^\mathbb{N}$ by

$$g_k(z) = \int_{\mathbb{T}} \|G_k(y, z, M)\| dm(y).$$

In view of equation (5.2.29) we have $\sup_k \mathbb{E}(g_k) < \infty$ and, by the submartingale convergence theorem, $\sup_k g_k(z) < \infty$ for almost every $z \in \mathbb{T}^{\mathbb{N}}$. Put $g(z) = \sup_k g_k(z)$ and choose $\Omega_1 \subseteq \mathbb{T}^{\mathbb{N}}$ with $\mathbb{P}(\Omega_1) = 1$ such that $g(z) < \infty$ for any $z \in \Omega_1$.

We next construct an integer sequence M, and an increasing sequence of scalars $r(k) \in (0, 1)$ with $r(k) \uparrow 1$ such that for fixed $z \in \Omega_1$, there exists $h(\cdot, z) \in H^1(\mathbb{D}, X)$, such that

$$\|G_k(y, z, M) - h(r(k)y, z)\| \leq 2 \cdot (4^{-k} g(z)), \tag{5.2.30}$$

for $y \in \mathbb{T}$ and $k \in \mathbb{N}$.

We inductively determine $M(k)$ and $r(k)$ together with integers $u(k), v(k)$ and coefficients $f(p) = f(p, z)$ with $f(p) \in X$ such that we have the equality

$$G_k(y, z, M) - G_{k-1}(y, z, M) = \sum_{p=u(k)}^{v(k)} f(p) y^p, \tag{5.2.31}$$

and, moreover, denoting the analytic polynomial on the right-hand side of (5.2.31) by $h_k(y) = h_k(y, z)$, we get

$$\|h_k(r(k)y) - h_k(y)\| \leq 2^{-k} g(z) \quad \text{and} \quad \|h_k(r(j)y)\| \leq 2^{-2k+j} g(z),$$

for any $j \leq k - 1$. Remark that the coefficients $f(p) = f(p, z)$ are given as Fourier coefficients by

$$f(p, z) = \int_{\mathbb{T}} G_k(y, z, M) y^{-p} dm(y), \tag{5.2.32}$$

for $u(k) \leq p \leq v(k)$. The induction begins with $u(1) = 1$ as input. Since $F_0 = 0$ and F_1 is a trigonometric polynomial with $M(1) = 1$ there exists an integer $v(1) > u(1)$ such that with $M = (M(1), 0, \dots)$ we have

$$G_1(y, z, M) = \sum_{p=u(1)}^{v(1)} f(p) y^p.$$

We let h_1 denote the right-hand side polynomial of the above equation. Next, we choose $0 < r(1) < 1$ close enough to 1 that $|r(1)^{v(1)} - 1| \leq 4^{-1}/v(1)$. Thereafter, we select an integer $u(2) > v(1)$ large enough that $r(1)^{u(2)} \leq 4^{-1}(1 - r(1))$.

At stage $k - 1$ we determined the construction up to $M(k-1)$, $r(k-1)$, $v(k-1)$, and $u(k)$. Next we select $M(k)$, $r(k)$, $v(k)$, and $u(k+1)$. Since f_k is a trigonometric polynomial, we may determine the integer $M(k)$ large enough and thereby $v(k) > u(k)$ such that with $M = (M(1), \dots, M(k), 0, \dots)$ we have,

$$G_k(y, z, M) - G_{k-1}(y, z, M) = \sum_{p=u(k)}^{v(k)} f(p)y^p. \tag{5.2.33}$$

We let h_k be the analytic polynomial on the right-hand side of the above equation. Next, we choose $r(k) \in (r(k-1), 1)$ close enough to 1 such that

$$\left|r(k)^{v(k)} - 1\right| \leq 4^{-k}/v(k). \tag{5.2.34}$$

Finally, we select $u(k+1) < v(k)$ such that

$$r(k)^{u(k+1)} \leq 4^{-k} \quad \text{and} \quad r(j)^{u(k+1)} \leq 4^{-2k+j}, \tag{5.2.35}$$

for $1 \leq j \leq k-1$. Exploiting (5.2.35) and (5.2.32), we get $\sum_{j=k+1}^{\infty} \|h_j(r_k y)\| \leq 4^{-k} g(z)$, for any $k \in \mathbb{N}$. Consequently, if $0 < r < 1$ and $y \in \mathbb{T}$ the power series on the right-hand side below is absolutely converging, and with

$$h(ry) = \sum_{j=1}^{\infty} h_j(ry),$$

the mapping h is analytic in $\mathbb{D}$. It remains to show that $h \in H^1(\mathbb{D}, X)$. To this end we observe that by evaluating the telescoping sum (5.2.33) and estimating with (5.2.34) and (5.2.32) we obtain

$$\left\| G_k(y, z, M) - \sum_{j=1}^{k} h_j(r(k)y) \right\| \leq 4^{-k} g(z),$$

which gives condition (5.2.30). By first applying triangle inequality to (5.2.30), then integrating over $y \in \mathbb{T}$, and finally taking the supremum over k we find that

$$\sup_k \int_{\mathbb{T}} \|h(r(k)y)\| dm(y) \leq \sup_k \int_{\mathbb{T}} \|G_k(y, z, M)\| dm(y) + 2g(z).$$

Clearly, the term on the right-hand side equals $3g(z) < \infty$, and hence $h \in H^1(\mathbb{D}, X)$.

Invoking that the Banach space X satisfies condition (ii), we deduce that the radial limits $\lim_{k\to\infty} h(r(k)y)$ exist, for almost every $y \in \mathbb{T}$. Hence by (5.2.30), for any $z \in \Omega_1$ the limit $\lim_{k\to\infty} G_k(y, z, M)$ exists for almost every $y \in \mathbb{T}$. In other words,

$$\mathbb{P} \otimes m \left\{ z \in \mathbb{T}^N, y \in \mathbb{T} : \lim_{k\to\infty} G_k(y, z, M) \text{ exists} \right\} = 1.$$

By Fubini's theorem, there exists a set of full measure $T_0 \subseteq \mathbb{T}$ such that for any $y \in T_0$ the limit $\lim_{k\to\infty} G_k(y, z, M)$ exists, for almost every $z \in \mathbb{T}^{\mathbb{N}}$. Since for any fixed $y \in \mathbb{T}$ and M, the sequences $(G_k(y, \cdot, M))_{k=1}^{\infty}$ and $(F_k(\cdot))_{k=1}^{\infty}$ are identically distributed, we finally proved that $F_k(\cdot)$ converges almost surely. □

5.2.4 Real Variable Characterization

Summing up the content of Sections 5.2.1 and 5.2.3, we have shown that the aRNP of a Banach space X is equivalent to the property that every L^1-valued Hardy martingale converges almost surely. See Theorem 5.2.15 and Theorem 5.2.2. We continue by showing that the condition

$$H^{\infty}(\mathbb{T}, X) = H^{\infty}(\mathbb{D}, X)$$

may be expressed entirely in real variable terms, using plurisubharmonic functions on X. We will prove that the (aRNP) is equivalent to each of the following properties,

(i) the plurisubharmonic variational principle (Theorem 5.2.16);
(ii) closed bounded subsets of X admit nonempty, plurisubharmonic slices of arbitrary small diameter (Theorem 5.2.21).

The Plurisubharmonic Variational Principle

Let X be a Banach space. Fix $C \subseteq X$ and $f\colon C \to \mathbb{R}$. We say that f assumes a strong maximum in C at $x_0 \in C$ if

(i) f is upper bounded, that is, $\sup\{f(y) : y \in C\} < \infty$,
(ii) $f(x_0) = \sup\{f(y) : y \in C\}$,
(iii) any sequence $z_n \in C$ maximizing f in C is norm convergent to x_0, that is, if $\lim_{n\to\infty} f(z_n) = f(x_0)$ then $\lim_{n\to\infty} z_n = x_0$.

Let X be a complex Banach space. Recall that $\varphi\colon X \to \mathbb{R}$ is upper semi-continuous if $\{x \in X\colon \varphi(x) < \alpha\}$ is open for any $\alpha \in \mathbb{R}$. An upper semi-continuous $\varphi\colon X \to \mathbb{R}$ is called plurisubharmonic if

$$\varphi(x) \le \int_{\mathbb{T}} \varphi(x + y\zeta)dm(\zeta), \tag{5.2.36}$$

for $x, y \in X$. We denote by $\mathrm{Psh}^{(1)}(X)$ the set of all plurisubharmonic φ satisfying the 1-Lipschitz estimate

$$|\varphi(x) - \varphi(y)| \le \|x - y\|_X,$$

for any $x, y \in X$.

We say that X satisfies the *plurisubharmonic variational principle* if, for every closed bounded subset $C \subset X$ and every upper-bounded and upper-semicontinuous $g\colon C \to \mathbb{R}$, there exists $\varphi \in \mathrm{Psh}^{(1)}(\mathrm{X})$ such that

$$g + \varphi \quad \text{assumes a strong maximum in } C.$$

(Here $g\colon C \to \mathbb{R}$ is upper semicontinuous if $\{x \in C : g(x) < \alpha\}$ is open for any $\alpha \in \mathbb{R}$.)

The first main theorem of this section shows that the aRNP and the plurisubharmonic variational principle are equivalent conditions on a Banach space X.

Theorem 5.2.16 *A complex Banach space X satisfies the plurisubharmonic variational principle if and only if X satisfies the aRNP.*

The proof of Theorem 5.2.16 will be given in two steps as outlined in Theorem 5.2.18 and Theorem 5.2.20. In both steps we utilize Proposition 5.2.17, which introduces a flexible condition, equivalent to plurisubharmonic variational principle.

Proposition 5.2.17 *For a complex Banach space X, the plurisubharmonic variational principle is equivalent to the following condition: For any $\epsilon > 0$, any set $\mathbb{K}$, and pair of functions,*

$$f\colon \mathbb{K} \to X \quad \textit{and} \quad v\colon \mathbb{K} \to \mathbb{R},$$

satisfying $\sup_{z\in\mathbb{K}} \|f(z)\|_X < \infty$ and $\sup_{z\in\mathbb{K}} v(z) < \infty$, there exists $\varphi \in \mathrm{Psh}^{(1)}(\mathrm{X})$ and $\tau > 0$ such that

$$\operatorname{diam} f(\mathbb{K}_\tau) < \epsilon,$$

where $\mathbb{K}_\tau = \{z \in \mathbb{K} : v(z) + \varphi(f(z)) > \rho - \tau\}$ and $\rho = \sup\{v(z) + \varphi(f(z)) : z \in \mathbb{K}\}$.

Proof We first introduce notation used throughout the proof. Given $\tau > 0$ and $f\colon C \to \mathbb{R}$ we let

$$S(C, f, \tau) = \left\{ x \in C : f(x) > \sup_{y\in C} f(y) - \tau \right\}.$$

We remark that if $C \subseteq X$ is closed and $f\colon C \to \mathbb{R}$ is upper bounded and upper semicontinuous, then f assumes a strong maximum in C if and only if

$$\lim_{n\to\infty} \operatorname{diam}\,(S(C, f, \tau_n)) = 0,$$

for some (or equivalently any) sequence $\tau_n > 0$ satisfying $\lim \tau_n = 0$.

Step 1 First, assume that X satisfies the plurisubharmonic variational principle. Fix a set $\mathbb{K}$, let $f\colon \mathbb{K} \to X$ be bounded, and $v\colon \mathbb{K} \to \mathbb{R}$ be upper bounded. Put $C_0 = f(\mathbb{K})$. Define $g_0\colon C_0 \to \mathbb{R}$ by

$$g_0(x) = \sup\{v(z) : f(z) = x,\ z \in \mathbb{K}\}. \tag{5.2.37}$$

Let C denote the norm closure of C_0 in X. Define $g\colon C \to \mathbb{R}$ by $g(x) = \limsup_{y\to x} g_0(x)$. Thus defined, g is upper semicontinuous. Invoking the plurisubharmonic variational principle we may choose $\varphi \in \mathrm{Psh}^{(1)}(\mathrm{X})$ and a uniquely defined $x_0 \in C$ such that

$$g + \varphi \text{ assumes a strong maximum in } C \text{ at } x_0 \in C.$$

Hence for any $\epsilon > 0$ there exists $\tau > 0$ such that $D_\tau = \{x \in C : (g+\varphi)(x) > (g+\varphi)(x_0) - \tau\}$ satisfies $\operatorname{diam} D_\tau < \epsilon$. Put $\rho = \sup\{v(z) + \varphi(f(z)) : z \in \mathbb{K}\}$ and let

$$E_\tau = \{x \in C_0 : (g_0 + \varphi)(x) > \rho - \tau\}.$$

Note that $E_\tau \subseteq D_\tau$ and hence $\operatorname{diam} E_\tau < \epsilon$ and, moreover,

$$\mathbb{K}_\tau = \{z \in \mathbb{K} : v(z) + \varphi(f(z)) > \rho - \tau\}$$

satisfies $f(\mathbb{K}_\tau) = E_\tau$. Hence again $\operatorname{diam} f(\mathbb{K}_\tau) < \epsilon$.

Step 2 Next we verify the plurisubharmonic variational principle using the condition of the Proposition. Let $C \subseteq X$ be closed bounded. Let $g\colon C \to \mathbb{R}$ be upper bounded and upper semicontinuous. Put $\mathbb{K} = C$ and define

$$f\colon \mathbb{K} \to X \quad \text{and} \quad v\colon \mathbb{K} \to \mathbb{R}$$

by putting $f(z) = z$ and $v(z) = g(z)$. Thus f is bounded and v is upper bounded. Given $\epsilon > 0$, there exists $\tau > 0$ and $\varphi \in \mathrm{Psh}^{(1)}(\mathrm{X})$ such that with $\rho = \sup\{v(z) + \varphi(f(z)) : z \in \mathbb{K}\}$, the set

$$\mathbb{K}_\tau = \{z \in \mathbb{K} : v(z) + \varphi(f(z)) > \rho - \tau\}$$

satisfies $\operatorname{diam} f(\mathbb{K}_\tau) < \epsilon$. Note that $f(\mathbb{K}_\tau) = S(C, g+\varphi, \tau)$. Summing up, we have shown that for any $\epsilon > 0$ there exists $0 < \tau < \epsilon$ and $\varphi_\epsilon \in \mathrm{Psh}^{(1)}(\mathrm{X})$ such that

$$\operatorname{diam} S(C, g + \varphi_\epsilon, \tau) < \epsilon.$$

Step 3 Next, define

$$O_n = \left\{\varphi \in \mathrm{Psh}^{(1)}(X) : \operatorname{diam} S(C, g+\varphi, \tau) < 1/n, \ \text{for some } t > 0\right\}.$$

Clearly O_n is an open subset of the complete metric space $\mathrm{Psh}^{(1)}(\mathrm{X})$. We will verify below that O_n is dense in $\mathrm{Psh}^{(1)}(\mathrm{X})$. Now apply Baire's theorem to conclude that

$$\bigcap_{n=1}^{\infty} O_n \quad \text{is dense in } \mathrm{Psh}^{(1)}(X).$$

Pick $\varphi_0 \in \bigcap_{n=1}^{\infty} O_n$. By definition, there exists a sequence $\tau_n > 0$ such that $\operatorname{diam} S(C, g + \varphi_0, \tau_n) \leq 1/n$. Taking into account that C is closed and that g is upper bounded and upper semicontinuous, we infer that $g + \varphi_0$ assumes a strong maximum in C.

Step 4 It remains to prove that O_n is dense in $\mathrm{Psh}^{(1)}(\mathrm{X})$. To this end we fix $\varphi_1 \in \mathrm{Psh}^{(1)}(\mathrm{X})$ and $\gamma > 0$. We will show that there exists $\varphi_2 \in O_n$ such that $\|\varphi_1 - \varphi_2\|_{\mathrm{Lip}} < 2\gamma$.

Clearly, $\gamma^{-1}(g+(1-\gamma)\varphi_1)$ is upper semicontinuous on C and upper bounded. By Step 2, we obtain $\sigma_n > 0$ and $\psi \in \mathrm{Psh}^{(1)}(\mathrm{X})$ such that

$$\operatorname{diam} S\left(C, \gamma^{-1}(g + (1-\gamma)\varphi_1) + \psi, \sigma_n\right) < 1/n.$$

Equivalently, we have $\operatorname{diam} S(C, g + (1-\gamma)\varphi_1 + \gamma\psi, \gamma\sigma_n) < 1/n$. Finally, set $\varphi_2 = (1-\gamma)\varphi_1 + \gamma\psi$. Clearly $\varphi_2 \in O_n$ and $\varphi_2 - \varphi_1 = \gamma(\varphi_1 - \psi)$. Hence $\|\varphi_1 - \varphi_2\|_{\mathrm{Lip}} < 2\gamma$. □

Theorem 5.2.18 *If a complex Banach space X satisfies the plurisubharmonic variational principle, then it satisfies the aRNP. That is,* $H^\infty(\mathbb{D}, X) = H^\infty(\mathbb{T}, X)$.

Proof Let $f \in H^\infty(\mathbb{D}, X)$. Assuming that X satisfies the plurisubharmonic variational principle, we show that the radial limit $\lim_{r\to 1} f(r\zeta)$ exists in X for almost every $\zeta \in \mathbb{T}$. We obtain this by proving that for any $\epsilon > 0$ and any measurable $A \subseteq \mathbb{T}$ with $m(A) > 0$ there exists a measurable $A_1 \subseteq A$ satisfying $m(A_1) > 0$, and that

$$\limsup_{r,s\to 1} \|f(r\zeta) - f(s\zeta)\|_X < \epsilon, \text{ for almost every } \zeta \in A_1. \tag{5.2.38}$$

In the next two steps of the proof we set up the variational problem that allows us to define the desired measurable set A_1 in Step 3 of the proof.

Step 1 Put $u(z) = \|f(z)\|_X$ and $\lambda = \sup_{z\in\mathbb{D}} u(z)$. Clearly, u is uniformly bounded and subharmonic in $\mathbb{D}$. Littlewood's theorem (Theorem 1.1.19) asserts that the radial limit $\lim_{r\to 1} u(r\zeta)$ exists in $\mathbb{R}$ for almost every $\zeta \in \mathbb{T}$. Fix $A \subseteq \mathbb{T}$ with $m(A) > 0$, define $v \in L^\infty(\mathbb{T})$ by

$$v = -1_{\mathbb{T}\setminus A}(u + \lambda + 1),$$

and let $v(z) = \int v p_z dm$ be its harmonic extension to the unit disk. Clearly, we have $\sup_{z\in\mathbb{D}} |v(z)| < 2\lambda + 1$ and

$$\lim_{r\to 1} v(r\zeta) = \begin{cases} -(u(\zeta) + \lambda + 1) & a.e.\, \zeta \in \mathbb{T}\backslash A, \\ 0 & a.e.\, \zeta \in A. \end{cases} \tag{5.2.39}$$

Fix $\epsilon > 0$, and apply Proposition 5.2.17 to $v\colon \mathbb{D} \to \mathbb{R}$ and $f\colon \mathbb{D} \to X$. This yields the existence of $\varphi \in \text{Psh}^{(1)}(\text{X})$ and $\tau > 0$ such that with

$$\rho = \sup\{v(z) + \varphi(f(z)) : z \in \mathbb{D}\},$$

the set $D_0 = \{z \in \mathbb{D} : v(z) + \varphi(f(z)) > \rho - \tau\}$ satisfies $\operatorname{diam} f(D_0) < \epsilon$ and $D_0 \neq \emptyset$.

Step 2 Since $f \in H^\infty(\mathbb{D}, X)$ and $\varphi \in \text{Psh}^{(1)}(\text{X})$, the compositon $w(z) = \varphi(f(z))$ is bounded and subharmonic in $\mathbb{D}$. Hence, by Littlewood's theorem, its radial limits, $w(\zeta) = \lim_{r\to 1} w(r\zeta)$, exist for almost every $\zeta \in \mathbb{T}$. For later reference we record the following estimates: First, since $\varphi \in \text{Psh}^{(1)}(\text{X})$, for almost every $\zeta, \zeta_0 \in \mathbb{T}$,

$$w(\zeta) - w(\zeta_0) \leq \lim_{r\to 1} \|f(r\zeta) - f(r\zeta_0)\|_X \leq u(\zeta) + \lambda. \tag{5.2.40}$$

Second, for almost every $\zeta_0 \in A$ we have $w(\zeta_0) = \lim_{r\to 1}(v(r\zeta_0) + \varphi(f(r\zeta_0)))$, by (5.2.39). Hence $w \leq \rho$, almost everywhere on A and

$$w \leq u + \lambda + \rho. \tag{5.2.41}$$

Step 3 Define $A_1 = \{\zeta \in \mathbb{T} : v(\zeta) + w(\zeta) > \rho - \tau\}$. We prove next that $m(A_1) > 0$, that $A_1 \subseteq A$, and that condition (5.2.38) holds almost surely on A_1.

We first verify that $m(A_1) > 0$. To this end recall that $D_0 \neq \emptyset$. Pick $z_1 \in D_0$ and integrate $v + w$ against the Poisson kernel p_{z_1}. This gives

$$\int_{\mathbb{T}} (v + w)(\zeta) p_{z_1}(\zeta) dm(\zeta) \geq v(z_1) + w(z_1) = v(z_1) + \varphi(f(z_1)) > \rho - \tau,$$

and hence $m(A_1) > 0$. Next, obtain the inclusion $A_1 \subseteq A$ by showing that almost everywhere on $\mathbb{T}\backslash A$ we have $v + w \leq \rho - \tau$. Indeed, by (5.2.39) and (5.2.41), for almost every $\zeta \in \mathbb{T}\backslash A$,

$$\begin{aligned} (v + w)(\zeta) &= -(u(\zeta) + \lambda + 1) + w(\zeta) \\ &\leq -u(\zeta) - \lambda - 1 + u(\zeta) + \lambda + \rho \\ &= \rho - 1. \end{aligned} \tag{5.2.42}$$

Since $0 < \tau < 1$, we have $\rho - 1 \leq \rho - \tau$, and hence $\zeta \notin A$ implies $\zeta \notin A_1$ or $A_1 \subseteq A$.

We finally turn to verifying condition (5.2.38). Since $m(A_1) > 0$, Lebesgue's differentiation theorem implies that for almost every $\zeta \in A_1$, there exists $r_0 = r_0(\zeta) < 1$ such that for $r_0 < r < 1$ we have $r\zeta \in D_0$ and, hence,

$$\limsup_{r,s\to 1} \|f(r\zeta) - f(s\zeta)\|_X \leq \operatorname{diam}(f(D_0)),$$

for almost every $\zeta \in A_1$. Since, by Step 1, $\operatorname{diam}(f(D_0)) \leq \epsilon$, we have completed the verification of (5.2.38). □

Next, we turn to proving Theorem 5.2.20, which asserts that the aRNP implies the plurisubharmonic variational principle. Since Theorem 5.2.18 is the converse of Theorem 5.2.20, this will imply Theorem 5.2.16, our first real-variable characterization of those Banach spaces X for which

$$H^\infty(\mathbb{D}, X) = H^\infty(\mathbb{T}, X).$$

The plurisubharmonic envelope: We prepare the proof of Theorem 5.2.20 by proving the Hardy martingale representation of the plurisubharmonic envelope, a result of independent interest.

Let X be a complex Banach space. As defined in (5.2.36), an upper semi-continuous $\varphi\colon X \to \mathbb{R}$ is called plurisubharmonic if

$$\varphi(x) \leq \int_{\mathbb{T}} \varphi(x + y\zeta)dm(\zeta),$$

for $x, y \in X$. We denote by $\mathrm{Psh}^{(1)}(\mathrm{X})$ the set of all plurisubharmonic φ satisfying the 1-Lipschitz estimate

$$|\varphi(x) - \varphi(y)| \leq \|x - y\|_X,$$

for any $x, y \in X$.

We associate to any Lipschitz continuous $g\colon X \to \mathbb{R}$ its plurisubharmonic envelope. Put $g_0(x) = g(x)$ and

$$g_n(x) = \inf\left\{\int_{\mathbb{T}} g_{n-1}(x + y\zeta)dm(\zeta) : \; y \in X\right\}. \tag{5.2.43}$$

By construction, $g_n(x) \leq g_{n-1}(x)$ for $x \in X$ and $n \in \mathbb{N}$, hence $\{g_n\}$ is convergent. We put $\varphi_g(x) = \lim_{n\to\infty} g_n(x)$. It is easy to see that φ_g is plurisubharmonic. Indeed, by equation (5.2.43) and the monotone convergence theorem, we have

$$\begin{aligned}\varphi_g(x) &= \lim_{n\to\infty} g_n(x)\\ &\leq \lim_{n\to\infty} \int_{\mathbb{T}} g_{n-1}(x+y\zeta)dm(\zeta)\\ &= \int_{\mathbb{T}} \varphi_g(x+y\zeta)dm(\zeta),\end{aligned}$$

for any $x, y \in X$. We next remark that φ_g is the largest plurisubharmonic function less than or equal to g. Specifically, $\varphi \leq \varphi_g$ for any plurisubharmonic $\varphi\colon X \to \mathbb{R}$ satisfying $\varphi \leq g$. To see this we first check that $\varphi \leq g_n$ for any $n \in \mathbb{N}$. We start using $\varphi(x) \leq g(x)$ for $x \in X$. Then,

$$\begin{aligned}g_1(x) &= \inf\left\{\int_{\mathbb{T}} g(x+y\zeta)dm(\zeta) : y \in X\right\}\\ &\geq \inf\left\{\int_{\mathbb{T}} \varphi(x+y\zeta)dm(\zeta) : y \in X\right\}\\ &\geq \varphi(x).\end{aligned}$$

Assuming that $\varphi \leq g_{n-1}$ and repeating the above calculation with g replaced by g_{n-1} gives $\varphi \leq g_n$. Passing to the limit we showed that $\varphi \leq g$ implies

$$\varphi_g(x) = \lim_{n\to\infty} g_n(x) \geq \varphi(x).$$

This justifies our calling φ_g the plurisubharmonic envelope of g.

The next result enables us to re-express plurisubharmonic envelopes in terms of simple Hardy martingales. We will use it below to transfer analytic statements about the aRNP to probabilistic ones.

Proposition 5.2.19 *Let X be a complex Banach space and let $g\colon X \to \mathbb{R}$ be Lipschitz continuous with*

$$|g(x) - g(y)| \leq \|x - y\|.$$

If g_n is defined by equation (5.2.43), then

$$|g_n(x) - g_n(y)| \leq \|x - y\|, \tag{5.2.44}$$

for any $x, y \in X$, and

$$g_n(x) = \inf\{\mathbb{E}g(F_n)\}, \tag{5.2.45}$$

where the infimum is taken over simple, X-valued Hardy martingales $(F_k)_{k=1}^n$ satisfying $\mathbb{E}(F_n) = x$.

Proof We prove condition (5.2.44) inductively, starting with $n = 1$. Fix $\epsilon > 0$ and $x, y \in X$. Determine $v \in X$ such that $g_1(y) \geq \int_{\mathbb{T}} g_0(y + v\zeta)dm(\zeta) - \epsilon$. Then we routinely estimate

$$\begin{aligned} g_1(x) - g_1(y) &\leq \int_{\mathbb{T}} g_0(x + v\zeta) - g_0(y + v\zeta)dm(\zeta) + \epsilon \\ &\leq \|x - y\| + \epsilon. \end{aligned}$$

Exchanging the role of x and y and taking into account that $\epsilon > 0$ was arbitrary we have (5.2.44) for $n = 1$. Assume now that $|g_{n-1}(x) - g_{n-1}(y)| \leq \|x - y\|$, for $x, y \in X$. Replacing g_0 by g_{n-1} and g_1 by g_n gives (5.2.44) by repeating the above estimates.

We now turn to proving (5.2.45). Let $(F_k)_{k=1}^n$ be a simple X-valued Hardy martingale represented as

$$F_k(w) = x + \sum_{j=1}^{k} h_j(w)w_j, \tag{5.2.46}$$

where $w \in \mathbb{T}^{\mathbb{N}}$ with $w = (w_i)$, and $h_j(w) = h_j(w_1, \ldots, w_{j-1})$ takes values in X. Then obviously,

$$g_n(x) \leq \mathbb{E}g(F_n). \tag{5.2.47}$$

We now proceed to verify the converse estimate to (5.2.47) in the following form. Given $x \in X$, $\epsilon > 0$ and $n \in \mathbb{N}$ we determine inductively a simple Hardy martingale represented as (5.2.46) such that

$$g_n(x) \geq \mathbb{E}g(F_n) - \epsilon. \tag{5.2.48}$$

Step 1 Let $w \in \mathbb{T}^{\mathbb{N}}$ with $w = (w_i)$ and $w_i \in \mathbb{T}$. Choose $y \in X$ such that

$$g_n(x) \geq \int_{\mathbb{T}} g_{n-1}(x + yw_1)dm(w_1) - \epsilon/2.$$

Put $h_1 = y$ and $F_1(w) = x + h_1 w_1$.

Step 2 Given $w_1 \in \mathbb{T}$, and using that g_{n-1} is Lipschitz continuous, we determine a measurable X-valued step function $h_2(w) = h_2(w_1)$ such that

$$g_{n-1}(F_1(w)) \geq \int_{\mathbb{T}} g_{n-2}(F_1(w) + h_2(w)w_2)dm(w_2) - \epsilon/4.$$

We put $F_2(w) = F_1(w) + h_2(w)w_2$.

Step k + 1 We continue in an obvious manner and assume we have a step function $h_k(w) = h_k(w_1, \ldots, w_{k-1})$ and $F_{k-1}(w)$ with $\mathbb{E}F_{k-1} = x$ such that $F_k(w) = F_{k-1}(w) + h_k(w)w_k$, satisfies

$$g_{n-(k-1)}(F_{k-1}(w)) \geq \int_{\mathbb{T}} g_{n-k}(F_k(w))dm(w_k) - \epsilon/2^k.$$

Using that $g_{n-(k+1)}$ is Lipschitz continous, we now determine an X-valued step function h_{k+1} with $h_{k+1}(w) = h_{k+1}(w_1, \ldots, w_k)$ such that $F_{k+1}(w) = F_k(w) + h_{k+1}(w)w_{k+1}$ satisfies

$$g_{n-k}(F_k(w)) \geq \int_{\mathbb{T}} g_{n-(k+1)}(F_{k+1}(w))dm(w_{k+1}) - \epsilon/2^{k+1}. \tag{5.2.49}$$

Summary: We carry out the above construction until $k + 1 = n$. Then we obtained a simple Hardy martingale $(F_k)_{k=1}^n$ with $\mathbb{E}(F_n) = x$ and representation (5.2.46). Successively applying estimate (5.2.49) gives

$$g_n(x) \geq \int_{\mathbb{T}} \cdots \int_{\mathbb{T}} g(F_n(w_1, \ldots, w_n))dm(w_1) \ldots dm(w_n) - \epsilon. \tag{5.2.50}$$

To see that conditon (5.2.48) holds true it remains to rewrite the right-hand side of (5.2.50) as $\mathbb{E}g(F_n) - \epsilon$. In view of (5.2.47), we thus obtain (5.2.45). □

Theorem 5.2.20 *For any Banach space, the aRNP implies the plurisubharmonic variational principle.*

Proof Let X be a complex Banach space satisfying the aRNP. By Theorem 5.2.15 and Theorem 5.2.2, every L^1-bounded, X-valued Hardy martingale converges almost everywhere. We will prove that X satisfies the plurisubharmonic variational principle. To this end, we fix $\mathbb{K}$, $v\colon \mathbb{K} \to \mathbb{R}$, and $f\colon \mathbb{K} \to X$ satisfying $\sup_{z\in\mathbb{K}} v(z) < \infty$ and $\sup_{z\in\mathbb{K}} \|f(z)\|_X < \infty$. We fix $\epsilon > 0$ and proceed to verify the condition of Proposition 5.2.17. The desired plurisubharmonic function φ is determined in the first step of the proof.

Step 1 Fix $z \in \mathbb{K}$. We define a 1-Lipschitz function $g(z, \cdot)\colon X \to \mathbb{R}$ by putting

$$g(z, y) = \inf_{w\in\mathbb{K}} \{v(z) - v(w) + \|y - f(w)\| \; : \; \|f(z) - f(w)\|_X > \epsilon/2\}.$$

Let $\varphi(z, \cdot)$ denote the plurisubharmonic envelope of $g(z, \cdot)$, which is again a 1-Lipschitz function on X, by Proposition 5.2.19. Exploiting that X satisfies the aRNP, we will prove below that there exists $z_0 \in \mathbb{K}$ such that

$$\varphi(z_0, f(z_0)) > 0. \tag{5.2.51}$$

Assume the existence of $z_0 \in \mathbb{K}$ satisfying condition (5.2.51). We abbreviate $\varphi(y) = \varphi(z_0, y)$, put $\rho = \sup\{v(z) + \varphi(f(z)) : z \in \mathbb{K}\}$, and let $\tau = \varphi(f(z_0))$. Then clearly, $0 < \tau < 1$ and

$$v(z_0) \leq \rho - \varphi(f(z_0)) = \rho - \tau.$$

We put

$$D_0 = \{z \in \mathbb{K} : v(z) + \varphi(f(z)) > \rho - \tau\}.$$

In order to prove that X satisfies the plurisubharmonic variational principle we will verify that $\operatorname{diam} f(D_0) < \epsilon$. To this end, assume that there exists $z_1 \in \mathbb{K}$ such that $\|f(z_0) - f(z_1)\|_X > \epsilon/2$. Then,

$$\varphi(f(z_1)) \leq g(z_0, f(z_1)) \leq v(z_0) - v(z_1),$$

or $v(z_1) + \varphi(f(z_1)) \leq v(z_0) \leq \rho - \tau$, and hence $z_1 \notin D_0$. Thus we have verified that $\operatorname{diam} f(D_0) < \epsilon$.

Step 2 We now turn to the main part of the proof, verifying our claim that there exists $z_0 \in \mathbb{K}$ satisfying $\varphi(z_0, f(z_0)) > 0$. Assume to the contrary that $\varphi(z, f(z)) \leq 0$ for any $z \in \mathbb{K}$. Under this hypothesis we inductively construct increasing integers $N_k \uparrow \infty$, a sequence of countably valued $G_k : \mathbb{T}^{N_k} \to \mathbb{K}$ and $M_k \in L^1\left(\mathbb{T}^{N_k}, X\right)$ such that (M_k) is a subsequence of an L^1-bounded, X-valued, simple Hardy martingale satisfying,

$$\mathbb{E} \sum_{k=1}^{\infty} \|f(G_{k+1}) - f(G_k) - (M_{k+1} - M_k)\|_X \leq 1 + 2 \sup_{z \in \mathbb{K}} v(z), \tag{5.2.52}$$

where the expectation is taken over $\mathbb{T}^{\mathbb{N}}$, and

$$\mathbb{P}\left\{w \in \mathbb{T}^{N_{k+1}} : \|f(G_{k+1}(w)) - f(G_k(w))\|_X > \epsilon/2\right\} \geq 1 - 2^{-k}. \tag{5.2.53}$$

Step 3 We start the construction by picking $x_0 \in \mathbb{K}$ and putting $G_0 = x_0$ and $M_0 = f(x_0)$. Now let $k \in \mathbb{N}$ and assume that we already determined, increasing integers N_m, for $m \leq k$, countably valued $G_k : \mathbb{T}^{N_k} \to \mathbb{K}$, and $M_m \in L^1(\mathbb{T}^{N_m}, X)$ satisfying (5.2.52) and (5.2.53).

Next fix $w \in \mathbb{T}^{N_{k+1}}$. Recall that we assume $\varphi(G_k(w), f(G_k(w))) \leq 0$. By Proposition 5.2.19, we obtain $n_k = n_k(w) \in \mathbb{N}$ and a simple X-valued Hardy martingale $(U_j(w, \cdot))_{j=1}^{n_k}$ with $U_0(w, \cdot) = 0$ such that

$$\mathbb{E}\left[g(G_k(w), f(G_k(w)) + U_{n_k}(w, w'))\right] < 2^{-k-2},$$

where – for fixed $w \in \mathbb{T}^{N_{k+1}}$ – the expectation is taken over $w' \in \mathbb{T}^{n_k(w)}$. Evaluating the definition of $g(z, y)$, we may now determine $H = H(w, w')$ countably valued such that

$$\operatorname{ess}\inf_{w' \in \mathbb{T}^{n_k(w)}} \|f(H(w,w')) - f(G_k(w))\| > \epsilon/2, \tag{5.2.54}$$

and

$$\mathbb{E}\left[v(G_k(w)) - v(H(w,\cdot)) + \|f(G_k(w) + U_{n_k}(w,\cdot)) - f(H(w,\cdot))\|\right] < 2^{-k-2}, \tag{5.2.55}$$

where again for $w \in \mathbb{T}^{N_{k+1}}$ fixed, the expectation is over $w' \in \mathbb{T}^{n_k(w)}$.

Step 4 Now since $n_k(w)$ depends measurably on $w \in \mathbb{T}^{N_{k+1}}$, there exists $L \in \mathbb{N}$ such that $A = \{w \in \mathbb{T}^{N_{k+1}} : n_k(w) \leq L\}$ satisfies $\mathbb{P}(A) > 1 - 2^{-k}$ and

$$\mathbb{E}1_{\mathbb{T}^{N_{k+1}} \setminus A}(w) 1_{\mathbb{T}^{n_k(w)}}(w') \|U_{n_k}(w,w')\| + \|M_k(w)\| < 2^{-k-2}. \tag{5.2.56}$$

We next define $G_{k+1} : \mathbb{T}^{N_k} \times \mathbb{T}^L \to \mathbb{K}$. We put

$$G_{k+1}(w,w') = \begin{cases} G_k(w) & w \notin A, \\ H(w,w') & w \in A, \quad w' \in \mathbb{T}^{n_k(w)}. \end{cases}$$

By estimate (5.2.54) and the lower bound $\mathbb{P}(A) > 1 - 2^{-k}$ we have (5.2.53). Next, we define $M_{k+1} \in L^1(\mathbb{T}^{N_k} \times \mathbb{T}^L)$ by

$$M_{k+1}(w,w') = \begin{cases} M_k(w) & w \notin A, \\ U_{n_k}(w,w') - M_k(w) & w \in A, \quad w' \in \mathbb{T}^{n_k(w)}. \end{cases}$$

In view of estimates (5.2.55) and (5.2.56), G_{k+1} and M_{k+1} satisfy the crucial estimate

$$\mathbb{E}\left(v(G_k) - v(G_{k+1}) + \|f(G_k) + M_k - f(G_{k+1}) - M_{k+1}\|\right) < 2^{-k-2}. \tag{5.2.57}$$

This finishes the inductive construction. We next verify the estimate (5.2.52). Fix $n \in \mathbb{N}$, take the sum of (5.2.57) over $0 \leq k \leq n$, and obtain

$$\begin{aligned} \mathbb{E}\sum_{k=0}^{n} \|f(G_k) - f(G_{k+1}) - M_{k+1} + M_k\| &\leq 1 + \mathbb{E}\sum_{k=0}^{n} v(G_{k+1}) - v(G_k) \\ &= 1 + \mathbb{E}v(G_{n+1}) - v(G_0). \end{aligned}$$

Since $\mathbb{E}v(G_{n+1}) - v(G_0) \leq 2\sup_{z\in\mathbb{K}} v(z)$, we have verified (5.2.52).

Step 5 Now, by (5.2.52) the sequence $(\|f(G_k) - M_k\|_X)_k$ is bounded in $L^1(\mathbb{T}^{\mathbb{N}})$ and since $\sup_{z\in\mathbb{K}} \|f(z)\| < \infty$, the sequence (M_k) is also bounded in $L^1(X)$. We constructed (M_k) to be a subsequence of an X-valued simple Hardy martingale, and we assumed at the beginning that the Banach space X satisfies the aRNP. Hence (M_k) is convergent almost surely and in $L^1(X)$. By estimate (5.2.52), this implies that $f(G_k)$ converges as well, which contradicts (5.2.53). □

Small Plurisubharmonic Slices

The second main theorem of this section characterizes the aRNP in terms of plurisubharmonic slices.

Theorem 5.2.21 *A complex Banach space X satisfies the aRNP, if and only if for any bounded closed subset $C \subset X$ and $\epsilon > 0$, there exists $\varphi \in \mathrm{Psh}^{(1)}(X)$ such that*

$$C \cap \{\varphi > 0\} \neq \emptyset \quad \textit{and} \quad \operatorname{diam} C \cap \{\varphi > 0\} < \epsilon.$$

The two implications forming the content of Theorem 5.2.21 will be proven in Proposition 5.2.22 and Proposition 5.2.23 below.

Proposition 5.2.22 *If a Banach space X has the aRNP, then for any bounded closed subset $C \subset X$ and $\epsilon > 0$, there exists $\varphi \in \mathrm{Psh}^{(1)}(X)$ such that*

$$C \cap \{\varphi > 0\} \neq \emptyset \quad \textit{and} \quad \operatorname{diam} C \cap \{\varphi > 0\} < \epsilon.$$

Proof By Theorem 5.2.20 the aRNP implies the plurisubharmonic variational principle. Fix $\epsilon >$ and let C be a closed bounded subset of X. Define $g: C \to \mathbb{R}$ by $g(x) = 0, x \in C$, By assumption, there exists $\varphi \in \mathrm{Psh}^{(1)}(X)$ such that

$$C \cap \{g + \varphi > 0\} \neq \emptyset \quad \text{and} \quad \operatorname{diam} C \cap \{g + \varphi > 0\} < \epsilon.$$

Since $g = 0$ this gives the result. □

We show next the converse of Proposition 5.2.22.

Proposition 5.2.23 *Let X be a complex Banach space. If, for any bounded closed subset $C \subset X$ and $\epsilon > 0$, there exists $\varphi \in \mathrm{Psh}^{(1)}(X)$ such that*

$$C \cap \{\varphi > 0\} \neq \emptyset, \quad \textit{and} \quad \operatorname{diam} C \cap \{\varphi > 0\} < \epsilon,$$

then X satisfies the aRNP.

Proof Assume to the contrary that X fails the aRNP. By Theorem 5.2.6 there exist $\eta > 0$, a sequence $r_n \in (0, 1)$ with $\lim r_n = 1$, and $g \in H^\infty(\mathbb{D}, X)$ such that

$$\limsup_{k,n\to\infty} \|g_{r_n}(\zeta) - g_{r_k}(\zeta)\|_X > \eta, \tag{5.2.58}$$

for almost every $\zeta \in \mathbb{T}$. Let C denote the closure in X of $\{g_{r_n}(\zeta) : \zeta \in \mathbb{T}, n \in \mathbb{N}\}$. Thus C is a closed and bounded subset of X. By assumption, there exists $\varphi \in \mathrm{Psh}^{(1)}(X)$ such that

$$C \cap \{\varphi > 0\} \neq \emptyset \quad \text{and} \quad \operatorname{diam} C \cap \{\varphi > 0\} < \eta.$$

Consequently there exists $z \in \mathbb{D}$ such that $g(z) \in \{\varphi > 0\}$.

Next, set $f(w) = \varphi(g(w))$ where $w \in \mathbb{D}$. Since $f\colon \mathbb{D} \to \mathbb{R}$ is bounded and sub-harmonic it has radial limits almost everywhere. That is, there exists $M_f \subseteq \mathbb{T}$ with $m(\mathbb{T}\setminus M_f) = 0$ such that $F(\zeta) = \lim_{r\to 1} f(r\zeta)$ exists for $\zeta \in M_f$. Moreover,

$$f(w) \leq \int_{\mathbb{T}} F(\zeta)P(w,\zeta)dm(\zeta), \tag{5.2.59}$$

for $w \in \mathbb{D}$, where $P(w,\zeta)$ denotes the Poisson kernel of the unit disk. Clearly, since $g(z) \in \{\varphi > 0\}$ we have $f(z) > 0$. Hence by (5.2.59) the set

$$B = \{\zeta \in M_f : F(\zeta) > 0\}$$

satisfies $m(B) > 0$. Next, recall that $F(\zeta) = \lim_{r\to 1} f(r\zeta)$ for $\zeta \in M_f$. Hence for any $\zeta \in B$, there exists $n_0 = n_0(\zeta)$ such that $f(r_n\zeta) > 0$ for any $n > n_0$. Since $m(B) > 0$, invoking condition (5.2.58) gives that

$$\operatorname{diam} C \cap \{\varphi > 0\} \geq \operatorname{diam} \{g_{r_n}(\zeta) : \zeta \in B, n \in \mathbb{N}\} \cap \{\varphi > 0\} > \eta, \tag{5.2.60}$$

a contradiction. □

5.2.5 Holomorphic Injections

Let X, Y be complex Banach spaces, and let $B = B_X$ denote the closed unit ball of X. Let

$$J\colon X \to Y$$

be continuous (not necessarily linear), injective, and holomorphic in the sense that

(i) $J \circ f \in H^\infty(\mathbb{D}, Y)$ for any $f \in H^\infty(\mathbb{D}, X)$, and
(ii) $J^{-1} \circ g\colon \mathbb{T} \to X$ is measurable, for any bounded measurable $g\colon \mathbb{T} \to J(X)$.

We say that such a J is a preferred holomorphic injection if there exist closed sets $K_n \subseteq Y$ and scalars $\alpha_n \in (0,1)$ such that

$$\overline{J(B)}\setminus J(B) = \bigcup K_n,$$

and such that any $u \in J(B)$ may be separated uniformly from any of the sets K_n by a plurisubharmonic mapping as follows: For any $n \in \mathbb{N}$ and $u \in J(B)$ there exists a plurisubharmonic mapping $h\colon Y \to \mathbb{R}$ such that $h(u) > \alpha_n$, $h(v) \leq 0$ for $v \in K_n$, and $h(w) \leq 1$ for any $w \in \overline{J(B)}$.

Recall that Y satisfies the aRNP if every $g \in H^\infty(\mathbb{D}, Y)$ admits radial limits $\lim_{r\to 1} g(r\zeta)$ for almost every $\zeta \in \mathbb{T}$. Preferred holomorphic injections are tailor made for pulling back radial limits from Y to X.

Theorem 5.2.24 *Let X be a complex Banach space and assume that Y has the aRNP. If there exists a preferred holomorphic injection $J\colon X \to Y$ then X also has the* aRNP.

Proof Fix $f \in H^\infty(\mathbb{D}, X)$ with $\|f(z)\| < 1$ for any $z \in \mathbb{T}$. We will prove the existence of radial limits for f. By hypothesis on J we have $J \circ f \in H^\infty(\mathbb{D}, Y)$. Since Y satisfies the aRNP, $g(\zeta) = \lim_{r\to 1}(Jf)(r\zeta)$ exists almost everywhere. As $(Jf)(r\zeta) \in J(B)$, the radial limit function g takes values in $\overline{J(B)}$.

We next improve the conclusion of the previous paragraph and verify that $g(\zeta) \in J(B)$, for almost every $\zeta \in \mathbb{T}$. Assume to the contrary that

$$m\left\{\zeta \in \mathbb{T} : g(\zeta) \in \overline{J(B)}\backslash J(B)\right\} > 0.$$

Hence there exists $n \in \mathbb{N}$ such that $m\{g \in K_n\} > 0$. Put $C = \{g \in K_n\}$, let $\epsilon = \alpha_n/2 > 0$, and select $z_0 \in \mathbb{D}$ such that

$$\int_C P(z_0, \zeta) dm(\zeta) > 1 - \epsilon. \tag{5.2.61}$$

Clearly, condition (5.2.61) gives $\int_{\mathbb{T}\backslash C} P(z_0, \zeta)dm(\zeta) < \epsilon$. Define $u = J(f(z_0))$. Since $u \in J(B)$, we may separate it from K_n by plurisubharmonic mappings. Indeed, by hypothesis on J there exists $h\colon Y \to \mathbb{R}$, plurisubharmonic and uniformly continuous on bounded sets of Y, such that $h(u) > \alpha_n$, $h(v) \le 0$ for $v \in K_n$, and $h(w) \le 1$ for any $w \in \overline{J(B)}$. Now by subharmonicity we have

$$\alpha_n < h(u) \le \int_{\mathbb{T}} h(g(\zeta))P(z_0, \zeta)dm(\zeta). \tag{5.2.62}$$

We rewrite the integral on the right-hand side by splitting $\mathbb{T} = C \cup (\mathbb{T}\backslash C)$ where $C = \{\zeta \in \mathbb{T} : g(\zeta) \in K_n\}$. Since $h(g(\zeta)) < 0$ for $\zeta \in C$, and $h(g(y)) \le 1$ for $y \in \mathbb{T}\backslash C$, we find

$$\int_{\mathbb{T}} h(g(\zeta))P(z_0, \zeta)dm(\zeta) \le \int_{\mathbb{T}\backslash C} h(g(y))P(z_0, y)dm(y) \le \int_{\mathbb{T}\backslash C} P(z_0, y)dm(y).$$

Invoking estimates (5.2.61) and (5.2.62) then gives $\alpha_n < h(u) \le \epsilon$. Since we have $\epsilon = \alpha_n/2$, this is the desired contradiction. In summary, $m\left\{\zeta : g(\zeta) \in \overline{J(B)}\backslash J(B)\right\} = 0$, or $g(\zeta) \in J(B)$ almost everywhere.

The composition $\varphi(\zeta) = J^{-1}(g(\zeta))$ gives a well-defined element in $L^\infty(\mathbb{T}, X)$. In the last part of the proof we will check that f is the Poisson integral of φ.

Fix $0 < r < 1$ and $y \in \mathbb{T}$. Then in view of Lebesgue's dominated convergence theorem, we get

$$\begin{aligned}\int_{\mathbb{T}} \varphi(\zeta)P(ry,\zeta)dm(\zeta) &= \int_{\mathbb{T}} J^{-1}\left(\lim_{s\to 1} J(f(s\zeta))\right)P(ry,\zeta)dm(\zeta)\\ &= \int_{\mathbb{T}} J^{-1}\left(J\left(\lim_{s\to 1} f(s\zeta)\right)\right)P(ry,\zeta)dm(\zeta)\\ &= \int_{\mathbb{T}} \lim_{s\to 1} f(s\zeta)P(ry,\zeta)dm(\zeta)\\ &= \lim_{s\to 1}\int_{\mathbb{T}} f(s\zeta)P(ry,\zeta)dm(\zeta)\\ &= \lim_{s\to 1} f(sry) = f(ry).\end{aligned}$$

Finally, by Theorem 5.2.4 the representation of f as Poisson integral implies the existence of its radial limits almost everywhere. □

5.2.6 H^1-Uniform Convexity

Let X be a complex Banach space. Let $2 \leq q < \infty$. We say that X is H^1-uniformly convex of order q if there exists $\delta > 0$ such that for any $f \in H^1(\mathbb{T}, X)$,

$$\|f_0\|_X^q + \delta\left(\int_{\mathbb{T}} \|f - f_0\|_X dm\right)^q \leq \left(\int_{\mathbb{T}} \|f\|_X dm\right)^q, \qquad (5.2.63)$$

where $f_0 = \int_{\mathbb{T}} f dm$. For instance, $\mathbb{C}$ and $L^1(\Omega, \mu)$ would be H^1-uniformly convex of order 2. The next result, due to Haagerup and Pisier (1989), shows that H^1-convexity implies the existence of radial limits. For $f \in H^1(\mathbb{D}, X)$ and $0 < r < 1$, we write $f_r(\zeta) = f(r\zeta)$ where $\zeta \in \mathbb{T}$. Thus $f_r \in H^1(\mathbb{T}, X)$ and, by subharmonicity, the sequence $\|f_{r_n}\|_1$ is bounded and increasing as $r_n \uparrow 1$.

As we will show, when X is uniformly H^1-convex, then surprisingly, the convergence of $\|f_{r_n}\|_1$ forces f_{r_n} itself to converge in $L^1(\mathbb{T}, X)$.

Theorem 5.2.25 *Let $2 \leq q < \infty$ and let X be H^1-uniformly convex of order q. Then, for $f \in H^1(\mathbb{D}, X)$ and any choice of $r_n \in (0,1)$ with $\lim r_n = 1$, the sequence f_{r_n} converges in $L^1(\mathbb{T})$ and almost everywhere.*

Proof Let $f \in H^1(\mathbb{D}, X)$. We fix $0 < s < 1$ and put $g = f_s$ where $f_s(\zeta) = f(s\zeta)$. First, note that by a conformal change of variables, the H^1-convexity condition implies that

$$\|g(z)\|^q + \delta\left(\int_{\mathbb{T}} \|g - g(z)\| p_z dm\right)^q \leq \left(\int_{\mathbb{T}} \|g\| p_z dm\right)^q, \qquad (5.2.64)$$

for $z \in \mathbb{D}$ and $p_z(\zeta) = P(z, \zeta)$. We claim that g satisfies

$$\left(\int_{\mathbb{T}} \|g_t\| dm\right)^q + \delta\left(\int_{\mathbb{T}} \|g - g_t\| dm\right)^q \le \left(\int_{\mathbb{T}} \|g\| dm\right)^q, \tag{5.2.65}$$

where $0 < t < 1$ and $\zeta \in \mathbb{T}$. To get started we let $r = \sqrt{t}$ and use the semigroup property of Poisson kernels to observe that

$$g(y) - g_t(y) = \int_{\mathbb{T}} (g(y) - g_r(w)) P(ry, w) dm(w). \tag{5.2.66}$$

Next, take norms on both sides of (5.2.66), apply the triangle inequality to the right-hand side term, and integrate over $y \in \mathbb{T}$. This gives

$$\int_{\mathbb{T}} \|g - g_t\| dm \le \int_{\mathbb{T}\times\mathbb{T}} \|g(y) - g_r(w)\| P(ry, w) dm(y) dm(w).$$

Hence, by the subharmonicity estimates,

$$\int_{\mathbb{T}} \|g_t\| dm \le \int_{\mathbb{T}} \|g_r\| dm, \quad 0 < t < r < 1,$$

the qth root of the left-hand side in (5.2.65) is bounded by

$$\left(\left(\int_{\mathbb{T}} \|g_r(w)\| dm(w)\right)^q + \delta\left(\int_{\mathbb{T}\times\mathbb{T}} \|g(y) - g_r(w)\| P(ry, w) dm(y) dm(w)\right)^q\right)^{1/q}.$$

Using the triangle inequality in the two-dimensional ℓ_q space, it follows that the above term is bounded by

$$\int_{\mathbb{T}} \left(\|g_r(w)\|^q + \delta\left(\int_{\mathbb{T}} \|g(y) - g_r(w)\| P(ry, w) dm(y)\right)^q\right)^{1/q} dm(w). \tag{5.2.67}$$

Next, recall that $g_r(w) = g(rw)$ and that Poisson kernels are symmetric such that $P(rw, y) = P(ry, w)$. We now apply estimate (5.2.64) with $z = rw$ and obtain that (5.2.67) is bounded by

$$\int_{\mathbb{T}} \int_{\mathbb{T}} \|g(y)\| P(rw, y) dm(y) dm(w).$$

Since $\int_{\mathbb{T}} P(rw, y) dm(w) = 1$, the above integral is just (the qth root of) the right-hand side in (5.2.65). This completes the verification of the estimate (5.2.65).

Next, let $0 < r_n < 1$ be increasing with $r_n \uparrow 1$, and put $f_n = f_{r_n}$. It remains to show that f_n is a Cauchy sequence in $L^1(\mathbb{T}, X)$. To this end, fix $n, k \in \mathbb{N}$ and select $0 < t < s < 1$ such that $st = r_n$ and $s = r_{n+k}$. Since then $g = f_{n+k}$ and $g_t = f_n$, the estimate (5.2.65) yields

$$\delta \|f_{n+k} - f_n\|_{L^1(X)}^q \le \|f_{n+k}\|_{L^1(X)}^q - \|f_n\|_{L^1(X)}^q. \tag{5.2.68}$$

As $\|f_n\|_1$ is increasing and convergent, (5.2.68) shows that the sequence f_n satisfies the Cauchy condition $L^1(\mathbb{T}, X)$. Moreover, if $\varphi = \lim f_n$ then $f(z) = \int_{\mathbb{T}} \varphi p_z dm$ for any $z \in \mathbb{T}$ and, by Theorem 5.2.4, the sequence $f_n(\zeta) = f(r_n\zeta)$ converges in X for almost every $\zeta \in \mathbb{T}$. □

5.3 Hardy Martingale Cotype

Here, we study basic quantitative concepts associated with Hardy martingale convergence in Banach spaces, including complex uniform convexity, H^1-trees in Banach spaces, and Hardy martingale cotype.

We begin with a discussion of H^1-trees in a complex Banach space X. For $n \in \mathbb{N}$, let

$$\alpha_n(X) = \sup\left\{\inf_{1\le k\le n} \mathbb{E}\|\Delta F_k\|_X\right\}, \tag{5.3.1}$$

where the supremum is taken over those X-valued Hardy martingales (F_k) for which $\mathbb{E}(F_1) = 0$ and $\mathbb{E}\|F_n\| \le 1$. Clearly, the sequence $(\alpha_n(X))$ is nonincreasing. Demanding that

$$\inf_{n\in\mathbb{N}} \alpha_n(X) = 0 \tag{5.3.2}$$

imposes a condition on the Banach space X. In that case we say that the Banach space X does not admit H^1-trees of arbitrary depth. One might think that stronger conditions would arise when the infimum in (5.3.1) is replaced by an ℓ^q sum. Let $2 \le q < \infty$. We say that a Banach space X is of Hardy martingale cotype q if there exists $C > 0$ such that for any X-valued Hardy martingale (F_k),

$$\left(\sum_{k=1}^{n} (\mathbb{E}\|\Delta F_k\|_X)^q\right)^{1/q} \le C\mathbb{E}\|F_n\|_X, \tag{5.3.3}$$

for any X-valued Hardy martingale (F_k) and $n \in \mathbb{N}$. The feasible and meaningful values of q satisfy $2 \le q \le \infty$.

The space L^1: Before we proceed with a closer study of this property we discuss the case $X = L^1(\Omega)$. We derive the fact that L^1 has Hardy martingale cotype 2 with constant $C = \sqrt{2}$. To see this, we first verify that

$$(\mathbb{E}\|F_{k-1}\|_{L^1})^2 + \frac{1}{2}(\mathbb{E}\|\Delta F_k\|_{L^1})^2 \le (\mathbb{E}\|F_k\|_{L^1})^2, \tag{5.3.4}$$

for any L^1-valued Hardy martingale (F_k). To check (5.3.4) we proceed as in the proof of Theorem 2.5.9 specialized to $p = 1$. First, we condition to $\mathcal{F}_{k-1}$ by fixing $x \in \mathbb{T}^{k-1}$. Next, let $f_0 = F_{k-1}(x)$ and $f(z) = F_k(x, z)$, where $z \in \mathbb{T}$. Fix $\omega \in \Omega$ and apply the Haagerup–Pisier inequality (2.5.22) of Proposition 2.5.7 to the scalar-valued H^1-function $\zeta \to f(\zeta, \omega)$. This gives

$$|f_0(\omega)|^2 + \frac{1}{2}\left(\int_{\mathbb{T}} |f(\zeta,\omega) - f_0(\omega)| dm(\zeta)\right)^2 \le \left(\int_{\mathbb{T}} |f(\zeta,\omega)| dm(\zeta)\right)^2.$$

Taking expectations with respect to $\omega \in \Omega$, and using Minkowski's inequality and Fubini's theorem gives

$$\|f_0\|_{L^1}^2 + \frac{1}{2}\left(\int_{\mathbb{T}} \|f - f_0\|_{L^1} dm\right)^2 \le \left(\int_{\mathbb{T}} \|f\|_{L^1} dm\right)^2, \tag{5.3.5}$$

which yields the estimate (5.3.4). Summing the telescoping series resulting from (5.3.4), we get

$$\sum_{k=1}^{n} (\mathbb{E}\|\Delta F_k\|_{L^1})^2 \le 2(\mathbb{E}\|F_n\|_{L^1})^2, \tag{5.3.6}$$

for any $n \in \mathbb{N}$. In summary, the Banach space L^1 is of Hardy martingale cotype 2 and the cotype 2 constant of L^1 equals $\sqrt{2}$.

In Section 7.2 we will consider the trace class S_1, and prove Theorem 7.2.3, due to Haagerup and Pisier, which asserts that S^1 satisfies condition (5.2.63) with $q = 2$ and $\delta = 1/2$.

The space L^1/H_0^1: By contrast, L^1/H_0^1 does not satisfy Hardy martingale cotype estimates for any $q < \infty$. Indeed, Pisier's L^1/H_0^1-valued Riesz products provide suitable examples. By Theorem 5.1.4 there exists an L^1/H_0^1-valued Hardy martingale (G_k) such that

$$\mathbb{E}\|G_n\|_{L^1/H_0^1} \le 2 \quad \text{and} \quad \mathbb{E}\|\Delta G_k\|_{L^1/H_0^1} \ge 1, \quad \text{for} \quad 1 \le k \le n, \quad n \in \mathbb{N}. \tag{5.3.7}$$

The finite-dimensional building blocks of L^1/H_0^1 define a Banach space Y satisfying the aRNP, without being of Hardy martingale cotype-q.

For $k \in \mathbb{Z}$, let $e_k : \mathbb{T} \to \mathbb{T}$ be given by $e_k(\zeta) = \zeta^k$. We define the polynomial $L^1(\mathbb{T})$ subspaces

$$L_n^1 = \operatorname{span}\{e_k : |k| \le n\} \quad \text{and} \quad H_n^1 = \operatorname{span}\{e_k : 1 \le k \le n\}.$$

For each $n \in \mathbb{N}$, let $X_n = L_n^1/H_n^1$ be equipped with the quotient norm

$$\|x\|_{X_n} = \inf\left\{\|x + y\|_{L^1} : y \in H_n^1\right\}, \quad x \in L_n^1,$$

and define $Y = \{(x_n) : x_n \in X_n\}$ with

$$\|(x_n)\|_Y = \sum \|x_n\|_{X_n}.$$

Thus defined, Y is a separable dual space and hence satisfies the aRNP. (See the remarks following Theorem 5.7.4.) On the other hand, the Hardy martingale G_n defined by (5.1.8) takes its values in the space X_{4^n}. Hence in view of conditions (5.3.7), Y fails to satisfy Hardy martingale cotype q estimates, for any $q < \infty$.

5.3.1 Dichotomy

Here, we prove the dichotomy theorem of Xu (1988). It relates H^1-trees to Hardy martingale cotype as follows: either X has nontrivial Hardy martingale cotype, or else X admits H^1-trees of arbitrary depth. Equivalently, H^1-trees are the only reason why a Banach space may fail to have nontrivial Hardy martingale cotype.

Theorem 5.3.1 (Xu's dichotomy theorem) *For any Banach space X, the following dichotomy holds true: Either*

$$\inf \alpha_n(X) > 0,$$

and X contains H^1-trees of arbitrary depth, or X is of nontrivial Hardy martingale cotype, and there exists $2 \leq q < \infty$ and $C > 0$ so that

$$\left(\sum_{k=1}^{\infty} (\mathbb{E}\|\Delta F_k\|_X)^q \right)^{1/q} \leq C \sup_{k \in \mathbb{N}} \mathbb{E}\|F_k\|_X,$$

for any X-valued Hardy martingale (F_k).

The first step towards Xu's dichotomy theorem consists of proving that the H^1-tree constants $(\alpha_n(X))$ form a submultiplicative sequence. That is, we will prove that

$$\alpha_{mn}(X) \leq \alpha_m(X)\alpha_n(X),$$

for any $m, n \in \mathbb{N}$. To this end we will verify that a subsequence of a Hardy martingale also behaves like a Hardy martingale. The following so-called subsequence theorem for Hardy martingales makes this precise.

Theorem 5.3.2 (subsequence theorem) *Let X be a complex Banach space and (F_k) a Hardy martingale with values in X. For $n \in \mathbb{N}$, $\epsilon > 0$, and any choice of strictly increasing integers $(m(k))$, there exists an X-valued Hardy martingale (f_k) such that*

$$(1 - \epsilon)\mathbb{E}(\|\Delta f_k\|_X) \leq \mathbb{E}(\|F_{m(k)} - F_{m(k-1)}\|_X) \leq (1 + \epsilon)\mathbb{E}(\|\Delta f_k\|_X), \qquad (5.3.8)$$

and

$$(1 - \epsilon)\mathbb{E}(\|f_k\|_X) \leq \mathbb{E}(\|F_{m(k)}\|_X) \leq (1 + \epsilon)\mathbb{E}(\|f_k\|_X), \qquad (5.3.9)$$

for any $k \leq n$.

Proof First, fix $d \in \mathbb{N}$ and $M \in \mathbb{N}^d$ with $M = (m(l))$. Then for $z \in \mathbb{T}$, we put $z_l = z^{m(l)}$ and form $z^M = (z_l)$ where $1 \le l \le d$.

Next, fix $n \in \mathbb{N}$ and $1 \le k \le n$. For $d(k) \in \mathbb{N}$ and $M(k) \in \mathbb{N}^{d(k)}$ we put $M = (M(1), \dots, M(k), \dots, M(n))$. Thus $M \in \mathbb{N}^N$, where $N = \sum d(k)$. Given $y \in \mathbb{T}^n$ with $y = (y_1, \dots, y_n)$, we define

$$y^M = \left(y_1^{M(1)}, \dots, y_n^{M(n)}\right). \tag{5.3.10}$$

We have thus defined $y^M \in \mathbb{T}^N$, with $N = \sum d(k)$.

Step 1 Given $A \in \mathbb{N}$, we recursively define $M = (m(l) : 1 \le l \le k)$ by $m(1) = A$ and

$$m(l) = l^2 A^l m(l-1).$$

Thus M is an integer vector of length k. Next, fix $z \in \mathbb{T}$, put $z_l = z^{m(l)}$, and define $z^M \in \mathbb{T}^k$ by $z^M = (z_l : 1 \le l \le k)$. Given a trigonometric polynomial $C \colon \mathbb{T}^k \to X$, we define a trigonometric polynomial $C(\cdot, M) \colon \mathbb{T} \to \mathbb{C}$ by

$$C(z, M) = C(z^M).$$

Recall that $m(1) = A$. In view of the recursion defining M, the following observations are easy to see: First, if the spectrum of C is contained in $[-A, A]^k$, then

$$\int_{\mathbb{T}} C(z, M) dm(z) = \int_{\mathbb{T}^k} C(w) d\mathbb{P}(w), \tag{5.3.11}$$

where $\mathbb{P}$ denotes the Haar measue on $\mathbb{T}^k$. Second, if $(w_1, \dots, w_k) \in \mathbb{T}^{k-1}$ and if for fixed $(w_1, \dots, w_{k-1}) \subset \mathbb{T}^{k-1}$ the mapping $w_k \to C(w_1, \dots, w_{k-1}, w_k)$ defines an analytic polynomial in w_k, then $z \to C(z, M)$ is analytic as well.

Step 2 We fix an X-valued Hardy martingale (F_k) and assume that each F_k is an X-valued trigonometric polynomial. Next, fix $n \in \mathbb{N}$ and an increasing sequence of integers $m(0) < \cdots < m(k) < \cdots < m(n)$. We proceed to construct a Hardy martingale (f_k) for which condition (5.3.8) and (5.3.9) hold true.

To get started, we choose scalar-valued trigonometric polymomials p_k, q_k such that

$$\left| p_k - \left\|F_{m(k)}\right\|_X \right| \le \epsilon \mathbb{E} \left\|F_{m(k)}\right\|, \tag{5.3.12}$$

$$|q_k - \| F_{m(k)} - F_{m(k-1)} \|_X| \le \epsilon \mathbb{E} \left\|F_{m(k)} - F_{m(k-1)}\right\|, \tag{5.3.13}$$

and

$$\mathbb{E}(p_k) = \mathbb{E} \left\|F_{m(k)}\right\|_X \quad \text{and} \quad \mathbb{E}(q_k) = \mathbb{E} \left\|F_{m(k)} - F_{m(k-1)}\right\|_X. \tag{5.3.14}$$

In the above equations, $\mathbb{E}$ denotes the expectation over $\mathbb{T}^{\mathbb{N}}$. Determine $A_k \in \mathbb{N}$ so that the spectra of the polynomials F_k, p_k, q_k are contained in $[-A_k, A_k]^k$.

Step 3 Put $d(k) = m(k) - m(k-1)$, select vectors $M(k) \in \mathbb{N}^{d(k)}$ consisting of rapidly increasing integers, and put $M = (M(1), \ldots, M(n))$. We fix $y \in \mathbb{T}^n$ and specify $y = (y_1, \cdots, y_n)$ with $y_i \in \mathbb{T}$. Recall our convention that $y^M = \left(y_1^{M(1)}, \ldots, y_n^{M(n)}\right)$. Finally, for $k \le n$ define

$$f_k(y) = F_{m(k)}(y^M). \tag{5.3.15}$$

By equation (5.3.11) the integer vector M may be selected such that

$$(\mathbb{E}_k f_n)(y) = \left(\mathbb{E}_{m(k)} F_{m(n)}\right)(y^M), \qquad k \le n.$$

Since $F_{m(k)}(w)$ is analytic with respect to the variable $w_{m(k)}$, Step 1 yields that $f_k(y)$ is analytic with respect to the variable y_k. Thus we have verified that (5.3.15) defines an X-valued Hardy martingale (f_k).

Parallel to (5.3.15), we now put

$$s_k(y) = q_k(y^M) \quad \text{and} \quad r_k(y) = p_k(y^M).$$

Then $\mathbb{E}(p_k) = \mathbb{E}(r_k)$ and $\mathbb{E}(q_k) = \mathbb{E}(s_k)$. Hence, invoking the pointwise estimates (5.3.12) and (5.3.13) gives

$$(1-\epsilon)\mathbb{E}\|f_k\|_X \le \mathbb{E}\|F_{m(k)}\|_X \le (1+\epsilon)\mathbb{E}\|f_k\|_X$$

and

$$(1-\epsilon)\mathbb{E}\|\Delta f_k\|_X \le \mathbb{E}\left\|F_{m(k)} - F_{m(k-1)}\right\|_X \le (1+\epsilon)\mathbb{E}\|\Delta f_k\|_X,$$

for $k \le n$. □

We use Theorem 5.3.2 to prove that $(\alpha_n(X))$, the sequence of H^1-tree constants is submultiplicative. For simplicity of notation, we suppress the dependence on X and write $\alpha_n(X) = \alpha(n)$.

Lemma 5.3.3

$$\alpha(mn) \le \alpha(m)\alpha(n), \tag{5.3.16}$$

for any $m, n \in \mathbb{N}$.

Proof We let (F_k) be an X-valued Hardy martingale with $\mathbb{E}(F_k) = 0$. In order to prove (5.3.16) it suffices to check that

$$\inf_{1 \le j \le mn} \mathbb{E}\|\Delta F_j\|_X \le \alpha(n)\alpha(m)\mathbb{E}\|F_{mn}\|_X.$$

First, we group the martingale into n blocks of length m, and rewrite the infimum on the left-hand side as

$$\inf_{1\le j\le mn} \mathbb{E}\|\Delta F_j\|_X = \inf_{1\le k\le n} \inf_{(k-1)m<i\le km} \mathbb{E}\|\Delta F_i\|_X.$$

Then for any $1 \le k \le n$, we estimate on each block of length m separately, obtaining

$$\inf_{(k-1)m<i\le km} \mathbb{E}(\|\Delta F_i\|_X) \le \alpha(m)\mathbb{E}\left(\|F_{km} - F_{(k-1)m}\|_X\right).$$

Now the subsequence theorem asserts that $(F_{km} - F_{(k-1)m})_{k=1}^n$ behaves like a Hardy martingale difference sequence of length n. Specifically, by Theorem 5.3.2,

$$\inf_{1\le k\le n} \mathbb{E}\left\|F_{km} - F_{(k-1)m}\right\|_X \le \alpha(n)\mathbb{E}\|F_{mn}\|_X.$$

This gives the estimate (5.3.16). □

Proof of Theorem 5.3.1: We will prove that the assumption $\inf_n \alpha_n = 0$ implies that the Banach space X has nontrivial Hardy martingale cotype. Specifically, there exist $2 \le q < \infty$ and $C > 0$, such that

$$\left(\sum_{k=1}^{\infty}(\mathbb{E}\|\Delta F_k\|_X)^q\right)^{1/q} \le C \sup_{k\in\mathbb{N}} \mathbb{E}\|F_k\|_X,$$

for any X-valued Hardy martingale (F_k).

Step 1 Assume that $\inf \alpha_n = 0$. Using estimate (5.3.16) we verify that there exist $\beta > 0$ and $m_0 \in \mathbb{N}$ such that

$$\alpha(n) \le m_0^{\beta} n^{-\beta}. \tag{5.3.17}$$

Select $m_0 > 1$ so that $\alpha(m_0) \le 1/2$. Next, choose $\beta > 0$ so that $1/2 \le m_0^{-\beta}$. Thus we have found m_0 and $\beta > 0$ such that

$$\alpha_{m_0} \le m_0^{-\beta}.$$

Recall that the sequence $\alpha(n)$ is submultiplicative. Iteratively applying condition (5.3.3) gives

$$\alpha\left(m_0^k\right) \le m_0^{-k\beta}, \qquad k \in \mathbb{N}.$$

Let $n \in \mathbb{N}$ and choose $k \in \mathbb{N}$ such that $m_0^k \le n \le m_0^{k+1}$. Since $\alpha(n)$ is a decreasing sequence, we have

$$\alpha(n) \le \alpha\left(m_0^k\right) \le m_0^{-k\beta} \le m_0^{\beta} n^{-\beta}.$$

Step 2 Next, we prove that for $q > \beta^{-1}$, the Banach space X has Hardy martingale cotype q. Fix an X-valued Hardy martingale (F_k) and fix $N \in \mathbb{N}$. We rearrange the martingale differences (ΔF_k) according to their size. Determine therefore the integers $\sigma(1), \dots, \sigma(N)$, so that

$$\mathbb{E}\left\|\Delta F_{\sigma(1)}\right\| \geq \cdots \geq \mathbb{E}\left\|\Delta F_{\sigma(j)}\right\| \geq \cdots \geq \mathbb{E}\left\|\Delta F_{\sigma(N)}\right\|. \tag{5.3.18}$$

Next, fix $j \leq N$. Let $\{m(1), \dots, m(j)\}$ be the increasing rearrangement of the set $\{\sigma(1), \dots, \sigma(j)\}$. We use the orderd set $m(1), \dots, m(j)$ to block the martingale difference sequence. Let

$$D_l = F_{m(l)} - F_{m(l-1)} = \sum_{k=m(l-1)+1}^{m(l)} \Delta F_k, \qquad 1 \leq l \leq j.$$

Theorem 5.3.2 asserts that $D_1, \dots, D_j$ behaves like the martingale difference sequence of length j of a Hardy martingale. Indeed, by Theorem 5.3.2,

$$\inf_{1\leq l\leq j} \mathbb{E}\|D_l\| \leq \alpha(j)\mathbb{E}\left\|F_{m(j)}\right\|. \tag{5.3.19}$$

By inspection, $\Delta F_{m(l)} = D_l - \mathbb{E}_{m(l)-1}(D_l)$. Hence

$$\mathbb{E}\left\|\Delta F_{m(l)}\right\| \leq 2\mathbb{E}\|D_l\|. \tag{5.3.20}$$

Next, recall that we chose $\sigma(j)$ so that (5.3.18) holds true. Taking into account the estimates (5.3.19) and (5.3.20) gives

$$\mathbb{E}\left\|\Delta F_{\sigma(j)}\right\| \leq \inf_{1\leq l\leq j} \mathbb{E}\left\|\Delta F_{m(l)}\right\| \leq 2 \inf_{1\leq l\leq j} \mathbb{E}\|D_l\| \leq 2\alpha(j)\mathbb{E}\left\|F_{m(j)}\right\|. \tag{5.3.21}$$

Step 3 Select $q > \beta^{-1}$ and put $C = 2^q m_0^{\beta q} \sum_{j=1}^N j^{-\beta q}$. With (5.3.21) and (5.3.17) we estimate the martingale differences,

$$\begin{aligned}\sum_{j=1}^{N} \left(\mathbb{E}\|\Delta F_{\sigma(j)}\|_X\right)^q &\leq 2^q \sum_{j=1}^{N} \left(\alpha(j)\mathbb{E}\|F_{m(j)}\|\right)^q \\ &\leq C \sup_{1\leq j\leq N} (\mathbb{E}\|F_j\|)^q.\end{aligned}$$

Since $\{\sigma(1), \dots, \sigma(N)\} = \{1, \dots, N\}$ the proof is complete.

5.3.2 Complex Uniform Convexity

In this section we examine the notion of H^1-uniformly convex Banach spaces X more closely. We show that it implies convergence of L^1-bounded, X-valued Hardy martingales and discuss Xu's renorming theorem. Separately, we prove vector-valued extensions of the Davis and Garsia inequality (equation (2.5.13) in Theorem 2.5.4) and the Lépingle inequality (Theorem 2.5.5).

Recall that X is an H^1-uniformly convex Banach space of order $q \geq 2$ if there exist $\delta > 0$ such that for any $h \in H^1(\mathbb{T}, X)$ with $h_0 = \int_{\mathbb{T}} h dm$,

$$\|h_0\|_X^q + \delta\left(\int_{\mathbb{T}} \|h - h_0\|_X dm\right)^q \leq \left(\int_{\mathbb{T}} \|h\|_X dm\right)^q. \tag{5.3.22}$$

By conditioning, this implies that for any X-valued Hardy martingale (f_k),

$$(\mathbb{E}\|f_{k-1}\|_X)^q + \delta(\mathbb{E}\|\Delta f_k\|_X)^q \leq (\mathbb{E}\|f_k\|_X)^q. \tag{5.3.23}$$

Summing the telescoping series resulting from (5.3.23), we see that X satisfies Hardy martingale cotype q estimates. Xu's renorming theorem (Xu, 1988) asserts that Hardy martingale cotype implies H^1-uniform convexity as follows.

Theorem 5.3.4 (Xu's renorming theorem) *Let X be a Banach space of Hardy martingale cotype q. If $r > q$, there exists a quasinorm $|\cdot|$ on X such that*

$$\|x\|/2 \leq |x| \leq \|x\|, \tag{5.3.24}$$

and

$$|x| \leq \int_{\mathbb{T}} |x + \zeta y| dm(\zeta), \tag{5.3.25}$$

for any $x, y \in X$, and, moreover, there exists $\delta = \delta(r, q)$ such that for any $h \in H^1(\mathbb{T}, X)$,

$$|h_0|^r + \delta\left(\int_{\mathbb{T}} |h - h_0| dm\right)^r \leq \left(\int_{\mathbb{T}} |h| dm\right)^r, \tag{5.3.26}$$

where $h_0 = \int_{\mathbb{T}} h dm$.

Proof Let $q \geq 2$ and let $C_q > 0$ be the constant so that the Hardy martingale cotype estimate (5.3.3) holds true, and let $F = (F_m)$ be an integrable Hardy martingale. By Hölder's inequality, for any finite set $A \subset \mathbb{N}$,

$$\sum_{m \in A} \mathbb{E}\|\Delta F_m\|_X \leq C_q |A|^{1/p} \mathbb{E}\|F\|_X, \tag{5.3.27}$$

where $1/p = 1 - 1/q$. Next, for $k \in \mathbb{N}$ we define an equivalent quasinorm on X by putting

$$|x|_k = \inf\left\{\mathbb{E}\|F\|_X - \left(2C_q|A|^{1/p}\right)^{-1} \sum_{m \in A} \mathbb{E}\|\Delta F_m\|_X\right\},$$

where the infimum is taken over all integrable Hardy martingales $F = (F_k)$, for which $\mathbb{E}(F) = x$, and over all subsets $A \subset \mathbb{N}$ satisfying $|A| \leq k$. In view of the factor $(2C_q|A|^{1/p})^{-1}$ in the definition of $|x|_k$ and by inequality (5.3.27) we have the equivalence $\|x\|/2 \leq |x|_k \leq \|x\|$. It is considerably harder to check that

$$|x|_k \leq \int_{\mathbb{T}} |x + \zeta y|_k dm(\zeta), \qquad x, y \in X.$$

The decisive property of the quasinorm $|x|_k$ is however that for distances of order $\geq 1/k$ it satisfies already the desired complex convexity estimates: If $h \in H^1(\mathbb{T}, X)$ and $h_0 = \int_{\mathbb{T}} h dm$, then

$$\int_{\mathbb{T}} |h|_k dm = 1 \quad \text{and} \quad \int_{\mathbb{T}} |h - h_0|_k dm > 4C_q k^{-1/q},$$

which implies that

$$|h_0|_k \leq 1 - 1/k.$$

Finally, putting $c_0 = (\sum n^{-2})^{-1}$ we define the equivalent quasinorm on X by averaging

$$|x| = c_0 \left(\sum n^{-2} |x|_{2^n} \right). \tag{5.3.28}$$

We refer to Xu (1988) for the proof that (5.3.28) satisfies conditions (5.3.24)–(5.3.26). □

Open Problem: In Section 7.1.2 we prove that for any reflexive subspace $R \subset L^1$, the quotient space $X = L^1/R$ admits an equivalent quasinorm for which inequality (5.2.63) holds with $q = 2 + \epsilon$, for any $\epsilon > 0$. This will be obtained by combining the Bourgain–Davis theorem (Theorem 7.1.4) with Xu's renorming theorem (Theorem 5.3.4). It is open however, whether or not L^1/R admits an equivalent quasinorm satisfying H^1-uniform convexity estimates of order 2.

Xu (1988) proves that if X is an H^1-uniformly convex Banach space, then every L^1-bounded, X-valued Hardy martingale converges almost surely. In view of Theorem 5.2.15, this yields a second proof of the Haagerup–Pisier theorem (Theorem 5.2.25), which asserts that H^1-uniformly convex Banach spaces satisfy the aRNP.

Theorem 5.3.5 *Assume that the Banach space X satisfies inequality* (5.3.23). *Let* (F_k) *be an X-valued Hardy martingale. Then, for any subsequence* $m(k) \uparrow \infty$,

$$\left(\mathbb{E}\left\|F_{m(k-1)}\right\|_X\right)^q + \delta\left(\mathbb{E}\left\|F_{m(k-1)} - F_{m(k)}\right\|_X\right)^q \leq \left(\mathbb{E}\left\|F_{m(k-1)}\right\|_X\right)^q \tag{5.3.29}$$

and

$$\left(\sum_{n=1}^{k} \left(\mathbb{E}\left\|F_{m(n-1)} - F_{m(n)}\right\|_X\right)^q\right)^{1/q} \leq \delta^{-1/q} \mathbb{E}\|F_{m(k)}\|_X. \tag{5.3.30}$$

Consequently, if $\sup_{m \in \mathbb{N}} \mathbb{E}\|F_m\|_X < \infty$ *then* (F_m) *converges in* $L^1(X)$.

Proof Choose $\epsilon > 0$, and fix a subsequence $m(k) \uparrow \infty$ and an X-valued Hardy martingale (F_k). Apply the subsequence theorem (Theorem 5.3.2) to obtain a Hardy martingale (f_k) so that for any k, inequalities (5.3.8) and (5.3.9) hold true. Now use (5.3.23) for (f_k) to obtain

$$\begin{aligned}\left(\mathbb{E}\left\|F_{m(k-1)}\right\|_X\right)^q &+ \delta\left(\mathbb{E}\left\|F_{m(k-1)} - F_{m(k)}\right\|_X\right)^q \\ &\leq (1+\epsilon)(\mathbb{E}\|f_{k-1}\|_X)^q + \delta(\mathbb{E}\|\Delta f_k\|_X)^q \\ &\leq (1+\epsilon)(\mathbb{E}\|f_k\|_X)^q \\ &\leq (1+\epsilon)(1-\epsilon)^{-1}\left(\mathbb{E}\left\|F_{m(k)}\right\|_X\right)^q.\end{aligned}$$

Since $\epsilon > 0$ was arbitrary we obtain inequality (5.3.29). Finally, iterating (5.3.29) gives (5.3.30). As a corollary to (5.3.30), if (5.3.23) holds then $\sup_{m\in\mathbb{N}} \mathbb{E}\|F_m\|_X < \infty$ implies that the Hardy martingale (F_m) is a Cauchy sequence in $L^1(X)$, and hence is convergent. □

We now turn to previsible projection estimates for X-valued Hardy martingales, provided that X is H^1-uniformly convex. In Theorem 5.3.7 we obtain a vector-valued version of Lépingle's inequality (Theorem 2.5.5). The first step towards vector-valued previsible projection estimates is given by Lemma 5.3.6, which contains the nonlinear telescoping tools used in proving Theorem 5.3.7.

Lemma 5.3.6 *Let $1 \leq q \leq \infty$ and $1/p + 1/q = 1$. Let $n \in \mathbb{N}$. Assume that the nonnegative random variables $M_1, \ldots, M_n$ and $v_1, \ldots, v_n$, satisfy*

$$\mathbb{E}\left(M_{k-1}^q + v_k^q\right)^{1/q} \leq \mathbb{E}M_k,$$

for $1 \leq k \leq n$. Then

$$\mathbb{E}\left(\sum_{k=1}^{n} v_k^q\right)^{1/q} \leq 2\,(\mathbb{E}M_n)^{1/q}\left(\mathbb{E}\max_{k\leq n} M_k\right)^{1/p}.$$

Proof Define $0 \leq \epsilon \leq 1$ by $\epsilon^p = (\mathbb{E}\max_{k\leq n} M_k)^{-1}(\mathbb{E}M_n)$. Choose nonnegative $s_k \in L^\infty$ so that $\sum_{k=1}^n s_k^p \leq \epsilon^p$. Then using arithmetic and Hölder's inequality – as in the proof of Lemma 2.5.3 – yields

$$\mathbb{E}(v_k s_k) \leq \mathbb{E}\left(s_k^p M_{k-1}\right) + \mathbb{E}\left(M_{k-1}^q + v_k^q\right)^{1/q} - \mathbb{E}M_{k-1}.$$

Since $\mathbb{E}\left(M_{k-1}^q + v_k^q\right)^{1/q} \leq \mathbb{E}M_k$ and $\sum_{k=1}^n s_k^p \leq \epsilon^p$, summing over $k \leq n$ and invoking duality gives

$$\mathbb{E}\left(\sum_{k=1}^{n} v_k^q\right)^{1/q} \leq \epsilon^{-1}\mathbb{E}M_n + \epsilon^{p-1}\mathbb{E}\max_{k\leq n} M_k.$$

Inserting the value of $\epsilon > 0$ specified above finishes the proof. □

Theorem 5.3.7 *If X is an H^1-uniformly convex Banach space of order q, $(2 \le q < \infty)$, then for any X-valued Hardy martingale (F_k)*

$$\mathbb{E}\left(\sum_{k=1}^{\infty}(\mathbb{E}_{k-1}\|\Delta F_k\|_X)^q\right)^{1/q} \le A \sup_k \mathbb{E}\|F_k\|_X,$$

where $A = A(q, \delta)$.

Proof Let (F_k) be an X-valued Hardy martingale. Since X is H^1-uniformly convex, (5.3.22) is satisfied. Hence by conditioning we obtain that

$$\mathbb{E}\left(\|F_{k-1}\|_X^q + \delta(\mathbb{E}_{k-1}\|\Delta F_k\|_X)^q\right)^{1/q} \le \mathbb{E}\|F_k\|_X. \tag{5.3.31}$$

Apply the nonlinear telescoping lemma (Lemma 5.3.6) with $v_k = \delta^{1/q}\mathbb{E}_{k-1}\|\Delta F_k\|_X$ and $M_k = \|F_k\|_X$. This gives

$$\mathbb{E}\left(\sum_{k=1}^{\infty}(\mathbb{E}_{k-1}\|\Delta F_k\|_X)^q\right)^{1/q} \le A\left(\sup_k \mathbb{E}\|F_k\|_X\right)^{1-1/q}\left(\mathbb{E}\sup_k \|F_k\|_X\right)^{1/q},$$

where $A = A(q, \delta)$. It remains to invoke the L^1-maximal function estimates of Theorem 2.5.1. □

In Chapter 2, Theorem 2.5.4, we obtained the Davis and Garsia inequalities for scalar-valued Hardy martingales. We now turn to proving their vector-valued counterparts. At this point we introduce a second complex convexity condition on the Banach space X.

Hypothesis $\mathcal{H}(q)$: Let $2 \le q < \infty$. We say that a complex Banach space X satisfies hypothesis $\mathcal{H}(q) <$ if for each $M > 1$ there exists $\delta = \delta(M) > 0$ such that for any $z \in X$ and $g \in H_0^\infty(\mathbb{T}, X)$ satisfying $\|g\|_{L^\infty} \le M\|z\|_X$ the following estimate holds true,

$$\left(\|z\|^q + \delta\int_{\mathbb{T}}\|g\|_X^q dm\right)^{1/q} \le \int_{\mathbb{T}}\|z+g\|_X dm. \tag{5.3.32}$$

It is known that the H^1-uniform convexity condition of order q is weaker than hypothesis $\mathcal{H}(q)$. See Müller (2017).

We will establish the Davis and Garsia inequality for X-valued Hardy martingales provided that X satisfies hypothesis $\mathcal{H}(q)$. Specifically, Theorem 5.3.11 asserts that any X-valued Hardy martingale $F = (F_k)$ can be decomposed into the sum of Hardy martingales G and B such that

$$\mathbb{E}\left(\sum_{k=1}^{\infty}\mathbb{E}_{k-1}\|\Delta G_k\|_X^q\right)^{1/q} + \mathbb{E}\left(\sum_{k=1}^{\infty}\|\Delta B_k\|_X\right) \le A_q \sup \mathbb{E}(\|F_k\|_X),$$

where $2 \le q < \infty$ is determined by hypothesis $\mathcal{H}(q)$. The first step in the proof is of independent interest: It consists in proving the Davis decomposition for vector-valued Hardy martingales, and will be obtained without imposing any condition on the Banach space X. We emphasize that the proof of the scalar-valued Davis decomposition (2.5.16), based on Theorem 2.5.4, does not extend to the vector-valued case. Instead, we adapt the proof of Theorem 3.1.1 that resulted in condition (3.1.4).

Theorem 5.3.8 (Davis decomposition theorem) *Let X be a Banach space. Any X-valued Hardy martingale $F = (F_k)$ can be decomposed into the sum of X-valued Hardy martingales $F = G + B$ such that*

$$\|\Delta G_k\|_X \le M\|F_{k-1}\|_X \quad \text{and} \quad \mathbb{E}\left(\sum_{k=1}^{\infty} \|\Delta B_k\|_X\right) \le 8\mathbb{E}(\|F\|_X),$$

where $M \le 24$.

Proof Let $k \in \mathbb{N}$ and condition on $\mathcal{F}_{k-1}$ by fixing $x \in \mathbb{T}^{k-1}$. For $y \in \mathbb{T}$, put $h(y) = \Delta F_k(x, y)$ and $z = F_{k-1}(x)$. We apply Lemma 5.3.9 to $h \in H_0^1(\mathbb{T}, X)$ and $z \in X$. This gives $g \in H_0^\infty(\mathbb{T}, X)$ such that

$$\|g\|_{L^\infty(X)} \le M\|z\|_X \quad \text{and} \quad \|z + h\|_{L^1(X)} \ge \|z\|_X + (1/8)\|h - g\|_{L^1(X)}.$$

Define the splitting of ΔF_k by putting

$$\Delta G_k(x, y) = g(y) \qquad \text{and} \quad \Delta B_k(x, y) = h(y) - g(y).$$

This gives $\Delta F_k = \Delta G_k + \Delta B_k$, with

$$\|\Delta G_k\|_X \le M\|F_{k-1}\|_X,$$

and

$$\|F_{k-1}\|_X + \tau\mathbb{E}_{k-1}(\|\Delta B_k\|_X) \le \mathbb{E}_{k-1}(\|F_k\|_X) \tag{5.3.33}$$

with $\tau = 1/8$. Taking expectations on both sides of (5.3.33) and summing the resulting telescoping series we obtain

$$\sum \mathbb{E}(\|\Delta B_k\|_X) \le 8 \sup \mathbb{E}(\|F_k\|_X). \qquad \square$$

The proof of the Davis decompositon theorem (Theorem 5.3.8) is thus reduced to the following decomposition for vector-valued analytic functions. Again, the following lemma works for any Banach space X.

Lemma 5.3.9 *For any $h \in H_0^1(\mathbb{T}, X)$ and $z \in X$ there exists $g \in H_0^\infty(\mathbb{T}, X)$ so that*

$$\|g\|_{L^\infty(X)} \le M\|z\|_X$$

and

$$\|z\|_X + (1/8)\|h - g\|_{L^1(X)} \le \|z + h\|_{L^1(X)}, \tag{5.3.34}$$

where the constant satisfies $M \le 24$.

Proof The proof begins with the definition of $g \in H_0^\infty(\mathbb{T}, X)$. Thereafter, we successively collect the lower estimates for the right-hand side of (5.3.34).

Step 1 Determine $g \in H_0^\infty(\mathbb{T}, X)$ by putting

$$\rho = \inf\{t < \tau : \|h(z_t)\|_X > M\|z\|_X\}, \qquad g = N(h(z_\rho)),$$

where N denotes Doob's projection operator. Recall that $h \in H_0^1(\mathbb{T}, X)$. Hence the stopped martingale also satisfies $\mathbb{E}(h(z_\rho)) = 0$. By definition of ρ we have the uniform estimate $\|h(z_\rho)\|_X \le M\|z\|_X$. Since Doob's projection is a contraction between L^∞ spaces we obtain

$$\|g\|_\infty \le M\|z\|_X.$$

Moreover, $g \in H_0^\infty(\mathbb{T}, X)$ since N maps a space of holomorphic martingales into the corresponding Hardy space.

Step 2 We now turn to proving the integral estimates (5.3.34). The idea is to find a lower estimate for the right-hand side by integrating it against a suitable testing function. Define $A = \{\rho < \tau\}$. The set A is measurable with respect to the stopping time sigma-algebra $\mathcal{F}_\rho$. Since conditional expectations are L^1 contractions we have

$$\mathbb{E}(\|h(z_\tau)1_A\|_X) \ge \mathbb{E}(\|h(z_\rho)1_A\|_X) \ge M\|z\|_X. \tag{5.3.35}$$

Next, define $p = N(1_A)/2$. Clearly $0 \le p \le 1/2$ and, by the covariance formula, we get $\int p\,dm = \mathbb{P}(A)/2$ and

$$\int_{\mathbb{T}} \|h\|_X p\,dm = \frac{1}{2}\mathbb{E}(\|h(z_\tau 1_A\|_X).$$

Combining this with (5.3.35), the triangle inequality gives

$$\int_{\mathbb{T}} \|z + h\|_X p\,dm \ge \int_{\mathbb{T}} \|h\|_X p\,dm - \frac{1}{2}\mathbb{P}(A)\|z\|_X \ge \left(\frac{1}{2} - \frac{1}{2M}\right)\mathbb{E}(\|h(z_\tau)1_A\|_X). \tag{5.3.36}$$

Step 3 Define the outer function $q = \exp(\ln(1 - p) + iH\ln(1 - p))$. By construction, $q \in H^\infty(\mathbb{T})$. Since $h \in H_0^1(\mathbb{T}, X)$ we have $\int_\mathbb{T} hqdm = 0$. Put $q_2 = \Im q$ and $q_1 = \Re q$. Then by inspection, $\int_\mathbb{T} q_2dm = 0$ and $\int_\mathbb{T}(z + h)qdm = z\int_\mathbb{T} q_1dm$. In step 5 we will verify that for $C_1 > 3$

$$\int_\mathbb{T} q_1dm > 1 - C_1\mathbb{P}(A). \tag{5.3.37}$$

Assuming the crucial estimate (5.3.37) for the moment we may summarize the observations above as follows,

$$\int_\mathbb{T} \|z + h\|_X|q|dm \geq \left\|\int_\mathbb{T}(z + h)qdm\right\|_X \geq \|z\|_X(1 - C_1\mathbb{P}(A)). \tag{5.3.38}$$

Since $p + |q| = 1$, taking the sum of (5.3.36) and (5.3.38) gives

$$\int_\mathbb{T} \|z + h\|_Xdm \geq \|z\|_X + \left(\frac{1}{2} - \frac{1}{2M} - \frac{C_1}{M}\right)\mathbb{E}(\|h(z_\tau)1_A\|_X). \tag{5.3.39}$$

Step 4 It remains to obtain the link between the term $\mathbb{E}(\|h(z_\tau)1_A\|_X)$ arising on the right-hand sides of (5.3.39), (5.3.36), and $\int_\mathbb{T}\|h - g\|_X$. Recall that $A = \{\rho < \tau\}$. Hence,

$$h(z_\tau) - h(z_\rho) = (h(z_\tau) - h(z_\rho))1_A. \tag{5.3.40}$$

Using that Doob's projection contracts L^1 norms, we derive from (5.3.40) and (5.3.35) that

$$\int_\mathbb{T} \|h - g\|_Xdm \leq \mathbb{E}(\|h(z_\tau) - h(z_\rho)\|_X 1_A) \leq 2\mathbb{E}(\|h(z_\tau)1_A\|_X). \tag{5.3.41}$$

Now choose $M \geq 8C_1$ so that $(1/2 - 1/(2M) - C_1/M) > 1/4$ and merge inequalities (5.3.39)–(5.3.41) to obtain

$$\int_\mathbb{T} \|z + h\|_Xdm \geq \|z\|_X + \frac{1}{8}\int_\mathbb{T} \|h - g\|_Xdm.$$

Step 5 Here, we prove that inequality (5.3.37) holds true. We show

$$\int_\mathbb{T} q_1dm \geq 1 - 3\int_\mathbb{T} pdm. \tag{5.3.42}$$

Recall that $q_1 = (1 - p)\cos(H(\ln(1 - p)))$. Use $\cos(x) \geq 1 - x^2/2$ to get the pointwise inequality

$$q_1 \geq (1 - p) - \frac{1}{2}(H(\ln(1 - p)))^2. \tag{5.3.43}$$

We have thus reduced the lower L^1 estimate for q_1 to an upper L^2 estimate for the Hilbert transform. Since $0 \le p \le 1/2$, we have $(\ln(1-p))^2 \le 2p$, and hence

$$\int_{\mathbb{T}} (H(\ln(1-p)))^2 dm \le 2 \int_{\mathbb{T}} (\ln(1-p))^2 dm \le 4 \int_{\mathbb{T}} p dm. \tag{5.3.44}$$

Combining inequalities (5.3.43) and (5.3.44) gives (5.3.42), as required. □

Next, we assume that the Banach space X satisfies hypothesis $\mathcal{H}(q)$ and draw further consequences of Lemma 5.3.9. The following theorem prepares for the application of the nonlinear telescoping lemma (Lemma 5.3.6). It provides a one-step estimate that yields the vector-valued Davis and Garsia inequalities in Theorem 5.3.11.

Theorem 5.3.10 *Let $q \ge 2$, and let X satisfy hypothesis $\mathcal{H}(q)$. For every $z \in X$ and $h \in H_0^1(\mathbb{T}, X)$, there exists $g \in H_0^\infty(\mathbb{T}, X)$ such that $\|g\|_\infty \le M\|z\|_X$ and*

$$\left(\|z\|_X^q + \delta(q)\|g\|_{L^q(X)}^q\right)^{1/q} + \frac{1}{16}\|h - g\|_{L^1(X)} \le \|z + h\|_{L^1(X)}, \tag{5.3.45}$$

with $M \le 24$ and $\delta(q) = \delta(18)^{-q}$ where $\delta > 0$ is given by estimate (5.3.32).

Proof Fix $M \le 24$ such that the conclusion of Lemma 5.3.9 holds true. Pick $h \in H_0^1(\mathbb{T}, X)$ and $z \in X$. By Lemma 5.3.9, there exists $g \in H_0^\infty(\mathbb{T}, X)$ satisfying $\|g\|_{L^\infty(X)} \le M\|z\|_X$ and

$$\|z + h\|_{L^1(X)} \ge \|z\|_X + (1/8)\|h - g\|_{L^1(X)}. \tag{5.3.46}$$

On the other hand, by the triangle inequality we simply get

$$\|z + h\|_{L^1(X)} \ge \|z + g\|_{L^1(X)} - \|h - g\|_{L^1(X)}. \tag{5.3.47}$$

We let $0 < \alpha < 1$, form

$$(1 - \alpha)(5.3.46) + \alpha(5.3.47),$$

and apply the estimates of hypothesis $\mathcal{H}(q)$ to $\int \|z + g\|_X dm$. This gives

$$\|z + h\|_1 \ge (1 - \alpha)\|z\|_X + \alpha\left(\|z\|_X^q + \delta\|g\|_q^q\right)^{1/q} + c_\alpha\|h - g\|_1,$$

where $c_\alpha = (1 - 9\alpha)/8$. By the triangle inequality, the term on the right-hand side of the above expression is larger than

$$\left(\|z\|_X^q + \delta\alpha^q\|g\|_q^q\right)^{1/q} + c_\alpha\|h - g\|_1.$$

Choosing $\alpha = 1/18$ gives (5.3.45). □

Now we are ready to prove that hypothesis $\mathcal{H}(q)$ implies Davis and Garsia estimates for X-valued Hardy martingales.

Theorem 5.3.11 *Let $q \geq 2$, and let X be a complex Banach space satisfying hypothesis $\mathcal{H}(q)$. Then any X-valued Hardy martingale $F = (F_k)$ can be decomposed into the sum of X-valued Hardy martingales G and B such that $\|\Delta G_k\|_X \leq M\|F_{k-1}\|_X$ and*

$$\mathbb{E}\left(\sum_{k=1}^{\infty} \mathbb{E}_{k-1}\|\Delta G_k\|_X^q\right)^{1/q} + \mathbb{E}\left(\sum_{k=1}^{\infty} \|\Delta B_k\|_X\right) \leq A_q \sup_k \mathbb{E}(\|F_k\|_X).$$

Proof Let $k \in \mathbb{N}$ and condition on $\mathcal{F}_{k-1}$. Fix $x \in \mathbb{T}^{k-1}$ and $y \in \mathbb{T}$, then define

$$h(y) = \Delta F_k(x, y) \quad \text{and} \quad z = F_{k-1}(x).$$

We apply Theorem 5.3.10 to $z \in X$ and $h \in H_0^1(\mathbb{T}, X)$ and obtain $g \in H_0^\infty(\mathbb{T}, X)$, satisfying condition (5.3.45). Substituting back gives

$$\Delta G_k(x, y) = g(y) \quad \text{and} \quad \Delta B_k(x, y) = h(y) - g(y),$$

satisfying $\|\Delta G_k\|_X \leq M\|F_{k-1}\|_X$ and

$$\mathbb{E}\left(\|F_{k-1}\|_X^q + \delta(q)\mathbb{E}_{k-1}\|\Delta G_k\|_X^q\right)^{1/q} + c\mathbb{E}(\|\Delta B_k\|_X) \leq \mathbb{E}(\|F_k\|_X), \qquad (5.3.48)$$

where $c = 1/16$. Apply Lemma 5.3.6 with $M_k = \|F_{k-1}\|_X$ and $v_k^q = \mathbb{E}_{k-1}\|\Delta G_k\|_X^q$ and invoke Theorem 2.5.1, the maximal function estimate for Hardy martingales. This gives

$$\mathbb{E}\left(\sum_{k=1}^{\infty} \mathbb{E}_{k-1}\|\Delta G_k\|_X^q\right)^{1/q} + \mathbb{E}\left(\sum_{k=1}^{\infty} \|\Delta B_k\|_X\right) \leq A \sup_{n\in\mathbb{N}} \mathbb{E}\|F_n\|_X,$$

where $A = A(q, \delta)$. □

Pisier's Problem (Pisier, 1992a, 2016) asks if nontrivial Hardy martingale cotype of a complex Banach space X is determined by *simple* X-valued Hardy martingales. Recall that $G = (G_j : \mathbb{T}^{\mathbb{N}} \to X)$ is a *simple* X-valued Hardy martingale if $\Delta G_j(z) = f_{j-1}(z)z_j$, $z \in \mathbb{T}^{\mathbb{N}}$, where $f_{j-1} : \mathbb{T}^{\mathbb{N}} \to X$ depends only on the first $j-1$ coordinates of $z \in \mathbb{T}^{\mathbb{N}}$. See Section 2.2.1.

Assume that there exists $2 \leq q < \infty$ and $C > 0$ such that

$$\left(\sum_{j=1}^{\infty} (\mathbb{E}\|\Delta G_j\|_X)^q\right)^{1/q} \leq C \sup_{j\in\mathbb{N}} \mathbb{E}\|G_j\|_X,$$

for any *simple* X-valued Hardy martingale (G_j). Does it follow that X is of nontrivial Hardy martingale cotype? Does it follow that X satisfies the aRNP?

5.4 Unconditionality (aUMD)

We study the unconditional convergence of X-valued Hardy martingales in L^p where $1 \leq p < \infty$. We say that a Banach space X is unconditional for Hardy martingale differences if there exists $C_p < \infty$ such that

$$\mathbb{E}\left\|\sum_{k=1}^{n} \varepsilon_k \Delta F_k\right\|_X^p \leq C_p^p \mathbb{E}\|F_n\|_X^p, \qquad n \in \mathbb{N},$$

for any L^p-bounded, X-valued Hardy martingale (F_k), and $(\varepsilon_k) \in \{-1, 1\}^{\mathbb{N}}$. In this case we say that the UMD estimates hold true for X-valued Hardy martingales, or equivalently, that the Banach space X satisfies the analytical UMD property (aUMD$_p$).

In Section 5.4.1 we prove Kwapień's chracterization of Banach spaces containing c_0 in terms of almost surely bounded Rademacher series, and exploit it to show that the aUMD$_1$ implies the aRNP. We prove that c_0 is not finitely represented in any Banach space satisfying the aUMD$_p$, $1 \leq p < \infty$.

In Section 5.4.2 we prove Garling's extrapolation theorems to the effect that for any Banach space X, the aUMD$_p$ property is equivalent to aUMD$_1$. Separately, we show that ℓ^1 (and hence L^1) satisfies aUMD$_1$.

In Section 5.4.2 we prove Bourgain–Stein–Qiu inequalities for Banach spaces with the aUMD. They play an important role in the proof of Qiu's theorem (Theorem 7.3.4), which asserts that the iterated $L^p(L^q)$ spaces fail to satisfy the aUMD property for $1 \leq p \neq q < \infty$.

5.4.1 The Spaces c_0 and ℓ^1

Let X be a Banach space. We say that X contains a copy of c_0 if there exists a closed subspace $Y \subseteq X$ and a linear bijection $T : c_0 \to Y$ such that

$$\|T\| \cdot \left\|T^{-1}\right\| < \infty.$$

Bessaga and Pełczyński (1958b) established a very useful criteron for a Banach space to contain a copy of c_0. We refer the reader to Wojtaszczyk's book for the classical proof (Wojtaszczyk, 1991).

Theorem 5.4.1 (Bessaga–Pełczyński theorem) *Let $\{v_n\}$ be a sequence in a Banach space X. If*

$$\inf_n \|v_n\| > 0$$

and if there exists a constant $C < \infty$ such that

$$\sup_n \left\|\sum_{i=1}^{n} a_i v_i\right\| \leq C \sup_n |a_n|,$$

for any bounded sequence of scalars (a_n), then X contains a copy of c_0.

We let $\{\sigma_n\}$ denote the sequence of independent Rademacher functions on the unit interval $[0, 1]$, and we let λ denote Lebesgue measure on $[0, 1]$. Given a bounded sequence $\{f_n\}$ in a Banach space X, we let

$$G_n(t) = \sum_{i=1}^{n} \sigma_i(t) f_i \quad \text{and} \quad M(t) = \sup_n \|G_n(t)\|,$$

for $t \in [0, 1]$. Hoffmann-Jorgensen (1974) and, respectively, Kahane (1968, 1985) present the following basic facts on X-valued Rademacher series. Those facts hold true for any Banach space X.

(i) If $\lambda\{t \in [0, 1] : M(t) = \infty\} = 0$, then $M \in L^1$.
(ii) If $\{G_n\}$ is a Cauchy sequence in $L^1(X)$, then $\{G_n(t)\}$ converges in X for almost every $t \in [0, 1]$. Since the sequence $\{G_n\}$ forms a martingale, this assertion follows from the convergence theorem for closed vector-valued martingales (Theorem 2.1.3).

We now turn to Kwapień's theorem (Kwapień, 1974). In short, it states that if the Banach space X does not contain a copy of c_0 and if $\lambda\{t \in [0, 1] : M(t) = \infty\} = 0$, then $\{G_n(t)\}$ converges in X for almost every $t \in [0, 1]$.

Theorem 5.4.2 (Kwapień's theorem) *Let $\{f_n\}$ be a bounded sequence in a Banach space X, and let*

$$G_n(t) = \sum_{i=1}^{n} \sigma_i(t) f_i \quad \textit{and} \quad M(t) = \sup_n \|G_n(t)\|.$$

If $\lambda\{t \in [0, 1] : M(t) = \infty\} = 0$, and if $\lambda\{t \in [0, 1] : G_n(t)$ converges in $X\} \neq 1$, then X contains a copy of c_0.

Proof We assume that there exists a bounded X-valued Rademacher series that fails to converge almost everywhere. Using basic facts on Rademacher series (see above) we first show that this assumption implies conditions (5.4.2) and (5.4.3). The main step of Kwapień's argument consists in proving that (5.4.2) and (5.4.3) imply the hypothesis of the Bessaga–Pełczyński theorem (Theorem 5.4.1).

Part 1 We assume that $\{G_n(t)\}$ fails to be convergent in X for $t \in B \subseteq [0, 1]$ with $\lambda(B) > 0$. Clearly, the sequence $\{G_n\}$ forms a martingale in $L^1(X)$. Recall that for any martingale, convergence in $L^1(X)$ implies convergence almost everywhere. Hence there exists $\eta > 0$ and an increasing sequence of integers $n(k) \to \infty$ such that the sequence $F_k = G_{n(k+1)} - G_{n(k)}$ satisfies

$$\int_0^1 \|F_k\| d\lambda > \eta, \tag{5.4.1}$$

for any $k \in \mathbb{N}$. Recall that $\lambda\{t \in [0,1] : M(t) = \infty\} = 0$ implies $M \in L^1$. Since $\|F_k\| \le 2M$, the sequence $\|F_k\|$ is uniformly integrable on $[0,1]$. Hence (5.4.1) implies that there exists $\delta > 0$ and a measurable set Ω_0 with $\lambda(\Omega_0) > 0$ such that

$$\limsup_{k\to\infty} \|F_k(t)\| > \delta,$$

for $t \in \Omega_0$. We next form

$$H_n(s,t) = \sum_{k=1}^{n} \sigma_k(s) F_k(t).$$

By assumption, for almost every $t \in [0,1]$, there exists $\Omega_1 = \Omega_1(t)$ with $\lambda(\Omega_1) = 1$ such that

$$\sup_n \|H_n(s,t)\| < \infty,$$

for $s \in \Omega_1$. Put $\Omega_2 = \{t : \sup_n \|H_n(s,t)\| < \infty,\ s \in \Omega_1(t)\}$, and consider

$$\Omega_3 = \Omega_2 \cap \Omega_0.$$

Since $\lambda(\Omega_0) > 0$ and $\lambda(\Omega_2) = 1$ we have $\lambda(\Omega_3) > 0$. For $t_0 \in \Omega_3$, we may select an increasing sequence of integers $k(i) \to \infty$ such that $x_i = F_{k(i)}(t_0)$ satisfies

$$\inf_i \|x_i\| > 0 \tag{5.4.2}$$

and

$$\lambda\left\{ s \in [0,1] : \sup_n \left\| \sum_{i=1}^{n} \sigma_i(s) x_i \right\| < \infty \right\} = 1. \tag{5.4.3}$$

Part 2 The main part of the proof consists in showing that conditions (5.4.2) and (5.4.3) imply that X contains a copy of c_0.

Pick $K < \infty$ large enough, such that

$$A = \left\{ s \in [0,1] : \sup_n \left\| \sum_{i=1}^{n} \sigma_i(s) x_i \right\| < K \right\}$$

satisfies $\lambda(A) > 1/2$. We next observe that

$$\lim_{i\to\infty} \lambda(A \cap \{\sigma_i = a\}) = (1/2)\lambda(A),$$

for $a \in \{-1, +1\}$. Exploiting this observation, we may inductively choose an increasing subsequence $n(i) \to \infty$ such that

$$\lambda\left(A \cap \bigcap_{i=1}^{k} \{\sigma_{n(i)} = a_i\} \right) > 2^{-k-1},$$

for any $k \in \mathbb{N}$ and any $a_i \in \{-1, +1\}$. Put $J = \{n(i) : i \in \mathbb{N}\}$. Define the sequence of Rademacher functions $\{\sigma'_i\}$ by letting $\sigma'_i = \sigma_i$ for $i \in J$, and $\sigma'_i = -\sigma_i$ for $i \notin J$. Now consider

$$A' = \left\{ s \in [0,1] : \sup_n \left\| \sum_{i=1}^{n} \sigma'_i(s) x_i \right\| < K \right\}.$$

Since $\{\sigma'_i\}$ and $\{\sigma_i\}$ are identically distributed, we have $\lambda(A \cap S) = \lambda(A' \cap S)$ where $S = \bigcap_{i=1}^{k} \{\sigma_{n(i)} = a_i\}$. We claim that for any $k \in \mathbb{N}$ and $a_i \in \{-1, +1\}$ fixed, we have

$$A \cap A' \cap S \neq \emptyset.$$

Indeed, assuming that $(A \cap S) \cap (A' \cap S) = \emptyset$ gives a contradiction as follows:

$$2^{-k} = \lambda(S) \geq \lambda(A \cap S) + \lambda(A' \cap S) > 2^{-k}.$$

Finally, pick $t_0 \in A \cap A' \cap S$. Then

$$\left\| \sum_{i=1}^{k} a_i x_{n(i)} \right\| = \frac{1}{2} \left\| \sum_{i=1}^{n(k)} \sigma_i(t_0) x_i + \sum_{i=1}^{n(k)} \sigma'_i(t_0) x_i \right\| \leq K. \tag{5.4.4}$$

Note that any vector $(b_1, \ldots, b_k)$ with $\max |b_i| \leq 1$ is contained in the convex hull of $\{-1, 1\}^k$. Hence (5.4.4) holds true with $(a_i) \in \{-1, 1\}^k$ replaced by (b_i) with $\max |b_i| \leq 1$. Since $\inf_i \|x_{n(i)}\| > \delta$, we have finally established the hypothesis of Theorem 5.4.1 for the sequence $x_{n(i)}$. Thus we have proved that X contains a copy of c_0. □

We now present Garling's theorem, which asserts that any L^1-bounded, X-valued Hardy martingale converges almost surely, whenever X-valued Hardy martingales differences are unconditionally bounded.

Theorem 5.4.3 (Garling's theorem) *Let X be a complex Banach space. Assume that there exists $C_1 > 0$ such that for any X-valued Hardy martingale (F_n) we have,*

$$\mathbb{E} \left\| \sum_{k=1}^{n} \varepsilon_k \Delta F_k \right\|_X \leq C_1 \mathbb{E} \|F_n\|_X, \tag{5.4.5}$$

for any $n \in \mathbb{N}$, and any choice of signs $\epsilon_k \in \{+1, -1\}$. Then every L^1-bounded, X-valued Hardy martingale (F_n) converges almost surely and in the norm of $L^1(X)$.

Proof Let (F_n) denote an L^1-bounded, X-valued Hardy martingale. Put $K = \sup_n \mathbb{E}\|F_n\|_X$ and let $\sigma_n(t) = \text{sign}(\cos(2\pi 2^n t))$ be the sequence of independent

Rademacher functions on the unit interval $[0, 1]$. By assumption (5.4.5) and the maximal function estimate for Hardy martingales, we obtain

$$\mathbb{E} \sup_{n \in \mathbb{N}} \left\| \sum_{k=1}^{n} \sigma_k(t) \Delta F_k \right\|_X \leq e C_1 K,$$

for any $t \in [0, 1]$. Integrating over $t \in [0, 1]$ and applying Fubini's theorem gives

$$\mathbb{E} \int_0^1 \sup_{n \in \mathbb{N}} \left\| \sum_{k=1}^{n} \sigma_k(t) \Delta F_k \right\|_X dt \leq e C_1 K. \tag{5.4.6}$$

Now put $G_n(\omega, t) = \sum_{k=1}^{n} \sigma_k(t) \Delta F_k(\omega)$. By (5.4.6), there exist $\Omega \subseteq \mathbb{T}^{\mathbb{N}}$ with $\mathbb{P}(\Omega) = 1$ such that

$$\int_0^1 \sup_{n \in \mathbb{N}} \|G_n(\omega, t)\|_X dt < \infty,$$

for any $\omega \in \Omega$. By Theorem 5.4.4 the space X does not contain a copy of c_0. Hence, by Kwapien's theorem (Theorem 5.4.2), for any $\omega \in \Omega$ there exists $T_0 \subseteq [0, 1]$ with $|T_0| = 1$ such that $\lim_{n \to \infty} G_n(\omega, t)$ exists in X, for any $t \in T_0$. Equivalently,

$$\mathbb{P} \otimes \lambda \left\{ (\omega, t) : \lim_{n \to \infty} G_n(\omega, t) \text{ exists in } X \right\} = 1,$$

where λ denotes the Lebesgue measure on $[0, 1]$. By Fubini's theorem, there exists $S_0 \subseteq [0, 1]$ with $|S_0| = 1$ such that for any $s \in S_0$ there exists $\Omega_1 \subseteq \mathbb{T}^{\mathbb{N}}$ with $\mathbb{P}(\Omega_1) = 1$, such that $\lim_{n \to \infty} G_n(\omega, s)$ exists in X for any $\omega \in \Omega_1$. Thus, for $s \in S_0$ fixed, the L^1-bounded Hardy martingale $G_n(\cdot, s)$ converges almost surely. In view of the maximal function estimate for Hardy martingales, the sequence $G_n(\cdot, s)$ converges in $L^1(X)$.

Now fix $s \in S_0$ and $m, n \in \mathbb{N}$ with $n > m$. Then in view of (5.4.6) we have

$$\begin{aligned} \mathbb{E}\|F_n - F_m\| &= \mathbb{E} \left\| \sum_{k=m+1}^{n} \Delta F_k \right\| \\ &\leq C_1 \mathbb{E} \left\| \sum_{k=m+1}^{n} \sigma_k(s) \Delta F_k \right\| \\ &= C_1 \mathbb{E} \|G_n(\cdot, s) - G_m(\cdot, s)\|. \end{aligned}$$

Since $(G_n(\cdot, s))$ is a Cauchy sequence in $L^1(X)$, the same holds true for (F_n). Summing up, the Hardy martingale (F_n) converges in $L^1(X)$ and hence almost surely. □

Garling (1988) proved that if Hardy martingales ranging in a Banach space X are unconditionally bounded, X does not contain copies of ℓ_n^∞ uniformly, and used his result to give a second proof that the aUMD implies the aRNP.

Theorem 5.4.4 *Let $N \in \mathbb{N}$, $w_0 = e^{i\pi/N}$, and let $S_N = \left\{w_0^k : 1 \le k \le 2N\right\}^N$ and*

$$X = L^\infty(S_N).$$

There exists an X-valued Hardy martingale $F = (F_m)_{m=1}^N$ such that

$$\mathbb{E}\|\Delta F_m\|_X \ge c_1, \qquad \mathbb{E}\|F_m\|_X^p \le C_p, \qquad m \le N, \tag{5.4.7}$$

and

$$\mathbb{E}\left\|\sum_{m=1}^{N}(-1)^m \Delta F_m\right\|_X^p \ge c_p N^p \mathbb{E}\|F_N\|_X^p, \tag{5.4.8}$$

where $1 \le p < \infty$ and $C_p, c_p > 0$ are independent of N.

Proof The martingales are obtained by simply stopping a Steinhaus series. Put $S = S_N$, and let $s \in S$. Thus $s = (w_1, \dots, w_N)$ with $w_m \in \left\{w_0^k : 1 \le k \le 2N\right\}$. Let $z \in \mathbb{T}^\mathbb{N}$ with $z = (z_m)$. For s fixed, define

$$\tau(z, s) = \inf\{j \le N : |z_j w_j - 1| \ge 2\sin(\pi/(2N))\}. \tag{5.4.9}$$

Since $\tau(\cdot, s)$ is a stopping time with respect to the product filtration, we obtain a collection of scalar Hardy martingales by putting

$$F_m(z, s) = \sum_{j=1}^{m\wedge\tau(z,s)} (-1)^j z_j, \qquad 1 \le m \le N, s \in S.$$

We define X-valued Hardy martingales by stringing together the scalar ones as follows,

$$F_m\colon \mathbb{T}^\mathbb{N} \to L^\infty(S), \quad w \to (s \to F_m(z, s)).$$

We first verify conditions (5.4.7). Fix $m < N$. Put $A(z, s) = \{k : 1 \le k \le \tau(z, s)\}$. Clearly, we have

$$\Delta F_m(z, s) = (-1)^m z_m 1_{A(z,s)}(m). \tag{5.4.10}$$

Next, observe that for any $z \in \mathbb{T}^\mathbb{N}$ there exists $s \in S$ such that

$$m \wedge \tau(z, s) \ge m. \tag{5.4.11}$$

By (5.4.10) and (5.4.11) we get

$$\inf_z \max_{s\in S} |\Delta F_m(z, s)| \ge c_0 \quad \text{and} \quad \sup_z \max_{s\in S} |\Delta F_m(z, s)| \le 1,$$

which implies conditions (5.4.7). Next, we turn to estimate (5.4.8). The definition (5.4.9) of the stopping time yields the pointwise upper bound

$$|F_N(z, s)| = \left|\sum_{m=1}^{N\wedge\tau(z,s)} (-1)^m z_m\right| \le 2 + 2N\sin(\pi/(2N)) \le 5.$$

On the other hand, we have

$$\left|\sum_{m=1}^{N}(-1)^m \Delta F_m(z,s)\right| = \left|\sum_{m=1}^{N\wedge\tau(z,s)} z_m\right| \geq N \wedge \tau(z,s).$$

Invoking (5.4.11) again gives the lower bound

$$\inf_{z}\max_{s\in S}\left|\sum_{m=1}^{N}(-1)^m \Delta F_m(z,s)\right| \geq c_0 N,$$

which implies estimate (5.4.8). □

By Theorem 5.4.4, if a Banach space X satisfies the aUMD property, then it can't contain copies of ℓ_n^∞ uniformly. By a theorem of Maurey and Pisier (1976), X is therefore of finite cotype, and, moreover, there exists $q \in [2,\infty)$ and $A_q < \infty$, such that

$$\left(\sum_{j=1}^{n}(\mathbb{E}\|\Delta F_j\|_X)^q\right)^{1/q} \leq A_q(\mathbb{E}\|F_n\|_X), \tag{5.4.12}$$

for any n and any X-valued Hardy martingale (F_n). By Theorem 5.3.2 and condition (5.4.12), every L^1-bounded, X-valued Hardy martingale (F_n) is convergent in $L^1(X)$ (and almost surely).

Now we turn to $X = \ell_1$. We prove that martingale differences of ℓ_1-valued Hardy martingales are unconditionally bounded.

Theorem 5.4.5 *Let $X = \ell_1(\mathbb{C})$. There exists $C_1 > 0$ such that for any X-valued Hardy martingale (F_k) we have the unconditional estimates*

$$\mathbb{E}\left\|\sum_{m=1}^{k}\varepsilon_m \Delta F_m\right\|_X \leq C_1 \mathbb{E}\|F_k\|_X, \tag{5.4.13}$$

where $(\varepsilon_m) \in \{-1,1\}^{\mathbb{N}}$ and $k \in \mathbb{N}$.

Proof We fix $(\varepsilon_m) \in \{-1,1\}^{\mathbb{N}}$, and let (F_k) be a Hardy martingale with values in $X = \ell_1(\mathbb{C})$. Then for fixed $k \in \mathbb{N}$, we have

$$F_k = \left(F_k^{(n)}\right)_{n=1}^{\infty},$$

where $F_k^{(n)} : \mathbb{T}^{\mathbb{N}} \to \mathbb{C}$ is such that for each $n \in \mathbb{N}$ fixed, $F^{(n)} = \left(F_k^{(n)}\right)_{k=1}^{\infty}$ is a scalar Hardy martingale. Hence by Theorem 2.5.4, asserting the unconditionality of L^1-bounded scalar-valued Hardy martingales we get

$$\mathbb{E}\left|\sum_{m=1}^{k}\varepsilon_m \Delta F_m^{(n)}\right| \leq C_1 \mathbb{E}\left|F_k^{(n)}\right|. \tag{5.4.14}$$

Summing the estimates (5.4.14) over n amounts to passing to X-valued Hardy martingales where $X = \ell_1(\mathbb{C})$. Indeed, by (5.4.14),

$$\sum_{n=1}^{\infty} \mathbb{E}\left|\sum_{m=1}^{k} \varepsilon_m \Delta F_m^{(n)}\right| \leq C_1 \sum_{n=1}^{\infty} \mathbb{E}\left|F_k^{(n)}\right|. \tag{5.4.15}$$

Observe that the left-hand side of (5.4.15) is just the left-hand side of (5.4.13), and that the same holds true for the respective right-hand sides. This finishes the proof. □

5.4.2 Extrapolation

Garling's extrapolation theorem (Theorem 5.4.6) asserts that if a Banach space X satisfies the aUMD$_q$ for some $1 \leq q < \infty$, then X satisfies the aUMD$_p$ for all $1 \leq p < \infty$. Thus the property is independent of p, and we express this by dropping the subindex p. Garling's extrapolation theorem directly implies that the aUMD property passes from a Banach space X to $L^p(X)$ for $1 \leq p < \infty$.

Theorem 5.4.6 (Garling's extrapolation theorem) *Let X be a Banach space. Assume that there exists $1 \leq q < \infty$ and $C_q < \infty$, so that for every Hardy martingale (F_k) in $L^q(X)$ and $(\varepsilon_k) \in \{-1, 1\}^{\mathbb{N}}$,*

$$\mathbb{E}\left\|\sum_{m=1}^{k} \varepsilon_m \Delta F_m\right\|_X^q \leq C_q \mathbb{E}\|F_k\|_X^q, \qquad k \in \mathbb{N}.$$

Then for every $1 \leq p < \infty$, there exists $C_p < \infty$ such that for every Hardy martingale (F_k) in $L^p(X)$ and $(\varepsilon_k) \in \{-1, 1\}^{\mathbb{N}}$,

$$\mathbb{E}\left\|\sum_{m=1}^{k} \varepsilon_m \Delta F_m\right\|_X^p \leq C_p \mathbb{E}\|F_k\|_X^p.$$

The proof of Theorem 5.4.6 is comprised of two separate parts. The first, Theorem 5.4.8, allows the extrapolation from the endpoint $q > 1$ to just $p = 1$. The second part, Theorem 5.4.9, extrapolates from $q = 1$ to $1 \leq p < \infty$.

To extrapolate from $q > 1$ to $p = 1$ we utilize a method introduced by Maurey in his study of the Haar system (Maurey, 1975b). Fix a Banach space X, and a sequence of signs $\varepsilon = (\varepsilon_k) \in \{-1, 1\}^{\mathbb{N}}$. For a finite, X-valued Hardy martingale $F = (F_k)$ we put

$$T_\varepsilon F = \sum \varepsilon_k \Delta F_k. \tag{5.4.16}$$

The main idea of Maurey's approach is contained in the following lemma, which provides us with extrapolation estimates under the additional assumption that the martingale differences of $F = (F_k)$ are bounded by a previsible and increasing process. The Davis decomposition in Theorem 5.3.8 permits us to eliminate the previsibility assumption later.

Lemma 5.4.7 *Assume that there exists* $1 < q < \infty$ *and* $C_q > 0$ *so that for any Hardy martingale* (F_k) *in* $L^q(X)$,

$$\mathbb{E}\|T_\varepsilon F_k\|_X^q \le C_q^q \mathbb{E}\|F_k\|_X^q, \qquad k \in \mathbb{N}.$$

Let (G_k) *be an X-valued integrable Hardy martingale. Assume the sequence* (w_k) *is increasing, adapted, and satisfies* $\max_{m\le k} \|G_m\|_X \le w_{k-1}$ *for any* $k \in \mathbb{N}$. *Then,*

$$\mathbb{E}(\|T_\varepsilon G_k\|_X) \le AC_q\mathbb{E}(w_{k-1}), \qquad k \in \mathbb{N},$$

where $A = A(q)$.

Proof Let (G_k) be a Hardy matingale, and let (w_k) be an increasing and adapted process satisfying $\max_{m\le k} \|G_m\|_X \le w_{k-1}$. Put $v_k = w_k^{1/p}$, where $1/p + 1/q = 1$, and define the transformed martingale

$$Z_n = \sum_{m=1}^{n} v_{m-1}^{-1} \Delta G_m.$$

We rewrite Z_n using summation by parts (Abel transformation) and obtain the identities

$$\begin{aligned} Z_n &= v_{n-1}^{-1} G_n + \sum_{m=1}^{n-1} G_m \left\{ v_{m-1}^{-1} - v_m^{-1} \right\} \\ &= v_{n-1}^{-1} G_n + \sum_{m=1}^{n-1} G_m (v_{m-1} v_m)^{-1} \{ v_m - v_{m-1} \}. \end{aligned}$$

Since v_m is increasing and $\|G_m\| \le v_{m-1}^p$, the triangle inequality and the above identities give the following pointwise estimates for $\|Z_n\|$,

$$\|Z_n\| \le v_{n-1}^{p-1} + \sum_{m=1}^{n-1} v_{m-1}^{p-2} \{ v_m - v_{m-1} \} \le \left(1 + \frac{1}{p-1} \right) v_{n-1}^{p-1}. \tag{5.4.17}$$

Next, we prove a second pointwise estimate, namely

$$\|T_\varepsilon G_k\|_X \le 2 \left(\max_{m\le k} \|T_\varepsilon Z_m\|_X \right) w_{k-1}^{1/p}. \tag{5.4.18}$$

We now make a useful general observation: If $y_m \in X$ and if $0 \le a_m \le M$ is an increasing sequence of scalars, then, we claim that

$$\left\| \sum_{m=1}^{n} a_m y_m \right\|_X \le 2M \max_{k\le n} \left\| \sum_{m=1}^{k} y_m \right\|_X . \tag{5.4.19}$$

To see this, put $S_m = \sum_{j=1}^{m} y_j$ and rewrite

$$\sum_{m=1}^{k} a_m y_m = a_k S_k + \sum_{m=2}^{k} S_{m-1}(a_{m-1} - a_m). \tag{5.4.20}$$

Since $a_m \leq M$ and $a_{m-1} - a_m \leq 0$, we have $\sum |a_{m-1} - a_m| \leq M$. Hence (5.4.20) gives (5.4.19). Returning to the proof of (5.4.18) we write

$$T_\varepsilon G_k = \sum_{m=1}^{k} w_{m-1}^{1/p} w_{m-1}^{-1/p} \varepsilon_m \Delta G_m,$$

and apply (5.4.19) with $a_m = w_{m-1}^{1/p}$, $y_m = a_m^{-1} \varepsilon_m \Delta G_m$, and $M = w_{k-1}^{1/p}$.

Now we finish the proof using estimates (5.4.17) and (5.4.18). By Hölder's inequality and Doob's maximal theorem we obtain from (5.4.18) that

$$\mathbb{E}\|T_\varepsilon G_k\| \leq A_q (\mathbb{E}\|T_\varepsilon Z_k\|^q)^{1/q} (\mathbb{E} w_{k-1})^{1/p}.$$

Recall that $1/p + 1/q = 1$ and hence $q(p-1) = p$. The hypothesis on T_ε and the pointwise bound (5.4.17) imply that

$$\mathbb{E}\|T_\varepsilon Z_k\|^q \leq C_q^q \mathbb{E}\|Z_k\|^q \leq A_p C_q^q \mathbb{E}\left(v_{k-1}^p\right).$$

To finish the proof it remains to recall that $v_{k-1}^p = w_{k-1}$. □

Theorem 5.4.8 *Let T_ε be defined by equation (5.4.16). Assume that there exists $1 < q < \infty$ and $C_q > 0$ so that for any Hardy martingale (F_k) in $L^q(X)$,*

$$\mathbb{E}\|T_\varepsilon F_k\|_X^q \leq C_q^q \mathbb{E}\|F_k\|_X^q, \qquad k \in \mathbb{N}.$$

Then for any Hardy martingale (F_k) in $L^1(X)$,

$$\mathbb{E}\|T_\varepsilon F_k\|_X \leq C_q A \mathbb{E}\|F_k\|_X, \qquad k \in \mathbb{N},$$

where $A = A(q)$.

Proof We assume that T_ε – when restricted to Hardy martingales – is bounded on $L^q(X)$ with $1 < q < \infty$, and we will conclude that T_ε is bounded acting on Hardy martingales in $L^1(X)$. To this end, we fix a Hardy martingale F and apply to it the Davis decomposition in Theorem 5.3.8. This yields a splitting of F into Hardy martingales as $F = G + B$, such that

$$\|\Delta G_k\|_X \leq C\|F_{k-1}\|_X \quad \text{and} \quad \mathbb{E}\left(\sum_{j=1}^{k} \|\Delta B_j\|_X\right) \leq C\mathbb{E}(\|F_k\|_X), \tag{5.4.21}$$

for $k \in \mathbb{N}$. We next estimate the contribution of G and B by separate arguments. We use Lemma 5.4.7 to control $T_\varepsilon(G)$. To this end, put

$$w_{k-1} = C \max_{m \leq k-1} \|F_m\| + \max_{m \leq k-1} \|G_m\|.$$

Then by the first estimate in (5.4.21), $\max_{m\le k}\|G_m\| \le w_{k-1}$. Hence Lemma 5.4.7 applies and gives

$$\mathbb{E}\|T_\varepsilon G_k\| \le C\mathbb{E}(w_{k-1}),$$

where $C = C(q)$. Theorem 2.5.1, the maximal function estimates for Hardy martingales, gives $\mathbb{E}(w_{k-1}) \le C_1\mathbb{E}\|F_k\|$. Next we turn to $T_\varepsilon B_k$. Here, we apply the triangle inequality and invoke the second estimate of (5.4.21) to obtain

$$\mathbb{E}\|T_\varepsilon B_k\| \le \mathbb{E}\left(\sum_{j=1}^{k}\|\Delta B_j\|_X\right) \le C\mathbb{E}\|F_k\|.$$

Summing up the estimates for $T_\varepsilon(G)$ and $T_\varepsilon(B)$ we finally get $\mathbb{E}\|T_\varepsilon F_k\| \le C\mathbb{E}\|F_k\|$. □

We next reverse the direction of extrapolation. We start with the assumption that T_ϵ acts boundedly on Hardy martingales in $L^1(X)$, and obtain boundedness in L^p for any $p > 1$. The proof is based on $L^p(X)$ estimates for martingale sharp functions. Let $F = (F_n)$ be a martingale, then the sharp function of F is defined as

$$F^\sharp = \sup_{n\in\mathbb{N}}(\mathbb{E}_n(\|F - F_{n-1}\|_X).$$

Note that – up to a factor of two – Doob's maximal function $\sup_n\|F_n\|$ is a pointwise upper bound for the sharp function. The Fefferman–Stein inequalities assert that for $p > 1$, nontrivial lower bounds hold true in $L^p(X)$: There exist $C_p, c_p > 0$ so that

$$c_p\|F\|_{L^p(X)} \le \left\|F^\sharp\right\|_p \le C_p\|F\|_{L^p(X)}, \tag{5.4.22}$$

where $c_p \to 0$ and $C_p \to \infty$ as $p \to 1$. The right-hand side estimate of (5.4.22) is a direct consequence of the Doob's maximal function estimate; the left-hand side will be verified below.

Theorem 5.4.9 *Assume that there exists $A > 0$ so that for any X-valued integrable Hardy martingale $F = (F_k)$,*

$$\|T_\varepsilon(F)\|_{L^1(X)} \le A\|F\|_{L^1(X)}. \tag{5.4.23}$$

Then for any $p > 1$, and any Hardy martingale $F = (F_k)$ in $L^p(X)$,

$$\|T_\varepsilon(F)\|_{L^p(X)} \le (C_p/c_p)A\|F\|_{L^p(X)}, \tag{5.4.24}$$

where c_p, C_p arose in the Fefferman–Stein sharp function inequality.

Proof We start from the fact that condition (5.4.23) holds true and extrapolate to the values $p > 1$.

Step 1 Given an X-valued Hardy Martingale $F = (F_m)$ and $(\varepsilon_m) \in \{-1, 1\}^{\mathbb{N}}$, we let $G = (G_m)$ be defined as

$$G_n = \sum_{m=1}^{n} \varepsilon_m \Delta F_m.$$

We next observe that the sharp function of G is bounded – pointwise – by (a multiple) of the sharp function of F. Since $F - F_{n-1} = \sum_{m=n}^{\infty} \Delta F_m$, we have

$$(\mathbb{E}_n \|F - F_{n-1}\|_X)(x) = \int_{\mathbb{T}^{\mathbb{N}}} \left\| \sum_{m=n}^{\infty} \Delta F_m(x, w) \right\|_X d\mathbb{P}(w), \tag{5.4.25}$$

for $x \in \mathbb{T}^{n-1}$, and, similarly,

$$(\mathbb{E}_n \|G - G_{n-1}\|_X)(x) = \int_{\mathbb{T}^{\mathbb{N}}} \left\| \sum_{m=n}^{\infty} \varepsilon_m \Delta F_m(x, w) \right\|_X d\mathbb{P}(w). \tag{5.4.26}$$

Apply condition (5.4.23) to the right-hand side integrals in (5.4.25) and (5.4.26). This gives the pointwise relation,

$$(\mathbb{E}_n \|F - F_{n-1}\|_X) \le C_1 (\mathbb{E}_n \|G - G_{n-1}\|_X), \qquad n \in \mathbb{N}.$$

It remains to take the supremum over n, and invoke the Fefferman–Stein inequalities (5.4.22) to obtain (5.4.24).

Step 2 We turn to proving estimate (5.4.22) by employing the method of "good λ inequalities." Let $M_F = \sup_k \|F_k\|$, and fix $\lambda > 0$, $A > 1$. Our central claim is that the following measure estimates hold true,

$$\mathbb{P}\left(\{M_F > \lambda\} \cap \{F^{\sharp} \le \lambda/A\}\right) \le (2/A)\mathbb{P}(\{M_F > \lambda/2\}). \tag{5.4.27}$$

Towards proving (5.4.27) we consider

$$R = \inf\{l \in \mathbb{N} : \|F_l\| > \lambda\} \quad \text{and} \quad S = \inf\{k \in \mathbb{N} : \|F_k\| > \lambda/2\}.$$

Clearly, R and S are stopping times satisfying $\{S < \infty\} = \{M_F > \lambda/2\}$ and $\{R < \infty\} = \{M_F > \lambda\}$. Let $K \subseteq \mathbb{N}$ be the union of those $k \in \mathbb{N}$ for which

$$\{S = k\} \cap \{F^{\sharp} \le \lambda/A\} \ne \emptyset.$$

Note that for proving (5.4.27) it suffices to check that

$$\mathbb{P}(\{S = k\} \cap \{R < \infty\}) \le (2/A)\mathbb{P}\{S = k\}, \tag{5.4.28}$$

for any $k \in K$. Observe that we have $\|F_{k-1}\| \le \lambda/2$ on the set $\{S = k\}$ and $\|F_l\| > \lambda$ on $\{R = l\}$. Since $R \ge S$, if $\{S = k\} \cap \{R = l\} \ne \emptyset$ then $l \ge k$. Hence,

$$\int_{\{S=k\}\cap\{R=l\}} \|F - F_{k-1}\| d\mathbb{P} \ge (\lambda/2)\mathbb{P}(\{S = k\} \cap \{R = l\}), \tag{5.4.29}$$

for any $k \in \mathbb{N}$ and $l \geq k$. On the other hand, for $k \in K$ we have the upper estimate

$$\int_{\{S=k\}} \|F - F_{k-1}\| d\mathbb{P} \leq (\lambda/A)\mathbb{P}(\{S = k\}). \tag{5.4.30}$$

Note that the sets $\{S = k\} \cap \{R = l\}$, where $l \geq k$, form a disjoint covering of $\{S = k\} \cap \{R < \infty\}$. Hence, taking the sum in (5.4.29) over $l \geq k$ and comparing with (5.4.30) gives (5.4.28).

Step 3 By the estimate (5.4.27), for any $\lambda > 0$ and $A > 1$ we have,

$$\begin{aligned}\mathbb{P}(\{M_F > \lambda\}) &\leq \mathbb{P}\left(\{M_F > \lambda\} \cap \{F^\sharp \leq \lambda/A\}\right) + \mathbb{P}\left(\{F^\sharp > \lambda/A\}\right) \\ &\leq (2/A)\mathbb{P}\left(\{M_F > \lambda/2\}\right) + \mathbb{P}\left(\{F^\sharp > \lambda/A\}\right).\end{aligned}$$

Multiplying both sides by $p\lambda^{p-1}$ and integrating over $0 < \lambda < \infty$ gives

$$\|M_F\|_p^p \leq (2/A)2^p\|M_F\|_p^p + A^p \left\|F^\sharp\right\|_p^p.$$

Finally, choosing $A = 4 \cdot 2^p$ gives $\|M_F\|_p \leq 8 \cdot 2^p \left\|F^\sharp\right\|_p$. □

As an immediate consequence of Garling's extrapolation theorem (Theorem 5.4.6) we prove that if X satisfies the aUMD, then $L^p(X)$ for $1 \leq p < \infty$. Thus the following theorem, due to Garling (1988), yields a second proof of the fact that L^1 satisfies the aUMD.

Theorem 5.4.10 *Let X be a complex Banach space. Assume that there exists $1 \leq q < \infty$ and $C_q > 0$ such that*

$$\mathbb{E}\|T_\varepsilon G_k\|_X^q \leq C_q^q \mathbb{E}\|G_k\|_X^q,$$

for any Hardy martingale (G_k) in $L^q(X)$, for $\varepsilon \in \{-1, 1\}^{\mathbb{N}}$, and for $k \in \mathbb{N}$. Then for $1 \leq p < \infty$ and $Y = \ell^p(X)$, there exists $\widetilde{C_{q,p}} < \infty$, such that

$$\mathbb{E}\|T_\varepsilon F_k\|_Y^q \leq \widetilde{C_{q,p}} \mathbb{E}\|F_k\|_Y^q,$$

for any Hardy martingale (F_m) in $L^q(Y)$ and $\varepsilon \in \{-1, 1\}^{\mathbb{N}}$. The same conclusion holds for the Banach space $Y = L^p(X)$.

Proof Let $1 \leq p < \infty$ and apply Theorem 5.4.6 to the Banach space X. Hence the hypothesis of our theorem holds true with q replaced by p. Let $F = (F_k)$ be a Hardy martingale with values in $Y = \ell^p(X)$. Hence for fixed $k \in \mathbb{N}$,

$$F_k = \left(F_k^{(n)}\right)_{n=1}^\infty \quad \text{where} \quad F_k^{(n)} : \mathbb{T}^{\mathbb{N}} \to X.$$

For each $n \in \mathbb{N}$ fixed, $F^{(n)} = \left(F_k^{(n)}\right)$ is an X-valued Hardy martingale. Hence,

$$\mathbb{E}\left\|T_\varepsilon\left(F_k^{(n)}\right)\right\|_X^p \leq C_p^p \mathbb{E}\left\|\left(F_k^{(n)}\right)\right\|_X^p,$$

for $k, n \in \mathbb{N}$, and we have

$$\sum_{n=1}^{\infty} \mathbb{E} \left\| T_\varepsilon \left(F_k^{(n)} \right) \right\|_X^p \leq C_p^p \sum_{n=1}^{\infty} \mathbb{E} \left\| F_k^{(n)} \right\|_X^p .$$

Expressing the inequality above in terms of $Y = \ell^p(X)$, we obtained that T_ε acts boundedly for Hardy martingales in $L^p(Y)$, and

$$\mathbb{E}\|T_\varepsilon F_k\|_Y^p \leq C_p^p \mathbb{E}\|F_k\|_Y^p .$$

Finally, we apply Theorem 5.4.6 to the Banach space Y. This allows us to replace p in the above estimate by q, which gives the desired estimates in $L^q(Y)$. □

The space L^1/H_0^1: Since L^1/H_0^1 is not of finite Hardy martingale cotype, it also fails to satisfy the aUMD. Moreover, Pisier's vector-valued Riesz products display the extent to which UMD estimates fail to hold for Hardy martingales ranging in L^1/H_0^1. Indeed, by Theorem 5.1.5, for every $n \in \mathbb{N}$ there exists an L^1/H_0^1-valued Hardy martingale $(G_k)_{k=1}^n$ such that

$$\mathbb{E}\|G_n\|_{L^1/H_0^1} \leq C \quad \text{and} \quad \int_{\mathbb{T}^{\mathbb{N}}} \mathbb{E} \left\| \sum_{k=1}^{n} z_{n-k} \Delta G_k \right\|_{L^1/H_0^1} d\mathbb{P}(z) \geq \sqrt{n}, \tag{5.4.31}$$

where $C < \infty$ is independent of n.

Bourgain–Stein–Qiu Inequalities

In his thesis, Yanqi Qiu (2012a,b) determines a Banach space with an unconditional basis and with finite Hardy martingale cotype, failing the aUMD property. The work of Qiu (2012a,b) exploits the following Hardy martingale estimate (analogous to the Bourgain–Stein inequality).

Theorem 5.4.11 *Let $1 < p < \infty$. Assume that there exists $A > 0$ such that for every X-valued Hardy martingale $F = (F_m)$ and any $(\zeta_k) \in \mathbb{T}^{\mathbb{N}}$,*

$$\left\| \sum_{m=1}^{\infty} \zeta_m \Delta F_m \right\|_{L^p(X)}^p \leq A^p \|F\|_{L^p(X)}^p .$$

Let $k \in \mathbb{N}$, and let $F^{(k)} = \left(F_m^{(k)}\right)$, be an L^p-bounded X-valued Hardy martingale. Then

$$\mathbb{E}_{(\zeta)} \left\| \sum_{k=1}^{K} \zeta_k F_k^{(k)} \right\|_{L^p(X)}^p \leq A^p \mathbb{E}_{(\zeta)} \left\| \sum_{k=1}^{K} \zeta_k F^{(k)} \right\|_{L^p(X)}^p , \tag{5.4.32}$$

for any $K \in \mathbb{N}$, where $\mathbb{E}_{(\zeta)}$ denotes integration with respect to $\zeta \in \mathbb{T}^{\mathbb{N}}$ and $F_k^{(k)} = \mathbb{E}_k(F^{(k)})$.

Proof Let $F^{(k)} = \left(F_m^{(k)}\right)$ be an L^p-convergent, X-valued Hardy martingale. Let $\zeta, w \in \mathbb{T}^{\mathbb{N}}$. Put

$$G(\zeta, w) = \sum_{m=1}^{\infty} \zeta_m F^{(m+1)}(w), \qquad G^{(T)}(\zeta, w) = \sum_{m=1}^{\infty} \zeta_m F_m^{(m+1)}(w).$$

Let H and $H^{(T)}$ be given by Lemma 5.4.12. The assumption on X in combination with Lemma 5.4.12 yields

$$\mathbb{E}_{(\zeta)} \left\|G^{(T)}(\zeta, \cdot)\right\|_{L^p(X)}^p = \left\|H^{(T)}\right\|_{L^p(X)}^p \leq A^p \|H\|_{L^p(X)}^p = A^p \mathbb{E}_{(\zeta)} \|G(\zeta, \cdot)\|_{L^p(X)}^p.$$

Combining the last three equations above finishes the proof. □

The essence of the proof is contained in the following combinatorial construction, relating the estimate (5.4.32) to martingale transforms.

Lemma 5.4.12 *Let $F^{(k)} = \left(F_m^{(k)}\right)$ be a sequence of finitely supported X-valued Hardy martingales, and let*

$$G(\zeta, w) = \sum_{m=1}^{\infty} \zeta_m F^{(m+1)}(w), \qquad G^{(T)}(\zeta, w) = \sum_{m=1}^{\infty} \zeta_m F_m^{(m+1)}(w).$$

Then there exists an X-valued Hardy martingale $H = (H_n)$ so that for any $t > 0$,

$$\mathbb{P} \otimes \mathbb{P}\{\|G\| > t\} = \mathbb{P}\{\|H\| > t\}, \quad \text{and} \quad \mathbb{P} \otimes \mathbb{P}\left\{\left\|G^{(T)}\right\| > t\right\} = \mathbb{P}\left\{\left\|H^{(T)}\right\| > t\right\},$$

where $H^{(T)} = \sum_{m=1}^{\infty} \Delta_{2m+1} H$.

Proof Fix $k \in \mathbb{N}$ and let $F^{(k)} = \left(F_m^{(k)}\right)$ be an X-valued Hardy martingale. For $z \in \mathbb{T}^{\mathbb{N}}$ with $z = (z_n)$ we define

$$g_{2m}^{(2k)}(z_2, z_4, \ldots z_{2m}) = F_m^{(k)}(z_1, z_2, \ldots z_m),$$

and put $g^{(2k)}(z) = \lim_{m \to \infty} g_{2m}^{(2k)}(z)$. Thus $\mathbb{E}_{2m+1}(g^{(2k)}) = \mathbb{E}_{2m}(g^{(2k)})$ and $g^{(2k)}(z)$ depends only on the even coordinates of z. Clearly, $\mathbb{E}_{2m}(g^{(2k)})$ and $\mathbb{E}_m(F^{(k)})$ are equi-distributed. Next, define

$$H(z) = \sum_{m=1}^{\infty} z_{2m+1} g^{(2m+2)}(z).$$

Then $\Delta_{2m+1} H(z) = z_{2m+1} g_{2m}^{(2m+2)}(z)$. Since $g^{(2k)}(z)$ depends on the even coordinates of z, the random variables H and G are equidistributed and the same holds for $H^{(T)}$ and $G^{(T)}$. □

5.5 Embedding

In Section 2.3 we introduced Maurey's embedding of $H^1(\mathbb{T})$ into Hardy martingales with uniformly small increments. By iterating Theorem 2.3.1 we will show that any Hardy martingale may be considered as a subsequence of a Hardy martingale with increments that are bounded by a small, predictable, increasing process. The basic interest in this result comes from the previsibility assumption in extrapolation theorems such as Lemma 5.4.7.

Let $1/2 > \epsilon > 0$ and $w \in \mathbb{T}^{\mathbb{N}}$ with $w = (w_k)$. We define, inductively, $\varphi_1(w) = \epsilon w_1$ and

$$\varphi_n(w) = \varphi_{n-1}(w) + \epsilon(1 - |\varphi_{n-1}(w)|)^2 w_n. \tag{5.5.1}$$

Since $\varphi = (\varphi_n)$ is a uniformly bounded Hardy martingale whose limit is uniformly distributed over $\mathbb{T}$, that is,

$$\mathbb{P}\left(\{w \in \mathbb{T}^{\mathbb{N}} : \varphi(w) \in B\}\right) = m(B), \qquad B \subseteq \mathbb{T},$$

where $m(B)$ denotes the normalized Haar measure on $\mathbb{T}$. For convenience, we next restate Maurey's embedding Theorem 2.3.1 in the vector-valued setting.

Theorem 5.5.1 (Maurey's theorem) *For any $f \in H^1(\mathbb{T}, X)$,*

$$F_n(w) = f(\varphi_n(w)), \qquad w \in \mathbb{T}^{\mathbb{N}},$$

defines an X-valued Hardy martingale for which

$$\sup_{n \in \mathbb{N}} \mathbb{E}(\|F_n\|_X) = \int_{\mathbb{T}} \|f\|_X dm \tag{5.5.2}$$

and

$$\|\Delta F_n\|_X \leq 2\epsilon \int_{\mathbb{T}} \|f\|_X dm. \tag{5.5.3}$$

Proof We sketch Maurey's proof, emphasizing that the argument for scalar-valued H^1 functions carries over to the vector-valued case. It is straightforward to see that (F_n) is indeed an integrable X-valued Hardy martingale and that (5.5.2) holds true. We now turn to the pointwise estimates (2.3.3). Fix $w \in \mathbb{T}^{\mathbb{N}}$ and $n \in \mathbb{N}$. Then,

$$\Delta F_n(w) = f(\varphi_n(w)) - f(\varphi_{n-1}(w)).$$

Next, put $z = \varphi_n(w), u = \varphi_{n-1}(w)$, and use the Cauchy integral formula to obtain

$$f(z) - f(u) = \int_{\mathbb{T}} \left\{ \frac{\zeta}{\zeta - z} - \frac{\zeta}{\zeta - u} \right\} f(\zeta) dm(\zeta).$$

By the triangle inequality, we get

$$\|f(z) - f(u)\|_X \le \frac{|z-u|}{(1-|u|)(1-|z|)} \int_{\mathbb{T}} \|f\|_X dm. \tag{5.5.4}$$

We use the defining recursion (5.5.1) to see that

$$|\varphi_n(w) - \varphi_{n-1}(w)| = \epsilon(1 - |\varphi_{n-1}(w)|)^2 \tag{5.5.5}$$

and

$$(1 - |\varphi_{n-1}(w)|) \le (1-\epsilon)^{-1}(1 - |\varphi_n(w)|). \tag{5.5.6}$$

Since we put $z = \varphi_n(w)$ and $u = \varphi_{n-1}(w)$, the relations (2.3.5) and (5.5.6) imply that

$$\frac{|z-u|}{(1-|u|)(1-|z|)} \le 2\epsilon. \tag{5.5.7}$$

Combining the estimates (5.5.4) and (5.5.7) yields (5.5.3). □

Applying Maurey's theorem (Theorem 5.5.1) repeatedly, we associate to an arbitrary Hardy martingale a subsequence of a Hardy martingale *with small predictable increments* and almost identical norms. As mentioned above, this makes it possible to apply standard extrapolation theorems *without performing a Davis decomposition.* In that sense, the following embedding provides an alternative to Hardy martingale decomposition.

Theorem 5.5.2 *Let $\eta > 0$ and $1 \le p < \infty$. For any X-valued Hardy martingale $g = (g_k)$ there exists an X-valued Hardy martingale $G = (G_k)$, an increasing sequence of integers*

$$m(0) < m(1) < \cdots < m(n) < \cdots,$$

and a nonnegative adapted process (β_k) such that

$$\mathbb{E}\left(\sup_{k\in\mathbb{N}} \beta_k\right) \le \sup_{k\in\mathbb{N}} \mathbb{E}(\|g_k\|_X), \tag{5.5.8}$$

and so that the following conditions hold:

(i) *Small and previsible increments,*

$$\|\Delta G_k\|_X \le \eta\beta_{k-1}. \tag{5.5.9}$$

(ii) *Almost identical L^p norms,*

$$(1-\eta)\mathbb{E}\left(\|g_k\|_X^p\right) \le \mathbb{E}\left(\|G_{m(k)}\|_X^p\right) \le (1+\eta)\mathbb{E}\left(\|g_k\|_X^p\right) \tag{5.5.10}$$

and

$$(1-\eta)\mathbb{E}\left(\|\Delta g_k\|_X^p\right) \le \mathbb{E}\left(\|G_{m(k)} - G_{m(k-1)}\|_X^p\right) \le (1+\eta)\mathbb{E}\left(\|\Delta g_k\|_X^p\right). \tag{5.5.11}$$

Proof The proof iterates Maurey's theorem (Theorem 5.5.1). First of all we may assume that the martingale $g = (g_k)$ is finite and that, moreover, each g_k is a trigonometric polynomial. To keep the notation simple we restrict the presentation to the case $p = 1$.

Step 1 (Preparation) Since the martingale (g_k) consists of trigonometric polynomials, we have $A = \sum_{k\in\mathbb{N}} \mathbb{E}(\|\Delta g_k\|_X) < \infty$, Hence we may select $0 < \epsilon < \eta$ such that

$$\epsilon/\eta = (\sup \mathbb{E}(\|g_k\|_X)/A. \tag{5.5.12}$$

Since $g = (g_k)$ is finite, we have in fact $0 < \epsilon$. Let $\varphi = (\varphi_n)$ be the Hardy martingale (5.5.1) defined by $\varphi_1(w) = \epsilon w_1$ and

$$\varphi_n(w) = \varphi_{n-1}(w) + \epsilon(1 - |\varphi_{n-1}(w)|)^2 w_n, \qquad w \in \mathbb{T}^{\mathbb{N}}.$$

We shorten the notation and put $\Omega = \mathbb{T}^{\mathbb{N}}$.

Step 2 (Substitution) For $k \in \mathbb{N}$ and $u \in \Omega^k$ we write $u = (u^{(1)}, \ldots, u^{(k)})$ where $u^{(1)} \in \Omega, \ldots, u^{(k)} \in \Omega$. Define

$$\Phi^k \colon \Omega^k \to \mathbb{T}^k, \qquad \Phi^k(u) = \left(\varphi(u^{(1)}), \ldots, \varphi(u^{(k)})\right), \tag{5.5.13}$$

and form the linear substitution operator

$$T \colon L^1(\mathbb{T}^k) \to L^1(\Omega^k), \qquad Tf(u) = f(\Phi^k(u)). \tag{5.5.14}$$

Clearly, T is a contraction between the L^1 spaces in (5.5.14).

Fix $k \in \mathbb{N}$, $v \in \Omega^{k-1}$, and $w \in \Omega$. Then clearly, $u = (v, w) \in \Omega^k$ and we have

$$(Tg_k)(v, w) = g_k\left(\Phi^{k-1}(v), \varphi(w)\right).$$

We view g_{k-1} as a function on $\mathbb{T}^k$ that does not depend on the last variable. Hence we may apply the substitution operator T to g_{k-1}, and, since $\mathbb{E}_{k-1}(g_k) = g_{k-1}$, we observe the following commutation relation between expectations,

$$(Tg_{k-1})(v) = \mathbb{E}_{(w)}(Tg_k(v, w)). \tag{5.5.15}$$

Step 3 (An intermediary Hardy martingale) We fix $v \in \Omega^{k-1}$ and form the Hardy martingale with respect to the last variable,

$$h_m(w) = g_k\left(\Phi^{k-1}(v), \varphi_m(w)\right), \qquad m \in \mathbb{N}, w \in \Omega.$$

Theorem 5.5.1 asserts that $h = (h_m)$ is an X-valued Hardy martingale, and that its increments are small and uniformly bounded. Specifically, if we put

$$\alpha_{k-1}(v) = \mathbb{E}_{(w)}(\|Tg_k(v, w) - Tg_{k-1}(v)\|_X),$$

then

$$\sup_{w\in\Omega} \|\Delta h_n(w)\|_X \le \epsilon\alpha_{k-1}(v)$$

and

$$\sup_{n\in\mathbb{N}} \mathbb{E}_{(w)}\|h_n\|_X = \mathbb{E}_{(w)}(\|Tg_k(v,w)\|_X),$$

where the integration is over $w \in \Omega$.

Step 4 (Bounding the active variables) Since $g = (g_k)$ is assumed to be a finite martingale we pick $n \in \mathbb{N}$ so that

$$g = (g_k)_{k=1}^n. \tag{5.5.16}$$

We approximate Tg_k by stopping the martingales $\varphi = (\varphi_m)$ used in the definition of the linear substitutions T. We will replace the limit φ by one of its approximation φ_m, thereby reducing the number of active variables.

For any $m \in \mathbb{N}$, define the substitutions

$$T_m f(v) = f\left(\Phi_m^n(v)\right), \qquad \Phi_m^n(v) = \left(\varphi_m(v^{(1)}), \ldots, \varphi_m(v^{(n)})\right). \qquad v \in \Omega^n.$$

Since $n \in \mathbb{N}$ and $\epsilon > 0$ are fixed, there exists $K \in \mathbb{N}$ so that

$$\sup_{k\le n} \mathbb{E}_{\Omega^n}(\|T_K(g_k) - T(g_k)\|_X) \le \epsilon \sup_{k\le n} \mathbb{E}_\Omega(\|g_k\|_X) \tag{5.5.17}$$

and

$$\sup_{k\le n} \mathbb{E}_{\Omega^n}(\|T_K(\Delta g_k) - T(\Delta g_k)\|_X) \le \epsilon \sup_{k\le n} \mathbb{E}_\Omega(\|g_k\|_X), \tag{5.5.18}$$

where the integration on the left-hand side is with respect to the normalized Haar measure of Ω^n, and on the right-hand side we integrate over $\Omega = \mathbb{T}^{\mathbb{N}}$.

By construction, for each $k \le n$, the dependence of $T_K(g_k)$ is only on the variables,

$$\left(v_1^{(1)}, \ldots, v_K^{(1)}, \ldots\ldots, v_1^{(k-1)}, \ldots, v_K^{(k-1)}, w_1, \ldots, w_K\right).$$

Thus $T_K(g_k)$ is unambiguously defined on the torus product

$$\mathbb{T}^{Kk} \subseteq \mathbb{T}^{Kn}.$$

Step 5 (Conclusion) Put $N = Kn$, where $n \in \mathbb{N}$ and $K \in \mathbb{N}$ are defined by (5.5.16) and (5.5.17), respectively. Finally, we define the Hardy martingale $G = (G_k)_{k=1}^N$. Put

$$\rho\colon \Omega^n \to \mathbb{T}^{Kn}, \qquad \rho(v) = \left(v_1^{(1)}, \ldots, v_K^{(1)}, \ldots, v_1^{(n)}, \ldots, v_K^{(n)}\right),$$

then

$$G : \mathbb{T}^N \to X$$

is defined without ambiguity by putting

$$G(z) = (T_K g_n)(v), \qquad z = \rho(v).$$

Let $m(k) = Kk$. Then by the commutation relation (5.5.15),

$$G_{m(k)} = \mathbb{E}_{m(k)}(G)(z) = (T_K g_k)(v), \qquad z = \rho(v),$$

and

$$\left(G_{m(k)} - G_{m(k-1)}\right)(z) = T_K(g_k - g_{k-1})(v), \qquad z = \rho(v).$$

Thus in view of (5.5.17) and (5.5.18) we have verified the norm estimates (5.5.10) and (5.5.11).

Finally, we let $\mathbb{E}_K$ denote the conditional expectation projecting onto the first K variables of $\Omega = \mathbb{T}^{\mathbb{N}}$. Let

$$\mathbb{F}_K = \mathbb{E}_K \otimes \cdots \otimes \mathbb{E}_K$$

be the conditional expectation on Ω^n given by the n-fold tensor product of $\mathbb{E}_K$. Theorem 5.5.1 asserts that for $m(k-1) \leq j < m(k)$, we have the pointwise estimate

$$\|\Delta G_j(z)\|_X \leq \epsilon \mathbb{F}_K(\alpha_{k-1}(v)), \qquad z = \rho(v). \tag{5.5.19}$$

Now define the adapted process

$$\beta_{k-1}(z) = (\epsilon/\eta)\mathbb{F}_K(\alpha_{k-1}(v)), \qquad z = \rho(v).$$

Clearly, we have $\mathbb{E}_{\Omega^n}(\sup_k \mathbb{F}_K(\alpha_{k-1})) \leq \mathbb{E}_{\Omega^n}(\sum \|T\Delta g_k\|_X)$. Taking into account (5.5.12), specifying the relation between ϵ and $\eta > 0$, we obtain

$$\mathbb{E}\left(\sup_{k\in\mathbb{N}} \beta_k\right) \leq \sup_{k\in\mathbb{N}} \mathbb{E}(\|g_k\|_X). \tag{5.5.20}$$

Thus the estimate (5.5.19) translates to (5.5.9), and (5.5.20) gives estimate (5.5.8). □

5.6 Interpolation

In this section, we present a Banach couple (X_0, X_1) such that, X_0 and X_1 are isometric to ℓ_1 such that, nevertheless, each of the complex interpolation spaces X_θ contains a complemented copy of c_0. (A corresponding statement holds true for the real interpolation spaces.) Hence the aRNP or the aUMD of a Banach space does not pass to interpolation spaces.

Let $\{r_n\}$ be an enumeration of the vectors in $\{+1,-1\}^N$, $N \in \mathbb{N}$. Let c_{00} denote the vector space of finitely supported sequences, and let Z be a Banach space. If $E\colon c_{00} \to Z$ is a linear operator such that

$$\|E(r_n)\|_Z \le C,$$

then E extends uniquely to a linear operator on c_0 with $\|E\colon c_0 \to Z\| \le C$.

We denote by $\{e_n\}$ the unit vectors in c_0 and (by a slight abuse of notation) also the ones in ℓ^1 and let $\delta_n = (1 + \|r_n\|_{\ell_1})^{-n}$. Define the spaces

$$X_0 = \left\{ \left(\sum_{n=1}^{\infty} a_n e_n + b_n r_n, \sum_{n=1}^{\infty} \delta_n b_n e_n, \sum_{n=1}^{\infty} c_n e_n \right) : (a_n), (b_n), (c_n) \in \ell^1 \right\}$$

$$X_1 = \left\{ \left(\sum_{n=1}^{\infty} c_n e_n + b_n r_n, \sum_{n=1}^{\infty} b_n e_n, \sum_{n=1}^{\infty} \delta_n c_n e_n \right) : (a_n), (b_n), (c_n) \in \ell^1 \right\}$$

equipped with norms

$$\left\| \left(\sum_{n=1}^{\infty} a_n e_n + b_n r_n, \sum_{n=1}^{\infty} \delta_n b_n e_n, \sum_{n=1}^{\infty} c_n e_n \right) \right\|_{X_0} = \sum_{n=1}^{\infty} |a_n| + |b_n| + |c_n|$$

and

$$\left\| \left(\sum_{n=1}^{\infty} a_n e_n + c_n r_n, \sum_{n=1}^{\infty} b_n e_n, \sum_{n=1}^{\infty} \delta_n c_n e_n \right) \right\|_{X_1} = \sum_{n=1}^{\infty} |a_n| + |b_n| + |c_n|.$$

The spaces X_0 and X_1 are isometric to ℓ_1. To see this, fix $(x, y, z) \in X_0$ with $x = (x_n), y = (y_n)$, and $z = (z_n)$, and put $a_n = x_n - \sum y_m \delta_m^{-1} r_m(n)$, $b_n = y_n \delta_m^{-1}$, and $c_n = z_n$. Then the map

$$R_0(x, y, z) = ((a_n), (b_n), (c_n))$$

defines the isometry between X_0 and $\ell_1 \oplus \ell_1 \oplus \ell_1$. Similarly, the space X_1 is isometric to $\ell_1 \oplus \ell_1 \oplus \ell_1$. The spaces (X_0, X_1) form a Banach couple. Let $Y = c_0 \oplus \ell_1 \oplus \ell_1$, and let $J_j\colon X_j \to Y$ be the formal inclusion maps defined by

$$J_j(x, y, z) = (x, y, z).$$

As $\|\sum b_n r_n\|_{c_0} \le \sum |b_n|$, the inclusions J_j are contractions, therefore (X_0, X_1) is a Banach couple with ambient space Y. We let X_θ denote the complex interpolation space of order θ for the Banach couple (X_0, X_1). Thus for $(x, y, z) \in X_\theta$, we have the interpolatory estimates

$$\|(x, y, z)\|_{X_\theta} \le \|(x, y, z)\|_{X_0}^{1-\theta} \|(x, y, z)\|_{X_1}^{\theta}.$$

Finally, define the embedding $E\colon c_{00} \to Y$ by

$$E(x) = (x, 0, 0).$$

The theorem of Garling and Montgomery-Smith (1991) asserts that E extends to a bounded operator $E: c_0 \to X_\theta$ with closed range.

Theorem 5.6.1 *For $0 < \theta < 1$, there exists $C(\theta) > 0$ such that $\|E(r_n)\|_{X_\theta} \le C(\theta)$, and E extends to a bounded linear operator $E: c_0 \to X_\theta$ satisfying*

$$\|x\|_{c_0} \le \|E(x)\|_{X_\theta} \le C(\theta)\|x\|_{c_0}.$$

Proof It is easy to read off the following identities from the equations defining the norms in the spaces X_0 and X_1:

$$\begin{aligned}\|(r_n, \delta_n e_n, 0)\|_{X_0} &= \|(r_n, \delta_n e_n, 0)\|_{X_1} = 1,\\ \|(0, 0, \delta_n e_n)\|_{X_0} &= \|(0, \delta_n e_n, 0)\|_{X_1} = \delta_n,\\ \|(r_n, 0, 0)\|_{X_0} &= \|(r_n, 0, 0)\|_{X_1} = \|r_n\|_{\ell_1}.\end{aligned}$$

Moreover, by the triangle inequality

$$\|(0, \delta_n e_n, 0)\|_{X_0} \le \|(r_n, \delta_n e_n, 0)\|_{X_0} + \|(r_n, 0, 0)\|_{X_0} = 1 + \|r_n\|_{\ell_1}.$$

Now we turn to estimating the norm of $E(r_n) = (r_n, 0, 0)$ in the space X_θ. Start by applying the triangle inequality

$$\|(r_n, 0, 0)\|_{X_\theta} \le \|(r_n, \delta_n e_n, \delta_n e_n)\|_{X_\theta} + \|(0, \delta_n e_n, 0)\|_{X_\theta} + \|(0, 0, \delta_n e_n)\|_{X_\theta}.$$

First, as X_θ is an interpolation space of order θ for the Banach couple (X_0, X_1) we get

$$\|(r_n, \delta_n e_n, \delta_n e_n)\|_{X_\theta} \le \|(r_n, \delta_n e_n, \delta_n e_n)\|_{X_0}^{1-\theta}\|(r_n, \delta_n e_n, \delta_n e_n)\|_{X_1}^{\theta} \le 1 + \delta_n.$$

Second, using the above identities we have

$$\|(0, \delta_n e_n, 0)\|_{X_\theta} \le \|(0, \delta_n e_n, 0)\|_{X_0}^{1-\theta}\|(0, \delta_n e_n, 0)\|_{X_1}^{\theta} \le (1 + \|r_n\|_1)^{1-\theta}\delta_n^{\theta}.$$

Third, we have

$$\|(0, 0, \delta_n e_n)\|_{X_\theta} \le \|(0, 0, \delta_n e_n)\|_{X_0}^{1-\theta}\|(0, 0, \delta_n e_n)\|_{X_1}^{\theta} \le \delta_n^{1-\theta}(1 + \|r_n\|_1)^{\theta}.$$

Collecting terms we find $\|(r_n, 0, 0)\|_{X_\theta} \le C(\theta)$. Thus, we have verified that E extends to a bounded operator $E: c_0 \to X_\theta$ with $\|E(x)\|_{X_\theta} \le C(\theta)\|x\|_{c_0}$.

It remains to prove the lower bound $\|x\|_{c_0} \le \|E(x)\|_{X_\theta}$. Let $j \in \{0, 1\}$. Define the projection $P: X_j \to c_0$ by the equation

$$P(x, y, z) = x.$$

Clearly, $\|P: X_j \to c_0\| \le 1$, and since X_θ is the complex interpolation space, $\|P: X_\theta \to c_0\| \le 1$. Finally, by definition, $\mathrm{Id}_{c_0} = PE$, and hence $\|x\|_{c_0} = \|PE(x)\|_{c_0} \le \|P\| \cdot \|E(x)\|_{X_\theta}$. □

5.7 Notes

Section 5.1 presents the isomorphic classification theorem (Theorem 5.1.1) for polydisk algebras by Bourgain (1984b). Our presentation incorporates the seminar notes by Wolniewicz (1991). The introduction uses the book by Pełczyński (1977), which presents the subject prior to Bourgain (1984b):

(i) For $d \geq 2$, the ball algebra $A(\mathbb{B}_d)$ is not isomorphic to $A(\mathbb{T})$ (Mitiagin and Pełczyński, 1976).
(ii) For $d \geq 2$, $A(\mathbb{B}_d)$, and $n \geq 2$ is not isomorphic to $A(\mathbb{T}^n)$ (Henkin, 1967, 1968, 1970).
(iii) Bourgain (1985) proves that $A(\mathbb{T}^2)$ is not isomorphic to $A(\mathbb{T}^3)$, solving a problem posed specifically by Henkin (1970),

Although polydisk algebras $A(\mathbb{T}^d)$ (and their duals $A^*(\mathbb{T}^d)$) are pairwise non-isomorphic Banach spaces, P. Wojtaszczyk (1979a,b) proved that each $A(\mathbb{T}^d)$ is isomorphic to its c_0 sum. Consequently, Pełczyński's decomposition method is applicable to the class of polydisk algebras.

Theorem 5.1.4 is due to Pisier. Its proof is presented in Davis et al. (1984). The proof of Theorem 5.1.5 is extracted from Bourgain (1984b). Theorem 5.1.12, Theorem 5.1.13, Theorem 5.1.14, and their respective proofs are contained in Bourgain (1984b). The original proof of Theorem 5.1.15 in Bourgain (1984b) is based on analytic partitions of unity. The bi-Hardy martingale lifting theorem (Theorem 5.1.16) was known to Bourgain (1984b). The Steinhaus coefficient estimates, Theorem 5.1.17, and Theorem 5.1.18, are taken from Bourgain (1984b).

A detailed description of the measures arising in $A(\mathbb{T}^n)^*$ (the dual space) is contained in Bourgain (1985).

Related open problems discussed by Bourgain (1983b, 1989) include the dimension conjecture for the ball algebras $A(\mathbb{B}^n)$, and for $H^\infty(\mathbb{B}^n)$ and $H^\infty(\mathbb{T}^n)$.

In section 5.2, we treat the analytic Radon–Nikodym property (aRNP) introduced by Bukhvalov and Danilevich (1982). Bukhvalov and Danilevich (1982) state Theorem 5.2.1 (without proof). In the text we present the proof given by Hensgen (1986, 1991). For Theorem 5.2.2 see Bukhvalov and Danilevich (1982). Theorem 5.2.3 is the classical Fatou theorem. Theorem 5.2.4 is due to Bukhvalov (1976, 1981), the 1981 publication is the translation into English of the Russion original (Bukhvalov, 1976). The proof in the text is taken from Pallares and Kouba (1986–1987), which are the notes of the seminar talks at Groupe de Travail sur les Espaces de Banach, directed by Maurey and Piser. We refer to Ryan (1962) for the analytic counterpart of Theorem 5.2.4.

Theorem 5.2.5, due to Bukhvalov and Danilevich (1982), presents a vector-valued version of Fatou's Theorem.

Theorem 5.2.6 is due to Bu (1988).

Theorem 5.2.8, the remarks thereafter, and Theorem 5.2.9 are due to Dowling (1985) and Hensgen (1986, 1987). Theorem 5.2.11 was obtained by Dowling and Edgar (1988), exploiting the theorems of Lewis and Stegall (1973) and Ghoussoub and Rosenthal (1983).

Theorem 5.2.12 and Theorem 5.2.13 are due to Pisier (unpublished). Pisier's results were documented by Ghoussoub et al. (1989, Section V). A thorough development of real variable and topological methods related to the aRNP is contained in Ghoussoub, Lindenstrauss, and Maurey (1989), Ghoussoub and Maurey (1986, 1989), and Ghoussoub, Maurey, and Schachermayer (1989).

Theorem 5.2.14 and Theorem 5.2.15 are due to Edgar (1985, 1986).

The real variable characterisations of the aRNP are due to Bu (1988), Ghoussoub, Lindenstrauss, and Maurey (1989), and Ghoussoub and Maurey (1989). Our presentation incorporates Bu (1993) and Ghoussoub (1993, Chapter 2).

Proposition 5.2.19 is due to Edgar (1985) and Ghoussoub and Maurey (1989). Proposition 5.2.17 is due to Ghoussoub, Lindenstrauss, and Maurey (1989) and Ghoussoub and Maurey (1989), Theorem 5.2.18 and Theorem 5.2.20 are due to Ghoussoub and Maurey (1989). Theorem 5.2.16 and Theorem 5.2.21 are due to Ghoussoub, Lindenstrauss, and Maurey (1989) and Ghoussoub and Maurey (1989). The text presents the proof in these sources. Proposition 5.2.23 is due to Bu (1988),

Preferred holomorphic injections transporting the aRNP from the range to the domain were introduced by Ghoussoub and Maurey (1989), who applied their method to prove that the predual of the James tree space has the aRNP. Section 7.4 presents their proof in detail.

Theorem 5.2.24 is due to Ghoussoub and Maurey (1989). The conceptual link between martingale convergence and variational principles is due to Stegall (1978). Stegall's proofs were simplified by Bourgain (1979). See also Benyamini and Lindenstrauss (2000) and Phelps (1989) for presentations.

Theorem 5.2.25 is due to Haagerup and Pisier (1989), who applied it to prove that any noncommutative L^1 space satisfies the aRNP. See Haagerup and Pisier (1989, Corollary 2.3.). Kalton (1986a,b) treats the aRNP for quasi-Banach spaces.

In Section 5.3, the dichotomy theorem (Theorem 5.3.1), the subsequence theorem (Theorem 5.3.2), and the renorming theorem (Theorem 5.3.4) are due to Xu (1988). We refer to Beauzamy (1982) for Maurey's proof of the renorming theorem for general martingales. Maurey's proof served as the model for Xu's proof of Theorem 5.3.4 (Xu, 1988). Theorem 5.3.5 asserting that complex uniform convexity implies Hardy martingale convergence is also due to Xu (1988).

Theorem 5.3.7, Theorem 5.3.8, Theorem 5.3.10, and Theorem 5.3.11 are contained in Müller (2012, 2017). Motivation for these comes from a problem posed by Pisier (1992a, 2016): Is it possible to detect Hardy martingale cotype by testing simple Hardy martingales only?

The study of Banach spaces with aUMD was initiated by Garling (1988), who gave two proofs that aUMD implies the aRNP. The first, presented in Theorem 5.4.3, is based on Kwapień's theorem (Theorem 5.4.2) (Kwapień, 1974). The second proof starts with Theorem 5.4.4, which asserts that in an aUMD space c_0 is not finitely represented and that, by a theorem of Maurey and Pisier (1976), the latter property implies nontrivial cotype. In the presence of aUMD, nontrivial cotype implies nontrivial Hardy martingale cotype. Theorem 5.4.4 and Theorem 5.4.5 are due to Garling (1988).

Section 5.4.2 presents Garling's extrapolation theorem (Theorem 5.4.6) (Garling, 1988). The proof of Lemma 5.4.7 is due to Maurey (1980). The proof of Theorem 5.4.9 uses the argument recorded in Qiu (2012b, Section 2.6.4). It exploits sharp function estimates for general discrete filtrations presented in Long (1993). See Jiao, Xie, and Zhou (2015) and Jiao, Wu, Yang, and Yi (2017) for recent developments of the stopping time techniques involved.

Theorem 5.4.10 is due to Garling (1988).

Blower (1990) obtains a characterization of the aUMD property in terms of Fourier multipliers acting on $H^1(X)$. Important links between Banach spaces satisfying aUMD and H^∞ functional calculus were uncovered by Kalton and Weis (2001). Theorem 5.4.11 is due to Qiu (2012b). We refer to the book by Hytönen et al. (2016) for an account of the real martingale versions of Qiu's theorem (Theorem 5.4.11).

Theorem 5.5.1 is due to Maurey (1980). Theorem 5.5.2 is taken from Müller (2017). Edgar (1985) obtained approximation theorems for Hardy martingales by subsequences of simple Hardy martingales. We refer to Ghoussoub and Maurey (1989) and Bu and Schachermayer (1992) for further extended and strengthened embedding theorems.

In Section 5.6, Theorem 5.6.1 is due to Garling and Montgomery-Smith (1991). For related examples see Dilworth (1986, Section 4), who attributes the result to Pisier.

Appendix: RNP, UMD, and M-Cotype

Fix a Banach space X and a finite measure space $(F, \mathcal{F}, \mu)$. Let $\mathcal{G} \subseteq \mathcal{F}$ be a sub-sigma-algebra such that $F \in \mathcal{G}$ and let $\mathbb{E}_{\mathcal{G}}$ denote the associated conditional expectation operator. By Theorem 2.1.1,

$$\left\|\mathbb{E}_{\mathcal{G}} \otimes \mathrm{Id}_X : L^p(X) \to L^p(X)\right\| = 1,$$

for $1 \leq p \leq \infty$. We let $(\mathcal{G}_k)$ be a sequence of increasing sub-sigma-algebras of $\mathcal{F}$. A sequence of μ-Bochner integrable functions $(f_k : F \to X)$ is a $(\mathcal{G}_k)$-martingale if

$$\mathbb{E}_{\mathcal{G}_k}(f_{k+1}) = f_k.$$

We review the almost sure, weak, weak-*, and norm convergence of L^1-bounded martingales, leading up to the definition of the Radon–Nikodym property of Banach spaces. We give a characterization in terms of bounded harmonic functions, and recall the intrinsic description on RNP by Stegall's variational principle. Thereafter, we review the notions of type, cotype, and martingale cotype, and the UMD property of Banach spaces.

Almost Sure $\sigma(X^*, X)$-Convergence

Let X^* be the dual of a separable Banach space X. We next consider $L^1(X^*)$, bounded martingales. For those we have $\sigma(X^*, X)$-convergence pointwise, almost surely. The limiting function satisfies a weak measurability condition.

We say that $f : F \to X^*$ is weak-$*$-measurable if $\langle x, f(\cdot)\rangle : F \to \mathbb{R}$ is measurable for $x \in X$. We refer to Pisier (2016, Section 2.2) for the following result.

Theorem 5.7.1 *Let X be a separable Banach space, and let X^* denote its dual space. Let $(f_k : F \to X^*)$ be an X^*-valued $(\mathcal{G}_k)$-martingale sequence satisfying*

$$\sup_{k \in \mathbb{N}} \|f_k\|_{L^1(X^*)} < \infty. \tag{5.7.1}$$

Then there exists a weak-$*$ measurable $f : F \to X^*$ and $E \subset F$ with $\mu(E) = 0$ such that

$$\lim_{k \to \infty} \langle x, f_k(\omega) - f(\omega)\rangle = 0, \qquad x \in X, \quad \omega \in F \backslash E. \tag{5.7.2}$$

If, moreover, $\{\|f_k\| : k \in \mathbb{N}\}$ is a uniformly integrable subset of $L^1(F)$, then

$$\langle x, f_k\rangle = \mathbb{E}(\langle x, f\rangle | \mathcal{G}_k),$$

for $x \in X$, and $k \in \mathbb{N}$.

Proof By Doob's maximal inequality (2.1.7), and condition (5.7.1), there exists a subset of full measure G such that $\sup \|f_k(\omega)\|_{X^*} < \infty$ for $\omega \in G$. Next fix $\omega \in G$. Since the closed unit ball B_{X^*} is metrizable, and compact with respect to the $\sigma(X^*, X)$-topology, there exists a $\sigma(X^*, X)$ cluster point, call it $f(\omega)$ of $\{f_k(\omega) : n \in \mathbb{N}\}$.

Let D be a countable norm dense subset of the closed unit ball B_X. For $d \in D$, $(\langle d, f_k(\cdot)\rangle)$ forms an L^1-bounded scalar martingale. By the martingale convergence theorem (Theorem 1.1.3), there exists a subset $G(d) \subseteq G$ of full measure, such that $\lim_{n\to\infty}\langle d, f_k(\omega)\rangle = \langle d, f(\omega)\rangle$ for $\omega \in G(d)$. Finally, since D is countable $G_0 = \bigcap G(d)$ is of full measure, and

$$\lim_{n\to\infty} \langle d, f_k(\omega)\rangle = \langle d, f(\omega)\rangle,$$

for $d \in D$ and $\omega \in G_0$. Since D is norm dense in B_X, we obtain (5.7.2) with $E = F \backslash G_0$. The final assertion follows from Theorem 1.1.2. □

Almost Sure $\sigma(X, X^*)$-Convergence

Let $(F, \mathcal{F}, \mu)$ be a finite measure space, let X be a Banach space, and let X^* denote its dual space. We say that $f : F \to X$ is weakly measurable if $\langle f(\cdot), x^*\rangle : F \to \mathbb{R}$ is measurable for $x^* \in X^*$. The theorem of Phillips (see Pisier [2016, Section 1.8]) relates this concept to Bochner measurability.

Theorem 5.7.2 (Phillips's theorem) *If $f : F \to X$ is weakly measurable and if X is a separable Banach space, then f is Bochner measurable.*

The following theorem asserts that almost sure weak-convergence of an $L^1(X)$-bounded martingale sequence to a Bochner measurable limit implies its almost sure norm convergence. We refer to Davis et al. (1990), Kwapień and Woyczyński (1992, Theorem 4.2.3), or Pisier (2016, Section 2.4) for the presentation of the following result.

Theorem 5.7.3 *Let X be a Banach space. Let $(F, \mathcal{F}, \mu)$ be a finite measure space and let $(f_k : F \to X)$ be an X-valued martingale sequence satisfying*

$$\sup_{k\in\mathbb{N}} \|f_k\|_{L^1(X)} < \infty. \tag{5.7.3}$$

Then the following conditions are equivalent:

(i) *There exists a Bochner measurable $f : F \to X$ and $E \subset F$ with $\mu(E) = 0$ such that*

$$\lim_{k\to\infty} \langle f_k(\omega) - f(\omega), x^*\rangle = 0, \tag{5.7.4}$$

for $\omega \in F \backslash E$, and $x^* \in X^*$.

(ii) *There exists* $\widetilde{E} \subset F$ *with* $\mu\left(\widetilde{E}\right) = 0$ *and* $f \in L^1(F, X)$ *such that*

$$\lim_{k\to\infty} \|f_k(\omega) - f(\omega)\|_X = 0, \tag{5.7.5}$$

for every $\omega \in F\backslash\widetilde{E}$.

Proof Let $E \subseteq F$, let $f, f_k \colon F \to X$ be Bochner measurable, and assume that conditions (5.7.3) and (5.7.4) hold true. We are going to show that (5.7.5) holds true for some $\widetilde{E} \subset F$ with $\mu(\widetilde{E}) = 0$.

Part 1 Fix $A > 1$ and $\omega \in F$, and set $\sigma_A(\omega) = \inf\{k \in \mathbb{N} : \|f_k(\omega)\|_X > A\}$. Since $\{\sigma_A < \infty\} = \{\sup_{k\in\mathbb{N}} \|f_k(\cdot)\|_X > A\}$, Doob's maximal inequality yields the measure estimate

$$\mu(\{\sigma_A < \infty\}) \le \frac{1}{A} \sup_{k\in\mathbb{N}} \|f_k\|_{L^1(X)}.$$

Next, consider a Bochner integrable martingale sequence $g_k = f_{k\wedge\sigma_A}, k \in \mathbb{N}$, and set $g = f_{\sigma_A}$. Clearly, for $\omega \in F\backslash E$ we have

$$\lim_{k\to\infty} \langle g_k(\omega), x^*\rangle = \langle g(\omega), x^*\rangle, \qquad x^* \in X^*. \tag{5.7.6}$$

The choice of σ_A yields the pointwise bound $\|g_k(\cdot)\|_X \le A$.

Next, fix $x^* \in X^*$, form the uniformly bounded scalar-valued martingale $\varphi_k = \langle g_k(\cdot), x^*\rangle$, and put $\varphi = \langle g(\cdot), x^*\rangle$. In view of Lebesgue's dominated convergence theorem and (5.7.6) we obtain $\varphi_k = \mathbb{E}(\varphi|\mathcal{G}_k)$, $k \in \mathbb{N}$. Hence,

$$\langle g_k(\cdot), x^*\rangle = \langle \mathbb{E}(g(\cdot)|\mathcal{G}_k), x^*\rangle. \tag{5.7.7}$$

By the Hahn–Banach theorem, (5.7.7) implies that $g_k = \mathbb{E}(g|\mathcal{G}_k)$, $k \in \mathbb{N}$. Taking into account that $\|g\|, \|g_k\|$ are pointwise bounded by A, applying Theorem 2.1.3 gives $E_1 \subset F$, with $\mu(E_1) = 0$ such that

$$\lim_{k\to\infty} \|g_k(\omega) - g(\omega)\|_X = 0, \qquad \omega \in F\backslash E_1. \tag{5.7.8}$$

Part 2 Assume that there exists $B \subset F$ with $\mu(B) > 0$ such that (5.7.5) fails to hold for $\omega \in B$. Set $\delta = \mu(B)$ and let $A_0 = 4\sup_{k\in\mathbb{N}} \|f_k\|_{L^1(X)}/\delta$. Next consider the martingale sequence $g_k = f_{k\wedge\sigma_{A_0}}$, $k \in \mathbb{N}$, and set $g = f_{\sigma_{A_0}}$. Clearly we have

$$g_k(\omega) = f_k(\omega), \qquad g(\omega) = f(\omega), \quad \text{for} \quad \omega \in \{\sigma_{A_0} = \infty\}. \tag{5.7.9}$$

Combining (5.7.7), obtained in Part 1, and (5.7.9) gives

$$\lim_{k\to\infty} \|f_k(\omega) - f(\omega)\|_X = 0, \qquad \omega \in \{\sigma_{A_0} = \infty\}\backslash E_1. \tag{5.7.10}$$

Since

$$\mu(\{\sigma_{A_0} = \infty\}) = \mu(F) - \mu(\{\sigma_{A_0} < \infty\}) \geq \mu(F) - \delta/4,$$

we obtain a contradiction. □

Almost Sure Norm-Convergence

Let $(F, \mathcal{F}, \mu)$ be a finite measure space. We say that a sequence of Bochner integrable $(f_k : F \to X)$ converges almost surely if there exists $E \subset F$ with $\mu(E) = 0$ and $f \in L^1(F, \mathcal{F}, \mu, X)$ such that

$$\lim_{k\to\infty} \|f_k(\omega) - f(\omega)\|_X = 0,$$

for every $\omega \in F \backslash E$.

The almost sure convergence of $L^1(X)$-bounded martingales does not hold independent of the underlying Banach space X, rather it defines an isomorphic invariant of X. A Banach space X satisfies the RNP if for every finite measure space $(F, \mathcal{F}, \mu)$, any X-valued martingale $(f_k : F \to X)$ satisfying

$$\sup_{k\in\mathbb{N}} \|f_k\|_{L^1(X)} < \infty,$$

converges almost surely, where $L^1(X) = L^1(F, \mathcal{F}, \mu, X)$.

The following result due to Chatterji (1968) relates the Radon–Nikodym property to the almost sure convergence of $L^\infty(X)$-bounded martingales.

Theorem 5.7.4 (Chatterji [1968]) *A Banach space X satisfies the Radon–Nikodym property if and only if every $L^\infty(F, \mathcal{F}, \mu, X)$-bounded martingale $(f_k : F \to X)$ converges almost surely.*

Results: In the following list we collect basic and well-known results on the Radon–Nikodym proptery.

(i) A separable dual space (for instance ℓ^1, $H^1(\mathbb{T})$, or $L^p[0,1]$, for $1 < p < \infty$) satisfies the RNP. (Theorem of Dunford and Pettis [2000], also see Benyamini and Lindenstrauss [2000, Cor. 5.12].)

(ii) Any reflexive space satisfies the RNP. (Theorem of Dunford and Pettis [2000] also see Benyamini and Lindenstrauss [2000, Cor. 5.12].)

(iii) The spaces c_0, $L^1[0,1]$ fail the RNP. (See Benyamini and Lindenstrauss [2000, Section 5.2].)

(iv) The RNP is preserved under the formation of ℓ^1-direct sums. If each of the Banach spaces E_n satisfies the RNP, then also $(\sum E_n)_{\ell^1}$. (See Benyamini and Lindenstrauss [2000, Section 5.2].)

(v) Recently, an intrinsic spline characterization of Banach spaces with the RNP property was obtained in two separate steps in Müller and Passenbrunner (2020) and Passenbrunner (2020).

Harmonic Functions and RNP

The book by Pisier (2016, Sections 3.1–3.5) serves as our basic reference for the following link between vector-valued harmonic functions, vector-valued martingales, and the RNP.

Let X be a Banach space and let $u\colon \mathbb{D} \to X$ be a harmonic function. By definition, u is a Poisson integral if there exists $f \in L^1(\mathbb{T}, X)$ such that

$$u(z) = \int_{\mathbb{T}} P(z, \zeta) f(\zeta) dm(\zeta), \quad \text{for} \quad z \in \mathbb{D}.$$

We say that $u\colon \mathbb{D} \to X$ admits (strong) radial limits almost everywhere if there exists $E \subset \mathbb{T}$ with $m(E) = 0$, and a Bochner integrable $f\colon \mathbb{T} \to X$ such that

$$\lim_{r \to 1} \|u(r\zeta) - f(\zeta)\|_X = 0, \tag{5.7.11}$$

for $\zeta \in \mathbb{T} \backslash E$.

We refer to Bukhvalov and Danilevich (1982) for the next theorem showing that Poisson integrals may replace the role of closed martingales.

Theorem 5.7.5 *Let $1 \le p < \infty$, and let X be a Banach space. A harmonic function $u\colon \mathbb{D} \to X$ is the Poisson integral of $f \in L^p(\mathbb{T})$ if and only if*

$$\lim_{r \to 1} \int_{\mathbb{T}} \|u(rw) - f(w)\|_X^p dm(w) = 0.$$

In that case, u admits radial limits almost everywhere.

Bukhvalov and Danilevich (1982) were the first to realize the connection between the existence of radial limits for vector-valued harmonic functions and the RNP property of the underlying Banach space X.

Theorem 5.7.6 *A Banach space X satisfies the Radon–Nikodym property (RNP) if and only if every harmonic function $u\colon \mathbb{D} \to X$ satisfying*

$$\sup_{z \in \mathbb{D}} \|u(z)\|_X < \infty \tag{5.7.12}$$

admits radial limits almost everywhere.

Remark: We point out that the RNP is equivalent to the condition that every harmonic function $u\colon \mathbb{D} \to X$ satisfying

$$\sup_{0<r<1} \int_{\mathbb{T}} \|u(r\zeta)\|_X dm(\zeta) < \infty, \tag{5.7.13}$$

admits radial limits almost everywhere, which in turn holds if and only if for $1 < p < \infty$, and every harmonic function $u\colon \mathbb{D} \to X$ satisfying

$$\sup_{0<r<1} \int_{\mathbb{T}} \|u(r\zeta)\|_X^p dm(\zeta) < \infty, \tag{5.7.14}$$

there exists a p-Bochner integrable $f\colon \mathbb{T} \to X$ such that

$$\lim_{r\to 1} \int_{\mathbb{T}} \|u(r\zeta) - f(\zeta)\|_X^p dm(\zeta) = 0. \tag{5.7.15}$$

In view of Theorem 5.7.5, equation (5.7.15) implies that u is the Poisson integral of f and admits radial limits almost everywhere. For the above equivalences we refer to Bukhvalov and Danilevich (1982) for the range $1 < p < \infty$ and to Hensgen (1986) (as published in Hensgen [1991]) for the proof of the case $p = 1$.

Almost Sure Weak and Weak-* Radial Convergence

We consider the weak and the weak-* boundary convergence of vector-valued harmonic functions. We first show that weak radial convergence implies the existence of strong radial limits, and second that bounded harmonic functions ranging in the dual of a separable Banach space admit weak-* radial limits almost everywhere.

Let X^* be the dual of a Banach space X. A continuous $u\colon \mathbb{D} \to X^*$ admits weak-* radial limits almost everywhere, if there exists $E \subset \mathbb{T}$ with $m(E) = 0$ and a function $f\colon \mathbb{T} \to X^*$ satisfying

$$\lim_{r\to 1}\langle x, u(r\zeta) - f(\zeta)\rangle = 0, \quad \text{for } x \in X \text{ and } \zeta \in \mathbb{T}\setminus E,$$

and where $\langle x, f(\cdot)\rangle\colon \mathbb{T} \to \mathbb{R}$ is measurable for $x \in X$. For martingales ranging in a separable space, Theorem 5.7.1 asserts the almost sure existence of weak-$*$ limits. The same holds true for vector-valued harmonic functions. See Grossetête (1971, 1972). For proofs we refer to Hensgen (1986, 1991).

Theorem 5.7.7 *Let X be a separable Banach space with dual space X^*. Then every harmonic $u\colon \mathbb{D} \to X^*$ satisfying*

$$\sup_{0<r<1} \int_{\mathbb{T}} \|u(r\zeta)\|_{X^*} dm(\zeta) < \infty \tag{5.7.16}$$

admits weak- radial limits almost everywhere.*

Let X be a Banach space, and let X^* denote its dual space. We say that a continuous $u\colon \mathbb{D} \to X$ admits weak radial limits almost everywhere if there exists $F \subset \mathbb{T}$ with $m(F) = 0$ and a Bochner measurable $f\colon \mathbb{T} \to X$ such that

$$\lim_{r\to 1}\langle u(r\zeta) - f(\zeta), x^*\rangle = 0, \qquad x^* \in X^*, \quad \zeta \in \mathbb{T}\backslash F.$$

The next result, obtained by Hensgen (1986, 1991) links almost sure weak radial limits and almost sure strong radial limits analogous to the case of vector-valued discrete martingales. Its content is analogous to Theorem 5.7.3.

Theorem 5.7.8 *Let X be a Banach space with dual X^* and let $u\colon \mathbb{D} \to X$ be harmonic. Assume that*

$$\sup_{0<r<1} \int_{\mathbb{T}} \|u(r\zeta)\|_X dm(\zeta) < \infty, \tag{5.7.17}$$

and that $u\colon \mathbb{D} \to X$ admits weak radial limits almost everywhere. Then $u\colon \mathbb{D} \to X$ admits strong radial limits almost everywhere.

Intrinsic Characterization of RNP

We review two characterizations of Banach spaces with the RNP. One by the existence of small affine slices, and a second one by Stegall's variational principle.

Small affine slices: Let X be a Banach space and $C \subset X$. We say that C admits affine slices of arbitrary small diameter if, for any $\epsilon > 0$, there exists $x^* \in X^*$, with $\|x^*\| \le 1$, and $A \in \mathbb{R}$ such that

$$C \cap \{x^* + A > 0\} \neq \emptyset \quad \text{and} \quad \operatorname{diam}(C \cap \{x^* + A > 0\}) < \epsilon.$$

Theorem 5.7.9 *A Banach space X satisfies the RNP if and only if every bounded closed convex subset $C \subset X$ admits affine slices of arbitrarily small diameter.*

Theorem 5.7.9 is the result of many contributing authors. Benyamini and Lindenstrauss (2000, Chapter 5, Sec. 2) present its proof and a detailed historical discussion.

Stegall's variational principle: Fundamental work by Stegall (1978) exhibits deep connections between the RNP and optimization in Banach spaces.

Let X be a Banach space. Fix $C \subseteq X$ and $f\colon C \to \mathbb{R}$. We say that f assumes a strong maximum in C at $x_0 \in C$ if

(i) f is upper-bounded, that is, $\sup\{f(y) : y \in C\} < \infty$,

(ii) $f(x_0) = \sup\{f(y) : y \in C\}$,
(iii) If $z_n \in C$ and $\lim_{n\to\infty} f(z_n) = f(x_0)$, then $\lim_{n\to\infty} z_n = x_0$.

We say that the *affine variational principle* holds true in the Banach space X if for every bounded closed convex subset $C \subset X$ and every upper-bounded and upper-semicontinuous $g\colon C \to \mathbb{R}$ there exists $x^* \in X^*$, with $\|x^*\| \leq 1$, such that

$$g + x^* \quad \text{assumes a strong maximum in} \quad C.$$

(Recall that $g\colon C \to \mathbb{R}$ is upper semicontinuous if $\{x \in C\colon g(x) < \alpha\}$ is open for any $\alpha \in \mathbb{R}$.)

Theorem 5.7.10 (Stegall's variational principle) *A Banach space X satisfies the Radon–Nikodym property if and only if the affine variational principle holds true.*

Theorem 5.7.10 states Stegall's variational principle in the form given later, by Bourgain (1979). We refer to Benyamini and Lindenstrauss (2000, Chapter 5, Sec. 2) for additional information.

Type and Cotype

Let X be a Banach space. Let $(r_n)_{n=1}^\infty$ denote the sequence of independent $\{-1, 1\}$-valued Rademacher functions. Let $1 \leq p \leq 2$ and $2 \leq q < \infty$. We say that the Banach space is of type p or cotype q if there exists a constant $T_p(X) < \infty$, respectively, $C_q(X) < \infty$ if

$$\int_0^1 \left\| \sum_{k=1}^n x_k r_k(t) \right\|_X dt \leq T_p(X) \left(\sum_{k=1}^n \|x_k\|_X^p \right)^{1/p},$$

or

$$\left(\sum_{k=1}^n \|x_k\|_X^q \right)^{1/q} \leq C_q(X) \int_0^1 \left\| \sum_{k=1}^n x_k r_k(t) \right\|_X dt,$$

respectively, for $x_1, \ldots, x_n \in X$ and $n \in \mathbb{N}$. We say that X has nontrivial type, or respectively cotype, if there exists $1 \leq p \leq 2$, respectively $2 \leq q < \infty$, such that $T_p(X) < \infty$ and $C_q(X) < \infty$, respectively.

Finite representation: Let X, Y be Banach spaces. We say that Y is finitely represented in X if there exists a constant $C < \infty$ such that for any finite-dimensional subspace $E \subset Y$ there exists a subspace $F \subset X$ and a bijective linear operator $T\colon E \to F$ such that

$$\|T\colon E \to F\| \cdot \left\|T^{-1}\colon E \to F\right\| < C,$$

where the norm on E is the one induced by Y and, similarly, the norm on F is induced by X.

We say that a Banach space X is super-reflexive, if a Banach space Y, which is finitely represented in X, is reflexive.

Theorem 5.7.11 (Maurey and Pisier, 1976, 1974) *Let X be a Banach space. The space ℓ^1 is not finitely represented in X if and only if X has nontrivial type. Also c_0 is not finitely represented in X if and only if X has nontrivial cotype.*

Martingale Cotype

Let X be a Banach space, and let $2 \le q < \infty$. We say that X is of martingale cotype q if, for every $1 < p \le q$, there exists $C = C(p, q, X)$ such that for any filtered probability space $(F, (\mathcal{G}_n), \mu)$, and any $L^p(X)$-bounded $(\mathcal{G}_n)$-martingale $(f_n : F \to X)$ the following estimates hold true,

$$\sum_{k=1}^{n} \left(\int_F \|\Delta f_k\|_X^p d\mu \right)^{q/p} \le C^q \left(\int_F \|f_n\|_X^p d\mu \right)^{q/p}, \tag{5.7.18}$$

for $n \in \mathbb{N}$.

Let $2 \le q < \infty$. We say that a Banach space X is q-uniformly convex, if for every $1 < p \le q$, there exists a constant $\delta = \delta(p, q)$ such that for every probability space $(F, \mathcal{F}, \mu)$ and every $f \in L^p(F, \mathcal{F}, \mu, X)$,

$$\|\mathbb{E}(f)\|_X^q + \delta \left(\int_F \|f - \mathbb{E}(f)\|_X^p d\mu \right)^{q/p} \le \left(\int_F \|f\|_X^p d\mu \right)^{q/p},$$

where $\mathbb{E}(f) = \int_F f d\mu$.

Theorem 5.7.12 (Pisier, 1975a, 1992a) *Let $2 \le q < \infty$. A Banach space X is of martingale cotype q if and only if there exists an equivalent norm on X for which X is q-uniformly convex.*

We say that X has nontrivial martingale cotype if there exists $2 \le q < \infty$ such that $C(p, q, X) < \infty$ for $1 < p \le q$. Basic geometrical characterizations of Banach spaces with nontrivial martingale cotype were obtained by Pisier. For instance:

(i) Nontrivial martingale cotype and superreflexivity are equivalent conditions (Pisier, 1975a).
(ii) In particular, ℓ^1 is not finitely represented in a Banach space with nontrivial martingale cotype (Pisier, 1975a).

Unconditional Martingale Convergence

Let X be a Banach space, and let $1 < p < \infty$. We say that X has the UMD$_p$ property if there exists $C_p < \infty$ such that for any filtered finite measure space $(F, (\mathcal{G}_n), \mu)$, and any $L^p(X)$-bounded $(\mathcal{G}_n)$-martingale $(f_k \colon F \to X)$ the following estimates hold true,

$$\int_F \left\| \sum_{k=1}^{n} \varepsilon_k \Delta f_k \right\|_X^p d\mu \le C_p^p \int_F \|f_n\|_X^p d\mu, \tag{5.7.19}$$

for $(\varepsilon_k)_{k=1}^n \in \{-1, 1\}^n$, and $n \in \mathbb{N}$.

Results: We summarize the most basic properties of UMD spaces, referring to Maurey (1975b) and Pisier (1975b, 1992a):

(i) If, for a given Banach space X, the UMD_p property is satisfied, for some value of $1 < p < \infty$, then it holds true for any $1 < p < \infty$. Thus the property is independent of p and we simply say that X is a UMD space (Maurey, 1975b; Pisier, 1992a).
(ii) The spaces L^p, ℓ^p for $1 < p < \infty$ satisfy the UMD property.
(iii) Any UMD space X is superreflexive (Maurey, 1975b). Separate proofs of this result were obtained by Aldous (1979) who shows that c_0 does not embed in a UMD space, by Bourgain (1983c) who exploited the Hilbert matrix to show superreflexivity, and by Garling (1986) who provides a proof based on stochastic integration.
(iv) A UMD space with cotype q is also of martingale cotype q. This observation was exploited by Pisier (1975b) to obtain a superreflexive Banach space failing the UMD property.

6

Operators on $L^p(L^1)$

In this chapter we obtain vector-valued estimates for singular integral operators, martingale transforms, and orthogonal projections onto homogeneous polynomials of fixed degree. More generally, we are interested in exhibiting classes of linear operators

$$R\colon L^p(\Sigma,\sigma) \to L^p(\Sigma,\sigma), \quad 1 < p < \infty,$$

satisfying the following property: For any reflexive subspace $Y \subset L^1(\Omega,\mu)$, the vector-valued extension $\mathcal{R} = R \otimes J_\Omega$, $J_\Omega(f) = f$, $f \in Y$, satisfies the following estimates

$$\left\|\mathcal{R}\colon L^p((\Sigma,\sigma),Y) \to L^p((\Sigma,\sigma),Y)\right\| \le C(p,Y). \tag{6.0.1}$$

In Section 6.2 we prove that if there exists $0 < \alpha < 1$ such that

$$\left\|\mathcal{R}\colon L^p\left((\Sigma,\sigma),L^1(\Omega,\mu)\right) \to L^p((\Sigma,\sigma),L^\alpha(\Omega,\mu))\right\| \le C(p,\alpha), \tag{6.0.2}$$

then estimate (6.0.1) holds true. We present Maurey's proof of Rosenthal's L^p-theorem, and apply it to show that estimate (6.0.1) holds true, provided that

$$\left\|\mathcal{R}\colon L^p((\Sigma,\sigma),L^q(\Omega,\mu)) \to L^p((\Sigma,\sigma),L^q(\Omega,\mu))\right\| \le C(p,q), \tag{6.0.3}$$

for $1 < p,q < \infty$. Classical results of Benedek, Calerón, and Panzone (1962) assert that the Hilbert transform satisfies condition (6.0.3). By Pisier's theorem (Theorem 2.4.15), the orthogonal projections onto homogeneous polynomials of fixed, finite degree satisfy condition (6.0.3).

In Section 6.1 we prove that condition (6.0.2) holds true for the Hilbert transform on $L^p(\mathbb{T})$, for its martingale counterpart on $L^p(\mathbb{T}^\mathbb{N})$, and for the stochastic Hilbert transform acting on Brownian martingales.

6.1 The Hilbert Transform

Let H denote the Hilbert transform, given as Fourier multiplier

$$H(y^n) = (-i)\mathrm{sign}(n)y^n,$$

for $y \in \mathbb{T}$. By the M. Riesz theorem for $1 < p < \infty$, H is bounded on L^p and Kolmogorov's theorem asserts that it maps L^1 to weak L^1,

$$\left\|H\colon L^p(\mathbb{T}) \to L^p(\mathbb{T})\right\| \leq A(p), \quad \left\|H\colon L^1(\mathbb{T}) \to L^{1,\infty}(\mathbb{T})\right\| \leq A,$$

where $A(p) \leq p^2A/(p-1)$. In this section we prove vector-valued extensions of these basic theorems. For a given probability space (Ω, μ) and $0 < \alpha < 1$ we let $J_\Omega\colon L^1(\Omega) \to L^\alpha(\Omega)$ denote the formal inclusion map $J_\Omega(f) = f$, and let

$$H \otimes J_\Omega$$

be defined initially on the algebraic tensor product $L^p(\mathbb{T}) \otimes L^1(\Omega, \mu)$. The main result in this section asserts that $H \otimes J_\Omega$ extends to a bounded operator such that

$$\left\|H \otimes J_\Omega\colon L^p(\mathbb{T}, L^1(\Omega)) \to L^p(\mathbb{T}, L^\alpha(\Omega))\right\| \leq C(p, \alpha), \tag{6.1.1}$$

where $C(p, \alpha) < \infty$, for $0 < \alpha < 1 < p < \infty$. The most important applications of estimate (6.1.1) are presented Section 7.5, where we prove Bourgain's lifting theorem (Theorem 7.5.1) for Steinhaus martingales with values in L^1/H_0^1, and Kislyakov's embedding theorem (Theorem 7.5.3).

6.1.1 Kislyakov's Good Lambda Inequality

We begin by reviewing the Lorentz spaces $L^{p,\infty}(\mathbb{T})$ and $L^{p,1}(\mathbb{T})$, where $1 \leq p < \infty$. The following compilation of basic facts and concepts is based on Chapter 2 in Bennett and Sharpley (1988). Given a measurable $f\colon \mathbb{T} \to \mathbb{R}$, we let $f^\circ\colon [0,1] \to \mathbb{R}^+$ denote its nondecreasing rearrangement, defined by

$$f^\circ(t) = \inf\{\lambda \in \mathbb{R}^+ : m(\{|f| > \lambda\}) \leq t\},$$

where m denotes the normalized Haar measure on $\mathbb{T}$. The Hardy–Littlewood inequality asserts that

$$\int_{\mathbb{T}} |fg| dm \leq \int_0^1 f^\circ(t) g^\circ(t) dt. \tag{6.1.2}$$

For the proof of inequality (6.1.2) we refer to Bennett and Sharpley (1988, Chapter 2). Let $1 \leq p < \infty$. We define $L^{p,\infty}(\mathbb{T})$ to consist of those (measurable) $f\colon \mathbb{T} \to \mathbb{R}$ for which

$$\|f\|_{p,\infty} = \sup_{0 \leq t \leq 1} t^{1/p} f^\circ(t) < \infty.$$

We have $\|f\|_{p,\infty} = \sup m(\{|f| > \lambda\})^{1/p}\lambda$, where the supremum is extended over $\lambda \in \mathbb{R}^+$. Next, define $L^{p,1}(\mathbb{T})$ to consist of those measurable $f\colon \mathbb{T} \to \mathbb{R}$ for which

$$\|f\|_{p,1} = \int_0^1 t^{1/p} f^\circ(t) \frac{dt}{t} < \infty.$$

The Hardy–Littlewood inequality (6.1.2) yields the following analogue of Hölder's inequality. If $1 < p < \infty$, and if $1/r = 1 - 1/p$, then

$$\int_{\mathbb{T}} |fg| dm \le \|f\|_{p,\infty} \|g\|_{r,1}. \tag{6.1.3}$$

Fix a measurable set $E \subseteq \mathbb{T}$. Taking into account that $\|1_E\|_{r,1} = rm(E)^{1/r}$, we obtain

$$\int_E |f| dm \le r\|f\|_{p,\infty} m(E)^{1/r}. \tag{6.1.4}$$

We next define vector-valued Lorentz spaces. Let (Ω, μ) be a probability space. Let $1 \le p < \infty$ and $0 < \alpha \le 1$. Given a measurable

$$F\colon \mathbb{T} \times \Omega \to \mathbb{C},$$

and $z \in \mathbb{T}$, define $F(z)\colon \Omega \to \mathbb{C}$ by $F(z)(\omega) = F(z, \omega)$. For $q = 1$ or $q = \infty$, we let

$$L^{p,q}(L^\alpha) = L^{p,q}(\mathbb{T}, L^\alpha(\Omega, \mu))$$

consist of those (measurable) F for which $f(z) = \|F(z)\|_{L^\alpha}$ satisfies $f \in L^{p,q}(\mathbb{T})$, and we put

$$\|F\|_{L^{p,q}(L^\alpha)} = \|f\|_{p,q}.$$

Similarly we let $L^1(L^1) = L^1(\mathbb{T}, L^1(\Omega, \mu))$, denote the space of measurable functions $F\colon \mathbb{T} \times \Omega \to \mathbb{C}$ for which

$$\|F\|_{L^1(L^1)} = \int_{\mathbb{T}} \int_\Omega |F| d\mu dm < \infty.$$

The first step in the proof of the extrapolation theorem consists in showing that the tensor product $H \otimes J_\Omega$ extends to a bounded operator

$$H \otimes J_\Omega\colon L^1(L^1) \to L^{1,\infty}(L^\alpha),$$

for $0 < \alpha < 1$, where $J_\Omega\colon L^1(\Omega) \to L^\alpha(\Omega)$ denotes the formal inclusion map.

Theorem 6.1.1 *Let $0 < \alpha < 1$. There exists $1 < A = A(\alpha) < \infty$ such that $H \otimes J_\Omega$ extends to an operator from $L^1(L^1)$ into $L^{1,\infty}(L^\alpha)$ with norm bounded by A.*

Proof Fix $F \in L^1(L^1)$, let $G = (H \otimes J_\Omega)(F)$, and put $f(y) = \|F(y)\|_{L^1}$ and $g(y) = \|G(y)\|_{L^\alpha}$. In order to prove the theorem, we will verify that

$$\sup_{\lambda>0} \lambda m\{y \in \mathbb{T} : g(y) > \lambda\} \leq A \int_{\mathbb{T}} f(y)dm(y). \tag{6.1.5}$$

Fix $\lambda > 0$ and let $B = \{g > \lambda\}$. By arithmetic and Fubini's theorem, we get

$$\begin{aligned} \lambda m(B) &\leq \lambda^{(1-\alpha)} \int_B g(y)^\alpha dm(y) \\ &\leq \lambda^{(1-\alpha)} \left\| \left(\int_B |G(y)|^\alpha dm(y) \right) \right\|_{L^1}. \end{aligned} \tag{6.1.6}$$

For $\omega \in \Omega$ fixed, Kolmogorov's estimate for the Hilbert transform and estimate (6.1.4), applied to $y \to |G(y, \omega)|^\alpha$ and $p = 1/\alpha$, yields

$$\int_B |G(y,\omega)|^\alpha dm(y) \leq C_\alpha \left(\int_{\mathbb{T}} |F(y,\omega)| dm(y) \right)^\alpha (m(B))^{1-\alpha}. \tag{6.1.7}$$

Using (6.1.7), we continue estimating the last line of (6.1.6) as follows,

$$\begin{aligned} &\lambda^{(1-\alpha)} \left\| \int_B |G(y)|^\alpha dm(y) \right\|_{L^1} \\ &\quad \leq C_\alpha (\lambda m(B))^{(1-\alpha)} \left(\int_{\mathbb{T}} \|F(y)\|_{L^1} dm(y) \right)^\alpha \\ &\quad = C_\alpha (\lambda m(B))^{(1-\alpha)} \left(\int_{\mathbb{T}} f(y) dm(y) \right)^\alpha. \end{aligned} \tag{6.1.8}$$

It remains to combine (6.1.6) and (6.1.8) to obtain (6.1.5). For later purposes we select A and fix it to > 1. □

Kislyakov's distributional estimate: Here, we present a crucial link between the (desired) estimates for $H \otimes J_\Omega$ to those of the Hardy–Littlewood maximal function. Recall that for $f \in L^1(\mathbb{T})$, the Hardy–Littlewood maximal function is defined by

$$Mf(y) = \sup \int_I |f(\zeta)| dm(\zeta)/|I|,$$

where the supremum is extended over those intervals in $\mathbb{T}$ that contain $y \in \mathbb{T}$.

Given $F \in L^1(L^1)$ and $G = H \otimes J_\Omega(F)$ let $f(y) = \|F(y)\|_{L^1}$ and $g(y) = \|G(y)\|_{L^\alpha}$. Let $A = A(\alpha)$ such that (6.1.5) holds true.

Theorem 6.1.2 *There exists $A_1 = A_1(\alpha, A)$ so that for any $\eta > 0$ and $\lambda > 0$,*

$$m\{g > A_1\lambda, Mf < \eta\lambda\} \leq \eta m\{Mg > \lambda\}, \tag{6.1.9}$$

where Mf and Mg denote the Hardy–Littlewood maximal function of f and g, respectively.

Proof First, note that Mg is lower semicontinuous. That is, $\{Mg > \lambda\}$ is an open subset of $\mathbb{T}$, for any $\lambda > 0$. Hence there exists a sequence of pairwise disjoint intervals (I_j), such that $\{Mg > \lambda\} = \bigcup I_j$. Fix j and put $I = I_j$. Clearly we may assume that $m(I) \leq 1/2$. We denote by $(2I)$ the interval with the same midpoint as I and twice its measure. Since the expanded interval $(2I)$ intersects the set $\{Mg \leq \lambda\}$, we have

$$\int_{2I} g dm \leq 2m(I) \inf_{x \in 2I} Mg(x) \leq 2m(I)\lambda. \tag{6.1.10}$$

We next claim that there exists $A_1 = A_1(\alpha, A)$ such that for any $\eta > 0$ and $\lambda > 0$, we have

$$m(I \cap \{g > A_1\lambda,\ Mf < \eta\lambda\}) \leq \eta m(I). \tag{6.1.11}$$

For $\eta > 1$ there is nothing to prove. We now consider the case where $0 < \eta < 1$. In order to show condition (6.1.11) we will verify that there exists $A_1 = A_1(\alpha, A)$ such that, for $0 < \eta < 1$ and $\lambda > 0$, the following implication holds true:

$$\text{If} \quad I \cap \{Mf < \eta\lambda\} \neq \emptyset \quad \text{then} \quad m(I \cap \{g > A_1\lambda\}) \leq \eta m(I). \tag{6.1.12}$$

Step 1 Let $F_2 = F 1_{\mathbb{T}\setminus 2I}$ and $G_2 = \mathcal{H}F_2$. For $x \in \mathbb{T}$ we put $g_2(x) = \|G_2(x)\|_{L^\alpha}$. We proceed by verifying the following implication: If there exists $x_0 \in I$ with $Mf(x_0) < \eta\lambda$, then there exists $x_1 \in I$ such that, with $C_1 = 6(1 + (A\eta)^\alpha)^{1/\alpha}$, we have

$$g_2(x_1) \leq C_1\lambda.$$

To this end, let $F_1 = F - F_2, G_1 = \mathcal{H}F_1$, and, for $x \in \mathbb{T}$, let

$$f_1(x) = \|F_1(x)\|_{L^1}, \quad g_1(x) = \|G_1(x)\|_{L^\alpha}.$$

Since $F_1 = F 1_{2I}$, we have

$$\int_{\mathbb{T}} f_1 dm = \int_{2I} f dm \leq 2m(I) \inf_{x \in 2I} Mf(x) \leq 2m(I)Mf(x_0) \leq 2m(I)\eta\lambda. \tag{6.1.13}$$

The weak type inequality (6.1.5) gives an estimate between the distribution of g_1 and the integral of f_1 as follows,

$$m(I \cap \{g_1 > 6A\eta\lambda\}) \leq \frac{A}{6A\eta\lambda} \int_{\mathbb{T}} f_1 dm \leq \frac{1}{3} m(I). \tag{6.1.14}$$

Consequently, inequalities (6.1.14) and (6.1.10) imply that

$m(I \cap \{g_1 \leq 6A\eta\lambda\}) \geq (2/3)m(I)$ and $m(I \cap \{g \leq 6\lambda\}) \geq (2/3)m(I)$, respectively.

Comparing the lower measure estimates shows that their intersection is non-empty. Pick $x_1 \in I \cap \{g \le 6\lambda\} \cap \{g_1 \le 6A\eta\lambda\} \ne \emptyset$. By the quasi-triangle inequality in $L^\alpha = L^\alpha(\Omega,\mu)$, we get

$$g_2^\alpha(x_1) \le g^\alpha(x_1) + g_1^\alpha(x_1) \le (6\lambda)^\alpha(1 + (A\eta)^\alpha),$$

and hence,

$$g_2(x_1) \le C_1\lambda, \quad \text{where} \quad C_1 = 6(1 + (A\eta)^\alpha)^{1/\alpha}. \tag{6.1.15}$$

Step 2 Here, we show that there exists $C_2 > 0$ so that for any $x \in I$,

$$g_2^\alpha(x) \le \lambda^\alpha(C_1^\alpha + C_2^\alpha). \tag{6.1.16}$$

We first use the kernel estimates for the Hilbert transform to show that there exists $C_2 > 0$ so that for any $x \in I$,

$$\|G_2(x) - G_2(x_1)\|_{L^\alpha} \le C_2\eta\lambda. \tag{6.1.17}$$

As $\alpha < 1$ the $L^\alpha(\Omega,\mu)$ norm on left-hand side of (6.1.17) is bounded by the $L^1(\Omega,\mu)$ norm. Using the triangle inequality in $L^1(\Omega,\mu)$, the kernel estimates for the Hilbert transform, and recalling that $f(y) = \|F(y)\|_{L^1(\Omega,\mu)}$, we obtain that

$$\begin{aligned}\|G_2(x) - G_2(x_1)\|_{L^1} &= \left\| \int_{\mathbb{T}\setminus 2I} \left(\frac{1}{x-y} - \frac{1}{x_1 - y} \right) F(y)dm(y) \right\|_{L^1} \\ &\le \int_{\mathbb{T}\setminus 2I} \frac{|x - x_1|}{|x-y||x_1-y|} \|F(y)\|_{L^1} dm(y) \\ &\le C_2 \inf_{x\in I} Mf(x).\end{aligned} \tag{6.1.18}$$

Since $\inf_{x\in I} Mf(x) \le Mf(x_0) \le \eta\lambda$, condition (6.1.17) follows directly from (6.1.18). In summary, for $x \in I$ we have

$$\|G_2(x)\|_{L^\alpha}^\alpha \le \|G_2(x_1)\|_{L^\alpha}^\alpha + \|G_2(x) - G_2(x_1)\|_{L^\alpha}^\alpha,$$

and thereby we obtain (6.1.16).

Step 3 Put $C_3 = (C_1^\alpha + C_2^\alpha + 2^\alpha A^\alpha)^{1/\alpha}$. Clearly $C_3 > 1$, because we fixed $A > 1$ in Theorem 6.1.1. By (6.1.16),

$$g^\alpha(x) \le g_1^\alpha(x) + g_2^\alpha(x) \le g_1^\alpha(x) + \lambda^\alpha C_3^\alpha, \tag{6.1.19}$$

for $x \in I$. Put $A_1 = 2^{1/\alpha}C_3$. As $A_1^\alpha - C_3^\alpha = C_3^\alpha$, the pointwise estimates (6.1.19) yield the inclusion

$$I \cap \{g > A_1\lambda\} \subseteq I \cap \{g_1^\alpha > C_3^\alpha\lambda^\alpha\}.$$

Applying the weak type (1:1) estimate (6.1.5) to g_1 gives

$$m(I \cap \{g > A_1\lambda\}) \leq m(I \cap \{g_1 > C_3\lambda\}) \leq \frac{A}{C_3\lambda}\int_{\mathbb{T}} f_1 dm.$$

Since $2A \leq C_3$, by estimating f_1 with (6.1.13) we obtain $m(I \cap \{g > A_1\lambda\}) \leq \eta\lambda$. This verifies the implication (6.1.12), and consequently that the crucial estimates (6.1.11) hold true for any interval I in the disjoint cover of the open set $\{Mg \leq \lambda\}$.

Step 4 (Summing up) It remains to string together the estimates (6.1.11). We begin this process by noting that $A_1 > 1$ and that $m\{Mg \leq \lambda, g > A_1\lambda\} = 0$. Hence,

$$m\{g > A_1\lambda, Mf < \eta\lambda\} = m\{g > A_1\lambda,\ Mf < \eta\lambda,\ Mg > \lambda\}. \tag{6.1.20}$$

Let (I_j) be the sequence of pairwise disjoint intervals satisfying $\{Mg > \lambda\} = \bigcup I_j$. In view of the identity (6.1.20) we obtain,

$$m\{g > A_1\lambda, Mf < \eta\lambda\} = \sum_{j=1}^{\infty} m(I_j \cap \{g > A_1\lambda, Mf < \eta\lambda\}). \tag{6.1.21}$$

By the estimates (6.1.11), for $A_1 = A_1(\alpha, A)$ chosen in the previous step we have

$$\sum_{j=1}^{\infty} m(I_j \cap \{g > A_1\lambda, Mf < \eta\lambda\}) \leq \eta \sum_{j=1}^{\infty} m(I_j) \leq \eta m\{Mg > \lambda\}, \tag{6.1.22}$$

for any $\eta > 0$ and $\lambda > 0$. Combining (6.1.20)–(6.1.22) gives (6.1.9). □

Extrapolation for the Hilbert transform H: We now exploit Theorem 6.1.2 to obtain the main theorem in this section. We show that for any probability space (Ω, μ) the tensor product

$$H \otimes J_\Omega \colon L^p(L^1) \to L^p(L^\alpha)$$

defines a bounded operator, for $0 < \alpha < 1$ and $1 < p < \infty$, where $J_\Omega \colon L^1(\Omega) \to L^\alpha(\Omega)$ denotes the formal inclusion map, and where

$$L^p(L^1) = L^p(\mathbb{T}, L^1(\Omega, \mu)), \qquad L^p(L^\alpha) = L^p(\mathbb{T}, L^\alpha(\Omega, \mu)).$$

Theorem 6.1.3 *Let $1 < p < \infty$ and $0 < \alpha < 1$. There exists $A = A(p, \alpha)$ such that $H \otimes J_\Omega$ extends to a bounded operator from $L^p(L^1)$ into $L^p(L^\alpha)$ with norm $\leq A$.*

Proof Given $F \in L^1(L^1)$ and $G = H \otimes J_\Omega(F)$, let $f(y) = \|F(y)\|_{L^1}$ and $g(y) = \|G(y)\|_{L^\alpha}$. By Kislyakov's distributional estimate (6.1.9), there exists $A_1 < \infty$ such that

$$m\{g > A_1\lambda, Mf < \eta\lambda\} \leq \eta m\{Mg > \lambda\},$$

for any $\eta, \lambda > 0$. This gives

$$m\{g > A_1\lambda\} \leq m\{Mf > \eta\lambda\} + \eta m\{Mg > \lambda\}, \tag{6.1.23}$$

for $\lambda > 0$ and $\eta > 0$. Multiplying both sides of (6.1.23) by $p\lambda^{p-1}$, and integrating over $0 < \lambda < \infty$ gives

$$\int_{\mathbb{T}} g^p dm \leq \frac{A_1^p}{\eta^p} \int_{\mathbb{T}} (Mf)^p dm + \eta A_1^p \int_{\mathbb{T}} (Mg)^p dm.$$

In view of the L^p bounds for the Hardy–Littlewood maximal function we obtain $A_2 = A_2(\alpha, p)$ such that

$$\int_{\mathbb{T}} g^p dm \leq \frac{A_2}{\eta^p} \int_{\mathbb{T}} f^p dm + A_2\eta \int_{\mathbb{T}} g^p dm.$$

The above estimates hold for $0 < \eta < 1$. Given $1 < p < \infty$, we select $\eta > 0$ so small that the factor in front of the second integral on the right-hand side is bounded by 1/2. Thus we obtain $A = A(p, \alpha)$ such that

$$\int_{\mathbb{T}} g^p dm \leq A^p \int_{\mathbb{T}} f^p dm.$$

□

Extrapolation for $\mathbb{H}$: Recall that in Section 2.4.1 we defined the Hilbert transform on the infinite torus product $\mathbb{T}^{\mathbb{N}}$, and denoted it by $\mathbb{H}$. Consider $J_\Omega\colon L^1(\Omega) \to L^\alpha(\Omega)$, given by $J_\Omega(f) = f$, and let $\mathbb{H} \otimes J_\Omega$ be defined initially on the algebraic tensor product $L^p(\mathbb{T}^{\mathbb{N}}) \otimes L^1(\Omega)$. We will now prove estimates for $\mathbb{H} \otimes J_\Omega$ analogous to those of Theorem 6.1.3. This will be done below by applying the transfer method to trigonometric polynomials on $\mathbb{T}^{\mathbb{N}}$ with values in $L^1(\Omega)$. In the next theorem we employ abbreviations as follows:

$$L^p(L^1) = L^p(\mathbb{T}^{\mathbb{N}}, L^1(\Omega)) \quad \text{and} \quad L^p(L^\alpha) = L^p(\mathbb{T}^{\mathbb{N}}, L^\alpha(\Omega)).$$

Theorem 6.1.4 *Let $0 < \alpha < 1 < p < \infty$. There exists $A = A(p, \alpha)$ such that $\mathbb{H} \otimes J_\Omega$ extends to a linear operator form $L^p(L^1)$ into $L^p(L^\alpha)$ with norm bounded by A.*

Proof Let $X = L^1(\Omega)$, and let F be an X-valued trigonometric polynomial on $\mathbb{T}^{\mathbb{N}}$. To specify notation used below we fix $z \in \mathbb{T}^{\mathbb{N}}$ and let $\mathbb{H}F(z)$ be defined by

$$(\mathbb{H}F)(z)(\omega) = (\mathbb{H} \otimes J_\Omega F)(z, \omega),$$

where $\omega \in \Omega$. We now prepare for the application of the transfer method by recalling convenient notation introduced earlier. If $z \in \mathbb{T}^{\mathbb{N}}$, $y \in \mathbb{T}$, and if $M = (M_k)$ is a sequence of integers, then we put $y_k = y^{M(k)}$ and define $(z \cdot y^M) = (z_k y_k)$. By Proposition 2.4.6, there exists a sequence of integers $M = (M_k)$ such that the X-valued trigonometric polynomial

$$G(y, z, M) = F(z \cdot y^M), \qquad z \in \mathbb{T}^{\mathbb{N}}, y \in \mathbb{T},$$

satisfies the identity

$$(\mathbb{H}F)(z \cdot w^M) = (H(G(., z, M)))(w), \tag{6.1.24}$$

where the Hilbert transform H acts on the first variable of G. Note that the above equation expresses equality between elements of X. By rotational invariance and by Fubini's theorem we have

$$\int_{\mathbb{T}^{\mathbb{N}}} \| \mathbb{H}F(z)\|_\alpha^p d\mathbb{P}(z) = \int_{\mathbb{T}^{\mathbb{N}}} \int_{\mathbb{T}} \| \mathbb{H}F(z \cdot w^M)\|_\alpha^p \, d\mathbb{P}(z).$$

Fix now $z \in \mathbb{T}^{\mathbb{N}}$, invoke the transfer identity (6.1.24) and apply Theorem 6.1.3. This gives

$$\int_{\mathbb{T}} \| \mathbb{H}F(z \cdot w^M)\|_\alpha^p \, dm(w) \leq C_p^p \int_{\mathbb{T}} \|G(y, z, M)\|_X^p dm(y)$$

Integrating over $z \in \mathbb{T}^{\mathbb{N}}$ and using rotational invatiance of the Haar measure on $\mathbb{T}$ gives

$$\int_{\mathbb{T}^{\mathbb{N}}} \| \mathbb{H}F(z)\|_\alpha^p d\mathbb{P}(z) \leq C \int_{\mathbb{T}^{\mathbb{N}}} \int_{\mathbb{T}} \|F(z \cdot y^M)\|_X^p \, dm(y) d\mathbb{P}(z) \leq C \int_{\mathbb{T}^{\mathbb{N}}} \|F(z)\|_X^p d\mathbb{P}(z).$$

□

Remark: Recall that the Riesz projection onto Hardy martingales is given by $P = (1/2)(\mathrm{Id} - i\mathbb{H})$. Hence the conclusion of Theorem 6.1.4 holds true with $\mathbb{H}$ replaced by P. Similarly, we may replace $\mathbb{H}$ by the projection onto simple Hardy martingales P_0.

6.1.2 Garling's Good Lambda Inequality

In this section we present the stochastic approach to Theorem 6.1.3, due to D. J. H. Garling, and obtain vector-valued estimates for the stochastic Hilbert transform.

Let $(\Sigma, (\mathcal{F}_t), \mathbb{P})$ denote Wiener's filtered probability measure. Let $R = (R_t)$ be a real-valued, square-integrable martingale on Wiener space with stochastic integral representation $R_t = R_0 + \int_0^t Y_s dz_s + \overline{Y_s} d\overline{z}_s$. We recall that in Section 1.4 the stochastic Hilbert transform of R is defined by

$$\mathcal{H}R = i\int_0^\infty \overline{Y_s} d\overline{z}_s - Y_s dz_s. \tag{6.1.25}$$

Thus $\mathcal{H}R$ is again real valued, and there holds the identity $(\mathcal{H}R)_t = \mathcal{H}(R_t)$ for $t > 0$. In fact, it holds for any stopping time T on Σ.

Inequality (1.4.3) in Theorem 1.4.2 provides the starting point for the work in this section. It asserts that the stochastic Hilbert transform satisfies the weak type $(1 : 1)$ estimate

$$\sup_{\lambda \geq 0} \lambda \mathbb{P}\left\{\sup_{t \geq 0} |\mathcal{H}(R)_t| > \lambda\right\} \leq C\mathbb{E}|R|.$$

We let (Ω, μ) be a fixed probability space and let $J_\Omega \colon L^1(\Omega) \to L^\alpha(\Omega)$ denote the inclusion map. In this section we prove the result of Garling (Garling, D. J. H. $L^1(\Omega)/H^1(\Omega)$H is a Grothendieck space of cotype 2. pages 1–18, 1992, unpublished.) that the tensor product

$$\mathcal{H} \otimes J_\Omega \colon L^p(L^1) \to L^p(L^\alpha)$$

defines a bounded operator, for $0 < \alpha < 1$ and $1 < p < \infty$, and where

$$L^p(L^1) = L^p(\Sigma, L^1(\Omega, \mu)), \qquad L^p(L^\alpha) = L^p(\Sigma, L^\alpha(\Omega, \mu)).$$

The norm of $\mathcal{H} \otimes J_\Omega$, while depending on α and p, is independent on the fixed probability space (Ω, μ).

We begin by fixing $F \in L^1(\Sigma) \otimes L^1(\Omega)$ and $G = \mathcal{H} \otimes J_\Omega(F)$. For $y \in \Sigma$ we define $g_*(y)$ and $f(y)$ by

$$g_*(y) = \left\|\sup_{t>0} |G_t(y, \cdot)|\right\|_{L^\alpha(\Omega)} \quad \text{and} \quad f(y) = \|F(y, \cdot)\|_{L^1(\Omega)}. \tag{6.1.26}$$

The first step consists of proving that $\mathcal{H} \otimes J_\Omega$ extends boundedly to an operator from $L^1(L^1)$ to $L^{1,\infty}(L^\alpha)$. This is accomplished by the following theorem.

Theorem 6.1.5 *For $0 < \alpha < 1$ there exists $A = A(\alpha) > 0$ such that*

$$\sup_{\lambda > 0} \lambda \mathbb{P}\{g_* > \lambda\} \leq A\mathbb{E}(f).$$

Proof Fix $\lambda > 0$ and let $B = \{\omega \in \Sigma : g_*(\omega) > \lambda\}$. By arithmetic and Fubini's theorem we get

$$\begin{aligned}\lambda\mathbb{P}(B) &\leq \lambda^{(1-\alpha)} \int_{\Sigma} (1_B(y) g_*^{\alpha}(y)) d\mathbb{P}(y) \\ &\leq \lambda^{(1-\alpha)} \int_{\Omega} \int_{\Sigma} 1_B(y) \sup_{t>0} |G_t(y, \omega)|^{\alpha} d\mathbb{P}(y) d\mu(\omega).\end{aligned} \tag{6.1.27}$$

For $\omega \in \Omega$ fixed, the weak type estimate (1.4.3) for the stochastic Hilbert transform and the estimate (6.1.4), applied to $y \to \sup_{t>0} |G_t(y, \omega)|^{\alpha}$ and $p = 1/\alpha$, yields

$$\int_B \sup_{t>0} |G_t(y, \omega)|^{\alpha} d\mathbb{P}(y) \leq C_{\alpha} \left(\int_{\Sigma} |F(y, \omega)| d\mathbb{P}(y) \right)^{\alpha} (\mathbb{P}(B))^{1-\alpha}. \tag{6.1.28}$$

Inserting (6.1.28) to the last line of equation (6.1.27) and applying Hölder's inequality gives

$$\lambda\mathbb{P}(B) \leq C_{\alpha} (\lambda\mathbb{P}(B))^{(1-\alpha)} \left(\int_{\Sigma} \int_{\Omega} |F(y, \omega)| d\mu(\omega) d\mathbb{P}(y) \right)^{\alpha}. \tag{6.1.29}$$

It remains to cancel the factor $(\lambda\mathbb{P}(B))^{(1-\alpha)}$ on both sides of (6.1.29) and to raise the resulting equation to the power $1/a$. For later purposes we select and fix $A > 1$. □

Next, we prove good λ-inequalities for the stochastic Hilbert transform. As above, we first fix $F \in L^1(L^1)$, and let $G = \mathcal{H} \otimes J_{\Omega}(F)$. We need to introduce further notation as employed in the proof of Garling's good λ-inequalities. For $t > 0$, take the conditional expectations

$$F_t = \mathbb{E}(F|\mathcal{F}_t) \quad \text{and} \quad G_s = \mathbb{E}(G|\mathcal{F}_s), \quad 0 < s \leq t,$$

and put

$$g_t(y) = \left\| \sup_{s \leq t} |G_s(y, \cdot)| \right\|_{L^{\alpha}(\Omega)}, \quad f_t(y) = \|F_t(y, \cdot)\|_{L^1(\Omega)}, \quad y \in \Sigma. \tag{6.1.30}$$

Thus $(g_t(\cdot))$ is increasing, $(\mathcal{F}_t)$-adapted, and $g_*(y) = \lim_{t \to \infty} g_t(y)$, for $y \in \Sigma$. Finally, for a given $R \colon \Sigma \to [0, \infty]$ we define the sampled process g_R by

$$g_R(y) = \begin{cases} g_{R(y)}(y) & \text{if} \quad R(y) < \infty, \\ g_*(y) & \text{if} \quad R(y) = \infty. \end{cases}$$

The following theorem links F and $\mathcal{H} \otimes J_{\Omega}(F)$ by good λ-inequalities analogous to Theorem 6.1.2.

Theorem 6.1.6 *For $0 < \alpha < 1$ there exists $A_1 = A(\alpha, A(\alpha)) > 0$, such that for any $\eta > 0$ and $\lambda > 0$*

$$\mathbb{P}\left\{ g_* > A_1 \lambda \text{ and } \sup_{t>0} f_t \leq \eta\lambda \right\} < \eta \mathbb{P}\{g_* > \lambda\}. \tag{6.1.31}$$

Proof Fix $0 < \alpha < 1$. Choose $A > 1$ such that the conclusion of Theorem 6.1.5 holds true. Determine $A_1 > 0$ so that $2A \leq (2^{-1/\alpha}A_1 - 1)$. For $\eta > 0$ and $\lambda > 0$ we define the $(\mathcal{F}_t)$-stopping times, S and T, by

$$T = \inf\{t > 0 : g_t > A_1\lambda \text{ or } f_t > \eta\lambda\}, \quad S = \inf\{t > 0 : g_t > \lambda \text{ or } f_t > \eta\lambda\}.$$

Step 1 Clearly we have $S \leq T$. Let $X = \{\sup_{t>0} f_t \leq \eta\lambda\}$, then for $y \in X$,

$$\{g_* > A_1\lambda\} \cap X \subseteq \{T < \infty\} \cap X \subseteq \{S < T < \infty\} \cap X.$$

Observe that $\{g_t \leq \lambda\} \subseteq \{S = \infty\} \subseteq \{S = T\}$, and hence $\{S < T\} \subseteq \{g_* > \lambda\}$.

Step 2 Let G_T, G_S denote the sampling of the process (G_t) at the stopping times T and S, respectively. We define $G^{(1)} = G_T - G_S$, fix $y \in \Sigma$, and set

$$g_*^{(1)}(y) = \left\| \sup_{t>0} |G_t^{(1)}(y, \cdot)| \right\|_{L^\alpha(\Omega)}.$$

The quasi-triangle inequality in L^α gives $g_T \leq 2^{1/\alpha}\left(g_S + g_*^{(1)}\right)$. Consequently, we obtain

$$g_*^{(1)}(y) > 2^{-1/\alpha} g_T(y) - g_S(y) \geq 2^{-1/\alpha}A_1\lambda - \lambda,$$

for any $y \in \{S < T < \infty\} \cap X$. Recording the information obtained so far we note that

$$\left\{g_* > A_1\lambda \text{ and } \sup_{t>0} f_t \leq \eta\lambda\right\} \subseteq \left\{g_*^{(1)} > \left(2^{-1/\alpha}A_1 - 1\right)\lambda\right\}. \tag{6.1.32}$$

Step 3 Let F_T, F_S denote the sampling of the process (F_t) at the stopping times T and S, respectively. Define $F^{(1)} = F_T - F_S$, and set $f^{(1)} = \left\|F^{(1)}\right\|_{L^1(\Omega)}$. We have $|F_T| \leq \eta\lambda$ and $|F_S| \leq \eta\lambda$. Hence $f^{(1)} \leq 2\eta\lambda$, and clearly $F^{(1)} = 0$ on $\{S = T\}$. Taking into account that $\{S < T\} \subseteq \{g_* > \lambda\}$, we obtain

$$\mathbb{E}\left(f^{(1)}\right) \leq 2\eta\lambda\mathbb{P}\{S < T\} \leq 2\eta\lambda\mathbb{P}\{g_* > \lambda\}. \tag{6.1.33}$$

Step 4 Finally, we combine Theorem 6.1.5 and estimate (6.1.33) to obtain

$$\mathbb{P}\left\{g_*^{(1)} > \left(2^{-1/\alpha}A_1 - 1\right)\lambda\right\} \leq \frac{A\mathbb{E}\left(f^{(1)}\right)}{(2^{-1/\alpha}A_1 - 1)\lambda} \leq \frac{2A\eta}{(2^{-1/\alpha}A_1 - 1)}\mathbb{P}\{g_* > \lambda\}.$$

By our choice of A_1 we have $2A \leq \left(2^{-1/\alpha}A_1 - 1\right)$. Hence in view of (6.1.32), we completed the proof of (6.1.31). □

Extrapolation for $\mathcal{H}$: (Recall that $\mathcal{H}$ denotes the stochastic Hilbert transform defined by (6.1.25).) As above, $(\Sigma, \mathbb{P})$ denotes Wiener's space, and (Ω, μ) a second, fixed probability space. Accordingly, we shorten our notation to

$$L^p(L^1) = L^p(\Sigma, L^1(\Omega, \mu)), \qquad L^p(L^\alpha) = L^p(\Sigma, L^\alpha(\Omega, \mu)).$$

We show now that Theorem 6.1.6 and Doob's maximal inequality (1.1.27) imply that $\mathcal{H} \otimes J_\Omega$ defines a bounded operator from $L^p(L^1)$ to $L^p(L^\alpha)$, for $0 < \alpha < 1 < p < \infty$.

Theorem 6.1.7 *Let $0 < \alpha < 1 < p < \infty$. There exists $A = A(p, \alpha)$ such that $\mathcal{H} \otimes J_\Omega$ extends to a bounded operator from $L^p(L^1)$ into $L^p(L^\alpha)$ with norm $\le A$.*

Proof Let $0 < \alpha < 1$, $F \in L^1(L^1)$, and $G = \mathcal{H} \otimes J_\Omega(F)$. We define f_t by (6.1.30), f and g_* by (6.1.26). By Theorem 6.1.6, there exists $A_1 = A_1(\alpha) > 0$, such that

$$\mathbb{P}\{g_* > A_1 \lambda\} \le \mathbb{P}\left\{\sup_{t>0} f_t \le \eta\lambda\right\} + \eta \mathbb{P}\{g_* > \lambda\}, \tag{6.1.34}$$

for $\lambda > 0$ and $\eta > 0$. Multiplying both sides of (6.1.34) by $p\lambda^{p-1}$, and integrating over $0 < \lambda < \infty$ gives

$$\mathbb{E}\left(g_*^p\right) \le \eta^{-p} A_1^p \mathbb{E}\left(\sup_{t>0} f_t^p\right) + \eta A_1^p \mathbb{E}\left(g_*^p\right).$$

By Doob's inequality (1.1.27) for the martingale maximal function, we obtain $A_2 = A_2(\alpha, p)$ such that

$$\mathbb{E}\left(g_*^p\right) \le \eta^{-p} A_2^p \mathbb{E}\left(f^p\right) + \eta A_1^p \mathbb{E}\left(g_*^p\right). \tag{6.1.35}$$

Next, we choose $\eta > 0$ so small that the factor in front of the second integral on the right-hand side of (6.1.35) is bounded by $1/2$. This gives $A = A(p, \alpha)$ such that $\mathbb{E}\left(g_*^p\right) \le A^p \mathbb{E}\left(f^p\right)$. □

Remark: Garling's theorem (Theorem 6.1.7) implies Bourgain's theorem (Theorem 6.1.3). To see this, it suffices to recall the identity $Hu(z_\tau) = \mathcal{H}(u(z_\tau))$, for $u \in L^2(\mathbb{T})$, which was obtained in the discussion at the beginning of Section 1.4.

6.1.3 More on Extrapolation

We now exploit the theory of A_p-weights to obtain a second proof of Theorem 6.1.3. We let H denote the Hilbert transform on $\mathbb{T}$, let (Ω, μ) be a probability space, and let $J_\Omega \colon L^1(\Omega) \to L^\alpha(\Omega)$ denote the formal inclusion map. We will show that

$$H \otimes J_\Omega \colon L^{p,1}(L^1) \to L^{p,\infty}(L^\alpha)$$

defines a bounded operator, for $0 < \alpha < 1$ and $1 \leq p < \infty$, where

$$L^{p,1}(L^1) = L^{p,1}(\mathbb{T}, L^1(\Omega)) \quad \text{and} \quad L^{p,\infty}(L^\alpha) = L^{p,\infty}(\mathbb{T}, L^\alpha(\Omega)).$$

Importantly, the value $p = 1$ is included in this statement. Indeed, the case $p = 1$ was proven in Theorem 6.1.1. The proof presented below rests on the crucial idea of Kislyakov (1981) of passing to A_1 weights, and invoking the theorem of Hunt, Muckenhoupt, and Wheeden (1973).

Weighted weak type (1:1) estimates: The Hardy–Littlewood maximal operator is defined as

$$Mw(y) = \sup \int_I |w| dm |I|^{-1},$$

where the supremum is extended over those intervals $I \subseteq \mathbb{T}$ for which $y \in I$. Recall that the Hardy–Littlewood estimate for the maximal operator asserts that

$$\lambda m\{Mw \geq \lambda\} \leq 2 \int_{\{Mw>\lambda\}} wdm, \tag{6.1.36}$$

for any $w \in L^1(\mathbb{T})$. We say that a nonnegative $w \in L^1(\mathbb{T})$ satisfies the A_1 condition if there exists $C > 0$ such that

$$Mw(y) \leq Cw(y),$$

for $y \in \mathbb{T}$. Hunt, Muckenhoupt, and Wheeden (1973) proved that Kolmogorov's weak type (1:1) estimate holds true for any $w \in A_1$ in the following form: For $0 < \alpha < 1$ there exists $C_\alpha > 0$ such that for

$$\int_B |Hh|^\alpha dm_w \leq C_\alpha \left(\int_{\mathbb{T}} |h| dm_w \right)^\alpha (m_w(B))^{1-\alpha},$$

for $h \in L^1(\mathbb{T}, m_w)$, and $B \subseteq \mathbb{T}$, where the measure m_w is defined by integration,

$$m_w(B) = \int_B w(y) dm(y), \qquad B \subseteq \mathbb{T}.$$

We begin with a short observation stating that the Hardy–Littlewood maximal function preserves the class of Lorentz spaces $L^{q,\infty}(\mathbb{T})$.

Lemma 6.1.8 *Let $1 < q < \infty$ and $g \in L^{q,\infty}(\mathbb{T})$, then*

$$\|Mg\|_{q,\infty} \leq 2q/(q-1)\|g\|_{q,\infty}.$$

Proof Let $\lambda > 0$, assume that $g > 0$, and put $E = \{Mg \geq \lambda\}$. Using the Hardy–Littlewood inequality (6.1.36), followed by the duality between the Lorentz spaces gives

$$\lambda m(E) \leq 2 \int_E gdm \leq 2\|1_E\|_{p,1}\|g\|_{q,\infty},$$

where $1/p+1/q = 1$. Taking into account that $\|1_E\|_{p,1} = pm(E)^{1/p}$ and invoking arithmetic we obtain

$$\lambda m(E)^{1-1/p} = (2p)\|g\|_{q,\infty}.$$

In summary, since $1 - 1/p = 1/q$ and $p = q/(q-1)$ we have

$$\|Mg\|_{q,\infty} = \sup_{\lambda>0} \lambda m\{Mg \geq \lambda\}^{1/q} \leq 2q/(q-1)\|g\|_{q,\infty},$$

as claimed. □

Theorem 6.1.9 *Let $1 \leq p < \infty$ [sic!] and $0 < \alpha < 1$. There exists $A = A(p, \alpha)$ so that for any probability space (Ω, μ) the tensor product*

$$H \otimes J_\Omega$$

extends to a bounded operator from $L^{p,1}(L^1)$ into $L^{p,\infty}(L^\alpha)$ with norm $\leq A$.

Proof Recall that the case $p = 1$ is already covered by Theorem 6.1.1. Now we consider $p > 1$. Fix $z \in \mathbb{T}$. We are given L^1-valued trigonometric polynomials F and $G = H \otimes J_\Omega F$. Define

$$F(z)(\omega) = F(z, \omega),$$

where $\omega \in \Omega$ and put $f(z) = \|F(z)\|_{L^1}$. Similarly, we let $g(z) = \|G(z)\|_{L^\alpha}$. We will prove that there exists $A = A(p, \alpha)$ so that

$$\|g\|_{p,\infty} \leq A\|f\|_{p,1}, \tag{6.1.37}$$

where the norms are taken in $L^{p,1}(\mathbb{T})$ and $L^{p,\infty}(\mathbb{T})$, respectively

Step 1 We start with the idea, due to Kislyakov (1981) of producing a weight on which the action of the Hardy–Littlewood maximal operator stabilized. Since we assumed $1 < p < \infty$, the conjugate exponent $q = p/(p-1)$ satisfies $1 < q < \infty$. Set $g_0 = g^{p-1}$, and put $g_{n+1} = M(g_n)$, for $n \in \mathbb{N}_0$. Note that

$$\|g_0\|_{q,\infty} = \|g\|_{p,\infty}^{p-1}.$$

Choose $0 < \delta < 1$ such that $2\delta q < 1/4$. Then

$$w = \sum_{n=0}^{\infty} \delta^n g_n \quad \text{satisfies} \quad g_0 \leq w, \quad \text{and} \quad Mw \leq Cw,$$

where $C = C(\delta) < \infty$. By Lemma 6.1.8, the Hardy–Littlewood maximal function M preserves the space $L^{q,\infty}(\mathbb{T})$ and

$$\|w\|_{q,\infty} \leq C\|g\|_{p,\infty}^{p-1}. \tag{6.1.38}$$

Fix $\lambda > 0$ and let $B = \{g > \lambda\}$. By Chebyshev's inequality and arithmetic we get

$$\lambda^p m(B) \leq \lambda^p(\lambda^{1-p}) \int_B g(y)^{p-1} dm(y) \leq \lambda \int_B w(y) dm(y).$$

Define the weighted measure $m_w(B) = \int_B w(y)dm(y)$. Thus we restate the above estimate as $\lambda^p m(B) \leq \lambda m_w(B)$.

Step 2 Next, we prove that the following weighted weak type 1:1 estimate holds true,

$$\lambda m_w(B) \leq C \int_{\mathbb{T}} f(y) dm_w(y), \qquad B = \{g > \lambda\}. \tag{6.1.39}$$

Recall the convention $L^1 = L^1(\Omega, \mu)$. Let $B = \{g > \lambda\}$. Again, by arithmetic and Fubini's theorem,

$$\begin{aligned} \lambda m_w(B) &\leq \lambda^{(1-\alpha)} \int_B g(y)^\alpha dm_w(y) \\ &\leq \lambda^{(1-\alpha)} \left\| \left(\int_B |G(y)|^\alpha dm_w(y) \right) \right\|_{L^1}. \end{aligned} \tag{6.1.40}$$

Now recall that we defined w to be an A_1 weight. Hence for $\omega \in \Omega$ fixed, by the theorem of Hunt, Muckenhoupt, and Wheeden (1973), we have the following weighted weak type estimate for the Hilbert transform,

$$\int_B |G(y,\omega)|^\alpha dm_w(y) \leq C_\alpha \left(\int_{\mathbb{T}} |F(y,\omega)| dm_w(y) \right)^\alpha (m_w(B))^{1-\alpha}. \tag{6.1.41}$$

Inserting (6.1.41) into (6.1.40) and applying Hölder's inequality gives

$$\begin{aligned} &\lambda^{(1-\alpha)} \left\| \int_B |G(y)|^\alpha dm_w(y) \right\|_{L^1} \\ &\quad \leq C_\alpha (\lambda m_w(B))^{(1-\alpha)} \left(\int_{\mathbb{T}} \|F(y)\|_{L^1} dm_w(y) \right)^\alpha \\ &\quad = C_\alpha (\lambda m_w(B))^{(1-\alpha)} \left(\int_{\mathbb{T}} f(y) dm_w(y) \right)^\alpha. \end{aligned} \tag{6.1.42}$$

It remains to combine (6.1.40) and (6.1.42) to obtain the estimate (6.1.39).

Step 3 Here we exploit (6.1.3), which relates Lorentz spaces $L^{p,1}(\mathbb{T})$ and $L^{q,\infty}(\mathbb{T})$. We claim that

$$\int_{\mathbb{T}} f(y) dm_w(y) \leq \|f\|_{p,1} \|g\|_{p,\infty}^{p-1},$$

where the norms are taken in $L^{p,1}(\mathbb{T})$ and $L^{p,\infty}(\mathbb{T})$, respectively. To see this, we apply the Hardy–Littlewood inequality (6.1.3) to obtain

$$\int_{\mathbb{T}} f(y)w(y)dm(y) \leq \|f\|_{p,1}\|w\|_{q,\infty}.$$

Recall in Step 1, equation (6.1.38), we proved $\|w\|_{q,\infty} \leq C\|g\|_{p,\infty}^{p-1}$.

Summary: The estimates obtained in Step 1 through Step 4 yield

$$\|g\|_{p,\infty}^{p} \leq C \sup \lambda m_w\{g > \lambda\} \leq C\|f\|_{p,1}\|g\|_{p,\infty}^{p-1}.$$

Cancelling $\|g\|_{p,\infty}^{p-1}$ gives the result $\|g\|_{p,\infty} \leq A\|f\|_{p,1}$, proving (6.1.37). □

6.2 Reflexive Subspaces of L^1

In this section we consider reflexive subspaces Y of L^1 and operators on $L^p(Y)$. We prove that reflexive subspaces of L^1 are characterized by weighted L^p-inequalities (Rosenthal's theorem (Theorem 6.2.3)). As a consequence, the vector-valued Steinhaus projection and Hilbert transform are bounded on $L^p(Y)$ for every reflexive subspace $Y \subset L^1$. Our work in Chapter 7 rests on the results established here.

Let (Σ, σ) and (Ω, μ) be finite measure spaces. Let

$$L^p(L^s) = L^p((\Sigma, \sigma), L^s(\Omega, \mu))$$

denote the mixed-norm space consisting of functions $F \colon \Sigma \times \Omega \to \mathbb{C}$ measurable with respect to the product sigma-algebra such that

$$\|F\|_{L^p(L^s)} = \left(\int_{\Sigma}\left(\int_{\Omega} |F(z,\omega)|^s d\mu(\omega)\right)^{p/s} dm(z)\right)^{1/p} < \infty.$$

Given a closed subspace $Y \subseteq L^1(\Omega, \mu)$, the space $L^p(Y)$ consists of those $F \in L^p(L^1)$ that satisfy $F(z) \in Y$, for σ almost all $z \in \Sigma$ where $F(z) \colon \Omega \to \mathbb{C}$ is defined by the equation $F(z)(\omega) = F(z, \omega)$. We put

$$\|F\|_{L^p(Y)} = \left(\int_{\Sigma}\left(\int_{\Omega} |F(z,\omega)| d\mu(\omega)\right)^{p} dm(z)\right)^{1/p}.$$

Now fix a bounded linear operator R on $L^p(\Sigma, \sigma)$ and let J_Ω denote the inclusion map $J_\Omega(x) = x$. We consider the following properties of $\mathcal{R} = R \otimes J_\Omega$:

Property A $\mathcal{R}$ extends to a bounded operator on mixed norm spaces,

$$\|\mathcal{R}\colon L^p(L^q) \to L^p(L^q)\| \le A(p,q), \tag{6.2.1}$$

where $A(p,q) < \infty$ for $1 < p, q < \infty$.

Property B $\mathcal{R}$ extends to a bounded operator on

$$\left\|\mathcal{R}\colon L^p(L^1) \to L^p\,(L^\alpha)\right\| \le A(p,\alpha), \tag{6.2.2}$$

where $A(p,\alpha) < \infty$ for $0 < \alpha < 1$ and $1 < p < \infty$.

The main theorems of this section imply that under each of the above assumptions, the restricted operator $\mathcal{R}\colon L^p(Y) \to L^p(Y)$ is bounded, for any reflexive subspace Y of L^1, with norm $\le A(Y,p)$.

6.2.1 Uniform Integrability

Let (Ω,μ) be a probability space and let V be a subset of $L^1 = L^1(\Omega,\mu)$. We say that V is uniformly integrable if

$$\lim_{t\to\infty} \sup_{y\in V} \int_\Omega 1_{\{|y|>t\}}|y|d\mu = 0.$$

By the Dunford–Pettis theorem (Theorem 1.1.22) a subset of L^1 is uniformly integrable if and only if it is relatively compact for the $\sigma(L^1, L^\infty)$ topology. Theorem 1.1.23, which is a basic result from the earliest period of functional analysis, asserts that any bounded subset of a reflexive Banach space Y is relatively compact for the $\sigma(Y, Y^*)$ topology. In summary, a bounded subset, contained in a reflexive subspace of L^1, is necessarily uniformly integrable.

Proposition 6.2.1 *Let Y be a reflexive subspace of $L^1(\Omega,\mu)$. Then for any bounded sequence $\{y_n\} \subset Y$, convergence in measure implies convergence in the norm of $L^1(\Omega,\mu)$. For any $0 < \alpha < 1$, there exists $C = C(Y,\alpha)$ so that*

$$\|y\|_{L^1} \le C\|y\|_{L^\alpha}, \qquad y \in Y. \tag{6.2.3}$$

Proof Uniform integrability generalizes Lebesgue's dominated convergence theorem as follows: If a sequence $\{y_n\}$ of integrable random variables on (Ω,μ) converges in measure, and if $\{y_n\}$ is a uniformly integrable subset of $L^1(\Omega,\mu)$, then $\{y_n\}$ converges in the norm of $L^1(\Omega,\mu)$. Hence if $y_n \to 0$ in measure, then $\|y_n\|_{L^1} \to 0$.

Next we prove that (6.2.3) holds true. If not, there would exist a sequence $y_n \in Y$ so that $\|y_n\|_{L^1} = 1$ and $\|y_n\|_{L^\alpha} \longrightarrow 0$, as $n \longrightarrow \infty$. For any $\epsilon > 0$ we have

$$\mu\{|y_n| > \epsilon\} \le \epsilon^{-\alpha}\|y_n\|_{L^\alpha}^\alpha.$$

Thus we obtained a bounded sequence in a reflexive subspace of L^1 that converges in measure to 0 but not in the norm of $L^1(\Omega)$. □

Extrapolation

Let (Σ, σ) and (Ω, μ) be measure spaces. Let R be a bounded operator on $L^p(\Sigma, \sigma)$. Let $J_\Omega \colon L^1(\Omega) \to L^\alpha(\Omega)$ be the inclusion map given by $J_\Omega(f) = f$. We assume that $\mathcal{R} = R \otimes J_\Omega$ extends to a bounded operator

$$\left\| \mathcal{R} \colon L^p(L^1) \to L^p(L^\alpha) \right\| \le A(\alpha, p), \tag{6.2.4}$$

where $A(\alpha, p) < \infty$ for $0 < \alpha < 1$ and $1 < p < \infty$. Given a subspace $Y \subseteq L^1(\Omega, \mu)$, the norm on $L^p(Y)$ is the one induced by $L^p(L^1)$.

Theorem 6.2.2 *Let Y be a reflexive subspace of $L^1(\Omega, \mu)$. Assume that $\mathcal{R}$ satisfies* (6.2.4). *There exists $A = A(Y, p)$ so that the restriction of $\mathcal{R}$ to $L^p(Y)$ is bounded as follows*

$$\left\| \mathcal{R}_{|L^p(Y)} \colon L^p(Y) \to L^p(L^1) \right\| \le A. \tag{6.2.5}$$

Proof Fix $g \in L^p \otimes Y$. Then $\mathcal{R}(g)(z) \in Y$ for any $z \in \mathbb{T}$. Applying the estimate (6.2.3) gives $C = C(Y)$ such that

$$\int_\Omega |\mathcal{R}(g)(z)| d\mu \le C \left(\int_\Omega |\mathcal{R}(g)(z)|^\alpha d\mu \right)^{1/\alpha}.$$

Raising both sides to the pth power, and averaging over $z \in \mathbb{T}$, finally gives

$$\|\mathcal{R}g\|_{L^p(L^1)} \le C \|\mathcal{R}g\|_{L^p(L^\alpha)}. \tag{6.2.6}$$

By (6.2.4), we have

$$\|\mathcal{R}g\|_{L^p(L^\alpha)} \le A \|g\|_{L^p(L^1)}. \tag{6.2.7}$$

Combining inequalities (6.2.6) and (6.2.7) gives (6.2.5), where $A(Y, p, \alpha) = C(Y) A(\alpha, p, \Omega)$, for any $g \in L^p \otimes Y$. To complete the proof it suffices to note that $L^p \otimes Y$ is a dense subspace of $L^p(Y)$. □

Remarks: We have obtained an extrapolation theorem for operators satisfying Property B. By Theorem 6.1.3, Theorem 6.1.4, and Theorem 6.1.7, respectively, the Hilbert transform, the Riesz projection onto Hardy martingales, and the stochastic Riesz transform satisfy the estimate (6.2.4) with

$$A(p, \alpha) \le C_p/(1 - \alpha), \quad C_p \le Cp^2/(p - 1).$$

It is important to point out that the Steinhaus projection does not satisfy (6.2.4).

6.2.2 Rosenthal's L^p Theorem

We now set out to obtain extrapolation theorems that are applicable to the vector-valued Steinhaus projection. (More generally to operators satisfying the assumptions called Property A at the beginning of Section 6.2.) We will rest the proof on Theorem 6.2.3, Rosenthal's characterization reflexive L^1-subspaces.

Reverse-Hölder Estimates

This section contains the proof of Rosenthal's L^p theorem. We begin by commenting on the title of the section and recall a simple consequence of Hölder's inequality. Let $1 < r < \infty$ and pick $g \neq 0$ in the unit ball of L^r. Fixing $x \in L^q$ with $1/q + 1/r = 1$ and writing $x = g \cdot (x/g)$, we obtain by Hölder's inequality that

$$\int_\Omega |x| d\mu \leq \left(\int_\Omega \left| \frac{x}{g} \right|^q d\mu \right)^{1/q}, \tag{6.2.8}$$

for any $x \in L^q(\Omega, \mu)$ and $g \in L^r(\Omega, \mu)$ with $\|g\|_r \leq 1$. Rosenthal's L^p theorem simply states that a reverse-Hölder estimate holds true for any reflexive subspace of L^1.

Theorem 6.2.3 (Rosenthal's theorem) *Let Y be a reflexive subspace of $L^1(\Omega, \mu)$. Then there exists $1 < r < \infty$ and g in the unit sphere of $L^r(\Omega, \mu)$ such that*

$$\left(\int_\Omega \left| \frac{x}{g} \right|^q d\mu \right)^{1/q} \leq C \int_\Omega |x| d\mu, \qquad x \in Y, \tag{6.2.9}$$

where $C = C(Y, r)$ and $1/q + 1/r = 1$.

Comment: We may paraphrase Rosenthal's L^p theorem by saying that for any reflexive subspace Y of L^1 there exists $1 < q < \infty$ such that Y is isomorphic to a subspace of L^q. To wit, by Theorem 6.2.3, there exists $g \in L^r(\Omega, \mu)$ such that putting $T_{(1/g)}(x) = x/g$ defines a bounded multiplication operator

$$T_{(1/g)} : Y \to L^q(\Omega, \mu). \tag{6.2.10}$$

By (6.2.8), $T_{(1/g)}$ is invertible on its range. Thus the isomorphism between Y and a closed subspace of $L^q(\Omega, \mu)$ is provided by $T_{(1/g)}$ restricted to Y, and T_g restricted to $T_{(1/g)}(Y)$.

Outline of the proof: We follow Maurey's approach to Theorem 6.2.3. The proof consists of two distinct components. First, seeking a contradiction, we fix $1 < q$ (close to 1), and assume that for any $N > 1$ and any g in the unit ball of L^r with $1/r + 1/q = 1$, there exists $y \in Y$ such that the reverse-Hölder estimate (6.2.9) fails. Lemma 6.2.5, which encapsulates the hard, quantitative aspects of the proof, shows that then there exist $y_1, \ldots, y_n \in Y$, with $\|y_i\|_{L^1} = 1$, such that

$$\int_\Omega \left| \sum_{i=1}^n c_i y_i \right| d\mu \geq 1/2 \left(\sum_{i=1}^n |c_i|^q \right)^{1/q},$$

for any $(c_i) \in \mathbb{R}^n$ and $n \in \mathbb{N}$. A separate argument is provided for the proof that the lower estimates above contradict the fact that any bounded subset of Y is uniformly integrable.

We begin with the Dunford–Pettis characterization of uniformly integrable subsets in L^1. The proof below uses an idea of Dor (1975).

Proposition 6.2.4 *Let $V \subset L^1(\Omega, \mu)$ and let $A \geq 1$ be fixed. Assume that, for any $n \in \mathbb{N}$, there exist $y_1, \ldots, y_n \in V$ such that*

$$\frac{1}{A} \sum_{i=1}^n |c_i| \leq \int_\Omega \left| \sum_{i=1}^n c_i y_i \right| d\mu \leq \sum_{i=1}^n |c_i|, \tag{6.2.11}$$

for any $c_i \in \mathbb{R}$. Then V is not a uniformly integrable subset of $L^1(\Omega, \mu)$.

Proof Assume to the contrary that V is uniformly integrable. This implies that there exists $\delta > 0$ such that

$$\int_B |y| d\mu < \frac{1}{4A^2}, \tag{6.2.12}$$

for any $y \in V$, and $B \subseteq \Omega$ with $\mu(B) < \delta$. We fix now $n \subset \mathbb{N}$ large enough such that $(2A^2 + 1)/n < \delta$. By assumption on V, there exist $y_i \in V$ with $\|y_i\|_{L^1} = 1$, for $i \leq n$, such that

$$\frac{n}{A} \leq \int_\Omega \left| \sum_{i=1}^n \varepsilon_i y_i \right| d\mu, \tag{6.2.13}$$

for any choice of $\varepsilon_i \in \{-1, 1\}$. Averaging (6.2.13) over $(\varepsilon_i) \in \{-1, 1\}^n$, using Khintchine's inequality and applying the Cauchy–Schwarz inequality gives

$$\frac{n}{A} \leq \int_\Omega \left(\sum_{i=1}^n |y_i|^2 \right)^{1/2} d\mu \leq \left(\int_\Omega \max_{i \leq n} |y_i| d\mu \right)^{1/2} \left(\int_\Omega \sum_{i=1}^n |y_i| d\mu \right)^{1/2}. \tag{6.2.14}$$

Recalling that $\int_\Omega |y_i| d\mu = 1$, and cancelling the factor $n^{1/2}$ yields the crucial inequality

$$\frac{n}{A^2} \leq \int_\Omega \max_{i \leq n} |y_i| d\mu \leq n. \tag{6.2.15}$$

Choose a partition of Ω into pairwise disjoint measurable sets $D_1, \ldots, D_n$ such that $\max_{j \le n} |y_j(t)| = |y_i(t)|$, for $t \in D_i$ and $i \le n$. In view of inequality (6.2.15), this gives

$$\frac{n}{A^2} \le \sum_{i=1}^{n} \int_{D_i} |y_i| d\mu \le n. \tag{6.2.16}$$

By (6.2.16) there exists a subset $J \subseteq \{1, \ldots, n\}$ with cardinality at least $n/(2A^2 + 1)$, such that

$$\int_{D_i} |y_i| d\mu \ge \frac{1}{2A^2},$$

for $j \in J$. Finally, since the (D_i) are pairwise disjoint, there exists $j_1 \in J$ such that $\mu(D_{j_1}) < (2A^2 + 1)/n$. In view of (6.2.12) we have arrived at a contradiction. □

Remark: The converse of Proposition (6.2.4) holds true as well. If V is not a uniformly integrable subset of $L^1(\Omega, \mu)$ then there exists $\delta > 0$, a sequence (f_i) in V, and a sequence of sets (A_i) satisfying $\mu(A_i) \le 1/i$ such that $\int_{A_i} |f_i| d\mu > \delta$. Passing to a subsequence, we may assume that putting $B_i = A_i \setminus \bigcup_{j=i+1}^{\infty} A_j$ gives

$$\int_{B_i} |f_i| d\mu > \delta.$$

Since (B_i) are pairwise disjoint measurable sets, $y_i = f_i / \|f_i\|_1$, $i \le n$, satisfies inequality (6.2.11) with $A = 1/\epsilon$, for any $n \in \mathbb{N}$. See Wojtaszczyk (1991).

Maurey's variational problem: Extending the scope of our considerations, we introduce the following variational problem. Let u be an operator on a Banach space X with values $L^1(\Omega, d\mu)$, and let $1 < q < \infty$. We define

$$C_{1,q}\left(u \colon X \to L^1(\Omega, \mu)\right)$$

to be the *infimum* of those constants C for which there exists $g \in L^r(\Omega, \mu)$ with $1/r + 1/p = 1$ and $\int_\Omega |g|^r d\mu = 1$ such that

$$\left(\int_\Omega \left| \frac{u(x)}{g} \right|^q d\mu \right)^{1/q} \le C \|x\|_X, \qquad x \in X. \tag{6.2.17}$$

The problem consists in showing that the infimum defining $C_{1,q}(u)$ is attained for some g in the unit sphere of $L^r(\Omega, \mu)$. Its solution by Maurey (1973a, 1974) is presented in Theorem 6.2.8 below.

Assuming that Theorem 6.2.8 holds true, we now turn to the central component in the proof of Theorem 6.2.3, which is Lemma 6.2.5, known as Rosenthal's lemma. Its proof provides an analysis of operators for which $C_{1,q}(u)$ is exceedingly large.

Here starts Maurey's proof of Rosenthal's lemma: Fix $n \in \mathbb{N}$, $1 < q < \infty$, and $\epsilon > 0$. Then specify the parameters arising in the proof of Lemma 6.2.5 in the following order:

(i) Choose $0 < \delta < 1$ (very close to 1) such that

$$\left((n-1)(1-\delta^q) + (1-\delta)^q\right)^{1/q} < \epsilon n^{-1+1/q}. \tag{6.2.18}$$

(ii) Choose $0 < \sigma < 1$ small enough such that

$$\delta < (1-\sigma)^{1-1/q}. \tag{6.2.19}$$

(iii) In view of (6.2.19) choose K large enough such that

$$(1-\sigma)^{q-1}\left(1-(n/\sigma K)^{q-1}\right) > \delta^q. \tag{6.2.20}$$

(iv) Finally, choose $\overline{N} = \overline{N}(n, \epsilon, q)$ very large such that

$$(1-\sigma)^{q-1}\left(1-(n/\sigma K)^{q-1}\right)N^q - K^{q-1} > \delta^q N^q, \tag{6.2.21}$$

for any $N \geq \overline{N}$.

Lemma 6.2.5 (Rosenthal's lemma) *Let $n \in \mathbb{N}$, $1 < q < \infty$, and $\epsilon > 0$. Let $\overline{N} = \overline{N}(n, \epsilon, q)$ be specified by* (6.2.21) *and let Y be a finite-dimensional normed space. Then, for any operator*

$$u\colon Y \to L^1(\Omega, \mu)$$

satisfying $\|u\| \leq 1$ and $C_{1,q}(u) > \overline{N}$, there exist $y_i \in Y$ with $\|y_i\|_Y = 1$, $i \leq n$, such that

$$\left\|\sum_{i=1}^{n} c_i x_i\right\|_Y \geq (1-\epsilon)\left(\sum_{i=1}^{n} |c_i|^q\right)^{1/q},$$

for $(c_i) \in \mathbb{R}^n$.

Proof Let $1 < q < \infty$. Fix an operator $u\colon Y \to L^1(\Omega, \mu)$ with $\|u\| = 1$ and $C_{1,q}(u) = N$. Clearly, $N < \infty$, since Y is finite dimensional. By Theorem 6.2.8 below, the infimum defining $C_{1,q}(u)$ is attained for some g in the unit sphere of $L^r(\Omega, \mu)$ $(1/r + 1/q = 1)$. That is,

$$\int_\Omega \left|\frac{u(y)}{g}\right|^q d\mu \leq N^q \|y\|_Y^q, \qquad y \in Y. \tag{6.2.22}$$

Step 1 Put $h = |g|^r$, form the probability measure $\nu = h\mu$, and define the operator $u_1 : Y \to L^1(\Omega, \nu)$ by

$$u_1(y)(t) = h^{-1}(t)u(y)(t),$$

for $y \in Y$ and $t \in \Omega$. Clearly, we have $\|u_1\| = \|u\| = 1$ and

$$\int_\Omega |u_1(y)|^q \, d\nu = \int_\Omega \left|\frac{u(y)}{g}\right|^q d\mu, \tag{6.2.23}$$

for any $y \in Y$, and

$$C_{1,q}\left(u_1 : Y \to L^1(\Omega, \nu)\right) = C_{1,q}\left(u\colon Y \to L^1(\Omega, \mu)\right).$$

In the second step, we will repeatedly use that $C_{1,q}(u_1) = N$ and that for any w in the unit sphere of $L^r(\Omega, \mu)$, there exists $y \in Y$, satisfying $\|y\| = 1$, such that

$$\int_\Omega |u_1(y)|^q \, d\nu \leq N^q \quad \text{and} \quad \int_\Omega \left|\frac{u_1(y)}{w}\right|^q d\nu \geq N^q. \tag{6.2.24}$$

Step 2 Here, we go to the heart of the matter and prove the following statement: For every $A \subset \Omega$ satisfying $\nu(A) < n/K$, there exist $B \subset \Omega\backslash A$ and $y \in Y$ such that, with $f = u_1(y)$,

$$\|y\|_Y = 1, \quad \nu(B) < 1/K, \quad \int_B |f|^q \, d\nu > \delta^q N^q, \quad \text{and} \quad \int_\Omega |f|^q d\nu \leq N^q. \tag{6.2.25}$$

Recall that $\delta > 0$ and $K > 0$ are specified by (6.2.18) and (6.2.20), respectively. To this end, fix $A \subset \Omega$ with $\nu(A) < n/K$. Recall that $\sigma > 0$ is selected by (6.2.19). Next, define

$$w = (\sigma K/n)^{1/r} 1_A + c^{1/r} 1_{\Omega\backslash A}, \tag{6.2.26}$$

where

$$c = (1 - (\sigma K/n)\nu(A))/(1 - \nu(A)).$$

We verify directly that $\int w^r d\nu = 1$. Since $\|u_1\| = 1$ by (6.2.22) and (6.2.24), we find $y \in Y$ with $\|y\| = 1$ such that $f = u_1(y)$ satisfies the following estimates:

$$\int_\Omega |f| \, d\nu \leq 1, \quad \int_\Omega |f|^q \, d\nu \leq N^q, \quad \text{and} \quad \int_\Omega \left|\frac{f}{w}\right|^q d\nu \geq N^q. \tag{6.2.27}$$

Define $B = \{|f| \geq K\}\backslash A$. To check that $\nu(B) < 1/K$ is easy. Indeed, since $B \subseteq \{|f| > K\}$, simply using the first estimate in (6.2.27) gives $\nu(B) < 1/K$.

We verify next that the third estimate in (6.2.25) holds true. We begin with

$$\int_{\{|f|<K\}} |f| d\nu \leq K^{q-1} \int_\Omega |f| d\nu \leq K^{q-1}. \tag{6.2.28}$$

Unwinding the definition of w, and invoking the second estimate in (6.2.27), gives

$$\int_A \left|\frac{f}{w}\right|^q d\nu = (\sigma K/n)^{1-q} \int_A |f|^q \, d\nu < (\sigma K/n)^{1-q} N^q. \tag{6.2.29}$$

In view of the third inequality in (6.2.27), the upper bound of (6.2.29) gives the following crucial lower estimate,

$$\int_{\Omega\setminus A} |f|^q \, d\nu = c^{q-1} \int_{\Omega\setminus A} \left|\frac{f}{w}\right|^q d\nu \geq (1-\sigma)^{q-1} N^q \left(1 - (n/\sigma K)^{q-1}\right). \tag{6.2.30}$$

We first note that $B = (\Omega\setminus A)\setminus\{|f| < K\}$. Hence, combining the lower estimate (6.2.30) and the upper estimate (6.2.28) yields (6.2.25) as follows,

$$\begin{aligned}\int_B |f|^q \, d\nu &\geq \int_{\Omega\setminus A} |f|^q \, d\nu - \int_{\{|f|<K\}} |f|^q \, d\nu \\ &\geq (1-\sigma)^{q-1} N^q \left(1 - (n/\sigma K)^{q-1}\right) - K^{q-1} \\ &\geq \delta^q N^q,\end{aligned} \tag{6.2.31}$$

where for the last inequality we invoked (6.2.21).

Step 3 Here, we iteratively apply the conclusion of the previous step. We claim that there exist pairwise disjoint sets $A_1, \ldots, A_n \subset \Omega$ and $y_1, \ldots, y_n \in Y$ such that, with $f_i = u_1(y_i)$,

$$\|y_i\| = 1, \quad \nu(A_i) < 1/K, \quad \int_{A_i} |f_i|^q d\nu > \delta^q N^q, \quad \text{and} \quad \int_\Omega |f_i|^q d\nu \leq N^q, \tag{6.2.32}$$

for $1 \leq i \leq n$.

We use induction to prove the claim and begin by defining A_1 and y_1 satisfying (6.2.32). As $\|u_1\| = 1$ and $C_{1,q}(u_1) = N$, by (6.2.24) applied to $w = 1$, there exists $y_1 \in Y$ with $\|y_1\| = 1$ such that $f = u_1(y_1)$ satisfies

$$\int_\Omega |f| d\nu \leq 1, \qquad \int_\Omega |f|^q d\nu = N^q. \tag{6.2.33}$$

Put $A_1 = \{|f| > K\}$. By Chebyshev's inequality and the first estimate in (6.2.33) we have $\nu(A_1) = 1/K$ and, moreover, $\int_{\Omega\setminus A_1} |f|^{q-1}|f| d\nu \leq K^{q-1}$. Hence by the second estimate in (6.2.33) and (6.2.21), we obtain

$$\int_{A_1} |f|^q d\nu \geq N^q - K^{q-1} > \delta^q N^q.$$

Thus A_1 and y_1 satisfy (6.2.32).

Let $j \in \mathbb{N}$ with $j \leq n-1$. Assume that we have already determined pairwise disjoint sets $A_1, \ldots, A_j$ and $y_1, \ldots, y_j \in Y$ such that for any $1 \leq i \leq j$, condition

(6.2.32) holds true. Then put $A = A_1 \cup \cdots \cup A_j$. The induction hypothesis gives $\nu(A) \leq j/K$ where $j \leq n-1$. By (6.2.25), there exists $A_{j+1} \subseteq \Omega \backslash A$ and $y_{j+1} \in Y$ such that (6.2.32) holds true with i replaced by $j+1$.

Step 4 Recall that by ℓ_n^p we denote the vector space $\mathbb{R}^n$ equipped with the norm

$$\|(c_i)\|_q = \left(\sum_{i=1}^{n} |c_i|^q \right)^{1/q}.$$

Fix $y_1, \ldots, y_n \in Y$ and pairwise disjoint sets $A_1, \ldots, A_n \subset \Omega$ such that (6.2.32) holds true. We will now show that

$$\left\| \sum_{i=1}^{n} c_i y_i \right\|_Y \geq (1-\epsilon)\|(c_i)\|_q,$$

for $(c_i) \in \mathbb{R}^n$. To this end put $f_i = u_1(y_i)$. By (6.2.32),

$$\int_{\Omega \backslash A_i} |f_i|^q < (1-\delta^q)N. \tag{6.2.34}$$

Moreover, by duality there exists φ_i in the unit sphere of $L^r(\Omega, \nu)$ satisfying $\operatorname{supp}\varphi_i \subseteq A_i$ and

$$\left| N - \int_\Omega \varphi_i f_i d\nu \right| \leq (1-\delta)N. \tag{6.2.35}$$

By (6.2.34), for $i \leq n$ fixed we have

$$\sum_{j \neq i} \left| \int_\Omega \varphi_j f_i d\nu \right|^q \leq (n-1)(1-\delta^q)N^q. \tag{6.2.36}$$

Let $e_i \in \mathbb{R}^n$ denote the ith coordinate unit vector and define the operator $v \colon L^q(\Omega, \nu) \to \ell_n^p$ by

$$v(f) = \frac{1}{N} \sum_{i=1}^{n} \left(\int_\Omega \varphi_i f d\nu \right) e_i.$$

Put $x_i = v(f_i)$. In view of (6.2.35) and (6.2.36) we have

$$\begin{aligned} \|x_i - e_i\|_q^q &= \frac{1}{N^q} \sum_{j \neq i} \left| \int_\Omega \varphi_j f_i d\nu \right|^q + \frac{1}{N^q} \left| N - \int_\Omega \varphi_i f_i d\nu \right|^q \\ &\leq \frac{1}{N^q} [(n-1)(1-\delta^q)N^q + (1-\delta)^q N^q] \\ &\leq \epsilon^q n^{-q/r}, \end{aligned} \tag{6.2.37}$$

where we used (6.2.18) in the last estimate. Consequently, by (6.2.37) and Hölder's inequality we obtain that

$$\left\|\sum_{i=1}^n c_i x_i\right\|_q \geq \left\|\sum_{i=1}^n c_i e_i\right\|_q - \sum_{i=1}^n |c_i| \|x_i - e_i\|_q \geq (1-\epsilon)\left(\sum_{i=1}^n |c_i|^q\right)^{1/q}. \quad (6.2.38)$$

Since $\operatorname{supp}\varphi_i \subseteq A_i$ and since φ_i is in the unit sphere of $L^r(\nu)$, Hölder's inequality gives

$$\|v(f)\|_q \leq (1/N)\left(\sum_{i=1}^n \int_{A_i} |f|^q d\nu\right)^{1/q} \leq (1/N)\|f\|_{L^q(\nu)}. \quad (6.2.39)$$

By (6.2.22) and (6.2.23), we have $\|u_1 : Y \to L^q(\nu)\| \leq N$. Let $w\colon Y \to \ell_n^p$ be the composition $w = vu_1$. Invoking (6.2.39) gives $\|w\| \leq 1$. Hence, by (6.2.38) and the fact that $w(y_i) = x_i$, we finally obtain

$$\left\|\sum_{i=1}^n c_i y_i\right\|_Y \geq \left\|\sum_{i=1}^n c_i x_i\right\|_q \geq (1-\epsilon)\left(\sum_{i=1}^n |c_i|^q\right)^{1/q},$$

for $(c_i) \subset \mathbb{R}^n$. This finishes the proof. □

Proof of Theorem 6.2.3: Let Y be a reflexive subspace of $L^1(\Omega,\mu)$. Let $\mathrm{id}\colon Y \to L^1(\Omega,\mu)$ be the inclusion map given by $\mathrm{id}(x) = x$. Clearly $\|\mathrm{id}\| = 1$. In terms of operators, we have to show that there exists $1 < q < \infty$ such that

$$C_{1,q}\left(\mathrm{id}\colon Y \to L^1(\Omega,\mu)\right) < \infty.$$

Assume to the contrary that $C_{1,q}(\mathrm{id}) = \infty$ for any $1 < q < \infty$. We will show below that then for any $n \in \mathbb{N}$ and $\epsilon > 0$ there exist $y_1, \ldots y_n \in Y$ such that

$$(1-2\epsilon)\sum_{i=1}^n |c_i| \leq \left\|\sum_{i=1}^n c_i y_i\right\|_Y \leq \sum_{i=1}^n |c_i|. \quad (6.2.40)$$

Thus by Lemma 6.2.4, we determine a bounded subset of Y that fails to be uniformly integrable. This contradicts the reflexivity of Y.

Now fix $n \in \mathbb{N}$ and $\epsilon > 0$. Determine $1 < q < \infty$ close enough to 1, such that $n^{1-1/q} \leq (1+\epsilon)$. By Hölder's inequality, this choice of q gives

$$\sum_{i=1}^n |c_i| \leq n^{1-1/q}\left(\sum_{i=1}^n |c_i|^q\right)^{1/q} \leq (1+\epsilon)\left(\sum_{i=1}^n |c_i|^q\right)^{1/q}, \quad (6.2.41)$$

for $(c_i) \in \mathbb{R}^n$. Next, determine $\overline{N} = \overline{N}(n,\epsilon,q)$ such that condition (6.2.21) holds true. Choose a finite-dimensional subspace $Y_1 \subset Y$ such that

$$C_{1,q}\left(\mathrm{id}\colon Y_1 \to L^1(\Omega,\mu)\right) > \overline{N}.$$

Rosenthal's lemma (Lemma 6.2.5) asserts that there exist $y_1, \ldots, y_n \in Y$ such that $\|y_i\| = 1$ and

$$\left\| \sum_{i=1}^{n} c_i y_i \right\|_Y \geq (1 + \epsilon) \left(\sum_{i=1}^{n} |c_i|^q \right)^{1/q},$$

for $(c_i) \in \mathbb{R}^n$. In view of (6.2.41) we have obtained (6.2.40), which finishes the proof of Rosenthal's theorem.

Extrapolation

Let (Σ, σ) and (Ω, μ) be measure spaces. Let R be a bounded operator on $L^p(\Sigma, \sigma)$. We assume that for $1 < p, q < \infty$, there exists $A(p, q) < \infty$, such that $\mathcal{R} = R \otimes J$ is bounded as

$$\left\| \mathcal{R} \colon L^p(L^q) \to L^p(L^q) \right\| \leq A(p, q), \tag{6.2.42}$$

where J denotes the identity on $L^q(\Omega, \mu)$.

Theorem 6.2.6 *Let Y be a reflexive subspace of $L^1(\Omega, \mu)$. Assume that $\mathcal{R}$ satisfies the hypothesis* (6.2.42)*. Then there exists $1 < q < \infty$ with $q = q(Y)$ and $C(Y) < \infty$ such that*

$$\|\mathcal{R}_{|L^p(Y)} \colon L^p(Y) \to L^p(Y)\| \leq A, \tag{6.2.43}$$

where $A = C(Y) A(p, q)$.

Proof Apply Rosenthal's theorem (Theorem 6.2.3) to the reflexive space Y. Hence there exists $1 < r < \infty$ and g in the unit sphere of $L^r(\Omega, \mu)$ such that (6.2.9) holds true with $1/r + 1/q = 1$. Put $h = |g|^r$. Then $h \in L^1(\Omega, \mu)$ with $\|h\|_{L^1} = 1$. Define the probability measure $\nu = h\mu$. Then by (6.2.9),

$$\left(\int_\Omega |y h^{-1}|^q d\nu \right)^{1/q} \leq C \int_\Omega |y| d\mu, \tag{6.2.44}$$

for $y \in Y$. Fix $f \in L^p(\Sigma) \otimes Y$, then $f(z) \in Y$. Hence $\left(f h^{-1}\right)(z) \in Y$ for $z \in \Sigma$. By (6.2.44), we get

$$\left(\int_\Omega \left| \left(f h^{-1}\right)(z) \right|^q d\nu \right)^{1/q} \leq C \int_\Omega |f(z)| d\mu, \qquad y \in Y. \tag{6.2.45}$$

We apply the hypothesis (6.2.42) to the probability space (Ω, ν) and obtain

$$\left\| \mathcal{R} \colon L^p(L^q(\nu)) \to L^p(L^q(\nu)) \right\| \leq A, \tag{6.2.46}$$

and evaluating the norm estimate (6.2.46) on $\left(f h^{-1}\right)$ gives

$$\left\| \mathcal{R}(f h^{-1}) \right\|_{L^p(L^q(\nu))} \leq A \left\| f h^{-1} \right\|_{L^p(L^q(\nu))}. \tag{6.2.47}$$

Since $1 < q < \infty$, Hölder's inequality gives

$$\int_\Sigma \left(\int_\Omega |\mathcal{R} f(z)| d\mu \right)^p d\sigma(z) \le \int_\Sigma \left(\int_\Omega |\mathcal{R}(fh^{-1})(z)|^q \, d\nu \right)^{p/q} d\sigma(z).$$

Hence by (6.2.47) and (6.2.45), we obtain

$$\|\mathcal{R} f\|_{L^p(Y)} \le \left\|\mathcal{R}(fh^{-1})\right\|_{L^p(L^q(\nu))} \le A \left\|fh^{-1}\right\|_{L^p(L^q(\nu))} \le CA \|f\|_{L^p(Y)}. \qquad \square$$

Remarks: In Chapter 7 we apply Theorem 6.2.6 to the following operators.

For $1 < q < \infty$, the spaces $L^q(\Omega, \mu)$ satisfy the UMD condition. Hence the assumption (6.2.42) holds true for the Hilbert transform on $\mathbb{T}$, the Riesz projection onto Hardy martingales, the orthogonal projection onto simple Hardy martingales on $\mathbb{T}^{\mathbb{N}}$, and the stochastic Hilbert transform. See, for example, Burkholder (2001) and Pisier (2016). Theorem 6.2.6 applies to projections onto homogeneous polynomials of fixed degree. Indeed, for $k \in \mathbb{N}$ we put

$$\mathcal{S}^{(k)}(f) = \sum_{S,\varepsilon} \langle f, p_S^\varepsilon \rangle p_S^\varepsilon, \tag{6.2.48}$$

where $p_S^\varepsilon(z) = \prod_{j\in S} z_j^{\varepsilon_j}$, and where the sum is taken over $S \subset \mathbb{N}$ of cardinality k and $\varepsilon \in \{+1, -1\}^S$. Put $L^p(L^q) = L^p(\mathbb{T}^{\mathbb{N}}, L^q(\Omega))$. Then by Theorem 2.4.15,

$$\left\|\mathcal{S}^{(k)} \colon L^p(L^q) \to L^p(L^q)\right\| \le C(p, q)^k, \tag{6.2.49}$$

where $C(p, q) < \infty$, for any $1 < p, q < \infty$.

The orthogonal projection onto Steinhaus martingales may be written as the composition of $\mathcal{S}^{(1)}$ (see Theorem 2.4.15) followed by the Riesz projection onto Hardy martingales. Hence for the Steinhaus projection, assumption (6.2.42) of Theorem 6.2.6 holds true.

Rosenthal's theorem (Theorem 6.2.3) admits many applications pertaining to the structure theory of classical Banach spaces. For instance, Theorem 6.2.3 yields that none of the reflexive spaces

$$Z_p = \left(\sum \ell_n^1\right)_p, \qquad p > 1,$$

is isomorphic to a subspace of L^1. Indeed, if Z_p would embed into L^1, then by the operator (6.2.10), Z_p would be isomorphic to a subspace of L^q for some $q > 1$. In particular, Rosenthal's theorem would imply that Z_p is a Banach space of non-trivial type. Clearly, this contradicts the fact that Z_p contains copies of ℓ_n^1, $n \in \mathbb{N}$, uniformly.

Reverse-Minkowski Inequalities

We begin by observing that reverse-Hölder estimates imply reverse-Minkowski inequalities. Let X be a closed linear subspace of $L^1(\Omega,\mu)$ and $1 < q < \infty$. Let $K < \infty$ and assume that for any $\eta > 0$ there exists $g \geq 0$ in the unit sphere of $L^r(\Omega,\mu)$ $(1 < r < \infty)$ such that

$$\int_\Omega \left|\frac{x}{g}\right|^q d\mu \leq (K+\eta)^q \left(\int_\Omega |x| d\mu\right)^q,$$

for $x \in X$. Then writing $|x_i| = |x_i/g|g$ and applying Hölder's inequality gives

$$\int \left(\sum_{i=1}^n |x_i|^q\right)^{1/q} d\mu \leq \left(\sum_{i=1}^n \int \left|\frac{x_i}{g}\right|^q d\mu\right)^{1/q}$$

$$\leq (K+\eta)\left(\sum_{i=1}^n \left(\int |x_i| d\mu\right)^q\right)^{1/q},$$

for any $n \in \mathbb{N}$ and $x_1,\ldots,x_n \in X$. Since $\eta > 0$ was arbitrary, we observed that reverse-Hölder implies reverse-Minkowski in the following sense. Let $\mathrm{id}\colon X \to L^1(\Omega,\mu)$, then

$$\int_\Omega \left(\sum_{i=1}^n |x_i|^q\right)^{1/q} d\mu \leq C_{1,q}(\mathrm{id}) \left(\sum_{i=1}^n \left(\int_\Omega |x_i| d\mu\right)^q\right)^{1/q},$$

for any $n \in \mathbb{N}$ and $x_1,\ldots,x_n \in X \subset L^1(\Omega,\mu)$.

Let X be a Banach space, and let u be an operator on X with values $L^1(\Omega,\mu)$. Recall that we defined $C_{1,q}(u)$ to be the *infimum* of those constants C for which there exists g in the unit ball of $L^r(\Omega,\mu)$ such that

$$\left(\int_\Omega \left|\frac{u(x)}{g}\right|^q d\mu\right)^{1/q} \leq C\|x\|_X,$$

for $x \in X$. Above we showed in effect that for any linear operator $u\colon X \to L^1(\Omega,\mu)$, reverse-Minkowski estimates hold true as follows

$$\int_\Omega \left(\sum_{i=1}^n |u(x_i)|^q\right)^{1/q} d\mu \leq C_{1,q}(u) \left(\sum_{i=1}^n \|x_i\|_X^q\right)^{1/q},$$

for any $n \in \mathbb{N}$ and $x_1,\ldots,x_n \in X$. We now turn to Maurey's theorem, which states that the infimum defining $C_{1,q}(u)$ is attained for some g in the unit ball of $L^r(\Omega,\mu)$ (Maurey, 1973a, 1974). The proof is based on the mountain pass lemma of Ky-Fan. See Schwartz (1973).

Lemma 6.2.7 *Let K be a compact convex subset of a locally convex topological Hausdorff space E. Let $\mathcal{F}$ be a collection of concave, upper semi-continuous and finite functions on K such that*

$$\max_{f \in K} F(f) \geq 0, \tag{6.2.50}$$

for $F \in \mathcal{F}$. Then there exists $g \in K$ such that $F(g) \geq 0$, for any $F \in \mathcal{F}$.

Proof In $C(K)$, consider the open cone $\mathcal{N} = \{G \in C(K) : G < 0\}$. By (6.2.50), $\mathcal{F} \cap \mathcal{N} = \emptyset$. By the Hahn–Banach theorem we may separate the cones $\mathcal{F}, \mathcal{N}$ by an element in the dual of $C(K)$. There exists a finite measure $\nu \in C(K)'$ so that $\int F d\nu \geq 0$, for $F \in \mathcal{F}$, and $\int G d\nu < 0$, for $G \in \mathcal{N}$. It follows from the second condition that the measure ν is nonnegative, and nonvanishing. Hence we may assume that ν is a probability measure. Let $g \in E''$ be the barycenter of ν, that is, for each $\zeta \in E'$ we have

$$\langle g, \zeta \rangle = \int_K \langle f, \zeta \rangle d\nu(f). \tag{6.2.51}$$

As K is compact and convex we may choose $g \in K$ (and not just in E'').

We now claim that $F(g) \geq 0$, for any $F \in \mathcal{F}$. To this end, let B_F be the collection of affine and continuous functions pointwise larger than F. Since F is concave, upper semicontinuous and finite we have the pointwise relation

$$F = \inf\{G : G \in B_F\}.$$

By (6.2.51) we have $G(g) = \int G d\nu$ for $G \in B_F$, and

$$F(g) = \inf_{G \in B_F} G(g) \geq \int_K \inf_{G \in B_F} G d\nu = \int_K F d\nu \geq 0. \qquad \square$$

The next theorem shows that reverse-Minkowski inequalities imply reverse-Hölder estimates, and that the infimum defining $C_{1,q}(u)$ is attained.

Theorem 6.2.8 *Let X be a Banach space. Let u be an operator on X with values $L^1(\Omega, \mu)$. Let $1 < q < \infty$. If there exists $K > 0$ such that*

$$\int_\Omega \left(\sum_{i=1}^n |u(x_i)|^q \right)^{1/q} d\mu \leq K \left(\sum_{i=1}^n \|x_i\|_X^q \right)^{1/q}, \tag{6.2.52}$$

for any $n \in \mathbb{N}$ and $x_1, \ldots, x_n \in X$, then there exists a nonnegative g in the unit ball of $L^r(\Omega, \mu)$ $(1/q + 1/r = 1)$, such that

$$\left(\int_\Omega \left| \frac{u(x)}{g} \right|^q d\mu \right)^{1/q} \leq K \|x\|_X, \qquad x \in X. \tag{6.2.53}$$

Proof The reduction to Lemma 6.2.7 is carried out in the first two steps of the proof, where we verify the inequalities required by the hypothesis of the Ky-Fan lemma and we check weak semicontinuity of the concave functionals (6.2.54). In the final step, we show that applying Lemma 6.2.7 produces $g \in L^r(\Omega, \mu)$ with the properties stated in (6.2.53).

Step 1 Let $F = \{h\colon \Omega \to \mathbb{R}^+ : \int h^r d\mu \le 1\}$. The set K is convex and compact for the weak-topology in $L^r(\Omega,\mu)$. Fix $n \in \mathbb{N}$, and a vector $x = (x_1, \dots x_n)$ of elements in X, and put $f_i = u(x_i)$. Define a measure μ_x and a testing function h_x by putting

$$\mu_x = \left(\sum_{i=1}^n |f_i|^q\right)\mu \qquad \text{and} \qquad h_x = C\left(\sum_{i=1}^n |f_i|^q\right)^{1/(rq)},$$

where the normalizing factor C is chosen so that $\int h_x^r d\mu = 1$. Hence $h_x \in F$. Next, consider the nonlinear functional $F_x\colon F \to \mathbb{R}$ by

$$F_x(h) = K^q \sum_{i=1}^n \|x_i\|_X^q - \int h^{-q} d\mu_x. \tag{6.2.54}$$

We claim that

$$F_x(h_x) \ge 0. \tag{6.2.55}$$

We start the verification of (6.2.55) by identifying $C^{-r} = \int \left(\sum_{i=1}^n |f_i|^q\right)^{1/q} d\mu$ and evaluating

$$\int h_x^{-q} d\mu_x = C^{-q} \int \left(\sum_{i=1}^n |f_i|^q\right)^{-1/r} \left(\sum_{i=1}^n |f_i|^q\right) d\mu = C^{-q} \int \left(\sum_{i=1}^n |f_i|^q\right)^{1/q} d\mu.$$

Since $q/r = q - 1$, using the explicit term for C gives

$$\int h_x^{-q} d\mu_x = \left(\int \left(\sum_{i=1}^n |f_i|^q\right)^{1/q} d\mu\right)^q.$$

Thus, invoking (6.2.52), we obtain

$$F_x(h_x) = K^q \sum_{i=1}^n \|x_i\|_X^q - \left(\int \left(\sum_{i=1}^n |f_i|^q\right)^{1/q} d\mu\right)^q \ge 0,$$

as claimed.

Step 2 We verify that the functionals F_x are concave, and weak upper semicontinuous. Taking into account the minus sign in front of the second summand in (6.2.54) this amounts to verifying that

$$I(f) = \int f^{-q} d\mu, \qquad f \in F,$$

is convex and weakly lower semicontinuous with respect to the $\sigma(L^r(\mu), L^q(\mu))$ topology on $F = \{h\colon \Omega \to \mathbb{R}^+ : \int h^r d\mu \le 1\}$. Note first that the map $t \to t^{-q}$ is convex on $[0, +\infty)$. Hence, by inspection, I is a convex functional on F. Moreover, $F \subseteq L^r(\mu)$ is weakly compact and convex. Hence, in order to prove that I is $\sigma(L^r(\mu), L^q(\mu))$ lower semicontinuous, it suffices to check lower semicontinuity with respect to the $L^r(\mu)$ norm (Mazur's theorem). Fix a sequence f, f_n in K and assume convergence in the norm of $L^r(\mu)$. By passing to a subsequence we may assume almost sure convergence of $f_n^{-q} \to f^{-q}$. Since $f_n \geq 0$, Fatou's lemma implies that

$$\int f^{-q} d\mu \leq \liminf_{n\to\infty} \int f_n^{-q} d\mu.$$

Step 3 As verified above, F_x is concave and upper semicontinuous on F. Moreover, $\{F_x : x \in X^n\}$ is a convex cone of functions. By claim (6.2.55)

$$F_x(h_x) \geq 0,$$

for $x \in X^n$. Invoking Lemma 6.2.7 (the Ky-Fan lemma) shows that there exists $g \in K$ such that

$$F_x(g) \geq 0, \quad \text{for all} \quad x \in X^n. \tag{6.2.56}$$

Since $g \in K$, we have $g \geq 0$ and $\int g^r d\mu \leq 1$. It remains to show that, for our choice of g, (6.2.53) holds true. To this end, fix $y \in X$ and consider $x = (y, 0, \ldots, 0)$. Then by (6.2.56), we obtain the crucial inequality

$$F_x(g) = K^q \|y\|_X^q - \int g^{-q} |u(y)|^q d\mu \geq 0,$$

which gives (6.2.53). □

Extrapolation

Let Σ be a measure space and let R be a bounded operator on $L^p(\Sigma)$. We assume that for any probability space (Ω, μ), the tensor $\mathcal{R} = R \otimes J_\Omega$ is bounded as

$$\left\| \mathcal{R}\colon L^p(L^1) \to L^p\,(L^\alpha) \right\| \leq A, \tag{6.2.57}$$

where $A = A(p, \alpha) < \infty$, for $0 < \alpha < 1 < p < \infty$. Recalling our convention that Y be a closed subspace of $L^1(\Omega, \mu)$, the norm on the space $L^p(Y)$ is the one induced by $L^p(L^1)$.

Theorem 6.2.9 *Let Y be a closed subspace of $L^1(\Omega, \mu)$. Assume that there exists $K > 0$ and $1 < q < \infty$ such that for any $n \in \mathbb{N}$ and $y_1, \ldots, y_n \in Y$,*

$$\int \left(\sum_{i=1}^n |y_i|^q \right)^{1/q} d\mu \leq K \left(\sum_{i=1}^n \|y_i\|_{L^1}^q \right)^{1/q}, \tag{6.2.58}$$

and that $\mathcal{R} = R \otimes J_\Omega$ satisfies the hypothesis (6.2.57). Then the restriction of $\mathcal{R}$ to $L^p(Y)$ satisfies

$$\left\| \mathcal{R}_{|L^p(Y)}\colon L^p(Y) \to L^p(Y) \right\| \leq \overline{A}, \tag{6.2.59}$$

where $\overline{A} = A(p, \alpha) L(K, q, \alpha)$.

Proof Fix $K > 0$ and $1 < q < \infty$ such that (6.2.52) holds true for any $n \in \mathbb{N}$ and $y_1, \ldots, y_n \in Y$. By Theorem 6.2.8, there exists g in the unit sphere of $L^r(\Omega, \mu)$ $(1/r+1/q = 1)$ such that inequality (6.2.53) holds true. Putting $h = |g|^r$ we form the probability space $(\Omega, h\mu)$. By (6.2.53), we have

$$\left(\int_\Omega |yh^{-1}|^q \, hd\mu\right)^{1/q} \le K \int_\Omega |y| d\mu, \qquad y \in Y. \tag{6.2.60}$$

Step 1 We fix $0 < \alpha < 1$ and $y \in Y$. Choose $0 < \theta < 1$ such that $\alpha = q(1-\theta)/(q-\theta)$. Write

$$|y| = |yh^{-1}|^\theta \, |yh^{-1}|^{1-\theta} \, h,$$

and note that $\theta/q + (q-\theta)/q = 1$. Hence, applying Hölder's inequality gives

$$\int |y| d\mu \le \left(\int |yh^{-1}|^q \, hd\mu\right)^{\theta/q} \left(\int |yh^{-1}|^{(1-\theta)q/(q-\theta)} \, hd\mu\right)^{(q-\theta)/q}. \tag{6.2.61}$$

Invoke (6.2.60) to estimate the first factor on the right-hand side of (6.2.61) and use arithmetic to find that

$$\int_\Omega |y| d\mu \le K^{\theta/(1-\theta)} \left(\int_\Omega |yh^{-1}|^\alpha \, hd\mu\right)^{1/\alpha}. \tag{6.2.62}$$

Since $\theta/(1-\theta) = [q(1-\alpha)]/[\alpha(q-1)]$, we write $K^{\theta/(1-\theta)} = K(q, \alpha)$.

Step 2 First, $f \in L^p \otimes Y$ implies $\mathcal{R}(f) \in L^p \otimes Y$ or, equivalently, $\mathcal{R}(f)(z) \in Y$ for $z \in \Sigma$. Hence fixing z and applying (6.2.62) to $\mathcal{R}(f)(z) \in Y$ gives

$$\int_\Omega |\mathcal{R}(f)(z)| d\mu = K(q, \alpha) \left(\int_\Omega |\mathcal{R}(fh^{-1})(z)|^\alpha \, hd\mu\right)^{1/\alpha}. \tag{6.2.63}$$

Taking the L^p norm in (6.2.63) with respect to the variable $z \in \Sigma$, gives

$$\|\mathcal{R}f\|_{L^p(L^1(\mu))} \le K(q, \alpha) \left\|\mathcal{R}(fh^{-1})\right\|_{L^p(L^\alpha(h\mu))}. \tag{6.2.64}$$

Invoke (6.2.57), for the probability space $(\Omega, hd\mu)$ to obtain

$$\left\|\mathcal{R}\colon L^p(L^1(h\mu)) \to L^p\,(L^\alpha(h\mu))\right\| \le A. \tag{6.2.65}$$

Fix $f \in L^p \otimes Y$ and apply (6.2.65) to $fh^{-1} \in L^p(L^1(h\mu))$. This gives

$$\left\|\mathcal{R}(fh^{-1})\right\|_{L^p(L^\alpha(h\mu))} \le A \left\|fh^{-1}\right\|_{L^p(L^1(h\mu))} = A\|f\|_{L^p(L^1(\mu))}. \tag{6.2.66}$$

Combining (6.2.64) and (6.2.66) finally gives us (6.2.59). □

Remarks: Assume that $A(\alpha, p)$ satisfies inequality (6.2.57) and that K and q are determined by (6.2.58). The proof above shows that

$$A(K, p, q, \alpha) = A(\alpha, p)L(K, q, \alpha),$$

where $L(Kq, \alpha) = K^{[q(1-\alpha)]/[\alpha(q-1)]}$.

6.3 Notes

In Section 6.1, Theorem 6.1.2 is a result of Kislyakov (1991). Theorem 6.1.3 and Theorem 6.1.9 (which includes Theorem 6.1.1) were obtained by J. Bourgain, whose proofs were presented by Dechamps-Gondim (1985) to the Seminar Bourbaki 1983/84. The method of proof in Dechamps-Gondim (1985) is rooted in the idea of generating A_1 weights due to Kislyakov (1981). In the text we use Kislyakov's proof of Theorem 6.1.3.

A survey of the results together with extensions and applications are discussed by Bourgain (1984c). The work Bourgain and Davis (1986) builds on Bourgain (1984c) and presents a proof of Theorem 6.1.9 different from the one in Dechamps-Gondim (1985). The presentation in this chapter incorporates joint work with J. Penteker (Müller and Penteker, 2015).

The proof of the transference theorem (Theorem 6.1.4) uses well-known arguments going back to Bonami (1970) and Meyer (1968). Also see Bourgain (1984a,b), Haagerup and Pisier (1989) and Kislyakov (1991) for closely related applications of the transference argument.

The vector-valued estimates for the stochastic Hilbert transform presented in Theorem 6.1.7, Theorem 6.1.6, and Theorem 6.1.5 were obtained by Garling (Garling, D. J. H. $L^1(\Omega)/H^1(\Omega)$H is a Grothendieck space of cotype 2. pages 1–18, 1992, unpublished.) Clearly, Theorem 6.1.7 directly yields Theorem 6.1.3.

In Section 6.2 we use Wojtaszczyk (1991) as source on uniform integrability, weak compactness, and the Dunford–Pettis Theorem. The proof of the extrapolation theorem (Theorem 6.2.2), based on Proposition 6.2.1, is due to Kislyakov (1991). It is important to note that Kislyakov's proof of Theorem 6.2.2 does not rely on Rosenthal's change of density theorem (Theorem 6.2.3) (Rosenthal, 1973).

The proof of Rosenthal's theorem follows the presentation of Maurey (1973b) verbatim. Wojtaszczyk (1991, Theorem II.C. 12) is our source for the proof of Proposition 6.2.4. The extrapolation theorem (Theorem 6.2.6) is proved following Pisier (1982b). Also see Pisier (1986). We included Theorem 6.2.6 with the intention of applying it to the Steinhaus projection.

Theorem 6.2.8 is due to Rosenthal (1973). Its proof in the text follows Maurey (1973a). See Schwartz (1973) for the proof of the Ky-Fan mountain pass lemma. The proof of the extrapolation theorem (Theorem 6.2.9) follows Bourgain and Davis (1986).

We refer to the books by Li and Queffélec (2004), and Wojtaszczyk (1991) for different proofs of Rosenthal's theorem, and to García-Cuerva and Rubio de Francia (1985) for related applications of the Maurey–Rosenthal methods.

7

Formative Examples

In this chapter we continue the investigation of vector-valued Hardy martingales, the analytic Radon–Nikodym property (aRNP), Hardy martingale cotype, and the aUMD property of complex Banach spaces. Chapter 5 included the result that L^1 satisfies the aUMD property and that the following implications hold true:

$$\text{(aUMD)} \Rightarrow \text{Hardy martingale cotype} \Rightarrow \text{(aRNP)}.$$

In this chapter we obtain that for any reflexive subspace $R \subset L^1$, the quotient space L^1/R satisfies the aUMD and give examples showing that

$$\text{(aUMD)} \not\Leftarrow \text{Hardy martingale cotype} \not\Leftarrow \text{(aRNP)}\,.$$

In Section 7.1 we prove lifting theorems for Steinhaus and Hardy martingales ranging in L^1/R, where R is a reflexive subspace of L^1. We obtain the Kislyakov–Pisier theorem (L^1/R is a cotype 2 space, satisfying Grothendieck's theorem) and the theorem of Bourgain and Davis, which shows that L^1/R satisfies the aUMD and is of Hardy martingale cotype 2.

In Section 7.2, where we investigate unconditionality for the trace class S^1, the main triangular projection plays a central role. We prove the theorem of Kwapień and Pełczyński (1970), which states that S^1 does not embed into Banach spaces with an unconditional basis, and two theorems of Haagerup and Pisier (1989), which assert that S^1 is H^1-uniformly convex of order 2, and does not satisfy the aUMD property.

Section 7.3 presents remarkable ideas from the dissertation of Y. Qiu (2012b), who exhibited a family of Banach spaces, the iterated $L^p(L^q)$-spaces, which are super-reflexive, have an unconditional basis, and fail the aUMD property.

Section 7.4 is devoted to the pre-dual of the James Tree space $\mathrm{J_*T}$. We prove that $\mathrm{J_*T}$ satisfies the aRNP. Two methods of proof are presented. The first is due to Ghoussoub, Maurey, and Schachermayer (1989), and exploits holomorphic separation. The second, due to Ghoussoub and Maurey (1989), is based on holomorphic injections.

In Section 7.5 we obtain Bourgain's lifting theorem for Steinhaus martingales with values in L^1/H^1, and prove Kislyakov's embedding theorem. We form a separable dual space that fails Hardy martingale cotype q-estimates for any $q < \infty$.

7.1 L^1 Quotients by Reflexive Spaces

In Chapter 5 we considered Hardy martingales with values in a Banach space X and focused on martingale inequalities that gave rise to isomorphic invariants on the underlying Banach space. We obtained that L^1 is of Hardy martingale cotype 2 and satisfies the aUMD property. Specifically, in Section 5.3 we presented the Haagerup–Pisier inequality,

$$\left(\sum_{m=1}^{k}(\mathbb{E}\|\Delta F_m\|_{L^1})^2\right)^{1/2} \leq \sqrt{2}\mathbb{E}\|F_k\|_{L^1}, \tag{7.1.1}$$

for any L^1-valued Hardy martingale $F = (F_n)$. Section 5.4 contains Garling's unconditionality estimate for an L^1-valued Hardy martingale (F_n),

$$\mathbb{E}\left\|\sum_{m=1}^{k}\varepsilon_m\Delta F_m\right\|_{L^1} \leq C\mathbb{E}\|F_k\|_{L^1}, \tag{7.1.2}$$

for $(\varepsilon_m) \in \{-1, 1\}^{\mathbb{N}}$ and $k \in \mathbb{N}$.

In this section we prove that for any reflexive subspace $Y \subset L^1$, the quotient space $X = L^1/Y$ is of Hardy martingale cotype 2 and satisfies the aUMD property. This is accomplished by proving martingale lifting theorems, replacing the target space $X = L^1/Y$ by L^1. The associated estimates rest on the results of Chapter 6. It is well known that for any infinite-dimensional reflexive subspace $Y \subset L^1$, the quotient space L^1/Y fails to be isomorphic to L^1; more generally L^1/Y is not a script-L^1 space in the sense of Lindenstrauss and Rosenthal (1969). We refer to the book by Diestel et al. (1995, p. 318) for presentation of the details.

In Section 7.1.1 we present the lifting theorem for Steinhaus martingales and obtain that $X = L^1/Y$ is a Banach space of cotype 2 whenever Y is reflexive. We exploit its reflexivity by applying the extrapolation theorem (Theorem 6.2.6) to projections onto homogeneous polynomials of fixed degree.

In Section 7.1.2 we obtain that $X = L^1/Y$ has the aUMD property provided that Y is reflexive. In the proof we apply Theorem 6.2.2 to the Riesz projection. We close the section by proving a Hardy martingale lifting theorem for L^1/Y, valid for any closed subspace $Y \subset L^1$. Theorem 7.1.5 is based on Theorem 6.1.4.

7.1.1 The Kislyakov–Pisier Theorem

Here we discuss the lifting problem for Steinhaus martingales. Let $Y \subseteq L^1$ be a closed subspace and let $q\colon L^1 \to L^1/Y$ denote the canonical quotient map. Given $f_i \in L^1/Y$ the Steinhaus martingale lifting problem for L^1/Y asks for the existence of $g_i \in L^1$ satisfying the lifting relation $q(g_i) = f_i$ and the estimates

$$\int_{\mathbb{T}^n} \left\| \sum_{i=1}^{n} z_i g_i \right\|_{L^1} dm_n(z) \leq C(Y) \int_{\mathbb{T}^n} \left\| \sum_{i=1}^{n} z_i f_i \right\|_{L^1/Y} dm_n(z), \tag{7.1.3}$$

where integration is with respect to the uniform measure on $\mathbb{T}^n$, and $C(Y)$ is independent of $n \in \mathbb{N}$ or $f_i \in L^1/Y$. In this section we present the solution to these problems in the case when Y is a reflexive space.

The Steinhaus martingale lifting problem for L^1/Y is equivalent to the following lifting problems for operators acting on the space of Steinhaus martingales with values in $Y^\perp$, the annihilator of Y in L^∞. Let S_n denote the space of scalar Steinhaus martingales of length n. Thus $\varphi \in S_n$ is given by

$$\varphi(z) = \sum_{i=1}^{n} \alpha_i z_i, \tag{7.1.4}$$

where $\alpha_i \in \mathbb{R}$, $z_i \in \mathbb{T}$, and $z \in \mathbb{T}^n$ with $z = (z_i)$. We view S_n as being endowed with the norm induced by $L^1(\mathbb{T}^n)$. A linear operator $u\colon S_n \to Y^\perp$ is determined by $x_i \in Y^\perp$ and the relation

$$u(\varphi) = \sum_{i=1}^{n} \alpha_i x_i,$$

where φ is given by (7.1.4). Equivalently, we say that u is defined by the equation $u(z_i) = x_i$. The lifting problem for the operator u asks for the existence of a bounded linear operator

$$\widetilde{u}\colon L^1(\mathbb{T}^n) \to Y^\perp, \tag{7.1.5}$$

satisfying the lifting conditions $\widetilde{u}(z_i) = u(z_i)$ and the norm estimates

$$\|\widetilde{u}\| \leq K(Y)\|u\|, \tag{7.1.6}$$

where $K(Y)$ is independent of n or $x_i \in Y$. Thus by letting $i\colon S_n \hookrightarrow L^1$ denote the inclusion map, the following diagram commutes.

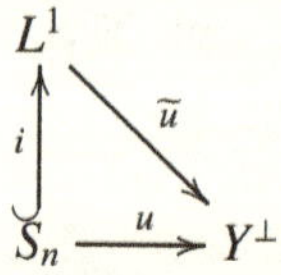

(7.1.7)

A classic theorem of Pisier (1978, 1983, 1986) asserts that the operator lifting problem above is equivalent to the Steinhaus martingale lifting problem for L^1/Y, and that the constants arising in inequalities (7.1.3) and (7.1.6) are related by

$$K(Y)/\sqrt{2} \leq C(Y) \leq \sqrt{2}K(Y).$$

The second lifting problem concerns operators defined on $Y \subseteq L^1$ with values in a cotype 2 Banach space. Let E be a Banach space of cotype 2 and let

$$u\colon Y \to E \tag{7.1.8}$$

be a bounded linear operator. Fix $K_1, K_2 \geq 1$. The lifting problem for the triple (E, u, Y) asks for the existence of a Banach space E_1, satisfying the cotype 2 estimate,

$$\text{Cotyp}_2(\text{E}_1) \leq \max\{\text{Cotyp}_2(\text{E}), K_2\},$$

an isometry $j\colon E \to E_1$, and a bounded linear operator $\widetilde{u}\colon L^1 \to E_1$ satisfying the lifting relation and norm estimates

$$\widetilde{u}_{|Y} = ju, \qquad \|\widetilde{u}\| \leq K_1\|u\|. \tag{7.1.9}$$

Thus letting $i\colon Y \hookrightarrow L^1$ denote the inclusion map, we obtain a triple $(E_1, \widetilde{u}, L^1)$ such that the following diagram commutes.

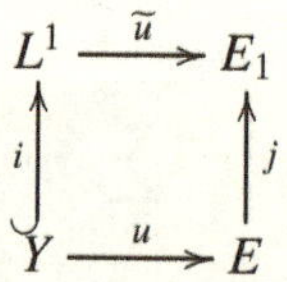

(7.1.10)

Pisier's work (1983, 1986) links martingale lifting and operator lifting: If the Steinhaus martingale lifting problem for L^1/Y is solvable with constant $C(Y)$, then the operator lifting problem for (E, u, Y) is solvable with constants

$$K_1 = 2(C(Y) + 1) \quad \text{and} \quad K_2 = 2\sqrt{2}C(Y),$$

for any Banach space E of cotype 2.

The construction of the Banach space E_1 and the extension operator $\widetilde{u}$ is provided by Kislyakov's lemma (Kislyakov, 1976) (also see Kalton and Pełczyński [1997]).

Lemma 7.1.1 *Let $Y \subseteq L^1$ be a closed linear subspace. For any Banach space E and any bounded linear operator $u\colon Y \to E$ there exists a Banach space E_1, an isometric embedding $j\colon E \to E_1$, and a bounded operator $\widetilde{u}\colon L^1 \to E_1$ such that the following conditions hold true:*

(i) $\widetilde{u}_{|Y} = ju$,
(ii) $\|\widetilde{u}\| = \|u\|$,
(iii) *L^1/Y is isometric to $E_1/j(E)$,*
(iv) *if u is an embedding, then so is $\widetilde{u}$.*

Proof We normalize $\|u\| = 1$. Let $Z = L^1 \times E$ with norm $\|(x_1, y)\| = \|x_1\| + \|y\|$, where $x_1 \in L^1$, $y \in E$. Define the linear subspace $W = \{(x, -u(x)) : x \in Y\} \subset Z$ and let $q\colon Z \to Z/W$ denote the canonical quotient mapping. We set

$$E_1 = Z/W \tag{7.1.11}$$

and define $j\colon E \to E_1$ by $j(y) = q(0, y)$, $y \in E$. Clearly, $\|j(y)\|_{E_1} \le \|y\|_E$ for $y \in E$. Conversely,

$$\begin{aligned}\|j(y)\|_{E_1} &= \inf\{\|(x, y - u(x))\|_Z : x \in Y\}\\ &= \inf\{\|x\|_{L^1} + \|y - u(x)\|_E : x \in Y\}\\ &\ge \|y\|_E + \inf\{\|x\|_{L^1} - \|u(x)\|_E : x \in Y\}\\ &= \|y\|_E.\end{aligned} \tag{7.1.12}$$

Thus, $\|j(y)\|_{E_1} = \|y\|_E$ for $y \in E$, proving that j is an isometric embedding. Next, observe that for $x \in Y$,

$$ju(x) = q(0, u(x)) = q((0, u(x)) + (x, -u(x)) = q(x, 0). \tag{7.1.13}$$

Next, define the linear operator

$$\widetilde{u}\colon L^1 \to E_1, \quad x_1 \to q(x_1, 0). \tag{7.1.14}$$

Clearly $\widetilde{u}$ is a contraction. The identity (7.1.13), yields that $\widetilde{u}_{|Y} = ju$, and in combination with (7.1.12) we find that $\|\widetilde{u}\| = \|u\|$.

Next, we exhibit the isometry between the quotient spaces L^1/Y and $E_1/j(E)$. Let $p_Y\colon L^1 \to L^1/Y$ and $p_E\colon E_1 \to E_1/j(E)$ denote the canonical quotient maps. For $x \in Y$ we have $\widetilde{u}(x) \in j(E)$. Hence $\widetilde{u}$ gives rise to a well-defined mapping $v\colon L^1/Y \to E_1/j(E)$ satisfying

$$p_E\widetilde{u} = vp_Y. \tag{7.1.15}$$

For $x_1 \in L^1, y \in E$, we have $\widetilde{u}(x_1) + j(y) = q(x_1, 0) + q(0, y) = q(x_1, y)$ and obviously $\inf\{\|y - u(x)\|_E : y \in E\} = 0$. Thus, by the identity (7.1.13),

$$\begin{aligned}\|v(p_Y(x_1))\|_{E_1/j(E)} &= \inf\{\|q(x_1,y)\|_{E_1} : y \in E\} \\ &= \inf\{\|(x_1 + x, y - u(x))\|_Z : y \in E, x \in Y\} \\ &= \inf\{\|x_1 + x\|_{L^1} + \inf\{\|y - u(x)\|_E : y \in E\} : x \in Y\} \\ &= \inf\{\|x_1 + x\|_{L^1} : x \in Y\} \\ &= \|p_y(x_1)\|_{L^1/Y}.\end{aligned} \tag{7.1.16}$$

In view of (7.1.16), $v\colon L^1/Y \to E_1/j(E)$ is an isometric injection. We show next that v is surjective. To this end fix $y_1 \in E_1$. Choose $z \in Z$ such that $y_1 = q(z)$. Write $z = (x_1, y)$ with $x_1 \in L^1$ and $y \in E$. Then,

$$v(p_Y(x_1)) = p_E(u_1(x_1)) = p_E q(x_1)) = p_E q(z) = p_E(y_1), \tag{7.1.17}$$

proving that v is surjective.

Now assume that u is an embedding. That is, assume that there exists $c > 0$ such that $\|u(x)\|_E \geq c\|u\|_Y$. We will show that $\widetilde{u}$ is an embedding as well. To this end, fix $x_1 \in L^1$ and estimate

$$\begin{aligned}\|\widetilde{u}(x_1)\|_{E_1} &= \inf\{\|x_1 - x\|_{L^1} + \|u(x)\|_E : x \in Y\} \\ &\geq \inf\{c\|x_1 - x\|_{L^1} + c\|x\|_Y : x \in Y\} \\ &\geq c\|x_1\|_{L^1}.\end{aligned} \tag{7.1.18}$$

□

The development of the above lifting problems and their interrelation began in the work of Kislyakov (1976) and Pisier (1978). (See also the presentation by Pełczyński (1980) and Kalton and Pełczyński (1997).) It played a fundamental role in the construction of Pisier's counterexample to Grothendieck's conjecture. See Pisier (1983, 1986). It is very instructive to review Section 4.5 (Talagrand's examples) in light of Kislyakov's lemma (Lemma 7.1.1).

We now turn to proving that for any reflexive subspace of $Y \subset L^1$, the Steinhaus lifting problem has a positive solution. Let $k \in \mathbb{N}$. Recall that in Section 2.2 we defined the projection $\mathcal{S}^{(k)}$ as

$$\mathcal{S}^{(k)}(f) = \sum_{S,\varepsilon} \langle f, p_S^\varepsilon \rangle p_S^\varepsilon,$$

where $p_S^\varepsilon(z) = \prod_{j \in S} z_j^{\varepsilon_j}$, and where the sum is taken over $S \subset \mathbb{N}$ of cardinality k and $\varepsilon \in \{+1, -1\}^S$. This formula applies to scalar-valued as well as vector-valued $f\colon \mathbb{T}^{\mathbb{N}} \to X$. For any X-valued trigonometric polynomial $A\colon \mathbb{T}^{\mathbb{N}} \to X$ the Steinhaus projection $\mathcal{S}(A)$, given by

$$\mathcal{S}(A)(w) = \sum_{k=1}^{\infty} \mathbb{E}(A(w)\overline{w_k})w_k,$$

is related to $\mathcal{S}^{(1)}$ by the identity,

$$\mathcal{S}^{(1)}(A) = \mathcal{S}(A) + \overline{\mathcal{S}(\overline{A})}.$$

The projections $\mathcal{S}^{(k)}$ act boundedly on the spaces $L^p(L^r) = L^p(\mathbb{T}^{\mathbb{N}}, L^r(\Omega, \mu))$. Specifically, by Theorem 2.4.15 there exists $C = C(p, r)$ such that

$$\left\| \mathcal{S}^{(k)} \colon L^q(L^r) \to L^p(L^r) \right\| \leq C^k, \tag{7.1.19}$$

for any $k \in \mathbb{N}$ and $1 < p, r < \infty$.

The next theorem, due to Kislyakov (1976) and Pisier (1978), respectively, solves the Steinhaus martingale lifting problem for any reflexive subspace of $Y \subset L^1$. The proof presented below is due to Pisier (1982b).

Theorem 7.1.2 (Kislyakov–Pisier theorem) *Let (Ω, μ) be a probability space and $1 \leq p < \infty$. Let Y be a reflexive space of $L^1(\Omega, \mu)$ and*

$$X = L^1(\Omega, \mu)/Y.$$

Let $n \in \mathbb{N}$, $x_1, \ldots, x_n \in L^1/Y$. There exist $g_1, \ldots, g_n \in L^1$, with $q(g_j) = x_j$, such that for every $1 \leq p < \infty$,

$$\mathbb{E}_{(w)} \left\| \sum_{j=1}^{n} g_j w_j \right\|_{L^1}^p \leq C(p) \mathbb{E}_{(w)} \left\| \sum_{j=1}^{n} x_j w_j \right\|_{L^1/Y}^p, \tag{7.1.20}$$

where the expectation is with respect to the Haar probability measure over $w \subset \mathbb{T}^n$ and hence,

$$\left(\sum_{j=1}^{n} \|x_j\|_{L^1/Y}^2 \right)^{1/2} \leq C\mathbb{E}_{(w)} \left\| \sum_{j=1}^{n} x_j w_j \right\|_{L^1/Y}. \tag{7.1.21}$$

Proof By Kahane's inequality (2.2.14), it suffices to prove (7.1.20) for $p = 2$. So fix a Steinhaus martingale $F = (F_k)$ in $L^2(X)$ with $F_n(w) = \sum_1^n x_j w_j$. Since $\mathcal{S}$ is a projection, we have

$$\mathcal{S}(F_k) = F_k. \tag{7.1.22}$$

Step 1 Recall that $q \colon L^1(\Omega) \to L^1(\Omega)/Y$ denotes the canonical quotient map. Let $A = (A_k)$ be a square-integrable martingale lifting of F, that is, $A_k \colon \mathbb{T}^{\mathbb{N}} \to L^1(\Omega)$, with $q(A_k) = F_k$ and

$$\|A_k\|_{L^2(L^1)} \leq 2\|F_k\|_{L^2(X)}. \tag{7.1.23}$$

Step 2 Next, we prove that for any $z \in \mathbb{T}^{\mathbb{N}}$ the following crucial inclusion holds true:

$$\mathcal{S}^{(1)}(A_k)(z) - A_k(z) \in Y. \tag{7.1.24}$$

To see this, we recall that $q(A_k) = F_k$ and

$$q\left(\mathcal{S}^{(1)}(A_k) - A_k\right) = \mathcal{S}^{(1)}(q(A_k)) - q(A_k) = \mathcal{S}^{(1)}(F_k) - F_k = 0.$$

Since the Steinhaus projection satisfies $\mathcal{S}(F_k) - F_k = 0$, we also have $\mathcal{S}(A_k)(z) - A_k(z) \in Y$.

Step 3 Let $0 < \theta < 1$. Convolve A with the Riesz product $J_\theta(w) = \prod\left\{1 + \frac{\theta}{2}\{w_j + \overline{w_j}\}\right\}$. Thus define

$$(\mathcal{J}_\theta A)(z) = \mathbb{E}\left(A(zw^{-1})J_\theta(w)\right).$$

Since $\mathbb{E}(A) = 0$, the identity in Proposition 2.4.9 yields

$$(\mathcal{J}_\theta A) = \sum_{k=1}^{\infty}\left(\frac{\theta}{2}\right)^k \mathcal{S}^{(k)}(A). \tag{7.1.25}$$

By orthogonality, we have $\mathcal{S}^{(k)}(A) = \mathcal{S}^{(k)}\left(A - \mathcal{S}^{(1)}(A)\right)$ for $k \geq 2$. Thus we rewrite (7.1.25) to obtain the second crucial identity of this proof,

$$\left(\frac{\theta}{2}\right)\mathcal{S}^{(1)}(A) = (\mathcal{J}_\theta A) - \sum_{k=2}^{\infty}\left(\frac{\theta}{2}\right)^k \mathcal{S}^{(k)}\left(A - \mathcal{S}^{(1)}(A)\right). \tag{7.1.26}$$

Taking the norm in $L^2\left(L^1\right)$, we therefore have

$$\left(\frac{\theta}{2}\right)\left\|\mathcal{S}^{(1)}(A)\right\| \leq \|(\mathcal{J}_\theta A)\| + \sum_{k=2}^{\infty}\left(\frac{\theta}{2}\right)^k \left\|\mathcal{S}^{(k)}\left(A - \mathcal{S}^{(1)}(A)\right)\right\|. \tag{7.1.27}$$

Now invoke that the operators $\mathcal{S}^{(k)}$ satisfy (7.1.19) and hence condition (6.2.42) holds true with $\mathcal{R} = \mathcal{S}^{(k)}$. In view of (7.1.24) and the reflexivity of Y, Theorem 6.2.6 shows that there exists a constant $C > 0$ such that for each $k \in \mathbb{N}$,

$$\left\|\mathcal{S}^{(k)}\left(A - \mathcal{S}^{(1)}(A)\right)\right\|_{L^2(L^1(\Omega))} \leq C^k \left\|A - \mathcal{S}^{(1)}(A)\right\|_{L^2(L^1(\Omega))}. \tag{7.1.28}$$

Since the Riesz product $J_\theta(w)$ is nonnegative and $\mathbb{E}(J_\theta) = 1$, the convolution by J_θ gives rise to a contraction such that $\|(\mathcal{J}_\theta A)\| \leq \|A\|$. Hence inequalities (7.1.27) and (7.1.28) imply that,

$$\left(\frac{\theta}{2}\right)\left\|\mathcal{S}^{(1)}(A)\right\| \leq \|A\| + \sum_{k=2}^{\infty}\left(\frac{C\theta}{2}\right)^k \left\|\left(A - \mathcal{S}^{(1)}(A)\right)\right\|. \tag{7.1.29}$$

Write $\epsilon = \theta/2$. By summing the geometric series, we infer from inequality (7.1.29) that

$$\left\|S^{(1)}(A)\right\|\left(\epsilon - \frac{(\epsilon C)^2}{1-\epsilon C}\right) \le \|A\|\left(1 + \frac{(\epsilon C)^2}{1-\epsilon C}\right).$$

Choosing $\epsilon < C^{-2}/8$, we obtain $C_1 > 0$ such that

$$\left\|S^{(1)}(A)\right\|_{L^2(L^1(\Omega))} \le C_1\|A\|_{L^2(L^1(\Omega))}. \tag{7.1.30}$$

Step 4 It remains to define the lifting Steinhaus martingale G. Fix $w \in \mathbb{T}^{\mathbb{N}}$, with $w = (w_j)$, put $\cos_j(w) = (w_j + \overline{w_j})/2$, and define $g_j = \mathbb{E}(A\cos_j(w))$, where integration is over $w \in \mathbb{T}^{\mathbb{N}}$. Then form

$$G(w) = 2\sum g_j w_j.$$

We have $q(g_j) = \mathbb{E}(F\cos_j(w)) = \mathbb{E}(F\overline{w_j})/2 = f_j$. Hence $q(G) = F$ and the Steinhaus martingale G is indeed a lifting of F. By the contraction principle (see Marcus and Pisier 1981, p. 46) we have,

$$\mathbb{E}\left\|\sum g_j w_j\right\|_{L^1}^2 \le C\mathbb{E}\left\|\sum g_j \cos_j(w)\right\|_{L^1}^2,$$

where integration is over $\mathbb{T}^{\mathbb{N}}$. Since $S^1(A) - \sum g_j \cos_j(w)$, the estimate (7.1.30) followed by (7.1.23) finally gives

$$\mathbb{E}\|G\|_{L^1}^2 \le C_2\mathbb{E}\|A\|_{L^1}^2 \le C_3\mathbb{E}\|F\|_X^2.$$

Step 5 In summary, given a Steinhaus martingale $F = (F_n)$ with $F_n(w) = \sum_1^n x_j w_j$, we have obtained a Steinhaus martingale $G = (G_n)$ with $G_n(w) = \sum_1^n g_j w_j$, where $g_j \in L^1$ is a lifting of $x_j \in L^1/Y$, and

$$\mathbb{E}\|G\|_1 \le C\mathbb{E}\|F\|_{L^1/Y}. \tag{7.1.31}$$

Using Khintchine's and Minkowski's inequalities, together with the fact that g_j is a lifting of x_j, gives

$$\frac{2}{\sqrt{\pi}}\mathbb{E}\|G\|_1 \ge \int_\Omega\left(\sum_{j=1}^n |g_j|^2\right)^{1/2} d\mu \ge \left(\sum_{j=1}^n \|x_j\|_{L^1/Y}^2\right)^{1/2}.$$

In view of (7.1.31), this gives inequality (7.1.21). □

7.1.2 The Bourgain–Davis Theorem

The theorem of Bourgain and Davis (1986) asserts that for any reflexive subspace $Y \subset L^1$, the quotient space L^1/Y is of Hardy martingale cotype 2 and satisfies the aUMD property. The central step of its proof consists in lifting L^1/Y-valued Hardy martingales.

Theorem 7.1.3 *Let (Ω, μ) be a probability space, and let Y be a reflexive subspace of $L^1(\Omega)$. We denote*

$$X = L^1(\Omega)/Y \quad \text{and} \quad L^p(L^1) = L^p(\mathbb{T}^{\mathbb{N}}, L^1(\Omega)).$$

For any Hardy martingale $F \in L^p(X)$, $1 \le p < \infty$, there exists a Hardy martingale $G \in L^p(L^1)$ satisfying

$$\|G\|_{L^p(L^1)} \le C\|F\|_{L^p(X)} \quad \text{and} \quad q(G_k) = F_k, \tag{7.1.32}$$

where $q\colon L^1(\Omega) \to X$ denotes the canonical projection and $C = C(p) < \infty$.

Proof For notational simplicity we give the proof of Theorem 7.1.3 only for the critical value $p = 1$. The remaining cases $p > 1$ have a simpler proof. See remarks below. We define $L^p(Y) \subseteq L^p(L^1)$ by the condition

$$F(z) \in Y, \qquad z \in \mathbb{T}^{\mathbb{N}}.$$

The norm on $L^p(Y)$ is the one induced by $L^p(L^1)$. Let P be the orthogonal Riesz projection onto the space of Hardy martingales, let $I_\Omega\colon L^1(\Omega) \to L^1(\Omega)$ be the formal identity, and set $\mathcal{P} = P \otimes I_\Omega$. Since, by assumption, Y is a reflexive space, Theorem 6.2.2 asserts that

$$\left\|\mathcal{P}_{|L^2(Y)}\colon L^2(Y) \to L^2(L^1)\right\| \le C.$$

Step 1 Let $A\colon \mathbb{T}^{\mathbb{N}} \to L^1(\Omega)$ be an integrable martingale lifting of F satisfying $q(A(z)) = F(z)$ and $\|A\|_{L^1(L^1)} \le 2\|F\|_{L^1(X)}$. For fixed $z \in \mathbb{T}^{\mathbb{N}}$, we put

$$\varphi(z) = \int_\Omega |A(z,\omega)| d\mu(\omega),$$

and form the outer function $g = \exp(\ln(\varphi + \epsilon) + i\mathbb{H}\ln(\varphi + \epsilon))$, where $\mathbb{H}$ is the Hilbert transform on $\mathbb{T}^{\mathbb{N}}$. Thus g defines an L^1-limit of an integrable Hardy martingale with $|g| = \varphi + \epsilon$ and $|g|^{-1}$ uniformly bounded. We define

$$G_k = g^{1/2}\mathcal{P}\left(A_k g^{-1/2}\right).$$

By construction, $G = (G_k)$ is a Hardy martingale. We prove in the following paragraphs that G is indeed a lifting of F, and that it satisfies the norm estimates (7.1.32).

Step 2 Let $z \in \mathbb{T}^{\mathbb{N}}$. We prove next that $A(z) - \mathcal{P}A(z) \in Y$. Specifically, we show that for $a_k = \Delta A_k$, we have

$$\mathbb{E}_{k-1}\left(a_k(z)\overline{z_k}^m\right) \in Y, \quad m \in \mathbb{Z}\backslash\mathbb{N}. \tag{7.1.33}$$

Indeed, since $q(a_k) = \Delta F_k$ and $F = (F_k)$ is a Hardy martingale, we infer that for $m \in \mathbb{Z}\backslash\mathbb{N}$,

$$q\left(\mathbb{E}_{k-1}\left(a_k(z)\overline{z_k}^m\right)\right) = \mathbb{E}_{k-1}\left(q(a_k(z))\overline{z_k}^m\right) = \mathbb{E}_{k-1}\left(\Delta F_k(z)\overline{z_k}^m\right) = 0. \tag{7.1.34}$$

For fixed $z \in \mathbb{T}^{\mathbb{N}}$, expand $a_k(z)$ in a Fourier series along the variable z_k,

$$a_k(z) = \sum_{m=-\infty}^{\infty} \mathbb{E}_{k-1}(a_k(z)\overline{z_k}^m)z_k{}^m,$$

and invoke (7.1.34) to obtain the crucial identity

$$q(a_k) = q\left(\mathcal{P}a_k + (\mathrm{Id} - \mathcal{P})a_k\right) = q(\mathcal{P}a_k). \tag{7.1.35}$$

Since, moreover, for $g_k = g^{1/2}\mathcal{P}\left(a_k g^{-1/2}\right)$ we have

$$q(g_k) - q(\mathcal{P}(a_k)), \tag{7.1.36}$$

summing the identities (7.1.35) and (7.1.36) gives $q(G_k) = q(A_k) = F_k$. Thus we obtain that G is a lifting of F.

Step 3 We next turn to proving the norm estimates (7.1.32). Fix $z \in \mathbb{T}^{\mathbb{N}}$ and recall that $A(z) - \mathcal{P}A(z) \in Y$. Hence

$$g(z)^{1/2}A(z) - \mathcal{P}\left(g(z)^{-1/2}A(z)\right) \in Y.$$

Since Y is reflexive, Theorem 6.2.2 implies that

$$\int_{\mathbb{T}^{\mathbb{N}}} \left\|\mathcal{P}\left(Ag^{-1/2}\right)(z)\right\|_{L^1(\Omega)}^2 d\mathbb{P}(z) \le C \int_{\mathbb{T}^{\mathbb{N}}} \left\|\left(Ag^{-1/2}\right)(z)\right\|_{L^1(\Omega)}^2 d\mathbb{P}(z). \tag{7.1.37}$$

Apply the Cauchy–Schwarz inequality to the product $G = g^{1/2}\mathcal{P}\left(Ag^{-1/2}\right)$. This gives

$$\int_{\mathbb{T}^{\mathbb{N}}} \|G(z)\|_{L^1} d\mathbb{P} \le \left(\int_{\mathbb{T}^{\mathbb{N}}} |g| d\mathbb{P}\right)^{1/2} \left(\int_{\mathbb{T}^{\mathbb{N}}} \left\|\mathcal{P}\left(Ag^{-1/2}\right)(z)\right\|_{L^1}^2 d\mathbb{P}\right)^{1/2}. \tag{7.1.38}$$

As $|g(z)| = \|A(z)\|_{L^1(\Omega)} + \epsilon$, the estimates (7.1.37) and (7.1.38) give

$$\int_{\mathbb{T}^{\mathbb{N}}} \|G(z)\|_{L^1(\Omega)} d\mathbb{P}(z) \le C \int_{\mathbb{T}^{\mathbb{N}}} \|A(z)\|_{L^1(\Omega)} d\mathbb{P}(z).$$

Finally, recall that we chose A to satisfy $\|A\|_{L^1(L^1)} \le 2\|F\|_{L^1(X)}$. □

Theorem 7.1.3 allows us to obtain the Bourgain–Davis theorem, by reduction to the L^1-estimates (7.1.1) and (7.1.2).

Theorem 7.1.4 *Let (Ω, μ) be a probability space. Let Y be a reflexive space of $L^1(\Omega)$. Then*

$$X = L^1(\Omega)/Y$$

satisfies the aUMD property and is of Hardy martingale cotype 2.

Proof We write $L^1 = L^1(\Omega)$. Let (F_j) be an X-valued Hardy martingale. By Theorem 7.1.3 there exists $C = C(Y)$ and an L^1-valued Hardy martingale (G_k) such that

$$\mathbb{E}\|G_k\|_{L^1} \leq C\mathbb{E}\|F_k\|_X \quad \text{and} \quad q(G_k) = F_k,$$

where $q\colon L^1 \to L^1/Y$ denotes the canonical quotient map. By estimates (7.1.1) and (7.1.2),

$$\mathbb{E}\left\|\sum_{j=1}^{k} \epsilon_j \Delta G_j\right\|_{L^1} \leq C\mathbb{E}\|G_k\|_{L^1} \quad \text{and} \quad \left(\sum_{j=1}^{k} (\mathbb{E}\|\Delta G_j\|_{L^1})^2\right)^{1/2} \leq C\mathbb{E}\|G_k\|_{L^1}$$

for every sequence $\epsilon_j \in \{-1, 1\}$. Since ΔG_j is a lifting of ΔF_j we obtain the estimates claimed in Theorem 7.1.4, that is

$$\mathbb{E}\left\|\sum_{j=1}^{k} \epsilon_j \Delta F_j\right\|_{X} \leq C\mathbb{E}\|F_k\|_X, \quad \text{and} \quad \left(\sum_{j=1}^{k} (\mathbb{E}\|\Delta F_j\|_X)^2\right)^{1/2} \leq C\mathbb{E}\|F_k\|_X. \qquad \square$$

We close this section by proving a Hardy martingale lifting theorem for L^1/Y where Y is a closed subspace of $L^1(\Omega)$, not necessarily relexive. Clearly the norm estimates of the lifted Hardy martingale will reflect the lack of reflexivity in Y.

Theorem 7.1.5 *Let Y be a closed subspace of $L^1(\Omega)$, let $0 < \alpha < 1$, and denote*

$$X = L^1(\Omega)/Y, \qquad L^p(L^\alpha) = L^p(\mathbb{T}^{\mathbb{N}}, L^\alpha(\Omega)).$$

For any $1 \leq p < \infty$, there exists $C = C(p, \alpha) < \infty$ such that for any Hardy martingale $F \in L^p(X)$ exists a Hardy martingale lifting $G \in L^p(L^\alpha)$ satisfying

$$\int_{\mathbb{T}^{\mathbb{N}}} \|G\|_{L^\alpha}^p d\mathbb{P} \leq C \int_{\mathbb{T}^{\mathbb{N}}} \|F\|_X^p d\mathbb{P} \quad \textit{and} \quad q(G_k) = F_k. \tag{7.1.39}$$

Proof The assertion of Theorem 7.1.5 includes the statement that for any fixed $0 < a < 1$,

$$\limsup_{p \to 1} C(p, \alpha) < \infty,$$

and that for any fixed $1 \le p < \infty$ we have $\limsup_{\alpha\to 1} C(p,\alpha) = \infty$. For notational simplicity we consider only the critical value $p = 1$. See remarks below. We let again P denote the orthogonal Riesz projection, and $J_\Omega : L^1(\Omega) \to L^\alpha(\Omega)$ be the formal inclusion map. Theorem 6.1.4 applied to $\mathcal{P} = P \otimes J_\Omega$ asserts that

$$\left\| \mathcal{P} \colon L^2(L^1) \to L^2(L^\alpha) \right\| \le C,$$

where $C = C_\alpha$. Fix $p = 1$ and let A be an integrable martingale lifting of F, satisfying $\|A\|_{L^p(L^1)} \le 2\|F\|_{L^p(X)}$. For fixed $z \in \mathbb{T}^{\mathbb{N}}$, we put

$$\varphi(z) = \int_\Omega |A(z,\omega)| d\mu(\omega),$$

and form the outer function $g = \exp(\ln(\varphi + \epsilon) + i\mathbb{H}\ln(\varphi + \epsilon))$. Thus g is an integrable Hardy martingale with $|g| = \varphi + \epsilon$, and $|g|^{-1}$ uniformly bounded. We define a Hardy martingale $G = (G_k)$ by putting

$$G_k = g^{1/2}\mathcal{P}\left(A_k g^{-1/2}\right).$$

Applying the second step of the proof of Theorem 7.1.5 shows that G is a lifting of F. We finally turn to proving the norm estimates (7.1.39). Apply the Cauchy–Schwarz inequality to the product $G = g^{1/2}\mathcal{P}\left(A g^{-1/2}\right)$. This gives

$$\int_{\mathbb{T}^{\mathbb{N}}} \|G(z)\|_{L^\alpha} d\mathbb{P} \le \left(\int_{\mathbb{T}^{\mathbb{N}}} |g| d\mathbb{P}\right)^{1/2} \left(\int_{\mathbb{T}^{\mathbb{N}}} \left\|\mathcal{P}\left(A g^{-1/2}\right)(z)\right\|_{L^\alpha}^2 d\mathbb{P}\right)^{1/2}.$$

By Theorem 6.1.4,

$$\int_{\mathbb{T}^{\mathbb{N}}} \left\|\mathcal{P}\left(A g^{-1/2}\right)(z)\right\|_{L^\alpha}^2 d\mathbb{P}(z) \le C_\alpha \int_{\mathbb{T}^{\mathbb{N}}} \|A(z)\|_{1,1}^2 |g(z)|^{-1} d\mathbb{P}(z).$$

Since $|g(z)| = \|A(z)\|_{L^1(\Omega)} + \epsilon$, taking $\epsilon \to 0$, gives (7.1.39). □

Remark: The proof above also shows that for general closed subspaces $Y \subseteq L^1(\Omega)$ and $X = L^1(\Omega)/Y$, the Hardy martingale $G = (G_k)$ given by

$$G_k = \mathcal{P}(A_k)$$

is a lifting of F and, for $1 < p < \infty$, it satisfies

$$\|G\|_{L^p(L^\alpha)} \le C_{p,\alpha} \|F\|_{L^p(X)}, \tag{7.1.40}$$

where $C_{p,\alpha} \le \frac{C_\alpha p^2}{p-1}$. In the special case when Y is reflexive we get

$$\|G\|_{L^p(L^1)} \le C_p \|F\|_{L^p(X)}, \tag{7.1.41}$$

where

$$L^p(X) = L^p(\mathbb{T}^{\mathbb{N}}, X),$$

and $C_p \leq \frac{p^2 C}{p-1}$. The simplified argument sketched here does not cover the value $p = 1$ in (7.1.40), respectively (7.1.41), for G.

7.2 The Trace Class

In Chapter 5 we derived Hardy martingale cotype 2 and the aUMD property for the space L^1, and we continued in Section 7.1.2 by proving that these properties pass over to the spaces L^1/Y, for every reflexive subspace $Y \subset L^1$.

We now turn to a noncommutative L^1 space. The trace class S^1 is defined in Section 1.1.4 where its most basic properties are listed. For a thorough discussion we refer to Wojtaszczyk (1991, Chapter III.G). We prove the theorem of Kwapien and Pełczyński which asserts that if K is infinite dimensional, then S^1 does not embed into a Banach space with an unconditional basis, and the theorem of Haagerup and Pisier that S^1 is of Hardy martingale cotype 2 and fails to satisfy the aUMD property.

Given a separable Hilbert space K we let $S^1 = S^1(K)$ denote the trace class of linear operators on K. Let $\{e_1, \ldots, e_n, \ldots\}$ be an orthonormal basis in K. Then

$$e_{i,j} = e_i \otimes e_j, \qquad i, j \in \mathbb{N},$$

ordered along the layers $\mathcal{L}_k = \{(i, j) : \max\{i, j\} = k\}$ forms a Schauder basis in $S^1(K)$. Given $A \in S^1(K)$ we write $A = (a_{i,j})$ if $A = \sum a_{i,j} e_{i,j}$, where the summation is taken along increasing $\mathcal{L}_k$. Recall that $\|A\|_{S^1} = \sup_{n \in \mathbb{N}} \left\| \sum_{i,j=1}^{n} a_{i,j} e_{i,j} \right\|_{S^1}$. On S^1, the main triangular projection is defined as

$$U^{(n)}(A) = \sum_{i=1}^{n} \sum_{j=i+1}^{n} a_{i,j} e_{i,j}. \tag{7.2.1}$$

With respect to trace duality $U = U^{(n)}$ is a self-adjoint operator. Indeed, we have

$$\operatorname{tr}(B^* U(A)) = \operatorname{tr}(U(B)^* A) = \overline{\operatorname{tr}(A^* U(B))} \tag{7.2.2}$$

and

$$\sup\{\|U(A)\|_1 : \|A\|_1 \leq 1\} = \sup\{\|U(B)\|_\infty : \|B\|_\infty \leq 1\}. \tag{7.2.3}$$

Let Y be a Banach space with an unconditional Schauder basis $(x_r)_{r=1}^\infty$, and biorthogonal functionals $(x_r^*)_{r=1}^\infty$ (defined by $\langle x_r^*, x_r \rangle = 1$ and $\langle x_r^*, x_s \rangle = 0$, if $r \neq s$). The unconditional basis constant of $(x_r)_{r=1}^\infty$ is defined by

$$\operatorname{ubc}(x_r) = \sup_{(\varepsilon_r)\in\{0,1\}^{\mathbb{N}}} \sup_{x\in Y,\, \|x\|\le 1} \sup_{n\in\mathbb{N}} \left\| \sum_{r=1}^{n} \varepsilon_r \langle x_r^*, x\rangle x_r \right\|_Y . \tag{7.2.4}$$

The Kwapien-Pełczyński theorem (Theorem 7.2.1) provides a deep connection between the main triangular projection and unconditionality in the trace class. First, if a Banach space Y contains a copy of S^1 then the unconditional basis constant of Y has to be larger than (a constant multiple of)

$$\sup_{n\in\mathbb{N}} \left\| U^{(n)} : S^1 \to S^1 \right\| .$$

Second, the operator norm of the main triangular projections satisfies the following lower bound,

$$\left\| U^{(n)} : S^1 \to S^1 \right\| \ge c \ln(n+1). \tag{7.2.5}$$

Theorem 7.2.1 is viewed as the noncommutative analoge to Pełczyński's theorem 4.2.12, which asserts asserting that $L^1(\Omega,\mu)$ does not embed in any Banach space with an unconditional basis unless (Ω,μ) is purely atomic, in which case $L^1(\Omega,\mu)$ is isometric to ℓ^1.

Theorem 7.2.1 (Kwapień–Pełczyński theorem) *If K is an infinite-dimensional separable Hilbert space, and $X = S^1(K)$, then for any Banach space Y, and any linear injection $T : X \to Y$,*

$$\|T\| \cdot \left\| T^{-1}_{|T(X)} \right\| < \infty \quad \textit{implies} \quad \operatorname{ubc}(Y) = \infty. \tag{7.2.6}$$

If $X = S^1(K)$, with $\dim K = n$, and if Y is a Banach space with an unconditional basis, then for any injective linear operator $T : X \to Y$,

$$\|T\| \cdot \left\| T^{-1}_{|T(X)} \right\| \cdot \operatorname{ubc}(Y) \ge c \ln(n+1). \tag{7.2.7}$$

Since the space S^1 is a separable dual space, it satisfies the RNP and the aRNP. In this section we show the results of Haagerup and Pisier (1989), that S^1 is of Hardy martingale cotype 2, and that it fails the aUMD.

Theorem 7.2.2 *If K is an infinite-dimensional separable Hilbert space and $X = S^1(K)$, then for every X-valued Hardy martingale $F = (F_m)$ and $n \in \mathbb{N}$,*

$$\sum_{m=1}^{n} (\mathbb{E}\|\Delta F_m\|_X)^2 \le 2(\mathbb{E}\|F_n\|_X)^2. \tag{7.2.8}$$

If $X = S^1(K)$, with $\dim K = n$, then there exists a Hardy martingale $F = (F_m)_{m=1}^{2n}$ of length $2n$ such that

$$\mathbb{E} \left\| \sum_{m=1}^{n} \Delta F_{2m} \right\|_X \ge c \ln(n+1)(\mathbb{E}\|F_{2n}\|_X). \tag{7.2.9}$$

In Section 7.2.1 we prove that the Banach space S^1 is H^1-uniformly convex, and derive from it the fact that it is of Hardy martingale cotype 2. In Section 7.2.2 we prove that S^1 lacks the aUMD property, and does not embed into a Banach space with an unconditional basis. In the proof of those results the main triangular projection plays a crucial role.

7.2.1 Complex Uniform Convexity

We turn to proving the first part of Theorem 7.2.2. The approach is based on the factorization theorem of Sarason (1967), which asserts asserting that for each $f \in H^1(\mathbb{T}, S^1)$ there exist $g, h \in H^2(\mathbb{T}, S^2)$ so that with $g(0) = \int g dm$, and $h(0) = \int h dm$,

$$f = g \cdot h, \quad f(0) = g(0)h(0), \quad \|f\|_{H^1(S^1)} = \|g\|_{H^2(S^2)}\|h\|_{H^2(S_2)}.$$

Theorem 7.2.3 *For every S^1-valued Hardy martingale (F_k),*

$$(\mathbb{E}\|F_{k-1}\|_{S^1})^2 + \frac{1}{2}(\mathbb{E}\|\Delta F_k\|_{S^1})^2 \leq (\mathbb{E}\|F_k\|_{S^1})^2, \tag{7.2.10}$$

for $k \in \mathbb{N}$.

By summing a telescoping series we obtain property (7.2.8) from Theorem 7.2.3. The essence of the argument consists in proving that S^1 is H^1-uniformly convex.

Proposition 7.2.4 *Let $f \in H^1(S^1)$, and $f_0 = \int_{\mathbb{T}} f dm$. Then*

$$\|f_0\|_{S^1}^2 + \frac{1}{2}\|f - f_0\|_{H^1(S^1)}^2 \leq \|f\|_{H^1(S^1)}^2.$$

Proof By Sarason's theorem there are $g, h \in H^2(S^2)$ so that with $g_0 = \int g dm$ and $h_0 = \int h dm$

$$f = g \cdot h, \quad f_0 = g_0 h_0, \quad \|f\|_{H^1(S^1)} = \|g\|_{H^2(S^2)}\|h\|_{H^2(S^2)}. \tag{7.2.11}$$

By the first two identities in (7.2.11) we get

$$f - f_0 = g(h - h_0) + (g - g_0)h_0. \tag{7.2.12}$$

The triangle inequality applied to (7.2.12) followed by the Cauchy–Schwarz inequality gives

$$\|f - f_0\|^2_{H^1(S^1)} \le 2\|g\|^2_{H^2(S^2)}\|h - h_0\|^2_{H^2(S^2)} + 2\|g - g_0\|^2_{H^2(S^2)}\|h_0\|^2_{H^2(S^2)}. \tag{7.2.13}$$

Next we use that $f_0 = g_0h_0$ to obtain

$$\|f_0\|^2_{S^1} \le \|g_0\|_{S^2}\|h_0\|_{S^2}. \tag{7.2.14}$$

Consider the weighted sum of (7.2.14) and (7.2.13) to see that

$$\|f_0\|^2_{S^1} + 1/2\|f - f_0\|^2_{H^1(S^1)} \le \left(\|g_0\|^2_{S^2} + \|g - g_0\|^2_{H^2(S^2)}\right)\|h_0\|^2_{S^2} + \|g\|^2_{H^2(S^2)}\|h - h_0\|^2_{H^2(S^2)}. \tag{7.2.15}$$

Note that, by Pythagoras's theorem,

$$\|g\|^2_{H^2(S^2)} = \|g_0\|^2_{S^2} + \|g - g_0\|^2_{H^2(S^2)}.$$

Hence the right-hand side term in (7.2.15) equals

$$\|g\|^2_{H^2(S^2)}\|h_0\|^2_{S^2} + \|g\|^2_{H^2(S^2)}\|h - h_0\|^2_{H^2(S^2)}.$$

Invoking Pythagoras's theorem again, and using (7.2.11) finally gives that

$$\|f_0\|^2_{S^1} + \frac{1}{2}\|f - f_0\|^2_{H^1(S^1)} \le \|g\|^2_{H^2(S^2)}\|h\|^2_{H^2(S^2)} = \|f\|^2_{H^1(S^1)}.$$ □

Proof of Theorem 7.2.3: We show that (7.2.10) holds true for $k \in \mathbb{N}$. To this end we condition to $\mathcal{F}_{k-1}$ by fixing $x \in \mathbb{T}^{k-1}$. Define $f \in H^1(S^1)$ by $f(w) = F_k(x, w)$. Then $f \in H^1(S^1)$ and $f_0 = F_{k-1}(x)$. Next, applying Proposition 7.2.4 gives the conditional estimate,

$$\|f_0\|^2_{S^1} + \frac{1}{2}\left(\int_{\mathbb{T}} \|f - f_0\|_{S^1} dm\right)^2 \le \left(\int_{\mathbb{T}} \|f\|_{S^1} dm\right)^2,$$

which we restate as

$$\|F_{k-1}\|^2_{S^1} + \frac{1}{2}(\mathbb{E}_{k-1}\|\Delta F_k\|_{S^1})^2 \le (\mathbb{E}_{k-1}\|F_k\|_{S^1})^2. \tag{7.2.16}$$

By Minkowski's inequality, we have

$$\left((\mathbb{E}\,\|F_{k-1}\|_{S^1})^2 + \frac{1}{2}\,(\mathbb{E}\,\|\Delta F_k\|_{S^1})^2\right)^{1/2} \le \mathbb{E}\left(\|F_{k-1}\|^2_{S^1} + \frac{1}{2}\,(\mathbb{E}_{k-1}\,\|\Delta F_k\|_{S^1})^2\right)^{1/2}.$$

Take the square root in (7.2.16) and integrate. Comparing, finally, the result with the last line gives (7.2.10).

7.2.2 Unconditional Convergence

In this section we prove the second part of Theorem 7.2.2. Let K be a complex Hilbert space with $\dim K = n$ and let $S^1 = S^1(K)$ denote the class of complex $n \times n$ matrices A, equipped with the norm $\|A\|_{S^1} = tr(A^*A)^{1/2}$. Recall that in (7.2.1) we defined the main triangular projection $U^{(n)}$ on S^1 by

$$U^{(n)}(A) = \sum_{i=1}^{n} \sum_{j=i+1}^{n} a_{i,j} e_{i,j}.$$

As we noted in (7.2.5), Gohberg and Krein (1970) and Kwapień and Pełczyński (1970) proved that the main triangular projections satisfy the lower bounds

$$\left\| U^{(n)} : S^1 \to S^1 \right\| \geq c \ln(n+1). \tag{7.2.17}$$

In Proposition 7.2.5 we re-express the operator $U^{(n)}$ by means of a martingale transform acting on Hardy martingales. Hence the norm of the main triangular projection provides an obstruction to proving martingale transform estimates for Hardy martingales. Thus (7.2.9) of Theorem 7.2.2 follows from the next proposition.

Proposition 7.2.5 *Let K be a complex Hilbert space with* $\dim K = n$ *and let* $S^1 = S^1(K)$. *For each* $A \in S^1$ *with* $A = (a_{i,j})$ *and* $z \in \mathbb{T}^{\mathbb{N}}$ *with* $z = (z_j)$ *we define the* S^1*-valued Hardy martingale* F *by setting*

$$F(z) = (a_{i,j} z_{2i+1} z_{2j})_{i,j}^{n}.$$

Then, for $z \in \mathbb{T}^{\mathbb{N}}$ *fixed,* $\|F(z)\|_{S^1} = \|A\|_{S^1}$ *and*

$$\left\| \sum_{j=1}^{n} \Delta F_{2j}(z) \right\|_{S^1} = \left\| U^{(n)}(A) \right\|_{S^1}. \tag{7.2.18}$$

Proof Let $A = (a_{i,j})_{i,j}^{n}$, and $(w_i), (v_j) \in \mathbb{T}^n$. Recall that the S^1 norm is invariant under multiplication by elementary products $\{w_i v_j\}$. Hence the matrix

$$B = (w_i v_j a_{i,j})_{i,j}^{n}$$

satisfies $\|A\| = \|B\|$ where the norm is taken in S^1. Clearly this gives $\|F(z)\| = \|A\|$. Next we turn to proving the identity (7.2.18). We expand $F(z)$ using the unit vector basis $\{e_{i,j}\}$ and obtain

$$F(z) = \sum_{j=1}^{n} \left[\sum_{i=1}^{j-1} a_{i,j} e_{i,j} z_{2i+1} \right] z_{2j} + \sum_{j=1}^{n} \left[\sum_{i=j}^{n} a_{i,j} e_{i,j} z_{2i+1} \right] z_{2j}.$$

Next, rewriting the second term on the right-hand side and leaving the first term unchanged we obtain

$$F(z) = \sum_{j=1}^{n} \left[\sum_{i=1}^{j-1} a_{i,j} e_{i,j} z_{2i+1} \right] z_{2j} + \sum_{j=1}^{n} \left[\sum_{i=1}^{j} a_{j,i} e_{j,i} z_{2i} \right] z_{2j+1}. \tag{7.2.19}$$

Since the conditional expectation $\mathbb{E}_j(F)$ on $\mathbb{T}^{\mathbb{N}}$ is given by integrating out the variables z_{j+1}, z_{j+2} etc., we have

$$\mathbb{E}_j F(z) = \mathbb{E}_w F(z_1, \ldots, z_j, w),$$

where $\mathbb{E}_w$ denotes integration with respect to $w \in \mathbb{T}^{\mathbb{N}}$. We now claim that

$$\mathbb{E}_{2j} F(z) = \mathbb{E}_{2j-1} F(z) + r_{2j}, \quad \text{where} \quad r_{2j} = \left[\sum_{i=1}^{j-1} a_{i,j} e_{i,j} z_{2i+1} \right] z_{2j}. \tag{7.2.20}$$

To see that (7.2.20) holds, we observe that $\mathbb{E}_{2j-1} F(z) = \sum_{k=1}^{j-1} d_k(z)$ where

$$d_k(z) = \left[\sum_{i=1}^{k-1} a_{i,k} e_{i,k} z_{2i+1} \right] z_{2k} + \left[\sum_{i=1}^{k} a_{k,i} e_{k,i} z_{2i} \right] z_{2k+1},$$

which together with (7.2.19) gives (7.2.20). Taking the sum over j in (7.2.20), we obtain

$$\sum_{j=1}^{n} \Delta F_{2j}(z) = \sum_{j=1}^{n} \left[\sum_{i=1}^{j-1} a_{i,j} e_{i,j} z_{2i+1} \right] z_{2j}.$$

Invoking once more that the S^1 norm is invariant under multiplication by elementary products such as $\{z_{2i+1} z_{2j}\}$ we finally obtain

$$\left\| \sum_{j=1}^{n} \left[\sum_{i=1}^{j-1} a_{i,j} e_{i,j} z_{2i+1} \right] z_{2j} \right\| = \left\| \sum_{j=1}^{n} \sum_{i=1}^{j-1} a_{i,j} e_{i,j} \right\| = \left\| U^{(n)} A \right\|,$$

where the norm is taken in the space S^1. □

For $A = (a(i, j)) \in S^1$, define the deletion operator $D^{(n)} : S^1 \to S^1$, by

$$D^{(n)}(A) = \sum_{i=1}^{n} \sum_{j=1}^{n} d(i, j) a(i, j) e_{i,j}, \tag{7.2.21}$$

where

$$d(i, j) = \begin{cases} 1 & \max(i, j) \in \{2k : k \in \mathbb{N}\}, \\ 0 & \max(i, j) \in \{2k - 1 : k \in \mathbb{N}\}. \end{cases} \tag{7.2.22}$$

Theorem 7.2.6 *Let Y be a Banach space with an unconditional Schauder basis $(x_r)_{r=1}^{\infty}$. Assume that there exists $0 < A_0 < \infty$ and a linear operator $T: S^1 \to Y$ such that*

$$A_0^{-1/2}\|y\|_{S^1} \leq \|T(y)\|_Y \leq A_0^{1/2}\|y\|_{S^1}, \quad y \in S^1. \tag{7.2.23}$$

Then

$$\left\|D^{(n)}: S^1 \to S^1\right\| \leq 4A_0\mathrm{ubc}(x_r), \quad n \in \mathbb{N}.$$

Proof The proof is composed of two distinct parts. First, the conclusion of the theorem is obtained for an embedding T that satisfies the additional condition (7.2.24). In the second part we provide the reduction to the special case treated in the first part.

Part 1 (Gliding humps) We first assume that in addition to (7.2.23), the operator $T: S^1 \to Y$ satisfies the conditions

$$\lim_{i\to\infty}\langle x_r^*, Te_{i,j}\rangle = \lim_{j\to\infty}\langle x_r^*, Te_{i,j}\rangle = 0, \quad r \in \mathbb{N}. \tag{7.2.24}$$

Fix $n \in \mathbb{N}$ and $\epsilon > 0$. Choose $A \in S^1$ with $A = (a(i,j))$ such that $\|A\|_{S^1} = 1$ and

$$\left\|D^{(n)}(A)\right\|_{S^1} \geq \left\|D^{(n)}: S^1 \to S^1\right\| - \epsilon. \tag{7.2.25}$$

In view of condition (7.2.24), a gliding hump argument yields the existence of strictly increasing integers $(p(k))_{k=1}^n$, $(q(k))_{k=1}^n$, and $(m(k))_{k=1}^{n+1}$, such that

$$b_k = \sum_{j=1}^{k} a(k,j)e_{p(k),q(j)} + \sum_{i=1}^{k-1} a(i,k)e_{p(i),q(k)}$$

satisfies the estimates

$$\left\|\sum_{r=m(k)+1}^{m(k+1)}\langle x_r^*, Tb_k\rangle x_r - Tb_k\right\|_Y + \left\|\sum_{r=m(k)+1}^{m(k+1)}\langle x_r^*, Tb_l\rangle x_r\right\|_Y \leq \epsilon, \tag{7.2.26}$$

for $1 \leq k \neq l \leq n$. Now specify $n = 2m+1$, and set

$$B = \sum_{k=1}^{n} b_k, \quad x = T(B), \quad C = \sum_{l=1}^{m} b_{2l}, \quad y = T(C).$$

In view of (7.2.26) we have

$$\left\|\sum_{l=1}^{m}\sum_{r=m(2l)+1}^{m(2l+1)}\langle x_r^*, x\rangle x_r - y\right\|_Y \leq m\epsilon. \tag{7.2.27}$$

By condition (7.2.23), $\|y\|_Y \geq A_0^{-1/2}\|C\|_{S^1}$. Since $\|C\|_{S^1} = \left\|D^{(n)}(A)\right\|_{S^1}$, (7.2.27) yields

$$\left\| \sum_{l=1}^{m} \sum_{r=m(2l)+1}^{m(2l+1)} \langle x_r^*, x\rangle x_r \right\|_Y + 2m\epsilon \geq \|y\|_Y \geq A_0^{-1/2} \left\|D^{(n)}(A)\right\|_{S^1}. \tag{7.2.28}$$

On the other hand, $\|x\|_Y \leq A_0^{1/2}\|B\|_{S^1}$ and $\|B\|_{S^1} \leq \|A\|_{S^1}$. By the unconditionality of the basis (x_r),

$$\left\| \sum_{l=1}^{m} \sum_{r=m(2l)+1}^{m(2l+1)} \langle x_r^*, x\rangle x_r \right\|_Y \leq \mathrm{ubc}(x_r)\|x\|_Y \leq \mathrm{ubc}(x_r)A_0^{1/2}\|A\|_{S^1}. \tag{7.2.29}$$

Combining (7.2.28) and (7.2.29), taking into account that $\epsilon > 0$ was arbitrary, we finally arrive at

$$\mathrm{ubc}(x_r)\|A\|_{S^1} \geq A_0^{-1} \left\|D^{(n)}(A)\right\|_{S^1}.$$

Part 2 (Reduction) Fix a linear operator $R\colon S^1 \to Y$ such that

$$A_0^{-1/2}\|y\|_{S^1} \leq \|R(y)\|_Y \leq A_0^{1/2}\|y\|_{S^1}, \quad y \in S^1. \tag{7.2.30}$$

Fix a biorthogonal functional x_r^*. Clearly, forming

$$(\langle x_r^*, R\,(e_{k,l})\rangle)_{k,l=1}^{\infty}$$

gives a (doubly indexed) bounded sequence of scalars. A Cantor diagonal argument yields subsequences $(k(i))$ and $(l(j))$ such that each of the sequences

$$i \to \left\langle x_r^*, R(e_{k(i),l(j)})\right\rangle \quad \text{and} \quad j \to \left\langle x_r^*, R(e_{k(i),l(j)})\right\rangle$$

are convergent. Next, put

$$E(i,j) = e_{k(2i),l(2j)} + e_{k(2i-1),l(2j-1)} - e_{k(2i),l(2j-1)} - e_{k(2i-1),l(2j)},$$

and define $V\colon S^1 \to S^1$ by setting

$$V(e_{i,j}) = E(i,j), \quad 1 \leq i,j \leq n. \tag{7.2.31}$$

Since $\mathrm{tr}E(i,j)E^*(i,j) = 2$ we obtain,

$$\frac{1}{2}\|A\|_{S^1} \leq \|V(A)\|_{S^1} \leq 4\|A\|_{S^1}, \quad A \in S^1. \tag{7.2.32}$$

The operator $T\colon S^1 \to S^1$ formed by $T = R \circ V$ satisfies the hypothesis (7.2.24) and (7.2.23) with A_0 replaced by $4A_0$. Hence (7.2.29) holds true with A_0 replaced by $4A_0$. □

Estimates for $\|U^{(n)}\|$ and $\|D^{(n)}\|$

We now present the link between the main triangular projection $U^{(n)}$ and the deletion operator $D^{(n)}$ (defined by (7.2.21) and (7.2.22)). Fix an odd integer n, and define the bijection $\tau\colon \{1,\dots,n\} \to \{1,\dots,n\}$ by

$$\tau(i) = \begin{cases} (n+2-i)/2 & \text{if} \quad 1 \le i \le n \quad \text{is odd,} \\ (n+1+i)/2 & \text{if} \quad 1 < i < n \quad \text{is even.} \end{cases} \tag{7.2.33}$$

Define the bijection $\sigma\colon \{1,\dots,n\} \to \{1,\dots,n\}$ by $\sigma(i) = n - \tau(i)$. Thus, if $1 \le i \le n$ is an odd integer and $1 \le j \le i$, then

$$\tau(i) \le \sigma(j) + 1 \cdot \quad \text{and} \quad \tau(j) \le \sigma(i). \tag{7.2.34}$$

If $1 < i < n$ is an even integer and $1 \le j \le i$, then

$$\tau(i) > \sigma(j) + 1 \quad \text{and} \quad \tau(j) > \sigma(i) + 1. \tag{7.2.35}$$

Define the permutation operator $P^{(n)}\colon S^1 \to S^1$ by

$$P^{(n)} e_{i,j} = e_{\tau(i),\sigma(j)}, \quad 1 \le i,j \le n. \tag{7.2.36}$$

It follows from (7.2.34) and (7.2.35) that

$$(P^{(n)})^{-1}[U^{(n)}]P^{(n)} = D^{(n)}. \tag{7.2.37}$$

The identity (7.2.37) yields the equality of operator norms

$$\left\|U^{(n)}\colon S^1 \to S^1\right\| = \left\|D^{(n)}\colon S^1 \to S^1\right\|. \tag{7.2.38}$$

We complete the respective proofs of Theorem 7.2.2 and Theorem 7.2.1 by showing that the sequence of main triangular projections is not uniformly bounded on S^1. In view of the trace duality formula, this is the same as saying that it is not uniformly bounded on S^∞. We obtain our lower bounds by testing $U^{(n)}$ on the Hilbert matrices

$$h_n(i,j) = (i - j + 1/2)^{-1}, \qquad 1 \le i,j \le n.$$

Hilbert's classical inequality gives uniform bounds for the bilinar forms associated to h_n. Thus there exists $C > 0$ such that

$$\sum_{i=1}^{n}\sum_{j=1}^{n} x(i)h_n(i,j)y(j) \le C\left(\sum_{i=1}^{n}|x(i)|^2\right)^{1/2}\left(\sum_{j=1}^{n}|y(j)|^2\right)^{1/2}.$$

This clearly gives $\sup_{n\in\mathbb{N}} \|h_n\|_\infty \le C$. Next, consider $k_n = U^{(n)}h_n$. We test the operator norm of k_n by choosing $x(i) = 1/\sqrt{n}$ and $y(j) = -1/\sqrt{n}$. With defintion (7.2.1) and by changing the order of summation we get

$$\sum_{i=1}^{n}\sum_{j=1}^{n} x(i)k_n(i,j)y(j) = n^{-1}\sum_{j=1}^{n}\sum_{i=1}^{j-1}(j-i-1/2)^{-1}$$
$$\geq n^{-1}c\sum_{j=1}^{n}\ln(j+1)$$
$$\geq c\ln(n+1).$$

Thus $\|k_n\|_\infty \geq c\ln(n+1)$. Since $k_n = U^{(n)}h_n$, we finally obtain (7.2.5). With trace duality (7.2.3), this gives (7.2.17), i.e.,

$$\left\|U^{(n)}: S^1 \to S^1\right\| \geq c\ln(n+1). \tag{7.2.39}$$

We next derive a corresponding upper bound for $\left\|U^{(n)}\right\|_{S^1}$. Let $m \in \mathbb{N}$, let $I(k), J(k)$ $(1 \leq k \leq m)$ be finite sets of integers satisfying

$$I(k) \cap I(l) = \emptyset, \quad J(k) \cap J(l) = \emptyset, \tag{7.2.40}$$

for $1 \leq k \neq l \leq m$. Given a set I we let 1_I denote its indicator function. Next, form the projection operator

$$P(A) = \sum_{k=1}^{m}\sum_{i=1}^{\infty}\sum_{j=1}^{\infty} 1_{I(k)}(i)1_{J(k)}(j)a(i,j)e_{i,j}, \quad A = (a(i,j)). \tag{7.2.41}$$

We claim that P is a contraction on S^1 (i.e., $\|P(A)\| \leq \|A\|$). To see this, consider first the matrix

$$B = \sum_{k=1}^{m}\sum_{l=1}^{m}\sum_{i=1}^{\infty}\sum_{j=1}^{\infty} 1_{I(k)}(i)1_{J(l)}(j)a(i,j)e_{i,j}. \tag{7.2.42}$$

Note that B is obtained from A by deleting rows and columns. Hence, clearly, $\|B\| \leq \|A\|$. Next, for $1 \leq k \leq m$ let $r_k: [0,1] \to \{-1,+1\}$ denote the independent Bernoulli random variables, and form the matrices

$$B(t) = \sum_{k=1}^{m}\sum_{l=1}^{m}\sum_{i=1}^{\infty}\sum_{j=1}^{\infty} 1_{I(k)}(i)1_{J(l)}(j)r_i(t)r_j(t)a(i,j)e_{i,j}. \tag{7.2.43}$$

Note that $\|B(t)\| = \|B\|$ for $t \in [0,1]$, and since $\int_0^1 r_k(t)r_l(t)dt = 0$ for $k \neq l$, we find

$$\int_0^1 B(t)dt = \sum_{k=1}^{m}\sum_{i=1}^{\infty}\sum_{j=1}^{\infty} 1_{I(k)}(i)1_{J(k)}(j)a(i,j)e_{i,j}. \tag{7.2.44}$$

Summing up, $\|P(A)\| \leq \|A\|$ as claimed.

The main triangular projection $U^{(n)}$ may be rewritten as the sum of approximately $\log n$ contractive projections, each satisfying the structural representation (7.2.40) and (7.2.41). This gives

$$\left\|U^{(n)}: S^1 \to S^1\right\| \leq C \ln(n+1). \tag{7.2.45}$$

We next specify $n = 2^m$ and construct the representation in $m = \log_2 n$ steps as follows:

Step 1 We are given integer intervals $I^0 = J^0 = [1, n]$, and form the triangle

$$T_1^0 = \left\{(i, j) \in I^0 \times J^0 : i + j \leq n + 1\right\}.$$

We set $I_1^1 = J_1^1 = [1, n/2]$, and define the square $S_1^0 = I_1^1 \times J_1^1$. Next, put $I_2^1 = J_2^1 = [n/2 + 1, n]$, and form the triangles of stage 1,

$$T_1^1 = T_1^0 \cap \left(I_1^1 \times J_2^1\right) \quad \text{and} \quad T_2^1 = T_1^0 \cap \left(I_2^1 \times J_1^1\right).$$

Step 2 We apply the construction of Step 1 to each of the stage 1 triangles T_1^1 and T_2^1, generating two squares, S_1^1 and S_2^1, together with the four triangles of stage 2,

$$T_1^2,\ T_2^2,\ T_3^2, \text{ and } T_4^2.$$

After step k, we have disjoint squares $S_1^l, \ldots, S_{2^l}^l$, for $0 \leq l \leq k - 1$. In view of (7.2.40) and (7.2.41), the union $S_l = S_1^l \cup \cdots \cup S_{2^l}^l$ determines a contractive projection

$$P_l(A) = \sum_{(i,j) \in S_l} a(i, j) e_{i,j}, \quad A = (a(i, j)).$$

After step m, we find $T_1^0 = \bigcup_{l=0}^{m-1} S_l$, which yields $U^{(n)} = \sum_{l=0}^{m-1} P_l$ and (7.2.45).

Open Problem: We let K be a separable complex Hilbert space, and we fix an orthonormal basis $\{e_k : k \in \mathbb{N}\}$ in K. Let A be a compact operator on K. We say that A is a triangular operator with respect to $\{e_k : k \in \mathbb{N}\}$ if

$$\langle A(e_k), e_j \rangle = 0, \quad \text{for } j > k.$$

The space $T^1 \subset S^1$ consists of the subspace of triangular operators. Pisier asked if the noncommutative version of the embedding $L^1 \hookrightarrow L^1/H^1$ holds true. Specifically, does there exist a linear subspace of S^1/T^1 isomorphic to S^1? The central tools used in Chapter 3 are available for the trace class. Most importantly, there exist Pisier's noncommutative versions of Jones's interpolation theorems for (H^1, H^∞) (Pisier, 1992b, 1993), noncommutative martingale inequalities (Jiao et al., 2020; Pisier and Xu, 1997; Randrianantoanina, 2007), and noncommutative martingale decomposition theorems (Parcet and Randrianantoanina, 2006; Perrin, 2009; Randrianantoanina, 2019).

7.3 Iterated $L^p(L^q)$ Spaces

In his thesis, Yanqi Qiu (2012a) constructed a family of Banach spaces with a 1-unconditional basis and with finite Hardy martingale cotype that nevertheless fail the analytic UMD property. This section presents this work.

We start by recalling a basic procedure for generating Banach spaces. Let E, F be Banach spaces and assume that E has an unconditional basis $\{e_i : i \in I\}$, where I is countable (or finite). Put $Y = \operatorname{span}\{e_i \otimes x_i : x_i \in F\}$. Define a norm on Y by putting

$$\|x\|_{E(F)} = \left\| \sum e_i \|x_i\|_F \right\|_E, \tag{7.3.1}$$

where $x = \sum e_i \otimes x_i$ is a finite sum. Finally, let $E(F)$ denote the completion of Y under the norm in (7.3.1). For instance, if we let $\{e_i : i \in \mathbb{N}\}$ denote the standard unit vector basis in ℓ_p then the norm on $\ell_p(\ell_q)$ is given by

$$\|(x_i)\|_{\ell_p(\ell_q)} = \left(\sum_{i=1}^{\infty} \|x_i\|_{\ell_q}^p \right)^{1/p},$$

where $\{x_i : i \in \mathbb{N}\}$ is a sequence of elements in ℓ_q. We write $\ell_p^{(2)}$ to denote the two-dimensional subspace of ℓ_p that is spanned by $\{e_1, e_2\}$. For fixed $1 \le p$, $q < \infty$ we write $E_1 = \ell_p^{(2)}\left(\ell_q^{(2)}\right)$, define inductively

$$E_n = \ell_p^{(2)}\left(\ell_q^{(2)}(E_{n-1})\right),$$

and let $X(p,q) = (\sum E_n)_{\ell^2}$. We list basic properties of $X(p,q)$ that follow routinely from its construction. For the details we refer to Qiu (2012a).

(i) In E_n, the n-fold tensor products of the unit vectors $\{e_1, e_2\}$ form an unconditional basis with basis constant $= 1$. Hence $X(p,q)$ is also a Banach space with a 1-unconditional basis. We may thus regard $X(p,q)$ as a Banach lattice. Since E_n is finite dimensional each of the lattices $X(p,q)$ $(1 \le p, q \le \infty)$ is reflexive.

(ii) The lattices $X(p,q)$ form a complex interpolation scale as follows: If $1 \le p_0, q_0, p_1, q_1 \le \infty$ and $0 < \theta < 1$, we let $[X(p_0,q_0), X(p_1,q_1)]_\theta$ denote the complex interpolation space between $X(p_0,q_0)$ and $X(p_1,q_1)$. See Calderón (1964). For $1/p = \theta/p_1 + (1-\theta)/p_1$ and $1/q = \theta/q_1 + (1-\theta)/q_1$ we have

$$X(p,q) = [X(p_0,q_0), X(p_1,q_1)]_\theta,$$

with equal norms.

(iii) Let $1 \le p,q \le \infty$, and let $0 < \theta < 1$ satisfy $\theta < 1 - 1/p$ and $\theta < 1 - 1/q$. Then putting $1/p_0 = (1/p - \theta/2)/(1-\theta)$ and $1/q_0 = (1/q - \theta/2)/(1-\theta)$, we have $1 \le p_0, q_0 \le \infty$ and

$$X(p,q) = [X(p_0,q_0), X(2,2)]_\theta.$$

Note that $X(2,2)$ is a Hilbert space. We have thus obtained $X(p,q)$ as complex interpolation space between a Hilbert space and an (otherwise arbitrary) Banach lattice. In Pisier (1979) and Pisier (2016, Chapter 11) one finds a thorough treatment of Banach spaces sharing this property – known as θ-Hilbertian. In particular, Pisier proves that θ-Hilbertian Banach spaces are super reflexive and of finite Hardy martingale cotype.

In this section we prove that if $1 < p \neq q < \infty$ then $X(p,q)$ fails to satisfy the aUMD property. This will be accomplished by showing that there exists $\kappa = \kappa(p,q) > 1$ such that for any $n \in \mathbb{N}$ there exists a Hardy martingale (F_k) with values in E_n satisfying $\mathbb{E}\|F_k\|_{E_n} \le 1$, and

$$\mathbb{E}\left\|\sum \epsilon_k \Delta F_k\right\|_{E_n} \ge \kappa^n, \tag{7.3.2}$$

where (ϵ_k) is a suitably chosen sequence of signs.

7.3.1 The Hardy Martingale S-Constant

Let $\{x_i\}$ be a sequence in a Banach space X. Assume that there exists a constant $C > 0$ such that for any sequence $F^{(k)} \in H^\infty(\mathbb{T}^\mathbb{N})$ and any set of integers $\{i_k\}$,

$$\int_{\mathbb{T}^\mathbb{N}} \left\|\sum \mathbb{E}_k\left(F^{(k)}\right)(z) x_{i_k}\right\|_X d\mathbb{P}(z) \le C \sup_{z\in\mathbb{T}^\mathbb{N}} \left\|\sum F^{(k)}(z) x_{i_k}\right\|_X. \tag{7.3.3}$$

We define the Hardy martingale S constant of the sequence $\{x_i\}$ to be

$$S(X,\{x_i\}) = \inf\{C\}, \tag{7.3.4}$$

where the infimum is extended over all $C > 0$ such that (7.3.3) holds for every sequence of bounded scalar-valued Hardy martingales. We now show that the S constant is a supermultiplicative quantity.

Theorem 7.3.1 *Let E be a Banach space with a 1-unconditional basis $\{e_i\}$. Let F be Banach space and let $\{f_j\}$ be any sequence in F. Then the S constants of $\{e_i \otimes f_j\}$ in $E(F)$ admit the lower estimate*

$$S(E(F),\{e_i \otimes f_j\}) \ge S(E,\{e_i\})S(F,\{f_j\}).$$

Proof We shorten the notation to $L^\infty(E) = L^\infty(\mathbb{T}^\mathbb{N}, E)$ and $L^1(E) = L^1(\mathbb{T}^\mathbb{N}, E)$. Fix a 1-unconditional basis $\{e_i\}$ in E, a sequence $\{f_j\}$ in F, and abbreviate $S(E) = S(E,\{e_i\})$ and $S(F) = S(F,\{f_j\})$.

Step 1 We resolve the infimum in (7.3.4) defining the constants $S(E)$ respectively $S(F)$. Fix $\epsilon > 0$ and select $K, M \in \mathbb{N}$, bounded scalar-valued Hardy martingales $\{F^{(k)}\}_{k=1}^K$, $\{G^{(m)}\}_{m=1}^M$, and increasing integers $\{i_k\}$, $\{j_m\}$ satisfying

$$\Big\|\sum F^{(k)} e_{i_k}\Big\|_{L^\infty(E)} \leq 1, \quad \Big\|\sum G^{(m)} f_{j_m}\Big\|_{L^\infty(F)} \leq 1, \tag{7.3.5}$$

and, simultaneously,

$$\Big\|\sum \mathbb{E}_k\left(F^{(k)}\right) e_{i_k}\Big\|_{L^1(E)} \geq S(E) - \epsilon, \quad \Big\|\sum \mathbb{E}_k\left(G^{(m)}\right) f_{j_m}\Big\|_{L^1(F)} \geq S(F) - \epsilon. \tag{7.3.6}$$

Step 2 Denote the product filtration on $\mathbb{T}^{KM}$ as $\mathcal{F}_1, \ldots, \mathcal{F}_{KM}$. We rewrite these as

$$\{\mathcal{F}_{(k,m)}\}_{k=1,m=1}^{K \quad M},$$

so that the lexicographic order on $\{\mathcal{F}_{(k,m)}\}$ coincides with the linear order on $\{\mathcal{F}_n\}$. Use $\mathbb{E}_{(k,m)}$ to denote the conditional expectation with respect to $\mathcal{F}_{k,m}$. Given $m \leq M$ and $k \leq K$ we select Hardy martingales $D^{(k)}(z)$ and $Z^{(k,m)}(z)$ such that for any $A \subset \mathbb{C}$,

$$\mathbb{P}\left\{z : D^{(k)}(z) \cdot Z^{(k,m)}(z) \in A\right\} = \mathbb{P} \otimes \mathbb{P}\left\{(v, w) : F^{(k)}(v)G^{(m)}(w) \in A\right\}.$$

In other words we demand that on the probability space $(\mathbb{T}^{\mathbb{N}}, \mathbb{P})$ the distribution of $D^{(k)}(z) \cdot Z^{(k,m)}(z)$ is identical to the one of $F^{(k)}(v)G^{(m)}(w)$ on the product probability space $(\mathbb{T}^{\mathbb{N}} \times \mathbb{T}^{\mathbb{N}}, \mathbb{P} \otimes \mathbb{P})$. Simultaneously, we arrange for the conditional expectations

$$\mathbb{E}_{(k,m)}\left(D^{(k)} \cdot Z^{(k,m)}\right)(z) \quad \text{and} \quad \mathbb{E}_k\left(F^{(k)}\right)(v)\mathbb{E}_m\left(G^{(m)}\right)(w)$$

to be identically distributed. For brevity we write $y^{(k,m)} = e_{i_k} \otimes f_{j_m}$, and define,

$$g(z) = \sum_{k=1}^K \sum_{m=1}^M g^{(k,m)}(z) y^{(k,m)},$$

where we put $g^{(k,m)}(z) = D^{(k)}(z)Z^{(k,m)}(z)$. By our choice of $g^{(k,m)}$, the Hardy martingale g and

$$h(v, w) = \sum_{k=1}^K \sum_{m=1}^M F^{(k)}(v)G^{(m)}(w)y^{(k,m)}$$

have identical L^∞ norms. That is, we have,

$$\|g\|_{L^\infty(\mathbb{T}^{\mathbb{N}}, E(F))} = \|h\|_{L^\infty(\mathbb{T}^{\mathbb{N}} \times \mathbb{T}^{\mathbb{N}}, E(F))}.$$

Taking into account the definition of h, it is easy to see that the product structure of $F^{(k)}(v)G^{(m)}(w)$ is matched by the tensor structure of the vectors $y^{(k,m)}$. Hence by (7.3.5) we have $\|h\|_{L^\infty(E(F))} \leq 1$.

Step 3 Next, consider the transformed Hardy martingales

$$T_1(g)(z) = \sum_{k=1}^{K}\sum_{m=1}^{M} \mathbb{E}_{(k,m)}\left(g^{(k,m)}\right)(z)y^{(k,m)}$$

and

$$T_2(h)(v,w) = \sum_{k=1}^{K}\sum_{m=1}^{M} \mathbb{E}_k\left(F^{(k)}\right)(v)\mathbb{E}_m\left(G^{(m)}\right)(w)y^{(k,m)}.$$

Now we exploit the choices made in the previous step to see that the respective L^1 norms are identical, namely that

$$\|T_1(g)\|_{L^1(\mathbb{T}^{\mathbb{N}},E(F))} = \|T_2(h)\|_{L^1(\mathbb{T}^{\mathbb{N}}\times\mathbb{T}^{\mathbb{N}},E(F))}. \tag{7.3.7}$$

Moreover, the right-hand side of (7.3.7) coincides with

$$\left\|\sum_{k=1}^{K} \mathbb{E}_k\left(F^{(k)}\right)e_{i_k}\left\|\sum_{m=1}^{M} \mathbb{E}_m\left(G^{(m)}\right)f_{j_m}\right\|_{L^1(\mathbb{T}^{\mathbb{N}},F)}\right\|_{L^1(\mathbb{T}^{\mathbb{N}},E)} . \tag{7.3.8}$$

By condition (7.3.6), the term in (7.3.8) is larger than $(S(E,\{e_i\}) - \epsilon)(S(F,\{f_j\}) - \epsilon)$. Since

$$\|T_1(g)\|_{L^1(E(F))} \geq (S(E,\{e_i\}) - \epsilon)(S(F,\{f_j\}) - \epsilon)$$

and $\|g\|_{L^\infty(E(F))} \leq 1$, we finally obtain that

$$S(E(F),\{e_i \otimes f_j\}) \geq (S(E,\{e_i\}) - \epsilon)(S(F,\{f_j\}) - \epsilon),$$

for any $\epsilon > 0$. □

Recall that $E_1 = \ell_p^{(2)}\left(\ell_q^{(2)}\right)$ forms basic building block of the sequence

$$E_n = \ell_p^{(2)}\left(\ell_q^{(2)}(E_{n-1})\right).$$

We next turn to presenting the main idea in Qiu's approach. We show that the Hardy martingale S constant of the unit vector basis in E_1 is strictly larger than 1. Hence, by Theorem 7.3.1, for E_n it can be made as large as we wish, by choosing $n >> 1$. On the other hand, as we show in Section 7.3.2, if a Banach space X has aUMD property and if $\{x_i\}$ is an unconditional basic sequence, then

$$S(X,\{x_i\}) < \infty.$$

Theorem 7.3.2 *Let $1 \leq p \neq q < \infty$. Let $E = \ell_p^{(2)}\left(\ell_q^{(2)}\right)$ and let $\{x_i\}_{i=1}^4$ denote the unit vector basis of E listed in the following order:*

$$x_1 = e_1 \otimes e_1, \quad x_2 = e_1 \otimes e_2, \quad x_3 = e_2 \otimes e_1, \quad x_4 = e_2 \otimes e_2. \tag{7.3.9}$$

Then there exists $\kappa = \kappa(p, q) > 1$ such that

$$\kappa \le S(E, \{x_i\}). \tag{7.3.10}$$

Proof Let $U\colon H^\infty(\mathbb{T}, E) \to H^1(\mathbb{T}, E)$ be defined by the relation

$$U\left(\sum_{i=1}^4 a_i x_i\right) = \left(\int_{\mathbb{T}} a_1 dm\right) x_1 + a_2 x_2 + \left(\int_{\mathbb{T}} a_3 dm\right) x_3 + a_4 x_4, \tag{7.3.11}$$

where $a_k \in H^\infty(\mathbb{T})$. We will prove that

$$\kappa \le \|U\| \le S(E, \{x_i\}), \tag{7.3.12}$$

where $\kappa = \kappa(p, q) > 1$ and $\|U\| = \left\|U\colon H^\infty(\mathbb{T}, E) \to H^1(\mathbb{T}, E)\right\|$.

Step 1 Here we prove that $\kappa \le \|U\|$ where $\kappa > 1$. We begin by preparing the construction of testing functions by which we will obtain the lower estimate for the norm of U. Fix disjoint intervals $I, J \subseteq \mathbb{T}$ with $m(I) = m(J) = 1/2$. Let $u, v, w, t > 0$ and define

$$f_1 = u 1_I + v 1_J, \quad f_2 = v 1_I + u 1_J, \quad f_3 = w 1_I + t 1_J, \quad f_4 = t 1_I + w 1_J.$$

Form the outer functions $a_k \in H^\infty(\mathbb{T})$ so that $|a_k| = f_k$, and define the E-valued testing function $g = \sum_{i=1}^4 a_i x_i$. First, note that $\|g(\zeta)\|_E$ does not depend on $\zeta \in \mathbb{T}$. Indeed, a direct evaluation gives

$$\|g(\zeta)\|_E = (u^q + w^q)^{p/q} + (v^q + t^q)^{p/q}. \tag{7.3.13}$$

Next, observe that $\int a_1 dm = \int a_2 dm = (uv)^{1/2}$, and hence

$$\|Ug(\zeta)\|_E = \left((uv)^{q/2} + w^q\right)^{p/q} + \left((uv)^{q/2} + t^q\right)^{p/q}. \tag{7.3.14}$$

Let $\alpha = p/q \neq 1$. Fix $\epsilon > 0$ and specialize u, v, w, and t as follows

$$u^q = \epsilon, \quad v^q = \epsilon/2, \quad w^{\alpha-1} = 1/4, \quad t = 1.$$

Denoting the associated E-valued testing functions by $g^{(\epsilon)}$, we obtain with (7.3.13) and (7.3.14), that for $\epsilon > 0$,

$$\left\|g^{(\epsilon)}\right\|_E = (\epsilon + w)^\alpha + (\epsilon/2 + 1)^\alpha, \quad \left\|Ug^{(\epsilon)}\right\|_E = \left(\epsilon/\sqrt{2} + w\right)^\alpha + \left(\epsilon/\sqrt{2} + 1\right)^\alpha. \tag{7.3.15}$$

Passing to the limit $\epsilon \to 0$ in (7.3.15) gives

$$\lim_{\epsilon \to 0} \left\|Ug^{(\epsilon)}\right\|_E = \lim_{\epsilon \to 0} \left\|g^{(\epsilon)}\right\|_E. \tag{7.3.16}$$

Differentiating (7.3.15) and evaluating at $\epsilon = 0$ we obtain

$$\left(\frac{d\left\|g^{(\epsilon)}\right\|_E}{d\epsilon}\right)_{\epsilon=0} = \alpha(1/4 + 1/2), \quad \left(\frac{d\left\|Ug^{(\epsilon)}\right\|_E}{d\epsilon}\right)_{\epsilon=0} = \alpha\left(1/\sqrt{32} + 1/\sqrt{2}\right). \tag{7.3.17}$$

If we would have $\|U\| \le 1$, then for any $\epsilon > 0$

$$\int_{\mathbb{T}} \left\| U g^{(\epsilon)} \right\|_E dm \le \sup_{\zeta \in \mathbb{T}} \left\| g^{(\epsilon)}(\zeta) \right\|_E .$$

With (7.3.17) and (7.3.16) this gives $\alpha \left(1/\sqrt{32} + 1/\sqrt{2}\right) \le \alpha(1/4 + 1/2)$, a contradiction. Hence there exists $\kappa > 1$ with $\kappa \le \|U\|$, which verifies the left-hand side of estimate (7.3.12).

Step 2 Here, we prove that $\|U\| \le S(E, \{x_i \colon 1 \le i \le 4\})$, where the x_i are specified in the unit vector basis (7.3.9). To this end we fix $a_i \in H^\infty(\mathbb{T})$ and form

$$g = a_1 x_1 + a_2 x_2 + a_3 x_3 + a_4 x_a.$$

Clearly we have $g \in H^\infty(\mathbb{T}, E)$. We next construct a representation of the mapping

$$g \to Ug$$

in terms of martingales and martingale transforms. Fix $z \in \mathbb{T}^{\mathbb{N}}$ and define

$$\varphi^{(1)}(z) = a_1(z_3), \quad \varphi^{(2)}(z) = a_3(z_3), \quad \varphi^{(3)}(z) = a_2(z_3), \quad \varphi^{(4)}(z) = a_4(z_3).$$

Attaching those to the unit vectors $\{x_i\}_{i=1}^4$ we obtain the E-valued Hardy martingale

$$f(z) = \varphi^{(1)}(z)x_1 + \varphi^{(2)}(z)x_3 + \varphi^{(3)}(z)x_2 + \varphi^{(4)}(z)x_4.$$

Note that f and g are related by the identity $f(z) = g(z_3)$. Next, we take progressive conditional expectations of the $\varphi^{(k)}$ and define the operator

$$T_1(f) = \mathbb{E}_1\left(\varphi^{(1)}\right)x_1 + \mathbb{E}_2\left(\varphi^{(2)}\right)x_3 + \mathbb{E}_3\left(\varphi^{(3)}\right)x_2 + \mathbb{E}_4\left(\varphi^{(4)}\right)x_4. \tag{7.3.18}$$

By definition of $S(E, \{x_i\})$ we have

$$\mathbb{E}\|T_1(f)\|_E \le S(E, \{x_i\}) \sup_{z \in \mathbb{T}^{\mathbb{N}}} \|f(z)\|_E. \tag{7.3.19}$$

Taking into account the position of the variable z_3, we find that

$$\mathbb{E}_1\left(\varphi^{(1)}\right) = \int_{\mathbb{T}} a_1 dm, \quad \mathbb{E}_2\left(\varphi^{(2)}\right) = \int_{\mathbb{T}} a_3 dm, \quad \mathbb{E}_3\left(\varphi^{(3)}\right) = a_2, \quad \mathbb{E}_3\left(\varphi^{(4)}\right) = a_4.$$

Inserting these terms into (7.3.18) and comparing with the relation (7.3.11) gives the crucial identities

$$\int_{\mathbb{T}} \|U(g)\|_E dm = \mathbb{E}\|T_1(f)\|_E \quad \text{and} \quad \sup_{z \in \mathbb{T}^{\mathbb{N}}} \|f(z)\|_E = \sup_{w \in \mathbb{T}} \|g(w)\|_E.$$

Thus by (7.3.19) we have $\|U\| \le S(E, \{x_i\})$, which is the right-hand side of estimate (7.3.12). □

7.3.2 Unconditional Convergence

We detail the relation between the aUMD property and the Hardy martingale S constant. The crucial link between these notions is provided by Theorem 5.4.11, the Bourgain–Stein–Qiu inequality for vector-valued Hardy martingales.

Theorem 7.3.3 *Let E be a Banach space. Assume that there exists $A > 0$ so that for any E-valued Hardy martingale $F = (F_k)$ and any $(\zeta_k) \in \mathbb{T}^{\mathbb{N}}$,*

$$\left\|\sum \zeta_k \Delta F_k\right\|_{L^p(\mathbb{T}^{\mathbb{N}},E)} \leq A\|F\|_{L^p(\mathbb{T}^{\mathbb{N}},E)}.$$

Then for any 1-unconditional basic sequence $\{e_i\}$ in E, the Hardy martingale S constant satisfies

$$S(E, \{e_i\}) \leq A.$$

Proof Let $\varphi^{(k)} = \left(\varphi_m^{(k)}\right)$ be a sequence of scalar-valued Hardy martingales, and let $i_k \uparrow \infty$ be an increasing sequence of integers. Applying Theorem 5.4.11, the Bourgain–Stein–Qiu inequality, to the sequence of E-valued Hardy martingales

$$F_m^{(k)}(w) = \varphi_m^{(k)}(w)e_{i_k},$$

where $m, k \in \mathbb{N}$ and $w \in \mathbb{T}^{\mathbb{N}}$, we obtain

$$\mathbb{E}_{(\zeta)}\left\|\sum_{m=1}^{\infty} \zeta_m \varphi_m^{(m)} e_{i_m}\right\|_{L^p(\mathbb{T}^{\mathbb{N}},E)}^p \leq A^p \mathbb{E}_{(\zeta)}\left\|\sum_{m=1}^{\infty} \zeta_m \varphi^{(m)} e_{i_m}\right\|_{L^p(\mathbb{T}^{\mathbb{N}},E)}^p. \tag{7.3.20}$$

As the basic sequence $\{e_i\}$ is 1-unconditional in E, the $L^p(E)$ norm in each of the terms in (7.3.20) is actually independent of $\zeta = (\zeta_m)$. Hence Hölder's inequality gives $S(E, \{e_i\}) \leq A$. □

Summing up the theorems of this section we obtain (7.3.2), and hence that $X(p, q)$ lacks the aUMD property for $1 < p \neq q < \infty$.

Theorem 7.3.4 *Let $1 \leq p \neq q < \infty$. There exists $\kappa = \kappa(p, q) > 1$ so that for any $n \in \mathbb{N}$ there exists an E_n-valued Hardy martingale (F_k) with $\mathbb{E}\|F_k\|_{E_n} \leq 1$, and a sequence of signs (ϵ_k) satisfying*

$$\mathbb{E}\left\|\sum \epsilon_k \Delta F_k\right\|_{E_n} \geq \kappa^n.$$

Proof In view of Theorem 7.3.2 and Theorem 7.3.1 there exists $\kappa > 1$ such that the Hardy martingale S constant of the tensor products basis in E_n is bounded from below by κ^n. Applying Theorem 7.3.3 with $E = E_n$ completes the proof. □

7.4 The James Tree

In this section we define the James tree (JT) and its (natural) predual, denoted J_*T. We prove that J_*T satisfies the aRNP. Thus every bounded analytic function on the unit disk, with values in J_*T, has radial limits almost everywhere and every L^1-bounded Hardy martingale with values in J_*T converges almost surely (and hence in L^1, by Corollary 2.5.2).

The James space: Recall that $I \subseteq [0, 1[$ is a dyadic interval if there exists $n \in \mathbb{N} \cup \{0\}$ and $1 \leq i \leq 2^n$ such that $I = [(i-1)/2^n, i/2^n[$. We let $\mathcal{I}$ denote the collection of all dyadic intervals contained in the unit interval $[0, 1[$. We fix $\mathcal{E} = \{K_n \in \mathcal{I} : n \in \mathbb{N}\}$ where

$$K_{n+1} \subset K_n \quad \text{and} \quad |K_{n+1}| = |K_n|/2,$$

for $n \in \mathbb{N}$. (In that case we say that $\mathcal{E}$ is an infinite segment of dyadic intervals.) Let $x = \{x(K) : K \in \mathcal{E}\}$ be a sequence of scalars indexed by the intervals in $\mathcal{E}$. We define the James norm

$$\|x\|_{J(\mathcal{E})} = \sup \left(\sum_{k=1}^{n} \left| \sum_{m=n_k+1}^{n_{k+1}} x(K_m) \right|^2 \right)^{1/2}, \tag{7.4.1}$$

where the supremum is taken over all sequences of increasing integers $\{n_k\}$ and $n \in \mathbb{N}$. We let $J(\mathcal{E})$ denote the space of sequences $x = \{x(K) : K \in \mathcal{E}\}$ satisfying

$$\|x\|_{J(\mathcal{E})} < \infty.$$

Note that $\|x\|_{J(\mathcal{E})} < \infty$ implies that $\sum_{K \in \mathcal{E}} x(K)$ exists. We denote by $J_0(\mathcal{E})$ the space of sequences $x = \{x(K) : K \in \mathcal{E}\} \in J(\mathcal{E})$ satisfying

$$\sum_{K \in \mathcal{E}} x(K) = 0.$$

The famous theorem of James (1950) asserts that $J_0(\mathcal{E})$ is of codimension one in its double dual space. More precisely, we have the equality

$$J_0(\mathcal{E})^{**} = J(\mathcal{E}).$$

The James Tree space: We define the James Tree to be a sequence space indexed by the collection of dyadic intervals $\mathcal{I}$. We say that $\mathcal{E}$ is a finite segment in $\mathcal{I}$ if there exist dyadic intervals I_0, I_1 with $I_1 \subset I_0$ such that $\mathcal{E} = \{J \in \mathcal{I} : I_1 \subseteq J \subseteq I_0\}$. Similarly, $\mathcal{E}$ is an infinite segment if there exists a dyadic interval I_0 and a point $t \in [0, 1[$, such that $\mathcal{E} = \{J \in \mathcal{I} : t \in J \subseteq I_0\}$.

Given a sequence of scalars $x = \{x(J) : J \in \mathcal{I}\}$, we define its James tree norm to be

$$\|x\|_{\mathrm{JT}} = \sup \left(\sum_{j=1}^{k} \left| \sum_{J \in \mathcal{E}_j} x(J) \right|^2 \right)^{1/2}, \tag{7.4.2}$$

where the supremum is taken over all $k \in \mathbb{N}$ and choices of pairwise disjoint segments $\mathcal{E}_1, \ldots, \mathcal{E}_k$. The James tree space then consists of those scalar sequences $x = \{x(J) : J \in \mathcal{I}\}$ for which $\|x\|_{\mathrm{JT}} < \infty$ and is denoted by JT.

Let $I \in \mathcal{I}$ and let $e_I \in \mathrm{JT}$ be defined by $e_I(K) = 1$ if $K = I$ and $e_I(K) = 0$ otherwise. Thus $\{e_I : I \in \mathcal{I}\}$ forms the unit vector basis of the space JT. When listed in lexicographic order, the unit vectors form a Schauder basis in JT. Let $(I(n) : n \in \mathbb{N})$ be a listing of the dyadic intervals in lexicographic order, let $e_n = e_{I(n)}$, and let (x_n) be a sequence of scalars such that

$$\sup_{m \in \mathbb{N}} \left\| \sum_{n=1}^{m} x_n e_n \right\|_{\mathrm{JT}} < \infty.$$

Then $\sum_{n=1}^{\infty} x_n e_n$ exists in JT. Thus $\{e_I : I \in \mathcal{I}\}$ is a boundedly complete Schauder basis. See (Lindenstrauss and Tzafriri, 1977, Definition 1.b.3). By (Lindenstrauss and Tzafriri, 1977, Proposition 1.b.4), JT is a dual space.

Dual and predual of JT: Given sequences of scalars $x = \{x(J) : J \in \mathcal{I}\}$ and $y = \{y(J) : J \in \mathcal{I}\}$ we let

$$\langle x, y \rangle = \sum_{J \in \mathcal{I}} x(J) y(J), \tag{7.4.3}$$

whenever the right-hand side series converges. We define $\|y\|_{\mathrm{JT}^*} = \sup\{\langle x, y \rangle\}$ where the supremum is extended over all finitely supported sequences x satisfying $\|x\|_{\mathrm{JT}} < 1$. Thus we may identify the dual space of JT as

$$\mathrm{JT}^* = \{y : \|y\|_{\mathrm{JT}^*} < \infty\}.$$

Let $\{f_I : I \in \mathcal{I}\}$ denote the biorthogonal functionals associated to the Schauder basis $\{e_I\}$. Specifically, $f_I \in \mathrm{JT}^*$ is defined by the equations $f_I(e_K) = 1$ if $I = K$, and $f_I(e_K) = 0$ otherwise.

We let $\mathrm{J_*T}$ be the norm closure of $\mathrm{span}\{f_I\}$ in JT^*. Since $\{e_I\}$ is a boundedly complete Schauder basis in JT, the dual space of $\mathrm{J_*T}$ is canonically identified with JT. See (Lindenstrauss and Tzafriri, 1977, Proposition 1.b.4). We denote this fact by

$$(\mathrm{J_*T})^* = \mathrm{JT}.$$

Basic projections on JT: Let $x \in \mathrm{JT}$ and $y \in \mathrm{JT}^*$ be given by scalar sequences $x = \{x(J) : J \in \mathcal{I}\}$ and $y = \{y(J) : J \in \mathcal{I}\}$. Any collection of dyadic intervals $\mathcal{U} \subseteq \mathcal{I}$ defines (formally) a self-adjoint and idempotent operation $P_{\mathcal{U}}$ on JT and JT^* by

$$P_{\mathcal{U}}(x) = \{x(J) : J \in \mathcal{U}\} \quad \text{and} \quad P_{\mathcal{U}}(y) = \{y(J) : J \in \mathcal{U}\}. \tag{7.4.4}$$

If $\mathcal{K}(I) = \{J \in \mathcal{I} : J \subseteq I\}$, $I \in \mathcal{I}$, then $P_{\mathcal{K}(I)}$ is a contraction on JT and JT^*. Next, fix a collection of pairwise dyadic intervals C and put $\mathcal{K} = \cup_{I \in C} \mathcal{K}(I)$. Then

$$\|P_C(x)\|_{\mathrm{JT}}^2 = \sum_{J \in C} |x(J)|^2 \quad \text{and} \quad \|P_{\mathcal{K}}(x)\|_{\mathrm{JT}}^2 = \sum_{J \in C} \|P_{\mathcal{K}(I)}(x)\|_{\mathrm{JT}}^2. \tag{7.4.5}$$

Since the basic projection operators are self adjoint, duality gives

$$\|P_C(y)\|_{\mathrm{JT}^*}^2 = \sum_{J \in C} |y(J)|^2 \quad \text{and} \quad \|P_{\mathcal{K}}(y)\|_{\mathrm{JT}^*}^2 = \sum_{J \in C} \|P_{\mathcal{K}(I)}(y)\|_{\mathrm{JT}^*}^2. \tag{7.4.6}$$

Based on the identities (7.4.6), Lindenstrauss and Stegall (1975) obtained the following pointwise decay estimates for the $P_{\mathcal{K}(I)}$ on functionals y converging to 0 along segments.

Proposition 7.4.1 *Fix $y \in \mathrm{JT}^*$ with $y = \{y(J) : J \in \mathcal{I}\}$ and assume that for any infinite segment of dyadic intervals $\{K_n : n \in \mathbb{N}\}$,*

$$\lim_{n \to \infty} y(K_n) = 0.$$

Then,

$$\lim_{n \to \infty} \max_{I \in \mathcal{I}_n} \|P_{\mathcal{K}(I)}(y)\|_{\mathrm{JT}^*} = 0,$$

where $\mathcal{I}_n = \{I \in \mathcal{I} : |I| = 2^{-n}\}$.

Lindenstrauss and Stegall (1975) give an important characterization of the difference set $\mathrm{JT}^* \backslash \mathrm{J}_*\mathrm{T}$ in terms of boundary limits along infinite segments of dyadic intervals, and identify the quotient space $\mathrm{JT}^*/\mathrm{J}_*\mathrm{T}$ with the nonseparable Hilbert space $\ell^2([0,1])$. Recall that the nonseparable Hilbert space $\ell^2([0,1])$ consists of those functions $f : [0,1] \to \mathbb{C}$ such that

$$\sup_F \sum_{\{t \in F\}} |f(t)|^2 < \infty,$$

where the supremum is extended over all *finite* subsets $F \subset [0,1]$.

Next, if $y \in \mathrm{JT}^*$ with $y = \{y(J) : J \in \mathcal{I}\}$, and if $\{K_n : n \in \mathbb{N}\}$ is an infinite segment of dyadic intervals, then the sequence of coefficients $\lim_{n\to\infty} y(K_n)$ exists in $\mathbb{C}$. By definition, if $y \in \mathrm{J}_*\mathrm{T}$ then $\lim_{n\to\infty} y(K_n) = 0$. Theorem 7.4.2 asserts that the limiting values determine $\mathrm{J}_*\mathrm{T}$ within JT^*.

Theorem 7.4.2 (Lindenstrauss–Stegall theorem) *For $y \in \mathrm{JT}^*\setminus \mathrm{J_*T}$ with $y = \{y(J) : J \in \mathcal{I}\}$, there exists an infinite segment of dyadic intervals $\{K_n : n \in \mathbb{N}\}$ such that*

$$\lim_{n\to\infty} y(K_n) \neq 0.$$

The identification of the quotient space is as follows:

$$\mathrm{JT}^*/\mathrm{J_*T} = \ell^2([0,1]).$$

Lindenstrauss and Stegall (1975) showed that $\mathrm{J_*T}$ fails to satisfy the Radon–Nikodym property. The main step in their proof is presented in Proposition 7.4.3.

Proposition 7.4.3 *There exist $\{y_I \in \mathrm{J_*T} : I \in \mathcal{I}\}$ satisfying the following conditions:*

(i) $\|y_I\|_{\mathrm{JT}^*} = 1$ *and* $\|y_I - y_J\|_{\mathrm{JT}^*} \geq 1/2$ *for* $I, J \in \mathcal{I}$.
(ii) $y_I = \frac{1}{2}\{y_{I_1} - y_{I_2}\}$, *where* $I_1, I_2 \in \mathcal{I}$ *denote the left and right half of* $I \in \mathcal{I}$, *respectively.*

Proof We put $y_I(K) = \frac{|I\cap K|}{|I|}$, where $I, K \in \mathcal{I}$, and define $y_I = \{y_I(K) : K \in \mathcal{I}\}$. First, we show that $\|y_I\|_{\mathrm{JT}^*} = 1$. To this end we fix $x \in \mathrm{JT}$ and evaluate,

$$\begin{aligned}
\langle y_I, x\rangle &= \sum_{K\in\mathcal{I}} y_I(K)x(K)\\
&= \sum_{K\in\mathcal{I}} \frac{|I\cap K|}{|I|}x(K)\\
&= \int_I \sum_{K\in\mathcal{I}} 1_K(t)x(K)\frac{dt}{|I|}\\
&\leq \sup_{t\in I}\left|\sum_{K\in\mathcal{I}} 1_K(t)x(K)\right|.
\end{aligned}$$

Note that for $t \in [0,1]$ fixed, the collection $\mathcal{E}_t = \{K \in \mathcal{I} : t \in K\}$ forms a segment of dyadic intervals. Next, by inspection, for any segment $\mathcal{E} \subset \mathcal{I}$ we obviously have the estimate,

$$\left|\sum_{K\in\mathcal{E}} 1_K(t)x(K)\right| \leq \|x\|_{\mathrm{JT}}.$$

Thus $|\langle y_I, x\rangle| \leq \|x\|_{\mathrm{JT}}$ for $x \in \mathrm{JT}$, and hence $\|y_I\|_{\mathrm{JT}^*} \leq 1$. Since $y_I(I) = 1$ we have also $\|y_I\|_{\mathrm{JT}^*} \geq 1$.

Next, we show that not only is $y_I \in \mathrm{JT}^*$ but actually that is $y_I \in \mathrm{J_*T}$. At this point we exploit Theorem 7.4.2. Fix an infinite segment of dyadic intervals $\{K_n : n \in \mathbb{N}\}$. Then clearly,

$$|I| \lim_{n\to\infty} y_I(K_n) = \lim_{n\to\infty} |I \cap K_n| = 0.$$

Thus in view of Theorem 7.4.2, we have $y_I \in \mathrm{J_*T}$.

Let $I, J \in \mathcal{I}$ with $I \neq J$. We show next that

$$\|y_I - y_J\|_{\mathrm{JT}^*} \geq 1/2.$$

We assume that $|J| \leq |I|$. If $|J| = |I|$ and $I \neq J$ then $I \cap J = \emptyset$. If, on the other hand, $|J| < |I|$ then clearly $|J| \leq |I|/2$ and hence $|I \cap J| \leq |J| \leq |I|/2$. Thus we observe that $|y_I(J)| \leq 1/2$ whenever $I \neq J$ and $|J| \leq |I|$. Clearly, since $y_J(J) = 1$ we obtain

$$\|y_I - y_J\|_{\mathrm{JT}^*} \geq |\langle y_I, e_J\rangle - \langle y_J, e_J\rangle| = |y_I(J) - y_J(J)| \geq 1/2.$$

Fix $I \in \mathcal{I}$ and let $I_1, I_2 \in \mathcal{I}$ denote the left and right half of $I \in \mathcal{I}$, respectively. We claim that

$$y_I(K) = \frac{1}{2}\left[y_{I_1}(K) - y_{I_2}(K)\right], \quad K \in \mathcal{I}.$$

Indeed, since $I_1 \cap I_2 = \emptyset$ and $I = I_1 \cup I_2$, we have,

$$|I| = 2|I_1| + 2|I_2| \quad \text{and} \quad |I \cap K| = |I_1 \cap K| + |I_2 \cap K|, \quad \text{for } K \in \mathcal{I}. \qquad \square$$

Theorem 7.4.4 *The space* $\mathrm{J_*T}$ *does not satisfy the Radon–Nikodym property.*

Proof Let $y_I \in \mathrm{J_*T}$ be a sequence satisfying the conclusion of Proposition 7.4.3. Put $\mathcal{I}_n = \{I \in \mathcal{I} : |I| = 2^{-n}\}$ and define $M_n \colon [0, 1] \to \mathrm{J_*T}$ by

$$M_n(t) = \sum_{I\in\mathcal{I}_n} y_I 1_I(t), \quad t \in [0, 1].$$

In view of Proposition 7.4.3, (M_n) is a $\mathrm{J_*T}$-valued martingale satisfying

$$\int_0^1 \|M_n(t)\|_{\mathrm{J_*T}} dt = 1 \quad \text{and} \quad \|M_n(s) - M_m(s)\|_{\mathrm{J_*T}} \geq 1/2,$$

for $m, n \in \mathbb{N}$, $m \neq n$, and almost every $s \in [0, 1]$. In summary, we have demonstrated the existence of an L^1 bounded martingale with values in $\mathrm{J_*T}$ failing to converge almost surely. $\square$

Molecules: If $\mathcal{E}$ is an infinite segment of dyadic intervals we define a linear functional $f(\mathcal{E})$ on JT by putting

$$\langle f(\mathcal{E}), x\rangle = \sum_{K\in\mathcal{E}} x(K) \quad \text{for} \quad x \in \mathrm{JT}.$$

By equation (7.4.2), which defines the norm in JT, the series is actually convergent and we have $|\langle f(\mathcal{E}), x\rangle| \le \|x\|_{\mathrm{JT}}$. If, more generally, $\{\mathcal{E}_j\}$ is a sequence of pairwise disjoint segments (finite or infinite) and if λ_j is a sequence of scalars with $\sum |\lambda_j|^2 \le 1$, then

$$y = \sum_{j=1}^{\infty} \lambda_j f(\mathcal{E}_j) \tag{7.4.7}$$

defines a bounded linear functional on JT such that $|y(x)| \le \|x\|$ for any $x \in \mathrm{JT}$. If in the series (7.4.7) only finitely many λ_j are nonvanishing, we say that y is a finite molecule in the unit ball of JT^*. We let M denote the set of finite molecules.

Due to the following theorem of Schachermayer, Sersouri, and Werner (1989), molecules are useful in the analysis of James tree spaces.

Theorem 7.4.5 *The convex hull of M is norm dense in the unit ball of* JT^*.

Open Problem: For the elements in unit ball of JT^*, find constructively a decomposition into absolutely convergent series of molecules.

7.4.1 Holomorphic Separation

In this section we prove that the predual of the James tree satisfies the aRNP. We present the proof by Ghoussoub, Maurey, and Schachermayer (1989) who based the approach on a refinement of the Lindenstrauss–Stegall theorem (Theorem 7.4.2).

We fix $I \in \mathcal{I}$ and let R_I denote the canonical restriction operator defined by $R_I(y) = \{y(J) : J \subseteq I\}$, where $y = \{y(J) : J \in \mathcal{I}\}$. Let y be in the closed unit ball of JT^*. Given $w \in \mathbb{D}$, say that y belongs to the class $\mathcal{K}_{w,I}$ if

$$\inf_{\mathcal{E}} \|R_I(y - wf(\mathcal{E}))\|_{\mathrm{JT}^*} \le (|w|/2)^2,$$

where the infimum is extended over all infinite segments $\mathcal{E}$ such that $I \in \mathcal{E}$. Theorem 7.4.6 is the first step towards proving the aRNP for the predual of the James tree.

Theorem 7.4.6 *For any countable dense set $C_d \subseteq \mathbb{D}$ we have*

$$\text{ball}(\text{JT}^*)\backslash \text{J}_*\text{T} = \bigcup_{w\in C_d}\bigcup_{I\in\mathcal{I}} \mathcal{K}_{w,I}.$$

If $w \in \mathbb{D}$ and $I \in \mathcal{I}$, there exists a sequence of quadratic polymomials h_n uniformly separating $\mathcal{K}_{w,I}$ and the predual J_*T *as follows: First,* $\lim_{n\to\infty} h_n(y) = 0$ *for any $y \in \text{J}_*\text{T}$, second,*

$$h_n(y) \le -|w|^2/2, \tag{7.4.8}$$

if $y \in \mathcal{K}_{w,I}$ and $2^{-n} \le |I|$, and third, $|h_n(y)| \le 1$ for any y in the unit ball of JT^*.

Proof First note that $\mathcal{K}_{w,I} \subseteq \text{JT}^*\backslash \text{J}_*\text{T}$, for $w \in \mathbb{D}$ and $I \in \mathcal{I}$. We fix $y \in \mathcal{K}_{w,I}$. Choose an infinite segment $\mathcal{E} \subset \mathcal{I}$ such that,

$$\|R_I(y - wf(\mathcal{E}))\|_{\text{JT}^*} \le 2(|w|/2)^2.$$

Since $\text{dist}(wf(\mathcal{E}), \text{J}_*\text{T}) = |w|$, we simply estimate as

$$\text{dist}(y, \text{J}_*\text{T}) \ge \text{dist}(wf(\mathcal{E}), \text{J}_*\text{T}) - \|R_I(y - wf(\mathcal{E}))\|_{\text{JT}^*} > 0.$$

Thus we prove that if $y \in \mathcal{K}_{w,I}$ then $y \in \text{JT}^*\backslash \text{J}_*\text{T}$.

Now we prove the converse inclusion. To this end we fix $y \in \text{ball}(\text{JT}^*)\backslash \text{J}_*\text{T}$. By Theorem 7.4.2, there exists $w_0 \in \mathbb{D}$ with $w_0 \neq 0$, and an infinite segment of dyadic intervals $\mathcal{E} = \{J_n : n \in \mathbb{N}\}$ such that

$$\lim_{n\to\infty} y(J_n) = w_0.$$

Fix $\epsilon > 0$ with $\epsilon < (|w_0|/4)^2$. By Theorem 7.4.5 there exists y_1 in the convex hull of M such that $\|y_1 - y\|_{\text{JT}^*} < \epsilon$. Accordingly, there exists $w_1 \in \mathbb{D}$ with $|w - w_1| < \epsilon$ such that $\lim_{n\to\infty} y_1(J_n) = w_1$. On the other hand, since y_1 is a finite linear combination of finite molecules this is possible only if there exists a dyadic interval $I \in \mathcal{E}$ such that

$$y_1(J_n) = w_1,$$

for any $J_n \in \mathcal{E}$ satisfying $J_n \subseteq I$ and $y_1(K) = 0$ for any $K \in \mathcal{I}$ satisfying $K \subseteq I$ and $K \notin \mathcal{E}$. In other words, y_1 satisfies $R_I(y_1 - w_1 f(\mathcal{E})) = 0$. In summary, if we select $w \in C_d$ such that $|w - w_1| < \epsilon$, then by the triangle inequality,

$$\|R_I(y - wf(\mathcal{E}))\|_{\text{JT}^*} \le \|y - y_1\|_{\text{JT}^*} + |w - w_1| \le 2\epsilon.$$

As $\epsilon < (|w_0|/4)^2$ we have $2\epsilon < (|w|/2)^2$, and hence $y \in \mathcal{K}_{w,I}$. This gives the desired inclusion

$$y \in \bigcup_{w\in C_d}\bigcup_{I\in\mathcal{I}} \mathcal{K}_{w,I}.$$

It remains to construct the sequence $\{h_n\}$ of quadratic polynomials that separate $\mathrm{J_*T}$ from a given set $\mathcal{K}_{w,I}$. To this end we let $\mathcal{E}(I,n) = \{J \in \mathcal{I} : J \subseteq I, |J| = 2^{-n}\}$, put $\zeta = |w|/w$, and define

$$h_n(y) = -\Re \sum_{J\in\mathcal{E}(I,n)} (y(J)\zeta)^2,$$

where $y \in \mathrm{JT}^*$ with $y = \{y(J) : J \in \mathcal{I}\}$. Note that for any $y \in \mathrm{JT}^*$ with $\|y\|_{\mathrm{JT}^*} \leq 1$ we have $|h_n(y)| \leq 1$. Similarly, by definition of the James tree predual, if $y \in \mathrm{J_*T}$ then $|h_n(y)| \to 0$ as $n \to \infty$. Finally, we fix any infinite segment of dyadic intervals $\mathcal{E}$ for which $I \in \mathcal{E}$. Then we have $h_n(wf(\mathcal{E})) = \Re(\zeta w)^2 = -|w|^2$, for any $n \in \mathbb{N}$ large enough. This then gives the estimate (7.4.8). □

Theorem 7.4.7 *Any $g \in H^\infty(\mathbb{D}, \mathrm{J_*T})$ has radial limits almost everywhere.*

Proof The proof begins with a general observation about weak-* radial limits. Let $g : \mathbb{D} \to \mathrm{J_*T}$ be analytic with $\|g(w)\|_{\mathrm{J_*T}} \leq 1$ for any $w \in \mathbb{D}$. Since JT is a separable Banach space, g has weak-* radial limits almost everywhere. Indeed, by Theorem 5.7.7 there exists $\varphi : \mathbb{T} \to \mathrm{JT}^*$ such that $\|\varphi(\zeta)\|_{\mathrm{JT}^*} \leq 1$ and for any $x \in \mathrm{JT}$ we have

$$\langle x, g(r\zeta)\rangle \to \langle x, \varphi(\zeta)\rangle,$$

as $r \to 1$, for almost every $\zeta \in \mathbb{T}$. The core idea of the proof consists in showing that the weak-* radial limits are actually contained in the predual $\mathrm{J_*T}$ and not in JT^*.

We turn to proving that $\varphi(\zeta) \in \mathrm{J_*T}$ almost everywhere. Assume to the contrary that $\varphi(\zeta) \in \mathrm{JT}^* \backslash \mathrm{J_*T}$ on a set of positive measure. By Theorem 7.4.6 there exists $w \in C_d$ and $I \in \mathcal{I}$ such that

$$A = \{\zeta \in \mathbb{T} : \varphi(\zeta) \in \mathcal{K}_{w,I}\}$$

satisfies $m(A) = \alpha > 0$. Fix $\epsilon > 0$ with $\epsilon < (|w|/4)^2$. By Lebesgue's differentiation theorem there exists $z \in \mathbb{D}$ such that

$$\int_A p_z(\zeta)dm(\zeta) > 1 - \epsilon,$$

where $p_z(\zeta) = (1 - |z|^2)/|\zeta - z|^2$ is the Poisson kernel. By Theorem 7.4.6 there exist quadratic polynomials h_n separating $g(z) \in \mathrm{J_*T}$ uniformly from $\mathcal{K}_{w,I}$. We have $h_n(g(z)) \to 0$ as $n \to \infty$, and if $2^{-n} \leq |I|$, then $h_n(y) \leq -|w|^2/2$ for $y \in \mathcal{K}_{w,I}$. Since, by construction, h_n is the real part of an analytic polynomial, we have the crucial identity

$$h_n(g(z)) = \int_{\mathbb{T}} (h_n \circ \varphi)(\zeta) p_z(\zeta) dm(\zeta).$$

Taking into account that $|h_n(y)| \leq 1$, for $\|y\|_{\mathrm{JT}^*} \leq 1$, our choice of the point z gives

$$\left|\int_{\mathbb{T}\setminus A} (h_n \circ \varphi) p_z dm\right| \leq \epsilon.$$

Conversely, since $\varphi(\zeta) \in \mathcal{K}_{w,I}$ for $\zeta \in A$, it follows that for any n large enough we have

$$\left|\int_A (h_n \circ \varphi) p_z dm\right| \geq (|w|^2/2)(1-\epsilon).$$

We may choose n still larger and obtain $|h_n(g(z))| \leq \epsilon$. Collecting terms we arrive at the inequalities

$$(|w|^2/2)(1-\epsilon) \leq |h_n(g(z))| \leq \epsilon.$$

Choosing $\epsilon > 0$ small enough, the above inequalities form the contradiction derived from the assumption that $A = \{\zeta \in \mathbb{T} : \varphi(\zeta) \in \mathcal{K}_{w,I}\}$ satisfies $m(A) = \alpha > 0$.

We have thus proved that $\varphi(\zeta) \in \mathrm{J}_*\mathrm{T}$ almost everywhere. Hence, φ is almost surely separably valued and $g(r\zeta)$ converges to $\varphi(\zeta)$ weakly. Theorem 5.7.8 asserts that the existence of weak radial limits implies that

$$\lim_{r\to 1} \|g(r\zeta) - \varphi(\zeta)\|_{J_*T} = 0,$$

for almost every $\zeta \in \mathbb{T}$. □

7.4.2 Holomorphic Injections

We present the method of Ghoussoub and Maurey (1989), which shows that the predual of the James tree has the aRNP by constructing a holomorphic injection from $\mathrm{J}_*\mathrm{T}$ into a predual of the *Ghoussoub–Maurey tree space* that, as it turns out, is a separable dual space.

The Ghoussoub–Maurey tree space: Recall that we let $\mathcal{I}$ denote the collection of all dyadic intervals contained in the unit interval $[0, 1)$. We say that $\mathcal{E} = \{K_n \in \mathcal{I} : n \in \mathbb{N}\}$ is an infinite segment of dyadic intervals if

$$K_{n+1} \subset K_n \quad \text{and} \quad |K_{n+1}| = |K_n|/2, \quad \text{for} \quad n \in \mathbb{N}.$$

In the opening paragraphs of the present section we introduced the James space $J(\mathcal{E})$ to consist of those sequences $\{x(K) : K \in \mathcal{E}\}$ for which

$$\|x\|_{J(\mathcal{E})} = \sup\left(\sum_{k=1}^{n}\left|\sum_{m=n_k+1}^{n_{k+1}} x(K_m)\right|^2\right)^{1/2} < \infty, \tag{7.4.9}$$

where the supremum is taken over all sequences of increasing integers $\{n_k\}$ and $n \in \mathbb{N}$. See definition (7.4.1). The Ghoussoub–Maurey tree space Y consists of those sequences $x = \{x(K) : K \in \mathcal{I}\}$ for which

$$\|x\|_Y = \sup_{\mathcal{E}} \|x\|_{J(\mathcal{E})} < \infty, \tag{7.4.10}$$

where the supremum is taken over all infinite segments for which $\bigcup_{J\in\mathcal{E}} J = [0, 1[$. Clearly, the norms in JT and Y are related by

$$\|x\|_Y \leq \|x\|_{\mathrm{JT}},$$

for every $x \in \mathrm{JT}$ and hence the formal inclusion map $i\colon \mathrm{JT} \to Y$ given by $i(x) = x$ is a bounded linear operator. Moreover, if $x \in Y$ with $x = \{x(K) : K \in \mathcal{I}\}$, then for every segment of dyadic intervals, $\mathcal{E}$, the right-hand side of

$$\langle f(\mathcal{E}), x\rangle = \sum_{K\in\mathcal{E}} x(K)$$

is convergent and defines a linear functional on Y satisfying $\|f_{\mathcal{E}}\|_{Y^*} = 1$. We let Y_1 be the Y-norm closure of the finitely supported sequences in Y, and put

$$Y_0 = \left\{x \in Y_1 : \langle f(\mathcal{E}), x\rangle = 0, \text{ for every infinite segment } \mathcal{E} \text{ with } \bigcup \mathcal{E} = [0, 1[.\right\}.$$

We let $Y_0^{\perp}$ denote the annihilator of Y_0 in Y^*. Thus $y \in Y_0^{\perp}$ if $y \in Y^*$ and $\langle x, y\rangle = 0$ for any $x \in Y_0$. By the Hahn–Banach theorem we have

$$Y_0^* = Y^*/Y_0^{\perp},$$

with equality of norms, that is,

$$\sup\{\langle y, x\rangle : x \in Y_0\} = \inf\{\|y + y_0\|_{Y^*} : y_0 \in Y_0^{\perp}\},$$

for any $y \in Y^*$.

The predual of Y: For a dyadic interval K we define $f_K \in Y^*$ by putting

$$\langle f_K, x\rangle = x(K),$$

for $x \in Y$ with $x = \{x(I)\colon I \in \mathcal{I}\}$. Clearly $\|f_K\|_{Y^*} = 1$. We let $Y_* \subseteq Y^*$ denote the norm closure of span$\{f_K\}$. Thus Y_* is a separable subspace of Y^*, and $\{f_K\colon K \in \mathcal{I}\}$ listed in lexicographic order is a Schauder basis of Y_*. Proposition 7.4.11 below asserts that the linear mapping

$$I\colon (Y_*)^* \to Y, \qquad I(\ell) = \{\langle \ell, f_K\rangle\colon K \in \mathcal{I}\}, \tag{7.4.11}$$

is well defined, bijective, and satisfies

$$\|I(\ell)\|_Y = \|\ell\|_{(Y_*)^*}, \quad \text{for} \quad \ell \in (Y_*)^*.$$

Thus Y_* is isometric to a predual of Y. Theorem 7.4.8 asserts that, moreover, Y_* is isomorphic to Y_0^*. Thus Y_* is isomorphic to a *separable dual space*.

Theorem 7.4.8 *The quotient space* $Y_0^* = Y^*/Y_0^\perp$ *is isomorphic to* Y_*. *The isomorphism is given by the restriction to* Y_* *of the canonical quotient map onto* $Y^*/Y_0^\perp$. *Moreover,*

$$(Y_0)^{**} = Y,$$

with equivalent norms.

Proof Let $Y_* \subseteq Y^*$ denote the norm closure of $\operatorname{span}\{f_K\}$. Let $q\colon Y^* \to Y^*/Y_0^\perp$ be defined by $q(y) = \{y + y_0 : y_0 \in Y_0^\perp\}$. We now prove the following assertions:

(i) For any $\alpha \in Y_0^*$ there exists $y \in Y_*$ such that $q(y) = \alpha$.

(ii) For any $y \in Y_*$,

$$\|y\|_{Y^*}/2 \le \|q(y)\|_{Y^*/Y_0^\perp} \le \|y\|_{Y^*}. \tag{7.4.12}$$

Thus we prove that the restriction to Y_* of the canonical quotient map is an isomorphism betwen $Y^*/Y_0^\perp$ and Y_*.

Step 1 Let $y \in Y_*$ with norm convergent expansion $y = \sum y(K) f_K$. First, assume that the series representing y is actually a finite linear combination. Hence there exists $n \in \mathbb{N}$ such that $y(K) = 0$ for $|K| \le 2^{-n}$. Fix $\epsilon > 0$ and choose $x \in Y$ with $\|x\|_Y = 1$ and

$$\langle y, x\rangle > \|y\|_{Y^*}(1 - \epsilon).$$

Define $x_1 \in Y$ by specifying

$$x_1(K) = \begin{cases} x(K) & \text{if } |K| > 2^{-n}, \\ -\sum_{L \supset K} x(L) & \text{if } |K| = 2^{-n}, \\ 0 & \text{if } |K| < 2^{-n}. \end{cases} \tag{7.4.13}$$

Thus $x_1 \in Y_0$ with $\|x_1\|_Y \le 2$ and $\langle y, x_1\rangle = \langle y, x\rangle$. Next, choose $y_0 \in Y_0^\perp$ such that $\|q(y)\|_{Y_0^*} \ge \|y + y_0\|_{Y^*}$. Then by simply evaluating the norm in Y^* we get $\|y + y_0\|_{Y^*} \ge (1 - \epsilon)\langle y - y_0, x_1/2\rangle$, and by our previous choices,

$$\langle y - y_0, x_1\rangle = \langle y, x_1\rangle = \langle y, x\rangle \ge (1 - \epsilon)\|y\|_{Y^*}.$$

Thus we have proved that $\|y\|_{Y^*}/2 \le \|q(y)\|_{Y^*/Y_0^\perp}$ for $y \in \operatorname{span}\{f_K\}$.

Step 2 It remains to get rid of the assumption that $y \in Y_*$ is a finite linear combination of the $\{f_K\}$. To this end we fix $y \in Y_*$ and choose $y_1 \in \operatorname{span}\{f_K\}$ such that $\|y - y_1\|_{Y^*} < \epsilon\|y\|_{Y^*}$ and $\|y_1\|_{Y^*} > (1-\epsilon)\|y\|_{Y^*}$. By the first step of the proof we get $\|y_1 + y_0\|_{Y^*} \geq \|y_1\|_{Y^*}/2$ for any $y_0 \in Y_0^\perp$. Then

$$\|y - y_1\|_{Y^*} \geq \|y_1 - y_0\|_{Y^*} - \|y - y_1\|_{Y^*} \geq \|y_1\|_{Y^*}(1/2 - \epsilon) \geq \|y\|_{Y^*}(1/2 - 2\epsilon).$$

Since $\epsilon > 0$ and $y_0 \in Y_0^\perp$ the above estimate gives (7.4.12).

Step 3 Let $\alpha \in Y_0^*$. Below, we construct $y \in Y_*$ such that $q(y) = \alpha$. For $K \in \mathcal{I}$ and $n \in \mathbb{N}$ let $\mathcal{I}(K,n) = \{L \in \mathcal{I} : L \subseteq K, |L| = 2^{-n}|K|\}$. We put

$$g_{K,n} = \sum_{L \in \mathcal{I}(K,n)} e_L$$

and define $d_{K,n} = e_K - g_{K,n}$. Thus $d_{K,n} \in Y_0$ and, moreover, for any choice of increasing integers $\{n_k\}$ – in Y_0 – the sequence $\{g_{K,n_k} - g_{K,n_{k+1}} : k \in \mathbb{N}\}$ is equivalent to the unit vector basis of ℓ^2. Consequently, $g_{K,n_k} - g_{K,n_{k+1}} \longrightarrow 0$ weakly in Y_0, and hence in Y_0, $d_{K,k}$ is a weakly convergent sequence as $k \longrightarrow \infty$. This allows us to define the coefficients

$$y(K) = \lim_{k\to\infty} \langle \alpha, d_{K,k} \rangle, \tag{7.4.14}$$

for any $K \in \mathcal{I}$. For any $n \in \mathbb{N}$ let $\mathcal{I}^{(n)} = \{K \in \mathcal{I} : |K| \geq 2^{-n}\}$, and put

$$y_n = \sum_{K \in \mathcal{I}^{(n)}} y(K) f_K.$$

First, we will show that for any $x \in Y_0$ with $x \in \operatorname{span}\{e_K\}$ we have $\langle a, x \rangle = \lim_{n\to\infty}\langle y_n, x\rangle$, second, that $\{y_n\}$ is norm convergent in Y^*, and third, that its limit $y \in Y_*$ satisfies $q(y) = \alpha$.

Step 4 Let $x \in Y_0$ with $x = \sum x(K)e_K$, and assume that only finitely many coefficients $x(K)$ are nonvanishing. Choose $n \in \mathbb{N}$ large enough such that $\langle y_n, x \rangle = \sum y(K)x(K)$. In view of (7.4.14), we have

$$\begin{aligned}
\sum y(K)x(K) &= \lim_{k\to\infty} \sum \langle \alpha, d_{K,k} \rangle x(K) \\
&= \lim_{k\to\infty} \Big\langle \alpha, \sum x(K) d_{K,k} \Big\rangle \\
&= \Big\langle \alpha, \sum x(K) e_K \Big\rangle - \lim_{k\to\infty} \Big\langle \alpha, \sum x(K) g_{K,k} \Big\rangle \\
&= \langle \alpha, x \rangle,
\end{aligned}$$

since $\lim_{k\to\infty}\langle \alpha, \sum x(K) g_{K,k}\rangle = 0$. We have thus verified that $\langle y_n, x \rangle = \langle \alpha, x \rangle$ for $n = n(x)$ large enough.

Step 5 Here, we prove that $\{y_n\}$ is norm convergent in Y^*. If not, there would exist $\delta > 0$ and a sequence of increasing integers $\{n_k\}$ such that $\|y_{n_k} - y_{n_{k+1}}\|_{Y^*} > \delta$. Since $y_{n_k} - y_{n_{k+1}} \in \operatorname{span}\{f_K\}$, by estimate (7.4.12), this implies that

$$\|y_{n_k} - y_{n_{k+1}}\|_{Y_0^*} > \delta/2.$$

Hence there exists $C < \infty$ and a sequence of testing functions $x_k \in Y_0$ such that $\langle y_{n_k} - y_{n_{k+1}}, x_k\rangle \geq 1$, $\|x_k\| \leq C$, and $x_k(K) = 0$ unless $K \in \mathcal{I}^{(n_{k+1})} \setminus \mathcal{I}^{(n_k)}$. By the Step 4 this gives

$$\langle \alpha, x_k\rangle = \langle y_{n_k} - y_{n_{k+1}}, x_k\rangle \geq 1.$$

On the other hand, $\|\sum d_k x_k\|_Y^2 \leq \sum |d_k|^2$ for any choice of $d_k \in \mathbb{C}$. Consequently, the sequence x_k converges weakly to 0 in Y_0, and hence $\langle \alpha, x_k\rangle \longrightarrow 0$ as $k \longrightarrow \infty$. This contradiction proves that $\{y_n\}$ is norm convergent in Y^*.

Step 6 Let $y = \lim y_n$ with convergence in the norm of Y^*. The last step in the proof consists in showing that $q(y) = \alpha$ or, equivalently, that $\langle y, x\rangle = \langle \alpha, x\rangle$ for any $x \in Y_0$.

Let $x \in Y_0$ and $m \in \mathbb{N}$. Choose $x_m \in \operatorname{span}\{e_K\}$ such that $x_m \in Y_0$ and $\|x - x_m\|_Y \leq 1/m$. Then by Step 2, there exists $n_0 = n_0(m)$ such that $\langle y_n, x_m\rangle = \langle \alpha, x_m\rangle$ for any $n \geq n_0$. Hence $\langle y, x_m\rangle = \langle \alpha, x_m\rangle$. Passing to the limit as $m \longrightarrow \infty$ yields $\langle y, x\rangle = \langle \alpha, x\rangle$. □

Let $y_1, y_2 \in \mathrm{JT}^*$ with $y_i = \{y_i(K)\}$. We define the pointwise product of y_1 and y_2 to be $y_1 \cdot y_2 = \{y_1(K) \cdot y_2(K)\}$. The metric properties of this operation provide an important link between the spaces JT^* and Y^*.

Theorem 7.4.9 *The pointwise product defines a bounded bilinear operator on* $\mathrm{JT}^* \oplus \mathrm{JT}^*$ *with values in* Y^* *such that*

$$\|y_1 \cdot y_2\|_{Y^*} \leq 2\|y_1\|_{\mathrm{JT}^*}\|y_2\|_{\mathrm{JT}^*}, \tag{7.4.15}$$

for any $y_1, y_2 \in \mathrm{JT}^*$.

Proof The proof exploits the molecular decomposition of the space JT^*. Let y_1, y_2 be molecules in JT^*. That is, there exist sequences of disjoint segments of dyadic intervals $\{\mathcal{E}_i\}$, respectively $\{\mathcal{G}_j\}$, and sequences of scalars $c_i, d_j \in \mathbb{C}$ satisfying $\sum |c_i|^2 \leq 1$ and $\sum |d_j|^2 \leq 1$, such that

$$y_1 = \sum c_i f(\mathcal{E}_i) \quad \text{and} \quad y_2 = \sum d_j f(\mathcal{G}_j),$$

where the representing series converge in the norm of JT^*. Recall that $f(\mathcal{E}) \in \mathrm{JT}^*$ is defined by $\langle f(\mathcal{E}), x\rangle = \sum_{K \in \mathcal{E}} x(K)$, for any $x \in \mathrm{JT}$.

Step 1 Let E_i, respectively G_i, denote the maximal dyadic interval in $\mathcal{E}_i$, respectively $\mathcal{G}_i$. For each $i \in \mathbb{N}$ fixed, let $J(i) = \{j \in \mathbb{N} : G_j \in \mathcal{E}_i, G_j \neq E_i\}$. Similarly, for $j \in \mathbb{N}$ fixed let $I(j) = \{i \in \mathbb{N} : E_i \in \mathcal{G}_j\}$. Given $i \in \mathbb{N}$ and $j \in J(i)$ we let $\mathcal{E}_{i,j} = \mathcal{E}_i \cap \mathcal{G}_j$, and for $j \in N$ and $i \in I(j)$ put $\mathcal{G}_{j,i} = \mathcal{G}_j \cap \mathcal{E}_i$. Then we have

$$y_1 \cdot y_2 = \sum_{i=1}^{\infty} c_i \sum_{j \in J(i)} d_j f(\mathcal{E}_{i,j}) + \sum_{j=1}^{\infty} d_j \sum_{i \in I(j)} c_i f(\mathcal{G}_{j,i}). \tag{7.4.16}$$

For $i \in \mathbb{N}$ and $j \in J(i)$ put $v_j = f(\mathcal{E}_{i,j})$. We claim that

$$\left\| \sum d_j v_j \right\|_{Y^*}^2 \leq \sum |d_j|^2.$$

To see this we use duality and fix $x \in Y$ with $\|x\|_Y \leq 1$. Then

$$\left\langle \sum d_j v_j, x \right\rangle = \sum d_j \langle v_j, x\rangle \leq \left(\sum |d_j|^2 \right)^{1/2} \left(\sum |\langle v_j, x\rangle|^2 \right)^{1/2}.$$

Next, observe that $\{\mathcal{E}_{i,j} : j \in J(i)\}$ forms a sequence of disjoint segments contained in $\mathcal{E}_i$. Hence,

$$\sum_{j \in J(i)} \left| \sum_{K \in \mathcal{E}_{i,j}} x(K) \right|^2 \leq \|x\|_Y^2.$$

Since $\langle v_j, x\rangle = \sum_{K \in \mathcal{E}_{i,j}} x(K)$, we obtain

$$\left\langle \sum d_j v_j, x \right\rangle \leq \left(\sum |d_j|^2 \right)^{1/2},$$

for any $x \in Y$ with $\|x\|_Y \leq 1$.

Clearly, the same proof shows that for $j \in N, i \in I(j)$ and $w_i = f(\mathcal{G}_{j,i})$ we have $\| \sum c_i w_i \|_{Y^*}^2 \leq \sum |c_i|^2$. Finally, we use the identity (7.4.16) and invoke the above estimates to conclude that

$$\begin{aligned} \|y_1 \cdot y_2\|_{Y^*} &\leq \sum_{i=1}^{\infty} |c_i| \left(\sum_{j \in J(i)} |d_j|^2 \right)^{1/2} + \sum_{j=1}^{\infty} |d_j| \left(\sum_{i \in I(j)} |c_i|^2 \right)^{1/2} \\ &\leq 2 \left(\sum_{i \in \mathbb{N}} |c_i|^2 \right)^{1/2} \left(\sum_{j \in \mathbb{N}} |d_j|^2 \right)^{1/2}. \end{aligned}$$

Step 2 Let $\epsilon > 0$ and $y_1, y_2 \in \mathrm{JT}^*$. Choose two sequences of molecules $\{m_i\}$, respectively $\{n_j\}$, and sequences of scalars $a_i, b_j \in \mathbb{C}$ satisfying $\sum |a_i| \leq \|y_1\|_{\mathrm{JT}^*}$ and $\sum |b_j| \leq \|y_2\|_{\mathrm{JT}^*}$, such that $y_1 = \sum a_i m_i$ and $y_2 = \sum b_j n_j$. Then the pointwise product admits the expansion

$$y_1 \cdot y_2 = \sum_{i,j} (a_i b_j)(m_i \cdot n_j).$$

Since by the previous step we have $\|m_i \cdot n_j\|_{Y^*} \leq 2$, we obtain inequality (7.4.15). □

Remarks: Comparing estimates (7.4.10) and (7.4.2) immediately gives $\|x\|_Y \leq \|x\|_{\mathrm{JT}}$, for every $x \in Y$, and $\|y\|_{Y^*} \geq \|y\|_{\mathrm{JT}^*}$, for every $y \in \mathrm{JT}^*$. Hence the assertion of Theorem 7.4.9 gives

$$\begin{aligned}\|y_1y_2y_3\|_{Y^*} &\leq 2\|y_1y_2\|_{\mathrm{JT}^*}\|y_3\|_{\mathrm{JT}^*}\\ &\leq 2\|y_1y_2\|_{Y^*}\|y_3\|_{\mathrm{JT}^*} \leq 4\|y_1\|_{\mathrm{JT}^*}\|y_2\|_{\mathrm{JT}^*}\|y_3\|_{\mathrm{JT}^*},\end{aligned} \tag{7.4.17}$$

for $y_1, y_2, y_3 \in \mathrm{JT}^*$. Moreover,

$$\left\|(y_1+y_2)^3 - y_2^3\right\|_{Y^*} \leq 4\|y_2\|_{\mathrm{JT}^*}\left(3\|y_1\|_{\mathrm{JT}^*} + 3\|y_1\|_{\mathrm{JT}^*}\|y_2\|_{\mathrm{JT}^*} + \|y_2\|_{\mathrm{JT}^*}^2\right). \tag{7.4.18}$$

To see that (7.4.18) holds true it suffices to note that

$$(y_1+y_2)^3 - y_2^3 = y_2\left(3y_1^2 + 3y_1y_2 + y_2^2\right)$$

and to apply (7.4.17) to each of the terms on the right-hand side.

Theorem 7.4.10 *The map $Sy = y \cdot y \cdot y$ defines an analytic injective function $S\colon \mathrm{J_*T} \to Y_*$, which sends bounded sets in $\mathrm{J_*T}$ to bounded sets in Y_*, such that $S(\{y \in \mathrm{J_*T} : \|y\|_{\mathrm{JT}^*} \leq 1\})$ is norm-closed in Y_*.*

Proof Let $y \in \mathrm{J_*T}$. Let $y_k \in \operatorname{span}\{f_K\}$ such that $\lim\|y - y_k\|_{\mathrm{JT}^*} = 0$. Clearly, $y_k^3 \in \operatorname{span}\{f_K\}$ and $y^3 - y_k^3 = (y - y_k)\left(y^2 + yy_k - y_k^2\right)$. By Theorem 7.4.9 and inequality (7.4.17),

$$\left\|y^3 - y_k^3\right\|_{Y^*} \leq C\|y - y_k\|_{\mathrm{JT}^*},$$

and hence $Sy = y^3$ is contained in $\overline{\operatorname{span}}\{f_K\}$, where the closure is taken with respect to the norm in Y^*. In other words, $Sy \in Y_*$. □

By Theorem 7.4.10, the map $Sy = y^3$ is a *preferred holomorphic injection*, as defined in Section 5.2.5. Since Y_* is a separable dual space it satisfies the aRNP. Hence, by Theorem 5.2.24, the space J_*T satisfies the aRNP.

We finally turn to proving that the map I, defined by (7.4.11), naturally identifies Y and $(Y_*)^*$. Recall that for $\ell \in (Y_*)^*$ we set $I(\ell) = \{\langle \ell, f_K\rangle : K \in \mathcal{I}\}$.

Proposition 7.4.11 *The linear mapping $I\colon (Y_*)^* \to Y$ given by $\ell \mapsto I(\ell)$ is well defined, bijective, and satisfies $\|I(\ell)\|_Y = \|\ell\|_{(Y_*)^*}$, for $\ell \in (Y_*)^*$.*

Proof Proposition 7.4.11 consists of three separate assertions. Accordingly, we show first that I is well defined and satisfies $\|I(\ell)\|_Y \leq \|\ell\|_{(Y_*)^*}$. Secondly, we prove that I is injective and that $\|I(\ell)\|_Y \geq \|\ell\|_{(Y_*)^*}$. The third part of the proof consists of showing that I is surjective.

Step 1 Fix $\ell \in (Y_*)^*$ and form $z = \{\langle \ell, f_K\rangle : K \in \mathcal{I}\}$. We will show that $\|z\|_Y \leq \|\ell\|$ and, hence, $z \in Y$.

We let $y_* \in Y_*$. Thus there exist scalars b_K such that $y_* = \sum b_K f_K$ where the series converges in the norm of Y^*, and hence,

$$\|y_*\|_{Y^*} = \sup_{\|y\|_Y \le 1} \left| \sum b_K y(K) \right|.$$

Next, we fix a segment of dyadic intervals $\mathcal{E}$, $k \in \mathbb{N}$, and let $(\mathcal{E}_j)_{j=1}^k$ be a decomposition of $\mathcal{E}$ into segments of dyadic intervals. For $1 \le j \le k$ we put $f(\mathcal{E}_j) = \sum_{L \in \mathcal{E}_j} f_L$. We have $f(\mathcal{E}_j) \in Y_*$ with $\|f(\mathcal{E}_j)\|_{Y^*} = 1$.

We let

$$N = \left(\sum_{j=1}^k |\langle \ell, f(\mathcal{E}_j) \rangle|^2 \right)^{1/2}, \tag{7.4.19}$$

and for $1 \le i \le k$ we define $a_i = \overline{\langle \ell, f(\mathcal{E}_i) \rangle} N^{-1}$ and set $z_* = \sum a_i f_{\mathcal{E}_i}$. Clearly z_* is contained in Y_*, since each of the $f(\mathcal{E}_i)$ is. It is now easy to see that $\|z_*\|_{Y^*} \le 1$. Indeed, we have

$$\begin{aligned} \|z_*\|_{Y^*} &= \sup_{\|y\|_Y \le 1} \sum_{j=1}^k N^{-1} \langle \ell, f(\mathcal{E}_j) \rangle \langle f(\mathcal{E}_j), y \rangle \\ &\le \sup_{\|y\|_Y \le 1} \left(\sum_{j=1}^k |\langle f(\mathcal{E}_j), y \rangle|^2 \right)^{1/2} \le 1. \end{aligned}$$

Next, we test $\ell \in (Y_*)^*$ on $z_* \in Y_*$ to obtain

$$\|\ell\|_{(Y_*)^*} \ge |\langle \ell, z_* \rangle| = \sum_{j=1}^k N^{-1} |\langle \ell, f(\mathcal{E}_j) \rangle|^2 = N. \tag{7.4.20}$$

Since the segment of dyadic intervals $\mathcal{E}$ and its decomposition $(\mathcal{E}_j)_{j=1}^k$ of $\mathcal{E}$ into segments of dyadic intervals were arbitrary, we obtain that $z \in Y$ and $\|z\|_Y \le \|\ell\|_{(Y_*)^*}$ by combining (7.4.20) and (7.4.19).

In summary, I is well defined and a contraction, $\|I(\ell)\|_Y \le \|\ell\|_{(Y_*)^*}$.

Step 2 We show that I is injective and that $\|I(\ell)\|_Y \ge \|\ell\|_{(Y_*)^*}$. Fix $\ell \in (Y_*)^*$ and form $z = \{\langle \ell, f_K \rangle : K \in \mathcal{I}\}$. For any $y_* \in Y_*$ with $\|y_*\| \le 1$, we will verify that

$$\langle \ell, y_* \rangle \le \|z\|_Y.$$

Fix $y_* \in Y_*$ with $\|y_*\| \le 1$, then there exist scalars b_K such that $y_* = \sum b_K f_K$, and where the series converges in the norm of Y^*. Hence,

$$\|z\|_Y \ge \langle y_*, z \rangle = \sum b_K \langle \ell, f_K \rangle = \left\langle \ell, \sum b_K f_K = \right\rangle.$$

By taking the supremum over $y_* \in Y_*$, with $\|y_*\| \le 1$, and recalling that $z = I(\ell)$ we obtain $\|I(\ell)\|_Y \ge \|\ell\|_{(Y_*)^*}$, as claimed.

Step 3 It remains to prove that I is surjective. To this end we fix $y \in Y$ and define the algebraic linear functional $\ell_y : Y_* \to \mathbb{R}$ by

$$\ell_y(z_*) = \sum_{L \in \mathcal{I}} \langle z_*, e_L \rangle y(L).$$

It is now straightforward to verify that $\ell_y \in (Y_*)^*$, that $\|\ell_y\| \leq \|y\|_Y$, and that $I(\ell_y) = y$. Indeed, fix $z_* \in Y_*$ with norm convergent series expansion $z_* = \sum b_K f_K$. Then $\langle z_*, e_L \rangle = b_L$ and, more generally,

$$\langle z_*, y \rangle = \sum b_L \langle f_L, y \rangle = \sum b_L y(L).$$

Thus $\ell_y(z_*) = \langle z_*, y \rangle$ and $\ell_y \in (Y_*)^*$ with $\|\ell_y\| \leq \|y\|_Y$. The identity $I(\ell_y) = y$ follows directly from the obvious fact that $\ell_y(f_L) = y(L)$. □

7.5 The Space L^1/H^1_0

In this section we discuss Steinhaus martingales in L^1/H^1_0. We present Bourgain's solution to the Steinhaus martingale lifting problem for L^1/H^1_0, and Kislyakov's embedding theorem.

7.5.1 Cotype 2

In Chapter 5 we noted that L^1/H^1_0 does not satisfy the aRNP and, consequently, fails Hardy martingale cotype and the aUMD. In view of the Bourgain–Davis theorem (Theorem 7.1.3), this sets L^1/H^1_0 apart from the quotient spaces L^1/Y, with $Y \subset L^1$ being reflexive.

In this section we obtain Bourgain's solution for the Steinhaus martingale lifting problem for L^1/H^1_0 proving for L^1/H^1_0 the exact analogue of the Kislyakov–Pisier theorem (Theorem 7.1.2).

Fixing notations, for $f \in L^1(\mathbb{T})$ we let $q(f) = \{f + H^1_0(\mathbb{T})\}$. Thus $q(f)$ is the element of L^1/H^1_0 that contains f and, conversely, f is a lifting of $q(f)$.

Theorem 7.5.1 *Let $n \in \mathbb{N}$, $x_1, \ldots, x_n \in L^1/H^1_0$. Then there exist $g_1, \ldots, g_n \in L^1$ such that $q(g_j) = x_j$ and*

$$\mathbb{E}_{(w)} \left\| \sum_{j=1}^{n} g_j w_j \right\|_{L^1} \leq C \mathbb{E}_{(w)} \left\| \sum_{j=1}^{n} x_j w_j \right\|_{L^1/H^1_0}, \tag{7.5.1}$$

where the expectation is with respect to the Haar measure on $\mathbb{T}^{\mathbb{N}}$, and hence,

$$\left(\sum_{j=1}^{n} \|x_j\|^2_{L^1/H^1_0} \right)^{1/2} \leq C \mathbb{E}_{(w)} \left\| \sum_{j=1}^{n} x_j w_j \right\|_{L^1/H^1_0}. \tag{7.5.2}$$

Below, we use the notational convention,

$$\Omega = \mathbb{T}^{\mathbb{N}}, \quad \mu = \mathbb{P}, \quad L^p(L^\alpha) = L^p(\mathbb{T}, L^\alpha(\Omega, \mu)), \quad 0 < \alpha \leq 1.$$

The main step in the proof of Theorem 7.5.1 is the following lifting result.

Theorem 7.5.2 *For $0 < \alpha < 1$, there exists $C_\alpha > 0$ such that the following holds. If $F = (F_n)$ is an integrable Steinhaus martingale with values in L^1/H_0^1, and $A = (A_n)$ an integrable martingale lifting of F with values in $L^1(\mathbb{T})$, then there exists a Steinhaus martingale $G = (G_n)$ so that*

$$q(G_n(w)) = q(A_n(w)), \quad w \in \Omega,$$

and

$$\left\|\widetilde{G}\right\|_{L^1(L^\alpha)} \leq C_\alpha \left\|\widetilde{A}\right\|_{L^1(L^1)}, \tag{7.5.3}$$

where

$$\widetilde{G}(t, w) = G(w, t), \qquad \widetilde{A}(t, w) = A(w, t), \qquad w \in \Omega, t \in \mathbb{T}.$$

Proof Let R denote the orthogonal projection onto $\overline{H_0^2}(\mathbb{T})$ with kernel $H_0^2(\mathbb{T})$. Thus $\mathrm{Id} - R$ coincides with the Riesz projection, and if $f_1, f_2 \in L^1(\mathbb{T})$ such that $q(f_1) = q(f_2)$, then $R(f_1) = R(f_2)$, and conversely. Moreover, if $f \in L^2(\mathbb{T})$ and $h, h^{-1} \in H^2(\mathbb{T})$ so that $fh^{-1} \in L^2(\mathbb{T})$, then there exists $g \in H^2(\mathbb{T})$ such that

$$hR\left(fh^{-1}\right) = f + g. \tag{7.5.4}$$

In other words, if $f \in L^1(\mathbb{T})$ is a lifting of some $x \in L^1(\mathbb{T})/H_0^1(\mathbb{T})$ then $hR(fh^{-1})$ is also a lifting of x. Next, recall Theorem 6.1.3, which asserts that for $0 < \alpha < 1$, there exists $C = C_\alpha$ so that

$$\left\|\mathcal{R}\colon L^2(L^1) \to L^2(L^\alpha)\right\| \leq C,$$

where $\mathcal{R} = R \otimes J_\Omega$ and $J_\Omega\colon L^1(\Omega, \mu) \to L^\alpha(\Omega, \mu)$ is the formal inclusion operator.

Step 1 Let $A = (A_n)$ be an integrable martingale lifting of F with values in $L^1(\mathbb{T})$. Hence for almost every $t \in \mathbb{T}$,

$$\widetilde{A}(t)\colon \Omega \to \mathbb{C}, \quad w \to \widetilde{A}(t, w),$$

is integrable. Put $\varphi(t) = \left\|\widetilde{A}(t)\right\|_{L^1(\Omega)}$ and let $g = \exp(\ln(\varphi + \epsilon) + iH(\ln(\varphi + \epsilon)))$, where H denotes the Hilbert transform on $\mathbb{T}$. Thus $g \in H^1(\mathbb{T})$ is the outer function satisfying $|g| = \varphi + \epsilon$. Finally, define $G = (G_n)$ by the equation

$$G_n(w, t) = g^{1/2}(t)\mathcal{R}\left(\widetilde{A_n} g^{-1/2}\right)(t, w), \qquad t \in \mathbb{T}, w \in \Omega.$$

Step 2 We show next that $G = (G_n)$ is a Steinhaus martingale and a lifting of $F = (F_n)$. Take into account that F was a Steinhaus martingale. Choose $x_j \in L^1/H^1_0$ such that

$$F_n(w) = \sum_{j=1}^{n} x_j w_j.$$

Let f_j be a lifting of x_j. By (7.5.4) we have $q(f_j) = q\left(g^{1/2}R\left(f_j g^{-1/2}\right)\right)$ and, moreover,

$$G_n(w,t) = \sum_{j=1}^{n} g^{1/2}(t)R\left(f_j g^{-1/2}\right)(t)w_j,$$

where $n \in \mathbb{N}, w \in \Omega$, and $t \in \mathbb{T}$. Hence $G = (G_n)$ is a Steinhaus martingale and a lifting of F.

Step 3 It remains to verify the norm estimates (7.5.3). Start by putting

$$\mathcal{R}\left(\widetilde{A_n}g^{-1/2}\right)(t) \colon \Omega \to \mathbb{C}, w \to \mathcal{R}\left(\widetilde{A_n}g^{-1/2}\right)(t,w).$$

Next, observe that with $L^\alpha = L^\alpha(\Omega)$, we have

$$\left\|\widetilde{G}(t)\right\|_{L^\alpha} = |g(t)|^{1/2}\left\|\mathcal{R}\left(\widetilde{A}g^{-1/2}\right)(t)\right\|_{L^\alpha}. \tag{7.5.5}$$

The Cauchy–Schwarz inequality applied to the right-hand side of (7.5.5) gives

$$\int_{\mathbb{T}}\left\|\widetilde{G}(t)\right\|_{L^\alpha} dm \le \left(\int_{\mathbb{T}}|g(t)|dm\right)^{1/2}\left(\int_{\mathbb{T}}\left\|\mathcal{R}\left(\widetilde{A}g^{-1/2}\right)(t)\right\|^2_{L^\alpha} dm\right)^{1/2}. \tag{7.5.6}$$

Theorem 6.1.3 provides the following upper bound for the right-hand side factor,

$$\int_{\mathbb{T}}\left\|\mathcal{R}\left(\widetilde{A}g^{-1/2}\right)(t)\right\|^2_{L^\alpha} dm \le C_a \int_{\mathbb{T}}|g(t)|^{-1}\left\|\widetilde{A}(t)\right\|^2_{L^1} dm. \tag{7.5.7}$$

Recall that $|g(t)| = \left\|\widetilde{A}(t)\right\|_{L^1} + \epsilon$. Letting $\epsilon \to 0$ and combining the estimates (7.5.6) and (7.5.7) gives

$$\int_{\mathbb{T}}\left\|\widetilde{G}(t)\right\|_{L^\alpha} dm \le C_\alpha \int_{\mathbb{T}}\left\|\widetilde{A}(t)\right\|_{L^1} dm. \qquad \square$$

Proof of Theorem 7.5.1: Let $w \in \Omega$ with $w = (w_j)$, let $a_j \in \mathbb{C}$ and $f(w) = \sum_{j=1}^{n} w_j a_j$. By Khintchine's inequality, for $0 < \alpha \le 1$,

$$\frac{\sqrt{\pi}}{2}\|f\|_{L^2(\Omega)} \le \|f\|_{L^1(\Omega)} \le C\|f\|_{L^\alpha(\Omega)}, \tag{7.5.8}$$

where $C = C_\alpha$. Let $F = (F_n)$ be an integrable, L^1/H_0^1-valued Steinhaus martingale, with $F_n(w) = \sum_{j=1}^n x_j w_j$ and $x_j \in L^1/H_0^1$. Let $A = (A_n)$, be an integrable martingale lifting of F with values in L^1, such that

$$\|\widetilde{A}\|_{L^1(L^1)} \leq 2 \int_\Omega \|F\|_{L^1/H_0^1} d\mu$$

By Theorem 7.5.2 there exists a Steinhaus martingale $G = (G_n)$, with $G_n(w) = \sum_{j=1}^n g_j w_j$ such that $g_j \in L^1(\mathbb{T})$ is a lifting of $x_j \in L^1/H_0^1$, satisfying

$$q(G_n(w)) = q(A_n(w)) \quad \text{and} \quad \|\widetilde{G}\|_{L^1(L^\alpha)} \leq C_\alpha \|\widetilde{A}\|_{L^1(L^1)},$$

where according to our convention

$$L^1(L^1) = L^1(\mathbb{T}, L^1(\Omega, d\mu)), \qquad L^1(L^\alpha) = L^1(\mathbb{T}, L^\alpha(\Omega, d\mu)).$$

Next we show that

$$\|\widetilde{G}\|_{L^1(L^1)} \leq C \|\widetilde{G}\|_{L^1(L^\alpha)}$$

Indeed Khintchine's inequality gives

$$\begin{aligned}
\int_\mathbb{T} \int_\Omega \left|\widetilde{G}(w,t)\right| d\mu(w) dm(t) &= \int_\mathbb{T} \int_\Omega \left|\sum_{j=1}^n g_j(t) w_j\right| d\mu(w) dm(t) \\
&\leq C_\alpha \int_\mathbb{T} \left[\int_\Omega \left|\sum_{j=1}^n g_j(t) w_j\right|^\alpha d\mu(w)\right]^{1/\alpha} dm(t) \\
&= C_\alpha \int_\mathbb{T} \left[\int_\Omega \left|\widetilde{G(w,t)}\right|^\alpha d\mu(w)\right]^{1/\alpha} dm(t).
\end{aligned}$$

Finally, using the left-hand side estimate in (7.5.8), Minkowski's inequality, and that g_j is a lifting of x_j we obtain

$$\begin{aligned}
\sqrt{2}\,\|\widetilde{G}\|_{L^1(L^1)} &\geq \int_\mathbb{T} \left(\sum_{j=1}^n |g_j(t)|^2\right)^{1/2} dm(t) \\
&\geq \left(\sum_{j=1}^n \|g_j\|_1^2\right)^{1/2} \qquad (7.5.9) \\
&\geq \left(\sum_{j=1}^n \|x_j\|_{L^1/H_0^1}^2\right)^{1/2}.
\end{aligned}$$

7.5.2 Kislyakov's Embedding Theorem

Let $S(\mathbb{T}^\mathbb{N})$ denote the space of L^1 convergent Steinhaus martingales equipped with the norm of $L^1(\mathbb{T}^\mathbb{N})$. By Khintchine's inequality, $S(\mathbb{T}^\mathbb{N})$ is isomorphic to a Hilbert space. In view of the discussion in Section 7.1.1, the Steinhaus martingale lifting theorem for L^1/H_0^1, Theorem 7.5.1 implies that any bounded operator

$$u\colon S(\mathbb{T}^{\mathbb{N}}) \to H^\infty(\mathbb{T})$$

admits a bounded linear extension $\widetilde{u}\colon L^1(\mathbb{T}^{\mathbb{N}}) \to H^\infty(\mathbb{T})$ such that $\widetilde{u}(G) = u(G)$, for $G \in S(\mathbb{T}^{\mathbb{N}})$ and $\|\widetilde{u}\| \leq C\|u\|$. Indeed, to see that Section 7.1.1 applies, it suffices to observe that the annihilator of $H_0^1(\mathbb{T})$ in $L^\infty(\mathbb{T})$ coincides with $H^\infty(\mathbb{T})$.

Bourgain's theorem (Theorem 7.5.4) generalizes the lifting property for operators u defined on Hilbert space to those $T\colon Y \to H^\infty$ where Y may be any reflexive subspace of L^1. Below, this will be obtained by duality from Kislyakov's embedding theorem (Theorem 7.5.3).

Let (Ω, μ) be a probability space and

$$L^1(L^1) = L^1(\mathbb{T}, L^1(\Omega, \mu)).$$

We fix a closed linear subspace Y of $L^1(\Omega, \mu)$. Let $L^1(Y)$ consist of those $F \in L^1(L^1)$ satisfying $F(z) \in Y$ for any $z \in \mathbb{T}$. Moreover, we define the subspace $H^1(L^1)$ of $L^1(L^1)$ by the condition

$$\int_{\mathbb{T}} F(z) z^n \, dm(z) = 0, \qquad n \in \mathbb{N},$$

and put $H^1(Y) = H^1(L^1) \cap L^1(Y)$. By definition $L^1(Y)$, respectively $H^1(Y)$, are closed subspaces in $L^1(L^1)$, respectively $H^1(L^1)$, and hence when equipped with the quotient norm the spaces

$$L^1(Y)/H^1(Y) \quad \text{and} \quad L^1(L^1)/H^1(L^1)$$

are indeed Banach spaces. Given $F_1 \in L^1(Y)$ and $F_2 \in L^1(L^1)$, the corresponding equivalence classes are defined by

$$[F_1]_1 = \left\{F_1 + H^1(Y)\right\}, \quad [F_2]_2 = \left\{F_2 + H^1(L^1)\right\}.$$

For any $F \in L^1(Y)$ we clearly have $[F]_1 \subseteq [F]_2$. Hence the inclusion map

$$I\colon L^1(Y)/H^1(Y) \to L^1(L^1)/H^1(L^1), \quad [F]_1 \to [F]_2$$

is a well-defined linear contraction for any closed subspace $Y \subseteq L^1(\Omega)$. In this section we show that under the additional hypothesis that Y is a reflexive subspace of L^1, the inclusion map I is invertible on its range (it is an embedding). We further ease the notation by putting

$$Z_1 = L^1(Y)/H^1(Y) \quad \text{and} \quad Z_2 = L^1(L^1)/H^1(L^1).$$

Theorem 7.5.3 *If Y is a reflexive subspace of $L^1(\Omega)$ then the inclusion operator $I\colon Z_1 \to Z_2$ is invertible on its range. There exists $C > 0$ such that*

$$\|[F]_1\|_{Z_1} \le C\|[F]_2\|_{Z_2}, \tag{7.5.10}$$

for $F \in L^1(Y)$. Consequently, for any $u_1 \in Z_1^*$ there exists $u_2 \in Z_2^*$ so that for any $F \in L^1(Y)$,

$$u_2([F]_2) = u_1([F]_1), \quad F \in L^1(Y), \quad \textit{and} \quad \|u_2\|_{Z_2^*} \le C\|u_1\|_{Z_1^*}. \tag{7.5.11}$$

Proof Let $\mathrm{Id} - R$ be the Riesz projection mapping $L^2(\mathbb{T})$ onto $H^2(\mathbb{T})$. Let J_Ω be the identity on $L^1(\Omega)$ and let Y be a reflexive subspace of $L^1(\mathbb{T})$. Theorem 6.2.2 applied to $\mathcal{R} = R \otimes J_\Omega$ gives

$$\left\|\mathcal{R}_{|L^1(Y)}\colon L^2(Y) \to L^2(L^1)\right\| \le C.$$

Step 1 Let $F \in L^1(Y)$. Since the algebraic tensor product $L^1(\mathbb{T}) \otimes Y$ is dense in $L^1(Y)$ we may assume that $F = \sum_{j=1}^n f_j x_j$, where $f_j \in L^1(\mathbb{T})$ and $x_j \in Y$. To $[F]_2$ select a lifting $A\colon \mathbb{T} \times \Omega \to \mathbb{C}$ such that

$$\|A\|_{L^1(L^1)} \le 2\|[F]_2\|_{Z_2}. \tag{7.5.12}$$

Fix $z \in \mathbb{T}$ and put $\varphi(z) = \|A(z)\|_{L^1}$. Let $\epsilon > 0$, let $g \in H^1(\mathbb{T})$ be the outer function satisfying $|g| = \varphi + \epsilon$, and put

$$G = g^{1/2}\mathcal{R}\left(Ag^{-1/2}\right). \tag{7.5.13}$$

We will show below that G is a lifting of $[F]_1$, and that there exists a constant $C = C(Y)$ so that

$$\|G\|_{L^1(L^1)} \le C\|A\|_{L^1(L^1)} \tag{7.5.14}$$

Once we verify that G is a lifting of $[F]_1$, the estimates (7.5.14) and (7.5.12) and the fact that A is a lifting of $[F]_2$ give the conclusion (7.5.10).

Step 2 Here we show that $[G]_1 = [F]_1$. Since A is a lifting of $[F]_2$ and $\mathcal{R}$ annihilates $H^1(L^1)$ we have

$$g^{1/2}\mathcal{R}\left(Ag^{-1/2}\right) = \sum_{j=1}^n g^{1/2}\mathcal{R}\left(f_j g^{-1/2}\right) x_j. \tag{7.5.15}$$

Moreover,

$$g^{1/2}\mathcal{R}\left(f_j g^{-1/2}\right) - f_j \in H^1(\mathbb{T}), \tag{7.5.16}$$

hence combining (7.5.16) with (7.5.15) and estimate (6.2.43) gives $G - F \in H^1(Y)$ or, equivalently, $[G]_1 = [F]_1$.

Step 3 Next we prove the $L^1(L^1)$ norm estimates for G. First, we fix $z \in \mathbb{T}$ and observe that by (7.5.13),

$$\|G(z)\|_{L^1} = \left|g^{1/2}(z)\right| \cdot \left\|\mathcal{R}\left(Ag^{-1/2}\right)(z)\right\|_{L^1}.$$

Integrating and applying the Cauchy–Schwarz inequality to the right-hand side we obtain the reduction to $L^2(L^1)$ estimates for $\mathcal{R}$,

$$\int_{\mathbb{T}} \|G(z)\|_{L^1} dm \le \left(\int_{\mathbb{T}} |g(z)| dm\right)^{1/2} \left(\int_{\mathbb{T}} \left\|\mathcal{R}\left(Ag^{-1/2}\right)(z)\right\|_{L^1}^2 dm\right)^{1/2}.$$

By (7.5.15), $\mathcal{R}\left(Ag^{-1/2}\right)(z) \in Y$ and since Y is reflexive, Theorem 6.2.2 implies that there exists a constant $C = C(Y)$ so that

$$\int_{\mathbb{T}} \left\|\mathcal{R}\left(Ag^{-1/2}\right)(z)\right\|_{L^1}^2 dm \le C \int_{\mathbb{T}} |g(z)|^{-1} \|A(z)\|_{L^1}^2 dm.$$

Recall that $|g(z)| = \|A(z)\|_{L^1} + \epsilon$. Letting $\epsilon \to 0$, the previous two inequalities yield (7.5.14). Since G is a lifting of $[F]_1$, the above estimate and (7.5.12) imply (7.5.10). Finally, by the Hahn–Banach theorem (7.5.10) yields (7.5.11). □

Recall that we view

$$H_0^\infty(\mathbb{T}) = \left\{g \in L^\infty(\mathbb{T}) : \widehat{g(n)} = 0,\ n \in \mathbb{Z}\backslash\mathbb{N}\right\},$$

as the dual space of $L^1(\mathbb{T})/H^1(\mathbb{T})$, and that the identification is provided by the bilinear form

$$\langle [f], g\rangle = \int_{\mathbb{T}} fg dm,$$

where $g \in H_0^\infty(\mathbb{T})$, $f \in L^1(\mathbb{T})$, and $[f] = \left\{f + H^1(\mathbb{T})\right\}$.

Theorem 7.5.4 *Let Y be a reflexive subspace of $L^1(\Omega)$. Any bounded linear operator $T\colon Y \to H^\infty(\mathbb{T})$ admits an extension to a bounded linear operator $T_1\colon L^1(\Omega) \to H^\infty(\mathbb{T})$ such that*

$$T_1(y) = T(y), \quad y \in Y \quad \text{and} \quad \|T_1\| \le C\|T\|,$$

where $C = C(Y)$ is given by Kislyakov's embedding theorem.

Proof Given $f \in L^1(\mathbb{T})$, $y \in Y$, and $z \in \mathbb{C}$, we define

$$u(f \otimes y)(z) = f(z)(Ty)(z).$$

By linearity, u has a well-defined extension to the algebraic tensor product $L^1(\mathbb{T}) \otimes Y$ and a uniquely defined and bounded extension to $L^1(Y)$ such that

$$\left\|u\colon L^1(\mathbb{T}, Y) \to L^1(\mathbb{T})\right\| \le \|T\|.$$

Next, define the linear functional

$$u_0 : L^1(\mathbb{T}, Y) \to \mathbb{C}, \quad F \to \int_{\mathbb{T}} u(F)(z)dm(z).$$

Clearly, $\|u_0\| \le \|u\|$ and $H^1(Y) \subseteq \ker(u_0)$. Hence u_0 induces a well-defined functional on the quotient space $Z_1 = L^1(Y)/H^1(Y)$,

$$u_1 : Z_1 \to \mathbb{C}, \quad [F]_1 \to u_0(F),$$

such that $\|u_1\| \le \|u_0\|$ and

$$u_1([f \otimes y]_1) = \langle [f], Ty \rangle, \quad f \in L^1(\mathbb{T}), \ y \in Y.$$

By Theorem 7.5.3 there exists an extension of u_1 to the quotient space $Z_2 = L^1(L^1)/H^1(L^1)$,

$$u_2 : Z_2 \to \mathbb{C},$$

satisfying conditions (7.5.11). In particular, u_2 inherits from u_1 the identities

$$u_2([f \otimes y]_2) = u_1([f \otimes y]_1) = \langle [f], Ty \rangle, \quad f \in L^1(\mathbb{T}), \ y \in Y.$$

Moreover, for $x \in L^1(\Omega)$ and $f \in L^1(\mathbb{T})$ we get

$$|u_2([f \otimes x]_2)| \le \|u_2\|_{Z_2^*} \|[f \otimes x]_2\|_{Z_2} \le C\|u_1\|_{Z_1^*} \|[f]\|_{L^1/H^1} \|x\|_{L^1(\Omega)}.$$

Finally, define the extension $T_1 : L^1(\Omega) \to H^\infty(\mathbb{T})$ by the duality relation,

$$\langle [f], T_1 x \rangle = u_2([f \otimes x]_2), \quad [f] \in L^1(\mathbb{T})/H^1(\mathbb{T}), \ x \in L^1(\Omega).$$

Thus defined, T_1 is an extension of T since

$$\langle [f], T_1 y \rangle = u_2([f \otimes y]_2) = u_1([f \otimes y]_1) = \langle [f], Ty \rangle, \quad y \in Y,$$

and bounded as follows,

$$\|T_1 x\| \le C\|u_1\|_{Z_1^*} \|x\|_{L^1(\Omega)} \le C\|T\| \cdot \|x\|_{L^1(\Omega)}, \quad x \in L^1(\Omega). \qquad \square$$

7.6 Notes

In Section 7.1, Theorem 7.1.2 is due to Kislyakov (1976) and Pisier (1978) separately. We present the proof due to Pisier (1982b). Pełczyński (1980) and Diestel, Jarchow, and Tonge (1995) contain detailed presentations of Kislyakov's approach (Kislyakov, 1976). The proof of Kislyakov's lemma (Lemma 7.1.1) given in the text, used Diestel, Jarchow, and Tonge (1995) and Kalton and Pełczyński (1997). The Hardy martingale lifting theorem

(Theorem 7.1.3), (and its interpretation in Theorem 7.1.4) were obtained by Bourgain and Davis (1986). Their approach also yields Theorem 7.1.5 and the remark thereafter.

In Section 7.2, we use Kwapień and Pełczyński (1970), Gohberg and Krein (1970), Pełczyński (1961), and Sarason (1967) in the introduction. Theorem 7.2.2 is due to Haagerup and Pisier (1989). We present their original proof using Theorem 7.2.3 and Proposition 7.2.5. For extensions and related results we refer to Blasco and Pełczyński (1991), Blower and Ransford (2004), Pisier (1992a), and Xu (1990, 1991). We presented the original proofs of Theorem 7.2.1 and Theorem 7.2.6 due to Kwapień and Pełczyński (1970).

Section 7.3 is devoted to the proof of Theorem 7.3.4, i.e., the main result of Qiu (2012a,b). We refer to Pisier (1979, 2016) for θ-Hilbertian Banach spaces and Bourgain (1983c) for the first super-reflexive Banach lattice that is not a UMD space. See also Bourgain (1984e). Garling (1988) observed that the example in Bourgain (1983c) may be adapted to show that there exists a super-reflexive Banach lattice without the aUMD condition. See Garling (1990) for the further development of Bourgain (1983c).

Hytönen, van Neerven, Veraar, and Weis (2016) present the real martingale versions of Qiu's theorem (Theorem 7.3.4).

In Section 7.4, the basic references for the James Tree space are James (1974) and Lindenstrauss and Stegall (1975). Proposition 7.4.1, Theorem 7.4.2, Proposition 7.4.3 and Theorem 7.4.4 are due to Lindenstrauss and Stegall (1975). Theorem 7.4.5 is due to Schachermayer, Sersouri, and Werner (1989). The book by Fetter and Gamboa de Buen (1997) contains a detailed presentation of the James Tree space. Theorem 7.4.6 is due to Ghoussoub, Maurey, and Schachermayer (1989) who used it to show Theorem 7.4.7, asserting that the predual of the James Tree space satisfies the aRNP. Theorem 7.4.8, Theorem 7.4.9, and Theorem 7.4.10 are due to Ghoussoub and Maurey (1989). In combination with their result on holomorphic injections (Theorem 5.2.24) they obtained a separate proof that the predual of the James Tree space satisfies the aRNP.

In Section 7.5, Theorem 7.5.1 is due to Bourgain (1984d). Bourgain and Davis (1986), Kislyakov (1991), and Pisier (1986) are the sources for our presentation of Theorem 7.5.2. Kislyakov's embedding theorem (Theorem 7.5.3), and its proof, is taken from Kislyakov (1991). The operator lifting theorem (Theorem 7.5.4) is due to Bourgain (1984a). The presentation in the text is based on Kislyakov (1991).

References

Acosta, G. and R. G. Durán. An optimal Poincaré inequality in L^1 for convex domains. *Proc. Amer. Math. Soc.*, 132(1):195–202, 2004.

Aldous, D. J. Unconditional bases and martingales in $L_p(F)$. *Math. Proc. Cambridge Philos. Soc.*, 85(1):117–123, 1979.

Aleksandrov, A. B. Spectral subspaces of the space L^p, $p < 1$. *Algebra i Analiz*, 19(3):1–75, 2007.

Anisimov, D. S. and S. V. Kislyakov. Double singular integrals: interpolation and correction. *Algebra i Analiz*, 16(5):1–33, 2004.

Arai, H. Measures of Carleson type on filtrated probability spaces and the corona theorem on complex Brownian spaces. *Proc. Amer. Math. Soc.*, 96(4):643–647, 1986a.

Arai, H. On the algebra of bounded holomorphic martingales. *Proc. Amer. Math. Soc.*, 97(4):616–620, 1986b.

Asmar, N. and E. Hewitt. Marcel Riesz's theorem on conjugate Fourier series and its descendants. In *Proceedings of the Analysis Conference*, Singapore 1986, volume 150 of North-Holland Math. Stud., pages 1–56. North-Holland, Amsterdam, 1988.

Bass, R. F. *Probabilistic Techniques in Analysis. Probability and its Applications.* Springer-Verlag, New York, 1995.

Beauzamy, B. *Introduction to Banach Spaces and Their Geometry*, volume 68 of North-Holland Math. Stud. North-Holland, Amsterdam-New York, 1982.

Bedrosian, E. A product theorem for Hilbert transforms. *Proceedings of the IEEE*, 51(5):868–869, May 1963.

Benedek, A., A.-P. Calerón, and R. Panzone. Convolution operators on Banach space valued functions. *Proc. Nat. Acad. Sci. USA*, 48:356–365, 1962.

Bennett, C. and R. Sharpley. *Interpolation of Operators*, volume 129 of *Pure and Applied Mathematics*. Academic Press, Inc., Boston, MA, 1988.

Benyamini, Y. and J. Lindenstrauss. *Geometric Nonlinear Functional Analysis. Vol. 1*, American Mathematical Society Colloquium Publications, vol. 48. American Mathematical Society, Providence, RI, 2000.

Bessaga, C. and A. Pełczyński. A generalization of results of R. C. James concerning absolute bases in Banach spaces. *Studia Math.*, 17:165–174, 1958a.

Bessaga, C. and A. Pełczyński. On bases and unconditional convergence of series in Banach spaces. *Studia Math.*, 17:151–164, 1958b.

Blasco, O. and A. Pełczyński. Theorems of Hardy and Paley for vector-valued analytic functions and related classes of Banach spaces. *Trans. Amer. Math. Soc.*, 323(1):335–367, 1991.

Blower, G. A multiplier characterization of analytic UMD spaces. *Studia Math.*, 96(2):117–124, 1990.

Blower, G. and T. Ransford. Complex uniform convexity and Riesz measures. *Canad. J. Math.*, 56(2):225–245, 2004.

Bochner, S. Additive set functions on groups. *Ann. Math. (2)*, 40:769–799, 1939.

Bochner, S. Generalized conjugate and analytic functions without expansions. *Proc. Nat. Acad. Sci. USA.*, 45:855–857, 1959.

Bogachev, V. I. *Measure theory. Vol. I, II.* Springer-Verlag, Berlin, 2007.

Bonami, A. Étude des coefficients de Fourier des fonctions de $L^p(G)$. *Ann. Inst. Fourier (Grenoble)*, 20(2):335–402, 1970.

Bourgain, J. *La Propriete de Radon–Nikodym.* Publications Mathematiques de l'Universite Pierre et Marie Curie No. 36, Paris, 1979.

Bourgain, J. Complémentation de sous-espaces L^1 dans les espaces L^1. In *Seminar on Functional Analysis, 1979–1980 (French)*, Exp. No. 27, 7. École Polytech., Palaiseau, 1980a.

Bourgain, J. Walsh subspaces of L^p-product spaces. In *Seminar on Functional Analysis, 1979–1980 (French)*, Exp. No. 4A, 9. École Polytech., Palaiseau, 1980b.

Bourgain, J. A counterexample to a complementation problem. *Compositio Math.*, 43(1):133–144, 1981.

Bourgain, J. Embedding L^1 in L^1/H^1. *Trans. Amer. Math. Soc.*, 278(2):689–702, 1983a.

Bourgain, J. On the primarity of H^∞-spaces. *Israel J. Math.*, 45(4):329–336, 1983b.

Bourgain, J. Some remarks on Banach spaces in which martingale difference sequences are unconditional. *Ark. Mat.*, 21(2):163–168, 1983c.

Bourgain, J. Bilinear forms on H^∞ and bounded bianalytic functions. *Trans. Amer. Math. Soc.*, 286(1):313–337, 1984a.

Bourgain, J. The dimension conjecture for polydisc algebras. *Israel J. Math.*, 48(4):289–304, 1984b.

Bourgain, J. Martingale transforms and geometry of Banach spaces. In *Israel Seminar on Geometrical Aspects of Functional Analysis (1983/84)*, pages XIV, 16. Tel Aviv University, Tel Aviv, 1984c.

Bourgain, J. New Banach space properties of the disc algebra and H^∞. *Acta Math.*, 152(1-2):1–48, 1984d.

Bourgain, J. On martingales transforms in finite-dimensional lattices with an appendix on the K-convexity constant. *Math. Nachr.*, 119:41–53, 1984e.

Bourgain, J. Some results on the bidisc algebra. *Astérisque*, 131:279–298, 1985.

Bourgain, J. Homogeneous polynomials on the ball and polynomial bases. *Israel J. Math.*, 68(3):327–347, 1989.

Bourgain, J. and W. J. Davis. Martingale transforms and complex uniform convexity. *Trans. Amer. Math. Soc.*, 294(2):501–515, 1986.

Bu, S. Q. Quelques remarques sur la propriété de Radon–Nikodým analytique. *C. R. Acad. Sci. Paris Sér. I Math.*, 306(18):757–760, 1988.

Bu, S. Q. On the analytic Radon–Nikodým property for bounded subsets in Banach spaces. *J. London Math. Soc. (2)*, 47(3):484–496, 1993.

Bu, S. Q. and B. Khaoulani. Une caractérisation de la propriété de Radon–Nikodým analytique pour les espaces de Banach isomorphes à leur carrés. *Math. Ann.*, 288(2):345–360, 1990.

Bu, S. Q. and W. Schachermayer. Approximation of Jensen measures by image measures under holomorphic functions and applications. *Trans. Amer. Math. Soc.*, 331(2):585–608, 1992.

Buhvalov, A. V. Hardy spaces of vector-valued functions. *Zap. Naučn. Sem. Leningrad. Otdel. Mat. Inst. Steklov. (LOMI)*, 65:5–16, 203, 1976.

Bukhvalov, A. V. Hardy spaces of vector-valued functions. *J. Sov. Math.*, 16:1051–1059, 1981.

Bukhvalov, A. V. and A. A. Danilevich. Boundary properties of analytic and harmonic functions with values in a Banach space. *Mat. Zametki*, 31(2):203–214, 317, 1982.

Burkholder, D. L. Distribution function inequalities for martingales. *Ann. Probability*, 1:19–42, 1973.

Burkholder, D. L. A sharp inequality for martingale transforms. *Ann. Probab.*, 7(5):858–863, 1979.

Burkholder, D. L. Martingale transforms and the geometry of Banach spaces. In *Probability in Banach Spaces, III* (Medford, Mass., 1980), volume 860 of *Lecture Notes in Math.*, pages 35–50. Springer, Berlin-New York, 1981.

Burkholder, D. L. A geometric condition that implies the existence of certain singular integrals of Banach-space-valued functions. In *Conference on Harmonic Analysis in Honor of Antoni Zygmund, Vol. I, II* (Chicago, Ill., 1981), Wadsworth Math. Ser., pages 270–286. Wadsworth, Belmont, CA, 1983.

Burkholder, D. L. Differential subordination of harmonic functions and martingales. In *Harmonic Analysis and Partial Differential Equations (El Escorial, 1987)*, volume 1384 of *Lecture Notes in Math.*, pages 1–23. Springer, Berlin, 1989.

Burkholder, D. L. Martingales and singular integrals in Banach spaces. In *Handbook of the Geometry of Banach Spaces, Vol. I*, pages 233–269. North-Holland, Amsterdam, 2001.

Burkholder, D. L., B. J. Davis, and R. F. Gundy. Integral inequalities for convex functions of operators on martingales. In *Proceedings of the Sixth Berkeley Symposium on Mathematical Statistics and Probability* (Univ. California, Berkeley, Calif., 1970/1971), *Vol. II: Probability Theory*, pages 223–240, 1972.

Burkholder, D. L. and R. F. Gundy. Extrapolation and interpolation of quasi-linear operators on martingales. *Acta Math.*, 124:249–304, 1970.

Burkholder, D. L., R. F. Gundy, and M. L. Silverstein. A maximal function characterization of the class H^p. *Trans. Amer. Math. Soc.*, 157:137–153, 1971.

Calderón, A.-P. Intermediate spaces and interpolation, the complex method. *Studia Math.*, 24:113–190, 1964.

Capon, M. Primarité de $l_p(L^1)$. *Math. Ann.*, 250(1):55–63, 1980.

Carleson, L. On convergence and growth of partial sums of Fourier series. *Acta Math.*, 116:135–157, 1966.

Carne, K. The algebra of bounded holomorphic martingales. *J. Funct. Anal.*, 45(1):95–108, 1982.

Chang, S.-Y. A., J. M. Wilson, and T. H. Wolff. Some weighted norm inequalities concerning the Schrödinger operators. *Comment. Math. Helv.*, 60(2):217–246, 1985.

Chatterji, S. D. Martingale convergence and the Radon–Nikodym theorem in Banach spaces. *Math. Scand.*, 22:21–41, 1968.

Coifman, R. R., P. W. Jones, and S. Semmes. Two elementary proofs of the L^2 boundedness of Cauchy integrals on Lipschitz curves. *J. Amer. Math. Soc.*, 2(3):553–564, 1989.

David, G. *Wavelets and Singular Integrals on Curves and Surfaces*, volume 1465 of *Lecture Notes in Mathematics*. Springer-Verlag, Berlin, 1991.

David, G. Unrectifiable 1-sets have vanishing analytic capacity. *Rev. Mat. Iberoamericana*, 14(2):369–479, 1998.

Davis, B. On the integrability of the martingale square function. *Israel J. Math.*, 8:187–190, 1970.

Davis, W. J., T. Figiel, W. B. Johnson, and A. Pelczynski. Factoring weakly compact operators. *J. Funct. Anal.*, 17:311–327, 1974.

Davis, W. J., D. J. H. Garling, and N. Tomczak-Jaegermann. The complex convexity of quasinormed linear spaces. *J. Funct. Anal.*, 55(1):110–150, 1984.

Davis, W. J., N. Ghoussoub, W. B. Johnson, S. Kwapień, and B. Maurey. Weak convergence of vector valued martingales. In *Probability in Banach Spaces 6* (Sandbjerg, 1986), volume 20 of *Progr. Probab.*, pages 41–50. Birkhäuser Boston, Boston, MA, 1990.

Dechamps-Gondim, M. Analyse harmonique, analyse complexe et géométrie des espaces de Banach (d'après Jean Bourgain). *Astérisque*, 1983(121–122):171–195, 1985. Seminar Bourbaki, Vol. 1983/84.

Delbaen, F. and W. Schachermayer. An inequality for the predictable projection of an adapted process. In *Séminaire de Probabilités, XXIX*, volume 1613 of *Lecture Notes in Math.*, pages 17–24. Springer, Berlin, 1995.

Dellacherie, C. and P.-A. Meyer. *Probabilités et potentiel*. Hermann, Paris, 1975. Chapitres I à IV, Édition entièrement refondue, Publications de l'Institut de Mathématique de l'Université de Strasbourg, No. XV, Actualités Scientifiques et Industrielles, No. 1372.

Dellacherie, C. and P.-A. Meyer. *Probabilities and Potential*, volume 29 of North-Holland Math. Stud. North-Holland, Amsterdam-New York, 1978.

Dellacherie, C. and P.-A. Meyer. *Probabilities and Potential B Theory of Martingales*, volume 72 of North-Holland Math. Stud. North-Holland, Amsterdam, 1982.

Diestel, J. *Geometry of Banach Spaces – Selected Topics*. Lecture Notes in Mathematics, Vol. 485. Springer-Verlag, Berlin-New York, 1975.

Diestel, J., H. Jarchow, and A. Tonge. *Absolutely Summing Operators*, volume 43 of *Camb. Stud. Adv. Math.* Cambridge University Press, Cambridge, 1995.

Diestel, J. and J. J. Jr. Uhl. *Vector Measures*. American Mathematical Society, Providence, R.I., 1977.

Dieudonné, J. *Éléments d'analyse. Tome I: Fondements de l'analyse moderne*. Traduit de l'anglais par D. Huet. Avant-propos de G. Julia. Nouvelle édition revue et corrigée. Cahiers Scientifiques, Fasc. XXVIII. Gauthier-Villars, Éditeur, Paris, 1968a.

Dieudonné, J. *Éléments d'analyse. Tome II: Chapitres XII à XV*. Cahiers Scientifiques, Fasc. XXXI. Gauthier-Villars, Éditeur, Paris, 1968b.

Dilworth, S. J. Complex convexity and the geometry of Banach spaces. *Math. Proc. Camb. Philos. Soc.*, 99(3):495–506, 1986.

Doob, J. L. *Stochastic Processes*. Wiley, New York, 1953.

Doob, J. L. Semimartingales and subharmonic functions. *Trans. Amer. Math. Soc.*, 77:86–121, 1954.

Dor, L. E. On projections in L_1. *Ann. Math. (2)*, 102(3):463–474, 1975.

Dosev, D., W. B. Johnson, and G. Schechtman. Commutators on L_p, $1 \le p < \infty$. *J. Am. Math. Soc.*, 26(1):101–127, 2013.

Dowling, P. N. Representable operators and the analytic Radon–Nikodým property in Banach spaces. *Proc. R. Ir. Acad. Sect. A*, 85(2):143–150, 1985.

Dowling, P. N. and G. A. Edgar. Some characterizations of the analytic Radon–Nikodým property in Banach spaces. *J. Funct. Anal.*, 80(2):349–357, 1988.

Dunford, N. and B. J. Pettis. Linear operations on summable functions. *Trans. Am. Math. Soc.*, 47:323–392, 1940.

Durrett, R. *Brownian Motion and Martingales in Analysis*. Wadsworth Mathematics Series. Wadsworth International Group, Belmont, CA, 1984.

Edgar, G. A. Disintegration of measures and the vector-valued Radon–Nikodým theorem. *Duke Math. J.*, 42(3):447–450, 1975.

Edgar, G. A. Complex martingale convergence. In Kalton N.J., Saab E. (eds) *Banach Spaces*. Lecture Notes in Mathematics, vol 1166. Springer, Berlin, Heidelberg, 1985.

Edgar, G. A. Analytic martingale convergence. *J. Funct. Anal.*, 69(2):268–280, 1986.

Enflo, P. and T. W. Starbird. Subspaces of L^1 containing L^1. *Studia Math.*, 65(2):203–225, 1979.

Fakhoury, H. Représentations d'opérateurs à valeurs dans $L^1(X, \Sigma, \mu)$. *Math. Ann.*, 240(3):203–212, 1979.

Fetter, H. and B. Gamboa de Buen. *The James Forest*, volume 236 of *London Mathematical Society Lecture Note Series*. Cambridge University Press, Cambridge, 1997.

Föllmer, H. Stochastic holomorphy. *Math. Ann.*, 207:245–255, 1974.

García-Cuerva, J. and J. L. Rubio de Francia. *Weighted Norm Inequalities and Related Topics*, volume 116 of North-Holland Math. Stud. North-Holland, Amsterdam, 1985.

Garling, D. J. H. Brownian motion and UMD-spaces. In *Probability and Banach Spaces* (Zaragoza, 1985), volume 1221 of *Lecture Notes in Math.*, pages 36–49. Springer, Berlin, 1986.

Garling, D. J. H On martingales with values in a complex Banach space. *Math. Proc. Camb. Philos. Soc.*, 104(2):399–406, 1988.

Garling, D. J. H. Random martingale transform inequalities. In *Probability in Banach Spaces 6* (Sandbjerg, 1986), volume 20 of *Progr. Probab.*, pages 101–119. Birkhäuser Boston, Boston, MA, 1990.

Garling, D. J. H. Hardy martingales and the unconditional convergence of martingales. *Bull. London Math. Soc.*, 23(2):190–192, 1991.

Garling, D. J. H. *Inequalities: A Journey Into Linear Analysis*. Cambridge University Press, Cambridge, 2007.

Garling, D. J. H. and S. J. Montgomery-Smith. Complemented subspaces of spaces obtained by interpolation. *J. London Math. Soc. (2)*, 44(3):503–513, 1991.

Garnett, J. B. *Bounded Analytic Functions*, volume 96 of *Pure and Applied Mathematics*. Academic Press Inc. [Harcourt Brace Jovanovich Publishers], New York, 1981.

Garnett, J. B. and P. W. Jones. The distance in BMO to L^∞. *Ann. of Math.*, 108(2):373–393, 1978.

Garnett, J. B. and P. W. Jones. BMO from dyadic BMO. *Pacific J. Math.*, 99(2):351–371, 1982.

Garsia, A. M. *Martingale Inequalities: Seminar Notes on Recent Progress*. W. A. Benjamin, Inc., Reading, Mass.-London-Amsterdam, 1973. Mathematics Lecture Notes Series.

Getoor, R. K. and M. J. Sharpe. Conformal martingales. *Invent. Math.*, 16:271–308, 1972.

Ghoussoub, N. *Duality and Perturbation Methods in Critical Point Theory*, volume 107 of *Cambridge Tracts in Mathematics*. Cambridge University Press, Cambridge, 1993.

Ghoussoub, N., J. Lindenstrauss, and B. Maurey. Analytic martingales and plurisubharmonic barriers in complex Banach spaces. In *Banach Space Theory* (Iowa City, IA, 1987), volume 85 of *Contemp. Math.*, pages 111–130. Amer. Math. Soc., Providence, RI, 1989.

Ghoussoub, N. and B. Maurey. H_δ-embedding in Hilbert space and optimization on G_δ-sets. *Mem. Amer. Math. Soc.*, 62(349):101, 1986.

Ghoussoub, N. and B. Maurey. Plurisubharmonic martingales and barriers in complex quasi-Banach spaces. *Ann. Inst. Fourier* (Grenoble), 39(4):1007–1060, 1989.

Ghoussoub, N., B. Maurey, and W. Schachermayer. Pluriharmonically dentable complex Banach spaces. *J. Reine Angew. Math.*, 402:76–127, 1989.

Ghoussoub, N. and H. P. Rosenthal. Martingales, G_δ-embeddings and quotients of L_1. *Math. Ann.*, 264(3):321–332, 1983.

Godefroy, G., N. J. Kalton, and D. Li. Operators between subspaces and quotients of L^1. *Indiana Univ. Math. J.*, 49(1):245–286, 2000.

Gohberg, I. C. and M. G. Krein. *Theory and Applications of Volterra Operators in Hilbert Space*. Translated from the Russian by A. Feinstein. Translations of Mathematical Monographs, Vol. 24. American Mathematical Society, Providence, R.I., 1970.

Grossetête, C. Sur certaines classes de fonctions harmoniques dans le disque à valeur dans un espace vectoriel topologique localement convexe. *C. R. Acad. Sci. Paris Sér. A-B*, 273:A1048–A1051, 1971.

Grossetête, C. Classes de Hardy et de Nevanlinna pour les fonctions holomorphes à valeurs vectorielles. *C. R. Acad. Sci. Paris Sér. A-B*, 274:A251–A253, 1972.

Haagerup, U. and G. Pisier. Factorization of analytic functions with values in noncommutative L_1-spaces and applications. *Canad. J. Math.*, 41(5):882–906, 1989.

Heins, M. *Hardy Classes on Riemann Surfaces*. Lecture Notes in Mathematics, No. 98. Springer-Verlag, Berlin-New York, 1969.

Helson, H. Conjugate series and a theorem of Paley. *Pacific J. Math.*, 8:437–446, 1958.

Helson, H. Conjugate series in several variables. *Pacific J. Math.*, 9:513–523, 1959.

Henkin, G. M. The nonisomorphy of certain spaces of functions of different numbers of variables. *Funkcional. Anal. i Priložen.*, 1(4):57–68, 1967.

Henkin, G. M. The Banach spaces of analytic functions in a ball and in a bicylinder are nonisomorphic. *Funkcional. Anal. i Priložen.*, 2(4):82–91, 1968.

Henkin, G. M. On non-isomorphism of Banach spaces of holomorphic functions. *Studia Math.*, 38:267–270, 1970.

Hensgen, W. Hardy–Räume vektorwertiger Funktionen. *Dissertation Ludwig-Maximilians-Universität München*, pages 1–151, 1986.

Hensgen, W. Operatoren $H^1 \to X$. *Manuscripta Math.*, 59(4):399–422, 1987.

Hensgen, W. Some remarks on boundary values of vector-valued harmonic and analytic functions. *Arch. Math. (Basel)*, 57(1):88–96, 1991.

Hoffmann-Jorgensen, J. Sums of independent Banach space valued random variables. *Studia Math.*, 52:159–186, 1974.

Hunt, R., B. Muckenhoupt, and R. Wheeden. Weighted norm inequalities for the conjugate function and Hilbert transform. *Trans. Amer. Math. Soc.*, 176:227–251, 1973.

Hunt, R. A. On the convergence of Fourier series. In *Orthogonal Expansions and their Continuous Analogues* (Proc. Conf., Edwardsville, Ill., 1967), pages 235–255. Southern Illinois Univ. Press, Carbondale, Ill., 1968.

Hytönen, T., J. van Neerven, M. Veraar, and L. Weis. *Analysis in Banach Spaces. Vol. I. Martingales and Littlewood-Paley Theory* Springer, Cham, 2016.

Itô, K., and McKean, H. P. Jr. *Diffusion processes and their sample paths*. Springer Verlag, Berlin-New York, 1965.

Ivanisvili, P., A. Lindenberger, P. F. X. Müller, and M. Schmuckenschläger. Hyper contractivity on the unit circle for ultraspherical measures: linear case. arXiv:2001.05567, DOI: 10.4171/RMI/1305, 2020.

James, R. C. Bases and reflexivity of Banach spaces. *Ann. of Math.*, 52:518–527, 1950.

James, R. C. A separable somewhat reflexive Banach space with nonseparable dual. *Bull. Am. Math. Soc.*, 80:738–743, 1974.

Jiao, Y., N. Randrianantoanina, L. Wu, and D. Zhou. Square functions for noncommutative differentially subordinate martingales. *Comm. Math. Phys.*, 374(2):975–1019, 2020.

Jiao, Y., G. Xie, and D. Zhou. Dual spaces and John–Nirenberg inequalities of martingale Hardy–Lorentz–Karamata spaces. *Q. J. Math.*, 66(2):605–623, 2015.

Jiao, Y., L. Wu, A. Yang, and R. Yi. The predual and John–Nirenberg inequalities on generalized BMO martingale spaces. *Trans. Am. Math. Soc.*, 369(1):537–553, 2017.

Johnson, W. B., B. Maurey, and G. Schechtman. Weakly null sequences in L_1. *J. Am. Math. Soc.*, 20(1):25–36, 2007.

Jones, P. W. L^∞ estimates for the $\bar{\partial}$ problem in a half-plane. *Acta Math.*, 150(1–2):137–152, 1983.

Jones, P. W. and P. F. X. Müller. Conditioned Brownian motion and multipliers into SL^∞. *Geom. Funct. Anal.*, 14(2):319–379, 2004.

Kadec, M. I. and A. Pełczyński. Bases, lacunary sequences and complemented subspaces in the spaces $L \subset p$. *Studia Math.*, 21:161–176, 1961/62.

Kahane, J.-P. *Some Random Series of Functions*. D. C. Heath and Co. Raytheon Education Co., Lexington, Mass., 1968.

Kahane, J.-P. *Some Random Series of Functions*, volume 5 of Cambridge Studies in Advanced Mathematics. Cambridge University Press, Cambridge, second edition, 1985.

Kalton, N. J. The endomorphisms of $L_p(0 \leq p \leq i)$. *Indiana Univ. Math. J.*, 27(3):353–381, 1978.

Kalton, N. J. Embedding L_1 in a Banach lattice. *Israel J. Math.*, 32(2-3):209–220, 1979.

Kalton, N. J. Linear operators on L_p for $0 < p < 1$. *Trans. Am. Math. Soc.*, 259(2): 319–355, 1980.

Kalton, N. J. Banach spaces embedding into L_0. *Israel J. Math.*, 52(4):305–319, 1985.

Kalton, N. J. Differentiability properties of vector valued functions. In *Probability and Banach Spaces* (Zaragoza, 1985), volume 1221 of Lecture Notes in Math., pages 141–181. Springer, Berlin, 1986a.

Kalton, N. J. Some applications of vector-valued analytic and harmonic functions. In *Probability and Banach spaces* (Zaragoza, 1985), volume 1221 of Lecture Notes in Math., pages 114–140. Springer, Berlin, 1986b.

Kalton, N. J. and A. Pełczyński. Kernels of surjections from L_1-spaces with an application to Sidon sets. *Math. Ann.*, 309(1):135–158, 1997.

Kalton, N. J. and L. Weis. The H^∞-calculus and sums of closed operators. *Math. Ann.*, 321(2):319–345, 2001.

Katznelson, Y. *An Introduction to Harmonic Analysis*. John Wiley & Sons, Inc., New York-London-Sydney, 1968.

Kazaniecki, K. and M. Wojciechowski. On the equivalence between the sets of the trigonometric polynomials. arXiv:1502.05994, 2014.

Kislyakov, S. V. What is needed for a 0-absolutely summing operator to be nuclear? In *Complex Analysis and Spectral Theory* (Leningrad, 1979/1980), volume 864 of *Lecture Notes in Math.*, pages 336–364. Springer, Berlin-New York, 1981.

Kislyakov, S. V. Spaces with a "small" annihilator. *Zap. Naučn. Sem. Leningrad. Otdel. Mat. Inst. Steklov. (LOMI)*, 65:192–195, 209, 1976.

Kislyakov, S. V. Absolutely summing operators on the disc algebra. *Algebra i Analiz*, 3(4):1–77, 1991.

König, H. On the best constants in the Khintchine inequality for Steinhaus variables. *Israel J. Math.*, 203(1):23–57, 2014.

Koosis, P. *Introduction to H_p Spaces*, volume 40 of *London Mathematical Society Lecture Note Series*. Cambridge University Press, Cambridge-New York, 1980.

Kuratowski, K. and C. Ryll-Nardzewski. A general theorem on selectors. *Bull. Acad. Polon. Sci. Sér. Sci. Math. Astronom. Phys.*, 13:397–403, 1965.

Kwapień, S. Unsolved problems (problem 3). *Studia Math.*, 38:467–483, 1970.

Kwapień, S. On the form of a linear operator in the space of all measurable functions. *Bull. Acad. Polon. Sci. Sér. Sci. Math. Astronom. Phys.*, 21:951–954, 1973.

Kwapień, S. On Banach spaces containing c_0. *Studia Math.*, 52:187–188, 1974. A supplement to the paper by J. Hoffmann-Jorgensen "Sums of independent Banach space valued random variables" (Studia Math. **52** (1974), 159–186).

Kwapień, S. and A. Pełczyński. The main triangle projection in matrix spaces and its applications. *Studia Math.*, 34:43–68, 1970.

Kwapien, S. and W. A. Woyczyński. *Random Series and Stochastic Integrals: Single and Multiple*. Probability and its Applications. Birkhäuser Boston Inc., Boston, MA, 1992.

Lechner, R. Factorization in SL^∞. *Israel J. Math.*, 226(2):957–991, 2018.

Lechner, R. Dimension dependence of factorization problems: Hardy spaces and SL_n^∞. *Israel J. Math.*, 232(2):677–693, 2019.

Lechner, R., P. Motakis, P. F. X. Müller, and T. Schlumprecht. The factorisation property of $l^\infty(X_k)$. *Math. Proc. Camb. Philos. Soc.*, 171(2):421–448, 2021

Lechner, R., P. Motakis, P. F. X. Müller, and T. Schlumprecht. Strategically reproducible bases and the factorization property. *Israel J. Math.*, 238(1):13–60, 2020.

Lépingle, D. Une inégalité de martingales. In *Séminaire de Probabilités, XII* (Univ. Strasbourg, Strasbourg, 1976/1977), volume 649 of *Lecture Notes in Math.*, pages 134–137. Springer, Berlin, 1978.

Lévy, P. *Processus stochastiques et mouvement Brownien. Suivi d'une note de M. Loève*. Gauthier-Villars, Paris, 1948.

Lewis, D. R. and C. Stegall. Banach spaces whose duals are isomorphic to $l_1(\Gamma)$. *J. Funct. Anal.*, 12:177–187, 1973.

Li, D. and H. Queffélec. *Introduction à l'étude des espaces de Banach*, volume 12 of *Cours Spécialisés*. Société Mathématique de France, Paris, 2004.

Li, D. and H. Queffélec. *Introduction to Banach Spaces: Analysis and Probability. Vol. 1*, volume 166 of *Cambridge Studies in Advanced Mathematics*. Cambridge University Press, Cambridge, 2018. Translated from the French by Danièle Gibbons and Greg Gibbons, For the French original see [MR2124356].

Lindenstrauss, J. and A. Pełczyński. Contributions to the theory of the classical Banach spaces. *J. Funct. Anal.*, 8:225–249, 1971.

Lindenstrauss, J. and H. P. Rosenthal. The $\mathcal{L}_p$ spaces. *Israel J. Math.*, 7:325–349, 1969.

Lindenstrauss, J. and C. Stegall. Examples of separable spaces which do not contain ℓ_1 and whose duals are non-separable. *Studia Math.*, 54(1):81–105, 1975.

Lindenstrauss, J. and L. Tzafriri. *Classical Banach Spaces. I*. Springer-Verlag, Berlin-New York, 1977.

Littlewood, J. E. Mathematical Notes (8); On Functions Subharmonic in a Circle (II). *Proc. London Math. Soc. (2)*, 28(5):383–394, 1928.

Liu, Z. A decomposition theorem for operators on L^1. *J. Operator Theory*, 40(1):3–34, 1998.

Long, R. L. *Martingale Spaces and Inequalities*. Peking University Press, Beijing; Friedr. Vieweg & Sohn, Braunschweig, 1993.

Marcus, M. B. and G. Pisier. *Random Fourier Series with Applications to Harmonic Analysis*, *Ann. Math. Stud.*, 101:1–150, 1981.

Maslyuchenko, O. V., V. V. Mykhaylyuk, and M. M. Popov. A lattice approach to narrow operators. *Positivity*, 13(3):459–495, 2009.

Maurey, B. Théorèmes de factorisation pour les opérateurs linéaires à valeurs dans un espace $L^p(U, \mu)$, $0 < p \leq +\infty$. In *Séminaire Maurey–Schwartz Année 1972–1973: Espaces L^p et applications radonifiantes, Exp. No. 15*, page 8. Centre de Math., École Polytech., Paris, 1973a.

Maurey, B. Un lemme de H. P. Rosenthal. In *Séminaire Maurey-Schwartz Année 1972–1973: Espaces L^p et applications radonifiantes, Exp. No. 21*, page 11. Centre de Math., École Polytech., Paris, 1973b.

Maurey, B. *Théorèmes de factorisation pour les opérateurs linéaires à valeurs dans les espaces L^p*. Société Mathématique de France, Paris, 1974.

Maurey, B. Sous-espaces complémentés de L^p, d'après P. Enflo. In *Séminaire Maurey-Schwartz 1974–1975: Espaces L^p, applications radonifiantes et géométrie des espaces de Banach, Exp. No. III*, pages 15 pp. (erratum, p. 1). 1975a.

Maurey, B. Système de Haar. In *Séminaire Maurey-Schwartz 1974–1975: Espaces L^p, applications radonifiantes et géométrie des espaces de Banach, Exp. Nos. I et II*, pages 26 pp. (erratum, p. 1). Centre Math., École Polytech., Paris, 1975b.

Maurey, B. Isomorphismes entre espaces H_1. *Acta Math.*, 145(1–2):79–120, 1980.

Maurey, B. and G. Pisier. Séries de variables aléatoires vectorielles indépendantes et propriétés géométriques des espaces de Banach. *Studia Math.*, 58(1):45–90, 1976.

Maurey, B. and H. P. Rosenthal. Normalized weakly null sequence with no unconditional subsequence. *Studia Math.*, 61(1):77–98, 1977.

Maurey, B. and G. Schechtman. Some remarks on symmetric basic sequences in L_1. *Compos. Math.*, 38(1):67–76, 1979.

Meyer, P. A. Démonstration probabiliste de certaines inégalités de Littlewood–Paley. IV. Semi-groupes de convolution symétriques. In *Séminaire de Probabilités, X (Première partie, Univ. Strasbourg, Strasbourg, année universitaire 1974/1975)*, pages 175–183. Lecture Notes in Math., Vol. 511. 1976.

Meyer, Y. Endomorphismes des ideaux fermes de L^1 (G), classes de Hardy et series de Fourier lacunaires. *Ann. Sci. École Norm. Sup. (4)*, 1:499–580, 1968.

Mitiagin, B. S. and A. Pełczyński. On the non-existence of linear isomorphisms between Banach spaces of analytic functions of one and several complex variables. *Studia Math.*, 56(2):175–186, 1976.

Montgomery-Smith, S. Concrete representation of martingales. *Electron. J. Probab.*, 3:No. 15, 15 pp. 1998.

Müller, P. F. X. Holomorphic martingales and interpolation between Hardy spaces. *J. Anal. Math.*, 61:327–337, 1993.

Müller, P. F. X. *Isomorphisms Between H^1 spaces*, volume 66 of *Instytut Matematyczny Polskiej Akademii Nauk. Monografie Matematyczne (New Series)* Birkhäuser Verlag, Basel, 2005.

Müller, P. F. X. A decomposition for Hardy martingales. *Indiana Univ. Math. J.*, 61(5):1801–1816, 2012.

Müller, P. F. X. A decomposition for Hardy martingales II. *Math. Proc. Camb. Philos. Soc.*, 157(2):189–207, 2014.

Müller, P. F. X. A decomposition for Hardy martingales III. *Math. Proc. Camb. Philos. Soc.*, 162(1):173–189, 2017.

Müller, P. F. X. and M. Passenbrunner. Almost everywhere convergence of spline sequences. *Israel J. Math.*, 240(1):149–177, 2020.

Müller, P. F. X. and J. Penteker. Lecture notes on singular integrals, projections, multipliers and rearrangements. In *IMPAN Lecture Notes, Winter School on Harmonic Analysis, Bedlewo*, www.impan.pl/swiat-matematyki/notatki-z-wyklado~/scriptsios_2.pdf, 2015.

Müller, P. F. X. and K. Riegler. Radial variation of Bloch functions on the unit ball of R^d. *Ark. Mat.*, 58(1):161–178, 2020.

Müller, P. F. X. and P. Yuditskii. Interpolation for Hardy spaces: Marcinkiewicz decomposition, complex interpolation and holomorphic martingales. *Colloq. Math.*, 158(1):141–155, 2019.

Neveu, J. *Discrete-Parameter Martingales*. North-Holland, Amsterdam-Oxford; American Elsevier Publishing Co., Inc., New York, revised edition, 1975. Translated from the French by T. P. Speed, North-Holland Mathematical Library, Vol. 10.

Novikov, I. and E. Semenov. *Haar Series and Linear Operators*, volume 367 of *Mathematics and its Applications*. Kluwer Academic Publishers Group, Dordrecht, 1997.

Odell, E. and B. Zheng. On the unconditional subsequence property. *J. Funct. Anal.*, 258(2):604–615, 2010.

Osekowski, A. *Sharp Martingale and Semimartingale Inequalities*, volume 72 of *Instytut Matematyczny Polskiej Akademii Nauk. Monografie Matematyczne (New Series)*. Birkhäuser/Springer Basel AG, Basel, 2012.

Pallares, A. and O. Kouba. *Seminar Notes*. Groupe de Travail sur les Espaces de Banach. Equipe d'Analyse, Universite Paris VI, 1986-1987.

Parcet, J. and N. Randrianantoanina. Gundy's decomposition for non-commutative martingales and applications. *Proc. London Math. Soc. (3)*, 93(1):227–252, 2006.

Passenbrunner, M. Martingale inequalities for spline sequences. *Positivity*, 24(1):95–115, 2020.

Passenbrunner, M. Spline characterizations of the Radon–Nikodým property. *Proc. Am. Math. Soc.*, 148(2):811–824, 2020.

Pełczyński, A. On the impossibility of embedding of the space L in certain Banach spaces. *Colloq. Math.*, 8:199–203, 1961.

Pełczyński, A. *Banach Spaces of Analytic Functions and Absolutely Summing Operators*. American Mathematical Society, Providence, R.I., 1977. Expository lectures from the CBMS Regional Conference held at Kent State University, Kent, Ohio, July 11–16, 1976, Conference Board of the Mathematical Sciences Regional Conference Series in Mathematics, No. 30.

Pełczyński, A. Geometry of finite-dimensional Banach spaces and operator ideals. In *Notes in Banach Spaces*, pages 81–181. Univ. Texas Press, Austin, TX, 1980.

Pełczyński, A. and M. Wojciechowski. Absolutely summing surjections from Sobolev spaces in the uniform norm. In *Progress in Functional Analysis* (Peñíscola, 1990), volume 170 of *North-Holland Math. Stud.*, pages 423–431. North-Holland, Amsterdam, 1992a.

Pełczyński, A. and M. Wojciechowski. Paley projections on anisotropic Sobolev spaces on tori. *Proc. London Math. Soc. (3)*, 65(2):405–422, 1992b.

Penteker, J. Postorder rearrangement operators. *Q. J. Math.*, 66(4):1103–1126, 2015.

Perrin, M. A noncommutative Davis' decomposition for martingales. *J. Lond. Math. Soc. (2)*, 80(3):627–648, 2009.

Petersen, K. E. *Brownian Motion, Hardy Spaces and Bounded Mean Oscillation*. Cambridge University Press, Cambridge-New York-Melbourne, 1977. London Mathematical Society Lecture Note Series, No. 28.

Phelps, R. R. *Convex Functions, Monotone Operators and Differentiability*, volume 1364 of *Lecture Notes in Mathematics*. Springer-Verlag, Berlin, 1989.

Pisier, G. Sur les espaces qui ne contiennent pas de l_n^1 uniformément. In *Séminaire Maurey-Schwartz (1973–1974), Espaces L^p, applications radonifiantes et géométrie des espaces de Banach, Exp. No. 7*, pages 19 pp. (errata, p. E.1). 1974.

Pisier, G. Martingales with values in uniformly convex spaces. *Israel J. Math.*, 20(3–4):326–350, 1975a.

Pisier, G. Un exemple concernant la super-réflexivité. In *Séminaire Maurey-Schwartz 1974–1975: Espaces L^p applications radonifiantes et géométrie des espaces de Banach, Annexe No. 2*, page 12. 1975b.

Pisier, G. Une nouvelle classe d'espaces de Banach vérifiant le théorème de Grothendieck. *Ann. Inst. Fourier*, 28(1):x, 69–90, 1978.

Pisier, G. Some applications of the complex interpolation method to Banach lattices. *J. Anal. Math.*, 35:264–281, 1979.

Pisier, G. Sur les espaces de Banach K-convexes. In *Seminar on Functional Analysis, 1979–1980 (French)*, pages Exp. No. 11, 15. École Polytech., Palaiseau, 1980.

Pisier, G. Holomorphic semigroups and the geometry of Banach spaces. *Ann. Math. (2)*, 115(2):375–392, 1982a.

Pisier, G. Quotients of Banach spaces of cotype q. *Proc. Am. Math. Soc.*, 85(1):32–36, 1982b.

Pisier, G. Counterexamples to a conjecture of Grothendieck. *Acta Math.*, 151(3–4):181–208, 1983.

Pisier, G. *Factorization of Linear Operators and Geometry of Banach Spaces*, volume 60 of *CBMS Regional Conference Series in Mathematics*. Published for the Conference Board of the Mathematical Sciences, Washington, DC; by the American Mathematical Society, Providence, RI, 1986.

Pisier, G. Factorization of operator valued analytic functions. *Adv. Math.*, 93(1):61–125, 1992a.

Pisier, G. Interpolation between H^p spaces and noncommutative generalizations. I. *Pacific J. Math.*, 155(2):341–368, 1992b.

Pisier, G. Interpolation between H^p spaces and noncommutative generalizations. II. *Rev. Mat. Iberoamericana*, 9(2):281–291, 1993.

Pisier, G. A polynomially bounded operator on Hilbert space which is not similar to a contraction. arXiv: arxiv.org/abs/math/9602207, 1996.

Pisier, G. A polynomially bounded operator on Hilbert space which is not similar to a contraction. *J. Am. Math. Soc.*, 10(2):351–369, 1997.

Pisier, G. *Martingales in Banach Spaces*, volume 155 of *Cambridge Studies in Advanced Mathematics*. Cambridge University Press, Cambridge, 2016.

Pisier, G. and Q. Xu. Non-commutative martingale inequalities. *Comm. Math. Phys.*, 189(3):667–698, 1997.

Popov, M. and B. Randrianantoanina. *Narrow Operators on Function Spaces and Vector Lattices*, volume 45 of *De Gruyter Studies in Mathematics*. Walter de Gruyter & Co., Berlin, 2013.

Qiu, Y. On the UMD constants for a class of iterated $L_p(L_q)$ spaces. *J. Funct. Anal.*, 263(8):2409–2429, 2012a.

Qiu, Y. Propriété UMD pour les espaces de Banach et d'opérateurs. *Thèse de Doctorat, Université Pierre et Marie Curie*, 2012b.

Randrianantoanina, N. Conditioned square functions for noncommutative martingales. *Ann. Probab.*, 35(3):1039–1070, 2007.

Randrianantoanina, N. Wu, L. and Xu, Q. Noncommutative Davis type decompositions and applications. *J. Lond. Math. Soc. (2)*, 99(1):97–126, 2019.

Ransford, T. *Potential Theory in the Complex Plane*, volume 28 of *London Mathematical Society Student Texts*. Cambridge University Press, Cambridge, 1995.

Revuz, D. and M. Yor. *Continuous Martingales and Brownian Motion*, volume 293 of *Grundlehren der Mathematischen Wissenschaften [Fundamental Principles of Mathematical Sciences]*. Springer-Verlag, Berlin, 1991.

Rosenthal, H. P. On subspaces of L^p. *Ann. Math. (2)*, 97:344–373, 1973.

Rosenthal, H. P. Sign-embeddings of L^1. In *Banach Spaces, Harmonic Analysis, and Probability Theory* (Storrs, Conn., 1980/1981), volume 995 of *Lecture Notes in Math.*, pages 155–165. Springer, Berlin, 1983.

Rosenthal, H. P. Embeddings of L^1 in L^1. In *Conference in Modern Analysis and Probability* (New Haven, Conn., 1982), volume 26 of *Contemp. Math.*, pages 335–349. Amer. Math. Soc., Providence, RI, 1984.

Rudin, W. *Real and Complex Analysis*. McGraw-Hill Book Co., New York-Düsseldorf-Johannesburg, second edition, 1974.

Rudin, W. *Functional Analysis*. International Series in Pure and Applied Mathematics. McGraw-Hill, Inc., New York, second edition, 1991.

Ryan, R. Boundary values of analytic vector valued functions. *Indag. Math.*, 65:558–572, 1962.

Ryll, J. and P. Wojtaszczyk. On homogeneous polynomials on a complex ball. *Trans. Am. Math. Soc.*, 276(1):107–116, 1983.

Rzeszut, M. and M. Wojciechowski. Independent sums of $H^1_n(\mathbb{T})$ and $H^1_n(\delta)$. *J. Funct. Anal.*, 273(2):836–873, 2017.

Sarason, D. Generalized interpolation in H^∞. *Trans. Am. Math. Soc.*, 127:179–203, 1967.

Sawa, J. The best constant in the Khintchine inequality for complex Steinhaus variables, the case $p = 1$. *Studia Math.*, 81(1):107–126, 1985.

Schachermayer, W., A. Sersouri, and E. Werner. Moduli of nondentability and the Radon–Nikodým property in Banach spaces. *Israel J. Math.*, 65(3):225–257, 1989.

Schwartz, L. Les applications p-sommantes. In *Séminaire Maurey-Schwartz Année 1972–1973: Espaces L^p et applications radonifiantes, Exp. No. 2*, page 19. Centre de Math., École Polytech., Paris, 1973.

Semenov, E. M. and S. N. Uksusov. Multipliers of series in the Haar system. *Sibirsk. Mat. Zh.*, 53(2):388–395, 2012.

Stegall, C. Optimization of functions on certain subsets of Banach spaces. *Math. Ann.*, 236(2):171–176, 1978.

Stegall, C. *Applications of Descriptive Topology in Functional Analysis*. Institutsbericht 289. J. Kepler Universität, Institut für Mathematik, Linz, 1985.

Szarek, S. J. On the best constants in the Khinchin inequality. *Studia Math.*, 58(2):197–208, 1976.

Talagrand, M. The three-space problem for L^1. *J. Am. Math. Soc.*, 3(1):9–29, 1990.

Tsuji, M. *Potential Theory in Modern Function Theory*. Chelsea Publishing Co., New York, 1975.

Varopoulos, N. T. The Helson–Szegő theorem and A_p-functions for Brownian motion and several variables. *J. Funct. Anal.*, 39(1):85–121, 1980.

Varopoulos, N. T. Probabilistic approach to some problems in complex analysis. *Bull. Sci. Math. (2)*, 105(2):181–224, 1981.

Wark, H. M. A remark on the multipliers of the Haar basis of $L^1[0, 1]$. *Studia Math.*, 227(2):141–148, 2015.

Wark, H. M. Operator-valued Fourier Haar multipliers on vector-valued L^1 spaces. *J. Math. Anal. Appl.*, 450(2):1148–1156, 2017.

Weis, L. On the representation of order continuous operators by random measures. *Trans. Am. Math. Soc.*, 285(2):535–563, 1984.

Weissler, F. B. Logarithmic Sobolev inequalities and hypercontractive estimates on the circle. *J. Funct. Anal.*, 37(2):218–234, 1980.

Weisz, F. *Martingale Hardy Spaces and their Applications in Fourier Analysis*, volume 1568 of *Lecture Notes in Mathematics*. Springer-Verlag, Berlin, 1994.

Wojtaszczyk, P. Decompositions of H_p spaces. *Duke Math. J.*, 46(3):635–644, 1979a.

Wojtaszczyk, P. On projections in spaces of bounded analytic functions with applications. *Studia Math.*, 65(2):147–173, 1979b.

Wojtaszczyk, P. Projections and isomorphisms of the ball algebra. *J. London Math. Soc. (2)*, 29(2):301–305, 1984.

Wojtaszczyk, P. *Banach Spaces for Analysts*, volume 25 of *Cambridge Studies in Advanced Mathematics*. Cambridge University Press, Cambridge, 1991.

Wolniewicz, T. *Seminar Notes*. Arbeitsgruppe Funkionalanalysis. Institut f. Mathematik, J. Kepler University, Summer 1991.

Xu, Q. Littlewood–Paley theory for functions with values in uniformly convex spaces. *J. Reine Angew. Math.*, 504:195–226, 1998.

Xu, Q. H. Inégalités pour les martingales de Hardy et renormage des espaces quasinormés. *C. R. Acad. Sci. Paris Sér. I Math.*, 306(14):601–604, 1988.

Xu, Q. H. Applications du théorème de factorisation pour des fonctions à valeurs opérateurs. *Studia Math.*, 95(3):273–292, 1990.

Xu, Q. H. Analytic functions with values in lattices and symmetric spaces of measurable operators. *Math. Proc. Camb. Philos. Soc.*, 109(3):541–563, 1991.

Yudin, V. A. On the Fourier sums in L_p. *Proc. Steklov Inst. Math.*, 180:279–280, 1989.

Zygmund, A. *Trigonometric Series. Vol. I, II*. Cambridge Mathematical Library. Cambridge University Press, Cambridge, 1988.

Notation Index

Subject Index

For EU product safety concerns, contact us at Calle de José Abascal, 56–1°,
28003 Madrid, Spain or eugpsr@cambridge.org.

www.ingramcontent.com/pod-product-compliance
Lightning Source LLC
LaVergne TN
LVHW102007210926
848483LV00009B/115
* 9 7 8 1 1 0 8 8 3 8 6 7 2 *